Alfred I. du Pont

Alfred Irénée du Pont, 1864–1935

Alfred I. du Pont

THE MAN AND HIS FAMILY

Joseph Frazier Wall

New York · Oxford
OXFORD UNIVERSITY PRESS · 1990

Oxford University Press

Oxford   New York   Toronto
Delhi   Bombay   Calcutta   Madras   Karachi
Petaling Jaya   Singapore   Hong Kong   Tokyo
Nairobi   Dar es Salaam   Cape Town
Melbourne   Auckland

and associated companies in
Berlin   Ibadan

Copyright © 1990 by Oxford University Press, Inc.

Published by Oxford University Press, Inc.,
200 Madison Avenue, New York, New York 10016

Oxford is a registered trademark of Oxford University Press

Library of Congress Cataloging-in-Publication Data
Wall, Joseph Frazier.
Alfred I. du Pont : the man and his family / Joseph Frazier Wall.
p. cm.   Includes bibliographical references.
ISBN 0-19-504349-9
1. Du Pont, Alfred Irénée, 1864–1935.   2. Du Pont family.
3. Businessmen—United States—Biography.   I. Title.
HC102.5.D78W35   1990
338.7′66′092—dc20   [B]   89-16141   CIP

9 8 7 6 5 4 3 2 1
Printed in the United States of America
on acid-free paper

For
Sheldon Meyer, my editor
Francis Wigdahl, my life-long friend
Mary Jane Wall, my sister
Three very special people in my life

Foreword

> The White Rabbit put on his spectacles, "Where
> shall I begin, please your Majesty?" he asked.
>
> "Begin at the beginning," the King said gravely,
> "and go on till you come to the end; then stop."
>
> Lewis Carroll, *Alice's Adventures in Wonderland*

THE WHITE RABBIT'S question, "Where shall I begin?" is also the historian's question. No problem is more basic or more troublesome than deciding just where one should jump into the stream of history in order to examine one particular segment of its flow. Those social and economic historians who have never tried their hand at writing biography have always envied the biographer, whose subject seemingly provides convenient terminal points. The recorded life of any individual must surely begin at the moment in which the umbilical cord is cut and the infant takes its first gasp of air, and the story must certainly end with the subject's last exhalation of breath and final contraction of the heart. There should be no difficulty here in following the sage advice given by Lewis Carroll's King of Hearts. How fortunate must be the biographer who knows just where to start and just when to stop, and then needs only to fill in what has happened to his or her subject between those two fixed moments of birth and death.

The task of beginning and ending a biography, however, as the experienced biographer knows, is never quite that simple, for the full story of any individual is never neatly concurrent with that person's life span. Alfred Irénée du Pont's physical life began on the morning of 12 May 1864 in a large stone house near the west bank of the Brandywine River in northern Delaware, and it ended seventy-one years later, on 29 April 1935, in an impressive mansion, called Epping Forest, on the banks of the Saint Johns River in northern Florida.

But Alfred's real story begins a century and a quarter earlier, with the birth of Pierre Samuel du Pont on 14 December 1739 in a small flat above a watchmaker's shop on the rue Harlay, near the Seine in the center of

Paris. And one may argue that it has not ended even yet, as the story of the control and disposition of his estate continues, as does the good work he hoped he had endowed forever.

America has had at least two families who, because their members have made major contributions to our history through more than two generations, are entitled to be termed dynasties: the Adams family of Massachusetts and the du Ponts of Delaware. Henry Adams, always acutely aware of the dynastic position in which he had been placed by the accident of birth, begins his autobiography, written in the third person, with the classic line, "Had he been born in Jerusalem under the shadow of the Temple and circumcised in the Synagogue by his uncle the high priest, under the name of Israel Cohen, he would scarcely have been more distinctly branded. . . ."[1]

Equally branded was Alfred Irénée du Pont, the great-great-grandson of Pierre Samuel du Pont.[2] What the First Unitarian Church, the Boston State House, the Charles River, Beacon Hill, the country house at Quincy, and the name Adams were to the boy Henry, so were Christ Church, the graining mill, the Brandywine River, Louviers, Swamp Hall, and the name du Pont to the boy Alfred. Marked by their respective dynastic heritages, both boys would bear their family names with great pride but also with no little pain. Because of their names, their boyhood expectations were high and their later disappointments unanticipated.

The study of either genetics or the impact of environment upon child development has failed to provide the historian with an explanation of how an ordinary family line can suddenly flare bright with extraordinary genius or—even more difficult—can sustain that remarkable talent through four or five generations. James Truslow Adams, in writing of the family having the same name but not related to him, addresses this problem but fails to provide an answer. He can only conclude that "without warning, like a 'fault' in the geologic record, there is a sudden and immense rise recorded in the physical energy of a family. . . . Fascinating as the problem is, it is insoluble."[3]

Even a cursory examination of the background of America's two great dynastic families, however, reveals some interesting similarities that might provide possible clues to the continued prominence of the Adams and du Pont families. Both, for as far back as we can trace the families, came of plebeian stock on the paternal side. They were within the lower echelons but not at the bottom of their societies' socioeconomic classes. Both families became dissenters from the established churches of their respective countries, and both produced young men with enough ambition to break from their families' long-established places of residence to search for another way of life in a new environment. The first Henry Adams moved from Barton St. David parish, England, to Braintree (later named Quincy),

Massachusetts, and Samuel du Pont migrated from Rouen, France, to Paris—a much shorter distance, to be sure, but a no less alien environment. Perhaps most interestingly, in both families the first genius to appear was a result of the father's marrying a woman of a distinctly higher social class than his own. It was the first John Adams's marriage to Susanna Boylston, daughter of Peter Boylston of Brookline, that produced *the* John Adams who would lift the Adams family from obscurity to lasting historical fame. It was Samuel du Pont's marriage to Anne Alexandrine Montchanin, the daughter of an ancient noble family who had lost their lands because of their adherence to the Huguenot faith, that produced Pierre Samuel du Pont.

But if there are similarities of background in the two dynastic families, there are also important differences. The Adams family came of the yeoman stock that best represented Britain's population long before that country became "a nation of shopkeepers." The du Ponts, however, had been townspeople of Rouen for generations prior to Samuel du Pont's seeking a larger stage in metropolitan Paris. The Adams family remained, even after their migration to America, what they had been in England—complacent tillers of the soil, untroubled by any higher ambition. Among the du Ponts, however, there had long been manifest the grasping avarice of the French *petite bourgeoisie*—no wide horizons of ambition, to be sure, but a grim determination to get, hold, and increase one's material possessions. Then, within the same decade, the 1730's, the lightning bolt of genius struck both families, and they were never the same again.

Surely neither the first John Adams nor Samuel du Pont expected to beget an offspring who would effect such a remarkable transformation within his family and for his nation. John and Susanna Adams did have a somewhat higher ambition for their first-born son, John, than they had for themselves. At a considerable sacrifice of the income they derived from their small farm in Braintree, they sent their sixteen-year-old son to Harvard, in the hope that he would successfully complete his studies there and become a village pastor like his uncle, thus bringing an additional measure of prestige to the Adams family in Puritan New England. Samuel du Pont's ambition for his second and only living son did not extend even that far; it was limited to the small watchmaking shop on the rue Harlay. If little Pierre could become as good, or perhaps even a better watchmaker than he, Samuel would be quite content. In planning their sons' futures, both families failed to take into account the problems that genius can create.

Acknowledgments

ONE OF THE LAST and certainly one of the most pleasant tasks that a writer undertakes in preparing a manuscript for publication is to add a brief—and always inadequate—note of appreciation for the many persons whose assistance proved essential in the creation of the book.

I have been twice blessed and doubly indebted in this instance, for my research of necessity had two centers of operations: the Eleutherian Historical Library in Wilmington, Delaware, which houses the papers of the du Pont family, except for those of the subject of this biography; and the offices of the Alfred I. du Pont Estate in Jacksonville, Florida, where Alfred I. du Pont's papers have been stored. During the many long years that this work was in slow progress, both Wilmington and Jacksonville became second homes to my wife and me, for there we found not only the basic resources for this book but also the personnel to unearth that raw material. In the process, many of these persons became our close friends.

At the risk of sounding like the usual recipient of an Oscar award who feels obliged to mention every other member of the movie's cast and crew, let me name those persons who have been important to this production.

In Wilmington: Walter Heacock, at that time Director of the Hagley Museum and the Eleutherian Library, and his successor, Glenn Porter; Richmond Williams, Eleutherian Librarian; Marjorie McNinch, Reference Archivist, and the current Curator of Manuscripts, Michael Nash, who has kindly given me permission to quote from the du Pont family papers; the late Dr. A. R. Shands, Jr., former director of the Nemours Children's Hospital, who along with his wife Polly and Mrs. A. B. Strange provided us with a most hospitable introduction to Wilmington and the Brandywine valley; the late Kennard Adams, manager of the Nemours grounds and mansion; three members of the du Pont family in particular: the late Maurice du Pont Lee, Alfred I. du Pont's nephew, whose initial help to me in

Wilmington was truly invaluable; A. I.'s grandson, Alfred du Pont Dent; and A. I.'s cousin, several times removed, Francis I. "Nick" du Pont. My very special and warm thanks go to John Beverley Riggs, former Curator of Manuscripts at the Eleutherian, and his assistant, Ruthanna Hines; to Ruth Linton, former Curator of the Nemours mansion; and a very special tribute to the late Mrs. B. Bright Low, an unfailing source of knowledge, understanding, and good humor to encourage and sustain me in this undertaking.

In Jacksonville: I am especially indebted to the former trustees of the Jessie Ball du Pont Religious, Charitable and Educational Fund: the late William B. Mills, the late Edward Ball, the Rev. Alexander D. Juhan, and especially to Hazel Williams, Executive Director of the Fund. I also wish to thank Miss Williams's successor, George Penick, for seeing me through the final stages of the book. These trustees have not only provided funds for my research, but have also given unstintingly their encouragement and their patience during the book's long gestation. My appreciation must also be expressed to other Jacksonvilleans: Helen Mills and Alice Juhan; Jo Ann Bennett, secretary to the director of the Jessie Ball du Pont Fund; the late Irene Walsh, personal secretary to both Alfred I. du Pont and Edward Ball; Mrs. W. T. Edwards, who very kindly gave me access to her husband's papers; and to Edward Ball's business associates, J. C. Belin and Tom S. Coldewey, who provided valuable insights into the Port St. Joe paper mill operations.

There were others outside the Wilmington-Jacksonville circuit whose interest and assistance I greatly appreciated: Mrs. Nadine Labbé of St. Germain-en-Laye, France, who did valuable research for me at the Bibliothèque National in Paris in the writings of Pierre Samuel du Pont de Nemours; Mrs. Ruth Morrell of St. Louis, Ed Ball's former wife; Rebecca Harding Adams, Jessie du Pont's cousin, who remembered the happy times A. I. experienced at Harding's Landing and Ball's Neck; and Mr. and Mrs. Robert Dubell, whose family were close friends of A. I. and Jessie.

Finally, I give thanks once again to my patient editor, Sheldon Meyer; to my typist, Lucille Boyd De Jong, who managed somehow to turn the illegible yellow pad pages into a legible manuscript; and above all, as always, to my wife, Beatrice Mills Wall, my chief research assistant, my most appreciative critic, and the first reader and editor of the slowly emerging manuscript.

Grinnel College   J. F. W.
July 1989

Contents

Alfred I. du Pont

SELECTED BRANCHES AND TWIGS ON THE DU PONT FAMILY TREE

THE TRUNK OF THE TREE

Samuel du Pont m. Anne Alexandrine de Montchanin
1708–76 — 1718–56

Pierre Samuel du Pont de Nemours m. (1) Nicole Marie Le Dèe de Rencourt
1739–1817   1743–84

m. (2) Widow Marie Françoise Robin Poivre
1748–1841

Victor Marie   Paul François   Eleuthère Irénée
1767–1827   1769–70   1771–1834

BRANCH I VICTOR MARIE du PONT de NEMOURS FAMILY

Victor Marie du Pont m. Gabrielle de La Fite de Pelleport
1767–1827   1770–1837

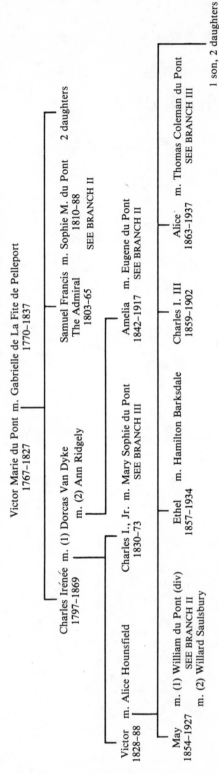

Samuel Francis m. Sophie M. du Pont   2 daughters
The Admiral   1810–88
1803–65   SEE BRANCH II

Amelia m. Eugene du Pont
1842–1917   SEE BRANCH II

Charles I. III   Alice   m. Thomas Coleman du Pont
1859–1902   1863–1937   SEE BRANCH III

1 son, 2 daughters

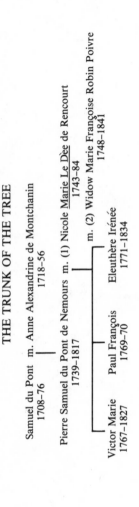

Charles Irénée m. (1) Dorcas Van Dyke
1797–1869
m. (2) Ann Ridgely

Charles I., Jr. m. Mary Sophie du Pont
1830–73   SEE BRANCH III

Ethel   m. Hamilton Barksdale
1857–1934

Victor m. Alice Hounsfield
1828–88

May   m. (1) William du Pont (div)
1854–1927   SEE BRANCH II
m. (2) Willard Saulsbury

Note: All of the branches on the du Pont family tree have been pruned to include only those persons who are relevant to the Alfred I. du Pont story.

BRANCH II ELEUTHERE IRENEE du PONT de NEMOURS FAMILY

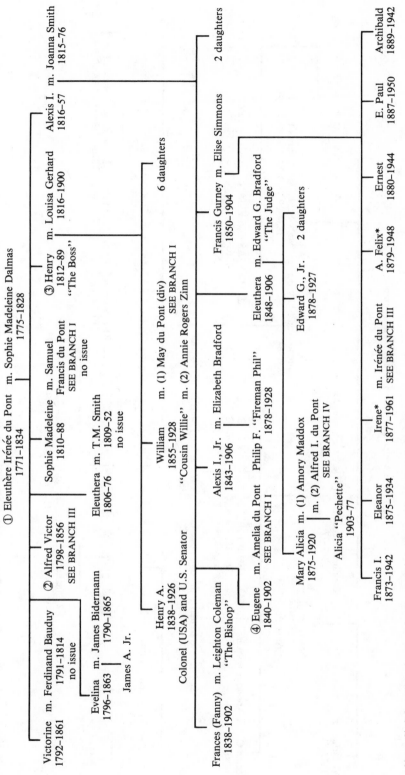

① Eleuthère Irénée du Pont 1771–1834 m. Sophie Madeleine Dalmas 1775–1828

Victorine 1792–1861 m. Ferdinand Bauduy 1791–1814 no issue

② Alfred Victor 1798–1856 SEE BRANCH III

Sophie Madeleine 1810–88 m. Samuel Francis du Pont SEE BRANCH I no issue

③ Henry 1812–89 "The Boss" m. Louisa Gerhard 1816–1900

Alexis I. 1816–57 m. Joanna Smith 1815–76

Evelina 1796–1863 m. James Bidermann 1790–1865

James A. Jr.

Eleuthera 1806–76 m. T.M. Smith 1809–52 no issue

Henry A. 1838–1926 Colonel (USA) and U.S. Senator

William 1855–1928 "Cousin Willie" m. (1) May du Pont (div) m. (2) Annie Rogers Zinn SEE BRANCH I

6 daughters

Francis Gurney 1850–1904 m. Elise Simmons

2 daughters

Frances (Fanny) 1838–1902 m. Leighton Coleman "The Bishop"

④ Eugene 1840–1902 m. Amelia du Pont SEE BRANCH I

Alexis I., Jr. 1843–1906 m. Elizabeth Bradford

Philip F. "Fireman Phil" 1878–1928

Eleuthera 1848–1906 m. Edward G. Bradford "The Judge"

Edward G., Jr. 1878–1927

2 daughters

Mary Alicia 1875–1920 m. (1) Amory Maddox m. (2) Alfred I. du Pont SEE BRANCH IV

Alicia "Pechette" 1903–77

Francis I. 1873–1942

Eleanor 1875–1934

Irene* 1877–1961 m. Irénée du Pont SEE BRANCH III

Ernest 1880–1944

A. Felix* 1879–1948

E. Paul 1887–1950

Archibald 1889–1942

Note: Circled numbers give the order of succession to the presidency of the du Pont Company.
*Designates the two children of Francis Gurney du Pont who sided with Cousin Pierre against A.I.

BRANCH III ALFRED VICTOR du PONT FAMILY

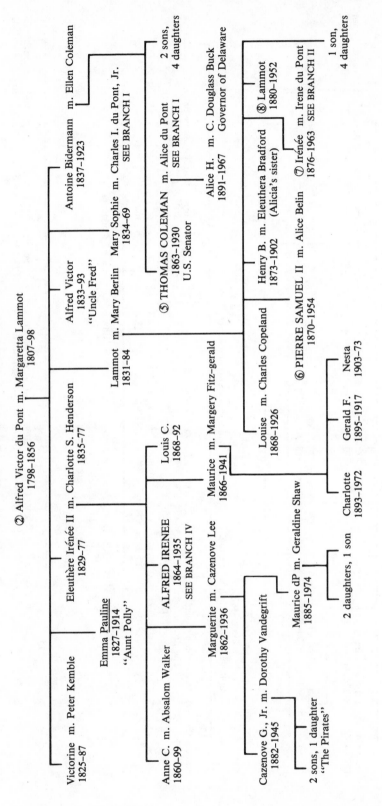

② Alfred Victor du Pont m. Margaretta Lammot
1798–1856   1807–98

Victorine m. Peter Kemble
1825–87

Eleuthère Irénée II m. Charlotte S. Henderson
1829–77   1835–77

Emma Pauline
1827–1914
"Aunt Polly"

Alfred Victor
1833–93
"Uncle Fred"

Antoine Bidermann m. Ellen Coleman
1837–1923

Mary Sophie m. Charles I. du Pont, Jr.
1834–69   SEE BRANCH I

2 sons,
4 daughters

Lammot m. Mary Berlin
1831–84

⑤ THOMAS COLEMAN m. Alice du Pont
1863–1930   SEE BRANCH I
U.S. Senator

Alice H. m. C. Douglass Buck
1891–1967   Governor of Delaware

Louis C.
1868–92

Maurice m. Margery Fitz-gerald
1866–1941

Louise m. Charles Copeland
1868–1926

Henry B. m. Eleuthera Bradford
1873–1902   (Alicia's sister)

⑥ PIERRE SAMUEL II m. Alice Belin
1870–1954

⑧ Lammot
1880–1952

⑦ Irénée m. Irene du Pont
1876–1963   SEE BRANCH II

1 son,
4 daughters

Anne C. m. Absalom Walker
1860–99

ALFRED IRENEE
1864–1935
SEE BRANCH IV

Marguerite m. Cazenove Lee
1862–1936

Maurice dP m. Geraldine Shaw
1885–1974

Charlotte
1893–1972

Gerald F.
1895–1917

Nesta
1903–73

Cazenove G., Jr. m. Dorothy Vandegrift
1882–1945

2 daughters, 1 son

2 sons, 1 daughter
"The Pirates"

Note: Names in capital letters are the three cousins who took over the company in 1902.
Circled numbers give the order of succession to the presidency of the du Pont Company.

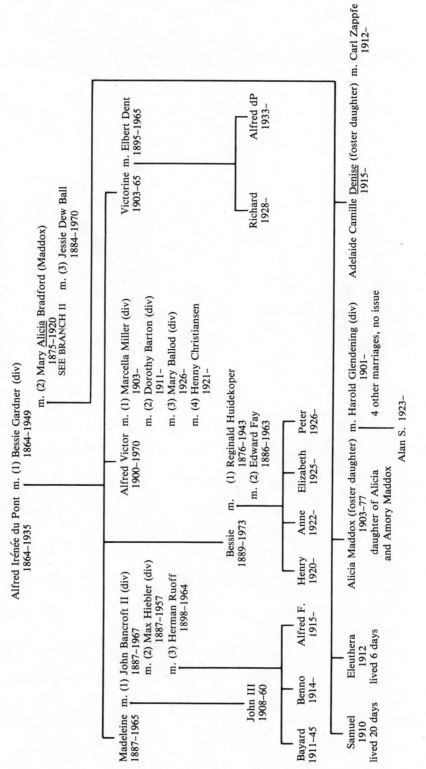

BRANCH IV ALFRED IRENEE du PONT FAMILY

The First Three Generations until Alfred I. du Pont's Death in 1935

Alfred Irénée du Pont m. (1) Bessie Gardner (div)
1864–1935   1864–1949

m. (2) Mary Alicia Bradford (Maddox)
1875–1920
SEE BRANCH II   m. (3) Jessie Dew Ball
1884–1970

Victorine m. Elbert Dent
1903–65   1895–1965

Richard   Alfred dP
1928–   1933–

Adelaide Camille Denise (foster daughter) m. Carl Zappfe
1915–   1912–

Alfred Victor m. (1) Marcella Miller (div)
1900–1970   1903–
m. (2) Dorothy Barton (div)
1911–
m. (3) Mary Ballod (div)
1926–
m. (4) Henny Christiansen
1921–

Bessie   m.   (1) Reginald Huidekoper
1889–1973   1876–1943
m. (2) Edward Fay
1886–1963

Henry   Anne   Elizabeth   Peter
1920–   1922–   1925–   1926–

Madeleine m. (1) John Bancroft II (div)
1887–1965   1887–1967
m. (2) Max Hiebler (div)
1887–1957
m. (3) Herman Ruoff
1898–1964

John III
1908–60

Bayard   Benno   Alfred F.
1911–45   1914–   1915–

Alicia Maddox (foster daughter) m. Harold Glendening (div)
1903–77   1901–
daughter of Alicia   4 other marriages, no issue
and Amory Maddox
Alan S. 1923–

Samuel   Eleuthera
1910   1912
lived 20 days   lived 6 days

Prologue
A Tale of Two Rivers and Two Men
1739–1834

PART I   The River Seine and Pierre Samuel du Pont,
Founder of the Family

WHEN SAMUEL DU PONT arrived in Paris in 1735, carrying in his satchel a few clothes and the tools of his trade, and in his fingers the dexterity that he had learned from his years of apprenticeship in Rouen, he headed straight for the rue Harlay, the street of fine watchmaking, where he opened his little shop directly across from that of the Montchanin brothers. It did not take long for Samuel to notice and to be attracted to his competitors' younger sister. During the day Samuel could observe her sitting behind the front window of the shop, where what daylight there was would be the brightest, painting numerals and designs on the watch dials of her brothers' making. In the evening, to his delight, Samuel discovered his small sitting room above the shop faced directly on Anne's bedroom window, behind which she frequently sat reading. Soon Samuel was posturing in front of his window like some turkey gobbler performing his ritual mating dance. Samuel would execute with his foil the elaborate fencing techniques of thrust and parry against an invisible foe with all of the grace of the most noble chevalier. Or he would bring out his flute and produce sounds of such amazing dissonance as to force Anne across the way to cover her ears and to laugh in both pleasure and pain.[1]

Obviously, Samuel had fallen in love with his young neighbor, but to him it must have seemed a hopeless suit. Anne was a Montchanin, and more than that, she had spent her childhood as a ward of her distant cousin, the Marquis de Jaucourt-Epenilles. Although she and her brothers were now on the same economic level as Samuel, the social caste system was still a greater determinant of status in eighteenth-century France than was family income. What chance did a du Pont from Rouen have of an alliance with this noble family from Charolais?

3

As it turned out, a very good chance, indeed, for Anne's recent move from the Jaucourt chateau to the small jewelry shop in Paris had been a far more radical and disruptive break in her life than the move from Rouen to Paris had been for her flute-playing suitor. Raised from infancy by the Jaucourts as a playmate for the Marquis's young daughter and regarding herself as a member of the family, she had been stunned when she was informed by the Marquise that, now that she was fifteen, she must consider her future. Of noble birth but penniless, she had to find a means of earning her dowry, so that within a few years she could marry into a station befitting her birth. The Marquise said that she had discussed this with her husband, and they were prepared to offer Anne a position as a domestic within the Jaucourt household. Since all of her living expenses would be taken care of, she would be able to save her entire monthly wage and within a few years would have a sufficient dowry that, along with her physical beauty, would attract some man of gentle birth to propose marriage to her.

With this announcement, Anne's secure world of luxury had crashed about her. Too proud even to consider serving as a maid to the cousins she had always regarded as her mother and sister, Anne had fled from the Jaucourt chateau in tears and had taken the first available transportation into Paris. There her brothers put her to work painting watch dials. Anne was grateful for this refuge, but she was also miserably unhappy and dreadfully bored. Samuel du Pont's posturing found a receptive audience. He was tall and handsome, and, even if he couldn't play the flute, he was a romantic figure to the lonely young girl. It would not be much of a change, to be sure, to move across the street into another watchmaking shop, but he was after two years' residency in Paris well established in business, and regarded even by her own brothers as one of the most skilled artisans in the trade. He would expect no dowry from her, and most important of all, he too was a Huguenot. Within two years of her arrival at her brothers' shop, she and Samuel du Pont were wed, being obliged to go across the border into the Netherlands so that they could be married properly in a Protestant church.

Anne very quickly discovered that the romantic suitor in the window across the way was not the same person as the uncommunicative husband bending over his workbench in the shop. The flute playing mercifully stopped, but so also did the wonderful make-believe fencing duels and much of the romance. As du Pont's *femme,* she was expected to do all of the menial tasks of the household, wait on customers in the shop, and have babies. Marriage had proved to be no escape after all.

Their first child, a son, was born within the first year of their marriage, but unfortunately the child lived only a few months. The following year, on 14 December 1739, Anne gave birth to a second son, whom Samuel named Pierre Samuel, after a favorite uncle in Rouen and himself. From birth the child was small and sickly—an unlikely and, indeed, very slender reed upon which to build a dynasty.

Determined not to lose this child within a few months of his birth, Anne persuaded her husband to allow her to send the infant into the country, where the fresh air and an abundant supply of milk from a strong peasant wet nurse might give the child a small chance for survival. The experiment proved to be a nearly fatal disaster. Visiting her baby a few weeks later, Anne discovered that the wet nurses's own more aggressive baby was getting most of her milk, and little Pierre Samuel was dying of malnutrition. The child was quickly transferred to another family, the local tavern keeper and his wife. Here Pierre Samuel survived the remaining months of his first year, but soon after he was weaned, the tavern keeper amused himself and delighted the infant by giving the child brandy whenever he cried. When Anne discovered this, the noble experiment in healthful country living was quickly abandoned. She brought the baby back to Paris and never again let the child out from under her supervision. The damage had been done, however, and Pierre Samuel, undersized, undernourished, and suffering from rickets, would bear the scars of his infancy for the rest of his life.

Even in Paris, under Anne's doting care, the child was not safe from further affliction. At the age of five, he fell from the stairs and broke his large nose, which was the distinguishing feature of the du Pont family. With his little round face and broken nose, he would always look like a small, battered, puglistic cherub, hardly a thing of beauty, but Anne loved him nevertheless. Although his small body had been stunted and warped by the trials of his infancy, nothing was deformed about his mind. Anne was surely the first to recognize that, somehow, out of their joyless and humdrum marriage she and Samuel had produced something truly remarkable—a child who had the brilliance of intellect to escape from the watchmaking future his father had planned for him and who would provide a vicarious exit for Anne as well. With the education she had received from the tutors in the Jaucourt household, Anne was able to give her son his first lessons in reading and writing at the age of three. By the time he was five, he was ready for more expert instruction than Anne could provide. After much cajoling, Anne finally persuaded her reluctant husband to provide the funds to engage a tutor. Her evaluation of her son's intellectual ability had not been based solely upon an overweening maternal pride. The instructor they employed was as impressed with his young charge's ability as was Anne. But Samuel du Pont was uneasy about the training his son was receiving—not at all the proper preparation for a child who must appreciate his station in life. Expectations would be raised that could not and should not be met. What point was there in filling his son's head with impractical classical learning and fanciful romantic rubbish? "Under no circumstances," he sternly warned the tutor, "is my son either to read or write poetry."[2] Perhaps as a consequence of this paternal decree, poetry became Pierre Samuel's favorite literary genre and provided him with the models for his own first literary exercises.

When Pierre Samuel was twelve, his proud tutor arranged for a public demonstration of his prize student's remarkable talents. The entire neighborhood of Parisian shopkeepers and day laborers attended. Young Pierre Samuel dazzled them all by his ability to "translate sight passages from the best Latin authors, answer questions in French and Latin grammar, on logic, rhetoric, fables, poetry, literary style—even civil law."[3] The neighbors were greatly impressed and Anne was ecstatic with pride, but Samuel was outraged that his dictum on his son's education had been so flagrantly violated. The boy had been turned into a damned poet after all. The lessons must stop immediately.

In desperation Anne turned to an even higher authority than the ruling pater familias. Shortly thereafter she announced to her husband that young Pierre Samuel had informed her that he had received a call from God to become a Protestant minister. When questioned by his father, the boy confirmed his mother's revelation. Not even Samuel was stubborn enough to stand against the combined force of God, Anne, and his only son. The boy was given a new tutor who could instruct him in Protestant theology as well as natural science and ancient philosophy. The only subject, however, for which Pierre Samuel showed no interest or talent was theology.

Within a few months, still another physical disaster struck the boy. Suffering from smallpox, he had a high fever for several days, and became so critically ill that the physician who came to call could not hear a heartbeat nor see a mist of breath on the mirror held before the child's mouth. The boy was pronounced dead. A neighbor woman laid out the small wasted body on an improvised funeral bier and prepared to sit by the corpse through the night. Anne, whose last and only hope appeared to be gone, had collapsed in grief in her bedchamber. Shortly afterwards, she heard a scream from the parlor. The corpse watcher was hysterical. The dead child had moved his arms and had mumbled a word. God had given her son back to Anne. Now more than ever, she was convinced of the miracle that was Pierre Samuel. Clearly, God did intend a special role for him on this earth.

Samuel, however, interpreted the miracle in a quite different way. The child had been left blind in one eye by the disease. Was this not a divine warning against further book learning? Enough of tutors with their fancy ways and even more fanciful ideas. Samuel himself would now take direct charge of his son's education. The long-forgotten foils were brought out, and Samuel gave his son daily lessons in the gentlemanly art of fencing, to build up his weakened body. So fascinated was the boy with the game that he began surreptitiously to seek out books on the history of weapons, which inevitably led him to a study of military science. Small and physically handicapped as he was, he began to dream of a career in the army. This he found much more attractive than his aborted study of theology.

When Samuel discovered his son's latest foray into the realm of book learning, he swept all of the hidden texts out of the house and firmly seated the boy in front of his watchmaking table. Now the young man would learn something useful, something that did not require books or great phsyical strength, and for which he would have the best teacher in all of Paris. The exit door out of the watchmaker's shop was apparently slammed shut, to both the boy and his mother. Pierre Samuel proved to be an amazingly apt pupil. Peering through the magnifying lens with his one good eye and moving his delicate little fingers with the dexterity of a fine lacemaker, he was now a joy for his father to behold. But the boy hated every minute of it.

In 1756, her spirit broken by Samuel's triumphant control of their son, and her body weakened by a succession of stillbirths, Anne Montchanin du Pont, dreamer of grand dreams, died—at the age of thirty-six. As she lay dying, she implored her husband and her son to love and to try to understand one another. Like many of her other hopes, this too would not be realized.

Soon after his wife's death, Samuel forced his son to move into a small attic room as a punishment for having dared to write a parody of the First Epistle of St. John, an act that Samuel considered nothing short of blasphemous. In despair, the boy attempted suicide with a hunting knife, but, discovered in the act, he was severely beaten by his father. In a rage, Pierre Samuel grabbed his few clothes and rushed out of the shop. There was an exit door after all, but it led only to another watchmaker's shop, for Pierre Samuel knew no other way to make a living. Having been apprenticed to one of the finest masters of the craft in the city, his father, young du Pont had no difficulty in finding employment in another shop. For five more years, he was to continue in his hated occupation.

In his new situation, however, his hands and eye might be fixed on the small tools of his trade, but his intellectual curiosity was no longer confined; it could now roam freely. When his roommate's severe illness forced Pierre Samuel to call in a physician to attend him, the young man met Doctor Barbon du Bourg, and the two quickly became close friends. Impressed as everyone except his father always was with Pierre Samuel's quickness of mind, du Bourg urged the youth to consider entering the field of medicine. Soon du Pont was spending his evenings reading medical texts and studying anatomical drawings. He would take an occasional day off from work to attend lectures on medicine, to observe surgical operations, and to visit clinics with his new-found friend. In those pre-licensing days, medicine was a simple skilled craft, like that of the watchmaker or barber. One needed no special degree. After his apprenticeship of a few months, Pierre Samuel could proclaim himself a physician, a title he would always keep even though he never practiced medicine until he was called upon to do so during the French Revolution when doctors were scarce and in great demand.

Young du Pont also found time in the evenings to write poetry and drama, and one of his plays was actually performed by an amateur theatrical group in Paris. These excursions into belles-lettres opened another door, into the world of writers and artists. Here young men could argue for hours whether French classicism was too confining to the artistic temperment, or on the meaning of a single line in Corneille's *Médée*. For the first time in his life, Pierre Samuel met people who were as intellectually alive as he and whose sharpness of mind could successfully challenge his own. Conversation, he discovered, could be as exhilarating as fencing, and he delighted in the quick thrusts and parry of ideas.

Frequently on pleasant Sunday mornings, while his literary companions were still sleeping off the heady hours of drinking and talking of the previous evening, Pierre Samuel would arise early and ride out into the country to visit former neighbors, the Dorés, who had recently moved from Paris to the small city of Nemours. Since the death of his own mother, Madame Doré had treated him like a son and had insisted upon his calling her Maman. It was a long trip for a day's outing—some forty miles from Paris—but well worth it. Pierre Samuel loved the ride along the Seine through the fertile lands of Ile-de-France. For a city boy who could remember nothing but the crowded, narrow streets of Paris, the countryside had a curious attraction. Perhaps those months of country living as an infant had given him something other than a stunted body—something as lasting, but far more beneficial.

The many hours in transit also gave him the opportunity to think in solitude, and as he looked out on the rich fields of grain, the fruit orchards, and tall stands of timber along the river, he slowly came to the realization that the wealth of France—the wealth of any nation—lay not in the palaces of the nobility, nor in the great cities with their grubby little shops and polluting factories, nor even in the seaports with their warehouses stocked with the world's goods, but within its tilled fields, and its true custodians were not the kings, princes, merchants, or manufacturers. They were the farmers. On these rides he began to develop a whole new thesis of political economy that would challenge the prevailing gospel of mercantilism, according to Jean Baptiste Colbert that had long been the only true economic faith for France and the entire Western world.

When he finally reached his destination at the spacious townhouse of the Dorés in Nemours, he was always assured of a warm reception. Maman

Doré would fuss over him and stuff him with her richest sauces and pastries; M. Doré would ask for the latest political news from Paris and listen to him with the respect due to an oracle; and their lively young cousin, who had the imposing name of Nichole Charlotte Marie-Louise Le de Rencourt, which the family had long ago shortened to Marie Le Dèe, would gaze upon him with adoring eyes. Small wonder that Pierre Samuel ever more frequently would give up his hours of needed Sabbath rest for this expedition into the country.

One Sunday, however, he found the usually gay Marie Le Dèe in tears when he arrived at the front gate. Between sobs, she told him that the Dorés had arranged for her to marry a local widower. Monsier des Naudières, who was incredibly old—fifty-five—and looked at least twenty years older than that. Pierre Samuel was outraged. He burst in upon Maman Doré and demanded to know how she could have possibly been a party to such an atrocity.

Maman then proceeded to give Pierre Samuel some facts of life. Marie was their ward. She had very little money of her own for a dowry, and she must find a husband while she was still of marriageable age. Monsieur des Naudières was a man of considerable property in Nemours and was eager to marry Marie. Did Pierre Samuel have any better arrangement in mind for Marie Le Dèe's future?

To his own great surprise, Pierre Samuel did. He found himself blurting out that he would marry Marie himself, although up to that moment he had never considered the girl anything other than the young ward of the Dorés who liked to listen to his poetry. But Maman Doré was not easily satisfied with this solution. How could he, an underpaid employee of a small watch shop, possibly consider supporting a wife like Marie, who was accustomed to the Dorés' standard of living? Again to his own surprise, Pierre Samuel had a quick answer. He had been planning to leave the watchmaking trade anyway. Now he would do so immediately. He was convinced that with his skill as a writer and with the many new contacts he had made among the intelligentsia of Paris—the men who would really count in this new enlightened era—he would soon make a fortune in comparison with which that of the old widower would pale into significance. It was a wild outburst of braggadocio, and it was difficult to imagine that Madame Doré would believe him—he could hardly believe it himself—but she agreed to give him two years in which to prove himself. Perhaps she had been waiting for many months to hear him say what he just had.

The ride back to Paris must have seemed longer than ever before to the suddenly sobered young man. Here he was practically betrothed to a young girl who had never before been an object of romantic interest to him, and not only was he committed to giving up the only occupation that had as yet provided him with even a bare subsistence, but also he must find a fortune within two years, in a still to be determined new field of endeavor. How could all of this have happened to him so unexpectedly, and by his own volition?

Well, at least the first step would be easy. He had always hated watch-making, and now that he had become skilled in the craft, no challenge was left for him at all—only a rote performance that bored him to distraction. He would be happy to leave, but before he did, there was one last task he must perform.

The following morning he set to work on a new watch, and for once this one received his full attention. It would be the finest watch he had ever made—perhaps the finest watch that anyone, including his father, had ever made. After several weeks of intense labor, the watch was finally finished. He carefully engraved on the back of the watch the inscription: *Du Pont filius composite, fecit, dedicavit patri suo.* The Latin inscription was pure inspiration, a final nose-thumbing gesture to his boorish father's distaste for classical learning. He wrapped the watch in gift paper, and on New Year's Day, 1763, he took his present to his father's new shop on the rue de Rich-elieu. We do not know what Samuel du Pont said, or what he thought, upon seeing his son again and receiving this unexpected gift. There is no record of what happened to the watch. Five generations later, du Pont's descendants would spend much time and energy and would offer a great deal of money to anyone who could produce this watch, but it was never found. Perhaps Samuel's affront at his former apprentice's hubris overcame his paternal pride in his son's accomplishment, and he simply destroyed the watch. Pierre Samuel could not have cared less. He was at last free to leave his father's trade forever. Whatever new occupation he undertook, how-ever, must be done with the greatest possible speed; the two-year deadline for saving Marie from Monsieur des Naudières hung over him heavily.

Where to go was the great problem. For a brief time he dreamed of the possibility of forming a regiment of stout Parisian companions and going south to Gibraltar, there to storm that vital fortress currently held by France's and Spain's mutual enemy, Great Britain. To return Gibraltar to its rightful owner would surely win him the fame and fortune he so des-perately needed.

There would always be that quixotic streak within du Pont's make-up that bordered upon the insane. With his vivid imagination, windmills could easily be converted into giants, and peasant girls into beautiful princesses who must be rescued. His eloquence allowed du Pont to persuade many otherwise quite rational people to accept his dreams at face value and to become his willing Sancho Panzas. He actually raised a small troop from among the hot-blooded gallants of Paris who were eager to follow him to

Gibraltar, but, happily for all concerned, in 1763 the Treaty of Paris ended hostilities between Britain and its Bourbon foes. There would be no liberation of Gibraltar after all.

With his dreams of military glory temporarily frustrated, Pierre Samuel turned to the only other pursuit that truly interested him. If it couldn't be the sword, it would have to be the pen. Again he had companions who were eager to help him. Some of the young poets and playwrights with whom he drank and argued knew other men who were older and higher up on the Parisian literary ladder. Pierre Samuel was brash enough and smart enough, once given a boost, to clamber up its rungs with alacrity. Very quickly, he found himself at the ethereal heights of the Encyclopédists—Diderot, d'Alembert, and the other lions of the French intelligentsia. They were intrigued by this curious little watchmaker with the broken nose—a youth who seemed to know something about everything and was not hesitant to tell it. For Pierre Samuel, coming from the rue Harlay, to find himself moving with familiar ease in the rococo salons of Montparnasse was a heady experience indeed, but he was not shy about reaching even higher. He sent off a packet of his poetry and one of his essays on the French financial situation to the man seated at the very pinnacle of France's literary Mount Olympus—the great Voltaire himself, now living in splendid exile in Switzerland. Voltaire actually read these unsolicited offerings from an unknown youth. "I see, Monsieur," he responded, "that you encompass two species one a little different from the other: finance and poetry. The waters of the Pactolus must be quite astonished at flowing with those of Permessus. You send me very pretty verses along with computations on 740 million."[4]

It would be from the waters of Pactolus, however, that du Pont would draw his fame and fortune, for "pretty verses" were no longer in fashion. Poetry and drama had become frozen in the deep ice of French classicism, and all of the best minds of France were now concentrated either upon satirical novels that made fun of all that was wrong with France or upon essays on political economy that offered proposals for reforming those defects. The age of Racine and Corneille had given way to that of Voltaire and the Economists. Realizing this, du Pont brushed aside his poems and plays and plunged into economic analysis.

His first pamphlet—and in some respects his most important—was written only a few weeks after he had given up his Gibraltar fantasy. *Reflections on the Wealth of the State* encapsulated all of those ideas about the value of land that had occurred to him as he rode down the country lanes to Nemours. It was a remarkable piece to come from a twenty-three-year-old youth. Du Pont anticipated by some thirteen years Adam Smith's basic thesis that money is not wealth, and by over a century Henry George's advocacy of a single tax on land to replace the countless nuisance taxes, which were regressive in principle, were difficult to collect, and brought in too little revenue over and above the cost of collection. This single tax, du Pont argued, would provide ample revenue for what should be the limited needs of government, the only legitimate function of the state being to provide security to property through its police force, army, and courts. The state should not otherwise interfere in the lives of its citizens. The people must be free to produce and to trade unhampered by state-supported monopolies, tariffs, or regulations. "Complete liberty of competition" would guarantee fair prices and all would prosper. Du Pont promised that his economic system would provide every nation with the only three basics necessary for a good society: property, security, and liberty.

The Economists were delighted. Here was the boldest statement yet against the restraints of mercantilism that had destroyed Spain economically and were rapidly doing the same thing to France. Overnight, *laissons faire*—let us tend to our own affairs—became the fashionable catchphrase in the salons and cafés of Montparnasse.

A copy of du Pont's pamphlet reached the court at Versailles and fell into the hands of Dr. François Quesnay, who as physician to the king's mistress, the Marquise de Pompadour, was for the moment one of the most powerful men in France. Quesnay read the work with amazed delight. Its author had lucidly spelled out the very ideas that Quesnay himself had been mulling over for a long time. Who was the author? Curiously enough, no name was given on the pamphlet. The word went out from Versailles: find us the person who wrote *Reflections*. A hasty inquiry brought back the incomplete information that the pamphlet had been written by a goldsmith or a watchmaker—somebody like that—anyway, a complete unknown whose initials were D. P. At Quesnay's further urgings, agents discovered a goldsmith named Du Ponchel, who was quickly brought to Versailles. The bewildered Du Ponchel was warmly greeted by Dr. Quesnay. No, the goldsmith knew nothing about political economy. No, he had never written a page of anything in his life, but if the noble sir was interested, he could make His Excellency a handsome pair of gold buckles for his shoes.

Du Ponchel was dismissed, and the search went on. Then the Marquis de Mirabeau recalled that he had recently received a rather interesting letter on France's financial difficulties from someone he had never heard of before—a Pierre Samuel du Pont. Perhaps he was the missing D. P. Pierre Samuel was found and was brought to court for a private audience. Du Pont would later recall the moment of meeting with Quesnay. "I was only a child when he held out his arms to me; it was he who made me a man."[5] Now the course had been set, and for du Pont, for France, and for America the future was to be greatly altered.

Once the doors of Versailles had been opened to du Pont, his political rise was swift. He became Quesnay's and, through him, Pompadour's, most ardent and loyal disciple—a disciple who could also serve as teacher and guide in the new realm of *laissez faire* economics. Articles, statistical reports, and philosophic essays poured from du Pont's pen. He was dubbed "our young agriculturist" by the great Pompadour herself, and his study, *De l'exportation et de l'importation des grains,* confirmed the appropriateness of her accolade. It also earned him a new appellation, "the friend of Turgot." Anne Robert Jacques Turgot was generally regarded by the Economists and those reform elements within the court at Versailles as having the most brilliant mind in France. Turgot's star, clearly in the ascendancy, would illuminate du Pont's march forward.

The eighteenth century may have been the Age of Reason, but French politics at its highest level still remained Byzantine in its irrational capriciousness. Very much depended upon who was able to capture the king's attention at the council table in the afternoon or who shared his bed at night. In Quesnay and Pompadour young du Pont had found able principals to take care of both fronts. They were also experts in the fine art of political high-wire acrobatics, but all three knew that a single false step or a sudden gust of whimsy from on high could send them all headlong into oblivion.

The unexpected death of the Marquise de Pompadour in April 1764 was more than just a strong crosswind—it was an earthquake that sent Quesnay's sycophantic followers scurrying back to the security of other support poles. Only du Pont dared to stand exposed alongside his master while the ground rocked beneath them. But Quesnay did not fall after all, and du Pont, having risked his political neck, was rewarded for his loyalty by being given the editorship of an important new supplement to the commercial gazette, the *Journal de l'agriculture, du commerce et des finances.*

His fame and fortune apparently secured, Pierre Samuel could now go to Nemours and claim his bride. He had met the two-year ultimatum, but just barely. When asked by Marie's father if he were a Protestant, Pierre Samuel could truthfully answer no. Of course he did not elaborate on this by explaining that he also was not a Catholic, but rather a Deist. Having met the family's religious test by a half-truth and having satisfied his own conscience, like Henry of Navarre before him, in the belief that "Marie was worth a mass," Pierre Samuel and Marie Le Dèe were married in Paris, in the church of Saint-Sulpice, on 28 January 1766. Although he had proposed almost unwittingly and had married somewhat deceptively under false religious colors, Pierre Samuel had nevertheless chosen wisely. Like his Massachusetts dynastic counterpart John Adams, who married Abigail Smith, du Pont found in Marie Le Dèe precisely the wife he needed, a woman who could manage their household affairs, at which he was hopelessly inept, while he attempted to manage the financial affairs of a nation. Marie gave the family practical grounding while Pierre Samuel's imagination soared over the rainbow. Marie Le Dèe was the first, but not the last, of the du Pont women who were to contribute greatly to the fortunes of the family. The women the du Pont men married and the daughters they sired have never received from the many historians of the du Pont dynasty the credit due to them.

Within the second year of their marriage, on 1 October 1767, a son was born to the du Ponts. He was christened Victor Marie, with the Marquis de Mirabeau as his godfather. In that same year, du Pont published his most ambitious work to date, *Physiocracy*. When asked to explain the title, he replied that he had drawn the word from classical Greek. It meant simply government based upon the natural, physical laws of the universe. At last the Economists—Quesnay, Turgot, and Mirabeau—had a name for their philosophy. Just as Newton a century before had revealed the perfect rationale of God's physical laws, which had brought order to the universe out of chaos, so now they, the Physiocrats, would apply these same scientific laws to man's chaotic political and economic systems and create a physiocratic paradise of peace and plenty for all. The former watchmaker was to be the herald of the new order.

The Physiocrats had their own journal, *Les Ephémérides du Citoyen,* whose title *Ephémérides,* again taken from a Greek word, *ephemeris,* meaning a record of astronomical observations, proclaimed their philosophy. This journal, however, was to be a diary of terrestrial observations. On 1 January 1769, du Pont was asked to take over its management.

One of the journal's more enthusiastic subscribers, Voltaire, wrote to du Pont to congratulate the periodical for having secured the services of his young friend and erstwhile poet as its editor. "No one has gone into detail of rural exploitation better than you have; no one has made us feel how dear those who work on the land should be to the State better than you have. I, too," Voltaire modestly added, "have the honor to be a tiller of the soil."[6] After all, had not Voltaire concluded his satirical masterpiece *Candide* with the injunction to everyone to tend quietly his own garden?

But du Pont, like Voltaire himself, could never do anything quietly. The new order may have been rushing to be born, but the old order, including state censorship of the press, was still in place. Du Pont was too noisily liberal. Within three years of his taking the editorship, *Ephémérides* was suppressed.

Du Pont was again unemployed and, with his family growing larger, his personal financial situation was nearly as precarious as that of the nation. A second son, Paul François, born in December 1769, lived less than a month, but a third son, born on 24 June 1771, gave every promise of being a healthy child. He was given the name Eleuthère Irénée, meaning freedom and peace, at the suggestion of his godfather, the great Turgot. To support his family, Pierre Samuel had to find some kind of lucrative employment, so he was most receptive to an offer from the Margrave of Baden to come to Karlsruhe as Counsellor of State to that small German province, with the further promise of becoming Director of Finance if things worked out well. Against the advice of Turgot, du Pont quickly accepted. He needed the money and, considering the disfavor with which he was viewed at Versailles after the suppression of his journal, he found it expedient to go into exile like Voltaire. As it turned out, however, within a few months du Pont realized he had made a mistake. The Margrave housed and dined him lavishly, but gave him almost nothing in the way of a salary. While at Karlsruhe, he did manage to write a comprehensive summary of the major principles of physiocracy, a book with the cumbersome title *Table raisonnée des principes de l'economie politique.* It was quickly translated into both German and English and gave him even greater international fame. Two Americans who read it with interest were Thomas Jefferson and Benjamin Franklin.

So also did a Polish nobleman, Prince Czartoryski. While in Paris, the prince sought out du Pont and offered him a position as tutor to his son. This time du Pont had the good sense to get the terms of employment in writing. All living expenses for him and his family would be defrayed. His yearly salary would be 10,000 francs, and in addition, at the end of ten years' service, du Pont would be given a bonus of 100,000 francs. As evidence of good faith, one-third of the bonus would be given to him immediately. It was indeed a princely offer, and this time both Turgot and Mirabeau urged him to accept. Du Pont needed no prompting from them. With 30,000 francs in hand, he suddenly had greater wealth than he had ever imagined he could possess.

The practical Marie had an immediate suggestion. Since her husband was always proclaiming that the only true wealth lay in land, why not buy a farm before they set out for Warsaw? Then they would at least have something to return to if his position did not last for the stipulated ten-year period. They quickly found the place they wanted, an estate called Bois des Fossés (Forest of the Trenches) because of the many ditches, reputedly dug by the Romans as fortifications, that ran across its fields. The farm lay just a few miles south of Marie's childhood home in Nemours, near the small village of Chevannes. It was a pastoral idyll, and Pierre Samuel was delighted. "I have taken," he boasted to his Uncle de Montchanin, "a preservative which I have been needing against the vapors and follies of ambition. Up to now, I have been strolling the earth, a sort of hero of romance, more noble than wise."[7]

Even with *pié d-à-terre,* du Pont was still a wanderer of the earth. On

18 July 1774, du Pont and Marie and their two small sons set off for Warsaw. Unfortunately, the summer of 1774 was a particularly inappropriate
moment for du Pont to be leaving Paris. France had a new king, the ami-
able young Louis XVI, and Turgot was back in favor at Versailles. As Controller General of Finance, he was in a position to make the physiocratic
paradise a reality. Within two months, du Pont left his tutoring position in
Warsaw and hurried back to France to accept the position of Inspector
General of Commerce. His short sojourn in Poland had been well worth
the trouble, however, for now the du Ponts had an estate to which Marie
and the boys could go while Pierre Samuel set about creating a new order
for France.

For the preceding one hundred years, western Europe had experienced
a greater measure of stability than it had known since the golden days of
the Roman empire. The Wars of Reformation had finally ended, and the
modern nation-states of France, Great Britain, Spain, and the Austrian
empire, born out of the travail of international and civil wars, had at last
consolidated and centralized their respective states under the Bourbon,
Hanover, and Hapsburg families. Boundaries between them had been fixed
and were generally respected by each of the other powers. Although the
seventeenth century had seen a long series of colonial wars to determine
which of the great powers would possess the North American continent,
that issue seemed to be finally settled in 1763 with the ouster of France
from the two great river valleys of the St. Lawrence and the Mississippi.

Then, suddenly, the entire fabric of western Europe, which had been
so laboriously knit together over the past millennium, began to unravel. It
would take thirty years of revolution and war before another *pax Europa*
could be established. In all of these events, Pierre Samuel du Pont would
be a conspicuous figure, first as one of the perpetrators of change and later
as one of the victims of the new order he had sought to effect.

The unraveling began not in Europe, however, but among the thirteen
British colonies on the continent of North America. What at first appeared
to be only a minor internal squabble over the question of taxation and local
autonomy soon became a major conflict that attracted the attention of all
of Europe. From the moment of the first protest over the stamp tax in 1765,
du Pont took a keen interest in the American cause. Philosophically, he was
a pacifist, and the name of his second son, Irénée—peace—meant more to
him than an obliging gesture to humor a powerful friend. But in practical
affairs of state, he was notably bellicose toward Great Britain. The old
Gibraltar dream was still with him, and the only wars he ever urged his
country to undertake were those that would pit France against its ancient
foe across the channel.

Toward the Americans, however, his affection was as great as was his animosity toward the Britons. Like his fellow countryman, Michel de Crèvecoeur, du Pont believed that Americans represented a new and enlightened breed of the human species, and in their vast land, which was largely unpolluted by cities and industry, they would have the best chance of any people to create a true physiocratic paradise. So du Pont cheered them when they declared their independence, not only from British colonial rule, but from the stultifying mortmain of their European heritage. Du Pont was in the forefront of those close to the throne in urging French support for the American revolutionaries. When Benjamin Franklin arrived in France in late December 1776 to begin the delicate negotiations that would lead to Franco-American treaties of amity and commerce, one of the first individuals he sought out was the author of *Physiocracy*. From then until the treaties were actually signed in 1778, du Pont was a frequent visitor to Franklin's residence in the village of Passy near Versailles.

The alliance proved to be of immense significance to both the United States and France. For the United States it brought needed money, matériel, troops, and the services of the French navy. For France, it brought a renewal of national pride and euphoric dreams of regaining the lost territories of Canada and Louisiana. When, in 1781, Lord Cornwallis's army surrendered to the combined French and American forces under General Washington at Yorktown, while the British regimental band played "The World Turned Upside Down," the ecstatic du Pont, in Paris, was convinced that the old order had indeed been turned on its head. Eighteenth-century enlightenment had won out over seventeenth-century mercantile imperialism. Although the sainted Quesnay and the noble Turgot, like Moses before them, had not lived to see the fulfillment of their dreams, their disciples, led by du Pont in France and Jefferson and Franklin in America, would now march triumphantly into the Promised Land.

Du Pont wanted and fully expected to be one of the chief negotiators at the treaty table in Paris in 1783. How pleasant it would be to find himself seated beside his friend Franklin at the head of the table, with the despicable British, no longer quite so proud, sitting in the losers' seats at the foot, begging for terms. But it was not to be. The Americans demonstrated that perfidy was not a monopoly of Albion after all. Unexpectedly and secretly, in violation of their alliance with France, they signed a separate peace treaty with Great Britain without bothering to inform their French allies. Du Pont never blamed Franklin for this act of betrayal. Clearly the fault lay with the other American negotiators, and most particularly with that notorious Francophobe, John Adams.

When the account sheets for this joint Franco-American venture were submitted, it was not difficult to discern that the assets were all on the Amer-

ican side, the debits all on the French side. The United States had its recognized independence and all of the former British colonial territory west
to the Mississippi. Great Britain still held Canada, and Spain kept Louisiana,
while France, which had financed this undertaking, had nothing for its efforts but a ruinous national deficit resulting from money borrowed at a high
rate of interest. But perhaps, the ever-optimistic du Pont thought, it was all
for the best. Now France would have to face up to the necessity for reform.

But in what way, and how soon? So much depended upon whose hands
would knead and mold the plastic Louis XVI. During these critical years of
indecision, Pierre Samuel du Pont, who wanted so much to be the sculptor
of a new order, found himself on the outer fringes of power now that Turgot was gone. He still held his position as Inspector General of Commerce,
and he stood in good favor with the king's chief and most trusted minister,
the Comte de Vergennes, but the job was not very demanding, and his
influence not very great.

During these years du Pont spent a great deal of time in transit between
Paris, Versailles, and his farm at Chevannes. The faithful and capable Marie
Le Dèe continued to manage Bois des Fossés and to raise the children. His
two sons were growing up, but they bore little resemblance to each other
in either physical appearance or in temperament. Victor Marie, the elder,
had inherited all of the exuberant gaiety, charm, and intellectual curiosity
of his father and all of the physical beauty of his mother and paternal grandmother, Anne Montchanin. The younger son, Eleuthère Irénée, had the
misfortune, it seemed to his father, of drawing most of his family inheritance from his grandfather, Samuel du Pont. Marred by a deep-red birth-
mark on his left cheek, he was not a child of beauty, and being quiet, intro-
spective, and solemn by nature, he attracted little attention from others.
Even as a young boy, however, he demonstrated a maturity and a sense of
responsibility that distinguished him from most children. What Victor
could obtain so easily by charm, little Irénée would have to work for and
get through industry and perseverance. Many years later, when Irénée had
proved his worth, Pierre Samuel was fond of saying that Victor was the
child of his and Marie's passion, while Irénée was the product of their more
sober reason. But that judgment came later. While the boy was growing up
and reaching manhood, his father thought him only stolid and dull like his
grandfather Samuel.

In the spring of 1784, in recognition for his services to the nation in
the peace negotiations with Great Britain, and at the prompting of Vergennes, Louis XVI bestowed upon du Pont and his descendants forever,
the patent of nobility. He was now Sieur du Pont, with his own coat of
arms—a shield with a blue background upon which stood a single Ionic
column surmounted by a helmet, and below, the motto: *Rectitudine Sto*

(Upright I stand). This motto would take on more significance in the years that lay ahead.

Shortly after this high moment of pride and self-satisfaction, Pierre Samuel was crushed by the sudden and unexpected death of Marie, at the age of forty-one, in September 1784. She was buried in the graveyard of the small church at Chevannes. No cross was put on her tombstone, only the inscription: "She constantly made the happiness of her husband and her children. She was an honor and an example to her relatives and her friends, the consolation, the blessing, the perpetual succor of the poor of this parish. Her soul must be in heaven."

But Pierre Samuel's soul was still very much on earth, preoccupied with affairs of state. He had just been appointed Councillor of State by the new Controller General, Charles Calonne, at a considerable increase in salary. Before taking his sons back with him to Paris to arrange for their education, du Pont, who regarded his newly acquired patent of nobility very seriously, arranged for a curiously anachronistic feudal ceremony in their home at Bois des Fossés. In the drawing room, seated beside the portrait bust of his wife, he called Victor, now seventeen, and Irénée, thirteen, to appear before him. He then solemnly handed his younger son his first sword. "I no longer have a wife," he solemnly intoned, "you have no mother. But here we have her likeness, and perhaps from Heaven she looks down upon us and is glad to see whatever we do that is worthy and good. . . . You must understand. I have told you often, that no privilege exists that is not inseparably bound to a duty." After handing Irénée his sword, he then said to both his sons, "Promise each other to be always firmly united, to comfort each other in every sorrow, to help each other in all efforts, to stand by each other in all difficulty and danger." The two boys so promised, and Irénée, who never entered into anything lightly, was to take this pledge very seriously indeed. He would have ample opportunity in the years ahead, not only to recall but to fulfill faithfully this promise of succor to his elder brother.[8]

Back in Paris, du Pont plunged headlong into the complicated negotiations of arranging a new commercial treaty with Great Britain. Much as he disliked Britain, he loved France and his physiocratic principles even more. He worked hard to lower the tariff duties on English manufactured goods and to get Britain to lower its import duties on French agricultural products. In these efforts, he had the strong support of America's new minister to France, Thomas Jefferson, who replaced Franklin in 1785. A close friendship developed between the two men that would endure for the remainder of du Pont's life.

Du Pont achieved a measure of success with the signing of a new Franco-British commercial treaty in September 1786, but it proved to be a victory of dubious value to France and to du Pont. The farmers of France had wanted more than they got, while the French textile manufacturers were outraged by the admission of British goods, which could undersell their own products. Calonne and du Pont were accused of having sacrificed French industry to "the deadly conceptions of Physiocracy." Furthermore, lower tariff revenues would in no way help ease France's mounting financial difficulties. Calonne was forced to propose to the king some drastic measure that would facilitate reforms. Drawing upon ancient historic precedent, Louis had two legislative options open to him. He could call for a meeting of the Estates-General, to which each of the three estates (the clergy, the nobility, and the commoners) would elect representatives, or he could bring together an Assembly of Notables, which he would appoint, representing the three estates, to propose reforms. The king chose the latter option as being less threatening to his absolute power. Du Pont was one of the thirty notables so appointed, and at the Assembly's first session he was elected secretary. That the Assembly of Notables failed in its mission was not because of a lack of proposals from the office of its energetic secretary. Having received no mandate from either the king or the people, the Assembly was notable only for the intensity of its internal dissension. Its only accomplishment was to effect the dismissal of Calonne and to send du Pont back into voluntary exile at Bois des Fossés.

Pierre Samuel left the government content at least in knowing that during his moment of power he had been able to secure promising positions for both of his sons. Victor, with his talent for foreign languages and his charmingly ingratiating manner, was a natural for the diplomatic corps. He was to go to the United States as secretary to the French legation in New York, while Irénée, who had always been interested in the natural sciences, was to study under the tutelage of the great French chemist Lavoisier at the national powder works at Essonnes.

Pierre Samuel himself was confident that his exile in the country would be short. The rapidly deteriorating financial situation would force the king and his new advisers to institute some reform measures. Once again his counsel would be needed. Meanwhile he was watching with keen interest, as always, events in America. Du Pont received with delight the news from his son Victor that the American Congress had called a constitutional convention to meet in Philadelphia in 1787. This step promised that a new frame of government would be created with a strong executive and a national legislative body having real powers.

Du Pont began to beat the drums for a similar constituent assembly in France to give his nation a much-needed written constitution. In this new government, the king would no longer be an absolute monarch, who apparently could do absolutely nothing in the present Byzantine order of intrigue and chaotic bureaucracy. The king would instead be transformed—elevated

in du Pont's opinion—to the position of Platonic philosopher king, something that du Pont and Jefferson had frequently discussed. In his new role,
the king would guide the national assembly and enforce the legislation that
it enacted within the carefully limited powers designated by the written
constitution. This political reform would ensure that all the necessary economic reforms in taxation, expenditures, and trade regulations would follow. It was the old physiocratic dream given new vitality by the Americans
meeting in Philadelphia.

It was also given new life by events in France. There were food riots in
Paris, Dijon, and Grenoble. The Queen, Marie Antoinette, was now openly
called Madame Déficit in the streets of Paris, and the mention of her name
would bring jeers from the crowd, for they attributed much of the government's financial difficulties to her reported extravagances at Versailles.
Louis, without the support of the Assembly of Notables, announced several
administrative and judicial reforms. He even recalled the Swiss banker
Jacques Necker to serve once again as minister of finance, but it was, as
Necker himself said, "too late." Necker had nothing new to offer to alleviate France's woes. As he had done during the American Revolution,
Necker proposed that France borrow more money. But who would give
loans at this moment—no matter how high the interest rates might be?
There remained one other alternative. In the summer of 1788 Louis
announced that he would call a meeting of the Estates-General on 1 May
1789. In the meantime, each district should hold its election, and the people should submit to their elected representatives a list of grievances for the
Estates-General to consider.

Pierre Samuel du Pont was overwhelmingly elected by the people of
Nemours district to be their representative to the Third Estate. Now the
tide was coming in and the surf was rising. Du Pont had already joined
forces with some old friends like Comte de Mirabeau and some new allies
like Charles Talleyrand, Bishop of Autun, and the Abbé Sièyes to form a
National Party. The party's task would be to dissolve the old division of the
three estates and turn the Estates-General into a true constituent assembly,
which would write a new constitution for France.

Within a month after the opening of the Estates-General at Versailles
in May 1789, their first objective had been achieved. The question that
Abbé Sièyes had posed in his little pamphlet that du Pont had published
that spring, "What is the Third Estate?" now had been answered as Sièyes
had wanted. The Third Estate was everything, and it would determine the
future of France. Representatives of the other two estates, clerics like Talleyrand and Sièyes, and noblemen like Mirabeau and Lafayette, might join
it, but the Third Estate would be the only representative body of France.
This did not mean democracy in the old Greek sense, nor did it mean

republicanism in the American manner. The king would stay, but he and the legislature itself would be guided by three principles that du Pont had provided:

> Every man has the right to do that which is not harmful to others.
>
> Every man has the right to the assistance of other men.
>
> Every man must be protected by the other men, and by the entire body of society, against all attacks on his liberty, on his property, on his security.

These principles had comprised the *cahier* that du Pont had written for Nemours when the king had asked each district to send in a memorial of its grievances. They would now provide the guidelines, du Pont was confident, for a French constitution.

For the next two years, du Pont and his liberal Nationalist friends were to ride the crest of the surf as it rushed toward an as yet unidentified shore. In spite of the king's rather feeble efforts to row against the tide, the forward movement was irresistible, and Mirabeau, du Pont, and the other liberals seemed to be in control. On 14 July 1789, the Bastille prison, that hated symbol of the autocracy, was stormed and razed. Then, in the early autumn of 1789, two stanchions of the *ancien régime* were swept away: the nobility renounced all of its feudal rights on 4 August, and in September du Pont rose in the Assembly and delivered a speech that was widly cheered by the representatives. He proposed the confiscation of all church lands. In that same memorable autumn, du Pont had the satisfaction of signing the Declaration of the Rights of Man, which beautifully amplified the three basic tenets of the good society that he had called for in his *cahier*. Progress was also being made in the writing of the new constitution, although the Assembly at the same time also had to concern itself with the governance of France. It was indeed an exhilarating time to be riding the crest of the surf.

But, as any boatman could have told du Pont, at the inevitable moment when the surf breaks, those who have been gliding forward with ease and seemingly so much in command of their vessel must be prepared to maneuver with consummate skill or be swamped and destroyed. With the king as their captain, whose only plan of action seemed to be to abandon ship, the Nationalist crew had no chance to maneuver. Their physiocratic craft was swamped, and of all those aboard the ill-fated vessel, only Talleyrand and du Pont would later be able to join Abbé Sièyes in his boast, "I have survived." But for du Pont, survival was to mean crawling to safety upon an alien shore.

There had been warnings that the crash was inevitable. In October 1789, at the same time that the king and queen were forced by a mob to leave Versailles and became virtual prisoners in the Tuileries Palace in Paris, du Pont spoke out in the Constituent Assembly against the further issuance of paper money, in an effort to check rising inflation. As a result, he was attacked by an angry mob outside the hall and was nearly thrown into the Seine. Still, he and his friends could not believe that they were losing control of the situation. In 1790 du Pont was elected president of the Assembly, and in that position of command he had the satisfaction of pushing through the Constitution of 1791, which provided for the kind of rational, representative government for which the Physiocrats had been working during the preceding twenty-five years.

Under the terms of the constitution, which barred members of the Constituent Assembly from being elected to the new Legislative Assembly, du Pont was no longer eligible to hold office. He was still one of the king's closest advisers, but that position of honor paid no salary. Du Pont desperately needed money and a new occupation. He turned to his son's mentor, Lavoisier, for help. The chemist lent du Pont 710,000 livres, secured by a 4 percent mortgage on Bois des Fossés. With this capital, du Pont bought a townhouse in one of the finest residential sections of Paris, on the Ile Saint Louis, across from Notre Dame. Here he installed printing presses on the ground floor, and on the rue de Richelieu, where his father had once had his watchmaking shop, Pierre Samuel opened a bookstore to sell the products of his press.

Irénée, who was at loose ends now that Lavoisier had left the national gunpowder factory at Essonnes to become commissioner of the national treasury, hurried back to Paris to assist his father in the print shop. He also enlisted in the recently established National Guard and was quickly elected commander of his company.

Irénée was happy to be back in Paris, for at age twenty, he had fallen deeply in love with a sixteen-year-old girl, Sophie Madeleine Dalmas, daughter of a shopkeeper in Paris. Pierre Samuel, however, was not at all happy with his son's choice. The recognition and honors that had been bestowed upon du Pont during the past twenty-five years had greatly expanded his easily inflatable self-esteem. Now he found it unthinkable that a son of Sieur du Pont de Nemours, Chevalier, and adviser to the King of France, should descend into the *petite bourgeoisie* out of which Pierre Samuel, by his own intelligence and wit, had so laboriously pulled himself and his family. He informed Irénée that he would never accept such a marriage, and should his son persist in his suit, Pierre Samuel would disavow him as his son. To his surprise—and perhaps to Irénée's as well—this boy, who had always been obedient and tractable, persisted, even fighting two duels with another suitor for Sophie's hand. Ultimately, Irénée's will proved stronger than his father's. On 26 October 1791, he and Sophie Madeleine were married, with Pierre Samuel's reluctant blessing.

Although Irénée's choice of wife was prompted only by love and not by reason—an affair of the heart and not the head—it proved to be one of the wisest decisions that the usually prudent and careful Irénée was ever to make. It was quickly apparent, even to Pierre Samuel, that Sophie Madeleine would assume the role in the family that had been left vacant by Marie Le Dèe's death. Like the mother-in-law she never knew, Sophie would prove to be the ballast for a father-in-law and a brother-in-law given to romantic flights into preposterous ventures. She would be the second of the strong du Pont women to hold a large family together through adversity, and as had been written on Marie's tombstone. "She constantly made the happiness of her husband and her children."

Nor was it a mistake politically for Irénée to descend into the *petite bourgeoisie* for which his father held such scorn. The immediate future belonged not to the intellectual elite, nor to the aristocracy of family and wealth, but to the small shopkeepers, the Dalmases of Paris—and even more, to the property-less *canaille* who roamed the streets mad with hunger and crying for blood. When Pierre Samuel's little world of rationalism foundered and sank, as surely it must, then all of his honors and titles, which he wore with such pride, would serve as weights, pulling him down and under.

The crash finally came on the hot afternoon of 10 August 1792. Pierre Samuel and Irénée were in the print shop when a messenger arrived with the news that a mob was forming with the intent of storming the Tuileries to get the royal family. For months Pierre Samuel had been preparing for just such an emergency. He and Irénée had raised their own private little army of sixty men from among friends and neighbors on Ile Saint Louis to protect their property against possible attack. Within minutes the troops were assembled, and the small band set off for the Tuileries in defense of their king. Du Pont and company arrived just as the royal family was making its hasty exit to seek sanctuary in the National Assembly. The king moved as if in a trance, but he recognized his old counsellor as he passed. "Ah, Monsieur du Pont," he murmured, "one always finds you where one has need of you."[9]

With the departure of the royal family, the National Guard, whose sympathies lay with the mob, also left, but du Pont's little band made the rash decision to stay alongside the Swiss Guard and defend the palace as if the king were still there. Perhaps in that way, the mob could be fooled and the royal family could reach the Assembly hall in safety. So they stayed, and most of them, along with the Swiss Guard, were slaughtered. Soon only

Pierre Samuel, Irénée, and six of their companions were left. *Rectitudine sto*—but for how much longer? Nor was there any reason left to maintain the ruse of protecting the king, for word reached them that the royal family had arrived safely at the National Assembly and Louis had given orders that all resistance at the Tuileries cease. Pierre Samuel and those few companions that remained alive hastily put on the *bonnets rouges,* red liberty caps taken from fallen *canaille,* and slipped out to mingle with the mob and manage their escape.

Irénée was eager to return immediately to their house and print shop, for he had left Sophie there alone, and their first child was due to be born at any moment. No one, he reassured his father, would suspect him of being at the Tuileries defending the king, since he was a member of the National Guard and an outspoken Jacobin. But for Pierre Samuel matters were different. Several times during the course of the fighting, individuals in the mob had recognized him and had yelled obscenities at him by name. He must go into hiding at once.

So Pierre Samuel fled through the streets of Paris in his ridiculous red cap and succeeded in reaching the house of Philippe Harmand, a cousin of Marie Le Dèe. Harmand, who was a student of astronomy, took du Pont by back streets to the astronomical observatory of the College of Four Nations, where du Pont's old Masonic friend, the noted astronomer Lalande, was in charge. Lalande hid him in a small attic space under the dome. Hiding in a structure that stood as a testament to man's scientific progress through rationality, du Pont must have remembered those halcyon days when he had edited a journal called *Les Ephémérides du Citoyen.* What naiveté to have ever thought that through the observation of the physical laws of the universe one could establish that same perfect order on earth. The Age of Reason had brought only the Reign of Terror, and the *citoyens* whom du Pont had addressed with such polished rhetoric and unimpeachable logic were not the same as the rabble who called themselves *citoyens* in the streets of Paris in 1792. Quesnay, Tugort, Mirabeau, and Madame Pompadour were all dead. France now belonged to Danton, Marat, Robespierre, and Madame Guillotine, whose blade was sharper and more effective than Madame Pompadour's wit and charm had ever been.

And Pierre Samuel du Pont de Nemours was in hiding behind a telescope and subsisting on the bread and water that the good Harmand managed to smuggle into him. In this refuge Pierre Samuel received the word that he had become a grandfather. Sophie had given birth on 30 August 1792 to a daughter, who was named Victorine. Victorine—what a strangely inappropriate name for this moment in French history, when all reason seemed to have fled and France was besieged by the armies of Austria and Prussia in a war that du Pont had vehemently argued against. But Pierre

Samuel liked the name Victorine. It gave promise for the future, and he still believed in progress.

Three days after the birth of Victorine, on 2 September, the mobs in the city went insane again. The prisons were stormed and most of the people awaiting trial and the guillotine were dragged out and massacred on the spot. In the terrible confusion of that September massacre, Pierre Samuel managed to escape from the observatory and slip out of the unguarded gates of Paris. By foot, he arrived in the small village of Cormeilles, a few kilometers west of Rouen, near the coast, where Harmand had a country house that du Pont could use as a hideout.

For several weeks Pierre Samuel stayed there, knowing it was unsafe to go anywhere else, as the Commune in Paris had issued a warrant for the arrest of "the traitor du Pont de Nemours." But, even in this darkest moment of adversity, du Pont was irrepressible. He began work on a major opus, *Philosophie de l'Univers,* in which he spelled out his deistic faith. "If I am a watch-maker, I, with the little that I have of wit, the immense clock of the Universe has also its clock-maker." Who is God? "I don't know, . . . [but] it would be quibbling with words for me to refuse to give to Intelligence the name God. . . . God is the intelligent one, the powerful one, the reasoning one, the motive force. Without him, Matter would be chaos." But there was something even above God—Nature. "Nature is not a being but a fact. It realizes the assembling of the essential properties of Intelligence and Matter." God had given man Pain as a goad to induce him to work and to teach him to avoid error. But along with Pain God bestowed a comforter, Hope, "who holds out to man in present misfortune the happiness that is to come."[10] So wrote Pierre Samuel in the dark days of 1792–93.

Not even this project, however, could keep du Pont from becoming bored in his isolated hideaway, far from family, friends, and news of current happenings. After a few months he decided to take the chance of returning to Bois des Fossés, where Sophie and the baby had gone as soon as they were strong enough to travel. Irénée, his revolutionary credentials being impeccable, had remained in Paris to run the printing presses so much in demand by the new republican government of France, with its constant output of proclamations, decrees, and pamphlets.

For the next year and a half, Pierre Samuel lived in his own home in apparent security. Various explanations have been offered as to why he was not arrested. His biographer, Pierre Jolly, hints that he was still being protected by someone within the inner circle of power, but that mysterious someone has never been identified, and it is difficult to imagine who in power in 1793 would serve as du Pont's protector. Another explanation offered is that the agents who searched for him went to the district of

Nemours to find Bois des Fossés, while actually the estate lies just across the line in another district. But their inability to find Bois des Fossés in that rural area of France is also difficult to comprehend. A more likely conjecture is that du Pont was reported to have been killed in the September massacres and his name was struck from the list of wanted enemies of the state. Even a dedicated revolutionary bureaucracy is subject to error.

But Pierre Samuel became overly confident of his security, and he foolishly made a few trips to Paris to see Irénée and to catch up on the latest news. There was much news to relate in the days of the Terror. Following the convening of a National Convention in September 1792 to replace the Constitution of 1791, which du Pont had believed would provide France with a model physiocratic government for all time, France was declared a republic. The king—now Citizen Louis Capet—was put on trial, found guilty of treason, and guillotined in January 1793. The hated Marie Antoinette followed her husband to the scaffold, as did all of the nobility who had not been able to flee the country and join France's enemies across the border. Then it was the turn of the Girondists, the moderate revolutionaries, and the intellectual elite—including Lavoisier, who had begged for a four-day reprieve from his execution in order to finish an experiment in which he was then engaged. Now the radicals were turning on each other. Danton was executed, Marat was assassinated in his bathtub, and Maximilien Robespierre emerged as dictator of France. But Irénée du Pont remained loyal to the aims of the Revolution. In spite of the violent means employed, he told his father, France had been reformed and the people in the streets were better off now than they had ever been under the *ancien régime*. Pierre Samuel, however, could only hope that the patient would survive the fever and would live to see a better day.

But the fever still raged, and Pierre Samuel had been foolhardy in exposing himself to the contagion by visiting Paris. His existence and whereabouts were discovered, and in early July, a small company of soldiers appeared at Bois des Fossés with orders to take du Pont to La Force prison in Paris, there to await trial and probable execution.

Even in prison, where so many of his friends had gone before him, du Pont was not subdued. Within a day or two, he was holding classes on political economy for the other inmates, and he wrote to Irénée asking for a chess board so that he could teach his fellow prisoners that ancient game. He also offered his services as a physician to those who were ill.

The first few days passed quickly enough; then, on the afternoon of 27 July, as du Pont was strolling in the prison yard with his old friend Admiral Treville and arguing, against the admiral's objections, for the abolition of slavery within the French West Indies colonies, the discussion was suddenly interrupted by the wild clamor of the tocsin outside the prison walls. Con-

vinced that there was to be a repetition of the September massacre of the
prisoners, du Pont urged the admiral and a few other friends to form a
barricade with their beds and arm themselves with the fire-irons taken from
their rooms. They resolved that if they were to die, they would die fighting.
A last valiant stand proved unnecessary, however, for the turnkey rushed up
with the news that the dictator Robespierre had just been arrested.

Although he had not known it at the time, du Pont had been scheduled
for execution on the following day, the 10th of Thermidor. Instead, it was
Robespierre and his general staff who climbed the steps to meet Madame
Guillotine on that day. The Reign of Terror was over. But Pierre Samuel
remained in prison. Both Victor, who had returned from the United States,
and Irénée made every effort through petitions and personal contacts to get
their father released. In the end it was Pierre Samuel himself who effected
his own liberation. When Citizen Voulland, the man who had signed the
order for his arrest and whose consent was essential to free du Pont, came
to La Force on a visit, Pierre Samuel managed to slip him a little pamphlet
on a national education system for France, which he had written while in
prison. A few days later, du Pont was set free. He hurried back to Bois des
Fossés.

Pierre Samuel's family was expanding. Victor, upon returning from
America, had found himself a bride. But there was no stepping down into
the *petite bourgeoisie* for Victor. He married Gabrielle Josephine de la Fite
de Pelleport, daughter of the late Marquis de Pelleport. During the Terror,
she had found refuge with her sister, who was a nun, in a convent—not
the safest place, to be sure, but better than being on the streets of Paris or
at the ancestral family estate. Soon she and Victor would leave for Charleston, South Carolina, where he was to be the French consul.

Irénée and Sophie had a second child, a girl named Lucille, who lived
only a few days.

Sophie and little Victorine went back to Paris to join Irénée, and Pierre
Samuel, alone at Boise des Fossés, decided it was time to consider another
marriage for himself. Long a secret admirer of Madame—now the widow—
Lavoisier, he proposed marriage. Madame Lavoisier was not interested in
his proposal, but she was interested in the 710,000-livre loan that her late
husband had made to du Pont. She was destitute, she told Pierre Samuel,
and she wanted either the money or Bois des Fossés, but not him. As du
Pont was in the same financial condition as she, and as he had no intention
of losing Bois des Fossés, he had to find another solution quickly for his
unmarried and impoverished state. Fortunately, a rich widow, Madame
Marie Françoise Poivre, whose late husband had been a friend of Pierre
Samuel, and who had her own pretensions to intellectualism, regarded the
philosopher of physiocracy as a genius, and warmly accepted the unex-

pected attention he suddenly bestowed upon her. They were married in the little town hall at Chevannes on 26 September 1795. For the moment at least, Pierre Samuel's financial difficulties were resolved.

But not his boredom. Du Pont, having so miraculously escaped the Terror, should have been content to settle down at Bois des Fossés with his adoring wife and lead the comfortable life of a country squire. Politics and Paris proved irresistible, however. France had still another constitution and its third government since 1789, and he was convinced that his country needed his services and advice. He had no difficulty in being elected to the upper legislative chamber, the Council of Ancients, of the new Directory government. With his official status secured, he hurried back to Paris to take his seat in the legislature and to begin publication of another political journal, *L'Historien*. Neither France nor du Pont, however, had learned anything about tolerance or discretion from their recent experiences. In a matter of months, du Pont was as disenchanted with the Directory as he had been with the Jacobin leadership of the National Convention, and the pages of the *L'Historien* contained anything but objective historical analysis. They were filled with angry diatribes against the corruption and ineffectiveness of the Directory. The government, in turn, proved to be no more tolerant of criticism, no more respectful of the freedom of the press, than had been Danton and Robespierre. *L'Historien* was suppressed on 4 September 1797, and the next day Pierre Samuel was once again an inmate of La Force prison. This time Irénée did not have the political immunity that he had enjoyed during the Terror. His printing office was closed by government order, and he joined his father in prison. After only one night, however, father and son were informed that they would be leaving La Force. Irénée was cleared of all charges, but Pierre Samuel was not to be let off so lightly. As a chronic troublemaker, he was to be deported to the newly established penal colony off the coast of French Guiana, which had already earned for itself the dread name "Devil's Island." It would have been more humane to be sentenced to execution, for the guillotine would have provided a speedier and less painful death. His rescue was effected by his old friend and frequent correspondent, Madame Germaine de Staël, the daughter of the Swiss financier, Jacques Necker. As one of Paris's most illustrious hostesses, whose famous salon attracted the most powerful people in Paris, including General Bonaparte and his wife Josephine, Madame de Staël had influence where it counted. She persuaded the Directors to pardon Pierre Samuel du Pont de Nemours on the grounds that as an old man of eighty years he could do no great harm to the Republic, and his deportation would mean his certain death. Du Pont, cleared of all charges, was told to return home. Pierre Samuel was grateful for Madame de Staël's intervention, but he never completely forgave her for the argument she had used in effecting his release. He was, he coldly informed her, not quite sixty, and he certainly wasn't ready to retire quietly and die in peace.

Indeed, he was not. He already had in mind an idea for what would be his grandest venture yet. The second trip to La Force prison had finally convinced him that France was politically hopeless. Liberty could be found only in the United States of America. Why not establish a colony there—he would call it Pontiana—which would be the Paradise of Physiocracy that he had never been able to create in France? There the du Pont family and other French refugees—indeed, lovers of liberty from every nation—would put into practice those principles of political economy that he had been advocating for the past three decades: every individual a free person, every family a landowner and tiller of the soil, and all of them collectively guaranteeing their mutual security and well-being.

To finance this undertaking, he and his two sons would form a company, Du Pont de Nemours Père, Fils & Cie., which would be capitalized at four million francs by shares to be sold at 10,000 francs per share. He quickly wrote up a prospectus, which promised: "No danger, no risk of loss of capital; the certainty of possessing for each share is a considerable property in a country where liberty, safety, independence really exist in a temperate climate, where the land is fertile and bountiful, in Virginia, and the western counties; the certainty of drawing interest—first at four per cent, then at six, then at eight; and finally after twelve years, the certainty of a capital that has been quadrupled, probably increased tenfold, possibly even twice tenfold."[1] As usual, Pierre Samuel was lifted by his own rhetoric and optimism, soaring far beyond reality into never-never land.

He showed the prospectus to his old friend Talleyrand, who, having spent much of the period of the Terror as an exile in America, knew all about speculating in American lands. Talleyrand liked the prospectus. He couldn't have made a better sales pitch himself. He also liked, now that he was Foreign Minister of the Directory, the idea of having du Pont in the United States as his own private political agent. Victor du Pont, too, as Consul General, would be a great help both to his father and to him. So Talleyrand gave the project his official blessing. He even arranged to have Pierre Samuel go to the United States in an official capacity, as traveling scholar for the National Institute of Sciences and Arts, ostensibly to make a survey of forestation in the United States. With Talleyrand's backing and with his prospectus in hand, du Pont enthusiastically peddled his stock, even going as far as Holland to find prospective investors.

But his plans began to go awry. First, the great rush to grab up the shares that du Pont had anticipated never happened. There were only three large investors: Lafayette, Necker, and the playwright Beaumarchais, who, as the largest investor, had the ill grace to die before he put in his money.

Of the four million francs needed, du Pont was able to get only 450,000 in hard cash.

A more serious threat to his plans arose in the worsening relations between France and the United States. France had placed an embargo on American shipping and had seized all American ships found in French ports, imprisoning their crews. French privateers were harassing American merchant vessels on the high seas, and in retaliation, President John Adams, the old Francophobe who had been du Pont's *bête noir* at the time of the peace negotiations in 1783, had refused to accept Victor's credentials as Consul General, declaring him *persona non grata*.

Victor du Pont returned to France on 24 June 1798. He was greatly disturbed about the imminence of war between France and the United States. He was certain the conflict would be a disaster for both countries, inasmuch as it would drive the United States into the arms of France's archenemy, Great Britain; he was so concerned about the situation that he immediately sent off a long report on the situation to Talleyrand. He admitted that the present American government under John Adams and the Federalists had shown itself "to be the most irreconcilable enemy of the [French] Republic," but he pointed out that current French policy had given the Federalists the public support they needed for a war with France.

> You ask me for facts, Citizen Minister: they present themselves en masse to my memory and I am troubled only in the matter of choice. One could make a volume of the collection of acts of violence, of brigandage, of piracy committed by French cruisers . . . directed principally against American commerce. . . . The war which is brewing between the two Republics will be disastrous for both. For the United States it will be the loss of their liberty and their independence; obliged to unite with England, they will no longer have the power to separate themselves from her.

For France it would mean losing the one ally she had and probably the loss of the colonies she still had in the Americas, as well.[12]

Talleyrand was impressed with Victor's account of the situation and was moved to action by the logic of Victor's prognostications. French policy toward America changed abruptly. France indicated it would receive any diplomatic envoy the United States might care to send. In the meantime, French harassment of American shipping on the high seas stopped. American vessels being held in French ports were released and American sailors were let out of prison. The war fever in both countries abated. Later historians, including Samuel Eliot Morison, would attribute the avoidance of a Franco-American war in 1798 largely to the efforts of two men: Victor du Pont, for his report and recommendations to Talleyrand, and John Adams, for his courage in defying the wishes of his own party by accepting

France's invitation to send a diplomatic representative to France.

Victor also had some good advice to offer his father. Pontiana, at least for the present moment, was an impossible dream. Du Pont *père* had no idea how inflated land values were in America. The capital they had on hand, 450,000 francs, would hardly buy enough land for the du Pont family. As for a colony of several hundred families in western Virginia or in Kentucky along the Ohio River—forget it.

Accepting that Pontiana would have to be postponed for a while, Pierre Samuel refused to forget it. Nor would he abandon his plans to emigrate to America. Pierre Samuel called a family council. His two sons and their families attended, as did the daughter of his wife by her first marriage and the daughter's husband, Jean Bureaux de Pusy. Pierre Samuel admitted that he had not raised the amount of capital he had anticipated, but he then ticked off the advantages they did have. With the support he had from Talleyrand, with his close friendship with Thomas Jefferson, whom Victor predicted would be the next President of the United States, with Victor's intimate knowledge of the United States and his valuable contacts there, and with Bureaux de Pusy's close associations with men of wealth in France, especially with the Marquis de Lafayette, they would make out all right. And what about Irénée? What could he contribute? Well, Irénée was hardworking and would always find something to do.

The old man had not lost his power of persuasion. He could still convert his auditors, even the practical and level-headed Sophie, into eager Sancho Panzas, ready to storm the coasts of America and tilt with the giants of that gigantic land. The plans were quickly drawn up. Madame Françoise du Pont and her son-in-law, Bureaux de Pusy, would depart on the first available transportation. Taking some of the company's capital with them, they would find an estate near, but preferably not in, New York City. The rest of the family—Pierre Samuel, the Victor du Ponts, the Irénée du Ponts, and Madame Bureaux de Pusy, who was expecting a child—would remain behind until after the baby was born and until Pierre and Victor could get the company's finances in order. They would then follow with what household furnishings they could bring and join the first two arrivals in the communal home, which by that time would have been purchased.[13]

So it was the two least willing conscripts to Pierre Samuel's venture, Madame Françoise du Pont and Bureaux de Pusy, who departed first. They were no sooner at sea than their ship was captured by a British frigate. They spent several weeks in England before they were able to continue their voyage. This inconvenience was trivial, however, in comparison with what the rest of the family would experience. Upon their arrival in New York, Madame du Pont and her son-in-law lost no time in securing a large, spacious home near Bergen Point, New Jersey, which they named Bon Sejour—Good Stay.

It took the remainder of the du Pont family considerably longer to settle their affairs in France than had been anticipated. The printing press in Paris had to be sold and a renter had to be found for Bois des Fossés. It was also no easy matter, with most of Europe still at war with France, to find a neutral ship able to accommodate the large family with its baggage and furniture.

The de Pusy baby was three and a half months old before the family was ready to depart. Pierre Samuel received word that one of the American ships that had been held in La Rochelle for the past two years had been recently issued its release papers. It would shortly be ready to depart for New York, and the captain would be honored to have the du Pont family as passengers. The name of the ship was the *American Eagle*. There was no opportunity to inspect the ship before signing on, but the du Ponts considered themselves fortunate to have found any accommodations. And Pierre Samuel found the ship's name charming. How appropriate it would be to soar to the Promised Land on the wings of an American eagle!

Having made the long overland trek from Chevannes to La Rochelle by carriage and wagon, the du Pont party assembled at the dock ready to board the *American Eagle* on 2 October 1799. It was quite a gathering. There was Pierre Samuel du Pont, his eyes feverishly bright with excitement; Victor and Gabrielle du Pont and their two children, Amelia Elizabeth, age three, and Charles Irénée, two; Eleuthère Irénée and Sophie with their three children, Victorine, now seven, Evelina Gabrielle, three, and Alfred Victor, one year old; Madame Île-de-France Bureaux de Pusy, with her three-month-old baby boy; and three nurses for the children. At the last moment, they also added Sophie's brother, Charles Joseph Dalmas, to the party. He had been conscripted for the French army but did not want to serve. So a substitute had been purchased for him, and he was included in the party. In addition, Pierre Samuel, still dreaming of Pontiana, had gone to speak to a group of prisoners who were scheduled to be deported to Devil's Island. He told them he had the authority to have released to his custody any able-bodied young man who would accompany the du Ponts to America. Two of the prisoners accepted the offer. There may have been moments on the high seas, however, when these two recruits regretted that they had not taken their chances with Devil's Island.

So here at the dockside they all gathered—twelve adults and six young children, along with a mountain of baggage and household goods, including Gabrielle's beautiful rosewood piano, which she couldn't bear to leave behind, as well as all of Pierre Samuel's books and voluminous notes for the biography of Turgot that he was planning to write. The ever-practical Sophie carried in her hamper a huge paté that she had made at the last moment and packed in straw. It would stand them in good stead.

One look at the *American Eagle* was enough to convince them that *American Turkey Buzzard* would have been a more appropriate name. Having lain in port for two years, the ship's hull was encrusted with barnacles and many of its sails appeared to be tattered and unmended. The captain had been so eager to depart when the release was given that he had not wasted time in getting the ship seaworthy. But he was delighted to have this large group of paying passengers, and he gave the du Pont family a hearty welcome aboard. There was no retreat now. For better or for worse, their lives were committed into his custody.

It proved to be for worse—far worse than any of them could have imagined. The captain and his officers were clearly incompetent, unable to maintain discipline among their crew, and their navigational skills and equipment, as soon became evident, were woefully inadequate. The crew, when it wasn't threatening mutiny, was busy trying to steal the du Pont's personal possessions. Pierre Samuel and his sons and recruits had to maintain a twenty-four-hour guard on their property. The other passengers, many of them Americans, resented the du Ponts, with their clannish, snobbish ways. Worst of all, the captain had not taken on adequate provisions for so large a company. Sophie's paté was soon gone, and the barrels of flour and salt pork were emptied. Stories would be told for generations among the du Ponts of how Pierre Samuel concocted delicious soups by scraping lard from the staves of the empty barrels and how Victor learned to catch and to prepare in a variety of tempting ways the numerous rats that shared their quarters. Three times they were saved from starvation only because the *American Eagle* was stopped on the high seas by British frigates, who quickly saw that this was no prize to capture but rather a ship in distress, which for the sake of its passengers must be given desperately needed provisions.

Nothing soared aboard the *American Eagle* except Pierre Samuel's spirits. He never gave up hope or his dream of the future. He was Voltaire's Dr. Pangloss and Charles Dickens's Mr. Micawber rolled into one. His ebullient optimism must have been more than a little trying to the other adults, but his grandchildren worshiped him. With his stories and songs and impromptu little theatre productions he could make them forget for hours the gnawing in their stomachs and their weariness with the ship's creaking and rocking.

The ship limped across the Atlantic, week after dreadful week. It took Columbus two months to cross the Atlantic on his first voyage to America. The *American Eagle* took three weeks longer than it had taken the *Santa*

Maria to make the same voyage three hundred years earlier. Somehow they all survived, even the baby, who was now twice as old as he had been at departure. The ship was scheduled to arrive in New York harbor, but their first sight of land, on 1 January 1800, was not Staten Island but Block Island off the coast of Rhode Island, some 150 miles off course. The ship made a brief stop at the dock on Block Island to allow the desperately weary passengers the opportunity to step on land after three months at sea. The du Pont family walked up the hill behind the port, and the first house they came to, with its door slightly ajar, its oil lamps glowing, and a fire blazing on the hearth, seemed to be offering them a cheery welcome. No one came to the door in response to their repeated knocks. Peering through the doorway, they could see a table set for dinner with bread and cheese, and there was a pot of soup bubbling over the fire. The temptation was too great. The famished du Ponts trooped in and in a matter of moments had consumed the prepared meal. As compensation, they left behind one gold louis coin, worth approximately five dollars, for their absent and unsuspecting hosts—generous payment for services rendered.

The du Ponts reluctantly returned to the *American Eagle* for the short leg of the voyage that remained. On 3 January 1800, they disembarked in Newport, Rhode Island. The trans-Atlantic ordeal was at last over. They were all safe on American soil, just as Pierre Samuel had promised they would be. It was the new year of a new century in a new land—the right time for a new beginning. It had been sixty-five years since Samuel du Pont had come to Paris from Rouen to open his watchmaking shop on the banks of the Seine. Now the Seine lay three thousand unbelievable miles behind his descendants. The du Ponts' future lay with Eleuthère Irénée du Pont, although none of them—least of all he—realized that. It lay along the banks of another river that none of them as yet had heard of—a little river called the Brandywine—an appropriate name for a settlement of French émigrés.

PART II   The River Brandywine and Eleuthère Irénée
du Pont, Founder of the Company

The du Pont family's first task, once they had recovered their land legs in
Newport after their three-month ordeal at sea, was to arrange for transportation to take them and their possessions to New York harbor and on to
their new home in New Jersey, where Madame du Pont and Bureaux de
Pusy eagerly awaited them. Arrangements were quickly made, and within
the month they were all finally reunited at Bon Sejour. What a welcoming
name that was. Equally appropriate, Pierre Samuel thought, was the
address—No. 91 Liberty Street—for the office they rented in New York as
city headquarters for the transplanted Du Pont de Nemours Père, Fils &
Cie., now renamed Du Pont de Nemours Father & Sons & Company of
New York.

With a lovely home in New Jersey, an office in Manhattan, and a letter
of welcome to "the ablest man of France" from his old friend, Thomas
Jefferson, Pierre Samuel's American life seemed complete. But he needed
something to do in order to start paying the handsome dividends that he
had promised to his French investors. Victor had been right—Jefferson confirmed that buying land at this time would be a foolish and losing proposition. There would be no Pontiana just yet, but there were many other
possibilities. "I have very good cards to play in this country," Pierre Samuel
wrote to Madame de Staël. "I have already seen enough here to judge that
the métier of the spider is worth more than that of the swallow. Here I am
spinning, and will spin my web with care, with precaution, and with all that
God has given me of intelligence."[1]

God had given Pierre Samuel a great deal of intelligence but very little
precaution. Neither he nor his elder son Victor could ever be a patient
spider, waiting for prey to come to his web. They were forever swallows,
flitting and skimming, catching some objects on the fly but mostly chasing
the unattainable. Pierre Samuel and Victor made up a list of enterprises
their new American company could undertake. They would enter the
import–export business; they would establish their own shipping lines for
freight and a regular packet-boat service for passengers and mail between
the United States and France. (God knew, and certainly the du Ponts knew
after their *American Eagle* experience, that there was need for reliable passenger service.) They would enter the brokerage business, and they would
invest other people's money in manufacturing and real estate opportunities.
More immediately, they would seek a concession from the French government to supply the French army in Santo Domingo, which was then
attempting to suppress a servile insurrection on that island colony. Seven—
at least seven—distinct and golden opportunities.

Pierre Samuel dashed off a new prospectus for the company. It was even more boastful and extravagant in its promises than the first one that he had written in France. His, Victor's, and Bureaux de Pusy's qualifications as economist, diplomat, and capitalist were extolled in rapturous terms, and one final sentence was added as an afterthought: "Eleuthère Irénée du Pont has also had much experience of business methods in France, in agriculture, manufactures, and the useful arts."

They had Thomas Jefferson's blessing, just as they had had Talleyrand's in France, but American politics in the election year of 1800 were a bit uncertain. To be on the safe side, Pierre Samuel engaged the former Secretary of the Treasury and the actual head of the Fedralist Party, Alexander Hamilton, as a lawyer for the company. The old lessons learned in France of political high-wire balancing had not been forgotten. Having thus secured support poles at both ends of the American political spectrum, the du Ponts, father and sons (but mostly first son, Victor), were ready for their entrepreneurial acrobatics.

It was then that second son, Irénée, somewhat diffidently approached his father with a proposal for Project No. 8. In the early autumn of 1800, while Pierre Samuel and Victor had been busy drawing up plans and attempting to raise more capital for their shipping line and other investments, Irénée, finding himself not essential to these activities, had accepted an invitation to go hunting with Victor's friend Colonel Louis Toussard, who had come to America from France to fight in the American Revolution and had settled on a farm near Wilmington, Delaware, at the conclusion of the hostilities. During the course of Toussard's and young du Pont's hunting expedition, they had run out of ammunition and had stopped in a country store to purchase gunpowder. Irénée had been shocked by the high price they had had to pay for the powder and later was even more dismayed by its poor quality. Toussard, however, had not been surprised. All American gunpowder was like this, he told Irénée. If one expected a gun to fire properly, one bought imported Englishpowder, but unfortunately here in the backwoods, it was not available.

Irénée had then asked to be taken on an inspection tour of an American powder plant, and Toussard had shown him the Lane-Decatur factory in Frankford, Pennsylvania. Irénée, after his years at Essonnes with Lavoisier, had no difficulty in spotting the problems in the American manufacturing process. The saltpeter, although of good quality, was poorly refined. The graining was not done correctly, reducing much of the powder to useless dust. The operation was hopelessly inefficient.

So Irénée had returned home with Project No. 8 to add to his father's and Victor's list. Pierre Samuel did not at first greet Irénée's proposal with much enthusiasm. The manufacture of gunpowder was, after all, a plebeian undertaking, lacking the scope and grandeur of colony building in Kentucky or brokerage houses and shipping lines in Manhattan. And it would be a strange trade for someone named Irénée, the son of a philosophical pacifist, to enter. But the more that Pierre Samuel thought about it, the more possibilities he saw. After all, there were other uses for gunpowder than the slaughter of fellow human beings on the battlefield or the high seas. This was especially true in America, where gunpowder was essential in taming the wilderness, clearing the land of wild animals, blasting out tree stumps, and cutting roads in the hills. Furthermore, Thomas Jefferson liked the idea. On further reflection, Pierre Samuel was even willing to sell Irénée's gunpowder to the United States military forces. After all, liberty must always be defended and philosophical pacificism could be supported only so far, especially if the potential enemy was Great Britain. Du Pont advised his son to find a location for his plant near the new Federal City of Washington, D.C., perhaps somewhere in Maryland or northern Virginia.

It was decided that Victor and Irénée should return to France as soon as possible to seek further financial backing for their many projects. Irénée could also acquaint himself with the most recent developments in the manufacture of gunpowder at Essonnes, and if at all possible, obtain from the French government the latest machinery for the refining and grinding processes. Pierre Samuel wrote a long letter to one of his original backers, the Swiss banker Jacques Bidermann, who lived in Paris. He went into great detail on each of the first seven projects, concluding with the statement, "My second son will explain to you the eighth plan that we have in view, and what we believe we can accomplish here by the manufacture of gunpowder—for which his skill in this art, the ignorance of it in America, the needs of government, those of the country, and even of the West Indies, give us not hope but a positive certainty of great profits."[2]

Early in February 1801, the two du Pont brothers arrived in France. Irénée was eager to accomplish his assigned tasks as quickly as possible and to get back to what he now regarded as home with Sophie and the children. In his usual methodical way, he had carefully computed the minimum capital he would need to get his gunpowder mill in operation—$36,000. Of this necessary funding, Pierre Samuel had promised to invest $22,000 from the parent Du Pont Company of New York. Irénée needed $14,000 more. With Papa's letter in hand, he was able to get from Bidermann and two French associates, Adrien Duquesnoy and Madame de Staël's uncle, Louis Necker, an additional $6,000. This still left $8,000 to raise, but Pierre Samuel was confident that there would be American subscribers for the remaining amount.

Having secured what he regarded as the necessary capital, Irénée registered in Paris the incorporation papers for E. I. du Pont de Nemours &

Cie., capitalized at $36,000, representing eighteen shares valued at $2,000 per share. Eleven shares were owned by the Du Pont Company of New York, one share each was allotted to Bidermann, Duquesnoy, and Necker, and the remaining four shares were to be sold to as yet undetermined American investors. Irénée, as director of the company, and the Du Pont Company of New York, as the major investor, would have full authority to determine when there should be a distribution of dividends, and all dividends paid would be allotted in thirty portions—eighteen parts would be distributed to the shareholders, nine parts would go to the director, and the remaining three parts would be reserved for expenses incurred in obtaining government contracts—a polite reference to the palm-soaping that every Frenchman knew was necessary in dealing with public officials.

Now that Irénée's project had legal existence on paper, the next and more difficult step was to give it physical reality. This meant getting the machinery, recruiting skilled workers, and finding the proper location in America for construction of his plant. One reason why Irénée had decided to incorporate his company in France was that he thought it might help him obtain the latest technical knowledge and even duplicates of the actual machinery used by the national powder works at Essonnes. He was surprised and delighted by the reception he was given by French officials both in the ordnance department of the government and at Essonnes. They were fully aware of the virtual monopoly that Great Britain had on the sale of quality gunpowder in America. It would be to France's distinct advantage to have the United States less dependent upon Britain. Now that Thomas Jefferson was President and the Federalists were out of power, the United States might once again be an ally of France, but only if the American bondage to the British munitions industry could be broken. So the red carpet was rolled out for Irénée, both in Paris and at Essonnes. The most advanced technological secrets of the trade were revealed to him in full detail. Processing machines were sold to him at cost. He was even allowed to recruit skilled French powdermen to be employed in his mill. This was a most unusual concession, for industrial nations in the late eighteenth and early nineteenth centuries guarded their highly trained personnel more zealously than they protected their royal treasuries. Irénée got all that he wanted and could afford to pay for.

His mission having been accomplished in less than five months, Irénée was eager to return to America, but Victor was not in as great a hurry. So far he had not been nearly as successful in achieving his objectives as was his brother. True, he had been fêted by Talleyrand, Madame de Staël, and Madame Lavoisier (who was now the wealthy Countess of Rumford but was still waiting for her first husband's loan to be repaid). He even had the opportunity to meet First Consul Napoleon Bonaparte, who since the coup d'état of November 1799 had become the virtual dictator of France. These were valuable contacts to establish, but raising additional capital for his and Papa's seven other projects took a great deal of time and effort and could not be rushed. Moreover, Victor, as always, thoroughly enjoyed mixing in high society. He was even planning a trip to Madrid, partly for business but also for pleasure.[3] His brother would have to return to America alone.

Irénée accepted this without question or argument. Swallows must always gracefully flit while spiders must patiently spin. Irénée returned to America in July 1801 and immediately started to look for a place to build his web. He knew just the kind of place he wanted. It must be close to a rapidly moving stream as a source of power. It could not be in a heavily populated area because of the danger of explosions, yet it could not be too isolated because of the need for mechanics and easily accessible transportation facilities. Papa du Pont had told his son to locate near Washington, but Irénée could find no site in that area that met all of these specifications. He found the ideal location in the very area in which, while hunting with Colonel Toussard, he first had the idea of building a gunpowder plant. After searching much of eastern Maryland and northern Virginia, he found his ideal location on the Brandywine, five miles north of Wilmington, near where Pennsylvania and Delaware share their curious semicircular boundary line. Here he would build his powder mill.

The Brandywine is one of America's shortest rivers. In fact, Delawareans have never considered it a river at all, just a creek, for it is not navigable, except by canoe, for more than two miles above its mouth. One of Wilmington's most illustrious native sons, the distinguished editor and author Henry Seidel Canby, has written lovingly of the river of his boyhood in *The Rivers of America* series, but even he has had to admit that the Brandywine "is no Thames or Hudson. It is a little river, only about sixty miles long from its double source to its mouth, with no world capital on its banks. It is not a famous boundary river, as is the Potomac or the Rio Grande; it is not a traffic river, as is the Mississippi; it was never a lumber river like the Kennebec or the Susquehanna; it has no Grand Canyon; it does not penetrate and open a continent like the great St. Lawrence. . . . Yet it does enter the Delaware at the exact point where the northern regions of industry and small-farm agriculture end and the plantation region of the South begins. It has a distinctive beauty, and an unusual grasp upon the imagination of those who know it well, and it has made history—military, social and economic—for three centuries. Thus it offers an excellent opportunity to study those subtle ways in which a river can influence events and men."[4]

The Brandywine has indeed made history—as a Revolutionary battlefield of note in the eighteenth century, as the home of the Brandywine school of artists in the twentieth century, and especially as a point of origin, along with Pawtucket, Lowell, and Troy, for America's industrial revolution. On this short stretch of river just above Wilmington there were established in the eighteenth century the first modern paper mill, the finest flour mills, and the most advanced cotton textile manufactory in America. And here at the beginning of the nineteenth century Irénée du Pont came to build America's gunpowder industry.

Although the Brandywine is short and lacks deep, navigable waters, it has other assets of inestimable value. Originating within the Welsh Hills of southeastern Pennsylvania, it flows quietly across the fertile Great Valley that lies between the Schuylkill and Susquehanna rivers, a pleasant gentle stream that complements the rich pastoral scenery of the many Quaker farms that border it. Then, just as it crosses the sate line into Delaware— and it is as if nature had intended a boundary line to be placed there—it strikes a range of igneous granite, called gabbro. It has dug a channel through this hard rock, and in less than four miles it descends 120 feet, in a mad rush to the Delaware River and on out to the sea. No industrial planner could have designed a better terrain for eighteenth-century industry, for the river has its fall line immediately above its broad, sluggish estuary. This felicitous union of fall-line power and tidewater transporation within a stretch of only four miles was highly attractive to the proto-industrialists of the eighteenth century.

In an age when the full potential of steam power had not yet been realized and when the even greater force of electricity was still the monopoly of nature's lightning bolts, the only natural sources of power, other than the strength of human and animal muscle, were wind, water, and heat. But in most places on the earth's surface wind is an uncertain, capricious force, and heat had only recently been successfully exploited by the Montgolfier brothers to lift their brightly colored balloons over the fields of rural France—a beautiful sight to behold but not considered to be of any great service to mankind. Water, however, was usually a constant force, and it was cheap and easy to capture. It had been the force of water rushing to find its lowest level that had for centuries turned the wheels of industry. It had ground grain, sawed wood, and in more recent times spun thread and woven cloth. For eight miles in Delaware water power was at its best, easily channeled into millraces and always ready to turn lathes, spin wheels, and revolve grindstones. Just below the so-called Great Falls, where the Brandywine met the sluggish Christina River to flow into the broad Delaware, lay the port of Wilmington, with its cargo vessels ready to take the barrels of flour, the bolts of cloth, and the rolls of paper to the markets of the world. Irénée had chosen his spot wisely.

Another attraction of the area for Irénée, in addition to the natural advantages of the Brandywine, was the fact that Wilmington had become a center of residence for French immigrants—those who had fled the Revolution in the 1790s and more recently those who had fled the even greater terror of the slave uprising in Santo Domingo. Among the latter emigrés was Pierre (now Anglicized to Peter) Bauduy, formerly a wealthy plantation owner in Haiti. A friend of Colonel Toussard, Bauduy had early on taken a keen interest in the du Ponts' Project No. 8. He had encouraged Irénée to build his gunpowder works in Delaware and had helped him find the land to purchase on the Brandywine. Even more helpful was his offer to buy two of the four shares of stock still outstanding. He also found a merchant in Philadelphia, Archibald McCall, who would buy the other two shares. So at last Irénée had his $36,000 capital in hand. He must now get the land.

Two possibilities were open—one parcel, lying almost within the city limits of Wilmington, was owned by a man named Harvey, the other, a ninety-six-acre tract farther up the Brandywine, lay some four miles outside of the city. Irénée much preferred the latter option for its isolated location. This land belonged to an old Quaker landowner named Jacob Broom. He had once tried his hand at industry himself, having built a cotton mill on the Brandywine in 1795. When the mill was destroyed by fire a few years later, Broom had given up on manufacturing, but he held his land, patiently waiting for societal demands to increase its value, since it was not of much use for agriculture. The Pennsylvania Quakers were gentle pacifists with respect to the battlefield, but they could be tough adversaries in the marketplace. Broom knew he had the land that du Pont wanted and needed, and he adjusted the prevailing price on land accordingly. He wanted $6,740 for the ninety-six acres on the west bank of the Brandywine, and after much haggling, he got it.

Then Irénée discovered that, as an alien, under Delaware law he was forbidden to own land. The company's lawyer, Alexander Hamilton, attempted to use his considerable political influence in Delaware to get the legislature to pass a special law exempting the du Ponts from this restriction, but to no avail. The prevailing prejudice against the French, which ironically Hamilton and his Federalist party had done so much to create, was too strong. It was necessary to have William Hamon, a naturalized Frenchman, hold the land in his name, with the understanding that it was the *de facto* property of Eleuthère Irénée du Pont.

Another problem quickly arose. Having done so well in his first land transaction with the du Ponts, the crafty old Broom built a dam across the Brandywine at a spot on his remaining land lying upstream from the du Pont purchase. He then offered this tract to Irénée at an even more inflated price. If du Pont refused to purchase it, then he and all the mill owners farther downstream would be at the mercy of Broom for an adequate flow of water. This time Alexander Hamilton did come to the rescue. He pointed out that the law gave a landholder riparian rights only to the center of any stream that bordered his land. For $100 Irénée bought the water rights that belonged to the landholder directly across the river from Broom and then destroyed the half of Broom's dam that extended into his water rights. There was no further trouble from Friend Broom. Irénée was discovering that in America it paid to have a good lawyer, even one whose politics did not agree with Irénée's.

In the summer of 1802, all was ready for Irénée and his family to make their move from Bon Sejour in New Jersey to the banks of the Brandywine. Sophie's brother, Charles Dalmas, went first, taking with him the family's household goods and livestock, including the prize Merino ram, Don Pedro, which Irénée had brought back with him from France. In July, Irénée, Sophie, and their three children followed overland by wagon, taking four days to travel the 130 miles over the miserable dirt roads that lay between Bergen Point and Wilmington.

On the land that Irénée had purchased from Broom there was a small log house that would serve the family until a larger home could be built. But the highest priority was building the powder works. With Bauduy's help, crews of masons, carpenters, and ditch diggers were rounded up. Millraces had to be dug and water gates installed along the plant's buildings. The local construction teams had never before been asked to build such a strangely designed factory. There would not be a single mill; instead, a series of strange, trapezoidal buildings would be scattered at safe distances from each other along the river bank. Each of these structures would consist of very heavy stone walls, much thicker than the masons deemed necessary, for the back and two side walls. The lower front side, facing the river, and the slanting roof of each structure had to be constructed of very thin wood, in which not one nail could be used. Indeed, there was not to be a single piece of metal anywhere in the structures. What was this crazy Frenchman up to? But once the workmen understood the function of the plant, the answer was obvious. Each part of the operation had to be a separate unit, so that a fire in one area would not create a chain reaction that would sweep through and blow up the whole process. Metal was taboo because a single fallen nail might strike a spark that would blow up the building. The powdermen who would later work here would also be allowed no piece of metal upon their persons, no coins in their pockets, not even a cobbler's tack in their shoes—only wooden sabots or soft moccasin-type slippers would be permitted.

Even if, with all these precautions, an explosion should occur in one building, it would spew its contents out of the thin, flared front side and roof, like a mortar aimed for the river water, and, it was hoped, would not rain destruction on the other buildings. The inhabitants of the Brandywine were getting a dangerous new neighbor, but at least its operation would be directed by a man of caution who knew his business and would take painstaking efforts to avoid unnecessary risk.

Irénée's construction demands in building his gunpowder factory proved to be far more expensive than he had anticipated. Labor costs were higher in America than in France, as was the cost of much of the building material. The amount allotted for plant construction was consumed long before the works were completed. Once again Peter Bauduy came to Irénée's assistance. In addition to the $4,000 he had already invested, Bauduy put in another $4,000, and when that was not enough, he was able to arrange a loan of $18,000 from a Philadelphia bank, secured by his own personal note. Bauduy now owned four shares of the company, while Irénée himself did not possess a single share in his own name. He began to feel that Bauduy had a greater claim on E. I. du Pont de Nemours & Cie. than he did, and so did Bauduy.

If Irénée was having financial difficulties in constructing his powder mill, they were minor compared to the problems that his father and brother were having with the other seven projects. Victor, in spite of his extended stay in France and his side-trip to Spain, had not been successful in attracting additional capital, and the French investors previously obtained were getting restless waiting for the promised dividends. The packet line they had hoped to establish for passenger service across the Atlantic had had the enthusiastic support of the American government but was vetoed by the French government as representing a threat to their own mercantile interests. They were successful in getting a commission to supply the French Army in Santo Domingo, but that turned out to be as much of a fiasco as everything else connected with France's ill-fated attempt to regain its lost colony in the West Indies. Two ships laden with du Pont supplies were lost at sea in the hurricane-ridden tropical waters, and only one managed to get to its destination. By that time, the French army, decimated by yellow fever and hopelessly outfought by the guerrilla tactics of the insurgents, gave up the struggle. In the meantime, Victor had made sizeable loans to Napoleon Bonaparte's young brother, who had settled in Baltimore and had married an American, Elizabeth Patterson. Victor was sure that Napoleon would stand by his brother, making good on the loans, and that the grateful First Consul would grant the du Ponts any concession they sought. But in the interim, Jerôme Bonaparte was unable to pay even the interest on the loans.

Pierre Samuel in the spring of 1802 decided he must return to France to try his hand at placating the French investors and, if at all possible, to raise more capital. He would transfer the parent company back to Paris.

Victor could remain at Bon Sejour, running his own company, V. du Pont de Nemours et Cie. of New York, and continue with the export–import business.

Whether this return was to be only a temporary visit or a permanent repatriation was uncertain—perhaps even to Pierre Samuel himself. Although he always insisted that he was at heart an American and that he had found his true home in the United States, the elder du Pont had never been able to transfer his loyalty totally to America. He still found the land alien and in many ways incomprehensible, and the English language struck him as barbaric and unusable. He also remained fascinated with French politics. As early as 1800, when he had just arrived at Bon Sejour with his heart full of hope and his head full of plans, he had expressed this ambivalence in a letter to Madame de Staël:

> I am writing to you in all confidence and even in all certainty at Paris or at Saint-Ouen. You have such a lively passion for great men and also for new things ... that surely you did not remain at Coppet [Switzerland] when France is being given its last hope, and even a reasonable hope, of finally attaining a government under which one can live, which knows the interests of the nation and the value of civil liberty.
>
> I do not need to confess to you that if the events which have taken place had preceded my American undertaking, I should not have exiled myself to the other end of the world; and I should probably be with you to talk about what we have done, what we can do, what we must do and what we shall do.
>
> But that is a happiness which is no longer reserved for me.[5]

For Madame Françoise du Pont there was no ambivalence whatsoever. She was desperately lonely in America, and once her daughter, son-in-law, and grandson had returned to France (where Bureaux de Pusy would become prefect of Lyons), she was more eager than ever to get back home. Although Pierre Samuel gave extravagant assurances to President Jefferson that this was only a business trip to Paris and that he would be back when circumstances permitted, as far as Madame du Pont was concerned, it was a one-way trip, and she would never return to America. As it turned out, both Françoise and Pierre Samuel du Pont were right.

When Jefferson got word of his old friend and trusted confidant's planned departure for France, he sent an urgent request to see du Pont before he left. As that proved impossible, Jefferson sent him a letter asking him to deliver some highly confidential correspondence to the American minister in Paris, Chancellor Robert Livingston. The cause of Jefferson's concern was the private intelligence that he had just received concerning the terms of the secret Treaty of San Ildefonso of 1 October 1800, under which Spain transferred all of the Louisiana territory to France.

The Mississippi River, with the great port of New Orleans at its estuary, was absolutely essential to the western territories of the United States. It was one thing to have this river and its seaport in the hands of a weak and compliant Spain from whom the United States had been able to obtain a treaty guaranteeing freedom of commercial traffic on the river and the right of deposit of goods in New Orleans. It was quite another matter to have Louisiana under the control of the most powerful military and political figure in Europe. It was obvious now why the French were fighting so vigorously to keep their small portion of the island of Santo Domingo. Haiti would provide the base for France's reassertion of colonial power in North America. The Mississippi would no longer be an artery of commerce, but a formidable military barrier to Americans pushing west across the Appalachians and down the Ohio. Under those circumstances, the states of Kentucky and Tennessee, as well as all of the old Northwest Territory, might secede from the Union and make their own arrangements with France. Vice-President Aaron Burr was already suspected of being involved in such a scheme. Therefore, Jefferson was eager to have du Pont serve as his private emissary to France.

In sending his dispatches to du Pont for delivery to Livingston, Jefferson left them unsealed and asked du Pont to read and comment upon them before departing. Du Pont was appalled by what he read. Jefferson was instructing his minister in France to inform the French government in no uncertain terms that the transfer of Louisiana from Spain to France would be considerd by the United States as nothing short of an act of war—a war that would, in Jefferson's words "annihilate her [France] on the ocean, and place that element under the despotism of two nations which I am not reconciled to the more because my own would be one of them."[6]

Du Pont hastily gave his alarmed comments. He told Jefferson that such a statement would be slapping a gauntlet in Napoleon's face. "'Give up that country or we shall take it' is not at all persuasive. 'We will defend it,' is the answer that comes naturally to any man," du Pont warned. He proposed that Jefferson instead say to Napoleon, "Give us Louisiana and at the first opportunity we shall restore Canada to you." But this was an argument that du Pont realized even while writing that Jefferson would find "not at all persuasive," so he tried another approach:

> What then are your means of acquiring and persuading France to an amicable cession of her property? Alas! Mr. President, the freedom of conventions, the natural taste of all peoples, of all individuals, for riches ... leave you with but one means when you have not an equal exchange to offer. It is acquisition, and it is payment in money. Consider what the most successful war with France and Spain would cost you. And contract for a part—a half, let us say. The two countries will have

made a good bargain. You will have Louisiana and probably the Floridas for the least expenditure possible; and this conquest will be neither envenomed by hatreds, nor sullied by human blood. France will ask of you as much as she can, and you will offer the least you can. But offer enough to bring her to a decision, if it can be done, before she takes possession.

We do not know what Jefferson's answer to this was, for he apparently asked du Pont to destroy his letter in response. Just prior to sailing, on 12 May, du Pont wrote once again, thanking the President for his letter and saying it had been burned as requested. Again he made a plea for the acquisition of Louisiana by purchase, not by war. Even if the American government did not have the full purchase price on hand at the moment, financial arrangements could certainly be made to extend payment over several years. After all, du Pont knew all about loans and promises to pay.[8]

Du Pont sailed for France not at all sure that he had persuaded Jefferson. But at least the seed of an idea had been planted, and Secretary of State James Madison found the future harvest proposed by du Pont most tempting. Independently of Jefferson, Madison urged Minister Livingston to move cautiously and not to appear to be too bellicose.

The situation in France was not much to du Pont's liking, either. Frustrated by failure in Haiti, Napoleon was not in a conciliatory mood. He was planning to make himself Emperor of the French people, and he wanted a French empire to go with the crown. Du Pont had been right. If Jefferson's ultimatum were presented to Napoleon, it would mean immediate war and the occupation of New Orleans by French troops. There was one shred of hope to which du Pont clung. Even more than he wanted the restoration of the Louisiana domain, Napoleon wanted money to finance the renewal of war with Great Britain. Du Pont advised Livingston of this fact and urged him not to present Jefferson's challenge at this time, but instead to play a waiting game. Livingston did not share the hope of a peaceful settlement, but he agreed to accept du Pont's advice for the moment, in accord with the instructions he had received from Madison.

Du Pont then proceeded to draft a treaty that he thought might prove acceptable to both parties. The terms he proposed would give to the United States the port of New Orleans and both east and west Florida for the sum of six million dollars. French vessels would continue to have the same rights of entry and of deposit in New Orleans and the Florida ports as American ships. In return, American vessels would have free navigation of the Mississippi. Du Pont sent this draft to Jefferson on 4 October 1802, but the President did not receive it until 31 December 1802.

When he finally got du Pont's letter, Jefferson was impressed with the proposal. He did not want a war with France, and it seemed to him that du Pont's plan offered a reasonable way out—giving New Orleans to the United States and guaranteeing the free use of the Mississippi, which should satisfy the westerners and preserve the American union. Although many in Congress were clamoring for war and for sending American troops to block French occupation of the trans-Mississippi region, Jefferson resolutely ignored the pressure. He announced that he was sending a second envoy, James Monroe, to France to join Livingston in the negotiations. Monroe's instructions were essentially to offer the French the terms drafted by du Pont, although no actual sum for the purchase of New Orleans and the Floridas was to be mentioned.

From then on, events moved rapidly toward the conclusion that du Pont had sought. On 3 March 1803, du Pont wrote to Jefferson,

> As a pacifistic philosopher, as a Frenchman and as a very warm friend of the United States, I thank you for having arranged a negotiation of the Louisiana affair. I think that it will be terminated to your satisfaction, perhaps before the arrival of Mr. Monroe. I have argued strongly in talks and by writing many times with M. de Talleyrand and most recently with Consul Le Brun.[9]

Although the affair was not settled until after Monroe's arrival, it was quickly resolved soon afterward, for Monroe was as popular with French officials as Livingston had been unpopular. The final agreement, however, went far beyond what Jefferson or even du Pont had dared to hope. Napoleon, suddenly weary of the whole North American project after the total defeat of General Leclerc's army in Haiti, authorized Ministers Marbois and Talleyrand to offer the whole territory of Louisiana for 60 million francs ($15 million). The treaty was concluded on 30 April 1803. A jubilant du Pont wrote to Jefferson: "Let me congratulate the United States and yourself on the wisdom through which, avoiding a war which would have thrown you into the arms of a redoubtable ally, you have acquired without shedding blood a territory ten times in extent and in fertility as the one you desired."[10]

Jefferson wrote in return, "For myself and my country, I thank you for the aids you have given . . . and I congratulate you on having lived to give those aids in a transaction replete with blessings to unborn millions of men, and which will mark the face of a portion of the globe so extensive as that which now composes the United States of America."[11]

Jefferson also expressed his gratitude in a more tangible manner. He wrote to the Secretary of War, Henry Dearborn:

> I enclose a letter from E. I. Dupont who has established a gunpowder manufactory in Wilmington. If the public avail themselves of his improvements in the art it would be to encourage improvement in one

of the most essential manufactures. I should be the more grateful by it as it would gratify his father, who has been a faithful and useful friend to this country. During my ministry in France he was at the head of the bureau of commerce with that country, and I was constantly indebted to his zealous exertions for all the ameliorations of our commerce with that country, which were obtained. On the late occasion, too, of Louisiana tho' he does not bring himself into view, I am satisfied that his just views of the subject have enabled him to make those energetic representations to Talleyrand, Marbois and others about the Consul, which his intimacy with them favored, and must have sensibly favored the result obtained.[12]

Of the many services that Pierre Samuel du Pont rendered to the United States, beginning with his efforts to secure a Franco-American alliance during the Revolution, none proved to be of greater value than his help in effecting the purchase of the Louisiana territory. He had prevented a disastrous war between France and America and had provided the solution by which the United States nearly doubled its territory, at a fraction of the cost of an uncertain war of conquest.

Irénée du Pont had reason to be personally gratified by the outcome as well. His gunpowder works were at last completed, and E. I. du Pont de Nemours & Company, was ready for business. Thanks to Jefferson's letter to Dearborn, one of the first orders the company received was for a shipment of black powder to be used by the marines against the Barbary Coast pirates, who for years had been the terror of commercial vessels in the Mediterranean. This was not one of the peaceful uses of gunpowder of which Irénée had boasted, but an order for 22,000 pounds of powder was most welcome.

Victor was still in New York, struggling with creditors who could not be silenced, and with debtors, including Jerôme Bonaparte, now King of Westphalia, who would not respond. Victor was only too happy to serve as the New York agent for the one du Pont project that still showed some promise of survival. The first barrel of gunpowder produced by E. I. du Pont de Nemours, early in the spring of 1804, was shipped off to Victor, who prepared an advertisement to appear in newspapers of the major cities:

E. I. DU PONT DE NEMOURS
GUN POWDER MANUFACTORY
Wilmington, Delaware

This new and extensive establishment is now in activity
and any quantity of powder, equal if not superior to any

manufactured in Europe will be delivered at the shortest
notice.

Samples to be seen at

V. DU PONT DE NEMOURS ET CIE.

New York[13]

Prospective buyers, however, would have had to hurry to see the samples in New York, for V. du Pont de Nemours et Cie. was not to have a much longer life. The company was mired in debt, and the only means available to Victor for extricating himself was declaration of bankruptcy. In 1805 his company was dissolved, and Victor's creditors took what few assets there were. Nothing was left of the first seven projects. Now the future of the family rested with Eleuthère Irénée and Project No. 8 on the banks of the Brandywine. Cashing in what little personal property he had that had escaped the bankruptcy judgment and adding to that the money that his wife had kept from her inheritance, Victor headed west to join Philip Church, Alexander Hamilton's nephew, who owned 100,000 acres of undeveloped land in the Genesee Valley of western New York. Victor planned to settle in the little town of Angelica and to operate a general store. It was a far cry from shipping lines and brokerage houses in New York, but Victor insisted that this latest venture could be the belated beginning of Papa's cherished Pontiana. Old dreams still flickered, and swallows still flitted. Gabrielle, who was expecting another child, would go with their three children to the Brandywine to stay with Irénée and Sophie until Victor had a place for them in Angelica.

If Irénée, to his regret, had lost his New York agent, he still had very much with him, to his increasing regret, his Wilmington agent, Peter Bauduy. There was a growing tension in the relationship between Irénée and the man who initially had been so helpful in providing both capital and counsel in getting the gunpowder operations underway. Bauduy, although genial and outgoing, was also aggerssive and ambitious. A born salesman, he could be invaluable to the new company in getting its product known to the general public and in bringing in orders. But as a former master of a large plantation, he found it exceedingly difficult to play the secondary role of overseer of sales and promotion. Irénée should continue to handle the production while he served as the front man, the spokesman for the company to the general public. Irénée, who had always been overshadowed within his own family as his father and brother strutted before the footlights to the plaudits of the crowd, was not ready to assume the same role within his own company. He was furious when he discovered that Bauduy was signing business letters as coming from "Du Pont, Bauduy & Company."

The name of the company, Irénée coldly informed his associate was E. I. du Pont de Nemours & Company. He, Bauduy, was simply a part of the "& Company." Bauduy had to yield on this point, but from that moment on, Irénée could not even make the pretense of liking his sales manager.

Bauduy did bring in orders, however, and the company began to show a profit. Forty-five thousand pounds of powder, which sold for $10,000, were turned out in 1804, the first year of operations, mostly due to purchases by the United States military forces. The following year sales rose to $33,000, and by 1807 they amounted to $43,000. Word spread that du Pont made the finest quality of gunpowder to be found anywhere—including Great Britain. Even Secretary of War Dearborn, who hated France and all Frenchmen and had been most reluctant to accede to the President's request that he place orders with the Du Pont Company, had had to agree after a series of tests were run that the Du Pont powder was best. After Mr. Jefferson's 1807 embargo shut off all trade with Europe, Du Pont powder proved to be a godsend for all those who previously had depended upon English powder. The embargo, in turn, proved to be a godsend to the Du Pont Company. Better than any tariff, it provided precisely the absoltue protection against foreign competition that small industries in America needed to make themselves into bigger industries. It may not have been good for shipping lines and the import–export business, but these enterprises no longer mattered to the du Ponts.

Unfortunately, increasing sales did not mean a great increase of cash in Irénée's pockets or of dividend checks in the hands of the shareholders. Almost all the profit was required for reinvestment in plant and materials, or far more irksome, to pay off old debts and interest on new debts. The $18,000 loan that Bauduy had negotiated had to be repaid, and then Irénée learned in 1807 that his father had borrowed another $20,000 from Talleyrand, ostensibly for capital improvement of the powder mill. Irénée received only $12,000 of this loan for his company. Papa kept the rest, but E. I. du Pont de Nemours & Company was charged for the entire loan.

Irénée never objected to his father about this matter. He was a man who kept his emotions tightly in check, yet he was often subject to periods of deep depression, when he would become even more taciturn and withdrawn. It was only to Sophie and to one or two close friends, such as to his brother-in-law, Charles Dalmas, that he would occasionally express his true feelings, as in the following letter, written in 1814, in answer to a suggestion that he give up the struggle and return to France:

> I do not see what I could do in France. I have spent my life here building up a very difficult industry and the disappointments I have had to bear have given me an habitual dullness and melancholy that would

be very out of place in society. Besides, how could I go? . . . I owe more
than sixty thousand dollars, chiefly in notes at the banks; so that my
debts amount to far more than my profits from the powder. I am forced
to stay here; the signatures that must be renewed every sixty days in bank
loans put me in exactly the situation of a prisoner on parole who must
show himself to the police every month.[14]

A major contributing factor to Irénée's debt load, in addition to the
debts incurred by his father and the demands of his own business, was the
obligation he assumed in assisting his brother. Victor had tried to make a
go of it in the Genesee wilderness, and in spite of his wife's loneliness and
unhappiness, he insisted for four years that things would improve. But Irénée had known from the beginning it had been one more flight into fantasy.
He wrote to his father in 1808 that Victor and his family would be "quite
ruined if they persist in staying on a farm of 500 acres covered with enormous forest trees, for which he has not paid, which he has no means of
cultivating, and by which he cannot possibly support his family, much less
give his children any kind of education."[15]

After receiving another pitiful letter from Gabrielle telling of Victor's
financial problems and of her and the children's fear of Indians, in this
Godforsaken wilderness, Irénée decided to take charge. He and the
accountant for his firm, Raphael Duplanty, made the long trek to Angelica
to bring Victor and his family back to civilization. Irénée promised to set
up his brother in business on the Brandywine. He and Duplanty, who had
a good head for business, could start a woolen mill directly across the river
from the powder mill. Irénée would furnish the initial capital for construction. Thanks to Don Pedro's numerous Merino progeny, the Brandywine
valley was now producing some of the best wool in the country. Woolen
manufacturing should prove to be a profitable venture.

It didn't take much arguing to persuade Victor that his Genesee venture
was a lost cause, and the family was reunited on the banks of the Brandywine. With Irénée's financial help, Victor and Gabrielle Josephine built a
somewhat smaller version of Irénée's and Sophie's home across the Brandywine—a spacious structure in the Southern style, with a two-story verandah facing the river. Gabrielle named their home Louviers, after her native
town in France.

Victor had at last found a nesting place, and Irénée had remained faithful to the oath he had given to stand by his brother "no matter what the
difficulty or the danger," but his burden had now been doubled, and he
would never be free of debt in his lifetime.

The wooden planks of the suspended footbridge across the Brandywine
that connected Louviers with the Eleutherian Mills estate of Irénée and
Sophie were soon worn smooth by Victor's four children, Amelia, Charles

Irénée, Samuel Francis, and little Julia Sophie Angelica, and Irénée's five, Victorine, now a beautiful young woman of nineteen, Evelina Gabrielle, Alfred Victor, Eleuthera, and one-year-old Sophie Madeleine. Within the next five years, there would be two more sons added to the family conclave: Henry, born in 1812, and Alexis Irénée, born on St. Valentine's Day, 1816. The du Pont children on both sides of the Brandywine would never lack for companionship, family ties were being drawn tight, and Irénée was content—no matter what the cost.

Shortly after the family's reunion Irénée was informed by Sophie that Victorine, whom he still regarded as a child, now had two suitors: Raphael Duplanty, Victor's partner in the woolen mill, and Ferdinand Bauduy, the son of Irénée's unwanted partner. Irénée was appalled by this news. Duplanty was a good man but much too old for Victorine, and Ferdinand was a pleasant enough youth, but much too young. When Sophie gently pointed out that Ferdinand was nearly as old as Irénée had been when they were married and Victorine was a year older than Sophie had been, Irénée's brusque answer was that he and Sophie had always been older and more mature than Ferdinand and Victorine. Irénée also suspected that this so-called infatuation on Ferdinand's part was only a plot contrived by his father "to marry the powder mill."

When Pierre Samuel was informed by letter of Victorine's two suitors, he had an even more eccentric suggestion. He said that he had always hoped Victorine might someday marry her uncle, Sophie's brother, Charles. *Grandpère* du Pont saw wisdom in the old French custom of marrying within the family. Other du Ponts would later adopt this practice, but not Victorine. She asserted that if she was to marry anyone, it would be Ferdinand, and no one else would even be considered. So Irénée gave his reluctant, qualified consent. Let Ferdinand go to France for two years as had been planned, and if upon his return the two were still interested in each other, Irénée would not prohibit the marriage.

Ferdinand departed for his two-year exile in 1811, and in 1813, he returned. Like Caesar, he came, he saw, and he totally conquered. He and Victorine were married on 9 November 1813, but his conquest proved short-lived. Eleven weeks after their marriage, on 21 January 1814, Ferdinand died of pneumonia. His grief-stricken father told Irénée that it had been his hope that the marriage of their children would return the two fathers to the closeness and friendship that they had once known. Perhaps now their shared grief would bind them even more firmly together. That was not to be. The stormiest period of their relationship still lay ahead. Within a few years, Peter Bauduy would become not only Irénée's competitor in a rival firm, but his implacable foe in the courts of law and within Irénée's own family. The very name Bauduy was to become anathema to

Irénée, but unfortunately he could never escape it, for his daughter bore the name and would remain Madame Bauduy for the remainder of her life.

The new difficulty between the two men had already begun to manifest itself prior to their children's marriage. It was a conflict common to partnerships of what should be done with the profits. Bauduy wanted them paid out to the shareholders as dividends. Irénée wanted to put the balance left after meeting interest payments on the outstanding loans back into the business. Bauduy wanted to enjoy the fruits of their labor. Irénée wanted to expand production. In 1810, anticipating that a war with either Great Britain or France—or perhaps both—was inevitable, Irénée, over Bauduy's violent objections, purchased additional land, the Hagley Farm, just down the river from the Eleutherian Mills, for $47,000. Here he intended to build a new and even more efficient plant in preparation for the increased business that he was sure would come.

Fortunately, Irénée's will prevailed, for war did come in 1812, as did the increased demand. E. I. du Pont de Nemours & Company was ready for it. Production jumped to 500,000 pounds of powder in 1812, with sales reaching the unprecedented figure of $148,597. The embargo of 1807 had given the company its boost for a take-off. War had sent it into an exhilarating orbit.

In Paris, Pierre Samuel could not contain his pride in his son's accomplishment on the Brandywine. His conversation was sprinkled with references to "my son, the great American industrialist," and to "Du Pont de Nemours and Company, the greatest gunpowder manufactory in the world." Pierre Samuel needed something to boast about, for his own financial situation was in total disarray, as had so often been true in the past. The old Du Pont de Nemours Father, Sons & Co., whose headquarters Pierre Samuel had transferred back to Paris when he and his wife had returned to France, had never yielded one franc to its disgruntled investors except in the eleven shares it owned in E. I. du Pont de Nemours & Company, and those dividends were not numerous. Du Pont *père* had kept his company alive by writing placating letters to the shareholders and by arranging additional loans from long-suffering friends like Talleyrand. In 1811, after nearly twenty years of war, even though there had been great victories on the battlefields of Europe, France was suffering an economic crisis nearly as severe as that which had precipitated the French Revolution of 1789. Banks were refusing further extensions and were demanding payment of principal on loans still outstanding. Pierre Samuel faced the same disaster that Victor had known in New York in 1805. His company could no longer maintain even a paper existence. Du Pont de Nemours Father, Sons & Co. was dissolved. All of its shares and all of its debts were transferred to E. I. du Pont de Nemours & Company. Irénée found himself, willy-nilly, the receiver of the old parent company. Its shareholders now became shareholders of the gunpowder factory on the Brandywine. This was not a welcome development for Irénée.

Papa had also, with the best intentions in the world, done another disservice to his son. He had bragged too often and too extravagantly about Irénée's success. The French investors now became insistent in their questioning. What is happening to all those great profits we have been hearing about? When are we going to get the dividends due us?

Even Pierre Samuel, who as always had fallen victim to his own hyperbole, began to ask himself why he and his now defunct company had not grown rich from their investment in E. I. du Pont de Nemours. Was it due to Irénée's mismanagement or—worse yet—was Irénée cheating him? Letter after patient letter from Irénée explaining the situation did little to improve Papa's understanding of the situation. His suspicions, at first only latent, became overt after the family in France received a letter from Peter Bauduy. Still angry over Irénée's decision to expand operations in 1810, Bauduy, without informing Irénée, wrote a vitriolic letter addressed to all the French investors charging his partner with bringing the company to the verge of ruin through overexpansion. "We now have," he wrote, "$140,000 invested in buildings, water power, and land on the Brandywine—a piece of madness that I was unable to prevent."[16]

The French investors threatened to bring legal action to force dissolution of the company and to liquidate its assets. Madame du Pont's daughter, now the Widow Bureaux de Pusy, was particularly outraged. She had inherited two shares from her late husband, and she wanted a return on her investment. She decided to come to America to assess the situation. Pierre Samuel and Madame du Pont were also outraged, an outrage that had to find expression in the written word. Pierre Samuel wrote one of the cruelest and most foolish diatribes he had ever composed in his long career of letter writing. In an epistle addressed to both Irénée and Victor, he accused them of being unnatural sons who were cheating their father and their stepsister, as well as their French friends who had trusted them. He disowned them as his sons, and he threatened to come to America to bring suit against them.

Pierre Samuel did not come, but Madame Bureaux de Pusy did. She was welcomed into the du Pont households on the Brandywine, since to Sophie and Gabrielle she would always be their dear sister Ile-de-France, who had shared with them the hardships of the trans-Atlantic crossing. But when she tried to get an accounting from Irénée, he made the tactical mistake of not spending the time necessary to explain in detail just what the situation was. He was frantically busy with wartime demands for increased production, and he didn't have the time or the energy to repeat to her what he had so frequently explained to his father. But Peter Bauduy was only too happy to spend hours talking with her about Irénée's mismanagement. He had a simple explanation for why Irénée would not go over the books of the company with her. It was because he was cheating her, along with the other investors, including Bauduy himself. Madame de Pusy coldly informed Irénée that she planned to file suit in the federal court in Philadelphia against him, and with that announcement, she departed for France.

Irénée was crushed by this declaration. His own stepsister had accused him of dishonesty and was planning to bring him into court as a swindler and a betrayer of family honor. For the first time it would be du Pont versus du Pont before the law. It would not, however, be the last time.

Happily for Irénée, at this dark moment another French investor decided to send his own personal emissary to inquire into the situation. The Paris-based Swiss banker, Jacques Bidermann, was also a victim of the economic crisis of 1811. He had closed his Paris bank, and one of the few assets he had left was his investment in E. I. du Pont de Nemours. He asked his young son, Jacques Antoine, to go to Wilmington to consult with Irénée. Perhaps Jacques could find out just what was going on.

Irénée did not make the same mistake with young Bidermann that he had made with his stepsister. Busy as he was, he put everything aside to confer with his visiting inspector. All company books were opened, and all notes on outstanding loans were thoroughly explained. Although Bidermann was only twenty-three, he was no novice in the field of finance. He quickly assessed the situation accurately. Everything that Irénée had written to his father and to the other French investors had been true, and the decision to put most of the profits back into the business had been correct. The Du Pont Company would not have been where it then was—the largest single manufactory of gunpowder in America, a plant capable of meeting the extraordinary demands of naval war with Great Britain—had Irénée not expanded his facilities at the very moment he did. All of the investors—indeed both America and France—owed a debt of gratitude to Irénée du Pont.

As he carefully scrutinized the books, Bidermann realized that Irénée had not made out as strong a case in his own defense as he might have done. Peter Bauduy, for all of his angry charges of being cheated out of what was due him, had over the past twelve years of the company's existence received over $100,000 in commissions, salary, and dividends, all for an investment of only $8,000, but Irénée's take, as director of the company, over the same period of time, had been only $16,500, far less than the $\frac{9}{30}$ths of net income to which he was entitled.

Bidermann was so impressed with what the company had already achieved under Irénée'smanagement and its potential for future growth that he asked to stay on as an employee. Irénée was overjoyed. At last he had an associate with whom he could work in harmony, a man who shared his views on management and plant development. Bidermann's report to his father and the other French investors did much to allay their fears. Pierre Samuel even wrote a letter of abject apology, retracting all that he had said in his earlier, impetuous indictment. The only two persons who were not silenced by Bidermann's stamp of approval were Peter Bauduy and Madame Bureaux de Pusy. Irénée's stepsister and her children continued with several suits, both in America and in France, against E. I. du Pont de Nemours and Company until 1824, when all charges were finally dropped and a settlement was reached, thanks in part to the aged Marquis de Lafayette's services as mediator.

Bauduy also remained a constant source of trouble for many years. When confronted with the indisputable evidence that Bidermann presented as to the very substantial returns he had realized on his investment, Bauduy dropped his suit and acceded to Irénée's request for his resignation from the company. He even agreed to sell his four shares. But with the knowledge he had acquired from his twelve years with Irénée, he opened up his own gunpowder factory in Eden Park, just south of the Christina River in Wilmington. He lured away some of Irénée's workers with promises of higher wages, and for the next several years he was one of the du Ponts' most active competitors in bidding for government contracts. At least he was gone, however, from both Irénée's powder factory and Victor's woolen mill, where he had also been an active partner. Bidermann took Bauduy's position as sales manager of the company. Bidermann also became Irénée's son-in-law upon his marriage to the du Ponts' second daughter, Evelina Gabrielle, in 1816. This was an event as pleasing to Irénée as Victorine's marriage to young Bauduy had been distressing. Ironically, the du Ponts were indebted to Peter Bauduy for having so alarmed Jacques Bidermann that he had sent his son to America to investigate. But that was a thought that probably never occurred to Irénée.

Among those present at Evelina's wedding was her grandfather, Pierre Samuel. After the promised "short business trip to Paris" that had lasted for thirteen years, the old man had returned to America in 1815, once again a political exile.

When he had first arrived in France in 1802, he had been extravagant in his praise of the Corsican adventurer who seemed to have become a second Augustus Caesar. Du Pont was convinced that France had at last found the philosopher-king whom the Physiocrats had sought. To his mind, Napoleon was a Solon, providing all of western continental Europe with a new, rational, and just Code of Laws; he was a Marcus Aurelius, dealing with the problems of empire and avoiding a conflict with the United States over Louisiana; and even in his many wars in Italy and Austria, he was a Washington or a Bolívar, extending the blessings of liberty, equality, and fraternity. But very soon disillusionment came, as it always did for Pierre Samuel in his assessment of French governments. Increasingly he saw Napoleon as a Nero in his tyrannical despotism, as an Attila in his constant, bloody conquests, as a Catherine de Medici in his use of nepotism to rule through relatives. Napoleon was destroying civil liberty in France, slaughtering French youths on the battlefields of Europe, and leading his nation toward inevitable disaster.

Believing these things, Pierre Samuel du Pont should have left France and returned to his family in Delaware. But he was preoccupied with his multi-volume edition of the works of Turgot, which would be the triumphant culmination of forty years of study and writing, and he was ensnared in financial difficulties that permitted no easy escape. Above all, Madame du Pont was adamant in her desire to remain in France. So the years passed and the pressure of du Pont's hatred for the Emperor Napoleon grew in its intensity because it had no outlet in the public media or in official duties. Censorship was more strict and more efficiently applied than it ever had been under the ancien régime. Every book and every pamphlet, no matter what the subject matter, was carefully scrutinized. If there was even a suggestion of criticism of the state, the work was suppressed. Even a new translation of the *Psalms of David* was put on the Napoleonic index of forbidden books. Pierre Samuel's old friend, Madame de Staël, wrote a book on German literature and culture that the censors deemed too complimentary of Teutonic culture and by inference, uncomplimentary to the French. The book's entire printing was destroyed and Madame de Staël once again fled into exile.

Du Pont wanted his nine-volume edition of Turgot's writings and philosophy also to be a veiled critique of contemporary French political economy. If it meant suppression of the work, so be it. At least he would have made a statement of principle. But he was to be denied even this satisfaction. The censors approved the *Oeuvres de Turgot* without a demurrer, and in some respects this was the cruelest action of all. So at the age of seventy-three, Pierre Samuel joined the underground conspiracy, led by Talleyrand, to work for the overthrow of the emperor.

In the summer of 1812, Napoleon overextended himself, as du Pont had been certain he eventually would, by invading Russia. The Russian winter of 1812–13 did what the combined forces of Britain, Austria, and Prussia had not been able to do during the previous twenty years. Napoleon's Grand Army was decimated, and by the spring of 1814 the Allies had crossed the Rhine and were marching on Paris. Now the conspiracy came out into the open. A provisional government was created and du Pont was elected Secretary General. Louis XVIII, brother of the ill-fated Louis XVI, was asked to return from his twenty-four-year exile and accept the crown.

Du Pont had no great love for the Bourbon family. He would have much preferred the restoration of the Republic under the Constitution of 1791, but if this was what the Allies insisted upon to preserve the integrity of the French nation, he would accept it. As Secretary General of the provisional government, du Pont had the pleasure of witnessing and notarizing the Act of Abdication signed by Napoleon at Fountainebleu on 11 April 1814, and placed his signautre directly below that of the deposed emperor.

Now du Pont was ready to start all over again, working for the restoration of civil liberties in France and converting the returning Bourbon relic of the past into a constitutional monarch if it were at all possible. But it was not to be. On 1 March 1815, Napoleon escaped from Elba and landed at Golfe Juan on the southern coast of France. As he headed north toward Paris, Napoleon was given a tumultuous welcome in each town that he passed through, and his small retinue grew into a formidable army. On 19 March 1815, Louis XVIII hastily fled into a second exile, but he wasn't quite as quick in departing France as was the Secretary General. Pierre Samuel du Pont was already aboard the ship *Fingal,* on his way to give America another try.

The du Pont family was once again complete, with the exception of Madame du Pont, whose state of health, happily for her, did not permit another Atlantic crossing. Pierre Samuel was overwhelmed with surprise and delight at what Irénée had accomplished on the banks of the Brandywine. In some respects, it exceeded even Pierre Samuel's wildest boasts. He wrote to Françoise du Pont, "I have seen the factory. It is gigantic, inconceivable that only one man was able to design and execute such things, especially this hydraulic machinery and these mechanical devices. . . . Irénée is a great man, with talent, courage, perseverance ten times greater than I had ever dared hope, although I have always held him in high esteem."[17] Irénée knew that the last line at least was not true, but he appreciated his father's sentiments.

Pierre Samuel seemed to be perfectly content to settle down at Eleutherian Mills with his children and grandchildren. Had the old man at last run down? Was this retirement? Pierre Samuel did not think so. "I do not believe it to be one. I look upon my voyage, my sojourn, only as the acquisition of a new, more peaceful place of study, where I can work to perfect myself, to ripen my ideas, to tie them together better, to place them with more order and skill before the eyes of the men whom God is calling upon, or will call upon, to propose and to draft Constitutions and Laws."[18] It sounded as if du Pont were still on call, waiting for a summons from France to return to write a new constitution and to set things in order again after the latest Napoleonic madness had ended.

In the meantime, he kept busy enough. He was delighted when the three South American states of Ecuador, Colombia, and Venezuela asked him to write constitutions for their newly established republics. He wrote a long letter (of book length) refuting the economic theories of J. B. Say, whose book, *A Treatise on Political Economy,* he had read while crossing the Atlantic. He carried on an active correspondence with Jefferson and went to visit the Sage of Monticello. He even returned to his first love, poetry, and wrote some simple verses for his grandchildren. As always, he kept a keen eye on events in France. He was not surprised by Waterloo, but he was also confident that the old order could never be restored by Count Metternich and the Congress of Vienna, "because people no longer believe in its promises, and because national pride will have been too much offended," he wrote Jefferson in 1815. "It seems highly probable that Germany, Italy, even England, will send away their Kings, and will renounce not only their Kings, but Royalty itself."[19]

Still the hopeful Republican to the end, he put his faith in the future:

> I feel no anxiety for my children. They have always been men of spirit, of probity and courage.... They have acquired a very distinguished capability. If they use it to procure for my Grandchildren an absolutely independent existence, they will be able to leave them among the most free and most enlightened of their enlightened and free fellow citizens.... Therefore, I have quite strong reasons for hoping that, under a Government where Nobility is not hereditary and has no influence in marriages, my family will become outstanding and will deserve so to be.[20]

It was a matter of great pride to Pierre Samuel that he could still be useful in promoting the advancement of his family. When he learned that Victor's second son, Samuel Francis, was interested in a military career, he wrote to Jefferson asking that the boy be granted a commission as a naval midshipman. Jefferson promptly wrote to President Madison to that effect, and at the age of fourteen, Midshipman Samuel Francis du Pont received his order to report for duty with the Mediterranean Squadron.

As the months passed, it became apparent that Madame du Pont would never join her husband in America and although he frequently spoke of his duty to go back to France to be by her side, there was always some task or family responsibility that kept him at Eleutherian Mills—the birth of his last grandchild, Alexis Irénée, or the marriage of his granddaughter Evelina.

Then, on the night of 16 July 1817, the old man was awakened by the sound of the fire gong down the hill by the mills. The charcoal-drying house was in flames. Every able-bodied man in the valley rushed to put out the fire before it reached the powder magazine. A bucket brigade was formed from the river up to the burning charcoal house, and among those passing buckets of water up the hill was old Pierre Samuel himself. Ignoring the pleas of his children and grandchildren to return to his bed, the old man worked on the line all night long. He felt the same exhilaration he had known when he had stood beside the Swiss Guard at the Tuileries many years before. By dawn the fire had been extinguished and the powder mills were saved. Wet and exhausted but flushed with excitement and a sense of accomplishment, the old man at last let Sophie lead him upstairs and get him settled in bed. He never got up again. On 7 August 1817, Pierre Samuel du Pont died, at the age of seventy-seven, the first du Pont to die as a result of an accident in the powder mills.

At the top of the incline behind Eleutherian Mills was a wooded area called Sand Hole Woods. There an acre was cleared of bush and leveled to provide a graveyard for the du Pont family. In the center of that area was placed the body of the cemetery's first occupant. A raised slab of sandstone was placed over the grave, with the inscription:

Sacred to the memory of

PIERRE SAMUEL DU PONT DE NEMOURS

Knight of the Order of Vasa, of the Legion of Honor and of the Order du Lis, Counsellor of State, Member of the first Constituent Assembly, President of the Council of Ancients and member of the Institute of France.

Born in Paris, December, A.D. 1739

Died at the Eleutherian Mills, August 7, A.D. 1817

The men of the Brandywine valley had saved the powder mills on the night that Pierre Samuel joined the bucket brigade, but such heroic efforts would not always prevail. Just eight months after the fire that had cost the elder du Pont his life, on the morning of 19 March 1818, there occurred one of the worst explosions in the history of the company. A glazing mill at Eleutherian suddenly exploded, for reasons never determined. It shot its contents out toward the river as it had been constructed to do, but some of the burning sparks were blown onto the roof of a drying house, which also caught fire and exploded. Then all hell erupted as the conflagration spread to the grand magazine where the finished powder was stored. There were three major explosions in all. They were felt as far away as Lancaster, Pennsylvania, forty-three miles to the north. Forty persons were killed, and bits of human flesh were found in the trees of Louviers across the river. Old Marshal Emmanuel de Grouchy, a veteran of most of the great Napoleonic battles, including Waterloo, happened to be visiting Victor at Louviers at the time. He was so sickened by the sight that he had to be escorted away from the scene of carnage and taken to a hotel in Wilmington. But the du Pont women, even Victor's fastidious and delicate wife, Gabrielle, did not flinch from the services demanded of them. Louviers was turned into a temporary hospital and morgue, and the women tended the wounded and laid out in shrouds what remained of the dead.

Irénée, who was in Philadelphia on one of his periodic note-signing missions, hurried back to the Brandywine. He found the original Eleutherian Mills leveled, and his own home was so damaged as to be uninhabitable for weeks. But none of his family had been killed, although Sophie had been struck in the side by a falling beam and had sustained an injury from which she never fully recovered.

It took a year for the Eleutherian Mills to be restored. In spite of what it meant to production and sales, Irénée insisted that the dependents of all the men killed in the accident be paid a monthly pension, although in those days there was no legal or generally accepted moral obligation to do so. Fortunately, the Hagley Mills farther downstream were undamaged and continued in full production. In a curious way, the accident eased somewhat the financial strain that Irénée had been laboring under. When news of the disaster reached France, Talleyrand, who had been insisting, after the death of Pierre Samuel, upon being repaid the loan of 100,000 francs he had made to the elder du Pont, now agreed to a ten-year extension. The French creditors of the bankrupt Du Pont Père, Fils & Cie. also agreed to accept new long-term notes at a lower rate of interest. So it could have been worse, Irénée thought, but God knows it had been bad enough, and no compensation would be adequate for the lives that had been lost. Surprisingly, only two of the surviving workmen refused to go back to their old jobs. All of the others accepted what had happened as an expected hazard of the trade. Security was tightened around the mills, however, for sabotage was suspected after one of the surviving workmen reported having seen an unknown man running from the glazing mill only moments before the explosion occurred.

Following the death of Pierre Samuel, Victor and Irénée had become the elder generation. They and their wives were the last remaining ties to France. Only the four oldest children had even the vaguest memories of Paris and Bois des Fossés, and French was no longer spoken in their homes except among the parents. The eldest sons of both brothers were now reaching the age where they could assume some of the business responsibilities. Victor was only too happy to turn the woolen mill over to his son Charles Irénée, who, although only twenty, seemed to have already settled into an adult maturity that his father had never fully achieved. With the woolen mill now under better management than ever before, Victor was freed from the daily routine he had never liked and could once again enjoy the life of the swallow. He entered state politics, and in spite of the handicap of being a French immigrant, by the charm of his personality and his gregariousness, he quickly achieved political prominence. Delaware had always been a curious enclave of conservative Federalist politics in the heart of Jeffersonian democracy. Along with New England, it had voted for the Federalist candidate for President as long as there was a Federalist party to vote for. After 1816, when the Federalist party had disappeared, Delaware supported the more conservative wing of the Democratic-Republican party, and looked first to John Quincy Adams and later to Henry Clay for leadership.

This was a political milieu in which Victor du Pont felt at home. In his advocacy of a high protective tariff and in his promotion of the political fortunes to his idol, Henry Clay, he was, of course, betraying the physiocratic faith of his father, but then Victor had never been a radical republican like his brother, Irénée. Victor helped lay the foundation for the creation of a Whig party in Delaware and was elected to the Delaware House of Representatives and then to the state Senate. He was as successful in playing the game of state politics as he had once been in international diplomacy, for the two fields of endeavor had much in common.

Even Irénée was no longer the radical he had been when he turned out propaganda for the radical Jacobins of Paris. Although he would never engage in political office-seeking like his brother, he was delighted and honored to be chosen a director of the United States Bank and to join his friend, Nicholas Biddle, in that citadel of financial conservatism. He regarded Andrew Jackson as a dangerous demagogue, much as his father had once viewed Danton and Robespierre. In 1832, at the height of the Nullification crisis, he had coldly refused to accept an order of several thousand pounds of gunpowder from the South Carolina government, which was threatening secession over the Tariff Act of 1832. Irénée and Victor were, each in his own way, setting a political pattern for the family that would endure for the generations to come.

For nearly ten years, Papa Pierre Samuel's was the only grave in Sand Hole Woods cemetery, lovingly tended by his children and grandchildren. Then, within less than two years, Irénée lost the two persons closest to him. Victor was the first. In seemingly perfect health, he was attending a political meeting in Philadelphia in January 1827 when he was suddenly stricken by a heart attack on the street, a block from his hotel. Carried to his room by friends, he died within a few hours. Although so different in every aspect of personality and talent, no two brothers were closer in fraternal love than were Victor and Irénée. For nearly thirty years, Irénée had carried Victor's financial burdens, but now that burden had been replaced by a grief that was much more difficult to bear.

Victor was buried in the place of honor, just to the right of his father, in the family cemetery. For many years Irénée had been the actual head of the family, but the protocol of primogeniture still prevailed in the grave-yard. Thereafter, Victor's descendants would claim their lots on the right of Pierre Samuel, and Irénée's would go to the left.

In November of the following year, 1828, it was necessary to open a place on the left half of the graveyard for Sophie Madeleine Dalmas du Pont. She had first complained of not feeling well in early July. The physicians were unable to diagnose her illness, which may have been cancer, and she grew steadily worse. Within a month she, who had always been so strong, was unable to take a step or feed herself. During the remaining twelve weeks of her life, Irénée hardly left her bedside. After Sophie's death, Irénée showed less and less interest in life. More and more, he turned the business details over to his faithful and competent son-in-law, Bider-mann, and to his own son, Alfred Victor. Irénée's periods of depression grew ever more frequent and of longer duration. He would disappear from the mills for hours at a time, and would be found sitting on a rock by the Brandywine, staring at nothing but his own grief.

But life still made its claims upon him. Victor's family and his own many children gave him a reason for living. In 1833, his youngest daughter, Sophie Madeleine, married her cousin from across the river, Victor's second son, Samuel Francis, now a lieutenant in the United States Navy. *Grandpère* du Pont, a strong believer in intramural family marriages, would have been pleased. The next year, Irénée's last unmarried daughter, Eleuthera, was wed to Thomas MacKie Smith, a young physician who had come to Wil-mington to open a practice. Victorine, the Widow Bauduy, continued to live at home with him and the two younger boys, Henry and Alexis. Both sons were keenly interested in the business—more so, it sometimes seemed to Irénée, than their older brother Alfred, who was expected to assume the directorship of the company.

By 1834, both the family and the company seemed to be in good order. The girls were all married and the boys had their careers already determined for them. The old debt to Talleyrand had at last been repaid, and Madame Bureaux de Pusy had been silenced. Negotiations were under way to pur-chase the few remaining shares still held by French investors. Soon the com-pany would be owned in its entirety by the family. This was as good a time as any for Irénée du Pont to make his exit from the stage.

Irénée did just that. On 30 October 1834, while in Philadelphia on business, he was stricken by a heart attack at almost exactly the same spot where Victor had fallen. Like his brother, he was carried to the hotel, where he died eleven hours later.

So ended the first two generations of the du Pont family saga. The two men, father and son, each in his own way, had made a particular contribution to the creation of an American dynasty. Pierre Samuel had provided a keen intelligence, an intellectual curiosity, a set of enlightened principles, an optimism, and above all, a *joie de vivre* that would be indispensable in maintaining the family's vitality. Eleuthère Irénée had provided the family with a solid base for its material fortunes. He had set precedents for industry, sober judgment, clear and realistic thinking, and above all, a sense of family loyalty and cohesion. These were valuable building blocks for the future.

There were, to be sure, debits as well as assets which the first two du Ponts bequeathed—impetuosity, quixotry, and glibness from Pierre Samuel; taciturnity, heavy sobriety, and melancholia from Eleuthère Irénée. The family would have to accept these debits as well. In varying combinations, the succeeding du Ponts would draw upon this mixed inheritance left to them by the two founders of their family.

I

Brandywine Heritage
1834–1864

ALFRED IRENEE DU PONT, the first son of Charlotte Henderson and Eleuthère Irénée du Pont II, was born on 12 May 1864 in the tall, three-story house that had been assigned to his parents by the ruling head of the family, Alfred's great-uncle Henry du Pont.

As the first son of the first son of the founder of the E. I. du Pont de Nemours and Company, the infant Alfred, according to the ancient Salic law that had prevailed in the land of the du Ponts' origin, should have been hailed as the heir apparent to the du Pont realm on the Brandywine. But power within the du Pont family had never descended in the Gallic-accepted line of primogeniture. It had instead gone to the male descendant who was the most talented, or at least the most determined to rule. The first Eleuthère Irénée du Pont, to whom they all owed their fortunes and their place in American society, had himself been the second and—for a long time—the underestimated son of an illustrious father. It was true that upon Eleuthère Irénée's death in 1834, the titular leadership of the family and the business had been bestowed upon his eldest son, Alfred Victor, but it became increasingly apparent to this second generation that the real power within the company and within the family should be vested in Alfred Victor's younger brother, Henry.

Now in 1864, although Henry was still in complete command of the du Pont dynasty, he and his sisters were already considering the line of succession into the third generation, in which the heir apparent was not Alfred Victor's eldest son, Eleuthère Irénée II, or even Henry's first son, Henry Algernon. The most likely successor to the throne of power was Alfred Victor's second son, Lammot, who had already demonstrated in many concrete ways that he had the talent, the industry, and the ambition to assume command of the family's fortunes whenever his uncle Henry relinquished the power.

If the newborn child, Alfred Irénée, should ever have designs upon the du Pont seat of authority in the fourth generation, he would not be able to claim it as his rightful birthright. He would have to fight for his right of succession. Only in the Sand Hole Woods cemetery did the protocol of undisputed primogeniture prevail. Although in the family burial ground Victor du Pont, the first son of Pierre Samuel, and Victor's descendants lay in the place of honor to the right of their distinguished forebear, the second son's honor lay spread out in the valley below, and it was his wealth that maintained the family's graves and sustained their living descendants.

Eleuthère Irénée du Pont had not lived as long as his father Pierre Samuel had. Irénée's sudden collapse on a street in Philadelphia in October 1834 and his death a few hours later came as a great shock to his family, his neighbors, and his employees. His obituary spoke for all of his associates when it noted that "a mysterious and inscrutable Providence has suddenly removed him from among us in the midst of his usefulness."[1] His children were now forced to assert the leadership of a new generation, for which they were quite unprepared.

Yet, in spite of its abrupt termination, Eleuthère Irénée's life had been far more complete and more satisfactorily fulfilled than his father's. Even on his deathbed, Pierre Samuel was still drawing up plans for a future that could never be realized. He sketched new constitutional frameworks for a republican France, he planned pastoral utopias for displaced peasants, he looked forward to a reunion with his wife and kinfolk in Paris, and he outlined in his head the many books that remained to be written. Seventy-eight years of hyperactivity had not been enough.

But sixty-three years were enough for Eleuthère Irénée du Pont to assure his immortality. The banks of the Brandywine supported his life's work—the largest gunpowder manufactory in America, firmly established and at last nearly debt-free. His family, including four daughters and three sons, along with their spouses and children, were securely ensconced where he had seated them. He left behind no unfinished business. The symphonic score was completed. Now all that his descendants need do was to perform in unison and harmony the work he had created.

It was, to be sure, not the composition that his father had expected or wanted the du Pont family to create. Pierre Samuel had wanted to build an agrarian utopia of small farms, a physiocratic dream to be realized in a Jeffersonian democracy that would stand in idyllic contrast to the ugly urban industrialism of monarchical Europe. Pierre Samuel could design, but he could not construct. That was the tragedy of his life. His son, Eleuthère Irénée, could construct, but not according to his father's specifications—and that would always be, even in his success, the source of his unhappiness for having failed his father. Pierre Samuel had wanted a land of self-sufficient farmers; Eleuthère Irénée had laid the foundation of a great industrial empire of wage-dependent laborers. Pierre Samuel had wanted economic equality and a redistribution of the old, inherited wealth; Eleuthère Irénée's venture would in time create a concentration of new wealth that would surpass the most avaricious hopes of Croesus. Pierre Samuel was a philosophic pacificst who dreamed of a world without war. Eleuthère Irénée's manufactory would produce terrible, new instruments of destruction. Eleuthère Irénée's reality seemed to be a mockery of Pierre Samuel's fantasy. Even the name du Pont was ironic, for there was no traversable bridge that could link the father's dreams to his son's deeds.

Yet in a curious way, Eleuthère Irénée had achieved a somewhat warped version of Pierre Samuel's utopian Pontiana. His manufactory was a community of mills, not of farmsteads, to be sure, but these mills were built in a woodland, not in the grimy center of an industrial city. They were powered not by coal-consuming steam boilers but by the natural flow of a country creek. There was no pollution of the air by coal smoke from chimneys or even by noise from the clatter of looms or the screech of saws. Birds still gathered in the ancient trees that crowded in upon the unobtrusive stone structures of the mills. Fish and young boys could still swim in the pure water that flowed down to the Delaware. It could even be argued that the natural order of things, so beloved by his physiocratic father, had been less disturbed by Irénée's mills than it would have been if Pierre Samuel had had his way and the land had been given over to the farmers, who would erode the hillsides by slash-and-burn clearing and by plowing up the thin topsoil. Irénée had brought a quiet, clean industry to the Brandywine, one that lived in harmony with its natural environment.

Occasionally, of course, the tranquility was suddenly shattered by a dreadful explosion, and the sky was reddened by hellish powder fires. Only then would the people of Wilmington to the south and the villagers of Chadds Ford to the north remember the true nature of the industry that lay sequestered in the wilderness. Fortunately, such accidents were rare. Most of the time, it was easy to forget that the Eleutherian Mills and the Hagley Yards were turning out a more deadly product than the cotton yard-goods or the stone-ground flour of their neighbors. The very isolation of the du Pont establishment helped to foster this tolerant forgetfulness in the adjacent towns.

The isolation also fostered a version of the communalism that Pierre Samuel had sought to create in his Pontiana. For Eleuthère Irénée, his family was his only meaningful world. He had created the E. I. du Pont de Nemours & Company for his father, his brother, his wife, and his children, and the nature of the company's product had necessitated its being built in a wilderness setting. This had meant creating a separate and independent community on the banks of Brandywine. Within this small world, the du Pont employees lived in houses built and owned by the company.

This community did not have the social and economic egalitarianism that Pierre Samuel had envisioned for his utopian community, but there was the egalitarianism of shared labor and shared danger. The du Pont sons and grandsons worked beside their day laborers, doing the same tasks and wearing the same rough work clothes as their employees, their faces smeared with the same black charcoal and their hands stained with the same yellow sulphur. The du Pont wives and daughters nursed the sick and taught the millworkers' children in the Brandywine Manufacturers' Sunday School, which Eleuthère Irénée had been instrumental in establishing in 1814. And when terrible moments of catastrophe did occur, the explosions shattered the windows of the workers' cottages and the managers' mansions alike. True egalitarianism was achieved in a death-style of accident if not in a lifestyle of privilege.

Within the du Pont family itself, there was a shared existence that went beyond communalism to something akin to familial communism. All of the du Pont homes, as well as their horses and carriages, were owned in common, the houses being assigned to each du Pont household according to the size and social needs of that family. The salaries paid to the du Pont sons who worked in the mills were modest and equal, and each of the seven du Pont children shared in the profits of the company, although the shares were not equal, but were divided according to a family-agreed-upon estimate of the contribution that each of the du Pont males employed by the company was making to the success of the business. Each of the du Pont daughters also received a share of the profits, although these shares were considerably smaller than those of their brothers and nephews and were divided equally among the sisters.[2]

There was a recognized head of the family who also served as the chief executive officer of the company. His authority was great, but it was not absolute. Like the British prime minister, he was considered to be *primus inter pares*, for on basic policy decisions that affected the welfare of both the family and the company, each of Eleuthère Irénée's children had a vote. The du Pont daughters' votes within these family councils were equal to those of their male counterparts, and not infrequently, especially for those issues involving family matters, their votes proved decisive.

This communal existence was an almost inevitable consequence of the émigré family's living in relative isolation and engaging in the manufacture of a product that required a non-urbanized environment. As long as Eleuthère Irénée had lived, the family had been patriarchal. He had been both *pater familias* and patron, and his supreme authority was neither shared nor disputed. But after his death, it was no longer possible or desirable to sustain the patriarchy among the siblings. Although it was tacitly assumed that Eleuthère Irénée's eldest son, Alfred Victor, was now the head of the family, he was insistent that at least for the present, his brother-in-law, James Antoine Bidermann, who had served as Eleuthère Irénée's right-hand man for twenty years, continue to act as head of the company. This was a wise decision on Alfred Victor's part, for during the transition from one generation to the next, Alfred had the opportunity to familiarize himself with all of the administrative details of the company under the able tutelage of Bidermann, but the latter had no intention of remaining at the head of the company. In the spring of 1837 he returned to France with his wife and child. He was now able to pay off the last debts still owed to the original stockholders. At the same time, he sold his own interest in the company and that of his wife, Evelina du Pont, back to the du Pont family. The remaining six children of Eleuthère Irénée were now on their own, in complete ownership and command of E. I. du Pont de Nemours & Company.

Although Alfred Victor shunned the formal title of president, the major responsibility for the continuing success of the company now fell upon him. He was as thoroughly familiar with every aspect of the business as only a du Pont child could be whose nursery playground had been the company yards and whose school vacations had been spent helping in the mills. He had had, moreover, two years of higher education at Dickinson College, where he had been privileged to study chemistry under the instruction of Thomas Cooper, the English immigrant who had become America's most eminent chemist.

Alfred Victor had returned from college in the spring of 1818, just in time to help rebuild the mills after the most devastating explosion the company had yet suffered. It was not a happy reintroduction to life on the Brandywine. Alfred Victor would have been content to have remained at Dickinson under the instruction of Cooper until he had received his bachelor of arts degree, and perhaps then to have sought an academic position himself where he could pursue his research in the quiet of an university laboratory. But he knew full well where both his duty and his destiny lay, and he returned to the Brandywine.

By 1837 his destiny had been fulfilled. He was head of both the company and the family. He took upon himself all of the company's administrative duties. He signed all letters. He kept the company books; all orders had to bear his signature, and he paid all of the bills. But he built a small research laboratory beside his office, and there he spent his happiest hours. The laboratory was his retreat from the practical affairs of business which he never enjoyed. He was more than eager to delegate the actual mill operations to his two younger brothers.

Henry du Pont, fourteen years Alfred Victor's junior, was a West Point graduate who had once dreamed of a career in the United States Army with the same longing that Alfred Victor had shown for academia. But he had resigned his commission and had returned to the Brandywine at the request of his father a few months before the latter's death. In contrast to that of his older brother, his return proved to be a true homecoming, and very quickly the company became the only real interest of his life.

Finally, the youngest brother, Alexis, who had not wanted to leave the Brandywine even for college, was never to have any career dreams except those that could be realized in the grinding and glazing sheds of the Eleutherian Mills.

Unlike his father, Alfred Victor did not have to search long for associates to aid him in the management of the concern. The three brothers formed a triangle of company operations, with Henry and Alexis forming the solid base, and Alfred Victor resignedly unhappy in his position at the apex.

The du Pont daughters also formed their own hierarchy of authority within the family, and among them the law of primogeniture had more significance than it had among their brothers. First in order of birth and in influence was Victorine. Born in 1792 in Paris at the moment that the last vestiges of the old order were being stomped under the feet of the Parisian mobs, Victorine was the only one of Irénée's children who had clear memories of France and who as an adult still felt more comfortable speaking French than she did English. Married at twenty-one to Ferdinand Bauduy, the son of her father's former partner, Victorine had known only two months of married life before her young husband succumbed to pneumonia. She remained in the family home, and after the death of her mother in 1828, she became her father's official hostess and the acknowledged head of the female side of the family. Although she was a child of the French Revolution, which her grandfather had been instrumental in initiating and her father had enthusiastically supported, she represented the antithesis of social change, and in her small way she succeeded in re-establishing on the banks of the Brandywine an *ancien régime* of social protocol and Old World manners. After her early bereavement, she never evinced the slightest interest in a second marriage. On the contrary, Madame Bauduy nurtured her precious widowhood with all of the tender and proud care she might have given to a child. Her younger siblings and even her father stood somewhat in awe of her pious good deeds and her unbending rectitude. She remained both the zealous guardian of the family's moral standards and its social arbiter throughout her long life.

The second daughter, Evelina Gabrielle, was born at Bois des Fossés in 1796. Only three years old when the family emigarated to America, she had but the dimmest memories of her native land. She had been fortunate in inheriting from her grandmother, Marie Le Dèe, great physical beauty and from her grandfather, Pierre Samuel du Pont, a vivacity and *joie de vivre*. As a young girl, she was the indisputable belle of the Brandywine. When Jacques Antoine Bidermann arrived from France in 1813 to inspect the Du Pont Company's books on behalf of his father and the other French stockholders and then stayed on to become Irénée's business partner, he fell an easy victim to Evelina's beauty and charm. Her marriage to Bidermann in 1816 greatly pleased her father, and unlike Victorine's tragically brief and unfulfilled marriage, Evelina's marriage was long and happy, and was terminated only by her death in 1863. Of the four du Pont daughters, only she produced a child—a son, James Irénée. In 1837, she and her husband and child went to France, where they stayed for several years. But the family ties were too strong. The Bidermanns returned to the Brandywine, although James no longer had any connection with the company. Eventually, their son returned to France. He established a permanent residence in Paris and he and his heirs became for his du Pont cousins on the Brandywine their one continuing "French connection."

Eleuthera du Pont, the third daughter, was the first of Irénée's children to be born in America. Gentle and sweet, she appeared to have no ambition other than to be Irénée's dutiful daughter and her sister Victorine's obedient shadow. Her world was narrowly circumscribed by the boundaries of the du Pont property, and she had little curiosity about exploring a larger scene. Even her marriage in 1834 to Thomas Mackie Smith, a young physician from Philadelphia, did little to broaden her horizons. Instead, her husband committed himself to her world by establishing their home within the du Pont enclave. Two years later, when Dr. Smith's younger sister, Joanna, married Eleuthera's younger brother, Alexis, the circle was drawn even closer, and the two Smith in-laws became an integral but undemanding part of the du Pont family.

Finally, there was Sophie Madeleine du Pont, the youngest of the four daughters and eighteen years the junior of her sister Victorine. Cute little Sophie was everyone's darling, and she would always be the baby of the family, even to her two younger brothers, Henry and Alexis. In 1833, Sophie, to the delight of her father, married her first cousin, Samuel Francis, Victor du Pont's younger son.

Marriage brought little change to Sophie's life—not even a change in her surname. To her family she remained the sweet, innocent child whom they had always known and loved. Her husband's long absences at sea, for months and even years, bestowed upon her a quasi-widowhood in which she could maintain the intimate relationships with her sisters and brothers that she had always known. She felt no conflicting loyalties and obligations to her husband, who was usually absent, or to children, since she had none.

Comfortably situated in Victor du Pont's old family home, Louviers, which from her earliest childhood had been almost as familiar to her as her father's house, she could look across the Brandywine to her birthplace and know all was well.

The three du Pont sons had no difficulty accommodating the husbands of the du Pont daughters within the power structure of the company. Evelina's husband, Jacques—or James, as he preferred to be called in America— had created his own place of importance within the company long before the du Pont sons had reached their maturity. Had he elected to remain with the company for longer than three years after the death of its founder, he might well have continued to be the dominating authority of E. I. du Pont de Nemours, but he had tactfully removed himself from the scene. Eleuthera's husband was content with his career as a physician. He wanted only to keep his wife and himself within the family circle. And Sophie's husband, Frank, as he was called by the family, wanted no part of the company except as a source of ammunition for his beloved ships-of-the-line. It was enough for him that his wife held a share of the interest in the company and could participate in the family councils.

But if the du Pont sons could largely ignore their sisters' husbands with respect to the management of the company, the du Pont daughters could not so easily dismiss the roles of their brothers' wives within the family circle. Accommodations were made for their sisters-in-law, but the places they occupied within the family largely depended upon their own individual personalities.

The wives of the two younger brothers presented no major problem of adjustment to the long-established family order. Louisa Gerhard, whom Henry married in 1837, was mostly ignored by the rest of the family. The daughter of a fairly well-to-do German merchant, Louisa always felt herself something of an alien in this tightly knit French family. As far as the family could determine, she was content to remain quietly in the background. She was never an active participant in the du Pont social life, and during a considerable part of her married life she was preoccupied with her eight children, and was often confined at home either in expectation of or in recuperation from giving birth. By the time her last child, William, was born in 1855, Louisa had become so accustomed to confinement that staying at home had become her way of life. Although her husband had, by that time, assumed full control of both the business and the family, Louisa made no effort to assert herself as First Lady of the du Pont realm. She was destined to survive all of the other du Ponts of her generation, living long enough to see the beginning of the twentieth century, but not even longevity gave her a position of command within the family.

Joanna Smith's position within the family was unique in that she had a double claim to the status of du Pont sister-in-law by virtue of being both the wife of Alexis du Pont and the sister of Eleuthera du Pont's husband. These ties alone would have ensured for her a far more integral place within the du Pont family than Louisa could ever achieve, but in addition she was far more sociable and outgoing than Louisa. She formed a particularly close companionship with both Eleuthera and Sophie. Her major influence within the family, however, was not social but religious. She and her brother were devout Episcopalians, and they brought to the du Pont family, with its long tradition of an ill-defined deistic rationalism, a new and unexpected fervor of belief. Joanna and her brother Thomas found a ready convert in Alexis, who was quickly transformed into a born-afresh Christian.

Alexis was never one to make a lukewarm commitment, whether to the company's business operations, which was his life, or to his newfound religion, which he hoped was to be his salvation. Alexis's conversion brought about the first serious rift within the family since Madame Bureaux de Pusy had charged her stepbrother Irénée with depriving her of her rightful share of the company's profits. Alexis, in his new religious zeal, was a far more active proselytizer than either his wife or brother-in-law. "We are in a business where every one of us should be prepared at any moment to meet our Maker," he repeatedly warned his family. Considering the frequency of accidental explosions in the plant, no one could deny the truth of that statement, but many in the family wanted to make their own preparations and did not accept Alexis's road map to salvation. When Alexis proposed that the family build an Episcopalian church on the company grounds to provide religious services for the family and their employees, his older brother's patience came to an end. For once Alfred Victor was prepared to assert his prerogative as head of the family. In an angry letter addressed to the entire family, he accused his brother of betraying the faith of his father, which was in truth no faith at all. There would never be, as long as Alfred Victor lived, any religious edifice on the du Pont lands. There would never be any attempt to prescribe or even provide a particular religious affiliation for the family members or for the company's employees. All persons connected with the du Pont establishment must feel free to worship as they pleased, or not to worship at all, if that pleased them. As for Alfred Victor himself, he warned that he would personally and quite happily "throw the first black-dressed, reversed collar man who appeared on the grounds into the Brandywine river."[3]

Although Alexis had allies, particularly his sisters Victorine and Eleuthera, the issue never came to a vote within the family council. Faced with Alfred Victor's uncharacteristic outburst of temper, the family quietly retreated. Alexis and his sisters went ahead with their plans to build Christ Church, but it was built with their own, not the company's funds, and it was carefully located on land that lay outside the du Pont communal property.

By converting her husband, Joanna Smith may perhaps be held indirectly responsible for creating this controversy, but it was the only disruption she would create in the familial harmony that had so long prevailed. She herself had no missionary ambitions among the family or its employees, and she had no social ambitions among her sisters-in-law.

The third female addition by marriage to the family, Alfred Victor's wife, Margaretta Lammot, was not of the same ilk as either Louisa or Joanna. She was the most assertive of the du Pont in-laws of either sex, and she ultimately prevailed as the dominant female figure among the du Ponts of her generation.

Margaretta's father, Daniel Lammot, a textile manufacturer in southern Pennsylvania, had early become acquainted with Irénée du Pont. The two men became close friends, and although Irénée had his own financial difficulties, he was never so strapped that he could not give his friend Daniel a helping hand in his times of even more pressing need.[4] Lammot's eldest daughter, Margaretta—or Meta, as the family called her—was approximately the same age as Eleuthera and was a frequent visitor in the du Pont household. Although she was nine years younger than Alfred Victor, in that isolated world on the Brandywine, attractive young teenagers from outside the family were rare commodities, and it was not long before she attracted the amorous attention of Irénée's eldest son. Her engagement at the age of seventeen to Alfred came as no surprise to either family. She and Alfred Victor were married in the fall of 1824. From that moment on, the entire community was aware of the powerful new mistress within the du Pont clan.

The marriage of Alfred Victor and Margaretta provided persuasive proof of the old maxim that opposites attract. Where Alfred Victor was quiet, diffident, and unassertive, Meta was bold, aggressive, and outspoken. Alfred Victor was not a strong and efficient manager of either his company or his family. He fussed over small details and lost sight of larger goals. When he took over the company, it was mostly debt-free, thanks to the long, unstinting efforts of his father and his brother-in-law, James Bidermann. He managed, during the thirteen years he served as its head, to incur a new and for that time substantial debt of $500,000, largely due to his generosity to friends and his inability to say no to those, including his father-in-law, who asked him to co-sign notes iof indebtedness. Meta, on the other hand, managed the family with a firmness and dispatch of which her children, her servants, and the local tradespeople could offer convincing testimony. The family's vegetable gardens, fruit orchard, and domestic live-stock kept for home consumption came under her direct surveillance, and productivity and thrift were her standing orders. Her household accounts balanced, and there was no place in her ledger for waste or for debt-incur-ring generosity.

Alfred Victor had never wanted or enjoyed the position of prominence given to him by the accident of order of birth, but Meta reveled in being the consort of the head of the du Pont establishment, a position she did not relinquish when her husband abdicated his in 1850. She was meant by nature to rule and could she and Alfred Victor have exchanged places, the du Pont company would have had a quite different history.

Meta never directly challenged Victorine's acknowledged position as the First Lady of the du Pont women. It was unnecessary, since each woman had her own sphere of interest and each avoided an overlap that could become a confrontation. Victorine's all-absorbing interest was the Brandywine Manufacturers' Sunday School, which her father, Irénée, had been instrumental in establishing in 1814. Irénée was motivated not only by a genuine concern for establishing a facility to provide the rudiments of education for his employees' children, but also, more selfishly, to give his beloved eldest daughter a new interest that might lift her out of the con-tinuing deep depression into which she had fallen after the death of her husband Ferdinand. He achieved both objectives. Starting as an informal school held on Sundays in the home of the textile manufacturer, John Sid-dall, the school soon outgrew its front parlor accommodations. In 1817, Irénée provided the land, and the Brandywine manufacturers and employ-ees, by subscription, provided the funds for a new building. From its incep-tion the school was nondenominational, "all sects joining together harmo-niously in the work." From 1817 until her death in 1861, Victorine served as superintendent of the school, with her sister Eleuthera as her assistant. The school remained the one absorbing interest of Victorine's life. Here she reigned supreme for forty-five years, providing the basics of a primary and secondary education to nearly 2,000 children, who ranged in age from five to eighteen years. Into this special world of Victorine, Meta had no wish to intrude.[5]

Meta's sphere of activity, aside from her own household, which she han-dled with expeditious efficiency, was the social life of the Brandywine valley. She served as a central intelligence agency for the every day as well as the extraordinary happenings in the various households, and in those pre-tele-phone days, the hired couriers of the du Pont establishment were kept busy delivering little notes of information from Meta, along with jars of jellies from her kitchen, baskets of produce from her gardens, and invitations to tea or supper at her home. No announcement of birth, engagement, illness, or death was official until it came via her.

Alfred and Meta produced almost as many children as Henry and Louisa, but unlike her sister-in-law, Meta did not undergo long confinements. Pregnancy and childbirth were brief, temporary inconveniences, which she put aside as quickly as possible. Enjoying remarkably good health herself, she found it difficult to understand her husband's proclivity for hypochondria, and she had little patience for other relatives and friends who were frequently ailing.

Joanna was not the only one to bring religion into the family. Meta came from a devout Swedenborgian family, and her faith was part of the dowry she brought to the du Pont family. Happily for Alfred Victor, Swedenborgianism, with its curious blend of scientific rationalism and ecumenical mysticism, was a religious expression that he was able to tolerate within his own immediate family without violating his own deistic conscience. Meta boldly held services in their home until such time as she could organize into a congregation the few scattered Swedenborgians in the Wilmington area and have a church built in town. Perhaps Swedenborgian ministers did not wear black suits or reversed collars. In any event, there is no record of Alfred's throwing the minister whom Meta had hired into the Brandywine River.

The great diversity of personality and style among Irénée du Pont's seven children and their six spouses is not surprising. What is surprising is how very well these thirteen people managed to live together in harmony and with apparent real affection for so many years. Irénée's own children had known little else outside their family, but for the men and women who married into the family, the adjustment could not have been easy. One of Margaretta's granddaughters, Marguerite du Pont, in reviewing John Winkler's unflattering portrayal of the du Pont family many years later, would succinctly state the problem. With a sharpness and acerbity worthy of her grandmother Meta, Marguerite criticized Winkler for the attention he devoted to the three du Pont brothers without giving proper due to their wives. "These three wives," Marguerite wrote her cousin Pierre, "whose names were not even mentioned in the history of the du Pont family, Margaretta Lammot, Joanna Smith and Louisa Gerhard, experienced in their own persons long years of poverty, monotony and hard work while the du Pont Powder Company was in the making. They were the back bone of Alfred, Henry and Alexis. They not only stood all Alfred, Henry and Alexis stood, but they stood Alfred, Henry and Alexis, too."[6]

Yet harmony prevailed among the du Ponts for many years prior to the

Civil War. On all of the basic issues they were in agreement: the welfare of the family, first; the success of the company, second; the propagation of the species, third. They were all staunch Henry Clay Whigs in a state that was then predominantly Democratic, and except for the one incident involving Alfred Victor and Alexis, there was no serious dispute over religion. Each of them had a place within the established scheme of things, and no one was so ambitious as to encroach upon another's recognized position.

This is not to imply that the du Ponts maintained a static society that permitted no individual development or communal change. Perhaps the greatest change, prior to the Civil War, in the lives of the du Ponts occurred in 1850, when at the age of fifty-two Alfred Victor decided that he had finally had enough of the job he had never really wanted in the first place. He resigned his position as chief executive officer of E. I. du Pont de Nemours. His stated reason was his failing health, but he had for so long complained of poor health that no one in the family could accept that this would necessitate his sudden departure from the company. The simple truth was that the company was not prospering under his administration. Its indebtedness was increasing at a disturbing rate, and not even the prosperity brought by the Mexican War had lasted long enough to reverse the trend. The company needed a far firmer hand than his if it was to survive its first half-century of existence. There was no debate over who was to succeed Alfred. There was only one candidate—his younger brother, Henry. Henry du Pont was thirty-eight when he assumed command, and immediately both the family and the company knew that the new master would be different in every respect from his immediate predecessor. Within a matter of weeks, Henry had learned for himself the sobriquet "Boss." Boss he was indeed, in every aspect of company operations and family arrangements, down to the smallest detail of pencil purchases for the office clerks or roof repair on the houses that he assigned to each du Pont household. And he remained Boss Henry for the next thirty-nine years, the longest tenure for a chief executive officer in the history of the company.

As one chronicler of the du Pont family has written, "If his older brother's [Alfred Victor's] motto had been 'Make it better; make it safer,' Henry's might be said to have been 'Make it cheaper; make it quicker.'"[7] Safety standards were not ignored—in his business they could not be—but the cost of safety could be lowered by regulations that were rigorously enforced and by new technology that would reduce the exposure to danger. Henry knew well what too few other entrepreneurs of that day realized—that profits were not dependent exclusively on the price put on the finished product. Profits were even more dependent upon the cost of production. Cut costs enough and prices can be cut to underbid competitors was his simple recipe for success. He knew that it is not what the market would bear in prices, but what the quality of the product would bear in the reduction of production costs that must concern the manufacturer.

With these precepts in mind, Henry proceeded to reorganize the company. The old generation of French immigrants and their sons who had worked in the du Pont mills since the beginning were now passing from the scene. It was an opportune moment to consider labor costs. Newly arrived Irish immigrants, refugees from the potato blight, could be hired cheaper than local American farm youths, so Henry brought in the Irish. Since transportation costs in getting the product to market were too high, he purchased his own Conestoga wagons to carry the powder kegs to the shipping terminals and built his own warehouses and wharves on the Delaware River, where the kegs could be picked up for shipment to the metropolitan markets.

The stories of Henry's thrift became family legends: the willow twigs he picked up each morning on his early walk to the office to add to the woodpile for the making of the essential charcoal, and the candles in the office that were used until the last flicker of the wick guttered out. Such small economies contributed little to cutting the overall cost of production, but as symbols they were powerful and constant reminders to employees and family alike of Henry's omnipresent concern. Unlike Alfred Victor, who had remained secluded in his office or in his private laboratory, Henry seemed to be able to be everywhere at once, appearing suddenly to scrutinize every detail of the grinding, glazing, or caking of the powder. With his fiery red beard framing his face, he looked as much like a son of Erin as his Irish employees, or in moments of anger, like the Devil himself. The quick nudge and the whispered warning, "The Boss is right behind you" proved to be the most effective safety monitoring device that the company had. Boss Henry's rule was despotic, but to most of his associates it was a benevolent despotism. The company began to prosper again, indebtedness was slowly whittled away, and the family did not complain, at least not audibly.

Death proved to be another agency for change, and during the first decade of Henry du Pont's regime, it visited this second generation of Brandywine du Ponts on three occasions, creating new widows and altering the established family structure. It first called unexpectedly one evening in January 1852, when Dr. Thomas Smith complained after dinner of a slight attack of indigestion. He asked his wife, Eleuthera, to bring him a spoonful of the medicine he used to treat his patients' stomach complaints. In her haste to comply, Eleuthera took the wrong bottle from the shelf and gave to her husband a tablespoon of the deadly alkaloid aconite, popularly known as monkshood or wolf's bane. As soon as he swallowed the dose,

Thomas sensed that a mistake had been made and when Eleuthera showed him the vial she still had in her hand, his fear was confirmed. He immediately took an emitic, but it proved futile. The fast-acting poison was in his system. He quickly lapsed into unconsciousness and died.[8]

It was cruelly ironic. Poor Eleuthera, who had never harmed a living creature in her life, who had frequently brought injured birds into her home, nursed them back to health, and then wept when they flew away— she , of all people, became the poisoner of her husband. Deprived of husband and never having had children, she had now only her old family, which she had in truth never really left. She became more than ever the quiet shadow of Victorine, with only the Brandywine Manufacturers' Sunday School to give some meaning to her life.

Four years later, Margaretta joined the growing sisterhood of widows when Alfred Victor died on 4 October 1856. His death was not unexpected, and it was a perverse vindication of his long years of complaining that his ailments were more serious than simple hypochondria. Alfred had so completely divorced himself from business affairs that his passing had little effect upon the company, nor did it materially affect the tenor of life for his immediate family. His seven children were nearly all grown, with only the youngest son, Antoine Bidermann, still in college. As for Margaretta, she continued to reign as the Grande Dame of the family, and her daily routine was as little affected by death as it had been by births in the family. Alfred exited from life as quietly as he had exited from the presidency of the company.

The third death in the family during the decade of the 1850s, however, had a far greater impact upon both the company and the family. Alexis du Pont, the youngest child of Irénée and Sophie du Pont, was in his own way as dynamic and forceful as his older brother Henry. Not even Henry dared tell Alexis anything about how to operate the mills to achieve higher quality or greater production. He was the greatest powdermaker the du Pont family was to produce, with the possible exception of his grand-nephew, Alfred Irénée. Among the workers, who had largely ignored Alfred Victor and who feared Henry, Alexis was not only respected but loved. If the DuPont Company had experienced little labor strife it was largely due to his presence in the mills and his ability to smooth over incipient disputes. Henry made visitations to the mills, but Alexis worked there, constantly sharing the hard labor and the same omnipresent danger. Therein lay the difference as far as the employees were concerned. Alexis also held the same favored position within the family, especially among his sisters. He was the youngest child in a large family and had been only twelve years old when his mother died. His older sisters had all served as surrogate mothers to him, but his natural charm and vivacity would have endeared him to the family regardless of the order of his birth. His enthusiastic conversion to orthodox Christianity had later caused a rift between him and Alfred Victor, but it brought him into ever closer relationship with Victorine and Eleuthera. Alfred Victor himself could not long remain at odds with Alexis, for the latter's growing piety did not stifle his natural wit. Of all the du Ponts of his generation, he had the liveliest sense of humor. He could make even the saturnine Alfred Victor and the bereaved Victorine laugh at his witticisms.

Alexis also seemed to have won the favor of the gods, for although he constantly exposed himself to danger and had on several occasions experienced the death of laborers working alongside him, he himself seemed to lead a charmed life as he escaped each accident unscathed. All the men working in the powder mills of necessity either became fatalists or they abandoned their jobs for less dangerous occupations. Workers who didn't sincerely believe that they wouldn't "go across the creek"—the workers' euphemism for being blown to bits—until their time was up had no business being powdermen, and the sooner they got out, the better for all concerned. But not even then were they absolutely sure of being safe if their number had already been called. One of the most frequently told stories was that of a worker who, having experienced several explosions after which he had been obliged to pick up the scattered heads, arms, and legs of his fellow workers, finally decided he had had enough and quit his job. The next week, as he was walking along a path on the far side of the Brandywine across from the mills, an explosion hurled a stone from the press room across the stream. With deadly accuracy, the projectile struck the man's head, killing him instantly. The moral of this probably apocryphal story was, "When your number is called, you can't escape by answering, 'Not here.'"[9]

Alexis's many seemingly miraculous escapes gave further credence to the workers' belief in fatalism. But then there was the one occasion when Alexis was not as fortunate as he had previously been which also served to confirm that faith.

Late Saturday afternoon, on the 22nd of August 1857, as the mills were being shut down for the weekend, Alexis came out of his house ready to drive into town to check on the progress being made on the building of the Episcopal Cathedral of St. John, whose construction he had been instrumental in initiating. Before departing for Wilmington, he suddenly decided to take one last tour around the grounds. He was particularly interested in seeing if everything was in readiness for the masons who were going to start needed stone repairs on one of the grinding mills early Monday morning. A quick look inside revealed that a large mixing box had been left in the mill. This would have to be removed before the masons could begin work. Alexis called seven of the men who were nearby to help him remove the box. The container was extremely heavy, the men were all in a hurry, and it proved to be exceedingly difficult to maneuver the box through the door. They nearly had it out when the rear corner of the box struck the stone door. There was a spark, the powder in the mixing box flared up, and the flame raced across the floor to a tub of waste material, which then exploded. Four of the movers were killed, and Alexis was blown thirty feet outside the mill. His clothes were ablaze but he was otherwise unhurt. His luck still held. Alexis jumped into the millrace to extinguish the flames on his shirt and pants, and while standing in the water noticed that the explosion in the grinding mill had hurled a burning piece of timber onto the roof of a nearby press room. This building, where the powder was pressed into cakes, was the most potentially dangerous mill of all. Should it explode, the entire works might go. Alexis leaped out of the water, ran toward the press mill and started to climb up on the roof to remove the burning ember. Just as he pulled himself up on the roof, the press mill exploded. He was hurled far out into the yard, where his eldest son, Eugene, found him, still alive but with his back broken and a main artery in his leg gushing forth blood. Eugene hastily made a tourniquet of his father's suspenders, and then he and several workers carried him on a window shutter up to his home. His back and leg injuries were not mortal wounds, but unfortunately his rib cage had also been crushed and both lungs had been punctured. For this, the doctors who were hastily summoned could do nothing. At last, Alexis's number had been called. It was his time to "go across the creek."[10]

Even so, Alexis did not succumb immediately to his wounds. He lived for another twenty-four hours, fully conscious but in great pain almost to the end. His was one of those long, protracted deathbed scenes which were so meaningful to the Victorians of the mid-nineteenth century. His last hours were fullly recorded for posterity by his sister Sophie:

> The first words spoken by our beloved Alexis after the men set him down in the entry were at the sight of his wife, uttered in his usual tone of voice, bright and calm, "Joanna, I am only hurt, don't come near me." Eleuthera knelt by him and threw her arms round him, kissing him on the forehead, the men exclaiming, "Don't Mrs. Smith, oh, don't." He directed his clothes to be cut off and said, "I am going to be stripped, send all the women away." We ran for blankets in which they wrapped him. Joanna, Fanny [his oldest daughter] and all then surrounded him and Eleuthera proposed his being carried to a couch to be put up in the dining room, but he said, "No, take me upstairs," and Joanna exclaimed, "Yes! lay him on his own bed."—the men carried him up in their arms, others supporting his feet, & placed him on his bed. As soon as he was laid down he looked up to Heaven & said, "Let me die the death of the righteous & let my last end be like His." The Doctors soon arrived. . . . The physicians did what they could to make him more comfortable. . . . Henry came as soon as the fire was extinguished in the yard. He had

seen Alexis in the yard who had told him "he only had a broken leg & to go on & put out the fire." Alexis said to him when he came, "Henry, I am badly hurt." Later he turned to Henry & said, "Harry, if I have not long to live I want you to see that St. John's Church is finished for I was bound to do it." Henry said, "Oh don't talk of that now." Joanna asked him if he would like her to pray. He answered Yes & she said the prayer for a person not likely to recover. He thanked her and at different times repeated, "Thank you, love, for that prayer, it was *such* a comfort." Mr. Parker came and had prayers. Alexis expressed the wish to receive the sacrament. The Doctors interposed & said it had to be deferred till the next morning. He asked if they thought he would live till then. They assured him there was no immediate danger, when he said to Mr. Parker, "Then I had rather wait till tomorrow." Mr. Parker then left & Drs. Jones & Bush arranged his leg, covering it with wet cloths, to stop the bleeding & c, during which he suffered much. He several times ejaculated, "Lord have mercy on me," but bore it heroically. . . . Joanna, Fanny, Eleuthera, Maria & the two Boys were near him: after one o'clock the Boys were told by their mother to go to bed: when they objected, and Alexis said, "Go Boys, you must go to bed."

The following morning Alexis was still alive but in great pain. He dictated his last will and testament, and then asked to see all of their employees. The workmen filed past, many of them weeping, and touched him gently on the shoulder. He urged each to get religion if they hadn't already done so. Then he spoke to each member of the family, greeting his brother-in-law Bidermann in the language of their parents, "Adieu, je m'envais mais ce n'est pas dire, je repose sur un Dieu tout puissant qui reçoit tous les pecherres." He gave special attention to Alfred's second son, Lammot, whom Alexis had already spotted as the likely heir apparent to the command of the company in the next generation. "Lammot, you're a noble fellow. You do wonders. I'd like to have seen Irénée [Alfred's oldest son, who was in Philadelphia on that weekend]. Say goodbye to him for me & tell him to take warning by me."

Irénée did arrive home, however, in time to bid farewell to his uncle. Alexis recognized him, and realizng that Irénée would want to know, described in detail how the accident occurred. "You have had a heavy loss," he told his junior partner, "you must go ahead and face it."

Sophie concluded her account by writing, "The last words we heard, in a solemn tone . . . were 'Heaven & Earth shall pass away, but thy words shall not pass away.' He then fell into a quiet sleep from which he never woke, but gently & peacefuly the spirit left this earth. Dr. Bush broke the silence by saying 'Alexis is with his God.' His sisters knelt down in prayer, rose, kissed him & left the room."[11]

For Sophie, and for her family who read this account, there was great comfort to be found in her recalling the manner of Alexis's death. There had been time enough for Alexis to tend to mundane matters like writing a will. There was time to bid loving farewells to all who were most dear to him. Above all, there had been time to reaffirm his faith in his God and in his hope for salvation. How his sisters must have treasured hs last words. As their beloved Bishop Alfred Lee would say, "It was a perfect dying." But this was a sentiment that neither their deistic forebears nor their more skeptical descendants would have been able fully to understand.[12]

Even if it was "a perfect dying" in a mid–nineteenth-century Christian sense, it was still death, with its shock and all of its finality. Alexis was only forty-one, and he left behind a wife of the same age and seven children, the youngest only three years old. He had held a central position in both the family and the company that could not be filled by anyone else. He was, furthermore, the first du Pont to be killed in a plant explosion, although there had been many in which their workers had died. His death had broken the magic charm that had seemed to make the du Pont family invulnerable to the constant danger to which they subjected themselves. That realization was a shock that struck each du Pont personally. They had lived with daily danger for over half a century. Now they fully realized they could die from it.

The family's sense of loss was great, but the full extent of that loss could never be accurately evaluated. What would have been the development and direction of the company had Alexis lived for another thirty years? He was already recognized not only within his own company but throughout the industry as the greatest powdermaker in America. How long would Alexis, with all of his energy, expertise, and drive, have been content to serve as a lieutenant under Henry's dictatorial command? Could the company continue for much longer to accommodate two such forceful figures, both bearing the name du Pont and belonging to the same generation? Would Alexis sooner or later have challenged Henry's position and in so doing, either defeated the Boss, destroyed himself, or worst of all, broken the company's and the family's unity beyond repair?

At the time of Alexis's death there were already indications of a rift between the two men that promised a power struggle. Francis Gurney Smith, the brother of Alexis's widow, Joanna, wrote to his sister two weeks after the accident that he was "glad to learn that you had had a conversation with Henry & now that the first step has been taken, I trust that the opportunities may be more frequent. I feel sure that his heart is much softened & that he desires to do all in his power to aid you, but he is naturally reserved & requires a little drawing out, but you know that confidence begets confidence, & the exhibition of it on your part will awaken it on his. Forgive me if I touch upon tender subjects, dearest Joanna, my desire to secure your happiness is my only plea."[13] Henry du Pont could risk a softened heart toward Alexis's widow. There was no one left within his generation of the family to challenge his authority. The company might be poorer for the loss of Alexis, but Henry was all the stronger.

Deaths in the family and resignations from the company were the major marking points by which the du Ponts measured the passage of time: "We moved into this house two years after Alfred left the Company," or "I remember well it was the year after Alexis's death that they left here and moved to Kentucky." Without these red circled calendar years which every family has, the past would quickly become a blur of undateable happenings. During most years, however, there was little to puncture the quiet, even tenor of life on the Brandywine save the normal events of growing up and growing older: going to school, leaving home for college, returning to the Brandywine, marrying, and giving birth to a new generation of du Ponts eager to repeat the cycle.

In the late summer of 1852, Henry du Pont hired a young man, R. Page Williamson, from Orange County, Virginia, to serve as a tutor for his own and several other children of the du Pont families. Young Williamson, in his weekly letters home to his mother, portrayed with considerable literary grace and candor what life was like among the du Ponts in the years 1852–53, a routine time during which there were no great events to give the period a special distinction. With the perspective and detachment that only an outsider can have, Williamson saw the du Ponts at school, at home, at work, and at play and dutifully reported on these activities to his inquisitive mother.

Lacking the financial resources for a college education, Williamson had as his only educational credentials the classical education he had received at a boy's academy in Louisa, Virginia. It had been precisely because of this limited education, however, that he had obtained the position of tutor. Henry du Pont, ever mindful of costs, offered him a much lower salary than a college graduate would have demanded. For Williamson, his position as tutor presented a frightening prospect. It is painfully apparent in his letters home that he was overawed by the responsibilities he had assumed in undertaking the education of ten children belonging to Henry, Alexis, and Charles I. du Pont, and to their mutual second cousin Gabrielle du Pont Breck:

> I have 10 scholars, 5 boys and 5 girls, among them Miss Fannie du Pont [Alexis's oldest child] . . . a young lady of 15. She is a fine Latin and Greek scholar, reads Horace with great eclat and at Greek she is hard to beat. *I hate to teach Gals.* The rest of my girls are from 12 to 8 years

of age. *Spoilt* bad children who always make me rejoice when night comes. The boys I manage well enough. They are quite well advanced. Henry du Pont Jr. is the best historian I ever met with; he is perfectly familiar with every king that ever reigned; i.e. in Europe and with almost all in Asia. He speaks of them as of acquaintances so familiar is he with their histories.

Because he was so inexperienced and unsure of his qualifications, Williamson found it particularly irksome to have Victorine Bauduy as a constant visitor in his classes. "I had almost forgotten to mention the most important character that figures in my school; it is a widowed sister of Mr. Du Pont's, who takes great interest in the school and pays me a daily visit. She is very fond of suggesting ways and means for which sugestions [sic] I would be more obliged to her for keeping to herself; tho' on the whole she is a right good sort of body, and if it were not that I don't like to have to please so many I have no doubt but that I would like her. It is very hard to please old women; but with Mr. Du Pont I would have no difficulty."[4] Victorine, because of her many years of teaching and supervising the Brandywine Manufacturers' Sunday School, made herself the self-appointed inspector general for all educational activities in the valley, and she must have been a trial indeed for the novice teacher.

Soon, however, Victorine left the young tutor to proceed on his own. Either she found he was doing an acceptable job or she became bored with his teaching. At any event, within a month he was able to write with considerable pleasure, "I have not for some time passed [sic] had much of Mrs. Beauday's [sic] company. She has been busily engaged nursing her sisters Mrs. Smith and Mrs. Henry du Pont. Mrs. Henry du Pont was taken sick upon the Saturday after I reached this place and I have not seen her since. . . . From what I saw of her I thought that I should like her very much."[5]

Williamson quickly settled into a daily routine that at first provided little variety:

You ask me to give you a full acct. of everything that concerns me of all I do and suffer; to be able to do this you must consent to spend the day with me and observe for yourself. In the morning at 5 minutes of six I get up and dress which takes about 20 minutes; at 15 after 6 I read my bible which brings me to 6½ when I begin my Greek and study till breakfast at a quarter after seven. I go down to quite a comfortable meal at which I meet with my host and hostess a young couple bearing the uncommon name of Peoples. They are a genteel and obliging couple and extremely solicitous about my comfort. [Peoples was a clerk in the Du Pont business office with whom Henry du Pont had made room and board arrangements for Williamson.] From breakfast till 8 o'clock, I usually walk about for exercise at 8 I resume my Greek and study until 5 minutes of nine when I resort to my school room and wring [sic] my bell where upon all my troubles begin. The bell being wrung [sic] in rush 10 of the worst children that it has evern been my lot to know; they have been spoiled by their parents and no authority taken with them by there [sic] teachers of whom in the last 7 years they have had 9. After wading thro an iniquity of little lessons, I come to 15 minutes of 11 when according to request and ancient usage, I give 15 minutes recess. At 11 the bell rings again and I am kept quite busy till 12 when I adjourn for dinner; I return immediately to my room and spend an hour on my Latin at two oclock I again return to my school and it passes off very much like the morning with an intermission of a few minutes of four at 5 oclock I break up for the day and walk about till 6 o'clock when I return to my tea. It is by request that my time is thus regulated at least that portion devoted to my school. The parents do not like to have their children much confined and of course I am very well disposed to comply with their wishes. After tea at 6½ oclock I retire to my room and study until 10 when I retire to rest and sleep sound until 6 again in the above you may see the whole view of my life for one day is so like another that it has become quite a routine.[16]

So the days passed for the du Pont children and their harried tutor, who had to cram furiously his study of Latin and Greek in order to stay at least abreast of his pupils while futilely trying to maintain order in the classroom.

In the Alfred du Pont home Page Williamson found his closest companions during the year he spent in the Brandywine valley, for Alfred's and Margaretta's children were the du Ponts nearest his own age. Their oldest daughter, named Victorine after her aunt, was already married and living in New York by the time that Williamson arrived on the scene, and their youngest child, Bidermann, was in preparatory school in Philadelphia, but their two younger daughters, Paulina and Sophie, became his close friends and provided him with his entry into Brandywine society. He described ir some detail his first visit to Alfred's and Margaretta's home:

> On Friday night last I took tea and spent the evening at Mr. Alfred Du Pont's by invitation. The family consists of Mr. and Mrs. Du Pont, two young ladies, the one about 20, the other 18 years of age and two sons of 21 and 23 years. From the style of living at Mr. H- du Ponts I had formed no idea of the elegant and sumptuous manner in which his brother lived. His daughters one of them particularly who is quite pretty, made themselves quite agreable [sic] and with chess and backgammon at both of which they were so polite as to allow me to beat I spent the evening very pleasantly; the mother is an elegant lady & that you may have a look at Miss Sophie Du Pont you have only to look at Geo.ᵃ Bankhead & imagine that she is pretty and young.[17]

With Lammot du Pont, Alfred's second son, who was only a year older than he, Williamson went to Whig political rallies and engaged in athletic contests. He was delighted when he was able to defeat Lammot in foot races, providing the distance did not exceed one hundred yards, for Lammot, at age 21, was the biggest and strongest youth in the valley. With Sophie and Paulina he went hunting and on nature trips. He also went to their home twice a week to work out in the gymnasium the girls had set up in one room of the house. "The attraction at Mr. du Ponts is not as you seem to insinuate his pretty daughters," he hastened to inform his mother, "but simply this that the young Misses du Pont are very fond of gymnastics and have a gymnasium fitted up . . . where we spend an hour or two twice a week. From the exercises I have derived an immense advantage."[18]

As much as Williamson enjoyed the company of Alfred Victor's children, it was Alfred Victor's youngest brother, Alexis, who was to have the greatest influence on his life. Although early in his tutoring Williamson had been impressed with the scholastic abilities of Alexis's daughter, Fanny, he did not get to know Alexis until late in October, when he met him at a Whig political rally. As he was able to do with everyone he met, Alexis immediately charmed Williamson, and the young tutor was delighted when Alexis invited him to accompany him the next day to attend church services in Wilmington. "We heard an exelent [sic] sermon from Mr. Franklin the Episcopal minister. Mr. Du Pont (Alexis) the only one of the family who is a member of the church (I mean the male portion) is one of the most thorough going Christians that I ever saw. On Wednesday evening last he came up to invite me to take tea at his house, which I did. I [sic] was the first time that I have visited him and I had a very nice time his wife is an elegant and exelent [sic] lady."[19]

From then on, Alexis was Williamson's mentor and spiritual guide. They spent many evenings together discussing religion. Alexis gave him sectarian tracts and loaned Williamson books from his personal library. The result was that at Christ Church one Sunday morning, which happened to be Communion Sunday, a day "never to be forgotten, I prayed fervently for direction from on high . . . [and] when the servants of the Most High were called to come and partake of the precious emblems of the Body and Blood of Christ, I seemed led by some unknown power to the Altar and there received the Holy emblems faithfully, prayerfully, & as I believe, to the everlasting benefit of my soul."[20] Alexis had made yet another, and in this case, quite easy, conversion to his faith.

With his newfound religious faith, his well-established friendships within the du Pont family, and his growing self-confidence, which led to greater success in the classroom, Williamson found his last few months on the Brandywine the happiest period he had yet experienced in his life. The du Ponts had received him into their exclusive circle, an acceptance that was officially recognized when he was asked in June 1853 to be the escort of both Paulina and Sophie du Pont to the most glittering social event of the season, a formal supper dance at the mansion of Joshua Gilpin, a wealthy paper and textile manufacturer in the Brandywine valley.[21]

Even this social acceptance by the du Ponts and the Gilpins, however, could never make Williamson their peer. An unsophisticated youth from the rural Piedmont region of Virginia, who prior to coming to Wilmington had never been more than thirty miles from his home, Williamson was always painfully aware of the great gulf that separated him from the du Ponts. He frequently wrote to his mother about the du Ponts' wealth:

> The du Ponts are descended from an ancient French family. Henry Du Pont's father raised a large family all of whom he settled in elegant mansions around him. . . . They form a community within themselves and live conjointly on the proceeds of an immense capital that they have never divided. You may have some light idea of their wealth when I tell you that in making the article of powder one of the materials viz. salt petre costs them at a time $100,000 this I learn is a small supply and only lasts 3 months.[22]

In stark contrast to his own humble origins, his limited education, and his even more limited financial resources, which had almost forced him to decline the Gilpin invitation because he felt he did not have clothing appropriate to the honor of escorting Alfred du Pont's daughters to the party, the months on the Brandywine were a rare and brief opportunity to experience a way of life that to him seemed to belong in a fairy tale. In July, when the term ended, the bell would strike midnight, and in place of the du Pont carriages there would be only the pumpkins that grew on his father's acreage in Orange County, Virginia. But his good friends Sophie, Paulina, and Lammot, their cousin Fanny, and, above all, their Uncle Alexis du Pont would remain very real persons who had once befriended Page, and had shaped his life.

The one other child of Alfred Victor and Margaretta who was living at home at this time, but whom Page Williamson never mentioned by name in his letters, was the oldest son, Eleuthère Irénée, who, like the grandfather for whom he was named, was always called Irénée. Williamson's failure to comment on Irénée in his weekly chronicle is not surprising. Most of the du Ponts as well seemed to forget or at least to ignore Irénée's existence. He was the eldest grandson of the founder of the company bearing the name du Pont and the first of his generation to be named a partner in the company and therefore he might have been expected to be the heir apparent to the du Pont throne of power. Irénée, however, was the least assuming and the least aggressive of Eleuthère Irénée du Pont's grandsons. As a small boy, he had been given the nickname "Bishop" by his brothers and sisters, not because of any proclivity to emulate the religiosity of his Uncle Alexis and Aunt Victorine, but because of his solemn and reserved demeanor. He seldom expressed any emotion, and his stoicism was viewed by his family as an indication of either a lack of feeling or a dullness of spirit. His father greatly underestimated the boy's intelligence and his mother found him sluggish and uninteresting in comparison to her other children. Neither of Irénée's parents understood or expected much from him.

But behind the wall of taciturnity that he had constructed around himself was a sensitive and passionate youth who, for whatever reason, could express his strong feelings only to himself. The volubility and extroverted exuberance of his mother, his siblings, and his cousins overwhelmed him, and he had withdrawn into himself, where he could dream his own dreams and where his loves and hopes were not du Pont fabricated. Irénée had never liked the world into which he had been born. Unlike his younger brother, Lammot, he had never enjoyed playing about the mills and following the workmen around the yards. Irénée's earliest nightmares were not of imaginary hobgoblins who lurked in the dark but of very real explosions, which broke the windows of his nursery in broad daylight.

The freest time he had ever known were the years he spent away from home in preparatory school and at the University of Pennsylvania. There he made friends who were not du Ponts and who accepted him for what he was and not for what he should be. Like his father, Irénée would have been happy never to return to the Brandywine, but he, too, knew full well what his destiny was, and so he had returned and entered the mills. The only analgesic he had to dull the pain he felt was withdrawal further into himself. He did his assigned tasks without complaint but without enthusiasm. His Uncle Henry could find nothing in his nephew's performance to criticize, his conscientious attention to detail was exemplary, but there was nothing that Boss Henry could enthusiastically praise either. Clearly, the second Eleuthère Irénée had not been born to command. By the time that Page Williamson came to the Brandywine to teach, Irénée had become an obedient and efficient automaton in the mills and a silent presence at his mother's dinner table. There had simply been nothing concerning Irénée for the young tutor to write home about.

It was not long after Williamson's sojourn in Delaware, however, that for the first time in his life Irénée provided a topic for conversation at dinner tables throughout the valley. This seemingly phlegmatic and dull young man, who had never before given cause to attract any attention to himself, fell in love—and with such passion and overtness as to cause the whole community to stare. That in itself was certainly noteworthy enough, but the object he had chosen for his affection was truly unbelievable. Irénée had fallen in love with Charlotte Shepard Henderson of Washington, D.C., the most fashionable, beautiful, and vivacious Southern belle ever to grace the Brandywine valley on her frequent visits to the Alfred du Pont home.

Charlotte was the daughter of Anne-Marie Cazenove and Archibald Henderson. Her mother's father, Antoine Charles Cazenove, had been born in Geneva but emigrated to America at the height of the French Reign of Terror in 1794. He had entered into a business partnership with another Swiss émigré, Albert Gallatin, who was later to win fame as Jefferson's Secretary of the Treasury. After his marriage to Anne Hogan in 1797, Cazenove settled in Alexandria, Virginia, where their daughter Anne-Marie was born. Through their mutual friends, Gallatin and Jefferson, the Cazenoves met Pierre Samuel du Pont and his two sons soon after the du Ponts arrived in Delaware, and the two families became close friends.[23]

Upon the recommendation of the du Ponts, the several Cazenove children attended the same academies in Philadelphia—Mt. Airy College for the boys and Mme. Rivardi's Seminary for the girls—that the du Pont children attended. Irénée du Pont's daughter, Sophie, would later recall, "The difficulties and inconveniences of travelling were such in those days that the young people from Alexandria could not go home for the shorter holidays, such as Christmas and Easter. . . . All such holidays were therefore spent by the Cazenove children at my father's—where they were loved and cherished as their own children by my parents, and where a bond of fraternal affection grew up among us all."[24] Anne-Marie Cazenove became a particularly close friend of Evelina and Eleuthera du Pont, and her younger sister Charlotte was the close companion of Sophie du Pont. The friendshps formed in these years continued after the young girls grew up and married.

In 1823, Anne-Marie Cazenove married General Archibald Henderson, commandant of the United States Marine Corps, a position that he was to hold longer than any other commandant in the history of the Corps. In the Commandant's House, popularly known as The Barracks, at 801 G Street, S.E., in Washington, D.C., on 25 September, 1835, Charlotte Henderson was born, the seventh of nine children.[25].

Fifteen years later, Charlotte repeated the pattern of holiday visits to the Brandywine and found the same warm welcome in Alfred du Pont's home that her mother had received from E. I. du Pont's family. Like some brilliantly plumed bird of passage, Charlotte would flit into the Brandywine valley on her way to or from Washington, bringing the latest tidbits of gossip from the nation's capital: what the ladies of the French and British ministries were wearing this season, whose political stars were in the ascendancy, what her father thought were the chances of war in Central America, or, more personally, which eligible bachelor she was currently in love with, for Charlotte was always falling in and out of love. By the time she was twenty, she had been engaged to be married on at least three separate occasions. When they weren't admiring her costumes, Margaretta du Pont and her daughters would sit and listen with fascination to the tales of Charlotte's latest conquest. Even Aunt Sophie would insist that dear little Charlotte must visit her in her home upon the hill, for after all, Charlotte was the namesake of her aunt Charlotte Cazenove Shepard, who had been "the closest and dearest friend" that Sophie had ever had. For that reason alone, Aunt Sophie, who was the stern moral arbiter of her many nieces and nephews, could forgive dear little Charlotte for her frivolity, fickleness, and too fancy dress.[26]

Alfred's four sons were equally fascinated with Charlotte's presence in their home. She was the most beautiful creature they had ever seen. Beside her, all of the young women in the Brandwyine valley paled into drabness. Their response to Charlotte's romantic conquests was, however, quite different from that of their sisters and mother. What chance had any of them if Charlotte could so casually dismiss handsome young lieutenants in her father's Marine Corps, dashing young Virginia cavaliers, or heirs to Philadelphia banking and mercantile fortunes? While his younger brother vociferously moaned over Charlotte's unattainable charms, Irénée, as usual, held his tongue and bottled up his emtions until that memorable day when he announced to his startled family that he was in love with Charlotte, and furthermore, he intended to have her. What possible chance did stolid and dull old "Bishop" have in winning the beautiful but elusive hand of Charlotte? If ever there was an instance of unrequited love, this was surely it.

If the family had been amazed by Irénée's uncharacteristic announcement, they were struck dumb by the growing realization that Charlotte was not summarily rejecting, but actually encouraging Irénée's advances. Why Irénée of all people? It could not be just his name or his family's wealth that attracted her. She could have had the pick of metropolitan society and could have laid claim to much greater fortunes. Irénée, to be sure, had the strong, masculine features of most of the du Pont men, which women found attractive, but he was certainly no Adonis. It could only be that Charlotte had breached the wall of reserve he had built around himself and saw an Irénée that no one in his family had ever seen. He was, moreover, an older man, six years Charlotte's senior, and far more mature than the youthful swains who had so assiduously but clumsily courted her in Washington and Virginia. Then, too, Charolotte thoroughly enjoyed the shock effect that this romance had. She would always find pleasure in doing the unexpected.

For whatever reason, Irénée's courtship of Charlotte was brief. Charlotte accepted his proposal, and they were married on 28 October 1858 in the Marine Commandant's mansion in Washington, D.C., with President Buchanan in attendance. For better and certainly also for worse, Irénée and Charlotte were wed. From this unlikely union Alfred Irénée du Pont was born six years later.

II

A Child Born Out of Civil War
1864–1877

T HE TWELFTH OF MAY 1864 was not a propitious moment for
either a nation or a child to be born. It had been three years and one
month to the day since the guns from the shore batteries of Charleston
harbor had opened fire upon the Federal island garrison of Fort Sumter, a
salvo that was intended to announce the birth of a new nation upon the
North American continent. But the birth had been far more difficult than
anticipated, and now after one thousand days the South was no nearer
achieving its independence than it had been on the twelfth of April 1861.
It was not the fresh scent of life but the fetid odor of death that hung over
the valleys of northern Virginia as General Grant's massive blue line beat
stubbornly against the thinner but unyielding gray line of General Lee's
troops in a prolonged engagement appropriately named the Battle of the
Wilderness. It was a long day of sorrow for two nations, one struggling to
be born, the other refusing to die.

Nor was it a time of celebration for the du Pont family, even though
on this day Charlotte Henderson du Pont, after five and a half years of
marriage and having given birth to two daughters, had at last given her
husband, Irénée, a son. It was not the day of rejoicing that it should have
been, for a dark shadow hung over this marriage, and none of the du Ponts
could have been particularly happy with the news that Charlotte had given
birth to yet another child.

For the newborn infant's grandmother, Margaretta, and for his numerous uncles and aunts, it did not seem six years but an eternity since Charlotte Henderson and Eleuthère Irénée du Pont had exchanged marriage
vows. The young couple's marriage had at the time seemed to be a romantic
fairy tale made real: the shy, reticent prince—certainly the least likely of all
her suitors—claiming as his own the beautiful, vivacious princess who might
have chosen any one of many eligible bachelors in Washington or Virginia

society. For the practical-minded members of all the families involved, this
union represented a matrimonial merger to gladden their hearts, for it
brought together the wealth and ancestral distinction of the du Ponts, the
Lammots, and the Cazenoves with the military prestige of Charlotte's
father, General Archibald Henderson.

The bride and groom themselves had also brought to this union their
own individual attributes. Charlotte Henderson apparently had everything
that a man and his family might want in a bride: beauty, wit, social poise,
and an unquenchable enthusiasm for life that should have made her a dazzling ornament on any family tree. Irénée, in turn, had precisely those qualities of character and temperament that would best complement and balance Charlotte's charms: seriousness of purpose, industry, a solemn
judiciousness, and a firmly established position within the du Pont hierarchy. It was, in short, to use a tired Victorian cliché, a marriage which
seemed to have been made in Heaven.

But by May 1864, it was apparent to all of the du Ponts, even to the
distracted Irénée and the distraught Charlotte, that this union was, instead,
a marriage spawned in Hell. For along with the golden sovereigns of beauty
and grace, Charlotte had brought as a part of her dowry the counterfeit
coin of mental instability inherited from her Grandmother Henderson's
family.[1]

If the Cazenove family ever knew of this dark stain in the Henderson
family into which Anne-Marie Cazenove had married, they had never given
the slightest intimation of the fact to the du Ponts. It was common knowledge, of course, that Charlotte's father was a bit peculiar. He was an autocrat who demanded blind obedience from both the officers and men under
his command and the wife and children under his roof, but his unreasonable
severity was attributed to his military background. His violent outbursts of
temper and equally difficult moods of sullen taciturnity were not regarded,
in the nineteenth-century patriarchal society, as indicating mental instability. He held a firm grasp upon the United States Marine Corps until the
day of his death, and neither his civilian superiors nor his military inferiors
were ever to question his competency. The President of the United States
himself, on the day of General Henderson's funeral, had paid him the signal
honor of following on foot as the funeral cortege made its long journey
from the Marine Barracks to the Congressional Cemetery.[2] When General
Henderson's last will and testament was read and it was discovered that in
a handwritten codicil he had bequeathed to his wife the Marine Commandant's mansion belonging to the United States government, this too was
dismissed by the probate court not as being an act of an unsound mind but
as a charmingly humorous bit of eccentricity.

Charlotte, however, did not escape as easily from the Henderson heritage. This beautiful young woman of light and laughter became a distraught creature of darkness and fury. Her descent into madness was coincidental with but not unrelated to the nation's descent into the maelstrom of civil war, and in time she personalized for the du Ponts the national tragedy of the 1860s. Her presence within the family caused its first searing division, a domestic civil war, and her insanity became in microcosm the nation's insanity.

In analyzing Charlotte du Pont's increasingly erratic progression into madness, it is difficult to separate heredity from environment and to assess the relative importance of each factor. Was Charlotte damned from the moment of her conception? Of her parents' nine children had she alone been singled out by some cruel Mendelian lottery to receive the full burden of an unwanted paternal inheritance? Or was it possible that, given different circumstances, she, like her father, might have been only brushed, not crushed, by heredity?

The du Pont family itself would endlessly debate and remain forever divided over these questions. Irénée's mother, the indomitable Margaretta du Pont, was convinced that Charlotte's disintegration into madness was inevitable and that the real tragedy—worse than that, the real crime—must be attributed to the Henderson family for not having revealed their family secret prior to Charlotte's marriage to Irénée. Margaretta, who was so strong herself that she found it difficult to tolerate weakness in others, saw nothing in the circumstances of Charlotte's married life that could possibly affect her mental stability. As evidence to support her position, Margaretta could and frequently did point out certain peculiarities in Charlotte's behavior that were manifest from the moment she had moved into the du Pont enclave as Irénée's bride. Such eccentricities as being particularly restless and unhappy at the time of the full moon must have been observed by her parents and siblings long before her marriage and should not have been kept secret.[3]

Charlotte's husband, Irénée, on the contrary, was equally convinced that there had been nothing wrong with Charlotte prior to her marriage. It was his own family, and particularly his mother and his Aunt Sophie, who had driven Charlotte into madness by constantly harassing her for being sympathetic to the Confederate cause. For this act of needless cruelty, Irénée never forgave his mother or his aunt. Charlotte's own family, the Hendersons and the Cazenoves, denying any duplicity in concealing the Henderson family secret, also blamed the du Ponts for Charlotte's fate. When she was first committed to an asylum, without any cause in the Hendersons' opinion, the du Ponts had doomed Charlotte to madness.

Whether or not heredity was the determining factor, one can hardly deny that the circumstances of Charlotte's life after she married Irénée gave her inherited tendencies full opportunity to flourish, like noxious weeds in a well-prepared soil. It was never easy for an outsider to marry into the du Pont family, for one perforce married the entire family as well, and one was expected to find social contentment and fulfillment within the du Ponts' small and isolated world on the banks of the Brandywine. For Margaretta Lammot and Joanna Smith, who had grown up in the nearby Chester Creek valley, the transition from one small manufacturing enclave to its neighboring settlement was not that difficult, but for Charlotte, who had been the belle of Washington and northern Virginia society, the confines of the Brandywine must have been stifling. She and Irénée had been settled in their small cottage on the Brandywine, assigned to them by Uncle Henry, for only two months when Charlotte received word on 6 January 1859 that her father had died. Two weeks later, her mother also died, and after that, there was no family base in Washington to which she could return for a brief renewal of her old life. The Brandywine community was now her only home. Marriage, then her parents' death—it had all been too abrupt and too permanently final.

She could, of course, make frequent one-day trips to Philadelphia for shopping and an occasional social event. She could even take an infrequent trip to New York City to visit Irénée's oldest sister, Victorine du Pont Kemble, which Charlotte delighted in, for although ten years older than Charlotte, Victorine treated her as an equal, as a person in her own right, not just a wifely appendage of Irénée. Aside from these family-approved outings, Charlotte was now expected to stay at home, manage the household for Irénée, and find all of the social life she needed in polite afternoon calls on her mother-in-law or teas with Aunt Sophie and Aunt Eleuthera. Irénée had his work, which kept him at the powder mill twelve to fourteen hours a day. Charlotte could have his babies, which should keep her happily employed at home. This was presumably God's (and most certainly it was the du Ponts') concept of the proper order of the universe.

Even under normal conditions, a woman as active and restless as Charlotte would have found life on the Brandywine tedious and irksome, but the Civil War made her existence within the du Pont family intolerable. The du Ponts were totally committed to the Union cause, and there were no more dedicated loyalists in the border state of Delaware than they. As munitions makers they were in a position to make their loyalty count. Even before the first shots were fired on Fort Sumter, Boss Henry had foreseen the possibility of an armed sectional conflict and had issued orders that no gunpowder was to be sold to Virginia or to any other state south of the Potomac. In so doing, he was following the example set by his father, who in 1832 had rejected a large and financially tempting order from the South Carolina state militia at a time when the Palmetto State was threatening to secede if the latest protective tariff act were enforced.

It is true, of course, that the Du Pont Company lost nothing financially by Boss Henry's resolute loyalist stand in 1861, for in the long war that followed, his mills were worked at full capacity and even then could not fully supply the insatiable demands of the Union armies for more ammunition. But even if this had not been true, the du Ponts would have preferred bankruptcy to allowing one keg of gunpowder to be used by Rebel forces.

Individual members of the du Pont family also directly participated in the conflict. Sophie du Pont's cousin-husband, Samuel Francis du Pont, was an admiral in the United States Navy and was commander of the squadron that captured Port Royal, South Carolina, in November 1861. He became the North's first great naval hero of the Civil War. Boss Henry's older son, Henry Algernon, a lieutenant colonel in the artillery, distinguished himself in the Virginia campaigns and was later awarded the Congressional Medal of Honor.

Lammot du Pont, Irénée's younger brother, at the outbreak of the war made a hurried trip to England and shrewdly bought up all of the British Indian saltpeter, an ingredient essential to the manufacture of gunpowder, that was available on the London exchange. When in retaliation for the Trent Affair in November 1861 the British government refused to issue a shipping permit for this cargo, Lammot was obliged to return to the United States without the saltpeter. He went directly to Washington to report to Secretary of State Seward the intransigence of the British government, which could have dire consequences for the war effort. Lammot was able to return to England on the very next passenger ship, carrying with him a letter from Lincoln to Prime Minister Henry Palmerston threatening the immediate recall of the American minister and the probable declaration of war by Congress against Great Britain if the shipment of saltpeter was not at once released.[4] Palmerston reluctantly yielded to this direct pressure from Washington. The release was signed and a triumphant Lammot promptly left port abroad the ship carrying his precious cargo.

Although this bold and successful venture by Lammot into high wartime diplomacy had to be kept a secret within the family, it gave him a position of prestige above all of the other du Ponts in Boss Henry's all-important evaluation. Later in the war Lammot would seek the more public glory of the battlefield by enlisting in and serving as adjutant for a volunteer company of Indiana infantrymen.[5]

Charlotte found herself isolated within this super-patriotic family. Although she had lived all of her life in the neutral territory of the District of Columbia, her parents had had a summer home in the Blue Ridge mountains, and she had always thought of herself as a Virginian. Most of her childhood girlfriends were now dedicated daughters of the Confederacy, and many of her former beaux as well as two of her brothers were serving in the Confederate army.[6] On the Brandywine, living with the du Ponts, Charlotte felt herself to be a pariah, an enemy alien, especially in the eyes of her mother-in-law and Aunt Sophie, whose zeal for the Union cause knew no bounds. With each Union victory, Charlotte had to endure the raucous hurrahs that rang out, and with each Confederate victory she found it even more difficult to bear the sullen silences and hostile stares directed toward her. It would have taken a far stronger and much less sensitive person than Charlotte to maintain equanimity under these conditions. One of Charlotte's living nightmares must have been the constant realization that her husband and his family were working day and night to grind and polish the powder that would shortly be directed toward her brothers, her cousins, and her friends. Small wonder that with each pasing month of this dreadful quadrennium of slaughter her inner tension grew, and her desire to escape the Brandywine, which she must perforce suppress, became a kind of madness in itself.

Charlotte's mental health had begun to deteriorate a year before the outbreak of the Civil War, however, and this gave support to Margaretta's assertion that the strained relations within the family produced by the war had nothing to do with Charlotte's instability. The first clear indication of difficulty followed the May 1860 birth of her first child, a daughter named Anne Cazenove after her recently deceased maternal grandmother. Charlotte apparently was particularly susceptible to postpartum depression, and with her it reached a frightening intensity. She took no pleasure in the newborn infant or in anything else, but sat alone in her darkened room quietly weeping. After a few days in which Charlotte seemed to grow ever more dejected, Margaretta took the baby into her home, an Irish wet nurse was found for the infant, and Charlotte was sent off to New York to stay with Victorine for an indefinite period of time. It was only with this move that Charlotte began to show a small flicker of interest in living. In October 1860, five months after the birth of the child, Margaretta wrote to Victorine to thank her for her report on Charlotte's condition:

> Certainly she is better, but she has been so before & I think until she is perfectly well she is as far off as ever from recovery & if she has a spell the 29th [the date of a full moon] it will even be discouraging at least. I want you to see the baby very much before she is sick or teething or anything to alter her. She is lovely nothing missing to make a perfect baby. I just doat [sic] on her & if her mother should recover it will be very hard for me give her up to her.[7]

Charlotte did recover, at least enough to yield to Irénée's entreaties to come back home to him and their little daughter. The outbreak of the Civil War and a second pregnancy were hardly conducive to her full recovery, however, and the tension between her and her mother-in-law increased. At least Charlotte had the sympathetic support of her husband in what was now an open feud with Margaretta. Irénée found unforgiveable his mother's lack of understanding of Charlotte's difficult position of being caught in a conflict of loyalties. When their second child, another girl, was born in December 1862, Irénée agreed with Charlotte that the child should not be named for her paternal grandmother, as was the custom. The baby was named Marguerite, which was close enough to Margaretta to make the intended insult obvious.

To the family's surprise, Charlotte survived the tensions of the war years and the births of Marguerite in 1862 and Alfred in 1864 in better health than she had known during the first two years of her marriage. Perhaps, Irénée hoped, her mental instability had been a temporary affliction—a terrible period of adjustment to marriage and motherhood that she now had successfully completed. Prior to Alfred's birth in 1864, Uncle Henry had been persuaded to assign the young couple new living quarters. In place of the small cottage they had first been given, Charlotte and Irénée now had a large, imposing house on Breck's Lane. Charlotte seemed to take a genuine delight in her new home, in her two young daughters and her infant son. Moreover, she had the loyal support of her husband in the continuing feud with his mother, for Margaretta was openly critical not only of Charlotte's political views but also of everything else that concerned her: how she kept house, how she spent her husband's money, how she raised her children, even how frequently the babies had colds, as if their childhood illnesses were willfully induced by their mother. It was not easy for Charlotte to bear all of this, but somehow she managed to keep from collapsing. It was as if she had determined that she would not break under the strain. She must have felt that, like her brothers, she was carrying the Confederate flag to battle, and no more than they could she retreat before the enemy.

She did have frequent tantrums, particularly when she was alone with Irénée, and there was an occasional flight to friends in Philadelphia or to her dear sister-in-law Victorine in New York. But she would quickly return to the comforting arms of Irénée, for their physical attraction was still there, and she found the same satisfaction in having Irénée's presence in her bed that she had known as a bride. Thus for four years she endured the seemingly unendurable.

The Civil War brought great prosperity to the Du Pont Company. Production reached levels that would have been considered impossible to achieve in 1861, but even with expanded facilities, the du Pont partners could not possibly meet the demands placed upon them by the United States government, and many small competitors sprang up throughout the country to meet the demands of a nation at war. The War of 1812 had given Eleuthère Irénée du Pont's small mill its first economic boost, but it was the Civil War that transformed the company into an industrial giant that would dominate gunpowder manufacturing for the next fifty years.

Individually, the du Ponts prospered beyond the wildest dreams of the company's founder and the most extravagant boasts of his father, Pierre Samuel. The major share of the wartime profits came to Henry du Pont, for in the reorganization of the company in 1857, after the deaths of Alfred Victor and Alexis, Henry received twenty of the forty-one shares of the company, which was now capitalized at $697,000. Alexis's estate was given fifteen shares, and Alfred Victor's two sons, Eleuthère Irénée II and Lammot, each received three shares. For the first time, the du Pont daughters were left out of sharing in the capital stock of the company. Later, in the midst of the Civil War, in 1863, a new partnership agreement was drawn up that gave Henry an even larger piece of the pie: 64.5 percent of the total 31 shares subscribed. Irénée and Lammot, representing the Alfred Victor family, each received five shares, and Eugene du Pont, Alexis's oldest son, one share, with the total capitalization of the partnership set at $620,000.[8] Federal income tax reports of 1862 showed Henry du Pont to have the largest income in the state of Delaware, reported at $270,000, which was six times as large as the second-highest income in the state. Irénée and Lammot, who each reported an income of $39,000, were tied for fifth place among the only nine persons in Delaware who had had incomes of over $20,000 during the previous year.[9]

If the war brought rich profits to the du Pont partners, it also had its debit side in anxiety and sorrow. Sabotage in the mills was a constant fear, for there were many Confederate sympathizers in the two southern counties of the small border state of Delaware. A more real possibility was an accidental explosion caused by carelessness of workmen in their haste to fill orders. There was even the remote possibility that the Brandywine valley might be invaded by Rebel troops, particularly in the summer of 1863, when General Lee made his bold foray into southern Pennsylvania, with Philadelphia as his reported objective. On 1 July, two days before the battle of Gettysburg, Joanna Smith du Pont wrote to her oldest daughter, Fanny, that all the families in the valley had packed up their valuables. "I *forced* Alexis to dig a hole down the hill and bury the coins. My silver is all in a bag ready to bury at the last moment, and all my jewelry is packed with my papers in boxes. I do not intend to leave here, if I can be allowed to say, but I may send away the two girls and keep the younger children with me. . . . It was reported yesterday they [the Rebels] were at Peach bottom, only *42 miles* from here, but it was not true. It has been discovered that the Rebels in Del. are sending arms to the prisoners in Fort Delaware!"[10]

The family also experienced anger, followed by sorrow, when Admiral du Pont, the hero of the capture of Port Royal, was repulsed in 1863 in his efforts to take Charleston, South Carolina. He blamed his failure on the Navy Department's not giving him adequate support. Because of this accurate but quite impolitic public accusation, du Pont was summarily dismissed from his command of the South Atlantic Blockade Squadron by a vindictive Secretary of the Navy, Gideon Welles. After nearly fifty years of distinguished service, du Pont resigned his commission and returned to the Brandywine a broken man. He died shortly thereafter, as much a casualty of the war as he would have been if shot in battle.

There was also, of course, the continuing friction that Charlotte's presence within the family created. It had been hoped that that too would end, along with all the other anxieties, once the rebellion was crushed and peace returned to a war-torn nation. So the du Pont family, particularly the women, joyously welcomed the end of the war, even though this meant a precipitous drop in the demand for their company's product and a greatly reduced income for them all. Even Charlotte, who mourned the demise of her beloved Confederacy, could find solace in the fact that her brothers had survived the conflict, and she herself would once again be able to visit her family in Virginia, North Carolina, and Florida, from whom she had been separated for four long years.

Unfortunately, the end of the war did not end the bitter internecine du Pont feud. It expanded and grew more intense as Margaretta next proceeded to alienate herself from her second son, Lammot. As soon as he was discharged from the service, he returned home and announced to the family that he was engaged to a local woman, Mary Belin. This came as a surprise—and a most unpleasant one, at that—to Margaretta. On 3 July 1865, she wrote to her youngest son, Bidermann, now in business partnership with his brother Fred in Louisville:

> I have a piece of news to communicate. Lammot is engaged to be married. Now guess & guess again. No one had the least idea—not a single individual & the lady, I think, was as much surprized as anyone. Even sharp sighted Polly had no notion of it,—Mary Belin!
>
> I have no objection but health. She is very delicate & certainly will go into a consumption—has had severe hemmorages [sic] & never without a cough. . . . But tis done & we must only hope for the best. He is happy & bright just now.[11]

Margaretta's real objection to this marriage, however, had nothing to do with Mary's health. Mary Belin was the daughter of Henry Belin, one

of the clerks employed by the Du Pont Company. He also had Jewish ante-
cedents; thus, on the bases of both current social standing and ancestry, his
daughter was simply not a suitable partner for the Lammot and du Pont
families. In many ways both subtle and bold, Margaretta made her view
obvious to both her son and prospective daughter-in-law.

Lammot did not have his brother Irénée's gentle patience. In neither
business nor family matters did he ever go out of this way to avoid confrontation and conflict. He always met an issue head on, never mind the
consequences. He soon let it be known to his mother that her social ostracism of his wife was not only perceived, but resented. If she did not want
to recognize his wife as a daughter then she no longer could claim him as
a son. He and Mary had their own friends in Wilmington, and the du Ponts
and their precious Brandywine circle could be damned. Lammot severed
relations with his mother. Margaretta could only write a plaintive little note
to her son Fred, the one son who never married and the one with whom
she still maintained a close relationshp, although—or perhaps *because*—he
lived in Louisville: "Lammot came home some days ago," she wrote, "but
I have not seen him & he passed [here] in the carriage this morning. . . .
How little he cares for me—'tis hard to have lived & loved him till 35 yrs,
& see how indifferent he is."[12]

If Lammot now regarded his mother as the enemy, he at least had one
friend and ally within the du Pont family. In the years that followed, Charlotte looked to Lammot more and more for support and sympathy, which
were always forthcoming. When during the economically depressed years
of the 1870s, Irénée attempted to put some restraints on Charlotte's expenditures, it was to Lammot she turned for information and advice:

> I have just received a few lines from Irénée which have considerably
> annoyed me, and as I am constantly worried in a similar way, I am going
> to trouble you to let me know if there are really any grounds for the
> uneasiness in relation to his financial concerns which he is constantly
> expressing.
>
> I sent the other day for a modest sum—one hundred dollars and he
> replied—"Please make this do, as my finances now admit of nothing but
> necessities, and I am now borrowing the moneys." Can you explain what
> this means as I know you could pretty well judge of his affairs by your
> own. . . . Since my return from abroad I have not had the slightest idea
> of what his income is. As this is a very painful & unsatisfactory position
> for me will you be good enough—in the strictest confidence on both
> sides, to let me know the true state of affairs.[13]

Lammot evidently supplied her with the information she sought. In a
second letter, written two weeks later, she thanked him "very sincerely for

telling me the whole truth, as it was that alone I was in search of. . . . I have not the slightest dread of stinting myself—indeed am particularly anxious to spend the smallest possible amount on myself. All the more that the sum I receive from my mother's estate is comparatively trifling. . . . My principal reason for consulting you in this matter is that I may know what sum I should spend upon the household and upon the children's education. . . . When I return next month I shall far better understand what course to take, for which I again acknowledge my indebtedness to you."[14]

But poor Charlotte was never be able to understand what course she should take, nor did she ever find it possible to "stint herself." Any effort upon Irénée's part to impose restrictions upon her led either to tantrums or to her flight from home.

Charlotte, who had found unexpected strength to withstand the tensions of the Civil War, now seemed to lack the determination necessary to cope with post-war life. In May 1866, her fourth child and second son, Maurice, was born. Charlotte seemed to bear the pregnancy and postnatal period well, and again Irénée had hopes that the worst was behind them. Even the relationship between his wife and his mother seemed to improve now that Margaretta had another daughter-in-law with whom to feud, and Charlotte's presence within the family no longer personified hated Rebel treason.

Irénée's hopes were crushed eighteen months after Maurice's birth, when the fifth and last child was born. This child, a son, was named Louis Cazenove after Charlotte's uncle. Then Charlotte's life quickly began to unravel. Charlotte, deeply sunk in postpartum depression, was sent off once again to stay with the long-suffering Victorine. After a few months she rallied enough to return home, but this time she did not experience the recovery she had managed eight years earlier after Anne's birth. Charlotte now directed all of her hostility not against her mother-in-law, but against her husband. The physical attraction she had once felt toward Irénée turned into physical revulsion. She could hardly stand the sight, not to mention the touch of him. Her condition rapidly deteriorated, and in March 1870 Irénée reluctantly yielded to the constant urging of his mother and the advice of her physician. He committed Charlotte to a private asylum in Philadelphia.

Margaretta, as always, served as the family's general reporter, keeping them informed about Charlotte's current condition. In April, she wrote to her son Fred:

> I cannot remember when I wrote to you, for I am in such an unhappy state that I pay no attention whether my last was to you or Binn [her son Bidermann]. . . . Charlotte still at the asylum, writing

every day to somebody saying she is quite well. At first furious that Irénée had taken her there, but in a letter written to Mary d P the last one
we know her to have written, she says it was right to have put here there
now she is well she ought to come out. . . . She is bitter against Irénée
which he knows, though most of the unkind things are not told him &
he is heartbroken. He sleeps badly eats less & has been in bed for three
days. . . . Charlotte spoke so much against Irénée at the Asylum that the
Dr. had to contradict it to the Drs. and attendants.

For two years C has not been right in her mind & took a kink against
Irénée, magnified and intensified every little thing he did, & wrote it all
to Victorine who thought her sane & believes all she said. Victorine has
been writing to Irénée that he ought to take her out & worries him
dreadfully. Crazy people can *write* very well tis their actions show that
reason and judgment are gone. . . . Emma says Charlotte has been very
unkind to Irénée for 13 months, & says Irénée never said an unkind
word in reply & gave her all she wanted & asked for. Tom Lytell told
Irénée she had been out of her mind for a long time he saw. Now as
this has been coming so long, I cannot imagine her well enough ever to
leave the place where she is well cared for with every amusement. She
wrote she had a donkey to ride every day. 300 acres are enclosed in
beautiful rides & walks.[5]

The three hundred acres and a donkey to ride were hardly adequate
compensation for being confined in an asylum, however, and in June 1870
Charlotte persuaded the doctors to release her. She stayed the summer with
Victorine, and then returned home, but there was no peace there for either
her or Irénée. The week after New Year's day 1871, she gathered up her
five children and their nurse and went off to live in the Franklin House in
Philadelphia. Margaretta wrote to her sister, Mary Lammot: "She goes to
Phila. to board with the 5 children & Emma & when persons have said, Oh!
how hard for Irénée, she says, well he can come up once a week & see the
children! . . . The children had a good teacher & were very happy at home,
but she has her own way in everything, & is *entirely selfish*. She is not fit to
have the children with her & still not crazy enough to put into an Asylum.
You may suppose how this worried me, & to see Irénée so unhappy & broken down." At the end of March, Margaretta again reported to Mary
Lammot:

Charlotte with the children have been three months in Phila but I
am glad to say, the children are to come home on Friday, they are crazy
to get back. She goes either to Alexandria or New York for a month or
so. Dislikes her home, husband, B/wine, & everything but excitement.
No one thinks her out of her mind, she is so gay & charming to strangers, but moody & does nothing, literally, not even dress herself when

alone. I don't think she will ever be different. Irénée gives her all she wants, lets her spend what she chooses, but she can't bear him. That is the point she is wrong on. Don't speak of this. I expect plenty of trouble hereafter about it.[16]

Three months had been long enough, at least to her own satisfaction, for Charlotte to prove the point that she was attempting to make: that she loved her children and that she was sane enough to take care of them. Never again would she find it necessary to leave home with her children. She, however, continued her ceaseless journeys throughout the rest of her short life, trying to find among strangers the peace and sanity she could never know at home with Irénée. Had she lived a half-century later, when divorce was not only accepted but almost expected within the du Pont family, she most certainly would have divorced her husband, and this might have been the salvation for both of them. But divorce was simply not an option in 1870, and for Charlotte there was no escape other than prolonged absences from home, with only an occasional hurried, brief visit to the Brandywine to embrace her children, to marvel over their growth during her absence, to distribute the many rich presents she brought with her, and then, after a short stay, a quick exit from the morass of Swamp Hall, the name that Irénée, half in jest, half in bitter truth, had given to their home on Breck's Lane.

In November 1871, Charlotte sailed for France, accompanied by her brother, Charles Cazenove. There she stayed with her aunt Pauline Cazenove Fowle. She remained in Europe for nearly two years. During her long absence, Irénée, at home with the children, both longed for and dreaded her return. During this time he began to show definite symptoms of tuberculosis. Margaretta wrote Fred in February 1872 that, upon her and her daughter Paulina's return from visiting Victorine in New York, they were met at the Wilmington station by the company's coachman, Tom Lytell, who was waiting to take them to Irénée's.

> Well, we found that the two weeks that we were away he had had a very bad cold & had increased it 3 times. He went to Cleveland quite sick came home worse & the next day they burnt the Press room, making alterations & he got his feet wet & stood in cold boots & c & was very sick. His cough is terrible. . . . I expect to find him in Pneumonia or typhoid, but there is nothing apparently the matter, but cough & tremendous expectoration. Annie & three of the children having the whooping cough . . . I almost think Irénée has it, or a sympathetic cough. He is very unhappy & the minute he speaks of Charlotte he begins to cough steadily. . . . Polly and I am staying. . . . Tomorrow I hope Irénée will get out of bed, but have no idea when we can get home

to stay. Emma is a most remarkable woman such powers as she has & such devotion to both Irénée & children. I cannot imagine what would become of them all without her.[17]

Although Margaretta had found it easy to announce that her daughter-in-law Mary Belin was suffering from consumption, she found it impossible to admit even to herself that her son was suffering from the same disease. But Irénée's trouble was no "sympathetic cough"—nor even whooping cough. Over the next five years, Irénée was frequently too weak to drag himself down to the mill. He would lie in bed, spitting blood and wishing he could die. These spells were followed by brief remissions in which he would return to work, trying to provide the same dedicated, undemanding service to the operations of the company that he had always given.

Distracted by worry over Charlotte and forced to spend much of his time on the job trying to mediate the increasingly frequent disputes between his brother Lammot and his Uncle Henry, the unhappy Irénée, who had always detested conflict of any sort, found his only happiness in the company of his children during those years. Miraculously, the five children, with marvelous resiliency, enjoyed an apparently happy and normal childhood in their abnormal household. Much of the credit for this must go to the oldest child, Anne, who had all of the poise, courage, and quiet managerial efficiency that had been the distinguishing characteristics of her great-grandmother, Sophie Dalmas du Pont. Although only eight years older than the youngest child, Louis, she had a maturity and balance far beyond her years, which enabled her to serve as a kind of surrogate mother to the younger four.

Marguerite, the second daughter, in turn, had the strength, the sharpness of wit, and the ready tongue of her grandmother Margaretta. Often impetuous and quick-tempered, she was prepared to do battle to protect her own interests and those of her siblings. Alfred, as the oldest of the three boys, was his younger brothers' model, their leader in their childhood games, and their valiant protector against the tough kids who lived in the workmen's cottages down the lane. Maurice, like his father Irénée, was quiet and unassertive, willing to accept Alfred's leadership, unwilling to cause any trouble within the family. It was only the youngest child, Louis, whom the older children tended to baby and overprotect, who was emotionally scarred by his childhood.

In later life the four older children never recalled anything especially bad about growing up in Swamp Hall. Alfred in particular remembered his childhood as being happy. For him, life on the Brandywine was not very different from what his father and uncles had known in their childhood. Swimming and fishing in the Brandywine, hunting squirrels and rabbits in the woods, poking around the mills to pick up the elementary skills of pow-dermaking from the busy workmen on the job, and trying to outwit the tutor—these were what Alfred would best remember.

Alfred greatly preferred the company of the millworkers' sons to that of his du Pont cousins who lived in the big houses scattered throughout the valley. The neighborhood boys taught him how to swim, how to take care of himself in a fight, how to swear, even how to blow smoke through his nostrils after a drag on a cigarette. Al, as he was called by his companions, belonged to the Down-the-Creek gang, and glorious were the days when the group engaged in hockey games, planned military maneuvers, and engaged in open combat with their hated rivals, the Up-the-Creekers.

Alfred managed to survive these wild and unregulated adventures of childhood with only one serious accident. This occurred one hot afternoon when he and his chums stripped off their clothes and Al, going first, dived headlong into the Brandywine, striking his head on a submerged rock. For-tunately, he did not break his neck, but he emerged from the water dripping blood from a broken nose. His injury did not seem serious enough to worry his father about it, so the accident went unreported, and the nose soon healed. But the ultimate consequences were far more serious than the cos-metic disfigurement that his great-great-grandfather, Pierre Samuel, suf-fered when he broke his nose as a child. Years later, deafness became a serious problem for Alfred, and he was told by a specialist that it was undoubtedly caused by damage done to his eustachian tubes from this untreated injury.[18]

Young Alfred could hardly be distinguished from any of the other boys in the gang; he was neither stronger nor weaker, neither cleaner nor better dressed, neither more cautious nor more bold than any of them. He differed from his fellow Down-the-Creekers in only two respects: first, he was a du Pont and the son of the boss of the other boys' fathers, but that was some-thing to be forgotten, not flaunted; and second, he truly loved music. The only thing that could bring him in willingly from play was his daily practice hour on the piano. In addition to the instruction he received on the piano, he taught himself to play the violin, the bass viol, and the flute. But his talent for music was also an attribute which was better to keep hidden from the gang.

As the oldest son, Alfred was the closest to Irénée. Nothing pleased him more than to be conducted through the mills, with his father serving as guide and instructor. Although Irénée's long hours of work and increas-ingly frequent bouts of illness did not allow him much time to spend with his children, the occasional good Sundays when Irénée was feeling well and Charlotte was far from home, he would take Alfred for long walks in the woods, identifying by their proper names the many wild plants and trees that grew there and telling the boy about his great-grandfather, Eleuthère Irénée, who was not only a chemist and a great powdermaker, but also an amateur botanist of considerable learning.

One of the high points of Alfred's youth was the weekend in 1876 when his father took him to the great Centennial Exposition in Philadelphia. Together they toured the exhibits and marveled at the newly invented telephone and the advanced mechanical technology displayed there. Father and son shared a love for machines: the beauty of their design, their functional practicality, the potential power they housed within their metal frames. Upon returning home, Irénée bought his son a Porter-Allen stationary steam engine with a twenty-horsepower capability. This was to be Alfred's machine shop laboratory, and he quickly learned how to disassemble, reassemble, and operate the engine with cool efficiency. It was his most prized possession.[19]

Even the occasional visits of their mother, which Irénée feared for their effect upon the children, seemed to have had little adverse consequences. The children, even little Louis, knew that their mother was not well and had to travel a great deal for reasons of health, but when she returned she always seemed so gay and happy and loving that it was difficult to think of her in any other way. And she arrived with an unbelievably large number of packages, so that it was like having several Christmases and birthdays all bunched together in one great feast of giving. There were, of course, the bad moments when they could hear their mother screaming in anger at their father, in spite of Irénée's best efforts to keep these disturbances subdued and isolated from the children's sight and hearing. But, after all, the children also got mad and yelled at each other. Was this so different? And their mother screamed at them, but always took their side in any argument with their nurse, the hated Emma, or their tutors.

The best time of all was when Charlotte would gather the children around her and tell them about her latest travels: of operas in New York or Paris, or flower festivals in South Carolina and hunts in Virginia. Alfred was particularly enthralled by the stories she brought back from her trip to Florida, then a land almost as alien and exotic as darkest Africa. The children's eyes grew large with wonder and longing as Charlotte, with all of the skill of a professional storyteller, described rivers teeming with alligators, the great Everglades swamp, the miles and miles of deserted white sand beaches, and the beautiful St. Johns River, flowing north into the sea. Here was a place where, Charlotte said, she would like to live permanently. Alfred thought he would, too.

But these moments were few, and all the more precious because of their rarity. Most of the time their mother was either a past memory to cherish or a future visitor to wait expectantly for. So, in spite of Irénée's fears and Margaretta's widely disseminated dire warnings, Charlotte's illness did not create a hostile environment for her children. Even under the critical scrutiny of their grandmother and their aunts it could be seen that the children were happy and prospering. Margaretta gave all the credit to the Alsatian nurse, Emma, whom she had personally found and employed for the family, but the children knew better. Emma was a tyrant who frequently beat them or punished them by depriving them of their meals. Under this strangely mixed administration of tyrannical despotism on the part of their nurse and benevolent neglect on the part of their mother, the five children were forced to find strength and order for themselves by banding together in mutual support. The ties that bound them to each other during these formative years were to last throughout their lives.

The truly bad moments, as far as the children were concerned, were certainly not times when Charlotte came home, nor even when Emma whipped them until angry red welts appeared on their backs. What the children truly dreaded were the nights when they could hear Papa's racking cough coming from his bedroom, and the following mornings when Granny Margaretta would descend upon the household to minister to their father and to join forces with Emma in ordering them about.

Unfortunately, these bad moments came ever more frequently, and by the spring of 1877 it was apparent to all of the children that Papa was dying. Mama came home, but this was not a good time either, for she seemed distracted and more restless than ever before. Now she could not, as she had before, find relief in flight. All through the hot summer, when the Brandywine valley steamed with heavy humidity and no relief was to be found even under the ceiling fans, Papa lay in his bed, gasping for breath between coughs, and Mama paced the hallways like a beautiful caged leopard constantly searching for a hidden and inaccessible exit to freedom. The only good thing to happen was that Mama fired Emma when she discovered the welts on Alfred's bare back.[20]

Then, on a particularly hot night in August, the storm, as it must do in such an electrically charged atmosphere, finally broke. The frightened children could hear their mother's screaming as two servants struggled to hold her down on her bed and to straighten out her convulsed body. Only a sedative administered by a hastily called physician brought a drugged tranquility to Charlotte's writhing. The next morning, when the sedation wore off, the spasms began again. For several days the doctor and nurse tried to quiet Charlotte, end the convulsions, and restore her sanity, but finally the doctor agreed with Margaretta: Charlotte must again be sent to the asylum in Philadelphia. Charlotte lived for only one week after arriving at the asylum. She returned to sanity briefly just before she died, but as Margaretta reported to her sister Mary, "she left no messages."[21] There was probably nothing more that Charlotte wanted to say to the du Pont family. At last she had found the exit she had been seeking.

Yet even in death, Charlotte could cause dissension in the family. Lammot, the only du Pont whom at the end Charlotte really trusted, went to Philadelphia to claim her body. He wanted to have the funeral in his home, believing that Charlotte would prefer that, inasmuch as she detested the du Pont family's Christ Church, but as Henry du Pont's wife, Louisa, reported to Sophie du Pont, "they thought the church was the right place & *no comments* could be made."[22] By "they," of course, Louisa meant Margaretta, who refused to enter her son Lammot's home.

A question also arose as to whether Irénée, who was desperately ill, should be informed of Charlotte's death. It was finally decided that he had to be told, although he was spared the full details of her final seizure. Louisa told Sophie that "Irénée has rallied wonderfully & has been spared all that could properly be withheld from him in relation to poor Charlotte. It was fortunate that he had his brothers at hand with their devotion and love to help him. . . . Willie [Louisa's and Henry's youngest son] brought word that Irénée was calm and composed but had been much overcome by the intelligence of dear Charlotte's death."[23]

Charlotte was buried in the du Pont family burial ground in Sand Hill Woods. Four weeks later, Irénée was buried beside her. Margaretta conveyed the news to her sister, "He had become a perfect skeleton, but did not suffer severely toward the last. Had been for a while wishing to die. Made all his arrangements & was as peaceful & resigned as an angel. No impatience or irritability & I may say he was a model for anyone. His life was sad & broke him down, but I trust he is happy now."[24]

Historians generally regard the year 1877 as the termination of Reconstruction, the last chapter marking the end of the Civil War period. For the ill-fated Charlotte and Irénée, the year 1877 also marked the end of their private civil war, but for them there had been no reconstruction, no reconciliation. They did finally find peace, but only the peace that prevailed in the cemetery. Perhaps Charlotte's spirit had managed finally to escape the bonds against which she had for so long struggled, but her body would lie beside her husband's on the banks of the Brandywine creek for all eternity.

Irénée was forty-nine years old when he died. Charlotte was forty-one. They left to the care of the du Pont family their five orphans, ranging in ages from seventeen to nine. This issue was the only fortunate legacy of their unhappy union.

III

The Way to a Du Pont Partnership
1877–1891

ELEUTHERE IRENEE DU PONT II, during his final lingering illness, had ample time as his mother said "to make all his arrangements" for dying. These arrangements included conversations with his five children. He gave particular attention to his eldest son. As Alfred would later remember the conversation, his dying father said to him, "Son, I am not going to be with you long. I doubt if your mother will. You must get an education and then come back, take off your coat and go ask your Uncle Henry for a job. I think the old company may need you sometime. I just have that feeling."[1]

These may not have been Irénée's exact words to his son, for undoubtedly Alfred's memory of that conversation had been enriched by later events, but the general substance of this testamentary statement is probably accurate. For some years Irénée had seen that his eldest son had both the talent for and interest in powdermaking to make a place of prominence for himself within the Du Pont Company to which Irénée had, with a keen sense of obligation but with no joy, given his life. Irénée once told his sister Polly that he had always hated the powdermaking business.[2] With young Alfred it was different. The boy had all of the enthusiasm and the magical feel for machinery that had marked with distinction both Irénée's uncle Alexis and his brother Lammot.

A long road lay ahead, however, before Irénée's last words to his son would be proved to be prophetic. The most immediately pressing problem was to determine what should become of Charlotte's and Irénée's five children. As soon as Irénée was laid in his grave beside the newly heaped mound that covered Charlotte's coffin, the children were sent by carriage back to the now strangely quiet and empty Swamp Hall. Their elders, both male and female, then retired to Henry's home for a family council to assess who could and would be willing to assume responsibility for each of the five children.

Swamp Hall must, of course, be vacated as soon as possible so that the home could be reassigned to some other du Pont family who had need for the additional space that the thirteen-bedroom house on Breck's Lane provided. The family quickly agreed on that point. The disposition of the five orphans among the several du Pont households required a considerably more lengthy discussion. The three older children did not present too great a problem. Annie and Marguerite could now continue their education at the finishing school in Philadelphia that du Pont girls had attended for thirty years. At thirteen, Alfred was now old enough to be sent off to the Reverend James C. Shinn's academy for boys in Waterford, New Jersey. During the school holidays the three older children could stay with their Grandmother Margaretta and their Aunt Polly. The two younger boys, Maurice, eleven, and Louis, nine, required more careful consideration. After much discussion, it was decided that the children's tutor would be dismissed, and the two young boys would also be sent off to a boarding school in Philadelphia. During vacations they could be passed around from house to house, so that no one family would have to assume full responsibility for them. Irénée had left his children an estate worth $500,000, which would provide an annual income of at least $25,000, certainly sufficient to pay for their schooling, in addition to providing them adequate funds for living expenses.[3] Uncle Fred, whom Irénée had designated in his will as the children's guardian, was appointed the family's representative to convey to the children, who must be anxiously awaiting a call, the details of the family's collective decision.

The children were, indeed, anxiously awaiting the decision, but not with hands-folded head-bowed quiet submission in the front parlor, as their elders envisioned. Immediately upon their arrival home from the cemetery, the children assembled their own rump session of a family council. They also reached immediate unanimity upon one point. Under no circumstances would they give up Swamp Hall. This was their home, their continuing link with the past, their base for all future operations. Nor would they be separated from each other. Annie would continue to be what she had long been in actuality—their surrogate mother, and Alfred would now have to assume the role of father, the protector of the family unit.

There was some discussion of how they should respond to the expected verdict from the family council being held at the Eleutherian Mills mansion. Alfred, aided and abetted by his two young brothers and by the fiercely impetuous Marguerite, urged military action. As Alfred sized up the situation, they were now facing a far more powerful and more frightening Up-the-Creek gang than he had ever encountered among his peers on the ice hockey field. With some hesitation, Annie finally gave her assent. After all, there was no orderly legal remedy available to them, for there was no court of appeal higher than that of Uncle Boss Henry. So the five raced off to arm themselves, Annie with a rolling pin, Marguerite with an axe, Alfred with a twelve-gauge shotgun Papa had recently given him, Maurice with a flint-lock heirloom pistol, and Louis with his toy bow and arrow.

It was this formidable company that confronted Uncle Fred when he rode into the driveway. The family had chosen wisely in selecting him as the emissary, for of all their uncles, he was the children's favorite. Upon catching sight of this gang of armed desperadoes, the startled Fred had difficulty in concealing his amusement. At a safe distance, he waved a white handkerchief and asked for a parley. The children agreed to talk but they held on to their weapons. Fred then presented the family's decision. The children were all to be sent off to boarding school, but he emphasized they would still remain a family. They would be together during the holidays. He pointed out that even if their parents had been living, they would soon all have left home for school anyway. The only thing that was being done now was to speed up the timetable a bit. But when asked if they would still have Swamp Hall to come home to, Fred had to admit that they would not. They would henceforth make their home either at Granny Margaretta's or at Uncle Lammot's.

As far as the children were concerned, that would not do at all. They quickly stated their conditions for laying down their arms. Being sent off to school would be all right (They were not about to die in defense of their present tutor!) But they would not give up their beloved Swamp Hall. They would fight to the last ditch for what had always been their home.

Uncle Fred was impressed. He said he would carry their terms back to his principals and would even plead their cause. He was as good as his word. Long a favorite of all of the du Ponts, especially his mother and his Uncle Henry—the two most powerful figures in the du Pont family council—Fred proved to be an effective advocate of the children's case. Both Margaretta and her daughter Polly could attest to the fact that Annie, even at seventeen, was a veteran and efficient household manager—a much better housekeeper and mother than the children's own mother had ever been. As for Uncle Henry, he was delighted when he heard about the axe, the rolling pin, the bow and arrow, and the guns. These were *enfants terribles* worthy of his respect. The case was reopened, and the family council did what it had rarely done before. It reversed a previous decision. The children could keep Swamp Hall, with Annie as the mistress of the household. She could continue her schooling at the female seminary in Wilmington, and the two younger boys would stay at home with their tutor and the household servants. Annie would be home except during school hours, and the family's faithful spinster cousin, Mary Van Dyke du Pont, who lived close by, would be happy to stop in every day to make sure everything was in order. Marguerite, who was by nature disputatious, would be sent off to school, perhaps to that new college that her mother's cousin, Pauline Fowle, and her husband, Henry Fowle Durant, had recently founded in Wellesley, Massachusetts. That would prevent any posible conflict between Annie and Marguerite over who was in authority at Swamp Hall. As for Alfred, he too, would be sent away to school, to Pastor Shinn's academy in New Jersey as had already been decided upon. At age thirteen, he had graduated beyond tutorial instruction at home and needed to begin a college preparatory course.

So it was quickly settled. Uncle Fred, with considerable relief, returned to Swamp Hall and told the children to lay down their arms. They had won a total victory. The family had accepted their terms unconditionally. Swamp Hall would remain theirs until such time as they desired to leave it for marriage and for homes of their own. The Battle of Swamp Hall was over without a single rolling pin hurled or arrow shot. There would not be, at least for the present, another civil war raging within its spacious domain.[4]

Because it was late in September when this settlement was reached, immediate preparations had to be made for sending Marguerite off to Wellesley College and Alfred to Waterford, New Jersey. These preparations received the full attention of all of the female members of the family. Many years later, Alfred, writing to a friend whose father had recently died, consoled him with the thought of "how fortunate you are to have been able to keep him with you for so many years and not to have lost him early in life as I did, when I was but thirteen years old, and thrown on the world with no one to advise or guide me."[5] The death of his father surely had been a grievous loss for Alfred, but the one thing he did not lack was someone "to advise or guide" him. He was surrounded by relatives who were more than eager to provide counsel, whether or not it was wanted or needed. Granny Margaretta, Aunt Polly, Cousin Mary Van Dyke, even Great Aunt Sophie, who, although unable to leave her own home, could still write frequent admonitory notes regarding proper dress, manners, penmanship, et cetera ad infinitum. They all clucked and fussed over the five children like a flock of mother hens trying to share a single brood of chicks. Never were orphans more attended to and nurtured than these five.

Somehow, in spite of all this attention, the children managed to get their own lives in order. Annie took over control of Swamp Hall, directing the servants with firm efficiency, supervising the little boys' tutorial instruction at home, and getting Marguerite and Alfred off to their respective schools in time for the opening of classes.

Neither Marguerite nor Alfred, however, was happy in that first year away from home. Marguerite suffered from an eye affliction that made reading very painful, and neither the doctors she consulted in Philadelphia nor those in Boston could find the source of the problem. She found the academic demands of Wellesley College more than she could meet, and at the end of the first year she was happy to transfer back to the less demanding finishing school in Philadelphia.[6]

Alfred's unhappiness at Pastor Shinn's school in Waterford was for reasons quite different from those of Marguerite. He found the good Reverend strong on piety but weak on pedagogy. The school offered nothing either to interest him or to challenge him academically. Furthermore, he was still too close to the home coop to escape the constant solicitous clucking of his relatives, as the following letter from Great Aunt Sophie to Granny Meta reveals:

> I send you another letter from dear little Alfred. I had written him (thro' my amanuensis) quite a long letter—beginning on inquiries about his rabbit traps—& telling him of the kind of traps the boys used when I was young & of those his grandfather Alfred taught Henry & me to make & C & C you see I interested him, as he replied by return mail. I wd not for the world make a comment on his letters lest it shd check his writing freely as he does & which is such a good thing for him & us. But I should think he ought to have a good writing master, for he don't [sic] write the beginning of a good hand. He ought to practice a large round hand. . . . I'm anxious to hear about Pauline & Annie's visit to the dear little fellow."[7]

The second year at the Shinn Academy proved to be even more tedious than the first. When Uncle Fred made one of his frequent trips to the Brandywine during a time when Alfred was home on vacation, the boy let his guardian know of his dissatisfaction with the school. His curriculum was mostly Bible study, no science or mechanical instruction, no laboratory, not even a workshop, just a lot of old, musty books and dull sermons. Young Alfred found a sympathetic listener in his uncle, for Fred du Pont regarded himself as being something of an expert on education. During the twenty-five years he had lived in Louisville he had made such a success of his paper pulp company and his investments in city railways and Kentucky coal fields as to become Louisville's wealthiest citizen. He was also the city's chief benefactor and had interested himself particularly in supporting secondary education. A few years later he built for his adopted town a high school for boys that gave special emphasis to vocational training and the physical sciences.

Although Fred du Pont knew that his mother and his aunts would probably be opposed to having young Alfred transferred from the good parson's school, which they themselves had selected because of its emphasis on Christian morality, he was determined to find another school for his nephew. The boy needed a curriculum that would prepare him for admission to the Massachusetts Institute of Technology. Fred had heard good things about the Phillips Academy in Andover, Massachusetts, reports that were enthusiastically confirmed by Charlotte's educator relatives, the Fowle Durants of Wellesley College.

Fred, as usual, was able to win over his mother to his way of thinking. Even Aunt Sophie, who had been the Reverend Shinn's chief supporter, was persuaded to accept the idea of a transfer to the Phillips Academy. Perhaps there the boy would find "the good writing master" who apparently was conspicuously absent at his present school.

In September 1879, Alfred took the long train ride from Wilmington to Boston, the farthest distance he had ever traveled from the Brandywine valley. Uncle Fred thought it necessary to accompany him, but at least this was better than having Granny Meta and Aunt Polly taking him by the hand to Parson Shinn's gloomy school.[8] Alfred knew he would miss his piano, his violin, and his bass viol, but he did find room in his portmanteau for his beloved flute. No matter how demanding this new school might be, Alfred was determined not to abandon his music entirely.

At the Phillips Academy, the boy was challenged intellectually for the first time in his life. He proved to be not a brilliant student but bright enough always to receive a passing grade in each subject and smart enough to do no more than was necessary to pass, thus leaving adequate time for athletics, the glee club, bicycling with his friends through the New England countryside, and in general having a good time.

In later years, Alfred liked to recall how very little money he had to spend for recreation or even for the necessities of life. Uncle Fred, in Alfred's recollections, was anything but generous in doling out monthly allowances. In answering a correspondent who had written to him complaining of the practice so common in private schools of expecting students to pay for a vast array of extracurricular activities and services over and above the established tuition and boarding costs, Alfred's comment was:

> There is nothing new in the attempt to extort money from pupils. It was so when I was a boy of fifteen and continued throughout college life. Fortunately, I was without money, so, with the exception of one class pin and one secret society pin my extraordinary expenditures were represented by zero. In addition I could not see that I lost cast [sic] perhaps for the reason that I didn't have any, but nevertheless I managed to get along, was happy and the boys all seemed to like me, poor as I was. I went through College on $30.00 a month—it is true, I darned my own socks and patched my own trousers, and the other boys cut my hair once or twice a year.[9]

Some of his fellow students in Andover, however, would remember Alfred's style of living as being quite different from what he described. In

1939, when Marquis James was writing a biography of Alfred, Jessie Ball du Pont, Alfred's wife, wrote to all of the almuni of Alfred's class she could find to ask for any useful information they might have regarding Alfred's school days at the Phillips Academy. One classmate responded: "I remember your late husband, a member of my class at Andover, only as the homeliest boy in the class and as one of the few boys who had a nickel plated high wheel bicycle.... Had I been an intimate friend of your husband, which was precluded by the difference in our financial standing, I might be of more help to you."[10]

Mrs. du Pont must not have been very pleased with this particular response to her inquiry for it did not appear in the James biography. Certainly, it was unfairly biased against Alfred in two respects. Photographs taken at the time belie the statement that he was the homeliest boy in the class. He was a handsome lad, large enough in both countenance and stature to bear with a regal style the large du Pont nose. Nor is it likely that he had selected his school friends on the basis of financial standing. Even as a small boy playing on the banks of the Brandywine, he had always preferred the companionshp of the children of the millworkers to that of the millowners' sons. One can be sure that Uncle Fred's monthly allowance did not enable Alfred to indulge in conspicuous consumption or to mark him among his school fellows as being a member of the financially elite.

On the other hand, it is doubtful that life at Andover was as penurious as Alfred chose to remember. In any event, the three years spent at the Phillips Academy were among the happiest he would know. As he wrote to Cousin Mary soon after his arrival, the academy "has a great many advantages in the way of learning ... the boys being very sotial [sic]. ... I think I will get along finely here at least it has all the appearances of it so far."[11]

And so he did. He even managed a minor declaration of independence from his overly solicitous family in his second year at Andover by taking a room in a different boarding house than the one his Granny Meta had selected for him at the Reverend Clough's residence. Even so, he found it necessary out of "a decent respect to the opinions" of his family to declare the reasons for this act of independence. "I received Grandma's telegram," Alfred wrote to Uncle Fred, "but not until I had engaged a room at Mrs. Tiltons whose house is nearly opposite to Mrs. Cloughs. There is but one boy here beside myself ... & I know him to be a good boy, and also very quiett [sic]. The meals here are good as Mrs. Cloughs and Mrs. Tilton is very social & kind. The only reason why I changed as I told Grandma was on account of the rooms. I think that you & Gran suppose my resasons for changing were because I would fall into the society of boys who were wild & perhaps dissipated, but, you need entertain no such idea, the facts are just as I represent them and I don't think that anyone will ever regret my changeing [sic]."[12] Even in distant Massachusetts he could feel the family peering over his shoulder.

Alfred completed his studies at the Phillips Academy in the summer of 1882. At eighteen he was now ready to enter the school that both he and his Uncle Fred had selected for him five years earlier—the Massachusetts Institute of Technology, which at that time was housed in a single building on Boylston Street, a block from the Boston Public Garden. His cousin, Thomas Coleman du Pont, the son of Bidermann du Pont, was a year older than Alfred and had already completed his freshman year at M.I.T. when Alfred arrived in the fall of 1882. Arrangements were made for the two du Pont cousins to room together in a boarding house near the college.

Alfred had not previously been well acquainted with his cousin Coleman, since the latter lived in Louisville, Kentucky, and had only occasionally made short visits to the Brandywine during the summer. The two now quickly became close friends. In Coleman, Alfred found sophisticated counsel on how to enjoy life at college. Coley knew all the best saloons in town. He had a list of female companions suitable for every occasion from cotillions in the Brahmin Back Bay district to more lively encounters in Haymarket Square. Coleman had an order of priorities in which having a good time held first place, with studying for his courses being very nearly at the bottom of the list. With his quick mind, breezy manner, ready wit, and facile tongue, Coleman had survived his first year at M.I.T. very well, and he was eager to instruct his young cousin on how to succed in college without really trying. Alfred would never be able to equal his instructor in mastering the fine art of hedonism, but in three short years he had come a long way from the Calvinistic austerity of Pastor Shinn's school in Waterford, New Jersey.

It was Coleman who advised Alfred not to bother with the formal curricular program required by the institute for obtaining a Bachelor of Science degree. Coleman himself had no intention of staying four years at M.I.T. just to obtain a degree. Two years would be sufficient to give him the M.I.T. label, which was all that was necessary. The important thing was not a piece of paper at the end of four years but a piece of the action in making money as quickly as possible. At the end of the current academic year, Coleman intended to return to Kentucky and seek his fortune in his Uncle Fred's and his father's coal mines. Alfred had a similar opportunity awaiting him on the Brandywine. Why worry about a degree? Take those courses that would have the most practical value for the immediate future and forget all that useless instruction in abstract, theoretical science and mathematics.

All of this nonofficial academic counseling made a great deal of sense to Alfred. Consequently, he also matriculated as a special student, and this status allowed him to ignore the requirements for a degree and to take the courses he wanted. During his first semester, he rather ambitiously registered for mathematics, chemistry, mechanical drawing, German, and forging, but as Coleman had undoubtedly warned him, this program proved to be far too demanding and quite unnecessary. For the second semester, Alfred dropped all the courses of the previous semester except chemistry and forging. A half-century later, the registrar of M.I.T., in reviewing Alfred's academic transcript at the request of Mrs. du Pont and Marquis James, remarked, with admirable tact and circumspection, that Mr. du Pont's record would indicate that here was a man "who knew what he wanted to do and did it."[3]

With less than half of what was considered a full course load, Alfred did very well indeed. Like his cousin Coleman, he now had ample time to devote to his social life. He was one of the few students at M.I.T. to join a Greek fraternal society, accepting a bid to pledge affiliation to Sigma Chi. He also had time for employment in off-campus jobs, since fraternity life and his other extracurricular activities required him to supplement his very limited financial resources. Uncle Fred had raised his monthly allowance from the five dollars which he had received at Andover to thirty dollars, but this advance in income from a dollar a week to a dollar a day was hardly commensurate with the young man's even more rapidly advancing social activities. Without the additional funds that he himself provided, Alfred could not have kept up with Coley, who had no such niggardly limitation placed upon him. After each snowfall Alfred was able to earn three dollars a day shoveling walks, but this was unfortunately seasonal work, and not even in Boston did the snow fall often enough to provide a supplement to his allowance sufficient to sustain his social life. It was then that his long years of dedication to music paid off. Occasionally he was able to substitute for an absent violinist in a theatre orchestra. This introduction to the stage led in turn to the opportunity to be a spear-bearing, sedan chair-carrying supernumerary in operatic productions at the Boston Theatre. But Grand Opera did not have as great an appeal for this young man with plebeian tastes as did the music hall revues at the Howard Atheneum in Scollay Square. As soon as the soprano had gasped out her last dying high note and the curtain had descended to a round of applause, Alfred would quickly change from his skimpy Egyptian slave costume into street clothes and hurry over to the Old Howard to gaze with delight upon the even more scantily costumed chorus girls.

It was while frequenting the night spots of those areas of Boston where the Cabots, Lowells, and Eliots seldom trod that young Alfred met and became a close friend of the great John L. Sullivan. Sullivan was to Irish Boston in the 1880s what Joe Louis would become to Black Detroit fifty years later. This brawling, wenching, hard-drinking folk hero of an oppressed minority had issued a standing offer of one thousand dollars to the man who could last four rounds with him in any boxing ring in the city. At first Alfred must have been tempted by the offer, for he had been the champion pugilist in his weight at Andover and had continued to demonstrate his prowess in the ring at M.I.T. One thousand dollars would have solved all of his financial problems for the remainder of his academic career in Boston. Fortunately, realism prevailed over fantasy, and he decided to win the great man's friendship rather than his purse. It proved surprisingly easy to do. One round of drinks at the bar, and Sullivan was won over. From that moment on the veteran fisticuffs champion and the green college kid were close friends. Alfred's earliest playmates on Breck's Lane had all been sons of Erin, and he now felt far more at home with the Sullivan entourage than he had ever felt with his Phillips classmates or his Sigma Chi brothers. Sullivan, in turn, was fascinated with this college boy with the aristocratic name who knew something about taking care of himself in the ring and could speak Sullivan's own language. So if Alfred couldn't beat Sullivan, he could and did join him.

John L. was delighted to give Alfred sparring lessons at William Muldoon's exhibition ring, and under Sullivan's expert tutelage, according to a cherished family legend, young du Pont, using an assumed name, engaged in several contests and usually emerged victorious. Almost every saloon or dance hall of any consequence in south Boston had a boxing ring in the back room where on Saturday nights young novices and tired, old, punchdrunk veterans of the ring squared off for purses that ranged from three to ten dollars, winner take all. Alfred is reputed to have won several of these purses. It was not a much more lucrative pursuit than shoveling snow, but it was a hell of a lot more fun. Alfred would always remember John L. Sullivan as the most distinguished person he met during his five years of residency in Massachusetts.[14]

At the end of his first year at M.I.T., Alfred accepted Coleman's invitation to go home with him to Louisville rather than to return to his own home. Coleman wanted to introduce his cousin to the social life of the Gateway City on the Ohio as well as to have one last fling himself before settling down to his job in the coal fields of eastern Kentucky. It was certainly the liveliest summer vacation that Alfred had ever enjoyed. Not only was Coley an even better guide to the social life of Louisville, his home town, than he had been in Boston, but best of all, he had no restrictions placed upon him even while residing in the parental home. Coleman's mother, Ellen Coleman du Pont, had died in 1876, the year before Alfred had lost both of this parents, and there were no Grannies or Aunts in Louisville to keep close watch on his comings and goings. There was only his wealthy and indulgent father, Bidermann, living in lonely splendor in the great mansion that Uncle Fred had built for himself but then had decided to present to his brother Bidermann, preferring for himself to continue living in the same shabby hotel suite he had rented when he first came to Louisville. Bidermann du Pont placed no limitations upon Coleman's expenditure of either time or money, for in the father's view, the son could do no wrong. The two boys revelled in the freedom that his summer offered them.

There was also an opportunity for Alfred to become better acquainted with his Uncle Fred on a one-to-one, man-to-man basis, which was a new experience for both of them. Alfred endeared himself to his old bachelor uncle by rigging up a mechanical fan for his office. What truly impressed Fred was not that his nephew could devise a fan that would work, but that he could do it with scrap pieces lying around the plant without costing Fred one cent.[15] The two got along together famously.

Fred was eager to discuss career possibilities with this mechanical genius of a nephew. In three years, Alfred would reach maturity and would inherit over $100,000 from his father's estate, an ample sum with which to launch a career in business for himself. Or his uncle would be happy to welcome him, along with his other nephew, Coleman, into his multifaceted business concerns: paper pulp mills, coal mines, street railway companies, or steel plants. In the West (for the family still spoke of Fred and Bidermann as having gone West to seek their fortunes) there was ample opportunity for a young man to rise quickly to the top if he was good enough, and Fred was convinced that young Alfred was good enough and ambitious enough to make the climb and that he would not find the same opportunity for rapid advancement within Boss Henry's domain.

Or, because Alfred had always had an interest in and a certain facile style for writing, Fred offered to introduce his nephew to the great Louisville *Courier Journal* editor, Henry Watterson. "Marse Henry" was personally indebted to Fred du Pont, since the latter had never pressed the gifted but somewhat erratic and improvident editor for prompt payment of his outstanding pulp paper bills. An introduction from Uncle Fred would be tantamount to a position on Watterson's paper.

To all of these tempting offers, however, Alfred politely but firmly said no. He was the eldest son, etc., etc., and his place was back on the Brandywine, making powder. He had never considered any other possibility. He belonged to the company of his ancestors and perhaps someday, if he was a good enough powderman, that company would belong to him. Fred did not press the issue, but let the youth know that as long as he and his Uncle Bidermann were still around, the offer would remain open.

In the fall of 1883, Alfred returned to M.I.T. for what he was sure would be his last year of formal education. Although the great John L. was there, awaiting him with open arms, Boston would not be the same place without Coley. Alfred again registered for a partial course load, although he now resumed his study of German. Coley's absence would permit at least one more academic course without unduly burdening him. In the spring term, he even signed up for a couple of courses in office management at a nearby business college. "Fortunately," he wrote his grandmother, "the Business college I attend is nearly opposite the Institute so I can skip from one to the other without losing much time but even then I can't accomplish all the work at all satisfactorily & I am afraid I will have to drop some of my Tech Studies to make scores for the rest." It is interesting to note that it was the M.I.T. courses, rather than the business college courses, that he intended to drop. Alfred evidently thought that he was learned enough to German and skilled enough in shop work to sacrifice them, but that if he was to win a place in Uncle Henry's organization, he should know something about business management.

Even in distant Kentucky, Coley managed to consume some of his cousin's time, for Alfred added the information in his letter to Margaretta that "I have been employing my spare time making a couple of telephones for Coleman who is going to rig them up at the mines thereby, he declares, saving him 7 miles walk each day. They are about finished & I am going to send them soon."[16] On his own, Alfred had learned a great deal more about Mr. Bell's marvelous invention, which he and his father had seen at the Philadelphia Exposition eight years earlier.

Two weeks after making this regular monthly report to his grandmother, Alfred was startled to read in the Boston *Globe* that there had been a terrible explosion at the Repauno Chemical Company, a subsidiary of the Du Pont Company in New Jersey that had been organized for the manufacture of dynamite. Among the six victims of the accident was Alfred's Uncle Lammot, the president of the company. Alfred took the next train from Boston back to Wilmington to attend his uncle's funeral. The only remaining son of Alfred Victor du Pont to be employed by the Du Pont Company was now gone. It was time, Alfred must have thought, as he stood by the open grave in Sand Hole Woods cemetery, for the third generation to report for duty.

The death of Lammot du Pont was the most damaging blow to its future that the Du Pont Company could possibly have suffered. During the century and a half in which members of the du Pont family were to control directly the affairs of the company, the one du Pont who most deserved the encomium "genius" was Lammot du Pont. All of the many talents of the other du Ponts—the chemical expertise of Eleuthère Irénée, the founder, and of Francis Gurney; the mechanical talent and inventiveness of Alexis I and of Alfred Irénée; the managerial skills of Henry and Pierre Samuel II; and the aggressive, venturesome spirit of Thomas Coleman du Pont—were all embodied in Lammot to a superlative degree. Alfred Victor's second son, to be sure, was not perfect. Lammot noticeably lacked the patience, the tolerance, and the modesty of his brother Irénée; the cautious prudence of Henry; or the personal concern and affection for the workers that won for Alexis I. and Alfred Irénée the devoted loyalty of their employees. Boss Henry, in particular, found it difficult to work with his gifted nephew, whose imperious nature was so like his own. From the first day of his employment, Lammot had found it almost impossible to adopt the subservient manner that Uncle Henry expected from those under his command. Margaretta used to say that whenever poor "Irénée was not in the powder yard he was busy trying to keep peace between Lammot and Uncle Henry."[17] After Irénée's death and Lammot's break with his mother over his marriage to Mary Belin, relations between Henry and his nephew became even more strained, since no one in the family now could or would attempt to mediate their quarrels and patch up their differences. Consequently, it was a great relief to both men when in 1882 Lammot took over the newly established dynamite and nitroglycerine plant at Repauno, New Jersey and moved his very large family to Philadelphia.

No one in the du Pont family, however, not even Henry, expected Lammot's move to be a permanent abandonment of the Brandywine valley. It had long been generally assumed by them all that when and if Uncle Henry ever relinquished his power and "shuffled off this mortal coil" (and the two events undoubtedly could only occur simultaneously), then Lammot would certainly become the next commander-in-chief of the family enterprise. While he was still a young man, Lammot's exploits had become as much a part of the family's legends as Pierre Samuel's escape from the guillotine in revolutionary France or Eleuthère Irénée's hunting trip with Colonel Toussard, which had led to the founding of E. I. du Pont de Nemours & Company. As his family later discovered, Lammot's distinction rested on a more solid base than such fabled exploits as his personally conveying munitions through the Dardanelles Strait to the beleaguered Russians during the Crimean War or getting the shipment of saltpeter away from the clutches of an unfriendly British government during the American Civil War.

Late in his life, Alfred I. got the idea of writing a history of the developing technology of gunpowder manufacturing. As a part of his research for what could have been a major contribution to the history of American technology, but which unfortunately never materialized, Alfred held a long interview with Henry Miller, an old-time employee of the Du Pont Company. Miller had first gone to work in the Hagley Yards in 1858 and his father before him had joined the company in 1822, so that together, father and son, in their employment, had encompassed nearly the entire history of the company. In the course of the interview Miller and Alfred du Pont touched upon every step in the process of powdermaking: how the grinding, pressing, and polishing were done in the early days, and what changes were made in both method and machinery during the ensuing years. Whenever Alfred would ask such questions as, "Who made the cooler for cooking the saltpeter in the refining process?" or "Who rebuilt the coal bin for the making of the charcoal?" or "Who invented the first cutting machine for the powder cakes?", inevitably the answer would be, "Why, Mr. Lammot did that; Mr. Lammot figured that one out."[8]

Lammot made two major contributions to the American explosives industry that were truly revolutionary. The first was his discovery in 1856 that Peruvian sodium nitrate could be used for the manufacture of blasting powder in place of the much more expensive potassium nitrate that composed the saltpeter of India. Not only was sodium nitrate far more plentiful and hence less expensive than Indian saltpeter but in containing a higher percentage of oxygen and nitrate, this "soda powder," as Lammot called it, was more powerful and produced a cleaner combustion than saltpeter. It could only be used for blasting, not for ammunition, but in times of peace, blasting powder was the Du Pont Company's most important product.[19]

Lammot's second contribution was to get the Du Pont Company involved, over Henry du Pont's vigorous objection, in the production of Alfred Nobel's invention of dynamite. In so doing, Lammot made major improvements in the manufacture of nitroglycerine products. Lammot had been fascinated with the amazing new explosive power that Nobel had unleashed as soon as he first read of it in 1863. Nitroglycerine, invented in 1845 by the Italian chemist Professor Ascanio Sobrero at the University of Turin, was regarded by its inventor as nothing more than an experiment of value only to the theoretical scientist. It was far too volatile and dangerous a substance to have any practical application. In the early 1860s, Alfred Nobel, working with his father, Emmanuel, was able to show the world what could be done with Professor Sobrero's powerful compound. By confining it within a tube, cushioning it with kieselguhr—a porous clay found in Germany—and then detonating it with a blasting cap of black powder fired by a lighted fuse, the Nobels had blasted a railway tunnel through solid granite near Stockholm quickly, cleanly, and for a fraction of the cost of black powder. Soon mining and railroad companies throughout the industrial world were clamoring for Herr Nobel's little sticks of dynamite.

If the majority of the gunpowder companies in America were impressed with and determined to get hold of Nobel's manufacturing process, either legally by obtaining franchise rights or illegally by stealing the patents, at least one powdermaker wanted no part of those satanic candles. "All hell will be to pay for this! Just you wait and see," Boss Henry roared at Lammont, who was eager for the Du Pont Company to get into the high explosive business.[20] Lammot waited, but not very patiently, for the Boss to be proved wrong.

Ultimately Henry's own desire for expansion and for the cartelization of the gunpowder industry trapped the old man and forced the Du Pont Company into becoming the country's largest producer of high explosive products. In 1873, after the financial panic which followed the collapse of the Jay Cooke banking concern in Philadelphia, the nation plunged into the most serious economic depression that it had yet experienced. But for the gunpowder industry, the depression had begun immediately after Lee surrendered to Grant in 1865. During the Civil War scores of small gunpowder plants had sprung up to meet the needs of the nation. Now the stillness that followed Appomattox was the stillness of death for many of these companies. For others, desperately trying to stay alive, the only way to delay the inevitable was to cut prices—below the cost of production if necessary. Boss Henry was a man who liked order. He viewed the chaos that now prevailed within his industry with dismay. The Du Pont Company itself was in no danger of going under. A whole new wonderful market for black powder had opened up as the nation turned from shooting young men to the more constructive task of blasting hillsides for railroad grades and mountains for gold and silver.

To maximize profits, however, it would be necessary to minimize competition. If this was not done, the du Ponts would still survive, but they would suffer. At the close of the Civil War, three giants dominated the field: E. I. du Pont de Nemours, which had 42 percent of the market; Laflin & Rand of Esopus, New York, which represented a consolidation of two small firms—Smith & Laflin, a pioneer in the powder business since the American Revolution, and the Rand Company; and the smallest of the trio, the Hazard Powder Company of Connecticut. It was time, Henry du Pont believed, for the Big Three to take action. The other two companies were interested, so in the spring of 1872, Henry made one of his rare trips to New York. At the Du Pont sales office at 70 Wall Street, officers from the other two companies met with Henry and Lammot. Four additional producers of black powder also accepted the invitation to send representatives. In this small gathering of what proved to be friendly rivals, agreement was quickly reached to form a Gun Powder Trade Association, which would establish fair prices for the industry, divide up marketing areas, and permit price cutting only within a specific locality in order to drive out nonmembers of the association. Du Pont, Hazard, and Laflin & Rand would each be given ten votes within the association; the Oriental Powder Company, located in Maine, would have six votes; and the remaining three companies—the American Powder Company of Massachusetts, and the Austin and the Miami Powder Companies, both located in Ohio—would each have four votes. Lammot du Pont was elected president of the Association, and A. E. Douglass of Hazard and Edward Greene of Laflin & Rand were chosen vice-president and secretary-treasurer, respectively. One big fish had been created to drive off the many little fish who were disturbing the feeding grounds.[21]

Henry du Pont was not satisfied with the creation of a cartel, or as he preferred to call it, a community of interest within the industry. His desire for domination and for profits led him to engage in the cannibalization of some of his competitors. Among the Big Three, the Hazard Powder Company was the smallest and most vulnerable to a takeover. Very quietly, the du Ponts bought out the Connecticut firm in 1875. It was important, however, to keep this transaction secret and to maintain the fiction of Hazard's remaining an independent rival firm. In this way, Hazard would continue to have its ten votes in the Association, but in reality, it meant that the Du Pont Company controlled twenty of the forty-eight votes among the directors of the Association. If Laflin & Rand had any knowledge of this transaction—and it is hard to believe that it did not—it preferred to go along with the game and to continue to regard Hazard as an independent company, for it was in Laflin & Rand's own best interest to maintain good relations with Du Pont. Having swallowed Hazard with the greatest of ease, Henry du Pont's appetite was not sated, but stimulated. Next to feel the bite was the California Powder Company, which had grown large enough on the Pacific Coast to dare to invade the Association's preserve east of the Rockies. Then the Du Pont Company turned its gaze southward. There was the tempting Sycamore Mills of Tennessee, which had been the major supplier of munitions to the South during the Civil War. Du Pont coveted its many Southern customers, who continued to purchase their powder from Sycamore Mills out of loyalty to the lost Confederate cause.

Then the Du Pont Company rapidly swept up eight small independent mills strategically located in the Pennsylvania coal fields, and also purchased one-third of the stock of the Austin Powder Company, which gave Du Pont control of Austin's four votes in the Association. In a moment of generosity, probably prompted by a desire to keep Laflin & Rand from asking awkward questions about Hazard's votes always being consistent with those of Du Pont, Henry du Pont invited his one remaining big rival to join him in the feast by consuming the Lake Superior Powder Company, which had an important market in the iron country of Michigan's upper peninsula. Together they also bought out the Oriental Powder Company of Maine, which had fallen heavily in debt during the depression of 1873.

By 1879, the Gun Powder Trade Association oligopoly had in reality become a dyarchy. Of the seven original films that had had voting rights in the Association, the Du Pont Company now owned outright or controlled three, and jointly owned a fourth company with Laflin & Rand. This left only the two Ohio companies independent of the Big Two's control, and they were small enough that they could be safely ignored. Boss Henry was now truly boss of the gunpowder industry. He had acquired the order and stability he had sought.[22]

But he had also acquired more than he had bargained for. To his consternation he found that he was deeply involved with something that he had vowed he would never be a part of. He was now a manufacturer of high explosives. The California Powder Company, which he had purchased in 1876 because it was the major producer of black powder on the West Coast, had another product that was rapidly becoming its best seller, a brand of dynamite that bore the trade name Hercules. One of the company's chemists, James Howden, had vastly improved Nobel's invention by substituting a mixture of sugar, magnesium carbonate, and Indian saltpeter for the inert kieselguhr cushioning material. The result was a dynamite stick that was 25 percent more effective than Nobel's kieselguhr dynamite. Because of this new feature containing an active rather than an inert base, the courts had ruled that Howden's patent was not an infringement of the Nobel patent, and the California Powder Company was authorized to produce its superior dynamite without having to hold a franchise right from Nobel or pay him a royalty.

The California Powder Company's position was enviable. The entire country was clamoring for its product. It had the patent for this product and the plant and the equipment to produce it. Once Henry du Pont acquired the California Powder Company, he was presented with a cruel dilemma. To allow the California Powder Company to continue the manufacture of its most profitable product would mean the violation of Henry's solemn vow that the Du Pont Company would never have anything to do with high explosives. But to shut down that production would in effect cause the bankruptcy of this valuable subsidiary, with a great loss of capital to the parent company. It was a question of principle versus profit. There was, of course, never any doubt as to which horn of the dilemma Henry would grab hold of. Given a choice between losing face and losing money, he would hide his face, admit he had been wrong, and run with the money. The du Ponts were in the dynamite business to stay. Having thus reluctantly put one foot into the water, Henry was drawn irresistibly into the main stream of the high explosives industry. Soon he was swimming merrily with the current.

Lammot du Pont had watched the old man's wheelings and dealings over the previous five years with an amused grin and an anticipatory gleam in his eye, for he had known full well what the purchase of the California Powder Company would inevitably mean. The company had the patent, and it had the facilities to produce the product but unfortunately the major portion of its market was in the eastern U.S., and what the company lacked was the transportation to get the product across the Rockies and the Great Plains to the Great Lakes and to the East Coast. The Union Pacific and the other major railroads refused to carry dynamite across the country as freight.

The only solution was to build a new plant in the Great Lakes region. There was no turning back. After Henry authorized the building of a Hercules plant near Cleveland, Lammot stepped forward and told the Boss that it was a fine idea, but what the company needed in addition was a third plant on the East Coast. Lammot assured his uncle that he could build a facility in New Jersey that in terms of efficient production would make all the other dynamite mills in America, if not in the world, seem obsolete by comparison. Four months after Henry gave his assent in January 1880, the Repauno Chemical Company was open for business. By July of 1880 it was producing a ton of dynamite daily. Lammot changed the brand name for his product from Hercules to Atlas. After all, Hercules had performed only twelve labors, while Atlas supported the world.

The Repauno Chemical Company was a joint venture of Du Pont, Hazard, and Laflin & Rand, which meant, of course, that Du Pont had two-thirds control of both management and profits. Lammot was made president of the new corporation and Henry's younger son, William du Pont, became secretary-treasurer.[23]

At last Lammot had his own plant, built according to his specifications, and run under his direction. He was his own boss, while Henry received the major share of the profits. Both men were satisfied. Lammot was determined that there would be no quill pens and candlelight at Repauno. By 1882, he had installed telephones throughout the plant. He was constantly making modifications and adaptations to the plant machinery, with the ultimate goal of building a facility that could be run with very little human labor, thus reducing both the cost of production and the danger to human life. Nor did Lammot remain satisfied with the product he had inherited from the California Powder Company. In place of the sugar that Howden had used as part of the base, Lammot substituted wood pulp, and he also found that he could use his Peruvian sodium nitrate instead of the expensive Indian saltpeter. Costs came down and sales and profits went up.

Having made important improvements on Howden's active base for dynamite, Lammot next turned his attention to the waste products that resulted from the manufacture of nitroglycerine. If these valuable acids could be recovered in their pure form, they could be recycled instead of being dumped, at great loss to the company, polluting the adjacent land and stream. He was working on this project in March 1884 when he lost his life.

On Friday afternoon, 28 March, in a workshop especially designed for testing of the recovery process, Lammot had drawn off into a lead-lined tank the last charge of the day. Because he had a business appointment in town, he left early, before the charge was properly disposed of, and his assistants mistakenly allowed the acids to remain in the tank. The next morning being a Saturday, Lammot would not normally have returned to the plant, but on this particular day he had arranged to meet a salesman from the Laflin & Rand St. Louis sales office. The two men were sitting in Lammot's office talking when a workman appeared to report that the acid left in the tank was "acting up very peculiar." Lammot rushed over to the workshop, where he found the mixture in the tank fuming and boiling like a witch's brew. Hurriedly, Lammot and the plant superintendent, Walter Hill, tried to turn the acid tank over to quench its contents in a water tank, but it was too late for such remedial action. The acid tank exploded. Lammot and Hill were buried under a bank of earth. Four other men, including the salesman standing outside, were killed by flying timbers.[24]

The du Ponts' crown prince was dead. The old king still reigned, but nothing would ever be the same again for either the company or the family. There was no obvious heir worthy to occupy the throne. It is, of course, impossible to evaluate fully what the Lammot's death meant. One can surmise, however, that had he lived for an additional twenty years, Lammot would have had the talent, the knowledge, and the imagination to move his company into a wide diversification of products and to effect in his lifetime the great chemical revolution that would be achieved at a much later date. His death may well have delayed for a generation the cellophane-wrapped, synthetic, plastic civilization of the future. "Better things for better living through chemistry" would now have to wait for the corporate researchers. The individual genius had been buried in Sand Hole Woods Cemetery too early to fulfill his promise.[25]

Returning to Boston after the funeral of his uncle, young Alfred could hardly wait for the school year to end. He was eager to get back to the

Brandywine and to take his place in the yards. He wanted to be a part of the transitional period as the old regime slowly yielded to a new but as yet undefined order. In May 1884, shortly after celebrating his twentieth birthday, Alfred decided he had had enough of classes and headed home.

Uncle Henry was not ready to receive him, however. Alfred was told that there would not be a place open to him until the end of September. The long summer stretched ahead with no specific assignment to occupy his time. Coleman was too busy in the Kentucky coalfields to find time to entertain him, and for Alfred, spending another summer in Louisville would mean subjecting himself to renewed pressure from Uncle Fred to become part of his organization. He would have to wait out the summer in Swamp Hall.

On his way back to Wilmington, Alfred had stopped off in Philadelphia to visit Lammot's widow and her nine living children. The oldest boy, Pierre Samuel II, was only fourteen, six years Alfred's junior, but he was mature beyond his years and took his new position as male head of the family very seriously. Already the younger children were calling him Dad, the form of address they would use in referring to him for the rest of his life. Mary Belin du Pont had just given birth to Lammot's eleventh child, a girl whom Mary named Margaretta, as a gesture of reconciliation toward her long-estranged mother-in-law.

While in Philadelphia, Alfred visited an exhibition on incandescent lighting that included models of all of Edison's devices as well as some foreign adaptations. Alfred had been fascinated with the subject of electric illumination ever since he and his father had seen their first electric arc lamp at the Philadelphia Exposition in 1876. In the intervening eight years, the arc lamp had made great inroads in replacing gas lamps for street lighting in Philadelphia and New York. There were even plans under way for its introduction in Wilmington. But the arc lamp was not suitable for indoor lighting. It gave too bright a light and it was noisy. Something else would have to be devised if electricity was to be used to light shops and homes. In 1879, after years of failure, Edison finally produced an incandescent light bulb with a filament that would not burn up as soon as the light was turned on. Three years later, he opened an electric station on Pearl Street in New York to supply electricity to several nearby businesses.

This development and its potential for the future fascinated Alfred in the summer of 1884. M.I.T. had introduced its first course in electrical engineering in 1882, the year that Alfred enrolled in the school, but he had not taken it. He did, however, do some experimenting on his own in the M.I.T. workshop and he had read every bit of information he could find on what Edison was atempting in his experiments at Menlo Park. Back at Swamp Hall after his careful inspection of the exhibit in Philadelphia, Alfred, assisted by two of his companions from the old Down-the-Creek gang, Gilly and Frank Mathewson, set up a German-made dynamo and a Shipman steam engine to run the dynamo. The icehouse in the back yard became Alfred's laboratory, and soon the three young men managed to produce enough electrical power to light a string of twenty lamps in the icehouse.

On a visit to Queen's Supply House in Philadelphia that summer to purchase additional electrical equipment, Alfred had the good fortune to meet Edison, who was there for the same reason. The Wizard of Menlo Park was greatly impressed by the young, self-taught enthusiast, who knew enough to ask the right questions and, better yet, knew enough to under-stand Edison's answers. Alfred was invited to visit Edison at Menlo Park, and he made several trips there that summer, which further fired his ambi-tion to illuminate the Brandywine valley with Mr. Edison's magic lamps. The summer passed much more quickly and profitably than Alfred had ever imagined it could.[26]

On 1 October 1884 a new name appeared on the Du Pont Company's payroll for that month: "Du Pont, Alfred I., $83.00." The name was du Pont, but his first tasks were no different than they would have been had his name been Gentieu or Dougherty or Miller. Du Ponts started at the bottom and worked their way up. The one big difference, of course, was that those named du Pont could work their way to a higher level than any-one else. Alfred's day in the yards began at 6:50 AM and lasted until 6:00 PM, with a one-hour break for dinner. During his ten hours of labor, the young man was expected to do whatever job he was told to perform, such as harnessing a team of horses in the stables and driving the team down to a car shed beside the narrow-gauge railroad track. There a railroad car was to be loaded with powder kegs, the horses hitched to the car, and the load hauled to the powder magazine in the Upper Yard. Then the empty car was taken to the soda house to be loaded with soda for the refinery. The soda had to be unloaded and a load of saltpeter picked up and taken to the rolling mills down in the Hagley Yards. When this round trip was com-pleted, he could begin all over again, and if it was not saltpeter or soda he was to load, carry, and unload, it was sulphur or charcoal. Smaller cars car-ried cargoes too dangerous to entrust to horsepower, such as "black dust" and soda from the composition houses. They had to be pushed by gangs of four to six men. On his first day Alfred learned how it felt to strain his back muscles in pushing forcefully but ever so carefully against a heavily loaded railway car. Powdermaking, viewed from the bottom looking up, was all sweat and ache, with no glamour attached. His fellow workers and even the foreman called him Mr. Alfred, to be sure, but nevertheless they called him to do the dirtiest and most menial of tasks.

Alfred survived his first day, but the sweetest music he had ever heard was the six o'clock whistle. He took the team of horses back to the stables, unhitched them and then rode his bicycle as fast as his tired leg muscles would permit along the riverfront road to Breck's Lane and the blessed sanctuary of Swamp Hall, where his sister Annie waited to greet him with a hot supper.

Alfred's apprenticeship as an unskilled laborer lasted until his cousin Francis Gurney du Pont, who was superintendent of the Hagley and the Lower Yards, promoted him to the rank of apprentice powderman. There was no increase in pay, but he was at last learning to be a real powdermaker, not just a teamster or a substitute for horsepower. He worked his way through a series of stints of duty, being assigned first to the composition house, where the charcoal and sulphur were mixed and ground together to make the black dust as the initial step in the manufacture of gunpowder. Then Alfred was sent to the rolling mills, one of the more dangerous stops in the operation. There the saltpeter, or if it was an order for blasting powder, the soda, was added to the black dust brought down from the composition house. The ingredients were combined and mixed by two gently revolving wheels. The slightest spark could mean a disastrous explosion. One entered this literal powder keg only to start or to stop the machinery, or to add water to the mixture once an hour. The remainder of the time one waited outside for the run to be completed, hoping all the while that no scrap of metal had found its way into the mixture.

The next assignment after the rolling mills was the press room where the black dust, now mixed with either saltpeter or soda, was pressed into two-foot-square slabs or cakes, which were then cut into smaller pieces about the size of a small lump of coal. In the press room the danger was even greater than in the rolling mill, for there was a far larger quantity of gunpowder on hand and the most serious explosions were those which started or reached the press room.

After the press room came the graining mill where the broken lumps of press cakes were crushed into grains of powder. The grains were then transported in fifty-pound bags to the glazing mill where they were smoothed and polished in revolving hollow cylinders and glazed with powdered graphite. Finally, the finished powder was ready to be sent to the packing house to be packed and sealed in kegs ready for shipment to market.

It took Alfred twenty-one months to complete his tour of duty, which took him through the entire process of powdermaking. This was a longer period of time than he had spent in the classrooms at M.I.T., and it was far more demanding and held much graver consequences if one failed. In this rigorous vocational school failure meant not just a red letter F on a report card but, at best, it meant dismissal from his life career before he had barely started. At worst, it meant a sudden and final trip "across the creek."

In July 1886 Alfred finally graduated from his training school. There was no elaborate commencement ceremony, only a formal handshake from Cousin Frank and a neat little handwritten note from Uncle Henry's office clerk informing him that he had been promoted to the position of Assistant Superintendent of the Lower Yards and that his salary would be increased from $1,000 to $1,500 per annum. This note meant far more to Alfred than any academic award. He was at last a true powderman, and that had always been his goal.[27]

During the two years that it had taken him to achieve this distinction, Alfred had also passed another important milestone. On 12 May 1885, he had reached his majority. He was no longer a ward of his Uncle Fred. His share of his father's estate was now his. With over $100,000 as his personal estate, with Swamp Hall as a home, and with the title of Assistant Superintendent of the Hagley and Lower Yards, he now certainly had the means, if he saw fit, to take a wife. And by the summer of 1886, he saw fit.

Alfred had never had any difficulty in attracting female companionship. He must have been high on the eligible bachelors list not only in the Brandywine valley but also in Boston and among his numerous distant cousins and their friends in Virginia society, but except for a brief flirtation with one of the Victor du Pont branch of cousins, Alice du Pont, whom everyone called Elsie, he had not given serious attention to any of the young women he escorted to dancing parties, concerts, and plays. The incipient romance with Elsie, however, came to an abrupt end. While he hesitated, his cousin Coleman, a man of action, proposed marriage to Elsie and won his suit. Not until May 1886 did Alfred at last meet the young woman whose appeal was such as to banish all hesitation and second thoughts.

Prior to attending the wedding of one of their Virginia cousins, Annie Cazenove Lee, Alfred and his sister Annie welcomed home their two younger brothers, Maurice, who was completing his second year at M.I.T., and Louis, who was a freshman at Yale University. After this brief reunion at Swamp Hall, the four du Ponts planned to go by train to Washington, D.C., and visit their sister Marguerite, who was now married to Cassius Lee. Her home would be the family's headquarters while they all participated in their cousin's wedding festivities in nearby Alexandria, Virginia.

Louis arrived at Swamp Hall accompanied by a young woman named Bessie Gardner. She was a distant cousin on their mother's side of the family, being a great-granddaughter of Eliza Cazenove Gardner, who was the eldest aunt of their mother, Charlotte. None of the Swamp Hall orphans had ever met members of this particular branch of the Cazenove family, but

Louis had discovered this heretofore unknown third cousin in New Haven, where she was living while her father, Dorsey Gardner, a distinguished linguist, was collaborating with the president of Yale University, Noah Porter, on the compilation of a new edition of Webster's dictionary.

Bessie Gardner was given a warm welcome at Swamp Hall and a cordial invitation to accompany the family to Washington and for the wedding of their mutual kinswoman, Annie Lee. Alfred, in particular, was delighted to have this unexpected addition to their wedding party, for he had never before met a young woman whom he found to be as attractive as Bessie. He had known other young women who were more beautiful, but never anyone with her style and her degree of sophistication. She did not play the role of a fluttering, empty-headed coquette, and he liked that. She had a ready wit and a quick tongue, and she could converse knowledgeably on important matters like politics, history, and current books. She also shared Alfred's love for the theatre and the opera. By the time cousin Annie Lee and her groom had said their "I do's" in the Episcopal Church in Alexandria, Alfred was more than eager to reenact the same ritual with Bessie and himself in the leading roles.

Consequently, Alfred suddenly displayed an unusual concern for his brother Louis's welfare at college and made several trips to Yale to see, as he told his sister Annie, "how Louis is getting along." Alfred did not do much to help Louis to improve his academic standing, but he did a great deal to promote his own standing with Bessie. In the late autumn of 1886, Alfred proposed marriage and Bessie promptly accepted. When Louis asked his older brother point-blank if he was going to marry Bessie Gardner and Alfred replied that he was, there was a quick expression on Louis's face of one who felt betrayed. Not until that moment had it occurred to Alfred that his brother, who was four years younger than Bessie, might also be in love with her. Louis said nothing more, and the incident could easily have been forgotten by Alfred but for the memory of Louis's hurt expression.

Bessie and Alfred were married on 4 January 1887 in St. James's Church in Philadelphia. After a brief honeymoon in Bermuda, they returned to Swamp Hall, which was now to be their home. In their absence, Gilly and Frank Mathewson had successfully carried out the detailed instructions given to them by Alfred for wiring the entire house for electric lights. No fenland ever glowed as brightly with miasmal ignis fatuus as did old Swamp Hall, with its newly installed incandescent lamps greeting the newlyweds upon their return.[28]

Annie du Pont had married her long persistent suitor, Absalom Waller, only two weeks before Alfred's wedding and had gone to live in Philadelphia. Except for the occasional visits of Maurice and Louis during college vacation, the five original orphans of Swamp Hall had now been reduced to one, and Alfred intended that there would soon be children again in the old family home who would not be orphans. Alfred's and Bessie's first child, a girl whom they named Madeleine, was born on 16 October 1887. A new era had begun in the old house on Breck's Lane.

Alfred I. du Pont may have entered a new phase of his life, but it was within a familiar old framework. As might be expected, Bessie had some major remodeling done in Swamp Hall, including the installation of four bathrooms and the addition of a billiard room on the ground floor, which quickly became Alfred's music room, but the basic form of the house remained the same, and Swamp Hall looked much the same as it had when Alfred's parents had moved in just prior to the Civil War.

Up at the mills also, the basic forms did not differ greatly from those of fifty years earlier. The first Eleuthère Irénée would have had no difficulty in finding his way around the yards had he returned to haunt his establishment. In the little frame house that served as the company office, the stubs of candles still flickered on Boss Henry's desk, and the scratch of his quill pen could still be heard as he painstakingly answered each incoming letter. Throughout the valley, in the company houses that Henry had assigned, the first, second, and third generations of E. I. du Pont's descendants still shared in common their horses and carriages, their profits and losses, their lives and the danger of imminent death.

Nothing changes and nothing will ever change, many of the increasingly disgruntled younger du Ponts complained, as long as Uncle Henry lives, and that may well be forever. No one seemed to have a stronger faith in his immortality than did old Henry du Pont himself. After Lammont's death, there was apparently no one being groomed to be Henry's successor. There were plenty of du Ponts available to be sure. There were his own two sons, Henry A., who had left a distinguished military career to head the sales division of the company, and William, who had succeeded Lammot as head of the Repauno Chemical Company; there were also Alexis's two sons: Eugene, who served as general manager directly under Henry, and Francis Gurney, a brilliant chemist, who seemed more interested in research than in his job as superintendent of both the Upper and Lower Yards; and, finally, in the third generation: Alfred, the assistant superintendent of the Lower Yards; Charles I., the one token representative of the Victor du Pont family, serving as assistant superintendent of the Upper Yards; and young Pierre Samuel II, Lammot's oldest boy, just finishing his studies at M.I.T. in preparation for returning to begin his apprenticeship at the mills. None of these men, however, had been designated heir apparent, least of all the three young men of the third generation, who had not even been granted a partnership in the company.

Henry's conservatism was a source of both humor and growing irritation among his numerous du Pont relatives. They laughed about his candles and his refusal to consider Alfred's offer to illuminate the mills with electric lights; they grumbled over the houses he so arbitrarily assigned to them; and they complained bitterly about the lack of opportunity for advancement and Henry's own greed in taking for himself and his two sons nearly 80 percent of the company's shares.

Much of this criticism was justified. Henry was dictatorial, selfish, and deaf to the suggestions of others. But a Bourbon reactionary who never learned anything new and never forgot anything old, he was not. The candles and the quill pens were not the true hallmarks of his administration. These were but idiosyncratic cosmetic symbols which have no real significance in the overall evaluation of his place in the history of the company. Indeed, it can be argued that during his thirty-nine years as head of the company he effected the fundamental changes in company policy and operations that made possible the revolutionary changes that occurred in the twentieth century. It was Henry du Pont who first extended the company's operations beyond the limited confines of the Brandywine valley by building a blasting powder plant on the Wapwallopen Creek in Pennsylvania in 1858. It was he who conceived of and forced his most important competitors to accept a Gun Powder Trade Association, thereby bringing an almost monopolistic order and stability to a fragmented and chaotic industry. He was to gunpowder what John D. Rockefeller was to oil and J. Pierpont Morgan would be to steel. And it was Henry du Pont, however reluctant he may initially have been, who took his company into the manufacture of nitroglycerine products, thus demonstrating by his success that E. I. du Pont de Nemours need not remain a single-product manufacturer. Even in the last year of his life, at the age of seventy-six, he was still dreaming big dreams, still making plans for expansion. In 1888, he initiated one of the largest undertakings in the history of the Du Pont Company during the nineteenth century. Under the supervision of his nephew Francis Gurney du Pont, in the little village of Mooar, near Keokuk, Iowa, he built the world's largest mills for the manufacture of black blasting powder. This plant was the stubborn old black powder man's counterargument to the craze for nitroglycerine. The Mooar mills remained an important producer of black powder long after the old mills on the Brandywine were silent and had fallen into ruins. Quite a record, all in all, for a quill pen pusher.[29]

Young Alfred was never one of Boss Henry's most vocal critics. He was disappointed when he could not persuade his uncle to install electric lights, but on most issues, Alfred had no complaints. Alfred got to keep the house he wanted. In general Alfred appreciated the strong, dependable leadership Henry provided. One always knew where one stood with the Boss, and young Alfred stood high in Henry's good graces. Alfred's father, Irénée, had always been Henry's favorite nephew, much preferred over the more gifted Lammot, and now Alfred appeared to be the fair-haired grand-nephew, for he alone was a true black powder man, the one du Pont of his generation who was most truly committed both to the company's heritage and to its future development under the du Pont banner.

In the early spring of 1889 Henry selected Alfred to go to France to gather information on recent progress that the French powder mills were reported to have made in the manufacture of brown prismatic powder. He was also entrusted with a more delicate mission that the United States Navy had asked the Du Pont Company to undertake. Alfred was commissioned to discover, if possible, the French method for making smokeless gunpowder. In 1886 a French chemist by the name of Vieille was reported to have developed what he called Poudre B, a single-base, pyro-nitrocellulose powder that upon combustion became wholly gaseous, burning fully, cleanly, and without smoke. As one historian of the gunpowder industry, William S. Dutton, has written, "Superiority of smokeless powder over black powder in firearms is as marked as the superiority of electricity over kerosene in illumination."[30] This was a comparison that Alfred du Pont would most certainly appreciate. Two years after Vieille's discovery, Nobel had come out with his version of smokeless powder, a double-base powder that was a compound of pyro-nitrocellulose and nitroglycerine. Nobel called his powder Ballistite. The British were now producing this form of smokeless powder under the name of Cordite. Clearly the United States was lagging far behind the great powers of Europe in this important area of ordnance development and the American Navy was understandably concerned. Alfred's task was to get hold of these trade secrets at as little cost to his company and his country as possible.

Armed with a letter of introduction from Secretary of State James G. Blaine to the American Minister to France, Alfred sailed for France on 20 March 1889, accompanied by Bessie, their infant daughter, and the child's nurse. It was Alfred's first trip abroad. At last he was to see the land of his forebears. For Bessie and Alfred the trip in its beginning had all of the festive joy of a second honeymoon.

Arriving in Paris, the young couple took rooms at the Hotel des Deux Mondes, an appropriately named residence for Alfred who during the next three months was indeed obliged to live in two worlds. There was the glittering world of Paris, which he explored with Bessie, dining in its restaurants, visiting its art galleries, and of taking a long coach ride out to Versailles to see the once glorious center of the Bourbon monarchy that Alfred's great-great-grandfather had futilely tried to reform and save.[31] There was also the other world of business, which was the reason for his trip and had its own excitement. Alfred's letter of introduction admitted him into the offices of the Ministry of War, which in turn led to the Director of Powders and Saltpeters, who graciously permitted Alfred to inspect the factories that produced brown prismatic powder. Although the inspection tour was hurried and brief, it was long enough for Alfred, who knew just what to look for, to see that the reports of French superiority in the manufacture of this product were not true. In fact, Alfred was surprised to see how obsolete some of the French machinery was. He would later recall how in France, England, and Belgium he saw the powder companies using technology and methods in the glazing mills that his great-grandfather and grandfather had used in the 1830s and 1840s but which the Du Pont Company had replaced with improved techniques before the Civil War.[32] As far as brown prismatic powder was concerned, the European manufacturers should be sending representatives to the Brandywine to learn how it really should be done.

When Alfred brought up the subject of smokeless powder, however, the welcome mat was abruptly yanked away. He was informed that not only was an inspection of the process out of the question, but even any discussion of the subject was forbidden. Poudre B for all but the highest French officials was in effect Poudre X—an unknown entity in a top-secret equation that France did not intend to give to a representative of the Du Pont Company an opportunity to solve.

Alfred's efforts to go directly to the French powder works and snoop around on his own proved equally fruitless. "I had thought of attempting to get into the upper yard, or its equivalent here and investigate the matter thoroughly," he wrote to Cousin Frank Gurney du Pont. "I find that our [security] system of picket fence, with wire on top, plus Mr. A. Burus, is publicity itself in comparison to the system employed here. Here they have a guard at the gate and one at each mill so you can see that nothing short of a pass printed by [Saint] Peter would be of any use. The mills are in charge of officers of high standing who would of course accept no bribe." Alfred even tried to buy from eight different French soldiers some of the cartridges containing smokeless powder, "offering as high as 1,000 francs per four or five shots but invariably met with the answer that money could not purchase them, as if found out, they were either sentenced to 20 years imprisonment or shot. This I afterwards ascertained to be true."[33]

Alfred had no alternative but to seek an answer elsewhere. He went to London in a futile attempt to get information about the British Cordite. Although rebuffed as forcefully there as he had been in France, Alfred did learn that the Coopal Company in Belgium also had the formula for smokeless powder. It was apparently a single-base powder, identical to the French product. A hurried trip to Wetteren, Belgium, just outside Ghent, to confer with the Coopal officials produced the interesting information that this company did indeed have the French formula for smokeless powder and that the Belgians, unlike the French, were willing to share this information with their good American friends, but for a price. That price included a very high royalty to be paid to Coopal for each pound of smokeless powder manufactured, and the condition that Du Pont agree to purchase the Coopal Company's machinery for making brown prismatic powder, although that machinery was inferior to what the Du Pont Company had already developed itself. This was the proposition. When Alfred asked if he might see a demonstration of the smokeless powder, he was told it was impossible. He could see the written reports of tests already conducted, but he would have to take it on faith that these reports were accurate. Believing he had no alternative if he was to carry out the mandate given him by his company and his government, Alfred signed a contract for options on both the Belgian brown powder machinery and the formula for smokeless powder.[34]

These arrangements were completed in the middle of June, and it was high time to return to the Brandywine. Both Alfred and Bessie were eager to leave. In letter after letter, Alfred complained of the wretched weather in both England and France, and of Bessie's and the baby's almost constant illness. "The baby has been sick since our arrival," he wrote his cousin Frank Gurney du Pont. "Bessie has been in bed a great deal of the time from a severe cold. I, as usual, interfered with my brilliancy and insisted on having the window opened at night for fresh air, not knowing the Parisians take advantage of the hours of darkness to inspect and repair their cesspools, the result was Bessie in bed next day with chills and fever." And again, "Don't give it away, but if I ever see Breck's Lane again I propose to stay there. Life at present is a burden. How anybody can live here is beyond my comprehension."[35]

At least his business mission, which was the real purpose of the trip, had been a success. Alfred believed he had fulfilled the mandate to get the formula for making smokeless powder. His superiors at home would not, however, view the contract with the same satisfaction that Alfred did. They would see it as an agreement to buy machinery that they did not need, and to pay an excessively high royalty for a promised formula that might not work. Fortunately, Alfred did not anticipate that reaction when he and Bessie departed for home.

When they arrived at Swamp Hall, they were shocked to learn that Uncle Henry was seriously ill—and was in fact dying. Even he was now prepared to accept his own mortality. Margaretta du Pont would later claim that Henry, finally accepting what he had for too long ignored, made a belated effort to prepare for a successor. She would tell her own family that she personally saw the letter he had secretly scratched out with his old quill pen and had sent to his nephew Fred in Louisville, asking him to come back to the Brandywine and take over as head of the company. Fred, according to Margaretta, sent this surprising letter to her for advice, and she had written back urging her son to stay where he was. Why should he kill himself, as his brothers Irénée and Lammot had done, working for the Du Pont Company only to put more money in the pockets of Henry's two sons. If this offer was actually made to Fred, as Margaretta claimed, then, as things turned out, his decision to follow his mother's advice proved to be unfortunate. Had Fred succeeded Henry as chief executive officer he very probably could have prevented the drift into the doldrums over the next ten years that very nearly brought an end to the Du Pont Company. And certainly this move would have prevented his own murder.[36] Fred, however, elected to stay with his profitable business in Louisville.

On 8 August 1889, his seventy-seventh birthday, the old Boss died without having designated his choice for a successor. The remaining and now leaderless du Pont partners met in executive session soon after the old man was buried to choose the new shepherd who would lead the du Pont flock into an uncertain future. Henry's two sons, Henry A. and William, now held four-fifths of the shares of the partnership as it had been reorganized in 1883 after Lammot had left the company to give his full attention to the Repauno Chemical Company. Either Henry A. or William could have taken control of the company had either one been supported by the other brother. Henry A. wanted and fully expected to be named president. He had interpreted his last conversation with his father as being a laying on of hands by which he had been anointed the new leader. Unfortunately, however, it had been a private conversation with no witnesses present. When Henry A. told his brother that their father wanted his eldest son to succeed him, William laughed. As Lammot's successor at Repauno, William did not want the presidency himself, but he was also damned sure he did not want Henry to have it either, for a deep-seated antagonism had existed between the two brothers for years. William told his brother that the other three partners did not want Henry either. Only the three sons of Alexis Sr. remained to be considered. Since Eugene was the eldest of the three, he was, almost by default, elected president. Henry A., sorely disappointed, never forgave William for his betrayal. The two brothers never spoke to each other again, and their mutual hostility became ever more bitter.

The next item on the agenda was to draw up a new partnership agreement. Since five du Pont partners remained, it was agreed that Henry A., William, and the three sons of Alexis Sr.—Eugene, Alexis Jr., and Francis Gurney du Pont—would each receive 20 percent of the new partnership. The three sons of Alexis Sr. would, of course, have to obtain their greatly increased shares in the partnership as a loan from Henry A. and William at 7 percent interest, to be paid back from future earnings of the company.

So the pie was neatly divided, but it left two du Ponts of the next generation, Alfred and Charles I., standing outside the door without a crumb.

The two assistant superintendents of the Upper and Lower Yards were
understandably outraged. Charles, who had always thought himself to be
something of an outsider, being a descendant of the Victor du Pont branch,
seemed resigned to this grossly unfair division of the spoils. Alfred was not.
He began to pound on the door and to demand entrance to the feast.

Alfred's elders were shocked. A share in the partnership was not something to be demanded. It was something that was granted at the benevolent
discretion of the ruling elders of the tribe. No du Pont had ever before
claimed a partnership in the company as his right. One simply waited and
hoped the gods would smile upon him. Alfred's behavior was judged as
gauche as if he had demanded the Order of the Garter from Queen Vic-
toria. But Alfred kept on pounding, and the whole family, particularly his
grandmother Meta, was aroused by the clamor he created.

To silence him, Eugene and his four partners agreed that each would
sacrifice one percent of his holdings, giving to Alfred a 5 percent share in
the partnership. That was not at all satisfactory to Alfred. It was not
enough, and moreover it still left Charles without a share. Alfred insisted
that he was demanding justice for Charles as much as for himself. The
deadlock was finally broken when Williams suddenly and unexpectedly
announced he would surrender his 20 percent, which could then be divided
equally between Alfred and Charles. William's magnanimous offer was not
born entirely out of altruism. William had never wanted to be a powderman. At Repauno, he lived in constant dread of meeting the same fate as
his cousin Lammot. Above all, his hostility toward his brother was such that
he did not wish to be a part of any organization that would bring him into
frequent association with the man who had always tried to dominate and
control his life. William preferred to take his money and escape the whole
du Pont enclave, which long stifled him both at work and at home, where
he had been entrapped in an unhappy marriage with a du Pont cousin.[37]

With William's departure from the company, a new partnership agreement was somewhat grudgingly drawn up. The partnership, dated 1 January
1891, had a total subscription value of $2,125,000. Each of the four senior
partners, Eugene, Francis G., Henry A., and Alexis, would keep his 20
percent of the partnership, with a subscription value of $425,000. The two
newly admitted partners, Charles I. and Alfred I. du Pont, would each have
a 10 percent share, with a subscription value of $212,500. It was necessary
for Charles to take his share as a loan from William. Alfred preferred to pay
William in cash, which he did by handing over the $100,000 estate he had
inherited from his father plus $25,000 he had received from Uncle Henry's
will as an unexpected token of the old man's affection for him. The remain-

ing $100,000 he borrowed from an obliging Uncle Fred, who could well understand why Alfred did not want to be indebted to any of the du Pont elders who had tried to keep him out of the partnership, not even to William, who had finally let him in.[38]

At the age of twenty-five, after five years of service to the company, Alfred I. du Pont was now a partner. It was much too early and much too easy for him to have achieved such an exalted position, his elders felt. It was much too late and much too difficult to have achieved what he justly deserved, Alfred felt.

IV

A Director Without Direction
1891–1902

EUGENE DU PONT was forty-nine years old when the mantle of leadership suddenly fell on his shoulders. In terms both of seniority of service and in general knowledge and experience in powdermaking, he was the logical successor to Henry. Eugene had worked in the Upper and Lower Yards since he was a teenage boy. He knew everything there was to know about the powdermaking business—both its triumphs and its disasters—everything, that is, except how to be a leader of the company. He had become a partner at the age of twenty-three, in 1863, one year before Alfred was born, when he was granted one share in the family partnership. During the ensuing twenty-six years, Eugene had slowly risen in the hierarchy until, at the time of Henry's death, he held the comprehensive but somewhat vague title of General Works Manager.

A generalist by experience and title, Eugene was by interest and talent a specialist. The only real satisfaction he ever found in his job was in working by himself in the chemistry laboratory, and his major contribution to the company had not been in operations or plant management, but in research. This research had led to two valuable patents for an improved type of charcoal, which had enabled the company to become the nation's leader in the production of brown prismatic powder.

An aloof and taciturn man, Eugene was only a silent presence in the mills, and to the du Pont family itself, he was little more. Aunt Sophie frequently urged her nephew Colonel Henry A. du Pont to return home more often to pay a visit to his father, Boss Henry. "To him your society is an exceeding enjoyment," Sophie wrote. "He is now for the most part surrounded by young men, none of whom sympathize with him. . . . Eugene is entirely shut up in himself, never speaking but to his wife and children."[1] It is doubtful that even Eugene's wife found her husband to be very lively company.

This loner, who had always had to live in the shadow of Uncle Henry and Cousin Lammot and who was not forceful or ambitious enough to cast a shadow of his own, suddenly found himself at the pinnacle of family power. His long years of service to the company had not prepared him for this position, and he found the mantle of authority not a comfortable garment, but a suffocating shroud.

He began boldly enough in trying to sweep away the dictatorial past. Although he had never taken part in the complaints directed against Henry's rule, he had certainly heard enough from his cousins and brothers over the years to know what they wanted done. There would be no more one-man rule within the company or within the family. The vast unitary domain that Henry had built up within the powdermaking industry would become a confederation of regional offices, each with its own semi-autonomous management.

And there would be no more foolish resistance to innovation. Cousin Alfred could install his electric arc lamps throughout the yards and his incandescent lights would resplendently illuminate the new home office building Eugene erected to replace Henry's one-room wooden shed. Each office and department within the plant would also be in direct communication with all the others by means of the telephone.

As real and welcome as Eugene's innovations were, they represented the true character of Eugene's regime as little as candles and quill pens had represented Henry's administration. In retrospect, Henry du Pont is clearly the true innovator who pushed the Du Pont Company forward, while Eugene, in spite of the electric lights and telephones and impressive organizational charts, sought only to hold the line. Eugene's administration meant stagnation, which, given the dynamics of late nineteenth-century capitalism, led inevitably to regression. Even the restructuring of the company into regional units proved to be a superficial show, for, as is often the case with someone who is unsure of his own ability to command, Eugene found it almost impossible to delegate authority. It was soon apparent that the new chief executive officer was as insistent as his predecessor had been in attending to all matters of policy and procedure himself. Unfortunately, Eugene lacked the aggressive vision that had given vitality and direction to Henry's dictatorial rule.

Alfred found that he had less authority as a junior partner under Eugene than he had had when he was only a salaried employee under Boss Henry. Alfred had never had difficulty in working with and presenting his views to Boss Henry, but the same was not true with Cousin Eugene. Alfred had aroused resentment among the senior partners at the beginning of Eugene's administration by his unrelenting demand for a place in the partnership.

Although Alfred had won that battle, it had been at the cost of whatever good relations he might have been able to establish with Eugene.

This initial chill in their relationship deepened into frigidity over the question of the implementation of the contract that Alfred had negotiated with the Coopal Company of Belgium. It was the first big policy issue to confront the new administration, and Eugene opposed honoring the contract, on the grounds that the Du Pont Company did not need the Belgians' machinery for the manufacture of brown prismatic powder. Eugene quite rightly considered himself to be the expert in this area, and it was apparent to him, from the drawings and specifications provided by Coopal, that the Du Pont Company already had developed a superior process. Alfred, who had seen the machinery in operation, had to concur with this judgment. Unfortunately, the purchase of the brown powder machinery was tied to the smokeless powder formula. No machinery, no formula. When questioned as to the evidence he had for the effectiveness of the formula, Alfred had to admit that he had nothing but the written reports that Coopal provided and these would have to be taken on their word alone as being true. Alfred had not been permitted to see any actual tests himself. Eugene vigorously protested the buying of a pig in the poke, sight unseen. It took the intervention of the United States Army, which like the Navy was becoming increasingly edgy over this dangerous new development in ordnance that some of their potential enemies now had, to carry the day for Alfred's contract. The Army insisted that the Du Pont partners buy the Coopal formula, along with the unwanted machinery, and the company reluctantly yielded to this demand.

Alfred's standing with his senor partners was hardly improved when, a few months later, tests revealed that the single-base powder produced by the Belgian formula was distinctly inferior to the double-base powder that the American inventor Hudson Maxim had recently developed—and which was available to the Du Pont Company.

Alfred's ill-fated contract did have one salutary consequence, however. It forced Eugene to take the one aggressively expansionist action of his administration. Convinced that the Du Pont Company could itself produce a better smokeless powder than Maxim's, Eugene ordered a new plant to be built at Carney's Point in New Jersey. He sent his brother, Francis Gurney du Pont, to this plant as manager, a job that Eugene himself would have welcomed. Frank du Pont was accompanied by Lammot's eldest son, Pierre Samuel II, who had just finished his apprenticeship in the company.

The transfer of Frank Gurney du Pont's main supervisory activities to the new smokeless powder plant at Carney's Point did not mean that Alfred lost a friend and ally in the Brandywine mills. Although Alfred had worked

more closely with Cousin Frank than with any other du Pont from the time
he had first reported for duty at the mills in October 1884, relations
between the two men had been distant and formal. Alfred, to be sure,
always held his elder second cousin in high esteem. Frank was a true powderman as well as a brilliant chemist. He was second only to Lammot du
Pont in scientific knowledge and in the practical application of that knowledge to product development. Unlike his older brother Eugene, Frank was
worthy of admiration, Alfred felt. Unfortunately, respect was not reciprocated. On the contrary, Frank seemed to take particular pleasure in finding
fault with Alfred. "I sit down to say in dispair [sic] that you did something
to the storage batteries at 10:20 o'clock this morning that has stopped the
second beating circuit, and what is worse, will not admit of its being connected with the minute circuit any more," Frank complained in one note
to Alfred. "If you are innocent of the charge, you will have to prove yourself
so, I cannot get the two circuits to run together, they make the most hor-
rible mess, mixing up together, and going any way except the way they are
expected to go. . . . I am grieved to the heart, as I just had the whole system
running together in the most perfect manner . . . now that is all over, and
all hope is lost."[2]

The two men had such different personalities and temperaments that it
is not surprising that their relations were less than cordial, in spite of their
close daily contact. Frank had much of his father's piety without Alexis Sr.'s
sense of humor and personal charm, and he was at least as conservative as
Eugene in giving younger men a chance for advancement. He was particularly irritated by Alfred's youthful cockiness and his brashness in demanding a partnership at too young an age. Cousin Frank took it upon himself
to instruct Alfred and young Pierre Samuel in proper respect for their elders
in the family.

Alfred, in turn, found Frank to be a study in paradox, more irritating
than it was fascinating. Alfred found it hard to understand how anyone
with Frank's talent and imaginative vision could at the same time be so
finical about unimportant details, writing Alfred notes such as: "Did I tell
you that I had set the Mill-wright clock back one minute? It was one minute
fast, and I put my keys into the switch and waited for one minute. . . . If
you will kindly set the other clocks you will oblige me"[3] Alfred enthusiastically applauded Frank's efforts to get better housing for their workers and
to convert the old Eleuthère Irénée du Pont mansion into a clubhouse for
their use,[4] but he was amazed that anyone who had such a concern for the
collective welfare of the company's labor force could be such an aloof
stranger to the workers as individuals. In contrast, it was said of Alfred by
his workers that "If Mr. Alfred knew you by name in the composing room,
he also knew you by name when you met him on Market Street." It was

doubtful that Frank knew his workers by name even in the composing room. So, although the two men worked together, signed their frequent letters to each other as "Your affectionate cousin," even shared ownership in a sailboat, they remained formal acquaintances, not close friends. Yet ironically, ten years after Frank's death, when Alfred faced the greatest crisis of his life, it was Frank's children who stood steadfastly loyal to Alfred, at the cost of their own careers within the company and their personal fortunes. The deep ties of friendship between the two families were not altered by the often strained relations between Alfred and Frank.

The decade during which Alfred served as an apprentice junior partner in the company proved to be far more difficult for him than the months he had served as an apprentice unskilled laborer. At the regular meetings of the partners, he and the other junior partner, Charles I. du Pont, were expected, like good Victorian children, to be seen but not heard. Although Alfred had become the expert in black powder, his opinion was never asked for, his approval never sought. Alfred suspected that there were frequent special meetings of the senior partners to which he was not invited.

Alfred, in turn, had little respect for the knowledge or ability of any of the senior partners, with the exception of Cousin Frank Gurney du Pont. Two of the senior partners were not even powdermen. Henry A. had had a distinguished career in the United States Army, but in the Du Pont Company he was concerned with selling, not making, gunpowder. Alexis Jr. had received his university degree not in chemistry or engineering but in medicine, and although he had only briefly been engaged in medical practice, he still bore the title of Doctor with great pride. As a young man Alexis had given up medicine to work with Fred du Pont in his cousin's coal and steel industries. Only in 1885, at Boss Henry's request, had Fred du Pont sent Alexis back to the Brandywine to serve in the home office. Alexis was a bright man and a capable office manager, but he was not a powderman. Nevertheless, it was to him that Eugene always turned for counsel.

Nor did Alfred find a congenial companion in the other junior partner, Charles I. du Pont. Charles held an equally isolated position within the company, but unlike Alfred, he raised no protest and kept his own counsel. The senior partners appreciated Charles's plodding industriousness and quiet subservience. Moreover, Charles became a hero in the yards during the great catastrophe of 7 October 1890, one of the worse disasters in the history of the company. Twelve men were killed outright and twenty more were seriously injured in the seven separate explosions that occurred within a few seconds of each other. The mills in the Upper Yards were almost totally razed, fifty homes in the vicinity were destroyed, and the shock was felt in Georgetown, Delaware, one hundred miles away. Alfred happened to be in Philadelphia on business at the time, but he felt the ground tremble even there. It was Charles, working in the laboratory in the Upper Yards, who rushed out and, at the peril of his own life, helped extinguish fires and rescue men trapped in the wrecked buildings. As a newspaper account of the time reported, "He conducted himself on this occasion, amid the danger which threatened the lives of everyone in the yard, with conspicuous bravery."[5] Charles remained as modest and unassuming in the unexpected limelight of publicity suddenly focused upon him as he was in performing the dull routine of his daily tasks. This, too, was noted with approval by the senior partners.

Alfred's strained relations with his superiors in the company must have been a subject of discussion in many du Pont households, but such talk paled into insignificance when over a period of four years the family was shaken by an equal number of major scandals, two concerned with marriage, two with death. For the first time in their nearly one hundred years of American residency, the du Ponts' private lives became a subject of notoriety in the metropolitan presses. The first scandal, in retrospect, was mild in comparison with its successors. It concerned Alfred's younger brother, Maurice, a most unlikely subject for scandal. The second son of Charlotte and Irénée du Pont had always been the most quiet and least obtrusive of their five children. As a child, he had been Alfred's obedient and devoted servant. Alfred had always been what Maurice could never be: the bold initiator of new games, the daring adventurer in all the many varieties of childhood perils, the staunch defender of the family's honor in fist fights and wrestling matches. Whenever his courage and strength permitted him to do so, Maurice followed Alfred's lead, but when his nerve or muscles failed him, Maurice would offer his brother his cheers and applause.

Maurice had also followed the academic trail blazed by his brother, attending Phillips Andover Academy and then M.I.T., but, unlike his brother, instead of simply dropping out of M.I.T. at the end of two years, he had transferred to Johns Hopkins University and had dutifully followed the prescribed curriculum, earning his degree in electrical engineering. For all of his efforts, however, he was not invited to join the Du Pont Company as an employee. Virtue, in this case, however, was to have its own peculiar rewards. Upon reaching his majority, Maurice received his share of his father's estate, and before settling down, he decided in 1889 to take his elderly spinster cousin, Mary Van Dyke du Pont, on a trip to Europe as a reward for her long years of faithful care of the orphans of Swamp Hall. It was Maurice's intent to join up, either in London or in Paris, with Alfred and Bessie, who were in Europe at that time on the smokeless powder mission. Before making contact with his brother, however, Maurice made another, quite unexpected contact. The liner on which he was traveling put into the port of Queenstown in Ireland. Having some time at his disposal, Maurice went ashore, stopped in a hotel pub for a drink, and there met a young hotel employee named Margery May Fitz-Gerald. One drink led to another, and to much spirited conversation. The young Irish lass was the most beautiful, witty, and charming woman Maurice had ever met. After one drink, he was madly in love. After several more, he proposed, and he threatened to stay in Queenstown until she consented to marry him. Margery consented, and Maurice, promising to return in the autumn to claim her as his bride, resumed his tour with Cousin Mary, and wisely kept his mouth shut about what he had seen and done in Queenstown.

In September, Maurice went back to Ireland. On the twelfth of October, Margery Fitz-Gerald and Maurice du Pont were married in St. Peter and Paul's Church in Cork.

At home during the preceding summer, Maurice had maintained a discreet silence about the only impetuous act of his life, and no one, not even Alfred, had any knowledge of Maurice's engagement and plans for marriage. With the wedding, however, Maurice's attachment became a matter of public record. Some weeks later, a reporter, idly scanning the marriage registry, noted the entry for 12 October: FITZ-GERALD, Margery May, of Queenstown to DU PONT, Maurice, of Wilmington, Delaware, USA. It took only a little sleuthing on the reporter's part to discover that DU PONT, Maurice, was indeed one of *the* du Ponts of America. The news quickly traveled via trans-Atlantic cable back to New York and to all points north, south, and west.

The du Pont family had its first inkling of Maurice's big news when they saw the headlines in the *Wilmington Evening Journal,* reprinted from the *Chicago Daily News:*

HE MARRIED A BARMAID

STORY OF MAURICE DU PONT'S ROMANTIC MATCH

BEAUTY AND VIRTUE HER ONLY DOWRY

The du Ponts were outraged, not so much by whom Maurice had married as by how they had been informed. (The du Ponts married with flowery notices on the society page, not with glaring headlines on the front page.) Only Alfred wrote to the address Maurice had left with him to offer his heartiest congratulations and to say that Swamp Hall was eager to welcome the newlyweds back to the Brandywine.[6]

Maurice and Margery did come back to Swamp Hall the following spring, and Alfred and Bessie persuaded some of the younger du Pont families to make their official calls. It was not long before Margery's charm and wit had won over even Granny Meta and Aunt Louisa. Soon Margery had been transformed, with Cinderella magic—at least within the du Pont fam-

ily—from "barmaid" into "the daughter of an ancient noble Irish family."
Margery herself rather enjoyed the more earthy designation given to her by
the press and would always sign her letters to her favorite in-law, Alfred,
"Your Irish barmaid sister."

Newspaper publicity, however distasteful, given to a marriage could be
forgiven and forgotten, but the publicity given to a divorce was quite
another matter. There had, of course, over the past century been
unhappy—even tragic—marriages in the du Pont family, most notably that
of Irénée and Charlotte du Pont, but there had never been a divorce. For
better or for worse, the du Ponts mated for life. This tradition was rudely
shattered in the spring of 1892, when William du Pont, Boss Henry's
younger son, went to Sioux Falls, South Dakota, and established residency
in that state, which at the time had the most liberal divorce laws in the
nation. Again, the du Ponts first received information of a momentous
event within the family through sensational headlines in the nation's press.
The news was not well-received.

The family, of course, had known for a long time that William and his
cousin-wife, May du Pont, Charles I.'s older sister, had been not happy in
their marriage. May had never wanted to be married to William. She had
been very much in love with a young lawyer in Wilmington, Willard Saulsbury. Saulsbury was, however, some seven years younger than May, and the
family regarded him as an ambitious youth who was eager to obtain May's
money and the prestige that came with her name. In the 1870s, the family
council under Uncle Henry's autocratic leadership was still powerful
enough to rule on matters of a personal nature within the family. The order
went out. May could not marry Saulsbury. She must marry Uncle Henry's
own son, William, who apparently had no choice in the matter either. William and May were joined in holy wedlock in 1878 and lived unhappily
afterward—but not forever. By 1892, they were in agreement on at least
one point. They had had enough. William had fallen in love with Annie
Rogers Zinn, a married woman living in Newcastle, Delaware, and Williard
Saulsbury was still waiting for May. So William became a six-month resident
of South Dakota, and the divorce was granted. William returned to Delaware to the arms of his beloved Annie, who had managed to free herself
from the encumbrance of her own unhappy marriage, and May was at last
free to marry her patient suitor. The two newly joined couples were now
free to live happily ever after.

The du Ponts, however, instead of blaming themselves for having
wrongfully interfered in the first place, attributed the turn of events to a
breakdown in the power of the family council. This scandal of divorce and
quick remarriage certainly would not have happened in the days of Boss
Henry. For most of the du Ponts, this scandal was one more piece of evi-

dence pointing up the ineffectiveness of Eugene's administration. Henry A., William's elder brother, was the most irate of all. Still smarting from what he regarded as his brother's betrayal in refusing to support him for the presidency of the company, Henry A. was utterly vindictive in his determination to make William a pariah in the family, not only during his lifetime but for all eternity. Not only would William be barred from the family's society, but Henry A. was determined that his brother should not even lie with the family in death. The word went out that under no circumstances was William to be buried in the consecrated ground of Sand Hole Woods Cemetery. Unfortunately for Henry A., however, William was seventeen years his junior, and not even the older brother's great social authority within the family could abrogate the natural operation of the actuarial tables. Henry A. died in 1926. When William died a few years later, he was placed, in accordance with his explicit instructions, in a lot adjoining his brother's.

William would ultimately triumph over his brother in the graveyard, but for the present, Henry A.'s decree of social ostracism for his brother was obediently, even eagerly observed by the family—with one notable exception. Alfred offered his friendship and support both to his wayward cousin and to the latter's former wife, May. He told both cousins that he would be happy to see either of them in any social gathering. The propensity of Alfred for collecting the du Pont social waifs under his wing did not go unnoticed among his elders.

As for William, he shed no tears over Henry's ban. Both he and his new wife had had enough of du Pont-dominated Wilmington society. William resigned as president of the Repauno Chemical Company, took his bride and fortune off to England for an extended honeymoon, and then returned to the United States, where he purchased the beautiful Virginia estate of Montpelier, the former residence of President James Madison.

May du Pont Saulsbury, also banned by Henry A. from du Pont society, could not so easily escape, but eventually she too would win out over what her new husband called "that God damned tribe."[7]

The family was still reeling from this game of matrimonial musical chairs, which had left only George Zinn, Annie Rogers du Pont's former consort, without a partner, when six months later, scandal again hit the family. This third scandal struck even closer to Alfred I. du Pont, for it concerned his second younger brother, Louis Cazenove du Pont.

Of Charlotte's and Irénée's five children, only Louis appeared to have been scarred by his parents unconventional and unhappy marriage. It had been soon after his birth that his mother had begun her final and irreversible decline into madness. His earliest childhood memories were of nights punctured by her screaming tantrums and of days during her long absences when he had cried for the want of her. He was only nine when his parents died. For a child as sensitive as Louis, probably no therapy could have alleviated his pain. Within the du Pont family, Louis was the child who most obviously could be characterized as being the beautiful and the damned. He was the most handsome of the five du Pont children. He also had one of the sharpest wits and liveliest minds in the family. Granny Meta, who was always ready to make a quick comparative judgment, frequently said that the most brilliant mind the du Pont family had produced since old Pierre Samuel was that of Louis Cazenove du Pont, her favorite grandson.[8]

Louis's great talent was to be tragically wasted, however. He had little motivation, even though he could do almost everything easily and well. School work presented no challenge and consequently was not worth the effort. Alfred was instrumental in having him placed in Phillips Andover Academy, but Louis barely managed to complete the course of study. After Phillips came Yale, which initially he liked for the challenge it presented, but too quickly, the familiar boredom set in. He fell in love with his cousin, Bessie Gardner, only to have Alfred take her away from him. It was then that Louis began to drink heavily. He cut classes and dropped courses, but the Yale authorities were reluctant to dismiss a du Pont. In 1891, after six years at Yale, he transferred to the Harvard Law School, which was willing to admit him even without a Yale degree. He had now reached his majority and could claim the last remaining portion of Irénée du Pont's estate. A year at Harvard convinced him that he had no interest in law, and he moved to New York to become, as he put it, "the only swell in the du Pont family."

Late in November 1892, at Alfred's urging, Louis came down to Swamp Hall for a visit. The years of dissipation had not marred his handsome face nor dulled his wit, but Louis's own hidden Dorian Gray portrait was not safely deposited in an attic storeroom. It hung heavy within him, becoming ever more ugly and more demanding. To be under the same roof with Bessie Gardner, now Mrs. Alfred I. du Pont, may have evoked old frustrations, for she had been the only prize that he had sought that he had not easily obtained.

On the afternoon of 2 December, Louis stopped by the Wilmington Club. He met a friend who asked if he would serve on the arrangements committee for the annual Christmas ball. Louis said that he was sorry he couldn't because he was leaving soon. He was going "a long distance away." Louis then went into the club library, sat down at a desk, and began to write a letter to a friend in New York. He didn't bother to finish the letter—another bit of uncompleted business in his unfulfilled life. A servant in the main lounge heard a shot. Rushing into the library, he found Louis slumped down in the desk chair, a bullet through his head and a pistol on the floor.

Louis was the family's first suicide. This in itself was scandalous enough to stun them all. To commit this mortal sin, however, not in the privacy of one's bedroom, as any gentleman, no matter how distraught, would have done, but in the sanctum sanctorum of the Wilmington Club, was indeed to die without a trace of sanctifying grace.[9]

If after these three scandals, the du Ponts believed they had become inured to all further shock, they were mistaken. The scandal that occurred five months after Louis's suicide was of such a nature as to shake even this benumbed family, for it concerned Alfred Victor—the family's Uncle Fred, the least likely du Pont to become involved in scandal. Good old, kindly Uncle Fred had devoted most of his sixty years tending to the demands and the needs of the rest of the family. As a young boy in school in Philadelphia, Freddie was the one whom his mother and his aunts always deputized for shopping errands: hymn books and religious tracts for Aunt Victorine's Sunday School, yard goods for his mother, and little indigo buntings from a pet shop for Aunt Eleuthera. As a young businessman in Louisville, Fred had provided a partnership for his younger brother, Bidermann, the least capable of the du Pont boys, and later, employment for his nephew, Thomas Coleman. It was Fred who had been designated by his older brother Irénée and his younger brother Lammot to be the guardian of their many children in the event of their deaths. Fred himself never married, and Irénée's five and Lammot's ten fatherless children became his responsibility. It was Uncle Fred who selected their schools, paid their allowances, and managed the shares of their fathers' estates until they reached their majority. It was Fred who never spent a cent more than necessary for himself from his own great wealth but who gave generously to provide public schools, parks, and jobs for the people of Louisville. A caring surrogate father, an enlightened employer, and Louisville's most generous public benefactor, Alfred Victor du Pont was widely respected, a model citizen. It is not surprising that when, on 16 May 1893, the press reported the unexpected news that he had been strickened by apoplexy and had died while calling at the home of his brother Bidermann, the shocked populations of both Louisville and Wilmington genuinely mourned the loss of this good man. The du Pont family could fine some solace in their grief in the panegyrics and laudatory editorials, which did much to offset the reams of sensational publicity the other members of the family had generated.

The du Ponts' comfort in this reflection was short-lived, however. Two days after the initial report of his death, which the *Louisville Commercial*[10] found to be a fitting conclusion to his life—"The quiet and simple character of his life unbroken in the scenes of death,"—the *Enquirer* in nearby Cincinnati published a very different version of Fred's death. His life may have been "quiet and simple," but it would seem that his death had been violent and complicated by blackmail and murder. Alfred Victor du Pont had not died quietly in his brother's arms. He had been shot through the heart while visiting Maggie Payne's house, Louisville's most expensive and notorious bordello, by one of Maggie's young "hostesses," who was angered by Fred's refusal to pay for the support of her child, who she claimed was also Fred's.

The Louisville papers flatly denied the *Enquirer* story and continued to maintain the validity of the original version of Fred du Pont's death. Many years later, however, Henry Watterson, the *Courier-Journal*'s great editor and Fred's close friend, admitted that the du Pont story was the only news story he had ever personally killed. In its place he had published a piece of pure fiction.[11] Nor did the du Pont family ever officially acknowledge the true cause of beloved Uncle Fred's death. They too clung to the respectability of the original story, no matter how synthetic its fabrication.

Fred's nephew, Coleman du Pont, who accompanied the body back to the Brandywine, told Alfred in great confidence, however, that he had been summoned to Maggie Payne's immediately after the shooting, for Coleman was no stranger in that particular residence. Coleman quickly arranged for a hearse to take his murdered uncle's body to Bidermann's house. Only then was Fred's death reported. An obliging coroner dutifully certified the death to have been caused by "effusive apoplexy." The casket was sealed and the false story was released.[12] It should be noted, however, that the family never brought suit for libel against the Cincinnati *Enquirer*. Nor was criminal charge made against Maggie Payne or any of her "hostesses" by the Commonwealth of Kentucky. With the connivance of officials and the family, it apparently had been possible to commit "the perfect murder" of a man whom all Louisville had called "the perfect citizen."

So preoccupied had the press been in covering the du Pont scandals that little attention was given to some far more significant activities in which the du Pont family was engaged. The major business of the du Ponts in the early 1890s was, after all, the creation of bigger and better chemical explosives, not more sensational and lurid journalistic explosions. Although the Du Pont Company's bread-and-butter business was still centered on the sportsman's black powder ammunition and the blasting powder produced in the Brandywine mills and at Mooar, Iowa, which represented Alfred's world, the Repauno Chemical Company and the new smokeless powder plant at Carney's Point in New Jersey held all the glamour of new research and the potential for future growth. From this world, Alfred felt himself to be largely isolated. He had few regrets about it since he was and always would be a black powder man. Like his Uncle Henry before him, he had to accept the reality of high explosives, but his mission in life was to produce the best black powder possible, and to improve both the efficiency and the safety of the methods for its production.

With the departure of William du Pont from the presidency of the Repauno Chemical Company, it was necessary to find new leadership for the company, which in less than a decade had become the nation's leading producer of nitroglycerine products. At the suggestion of Laflin & Rand, which held an important share of this company, J. Amory Haskell, the general manager of the Rochester & Pittsburgh Coal and Iron Company, was brought in to head Repauno. He in turn hired Hamilton Barksdale, a civil engineer who had helped construct a new railroad line for the Baltimore & Ohio Railroad between Baltimore and Philadelphia. Prior to their coming to the Repauno Chemical Company, neither Haskell nor Barksdale had had any experience with high explosives except as consumers of the product, but both men were superbly qualified as business managers and both were fast learners. Alfred could not have failed to note also that both Haskell and Barksdale were only thirty-one years old, just three years older than he, yet they were given complete authority by their two parent companies to run Repauno as they saw fit. Apparently it was possible to get to the top at a very young age, especially in a company that was co-owned by Laflin & Rand.

In fewer than three years, Haskell completely reorganized the high-explosives industry. Under his direction, a new company, the Eastern Dynamite Company, was formed, capitalized at $2,000,000. Seventy percent of its stock was exchanged for the capital stock of the Repauno and Hercules companies, and 30 percent for the Atlantic Dynamite Company, which had the great attraction of being the holder of all of Nobel's American patents. Atlantic was then reincorporated as a new subsidiary completely owned by the Eastern Dynamite Company.[13] This elaborate reshuffling of corporate structures meant simply that Du Pont, Laflin & Rand, and Hazard (which was, of course, secretly owned by Du Pont) gained a virtual monopoly over dynamite production in America.

Haskell was interested in more than using corporate wizardry to monopolize the industry. He was also interested in producing a superior product that justified that concentration of power. Under Lammot, Repauno had largely been a one-man show, with the office of research and development housed within his own ingenious head. Under Haskell, research became as organized and rational as the factory line itself. He employed Oscar R. Jackson as plant superintendent. Jackson, the son of Dr. Charles T. Jackson, the discoverer of ether as an anesthetic, quickly lived up to Haskell's claim that he was the most brilliant chemist in the high-explosives industry. Haskell also brought into the company his younger brother, Harry Haskell, who gave systematic order to office management.

For what he had been able to accomplish in three short years at Repauno, Amory Haskell was the logical choice in 1895 to succeed Soloman Turck, the long-time president of Laflin & Rand. Hamilton Barksdale was left in charge of the Repauno Chemical Company. Barksdale, who was able to recognize and reward ability in others younger than himself even if Eugene and his fellow senior partners back on the Brandywine could not, asked Alfred to joint the Repauno organization as vice-president, an office that had not existed under Haskell. This position provided Alfred with an entry into the high-explosives industry. It proved to be a valuable experience, but Alfred's main interest and real value to the Du Pont Company remained with the production of black powder.

Barksdale further sealed his alliance with the junior partners of the Du Pont Company by marrying Ethel, the elder sister of Charles I. du Pont. Not since James Antoine Bidermann had married E. I. du Pont's daughter Evelina in 1816 had the du Ponts found an in-law who would prove to be as able and influential within the company. With Haskell at Laflin & Rand and Barksdale at Repauno, the two former rival companies were now comfortably and securely intertwined. The stability and order that Boss Henry had set out to establish within the industry in 1870 had been fully realized.

Repauno, first under Lammot, but especially later under Haskell and Barksdale, represented a new frontier within the explosives industry that had been successfully invaded, exploited, and then settled. The du Ponts, assisted by their friendly rival, Laflin & Rand, could now regard dynamite, a territory initially explored by Alfred Nobel, as their established and firmly held domain in America. With the acquisition of the Atlantic Dynamite Company, the holder of the Nobel patents, and the formation of the Eastern Dynamite Company, the du Ponts had in effect declared a new Monroe Doctrine for the high-explosives industry. America was now Du Pont country. All Europeans, including the great Nobel himself, were expected to keep out. When, however, the ambitious du Ponts in 1896 attempted to blast their way into the lucrative market for dynamite in the mining operations in South Africa, they quickly discovered that their Monroe Doctrine was a two-way street. If they intended to invade the Eastern Hemisphere, then Nobel and his European affiliates, the Nobel-Dynamite Trust Company, Ltd., of London and the Vereinigte Köln-Rottweiler Pulverfabriken of Germany would retaliate by entering the American market. Nobel actually began construction of a dynamite plant at Jamesburg, New Jersey. A hastily called conference between Du Pont representatives and their British and German counterparts resulted in the so-called Jamesburg Agreement of

1897. Under the terms of this pact, the Europeans agreed to stay out of the American market, and the Gun Powder Trade Association in return promised not to encroach upon Europe or Africa. Henry du Pont's little "community" of East Coast and Midwestern gunpowder makers had now become an international cartel. The stability Boss Henry had sought for his industry within the United States had been extended globally.[14] The frontier days of dynamite production ended with this settlement.

But at the Du Pont Company's second New Jersey plant, located at Carney's Point, its newly appointed director, Francis Gurney du Pont, was still on the edge of a new frontier in high explosives. His task was to perfect an improved smokeless powder that could be produced in the same mass quantities that black powder and dynamite manufacturing had already achieved. After having proved that the Belgian firm's single-base formula was ineffective, Frank du Pont had turned his considerable research talent to adapting and improving the double-base formula that Nobel had first developed and which British firms were now utilizing in the production of Cordite.

In his efforts to achieve a smokeless gunpowder that would be superior to both the Nobel and the Hudson Maxim patents, Frank du Pont was ably assisted by his young cousin, Pierre Samuel II, Lammot's oldest son. Later, after they finished college, Frank's own sons, Francis I. and A. Felix, as well as Eugene du Pont's son, Alexis Irénée III, were added to the staff. Unlike Repauno, Carney's Point was strictly a du Pont family show. Within one year of the laboratory's construction at Carney's Point, the first gun cotton for naval mines and torpedoes had been produced, and by the end of 1892, fifty tons of gun cotton had been purchased by the U.S. Naval Bureau of Ordnance.

If the Navy was pleased with this beginning at Carney's Point, the Du Pont Company was not. Gun cotton was not the smokeless powder the company was seeking, and the United States Navy was not the customer the du Ponts were most eager to serve and to satisfy in the early 1890s. Frank was seeking not a high-explosive cotton for mines, torpedoes, and cannons, but a high-explosive powder that the ordinary weekend hunter could load into his shotgun to kill ducks and geese, and, moreover, a powder that would not cause the shotgun to explode, killing the hunter. The Du Pont Company had little interest in or concern for the needs of America's armed services. In spite of the efforts of President Chester Arthur's administration to replace the antiquated Civil War Navy with a modern fleet that could rate at least a third place standing to the far superior armadas of Great Britain and France, the United States was not a military power of any significance on either land or sea. The last serious Indian uprising had been put down in 1886 with the capture of the Apache Chief Geron-

imo, and from that time on there was little need even for the small garrisons
of troops that had been maintained in forts scattered across the Plains. Symbolic of these halcyon days of American isolationism and pacificism was the
military review that President Benjamin Harrison presented in 1890 for the
edification of the representatives of seventeen Latin American republics
assembled in Washington, D.C. for the first Pan American Congress. The
review's purpose, the President proudly announced to the visiting diplomats, was to show not how much military might the United States possessed, but rather how little. To most Americans war seemed an outdated
a form of barbarism, a practice confined to quarreling Balkan states or colonial Africa. With two vast oceans to separate it from Europe and Asia, with
a friendly nation to the north and a weak nation to the south, the United
States enjoyed natural security. Why then buy something that geography
had so generously provided free of charge?

If Congress and the general public had little interest in providing appropriations for defense expenditures, this did not mean that individual Americans were uninterested in ammunition. They had had a long-standing love
affair, dating back to the first English settlements, with guns—not guns on
battleships or foreign battlefields, but guns in their own hands to shoot the
still plentiful buffalo, the wild ducks, the migrating Canada geese, and the
passenger pigeons,—or much too often, either by accident or by design—
to shoot each other. The du Ponts did not need a market analysis to tell
them who their best and most reliable customers were.

Increasingly, those customers were inquiring about smokeless powder,
which was rumored to be a great improvement over the familiar sporting
black powder. It was reported to have greater firepower, with much less
recoil. Being smokeless, it left little residue within the barrel. The result
was much more rapid firing without fouling the weapon or tiring the
hunter. Frank du Pont's job at Carney's Point was to provide the customers
with what they wanted. His task was not simple, for smokeless powder as
so far developed in this country was not yet an appropriate ammunition for
the novice sportsman carelessly loading and firing from a duck blind.

The development of a smokeless powder that could meet the high standards set by black powder for velocity and pattern of shot was a major part
of the problem presented to Frank du Pont. There were also many other
details to be worked out, such as determining the proper wads and cartridge
cases to be used and the amount of powder to put in each shell, before the
Du Pont Company could guarantee its smokeless product as being as safe
to store and to use as its old reliable Eagle brand of black sporting powder.

These problems could not be researched and solved at leisure. In the
depression years of the early 1890s, the Du Pont Company could not afford
to lose customers to domestic and foreign competitors producing the cov-

eted smokeless powder shotgun shells. As Cousin Frank wrote to Alfred in the summer of 1893 regarding black powder production at the Brandywine mills, "I have been anxious to see you for a few days because owing to the dullness of business, we shall have to shut down. . . . We are in a very bad position from the outlook with prices reduced and dull trade, but the best we can do is put things in shape so that if we do have any trade, we can do it justice."[5] Nor did the black powder business improve during the ensuing twelve months. In May 1894, Frank wrote to his son, Francis I., then a student at Yale, "Business is worse than ever. Here we are shut down three days in the week and running three days, the men of course being paid for the time they work. . . . It is very bad and the outlook worse. It is the first time we have had to do anything of the kind on the Brandywine."[16] No one in the country awaited the Du Pont smokeless powder ammunition more eagerly than the du Ponts themselves. The pressure was on Frank du Pont to produce. Alfred, meanwhile, continued to turn out the best black powder on the market, just as his father, his grandfather, and his great-grandfather had before him. Alfred could only hope that the sales department would find the customers to allow him to keep on producing.

Black powder has as its basic components saltpeter, sulphur, and charcoal, but smokeless powder is predicated upon an entirely different concept. Its basic ingredient is nitrocellulose, commonly called gun cotton, which is a highly volatile material that can safely be worked on only by having the finely divided gun cotton immersed in a liquid and kept in suspension by the agitation of the liquid. The use of pure water, however, did not result in the desired granulation necessary to achieve grains of powder. The first real breakthrough in achieving a successful granulation process came when young Pierre Samuel, while still at the Brandywine laboratory prior to his transfer to Carney's Point, made the serendipitous discovery that the addition of nitro-benzol to the water caused a granulation of the mixture into a mass of separate grains suspended in the clear liquid. Although nitro-benzol produced granulation, it did not produce useable grains of smokeless powder. Nevertheless, Pierre Samuel's experiment of adding a compound to water to produce granulation opened up a whole new field of possibilities. After much more experimentation, it was discovered that gun cotton immersed in purified water to which acetone oil had been added would produce the desired grains of smokeless powder.[17] Two days before Christmas in 1893, E. M. Marble, the Du Pont Company's patent lawyer in Washington, D.C., wrote to Francis Gurney:

> At this season when all the world is given up to gayety, it gives me special pleasure to announce to you the Examiner [of the U.S. Patent Office] having charge of your smokeless powder applications, has for-

mally passed the acetone-oil application to issue, and that I will receive
the notice of allowance therefore on Tuesday next. . . .

The applications which I hold are as follows, (1) the improved process for dripping nitro-cellulose, allowed Aug. 22, 1893, (2) the nitro-
benzole-benzine process, allowed Oct. 23, 1893, and finally (3) the acetone-oil process, the allowance of which I will receive next week. . . .

Permit me to wish you a Merry Christmas and a happy and prosperous New Year.[18]

In preparation for testing the Du Pont smokeless powder against other
powders, both domestic and foreign, which were coming onto the market,
Frank du Pont decided to utilize the services of a German ordnance expert,
Armin Tenner, who had recently arrived in New York from Prussia to set
up an institute for the testing of gunpowder. Elliott S. Rice, the Du Pont
Company's general agent in Chicago, had first introduced Tenner to Frank
du Pont. Rice had met Tenner in Germany in the summer of 1892 and was
greatly impressed with the Prussian's background and expertise in all areas
of powdermaking. He urged the du Ponts to support Tenner in the establishment of his institute.

It is not clear whether or not Frank du Pont knew that Tenner was a
former employee of Wolff & Co., one of the Big Four of German gunpowder manufacturers. This company had recently developed the Walsrode
smokeless powder, which, it was rumored within the industry, would prove
to be the best powder yet produced. What is known, however, is that Rice,
upon returning to Wilmington in August 1892, gave Frank du Pont a sample of the Walsrode powder that Tenner had presented to him, along with
Tenner's report on the effectiveness of this powder.

Tenner himself did not arrive in the United States until a year later. By
this time Frank du Pont had developed his acetone-oil–processed smokeless
powder, which was remarkably similar to the Walsrode powder and for
which the Du Pont Company was awaiting a patent. Tenner at once
entered into negotiations with Frank du Pont to serve as the agent for
testing the Du Pont's newly developed smokeless powder. As might be
expected from a man who was wooing a potential client, Tenner, after
examining the Du Pont powder, was extravagant in his praise for its quality.
On 26 September 1893, Tenner wrote to Rice, "I would be surprised if Du
Pont should not prove the best or at least one of the best two smokeless
powders made in America, and it certainly will be nearly on an equal footing with Walsrode, should the matter not fall short in some of the points
yet to be determined."[19]

With this kind of promise, Frank du Pont could see the advertising
value that Tenner's tests could have for the company, so Tenner was commissioned to conduct the tests, for which service the Du Pont Company

gave $500 in support of the Tenner Testing Institute. Seven different powders, including both the Du Pont smokeless and the Du Pont black powder F.F.F., were tested for speed of firing, velocity, and pattern of shot. Unfortunately for Frank du Pont's dream of achieving a test result that could be proudly proclaimed in every sporting journal in the country, Du Pont Smokeless Powder rated lowest, earning only 150 points on the rating scale designed by Tenner. The Walsrode powder ranked first, with a score of 180, and Alfred must have taken some quiet satisfaction in the fact that his cherished black powder did better than Frank's smokeless powder, earning a score of 159 and placing fourth in the testing. Tenner added an apologetic little note to these test results, "I am sorry that your smokeless does not take the place as I would like to see it take, but I have called attention to the fact that it is not yet in the market and, no doubt, before it is introduced, will be improved in every respect sufficiently to make it as nearly perfect as any gun powder can be made."[20] This expression of regret did little to ease Frank du Pont's disappointment. Herr Tenner's tests proved not to be such stuff as advertisements are made of, and there were no further contributions from Du Pont to the Tenner Testing Institute. When Tenner's own dreams of a close association with the powerful Du Pont Company proved to be illusory, he quietly returned to his native Prussia. Frank du Pont probably would never have thought of him again had it not been for the patent suit that the Wolff Company brought against the Du Pont Company several years later.

At least it could be said that Tenner's tests were honest. Tenner had not tried to curry favor by falsifying the results. He had, moreover, been able to point out the problems with Du Pont Smokeless Powder. The powder itself was undoubtedly as good as the Walsrode product. The problem lay in the delivery system. Further experimentation was needed to determine the proper cartridge case, the size of wad used in the case, and the correct gun pressure to be used in firing. These were important questions, and it was some time before Frank du Pont and his associates arrived at the answers. For these suggestions alone, Tenner had more than earned his $500.

Had Tenner, however, provided the du Ponts with more than his invaluable criticism of their product? Had he perhaps given Frank du Pont an answer to the most basic questions of all, how to produce the grains of smokeless powder, by providing a sample of the Walsrode powder in the summer of 1892? These were the questions raised by the Wolff Company patent suit against the du Pont Company in 1901. In preparing his case for the defense, the Du Pont Company's lawyer, Charles Neave, wrote to Frank du Pont on 11 September 1901 pointing out that on 29 August 1892, Rice had written to Tenner that he had "paid Messrs. du Pont & Co., a visit, presenting for their consideration a sample of smokeless powder, together with your letter and your reports as to its successful use. They were, of course, interested and while they had heard of the powder, they promised to take up and test this sample." Neave expressed concern that Frank du Pont's application for a patent was filed in the Patent Office on 21 December 1892, four months after he received the Walsrode sample. Quite understandably, the Du Pont lawyer wanted to know if "you had been working on your process prior to August 1892. . . . It seems to me important that we should know what there was in Arnim [sic] Tenner's letter and report, which Mr. Rice handed to you with the sample of Walsrode Powder. . . . Was it possible from the examination which you made of the Walsrode Powder in August 1892, to determine what process was used in producing the granulation of that powder?"[21]

In his response, Frank du Pont flatly denied that he had known that Tenner was at that time an employee of Wolff & Co. He said that he had begun experiments on his own process in February or March of 1892. He also assured Neave that there was "nothing extraordinary" in having obtained a sample of the Walsrode powder. "Even now we get samples of every new powder that comes out, that we may know how it compares with our own. . . . No, it was not possible for us to have determined at that time what process was used to make the Walsrode powder. I am absolutely sure that Armin Tenner who arrived here in late 1893 did not tell us anything new."[22]

The last sentence of Frank du Pont's statement can be accepted as being true, but inasmuch as Tenner had arrived in the United States a year after the Carney's Point laboratory had already received a sample of the Walsrode powder, it is hardly relevant to the case. What can be questioned is Frank du Pont's assertion that it would have been impossible for him to have determined the process that the Wolff Company had used in producing the sample of the Walsrode powder which Tenner had so obligingly provided for him the previous year. A disinterested evaluator of this incident might well be as skeptical of this assertion as was the Wolff Company, especially in light of the desperate and futile efforts of Alfred du Pont in 1889 to bribe soldiers and to make surreptitious entries into the French powder mills in order to obtain a sample of their smokeless powder.

Judge George Gray, presiding over the Third Judicial Federal Circuit Court, however, apparently did not share this skepticism. On 29 May 1903, nearly two years after the suit had been brought by the Wolff Company against the Du Pont Company for an infringement of its Von Freeden patent, Judge Gray, a native of the Wilmington area and former attorney general of the state of Delaware, found for the defendant. The Du Pont patent for making smokeless powder through the acetone-oil process was held to be an original invention, independently arrived at, and therefore valid.[23] Judge Gray had again demonstrated the truth of the conventional wisdom held by most lawyers that the only suit more difficult than a libel suit for a plaintiff to win is a patent suit, particularly if the plaintiff happens to be a foreigner.

Alfred I. du Pont, from the perspective offered at the Hagley Yards on the Brandywine, could view the feverish activities at nearby Carney's Point with a certain aloof, amused detachment. Alfred had no radically new processes to discover or capture, no scientific tests to fail, no patent suits to defend. There for him was only the tried and true formula for making the best black powder anywhere in the world.

This is not to say that Alfred was entirely satisfied with his lot. His somewhat peripheral attachment to the Repauno Chemical Company, which he served as vice-president, had revealed to him what efficient management and the latest technology could accomplish under leadership as able as Hamilton Barksdale's. At Repauno, management did not eschew new methods simply because they were new nor ignore young men simply because they were young. In contrast, at the Brandywine mills the dead hand of the past lay heavily upon all operations, and it was there that Alfred was obliged by both choice and assignment to spend most of his working hours. Although Alfred was responsible for the production of black powder, which was still the Du Pont Company's basic product, he was not even given the courtesy of being designated superintendent of the yards. Frank du Pont held firmly to that title, although he spent nearly all of his time across the Delaware River at Carney's Point. Several times a week, prior to his departure for Carney's Point, Frank would meet with Alfred at seven o'clock in the morning to give his junior partner instructions for the day and the weekly production goals that Alfred was expected to meet.

Alfred, making his daily rounds through the mills, could see much that needed improvement: new machinery, new building designs, and above all a more efficient and rational movement of men and materials throughout the yards. There was far too much unnecessary backtracking and zigzagging in the movement of materials from the far end of the Upper Yards to the Hagley Yards a mile away and then back again. Lammot had protested this irrational arrangement thirty years before, but nothing had changed. Eugene and Frank were as deaf to Alfred's renewed suggestions for improvements as Boss Henry had been to Lammot's. Marquis James wrote in his biography of Alfred:

> Under these conditions Alfred turned out as good powder in the same time and at no more cost than was produced in up-to-date plants,

including those owned by Du Pont elsewhere. The secret was in his skill as a powder man, his incessant energy and above all his mastery of the variable human equation. An up-to-date yard indifferently manned can turn out no more work than an out-of-date yard expertly manned. Alfred fired the Brandywine working force, competent to begin with, with an *esprit de corps* which generated a positive enthusiasm for its task and for its boss.[24]

Although Alfred was not able to get the kind of time–motion efficiency that would have won the approval of America's great expert in scientific management, Frederick Winslow Taylor, he was able to make mechanical improvements that increased the efficiency and safety of the powdermaking business. Alfred made many of these innovations without either the authorization or indeed the knowledge of Eugene and Frank, and rarely did he bother to seek a patent for his inventions. Many years later, when he was asked which of all the mechanical improvements he had introduced into the powdermaking process he considered the most important, he answered that it was his design for a graining or corning mill that could be operated automatically without any workers' being present in what he considered to be the most dangerous single process in powdermaking. He added the comment, "Most of my inventions were powder making machinery which led to greater safety, the elimination of men from the mills and reducing the number of accidents, handling large amounts of powder at one time."[25]

His mechanical achievements were gratifying to Alfred's frequently bruised ego. His will could prevail in the area of machines, even if it did not in the head office, which he seldom visited, or at the weekly partners' meetings, which he seldom attended.

Another source of satisfaction for Alfred was his extension of the benefits of electricity beyond the plant to the du Pont clan. With the aid of the Mathewson brothers, Alfred wired the houses of Eugene, Frank, Henry A., Lammot's widow, Mary, and most surprisingly of all, his Aunt Louisa du Pont, Boss Henry's widow. Even Christ Church, where the du Ponts worshipped, and St. Joseph's Roman Catholic Church, where their Irish and Italian laborers went to Mass, were wired for Mr. Edison's incandescent lamps. The candlelit dimness that Uncle Henry had insisted upon for the Brandywine valley was gone forever. Even if Alfred could not enlighten Cousins Eugene and Frank on the advantages of modern production techniques in the mills, he could say, "Let there be light" in their front parlors.[26]

These minor triumphs over the entrenched establishment were not enough to satisfy Alfred's restless and innovative spirit. He had drafted his young cousin, Pierre Samuel, to play the piano in the small band he had organized, and after a concert the two young men would frequently walk home together, discussing their present frustrations and their bleak futures under the present regime.

Neither Alfred nor Pierre, to be sure, had any financial worries. When their Uncle Fred died in 1893, he left an estate of $2,000,000. Unfortunately he had not anticipated so sudden a demise, and had left no will. Under the Kentucky law of inheritance, one-half of his estate went to his mother, Margaretta du Pont, and the remaining half was divided into five equal shares, one share each to his two living siblings, Paulina and Bidermann, and three shares divided among the heirs of his one deceased sister, Victorine, and his two deceased brothers, Irénée and Lammot. From this division, Alfred and Pierre, along with their brothers and sisters, had each received $20,000.

Fred had once told his mother that he intended to leave a substantial sum to Alfred, Pierre, and Coleman, the eldest sons of his three brothers, Irénée, Lammot, and Bidermann. Margaretta was determined to carry out this expressed wish even though she had no legal obligation to do so. Accordingly, from the inheritance she received from her son's estate she took four thousand shares of stock in the Johnson Company, a steel rail company, and gave one thousand shares to each of these three grandsons and one thousand shares to her only living son, Bidermann. Each of the four portions was valued at $100,000. In addition, Margaretta cancelled the outstanding note for $100,000 held by Fred for the loan he had made to Alfred.[27]

Alfred thus received the largest share of his uncle's largesse, approximately $250,000. This footnote in the family's financial history was one that Alfred's sister, Marguerite, was never inclined to let her brother forget. Indirectly, Uncle Fred, through the generosity of Granny Meta, had provided his nephew Alfred with a substantial foundation upon which to build his fortune. It was not by reason of their small salaries, therefore, that these two young cousins were dissatisfied employees of the Du Pont Company.

In their long conversations Alfred and Pierre often compared their situations with that of their cousin Coleman, who from the moment he had started work in his Uncle Fred's coal mines in Kentucky had been given every opportunity for advancement. His ideas were listened to, his proposed innovations were accepted, and now that Fred du Pont was dead, Coleman at the age of thirty was in compelte control of his own destiny and was doing very well indeed in both steel and the city railway industry. There was no Cousin Frank to criticize everything he did, no Cousin Eugene to ignore everything he proposed.

Pierre, a precise, neat, and orderly young man, was mature far beyond his twenty-two years. He was appalled by the disorder and lack of planning that prevailed at Carney's Point. Cousin Frank was exceedingly finical about small details, as Alfred knew only too well, but at the same time he could be totally indifferent to keeping adequate supplies on hand or to making even the most basic capital improvements necessary to promote the research on smokeless powder. During one particularly cold winter at Carney's Point, Pierre described the chaos that prevailed there to his younger brother Henry Belin:

> The prospect is not very encouraging . . . the plant being practically shut down owing principally to F. G.'s methods of doing things. The gun cotton dripping rooms are hardly heated at all and it is impossible for the men to do anything as they can not wrap up much and must wear rubber gloves. The ice is about one or two inches thick all over the floors as all the splash and drip of water freezes immediately. . . . It is all the same old trouble of not looking ahead and making provisions for things that are bound to occur sooner or later.[28]

Pierre must have expressed his envy of Alfred, who only had to deal with Cousin Frank for an hour during their occasional seven o'clock morning conferences. Pierre had to live with Frank twelve or fourteen hours a day, week in, week out.

Alfred, however, saw his situation as being even more humiliating than Pierre's, for Alfred's status in the company should have made him less vulnerable to abasement. After all, he was six years older than Pierre and had given valuable service to the company for more than a decade. He was, moreover, one of only six partners, but for all of the influence he had on the company's policy, he might as well have been a hired hand.

So the two young men, on their long walks together, talked and complained about their lot. Frequently they spoke of leaving the company and striking out on their own. Pierre looked west to Louisville and Cousin Coleman, or north to Cleveland and their Uncle Fred's former close business associate, Tom Johnson, as possible opportunities for employment. Alfred was very excited about the recent development of a workable horseless carriage, powered by an internal combustion engine and fueled by gasoline. Alfred had early developed a love for motor cars. He had purchased his first automobile, a one-cylinder Benz, in 1897, and in keeping it running, he had demonstrated the same talent for auto mechanics as he had for all other machinery.[29] He considered entering the automobile industry, for surely it represented the future, being an invention that more than anything else could revolutionize American life.

Although the young men talked of flight, they stayed on, giving their best talents to the mills on the Brandywine and the laboratory at Carney's

Point. Pierre felt he could not leave his young brothers and sisters, for as their surrogate father, he had undertaken the management of the estates left to them by their father and Uncle Fred. For the immediate future at least, he could not head north to Cleveland and leave the entire burden of the family to his mother.

For Alfred, flight from the Du Pont Company did not present as many difficulties, since he could take his family responsibilities with him. In many subtle ways, however, his situation was more complicated. Alfred carried a heavier burden of inherited seniority within the family than his younger cousin and he had a far deeper love for powdermaking. It was to him more a vocation than an occupation; it was an art whose muse demanded total commitment. And Alfred could never erase from his conscience his father's haunting final words. The old company might still need him, although his superiors seemed unable to imagine, let alone recognize, this eventuality.

Then quite suddenly and unexpectedly, in the early spring of 1898, Alfred was presented with a major responsibility that his senior partners could not deny him, but instead had to urge him to undertake. In April 1898, the United States declared war on Spain. It was an unnecessary war that neither President McKinley, the majority of business interests in the country, the military commands, nor the diplomatic corps wanted or were prepared for. The Spanish-American War is a classic example of how public hysteria, carefully aroused and orchestrated by the mass media, can override pacificism, prudent judiciousness, and patient negotiations to the detriment of the public's true interests and general welfare. This war, in the words of historian Barbara Tuchman, was one more example of a nation's "march of folly." President McKinley, coerced by political pressures he was not strong enough to resist, asked for and received from Congress a declaration of war against a nation that had already yielded to every diplomatic demand made upon it.

William Randolph Hearst, the jingoistic editor of the *New York American,* is reported to have said, when he sent the artist Frederic Remington to Cuba in 1897 to illustrate Spanish atrocities, "You furnish the pictures and I'll furnish the war." Hearst and his equally bellicose rival publisher, Joseph Pulitzer of the *New York World,* furnished the nation with an unnecessary war, but unfortunately they could not provide the nation with the necessary material to fight that war. The United States in 1898 was woefully unprepared for combat. It had virtually no standing army; its antiquated navy was a one-ocean fleet that at the time war was declared was mostly concentrated in the far Pacific. There were no adequate supplies of rations, uniforms, or transportation facilities to feed, clothe, and move an army into Florida and on to the Caribbean islands. Above all, the country lacked weapons and munitions. All that were in abundant supply were a wildly

enthusiastic martial spirit, an untrained mob of eager volunteers, and a stirring, rallying cry: "Remember the Maine. To Hell with Spain." It is of such
stuff that chaos, if not catastrophe, is made.

An eager volunteer army can be given at least the rudiments of basic
training in a relatively short time. It can be clothed in old army surplus
uniforms, even if those heavy woolen suits had been meant for winter duty
on the Dakota plains and not for summer fighting in the tropics. An army
can be fed badly preserved, semi-rotten meat, even if this "embalmed beef,"
as the troops called it, was to bring about more deaths from diarrhea than
the enemy could produce with bullets in the hills of Cuba. The one thing,
however, for which there could be no hastily assembled substitute was ordnance. The Army and the Navy, brass hats in hand, called upon the du
Ponts to ask for gunpowder in vast quantities.

The Du Pont Company was happy to receive these requests from customers whom for the past two decades it had largely ignored as being of
little importance. Once again, as had been true in 1812, 1846, and 1861,
the armed services became the preferred customers whose demands upon
the company's production facilities would receive top priority. Once again,
war proved to be the escalator that lifted the company from the depression
of half-time shifts to the high plateau of around-the-clock, seven-days-a-
week production.

The military, to be sure, exacted a heavy price for the prosperity it
brought. If the military's demand for munitions was high, so also were its
demands for innovation in technology and quality. Dealing with the armed
services as customers was far more difficult than dealing with duck hunters
and coal operators, and the Army and the Navy had a flag-waving American
populace standing behind them to back them up. Their first demand was
for smokeless powder, because it was well known that the Spaniards had
this powder available to them for use in combat. This was an order that the
Du Pont Company could not possibly meet, however. Smokeless powder
was still in the experimental stage at Carney's Point. Nor were any of the
other powdermakers in America any further advanced in their production
of high explosives for military weapons. As smokeless powder was not available, the Navy and War Departments insisted upon brown prismatic powder for their large pieces of artillery. This the du Ponts could supply, but
hardly in the incredibly large quantity that the military demanded. The Du
Pont Company was curtly informed that it must produce the amount
needed and must do so in record time.

Faced with the seemingly impossible demand of twenty thousand
pounds of brown powder a day, the senior partners had to turn to their
much neglected junior partner, Alfred du Pont, to meet this challenge. It
was not immediately certain that Alfred would be available for this extraor-

dinarily demanding service to his company and his country, however, for immediately after the declaration of war on 25 April, Alfred had sought and obtained an officer's commission in the Army and had submitted his resignation to the Du Pont Company.

Given his and Pierre's growing dissatisfaction with their positions within the company, Alfred's precipitous patriotic action could be interpreted as an escape from what he had come to regard as an intolerable situation. It is doubtful, however, that Alfred regarded his enlistment as a final and irrevocable break with the company. Much more likely is his later assertion that he had signed up for military service, as he had once taken up the challenge to do battle with the Up-the-Creek gang, in order to defend the family honor and to shame his cousins who did not choose to enter into the fray. It did not prove too difficult, however, for the high-ranking military officers who had come down to the Brandywine to consult with Eugene and Frank du Pont to persuade Alfred to abandon his dreams of battlefield glory. The ordnance officers wanted Alfred du Pont. They assured him that as the best brown powder maker in America, he could contribute far more to the war effort by staying in Delaware than he could by charging the ramparts of San Juan hill. With this assessment, Eugene and Frank had to concur. Sensing that for once he was in a position to dictate his own terms, Alfred agreed to give up his commission and to remain at the Brandywine mills, providing he was given complete command of production. No more seven o'clock morning sessions with Cousin Frank to receive his orders directed from on high, no more nonsense about being the *assistant* superintendent of the yards. If the nation needed twenty thousand pounds of brown powder a day, he would produce that amount, but he had to be given full authority. Thus confronted by Alfred, who was strongly backed by the visiting military brass, Eugene and Frank could only acquiesce. Brown powder would be the only product made at the Brandywine yards for the duration of the war, and Alfred I. du Pont would be the producer. After fourteen years of humiliating subservience, he was at last his own boss.[30]

Alfred immediately set to work to accomplish his Herculean task. All production of sporting black powder ceased. The successful production of brown powder was largely dependent upon the quality of the charcoal. Five hundred pounds of wood, carbonized by steam, must produce exactly three hundred pounds of charcoal, no more and no less. For this exacting task, Alfred turned to John Thompson, the best charcoal burner in the business. Thompson worked eighteen hours a day for eighteen cents an hour. Each day, he managed to meet Alfred's production goal, only to find on the following day that the ante had been raised. Only the respected "Mr. Alfred" would have dared to demand and could have received from the workers such total dedication. And Mr. Alfred earned such loyalty only by subjecting himself to the same harsh discipline that he imposed. He worked the same eighteen hours as his charcoal burner did, and frequently he did not get home at all. He had the rare ability of being able to fall asleep in an instant—while tilted back in a chair or stretched out on top of a row of powder kegs—and then after twenty minutes to awaken feeling fully refreshed. As Marquis James has written about these hectic months in the spring and summer of 1898:

> [Alfred] had spent forty hours at a stretch in the Yards. He had evoked from his men a loyalty that drove them to deeds beyond their strength. He had felt the buoying force of the ambition that fires a slighted subordinate who means to show his superiors that he can do what they think no man can do. Much as these things count for any undertaking, they do not wholly explain the achievement which, so far as the record discloses, remains unequaled in the annals of the powder industry in America. Without ability of the highest order, Alfred could not have done what he did regardless of long hours, co-operative help, or impulse to vindication.[31]

No one, not even Eugene nor Frank, could deny that Alfred had done the impossible. Beginning with only one pound of brown prismatic powder in stock when he took over the Yards, Alfred within two months was producing the twenty thousand pounds a day that the ordnance officers had demanded—and an additional five thousand pounds for good measure.

Then, just as unexpectedly as it had begun, the war ended. In August Spain sued for peace unconditionally. On 15 August, Rear Admiral Charles O'Neil, Chief of the Naval Ordnance Bureau, wrote Eugene du Pont a letter quite different in tone from the peremptory demand he had made four months earlier. O'Neil now asked somewhat deferentially that the Du Pont Company cancel the Navy's outstanding order for one million additional pounds of brown powder and substitute for it a like amount of smokeless powder whenever it could be made available and "at a price which will compensate you for the expense to which you have been put in increasing your output of Brown Prismatic." To this request, Eugene replied:

> For the last four months we have devoted the whole output of our mills to the [Navy] Bureau and [Army] Ordance Department. Our customers who have been with us for years have been scantily supplied. We have jeopardized our business to a great extent. As to compensation, we fully believe that our business will be benefited to a much greater extent by turning our mills onto the regular manufacture than it would to continue to make Brown Prismatic powder. We, therefore, think that no

compensation is required for the money expended in increasing the out-put of Brown Prismatic powder.[32]

Eugene du Pont was not, of course, giving a direct answer to the request made by Rear Admiral O'Neil. The Navy did not want any more brown prismatic powder. It wanted smokeless powder for future contingencies in defense of the vast new empire seized from Spain, which extended from Puerto Rico to the Philippines. Eugene was not prepared to make any promises for smokeless powder. Profitable as this "splendid little war" had been for the Du Pont Company, Eugene was not particularly eager to keep a customer like the Navy, who demanded total control of the company's facilities. He was eager to return the Brandywine mills to their usual "use-ful, orderly business," for as he later said, "the smokeless military powder business has seen its maximum."[33] The Navy had to content itself with the thought that although it could not persuade Eugene to produce smokeless powder at least it did not have to pay for a million pounds of unwanted brown prismatic powder.

If Eugene du Pont was happy to return to the "useful, orderly business" of making premium black powder for the sportsmen and blasting powder for the mine operators and railway engineers, Alfred's reaction was more mixed. He too preferred making black powder, and it was a tremendous relief to return to a normal workday. Once again he had time available to take up his interests in music, automobiles, and electrical engineering, and to try to re-establish a normal family life with his wife and two young daughters at Swamp Hall. Alfred, however, knew what Eugene meant by an "orderly business." It meant the return to the old, leisurely pace of doing business under Eugene's dictation, strongly supported by the senior partners meeting in committee. Alfred would once again be reduced to the status of hired hand instead of production boss, someone who was expected to carry out orders without questioning them and to oversee the routine methods without having the authority to change them. "We will do it this way because we have always done it this way," might well have served as the company's slogan in the last years of Eugene's administration.

Eugene did make one major change in organization in the year following the Spanish-American War. The old partnership arrangement that had existed for nearly a century was dissolved, and on 23 October 1899, E. I. du Pont de Nemours & Company was incorporated in the state of Delaware with a capitalization of $2,000,000 which, of course, was simply an arbitrary figure that had no real relationship to the true value of the company. Each of the six partners received the same proportional amount of 20,000 shares, valued at $100 a share, as had been his interest in the old partnership. This meant that the four senior partners each received 4,000 shares in the new corporation and the two junior partners, Alfred and Charles, each received 2,000 shares. No additional stock was issued to any other persons.[34]

Years later, Alfred took credit for this change in organizational structure. Writing in 1931 to E. H. Batson, an official in the Internal Revenue Service, Alfred stated, "In 1890, I became a partner in the Company, which was then a co-partnership, and which at my request and instigation, was changed to a corporation in 1900."[35] Alfred, however, was as mistaken about the identity of the instigator of incorporation as he was about the date of incorporation. The man who pushed for incorporation in 1899, over the initial protests of Eugene, was Henry A. du Pont. Henry was not unaware of the problems existing within the partnership that had Eugene as its president and sole executive officer. By law, incorporation required a set of officers, and Henry hoped thus to weaken Eugene's ineffective one-man rule by giving company leadership a broader base. Eugene was intelligent enough to understand what incorporation meant, and at first he stubbornly refused to consider Henry's proposal. He finally yielded when he realized that incorporation might serve as a useful transition to that time when he was no longer on the scene to rule the company. He did, however, demand and receive from Henry and the other two senior partners the assurance that he would have their votes to elect him president of the new corporation.[36]

At the first meeting of the six stockholders after the formal act of incorporation, Eugene was duly elected president. The three other major stockholders, Henry A., Alexis Jr., and Francis Gurney, were each named vice-president, and Charles I. was elected secretary and treasurer of the company. Alfred was simply designated a director without any official position—a director without portfolio and certainly without any authority. To give Charles two offices and Alfred none was a gratuitous insult that would never be forgiven nor forgotten.

The three vice-presidents quickly discovered that their titles did not mean much either. Henry's hopes for a wider distribution of executive power were never realized. Eugene was as unwilling to delegate authority under the new organization as he had been in the partnership. He consulted with his three vice-presidents, but he alone executed, or more frequently failed to execute, company policy.

In this same year in which the old du Pont partnership became an incorporated company, Pierre Samuel du Pont decided to make his long-contemplated break with the company. His relations with Frank Gurney du Pont at Carney's Point, exacerbated by the war, had become more strained than ever. Carney's Point had not been able to meet the military's demands for smokeless powder, and Eugene's cavalier dismissal of Rear Admiral O'Neil's request for smokeless powder for delivery at some future date was a clear indication to Pierre that his and Cousin Frank's work had very low priority with Eugene. Alfred, at least, had had the excitement and satisfaction of accomplishment during the Spanish-American War, and now in peacetime, his black powder operations would once again receive top billing, even if Alfred himself was denied the status he deserved. But Pierre was faced with the dismal prospect of enduring the same muddled inefficiency at Carney's Point while working on a product that the president of the company regarded as unimportant except as it pertained to the sportsman's market. Even in this work, Pierre now found himself playing not second, but third, fiddle. After Frank's son, Francis I., who had recently graduated from Yale, joined his father at Carney's Point it soon became clear to Pierre that Francis I. was being groomed to take over the plant whenever his father stepped aside. It is hardly surprising that Pierre was particularly receptive to the query received in January 1899 from his Uncle Fred's former business associate, Tom Johnson, asking if he would be interested in succeeding Arthur Moxham as president of the Johnson Company. The company had recently sold its steel mills to the Federal Steel Company, which had prompted Moxham's move, but Tom Johnson needed an able young man who could handle the long-term liquidation of the company's other assets in land and the interurban railway in the Lorain, Ohio region. Johnson himself, as the newly elected reform mayor of Cleveland, was too preoccupied with politics to give attention to his business interests. Upon Coleman du Pont's and Moxham's recommendation he offered the position of the presidency to Pierre at a salary of $10,000 a year, two-and-a-half times the salary Pierre was receiving at Carney's Point.

Tempting as this offer was, there were still ties that bound Pierre to the Brandywine, and he decided to give the Du Pont Company one last chance to retain his services. He wrote to Eugene on 20 January 1899 telling him of the Johnson offer and asking if he could meet with the executive officers to discuss his future. In his reply Eugene made no counter-offer but asked Pierre to meet with him and the other directors on 26 January.

On the day following the meeting, Pierre wrote to his brother Henry Belin, who was in Asheville, North Carolina seeking treatment for tuberculosis, to inform him that he was "going to leave the Brandywine and go to Lorain." Pierre recounted his meeting with the Du Pont powers:

> I went hardly hoping that they would have any definite offer. I was not disappointed for Cousin Eugene simply said that when things were organized there would be several good places to be filled and thought that I could fill any of them. No special place was mentioned and no salary offered or any mention of giving or selling me an interest in the Company. I remarked that I hoped for something more definite and Alfred, for the whole partnership was lined up against the wall, spoke

up and said he thought as I had made a definite proposition that I was entitled to a definite answer. Nothing more was forthcoming so I got up with the remark that I thought that I understood the position and cleared out.[37]

Following this unsatisfactory meeting, Alfred and Charles tried to persuade Pierre to stay on. The present situation could not continue much longer, they assured him, and surely in a short time, Pierre would have an interest and an office in the company. Alfred could hardly have been very convincing in these hopeful expressions, and Pierre was tired of subsisting on great expectations. He had, after all, been with the company nearly twice as long as Alfred had been when he got his partnership interest. It made no sense now to give up a bird in hand at Lorain for a flock of elusive birds at the Brandywine. When Pierre the following day officially submitted his resignation to Cousin Frank, "'F. G.' . . . did not burst into tears or anything of the kind. I thought before that he was not anxious to have me stay and I am sure of it now."[38] Pierre could take a grim satisfaction in learning from Henry Belin shortly afterward that Carney's Point had lost $75,000 during 1898, which being a war year, should have given the plant the financial boost it needed. "It is not always agreeable to say 'I told you so' especially for those who are told, but I think the phrase would apply very well to some of our friends," was Pierre's smug response.[39]

If misery does indeed love company, Alfred had few to love after Pierre's departure. He was left without anyone with whom to exchange stories of plant inefficiency and administrative arrogance, and no one to share unrealized aspirations and possible routes of escape. Pierre had made his escape, but Alfred stayed on, although he was no longer sure why.

Coleman, who had left the Johnson Company at the same time as his good friend Arthur Moxham, quite surprisingly had moved from Louisville to Wilmington, for reasons that at the time were not apparent to anyone else, but Alfred hardly ever saw his former roommate and party companion. Coleman was seldom in Wilmington, for his job as interurban railway consultant and city bonding expert necessitated extensive travel. Alfred's grandmother, Margaretta Lamont du Pont, who had been Alfred's most loyal champion within the family, died in 1898 at the age of ninety-one, and Aunt Louisa, Boss Henry's widow, died two years later. The last ties with the du Pont second generation, which Alfred had always cherished far more than his associations with the third generation as represented by his four senior partner cousins, had now been broken.

Even at home, Alfred found little warmth and congenial companionship except with his two young daughters Madeleine and Bessie. Relations with his wife became increasingly strained, and as with so many other puzzles in his life, Alfred was not sure why. His increasing deafness made conversation ever more difficult, but his and Bessie's marital difficulties could not be explained simply by this physiological handicap. Bessie seemed to find deafness synonymous with dumbness, and she often treated her husband with the ill-concealed condescension that an unfeeling person would show toward a slow-witted dullard. She openly belittled him in his presence before her friends, his family, and even their children. She complained about his amateur musical performances and forced him to find some place other than Swamp Hall for him and his companions to practice. She disliked the easy camaraderie he enjoyed with his workers, and she felt obliged to remind him frequently that he was a du Pont, who should look and act like one. Many times Alfred was thankful for his deafness, which made it difficult for him to hear what she said to him—and about him to others.

The years immediately following the Spanish-American War were the most frustratingly difficult ones that Alfred was ever to experience. There would be times of greater crisis in his life, but none during which he felt as trapped, both at work and at home, as during the period from 1899 to 1902. The only bright moment to occur in his domestic life during these three years was the birth of his third child, a son, on 17 March 1900. It had been eleven years since Alfred's and Bessie's second child, little Bessie, whom every one called Bep, had been born. The arrival of Alfred Victor II after this long an interval seemed nothing short of a miracle to Alfred. That this happy event occurred on St. Patrick's Day made it seem especially propitious to Alfred and his many Irish employees, who celebrated the birth with fireworks and many rounds of bock beer. Perhaps now, Alfred thought, things could be different between him and Bessie, for Alfred's love for his wife had been instantly renewed in gratitude for the gift she had given him. But after the last skyrocket had exploded and paled, after the last keg had been tapped and drunk, he faced the sobering realization that nothing had changed, either at home or in the yards. There was his work of turning out in obsolete mills the best black powder made in America. There was his Tankopanicum Band that provided him with his only real moments of pleasure while playing the stirring marches of his own and his good friend John Philip Sousa's composition at the Rising Sun Tavern. And there was Swamp Hall, where he would continue to spend as little time as possible.

From this dreary pattern of existence, Alfred found few breaks except in music, and with his increasing deafness, he greatly feared that he would soon be deprived of this source of solace. In the autumn of 1899 Alfred suddenly decided he had to get away. For a few days he wanted to be a consumer instead of a producer of black powder. Alfred had been an avid hunter since his father gave him his first rifle at the age of ten, but it had been years since he had been able to take an extended hunting trip. He took his brother Maurice with him on this expedition as well as his boyhood chum and fellow electrician, Frank Mathewson.

There was good duck hunting to be had in the late fall along the Maryland side of the Chesapeake Bay, and Alfred had been told of a widow, Mrs. Fannie Harding, living at Harding's Landing on Ball's Neck, who rented rooms and provided board for hunters. The three musketeers settled in at Mrs. Harding's. At the Harding home and in his frequent visits to the nearby Thomas Ball plantation, Alfred found the peace and contentment he desperately needed. Neither the Widow Harding, with her five children, nor the Balls, with their five, had much money, but they had what Alfred with his wealth did not have—laughter on the front porch, good conversation in the dining room, and companionship while tramping through the marsh lands.

Captain Thomas Ball was this isolated region's most distinguished citizen. A descendant of the same Ball family that included Mary Ball, George Washington's mother, Ball had earned his military title by serving under the command of Robert E. Lee in the Civil War. After the war, Bell had gone to Texas to practice law, and there he had been elected to the Texas Senate. During President Grover Cleveland's first term, he had been appointed an assistant attorney general of the United States, but with the return of the Republicans to power in 1889, Ball had moved with his wife and five children to the old family plantation on Ball's Neck. Here the Balls eked out a modest subsistence with a little farming and a great deal of fishing, hunting, and clam digging. Theirs was a life without outside pressure from demanding supervisors, for there were no supervisors, and without inner pressure for upward mobility, for there was no gnawing ambition for the acquisition of more power and wealth. Alfred was enchanted with this little peninsula of sanity and vowed he would come again and often. It could never be his world, but it could be a brief release from all he knew and had to endure.[40]

Upon his return to the Brandywine, Alfred was saddled with the responsibility for the logistics of preparing for the great centennial celebration of the du Ponts' arrival in America. Every living descendant of the first Pierre Samuel had been invited to attend a great banquet, to be held on the banks of the Brandywine on the first day of the new century. At each setting would be placed a gold coin, a symbol of the legendary offering the starving du Ponts had left with their unknown and involuntary hosts on Block Island on 1 January 1800 in compensation for their meal.

More than one hundred du Ponts would assemble in commemoration of the event. Spacious as the du Pont mansions along the Brandywine were, they could not accommodate such a gathering. As Alfred would write to his recent hostess, Fannie Harding, on 14 January 1900:

I have had so many things crowded on me since the first of the year that I have really not had time to decently acknowledge the receipt of your nice letter. . . . On the first day of the year we celebrated the one hundredth anniversary of the landings of our ancestors in this country, and it was made the occasion of a family dinner at which congregated over a hundred. . . . As there was no room large enough, one had to be built, decorated, lighted and heated. . . . It is over thank goodness and it was altogether a success but I am glad I won't be alive to see the next one.[41]

Seated at the head of one of the long banquet tables and looking around the improvised dining hall where the du Ponts had gathered on the first day of the twentieth century, Alfred must have had much to reflect upon. The gold pieces lying on the tables to serve as place card tokens could have provided the Hardings and Balls with an income that would suffice for a year. At the head table sat Alfred's four senior associates, old men of the third generation, too tired now to manage the company their grandfather had founded, but too proud and stubborn to yield their power to the future. Seated at the lower tables were Alfred's cousins representing the fourth generation: Coleman, who had never been held in subservience by the heavy hand of the past; Pierre, who had managed to escape the restraining leash; and poor, sickly Charles I. du Pont—and even he was to be envied for having apparently escaped the anguish of frustrated ambition. Finally, there was Alfred himself, the builder of this festival hall, separated from his wife by more than the length of the long dining table, and still seated as far below the salt at the company's table of power as he had been when with the wild enthusiasm of youth he had rushed away from M.I.T. sixteen years earlier to look for his place at the family board.

The past was very much present in the room that night. Uncle Henry's widow, Aunt Louisa, the only living representative of the second generation, was too feeble to attend, but Alfred's Aunt Polly was there, and she was old enough to have known Eleuthère Irénée du Pont, the founder of the company. Several du Ponts present had known Alfred Victor, the second president, and all but the very youngest had memories—in some cases, too vivid memories—of Boss Henry. That night the du Pont family bade farewell to the nineteenth century and to the first century of their life in America. Alfred must have found all of this auld lang syne stuff a bit heavy. Less attention should be given to the past, he thought, and more attention to the future. The world was entering into a new century and in two years the Du Pont Company would enter its second century. The world would surely survive its transition into a new age, but for Alfred the more immediate question was whether the company could survive the new century. Unless radical changes came soon, it did not seem very probable.

V

The Way to Ownership
of the Company   1902

THE PERSISTENTLY gnawing question as to whether or not the Du Pont Company could long survive the slow rot of inefficiency kept troubling Alfred du Pont in the period immediately after the Spanish American War. Eventually, of course, the present senior partners belonging to the third generation of company management would pass from the scene, but Eugene and Alexis Jr. were only in their late fifties and Frances Gurney du Pont was still in his forties. Boss Henry had ruled with an iron hand until he was seventy-seven. If the present administration continued as long as its predecessor, Eugene and his associates would rule with ever more faltering hands for another fifteen to twenty-five years. Alfred very much doubted that the company could last that long. Of one thing he was certain: he could not endure his present situation that long.

Suddenly, however, all of Alfred's gloomy speculations were unexpectedly resolved. On 28 January 1902, after a one-week illness, Eugene du Pont died, at the age of sixty-one. The issue of the Du Pont Company's future could no longer be postponed. A successor to Eugene had to be designated immediately.

Those most centrally concerned have given many versions and interpretations of the events that followed Eugene's death. Their views of the succession to power within the company were to be so warped by later, unrelated events that it is difficult for historians to separate fact from prejudice. Any chronicler trying to distinguish between reality and fantasy must consider carefully all of these versions and then select those details from each account that produce a version that has the ring of verisimilitude.[1] The accounts as later given by Alfred, by his wife Bessie in her history of the Du Pont Company, and by his cousins, Henry A. and Pierre S. du Pont, agree on one point, that at the meeting of the company's officers held ten days after Eugene's death, each of the three surviving major stockholders were given the opportunity to take the presidency of the company.

As these sober-faced senior officers filed into Eugene's office to consider the question of a successor, both Henry A. and Alexis I. du Pont assumed that Francis Gurney du Pont would be the next president of E. I. du Pont de Nemours and Company. He was the logical choice in terms of both seniority of service and general knowledge of the powder-making business. He had been a partner in the company since 1873, nearly thirty years. He had served as superintendent of the Upper, the Lower, and the Hagley Yards and as general manager of the high-explosives plant at Carney's Point. Only in his fifty-second year, Frank was the youngest of the three vice-presidents, but this was in his favor, since it gave promise of providing longevity of service to the company. But before Henry A. du Pont could make a motion formally offering the presidency to Francis Gurney du Pont, Frank, to the amazement of the others, flatly and unequivocally rejected the presidency. He gave no reason, but simply stated that under no circumstances could he accept this position. Only later would the family learn that he was seriously ill and knew at the time that he had only a short time to live.[2]

Frank proposed that Henry A. du Pont take the position. As Boss Henry's eldest son, he at one time had held the largest share in the partnership. Henry A. had also inherited from his father a position of authority within the family that the other du Ponts accepted with the same respect that his subordinates in the army had once given him. In 1889, following Boss Henry's death, there had been nothing that Henry A. had wanted more than to succeed his father as president, and he had never forgiven his younger brother William for blocking his election. But thirteen years later, when the presidency was now his for the taking, Henry A. did not want it. He was sixty-four years old and no longer felt equal to the challenge of managing a company in serious need of a new direction. The principal reason for his refusal, however, was his involvement in Republican politics. Henry had his eye on a seat in the United States Senate. He anticipated having to expend much time and energy offsetting the efforts of the notorious utility magnate, John "Gas" Addicks, to buy the Republican party in Delaware if he was to emerge victorious over Addicks in the bloody contest for the possession of the Senate seat. Politely but firmly, Henry rejected the offer of the presidency of the Du Pont Company.[3]

Among the senior officers, that left only Alexis I. du Pont. He was the last of Alexis Sr.'s three sons to join the company. He had never served an apprenticeship in the mills but had come directly from a position of management in Fred du Pont's Louisville business firm to a partnership in the Du Pont Company. He may never have dirtied his hands with black charcoal and yellow sulphur, but in a little over a decade of service as an officer of the company he had proved himself a capable office manager. Of all of the partners, he also had had the greatest influence on his brother, Eugene, and his counsel was usually sound and beneficial. Alfred had never had a great deal of respect for Alexis because his cousin was no powdermaker and had never gone through the proper initiation rite of apprenticeship in the yards. Nevertheless, it is quite probable that Alexis might have proved to be a more able president than either his brother, Francis Gurney, or his cousin, Henry A. du Pont. He most certainly would have been a more effective chief executive than his brother Eugene had been. Alexis, however, also rejected the offer, although he was more candid than his brother in explaining why. He was in such poor health that the acceptance of such a responsibility was impossible.[4]

The short list of senior officers had now been exhausted without obtaining a successor to Eugene. There remained, to be sure, the two junior directors, Alfred I. and Charles I. du Pont. One look at the sickly Charles, sitting quietly at his secretarial table scribbling down the minutes of the meeting, was enough to convince his seniors that there was no point in proposing him as a candidate. Even if there were no other compelling reasons—and there were many—his obvious poor health would exclude him from consideration. His name was not even mentioned. Only one other name on the list of stockholders remained, that of Alfred I. du Pont.

Alfred had rarely attended meetings of the directors, believing that they were a waste of time, inasmuch as he was never asked for his opinion. This important meeting proved to be no exception. Alfred had not bothered to attend. Perhaps Alfred felt that it was a foregone conclusion that Francis Gurney du Pont would be offered and would accept the presidency. Under those assumed circumstances, any dissenting opinion he might offer would be entirely useless.

Alfred's absence permitted a candid discussion of his candidacy for the presidency. The most outspoken opponent of Alfred was his long-time immediate supervisor, Francis Gurney du Pont. Frank was unusually articulate in his denigration of Alfred. As Henry A. du Pont would much later recall the discussion, Frank "had formed an exceedingly low estimate of Alfred Irénée's good judgment and business ability, which he did not hesitate to express." Somewhat more moderately, Alexis expressed the same opinion. Henry himself was "disposed to be more lenient," but offered no strong support for Alfred's being offered the presidency, which everyone present knew he would be happy to accept.[5]

An impasse had been reached. No viable candidates remained among the officers of the company. The four stockholders present at this meeting must have looked at each other in dumb and desperate puzzlement. It was then that Francis Gurney dared to offer a most distasteful solution. The Du

Pont Company, which had existed for one hundred years and which now dominated the explosives industry as completely as the Carnegie Company dominated steel or Rockefeller dominated the oil industry, must courageously face up to the ignominy of selling out to a rival concern. The one possibility was that old and now friendly rival, Laflin & Rand. Frank spoke ardently and convincingly for his proposal. His colleagues listened and somewhat grudgingly found wisdom in his arguments. But how should they approach Laflin & Rand? And what would be the minimum amount that the du Pont family would accept as a purchase price? So closely had Eugene, like Boss Henry before him, kept to himself the assets of the company that the surviving Du Pont officers had only the vaguest idea of the true value of the property they were ready to offer for sale.

After considerable discussion, the stockholders present were able to agree upon both a procedure and a price. Hamilton Barksdale would be asked to serve as acting president of the company in order to negotiate a sale to Laflin & Rand. Barksdale was a logical choice for this delicate task. As president of the Repauno Chemical Company, he had proved himself to be an exceedingly able manager and a skillful decision maker. He was a close friend and former associate of the current president of Laflin & Rand, J. Amory Haskell, who had been Barkdale's predecessor as president of Repauno. Barksdale, moreover, was married to Ethel du Pont, Charles I.'s sister, which made him a member of the clan. If a sale could not be negotiated, Barksdale himself would be an acceptable choice to continue as president of the Du Pont Company. The directors also agreed that the minimum acceptable price for the company would be $12,000,000. With these major points agreed upon, Alexis was chosen as emissary to contact Barksdale and to secure his assent to the plan. Frank was given the more difficult assignment of telling Alfred what had taken place at the meeting.[6]

Considering his hostile reaction to the possibility of Alfred's being selected as president, Frank was a curious choice to be the messenger of bad tidings to Alfred. Perhaps Frank wanted this assignment and found a perverse pleasure in telling his cousin that he had been considered and found wanting. A more difficult question to answer is why Frank was so bitterly opposed to Alfred's candidacy. There were, to be sure, reasonable arguments that could be offered in opposition to Alfred's becoming the chief executive. No one questioned his competency as a production manager. He was the best powderman the company had known since Lammot, a true genius in mechanics, and no one since the first Alexis du Pont had had better relations with the millworkers. Frank, moreover, better than the other directors, could appreciate the production miracle that Alfred had performed during the Spanish American War. Yet having granted to Alfred all of these assets, Frank could still have raised valid questions as to Alfred's competency to serve as president. Alfred was not a committee man and was a notorious absentee from those board meetings to which he was invited. He had little understanding of some of the broader issues of company policy in the areas of sales, finance, and advanced research for which the president of a company is ultimately responsible.

Frank's vitriolic attack on Alfred, however, went beyond such rational arguments, and one can only conjecture as to the underlying reasons for the intensity of Frank's feelings toward his cousin. Alfred believed that he was himself largely to blame. He was, as he would later explain, "young, aggressive and not popular. I suppose I was impulsive—and not always polite."[7] His abrupt and brash manner must have galled Frank, whose sensitive *amour-propre* was easily bruised. Frank's antagonism, however, was more than a reaction to bad manners. It suggests a jealousy so deep-seated that Frank himself may not have been fully conscious of it. Frank was the youngest son of the youngest son of the founder of the company, and in a family keenly aware of birth order, be it expressed in seating arrangements at the dining table or in bequests from Uncle Fred's will, or even in the assignment of lots in the family cemetery, to be the last of the last was an unenviable position, especially when confronting Alfred, who was the first son of the first son of the first son of the founder. For all of his protests about being a mere hired hand, Alfred could never appreciate what it meant to have status denied one on the basis of birth order. Frank knew his own worth to the company, and he knew that others appreciated it as well, yet all of his life he had been outranked by Alfred's father. Irénée, by Uncle Lammot, by his cousin Henry A., even by his considerably less competent brother, Eugene. Now at last, when the opportunity was presented to him to assume command, the weakness of his own body had forced him to decline the offer. In the bitter frustration which consumed him, it was not too surprising that he should lash out against Alfred as he did.

That Frank was the one who first suggested and then argued forcefully for selling the Du Pont Company to an outside concern is also understandable. Fiercely loyal to his own immediate family and extraordinarily proud of the capabilities of his eldest son, Francis I., in whose interest he had eased Pierre du Pont out of the company, Frank knew that if one of the younger du Pont cousins—Alfred, Pierre, or even Coleman du Pont—should get the presidency, his own sons would never have a chance to reach the top. Far better to end the du Pont family's control completely than to have his sons live out their lives in subordinate positions within the company as he had done.

At the conclusion of the meeting, Francis Gurney must have felt that he had accomplished his objective. Although he was unable to accept the presidency himself, he had effectively destroyed Alfred's chances. He had also suggested a plan that would eliminate all of the other du Ponts from positions of power, for Frank had not the slightest doubt that Laflin & Rand would leap at the chance to buy E. I. du Pont de Nemours and Company.

Alfred is the only source we have for the meeting between him and his cousin. According to Alfred, Frank coldly and succinctly informed him that the question of succession to the presidency had been fully discussed at the meeting of the board of directors the previous afternoon. Neither Frank nor either of the other two senior partners had been able or willing to accept the presidency, nor had they found either Alfred or Charles to be an acceptable candidate for the position. Since the selection of Eugene's successor from among the current directors had proved to be impossible, efforts were now under way to explore the possibility of selling the company to another powdermaking concern.[8]

Frank undoubtedly expected Alfred to explode in anger at this news, given Alfred's temperament and his deep personal commitment to the company. Evidently Marquis James, Alfred's biographer, was also surprised by his subject's controlled reaction to this momentous news, for James wrote: "For once Alfred curbed his ready tongue, though his brain was reeling. Confess to the world the utter defeat and humiliation of the first industrial family of the United States! Sell the Company! . . . Though for years he had been steeling himself to face this very debacle, with difficulty Alfred kept his thoughts to himself."[9]

In their amazement at Alfred's self-control, both his cousin, Frank, and his biographer, James, underestimated Alfred's perceptivity. Although he found the news surprising, Alfred did not find it unwelcome. He had expected to hear that Frank Gurney was the new president, and if not he, then surely either Henry A. or Alexis. To Alfred any of these options would have been, to use James's word, the true debacle, for under the leadership of any one of these three representatives of the third generation, the company would continue to waste away from the same inertia that it had for too long suffered under Eugene's management. But with Frank, Henry A., and Alexis miraculously removed from the field, the game was not yet lost. The chance remained, however remote, that a younger du Pont could take control and lead the company into a brave new future. Alfred found he agreed with Cousin Frank on one point. It would be better if the company they both loved, each in his own way, was sold to an outsider than to allow it to go under while still in the du Pont family's possession because of a lack of proper management. The two men, to be sure, had substantially different definitions for the term "proper management." Frank had already failed in his efforts to find a du Pont whom he would accept as a proper manager. Now, although little time remained in the game, Alfred had his chance to score.

Sustained by this hope, Alfred could maintain with little difficulty a posture of calculated poise. He thanked Frank for the information and agreed that under the circumstances as outlined "a disposal of the assets of the Company did seem advisable."[10] On this subdued, unemotional note of seeming agreement, the conversation ended. A vaguely disturbed Frank, who had probably hoped for an outburst from Alfred that would further demonstrate the latter's instability, was left to his own puzzled conjectures, while Alfred hurried off to plan his strategy.

The only source of information for Alfred's next move is what Jessie Ball du Pont provided to Marquis James. According to Mrs. du Pont, Alfred left for New York the next morning to attempt to arrange a loan that would provide him with the capital necessary to purchase the company outright. Without any collateral except the hope of the future possession of the Du Pont Company providing he could persuade the other stock holders to sell their interests to him, Alfred's impetuous attempt to borrow $12,000,000 failed, if indeed it ever actually occurred.[11] New York bankers are seldom receptive to young, would-be borrowers, backed not by property but only with chimerical hopes, even if their names are du Pont.

Alfred then turned to his former M.I.T. companion, Coleman du Pont. Cousin Coley had always been ready to join Alfred—indeed, had usually been the instigator of the wildly improvised college capers. Even more than Alfred, Coleman had been eager to take a gamble if the stakes were high enough, no matter how remote the chances were of winning. Alfred now had a proposition that should rouse his cousin's sporting instincts, for there was little to lose if they should fail and an incredibly rich prize to be had if they succeeded. Coleman did not have enough funds at his own disposal to purchase the company, but Alfred would not have wanted Coleman to be the purchaser even if his cousin had been affluent enough to do so. What Alfred wanted was Coleman's expertise in making the appropriate financial arrangements. Alfred, like the rest of the family, had long been greatly impressed with Coleman's ability to produce, with a magician's legerdemain, money out of thin air to finance the many city railway ventures in which he had been involved.

Alfred's visit to Coleman's home on Delaware Avenue in Wilmington did not catch Coleman by surprise, for he too had been giving a great deal of thought to the question of succession ever since Cousin Eugene's death. He had even spoken with Pierre about it when Pierre had come down from Lorain for the funeral. Coleman was fully briefed on all of Pierre's and Alfred's complaints against the inefficient rule of the elders and the gloomy prospects of the company's future should that rule be continued. He, there-fore, gave a hearty welcome to Alfred and listened with great interest to his cousin's report on his conversation with Frank. Coleman was ready to join forces with Alfred, but he felt obliged first to consult with his wife, Elsie, for in giving up his present business Coleman had more to lose than did Alfred. Elsie was not overly enthusiastic. "Well, Coley," she is reported to have said, "you know what it is to be in business with your relatives." Taking this laconic response as assent, Coleman returned to his study, where he had left Alfred waiting, and told him that if the opportunity pre-sented itself after the next board meeting, he was ready to consider joining forces with Alfred to control the company.[12]

Everything now depended upon the success or failure of Alexis's mis-sion to Barksdale. If the president of Repauno should agree to serve as acting president of the Du Pont Company in order to negotiate a sale to Laflin & Rand, then the game was over. Alfred decided that he would not attend the next board meeting, at which time Alexis was expected to tell the board the results of his meeting with Barksdale, since the content of Alexis's report would become known soon enough. If Alexis had succeeded, nothing more was to be said or done. If he had failed to persuade Barksdale to take over, Alfred saw no point in tipping his hand prematurely. He would be in a better position to speak out when the board reassembled to consider a formal motion for the future disposition of the company.

Alexis's fateful meeting with Barksdale did not turn out as he and his brother Frank had hoped. Barksdale, surprised by the offer made to him, was too shrewd and cautious to leap at the bait, no matter how tempting it might seem. Instead, he asked Alexis to give him a few days to consider the proposition. This delay enabled Coleman to take positive action to affect the outcome. He and Barksdale were close friends and, by virtue of having married sisters of Charles I. du Pont, were also brothers-in-law. Coleman went directly to Barksdale, gave a frank account of his meeting with Alfred, and persuaded his brother-in-law not to accept Alexis's offer. When Alexis returned a couple of days later for an answer to the board's proposition, Barksdale told him that as long as there were other men bear-ing the du Pont name available and until the board had "exhausted all efforts to secure such a man to take the helm, I would not accept it."[13]

Alexis was obliged to report this unexpected response to the board on Thursday, 13 February. Frank was the only other board member visibly upset by the report. Charles and Colonel Henry A. du Pont kept silent. They obviously were not as eager to sell as were Frank and Alexis. Frank argued for going ahead with the plan of offering the company for sale even without Barksdale's cooperation. He was confident Laflin & Rand would not hesitate to purchase the company, but he was unable to get support for taking such action. The board decided to meet again the following day at which time Alfred had promised he would attend.

The next afternoon, St. Valentine's Day, the board reassembled. Alfred came to the meeting directly from the mills, dressed in his soiled work clothes. Leaning back against the wall, he closed his eyes and seemed to fall asleep while Frank slowly droned out a recapitulation of the previous board discussions. Then Frank offered a formal motion to sell the company to Laflin & Rand. Before the motion could be seconded by Alexis, Alfred opened his eyes and quietly offered an amendment. Instead of specifying Laflin & Rand, he suggested that the motion be changed to read the sale of the company "to the highest bidder." Frank, in looking around the room for guidance, saw nods of agreement from the other directors. The motion seemed innocuous enough and was a small price to pay for Alfred's assent to the main motion. So it was accepted by the chair as a friendly amendment and the main motion then carried unanimously.[14]

The issue appeared to be settled. Frank was ready to call for an adjournment when Alfred rose from his chair and quietly announced that he was now prepared to make a bid to purchase E. I. du Pont de Nemours and Company. For a moment, no one spoke. Colonel Henry and Charles du Pont, who had both been in communication with Coleman, may have been expecting some kind of statement from Alfred, but Frank and Alexis were obviously stunned by this bold announcement. Frank, nevertheless, was the first to speak up. In a voice choked with rage, he shouted in Alfred's face, "You can't have it. Besides, it's cash we want, you know."

Only then did Alfred's carefully maintained self-control finally snap. Alfred yelled back at his cousin, "Why not? Why can't I buy it? If you can't run the Company, sell it to someone who can!"[15] As he would later testify in a federal court, Alfred then pointed out to Frank "the fact that the business was mine by all rights of heritage; that it was my birthright. I told him that I would pay as much for the business as anybody else; that furthermore I proposed to have it. I told him I would require one week in which to perfect the necessary arrangements looking toward the purchase of the business and asked for that length of time."[16]

Alfred's reminder that he was the eldest son of the eldest son of the eldest son of the founder of the company was hardly the kind of tactful argument that would mollify and persuade Cousin Frank, who was so consumed with anger at this hated birthright claim that he could not even give an answer. Colonel Henry A. du Pont, the acknowledged head of the family, broke the silence. With the commanding dignity of person that was his greatest asset, the Colonel turned to the others and said, "Gentlemen, I think I understand Alfred's sentiment in desiring to purchase the business.

I wish to say that it has my hearty approval. I shall insist that he be given the first opportunity to acquire the property.'"[7]

Of all of the leading actors in this high drama of corporate disposition, Henry's role is the most interesting because it raises the most questions. Alfred, Frank, Alexis, and Charles acted out their parts according to a predictable script. But why did Colonel du Pont, to the amazement of his fellow players, suddenly become the champion of Alfred, a man to whom he had never been close or even particularly friendly? The explanation for Henry's support probably lies in a complex mixture of sentimental family pride and calculated business acumen. Alfred's heritage argument infuriated Frank but it aroused a strongly positive response from Henry. As the eldest son of the only genuine "Boss" the company had ever had, Henry could indeed "understand Alfred's sentiment in desiring to purchase the business." Fifteen years earlier, Henry himself had been consumed with the same desire. More than anyone else at this meeting, the Colonel was the proud bearer of the du Pont escutcheon. The family's coat of arms bears the bold motto, *Rectitudine Sto*—I stand upright. Were not the heirs about to debase this brave statement into an ignoble advertising slogan, *Venum Do*—I offer for sale the family heritage to the highest bidder? Faced with this prospect, Henry du Pont must have finally realized the gross enormity of Frank's proposal. Imagine having the name E. I. du Pont de Nemours replaced with Laflin & Rand!

A more practical consideration must also have strengthened Henry du Pont's resolve to speak in Alfred's defense. If the company should be sold outright for cash to an outside firm, neither he nor any other du Pont would ever again have a claim on the company's assets. Henry had no clearer idea of the true value of the company then did Frank or Alexis, but he was a sharp enough appraiser of business values to believe that the company's actual value must be considerably above the sale price they had so arbitrarily fixed. Furthermore, this figure did not take into account the potential future value of the company, providing it acquired the new expert management it so badly needed. If Alfred was given the full authority he had had during the Spanish American War, he could play a significant part in providing new direction to the company. Henry shared some of Frank's doubts about Alfred's ability to serve as the chief executive officer, but if Alfred could be persuaded to invite his cousins, Coleman and Pierre, to join him in a triumviral management, who knew what the Du Pont Company might become and what it might, within Henry's own lifetime, be worth? And if a financial arrangement could be devised with these three cousins that would allow Henry and the other senior partners a continuing interest in the company and some claim on its future profits, this would be much better than for each of the senior directors now to take his two and a half million dollars and run.

So Henry, at the critical moment when it had seemed that Frank had carried the day, spoke up in support of Alfred. Only he could have changed the ending of this drama, only he could have brought Alexis and Charles along with him. Frank stood alone, and Alfred was granted the week he said he needed to make the necessary arrangements.

Henry, however, wanted to make sure that the arrangements Alfred made would conform to his design. When Alfred left the room, elated at having received the best Valentine's gift of his life, the Colonel hurried after him. As Henry later remembered their conversation, he caught up with Alfred and said:

> "Referring to the offer which you made at the meeting to buy personally the whole stock of the Company, I assume of course, although you said nothing about it, that Thomas Coleman and Pierre are, or will be, associated with you in the proposed purchase." He [Alfred] said, "Yes, although as a matter of fact I have not heard from Pierre as yet, to whom I have written (or am about to write)." I replied, "With the understanding that Coleman and Pierre are associated with you in this proposition I assent to it most cordially and will do everything in my power to bring it about."[18]

With a handshake, these two unlikely allies parted company, Alfred to carry the good news to Coleman, Henry to meet with Alexis and Charles to lay plans for the delicate task of dealing with Frank.

Coleman was as excited as Alfred when he heard the report on the outcome of the board meeting. Alfred quickly discovered, however, that the price of Coleman's cooperation in this venture came high. Coleman made three demands that had to be accepted before he would agree to join Alfred in any attempted takeover. 1) Coleman, in total agreement with the Colonel, also demanded that Pierre also be brought into the management of the company; 2) Coleman must have 50 percent of the shares of the newly organized company—an amount equal to the combined shares of Alfred and Pierre; and 3) Coleman himself must be chosen as the president and chief executive officer of the company.[19]

It is doubtful that Alfred had anticipated so high a cost for Coleman's cooperation, but his cousin made it patently clear that these were non-negotiable demands. Why did Alfred so readily agree to the extortionary demands of someone who had never spent a day of labor in the yards, knew nothing about powdermaking, and was not prepared to invest any part of his own capital in this venture? This was a question which Alfred at a later date would frequently and bitterly ask himself and for which he would then have no answer. At the time that these demands were made, however, Alfred already knew the answer without even asking himself the question. Coleman held the upper hand, and both he and Alfred knew it. Without Coleman as an associate, Henry would not cooperate, and without the Colonel's active support, there was no chance of getting an agreement from the other shareholders to sell their stock to Alfred. An equally compelling reason for Alfred's quick assent to Coleman's demands was the necessity of devising a financial scheme that would satisfy the senior stockholders in lieu of twelve million dollars in hard cash. Alfred desperately needed Coleman's expertise in high finance, his ability to make promissory notes look like greenbacks.

Finally, it should be recognized that Coleman's terms did not seem as exorbitant to Alfred at the time as they would appear to him to be in retrospect. Alfred was as eager as Henry and Coleman to have Pierre brought into the venture. During the nine years that Pierre had been an employee of the company, he and Alfred had become very close companions. Alfred had great respect for his young cousin as a powdermaker, as an astute critic of Eugene's administration, and as a wise and competent administrator of his own family's funds. Pierre's subsequent career as president of the Johnson Company had further enhanced Alfred's respect for him. Together he and his cousin could provide the technical knowledge of the explosive industry that Coleman lacked. Nor did Alfred have any great difficulty in agreeing to Coleman's demand for the presidency. That office held no particular attraction for Alfred. He much preferred pushing production to pushing paper, committee meetings bored him, and he realized the difficulty his increasing deafness would pose in trying to conduct meetings. As far as Alfred was concerned, Coleman could expand his already capacious ego with whatever title he wanted to have posted on his office door.

It was only Coleman's request for 50 percent of the stock that may have seemed excessive to Alfred, but Alfred was willing to allow even this greedy claim if it was necessary to save the company. Apprently, Alfred did not attempt to negotiate even this point by making a counter-offer of a forty–thirty-thirty division. Unlike many other du Ponts, Alfred would never make the accumulation of great wealth his main pursuit in life. Nor was he envious of others, such as his great-uncle Henry, who managed to grab the biggest piece of the pie.

So with a minimum of discussion, the matter was settled. Alfred said he would write Pierre immediately to see if he would join the venture, but Coleman was a man of action, and time was short. Coleman went to the telephone in the hall and put in a call to Pierre in Lorain, Ohio. Standing in the hallway with Alfred at his side, Coleman shouted out to Pierre a brief summary of recent events in the Brandywine. As Pierre later recalled

> The conversation was brief—very brief—Coleman merely stated
> that Alfred had approached him on the plan of purchasing the properties
> of E. I. du Pont de Nemours and Company and reorganizing the busi-
> ness. The owners were willing to sell. He Coleman was willing to throw
> aside other personal affairs and to assume leadership in the reorganiza-
> tion if I was willing to go with him and undertake the financial part of
> the business. Would I do it? Yes.—Time less than three minutes and
> the plans and expectations of two lives had been completely changed.
> I made a hasty departure for Wilmington.[20]

Henry A. du Pont wasted no time either in fulfilling his part of the
agreement with Alfred. He met with Alexis and Charles to discuss how they
could mollify Frank's intransigence. They decided that only Alexis, acting
on his own, could possibly do this. The right emissary had been selected.
Although neither of the two principals involved left a record of this conversation, Alexis must have used convincing arguments to persuade his
brother, for shortly thereafter Henry received the gratifying news that
although Frank still preferred selling the company to Laflin & Rand, he
would no longer oppose the alternative disposition, providing acceptable
financial arrangements could be made, and also providing that under no
circumstances would Alfred serve as chief executive officer of the new
company.

Pierre left Lorain for Wilmington on Saturday evening, 15 February,
after having informed Tom Johnson of the opportunity that had suddenly
presented itself and receiving Johnson's blessing to go to the Brandywine
to explore the possibilities. Pierre arrived in Wilmington on Sunday, in time
to be greeted by the worst blizzard of the season. He was fortunate to be
able to get from the station to Coleman's house on Delaware Avenue. It
was not until Tuesday, 18 February, that the roads had been sufficiently
cleared for Pierre and Coleman to get to Swamp Hall. There, in Alfred's
billiard room, the three cousins met to design a financial plan that would
meet the approval of the senior stockholders.[21]

Having sold themselves as acceptable buyers to Henry, Alexis, Charles,
and the reluctant Frank, the three cousins now had to sell their credit rat-
ing. This was no easy task. If they had been willing and able to pool all of
their own assets, they could not have come up with one million dollars, not
to mention the expected twelve million. It was then that Coleman demonstrated not only his financial wizardry but also his salesmanship. The
three men agreed that they would not offer the shareholders anything in
cash. Instead they would form a new corporation, the E. I. du Pont de
Nemours Company, which would issue $12,000,000 worth of thirty-year

bonds, carrying an annual interest rate of 4 percent, which was precisely the rate of dividend the shareholders of E. I. du Pont de Nemours *and* Company were currently receiving. These bonds would be distributed among the present six shareholders in accordance with their percentage of the stock. Henry, Alexis, Frank, and the estate of Eugene would each receive 20 percent of the bonds; Alfred and Charles would each receive 10 percent. These would be mortgage bonds having first lien on all the property and assets of the company. In addition, the new company (which differed in name from the old company only by the omission of the conjunction "and") would issue capital stock with a par value of $20,000,000. This stock would be divided among the three cousins, with Coleman receiving stock at a par value of $10,000,000, and Alfred and Pierre each receiving $5,000,000 worth of stock.[22]

Coleman hurried off to sell the proposal to Henry, Alexis, and Charles. It was an easy sale, and Alexis was once again sent to secure Frank's approval. Upon further reflection, however, Coleman had misgivings about the proposal which he had just succeeded in selling to the ruling elders. The old Du Pont Company was entering into a new era with young, aggressive leaders. He was sure that his associates, Alfred and Pierre, had their own ideas of what innovations were needed, and although Coleman himself knew nothing about powdermaking he was, as always, full of expansive—and expensive—plans for the future. If all of the property they hoped to purchase was held as collateral for the first mortgage bonds, how could the new company possibly borrow additional funds that might shortly be needed? No financial institution would be willing to accept a second mortgage on property when no existing equity was held by the borrowers in that property. In another hurried meeting with his partners at which Coleman explained his concern, a new proposal was drafted in which Coleman would ask the sellers for a waiver on their holding a first lien on the property. In exchange for the waiver the sellers would receive 25 percent of the $20,000,000 of stock which was to be issued by the new company. Coleman was asking a great deal, for if the sellers accepted this proposal their $12,000,000 in bonds would be secured by nothing but their faith in the new owners' ability to perform. In return for this gamble, however, the former owners would share in the future profits of the new company—providing, of course, that it prospered.[23]

Alfred could only marvel at Coleman's audacity in proposing this scheme. Only a man as self-confident as he would dare approach those conservative old men with such a proposal. To Alfred's amazement, and probably to Coleman's as well, Henry leaped at the offer. This proposal was precisely what the Colonel wanted, for it gave him the opportunity to continue to share in the company's profits. Henry had great faith in the Du

Pont Company's future because, as a former military man, he well knew the human propensity to wage war. Munitions would always be needed, and the Du Pont Company would always stand ready to provide bigger and better explosives. Under Coleman's new arrangement, neither Henry nor his heirs would be excluded from the riches that the predictable future promised. Never had Coleman made a quicker sale. Henry's assent necessitated one more trip by Alexis to persuade his brother to accept still another proposal. Poor Frank must have felt that with each of Alexis's calls he sank deeper into the quagmire he had hoped to avoid. Further resistance was useless. Frank could do nothing but brusquely nod his acceptance.

On Thursday afternoon, 20 February, just six days after Alfred had made his unexpected offer and had asked for a week to make the necessary arrangements, the board of directors of E. I. du Pont de Nemours and Company met in formal session to vote on the sale of the company to Alfred I., Thomas Coleman, and Pierre S. du Pont. Except for Alexis, who had apparently worn himself out with his frequent trips to Frank's house and was too ill to attend, all of the directors of the old firm were present. The eldest son of the deceased Eugene was also present to vote the share of his father's estate. Colonel Henry presided and presented the three cousins' offer. It was accepted unanimously. Even Frank, who looked as if the anticipated Angel of Death had already visited him, voted yes. The matter was settled in principle. Only the proper legal forms had now to be drawn up and an act of incorporation for the new company registered with the state of Delaware.[24]

That same evening, Pierre dashed off letters to his three borthers, William K., Lammot, and Irénée. His letter to Irénée clearly conveys his jubilation, although it severely minimizes the difficulties that had to be surmounted to reach this moment of triumph:

Dear Irénée:—
Just after Cousin Eugene's death the stockholders of E. I. du Pont & Co. had a meeting and decided to sell out the business to Laflin & Rand as no one was ready to take up the management. Alfred made a request that he be permitted to take the property which request was granted. He came to Coleman and me and we decided to make an offer to the old stockholders to buy them out. This has been done and the proposition, as I am informed by Eugene [son of Eugene du Pont] this afternoon has been accepted. It is now only necessary to reduce it to writing . . .

This will doubtless come as a great surprise to you; nevertheless, it seems to be a "go." I think that there is going to be some hustling to get everything reorganized. We have not the slightest idea of what we are buying, but in that we are probably not at a disadvantage as I think that the members of the old company have a very slim idea of the property they possess. . . .

You may talk with Irene [Irénée's wife, who was a daughter of Francis Gurney du Pont] about this, but do not mention it outside until I advise you as the du Ponts are cranky (Aren't we all) and may not like to have the news announced before they do it themselves.

Your affectionate brother,
Pierre du Pont[25]

On Wednesday, 26 February, the stockholders of the old company met for the last time and formally accepted and signed the legal papers authorizing the transfer of the property to Alfred I., Thomas Coleman, and Pierre S. du Pont. The contract of sale, postdated 1 March 1902, embodied the financial arrangement as presented by Coleman the previous week with only a few minor changes. E. I. du Pont de Nemours and Company, as party of the first part, agreed to sell all of its property, assets, and good will to E. I. du Pont de Nemours Company, the party of the second part for $12,000,000 in thirty-year bonds bearing a 4 percent annual interest rate plus 28 percent of the $12,000,000 worth of stock that the new company would issue. Coleman had been obliged to accept the demands of the sellers that the new company should issue only $12,000,000 dollars worth of stock instead of the $20,000,000 he had proposed and provide the sellers with 28 percent of the share as a bonus instead of the 25 percent he had originally offered.[26]

The remaining 72 percent of the stock would be divided among the three buyers according to the proportions that Coleman had insisted upon. He would receive 36 percent of the total stock issue of the company, amounting to 43,200 shares. Alfred and Pierre would each receive 18 percent of the total issue, amounting to 21,600 shares each. Since Alfred was the only one of the three cousins who had been a stockholder in the old company, his holdings in the new company would be slightly larger than Pierre's. He received an additional 2,833 shares plus $1,200,000 in bonds—a small enough reward considering that he had been the initiator of the sale.[27]

The transaction was now completed. On Friday, 28 February 1902, one month to the day after Eugene's death, the papers were filed with the two parties' lawyers. The three cousins must have stared at each other in self-congratulatory wonder at the magnitude of their triumph. They were in possession of a company that over the past century had achieved a major share of the explosives industry in America and was now conservatively estimated to have a real value of $12,000,000. In addition, these three young men together owned outright 86,400 shares of the new company, which had a par value of $8,640,000. And what had each of them paid in hard cash from his own pocket for this bonanza? Exactly $700—the cost of the lawyers' services for drawing up the legal documents! Seldom in the history of American business enterprise had so much been purchased for so small an outlay of cash.[28] The three cousins, to be sure, had put in escrow their lives, their fortunes, and their sacred honor, but to them, with their youthful enthusiasm and vigor, this heavy indebtedness which they had assumed represented not involuntary servitude, but rather their declaration of independence. It was a testament of faith in their own abilities and in the continuing good will of the goddess Fortuna.

The three parted company temporarily to attend to those matters that each of them, quite typically, considered to be the most urgent: Coleman boarded a train for Louisville, where he would try to unravel and to terminate his many interlocking business affairs in Kentucky and Ohio, Pierre went to the treasurer's office to inspect with a bank examiner's critical eye the inventory of property newly acquired, and Alfred rushed back to the yards to discover what had happened to black powder production during the past two weeks while he had been preoccupied with the arduous but rewarding task of buying a company.

VI

The Triumphant Triumvirate
1902–1904

WHEN PIERRE DU PONT wrote to his brother Irénée on 20 February 1902 to announce that he and his two cousins had purchased E. I. du Pont de Nemours and Company, he closed his letter with the breezily cavalier admission that "we have not the slightest idea of what we are buying." Irénée must surely have been shocked by this seemingly irresponsible action so uncharacteristic of his cautious and prudent brother, but Pierre's younger siblings were not accustomed to questioning their older brother, whom they called Dad. Some years later, however, when Pierre was required to testify in court regarding the reorganization of the Du Pont Company in 1902, the U.S. district attorney was not fettered by any such filial inhibitions. He asked Pierre point-blank why he and his two cousins had been willing to purchase for $12,000,000 a business concern for which they had no appraised value nor even an inventory of that company's property, equipment, invested assets, and outstanding debits. Pierre's answer was equally blunt. "The character of the sellers," Pierre smugly assured the plaintiff's attorney. That had been enough of a guarantee for the purchasers.[1]

Pierre's faith in the moral integrity of his family elders was, to be sure, reinforced by his and Alfred's intimate acquaintance with the Du Pont mills and products. As Pierre would later write with greater candor in his unpublished manuscript history of the company, "Due to the hazards of the 'powder business' I knew it to be a custom to place little or no 'book value' on the manufacturing plants. As a demonstration of the wisdom of this procedure I had witnessed the conversion of nearly the entire 'Upper Yard' of the Brandywine Mills from an asset to a liability in the twinkling of an eye by the explosion of October 7th 1890. I was therefore prepared to find, as I actually did, little of plant value written on the books."[2]

Even a cursory glance at the company books, which Frank turned over to Pierre on 1 March, was enough to convince Pierre that he and his cousins had indeed received full value—and even more—for the $12,000,000 in notes which they had given to the sellers. The Du Pont Company owned and directly controlled five separate powder plants: the original mills on the Brandywine; the world's largest black powder plant at Mooar, Iowa; the somewhat smaller black powder mill at Wapwallopen, near Scranton, Pennsylvania; the Sycamore plant near Nashville, Tennessee; and the smokeless powder works at Carney's Point, New Jersey. As Pierre had correctly assumed, these five plants were greatly undervalued on the company books. They had a collective book value of only slightly over $2,000,000, although their replacement value would be several times that amount. The Mooar plant, as the largest, and most modern and efficient black powder company in the country, alone had a value well over $2,000,000.

In addition to the five production units, which formed the core of its assets, the company held stock and investments in other powder companies conservatively listed at a value of $4,000,000. And there was the quite unexpected little bonus of over $1,000,000 in cash as undistributed profits. Even with the assistance of his personal secretary, John J. Raskob, who had come from Lorain to help him, Pierre took several months to complete the inventory. His greatest problem was to keep the estimated value of the purchase low enough so that he and his two associates would not seem to have taken unfair advantage of the former owners. Pierre finally arrived at a statement of assets valued at $12,214,332.42, from which he subtracted a largely fictitious liability figure of $214,292.42, giving a final company value of $12,000,400.00. Oscar E. Morton, a certified public accountant and an expert auditor of company books in the explosives industry, was brought in to evaluate Pierre's inventory. He also tried to give as low an evaluation as his conscience would allow. He nevertheless arrived at a figure somewhat higher than Pierre's—$14,397,924.26. Even this evaluation could only be achieved by giving a very low book value, $2,338,368.98, for the company's five powder plants. Clearly, Coleman, Alfred, and Pierre had obtained a great deal for the $2,100 they had paid in hard cash for legal services.

Gratifying as these extremely conservative inventory figures must have been to the three cousins, they were still $12,000,000 in debt to the former owners. The company was theirs to own and operate, but they would have to make a minimum net annual profit of $500,000 simply to meet the interest due on the notes. In addition, they and their fellow shareholders would expect quarterly dividends on $12 million worth of stock. Pierre was not exaggerating when he wrote his brothers, "I think there is to be some tall hustling to get everything reorganized." Fortunately, each of the three cousins would prove himself to be an expert hustler in his own special field.

Official sanction for their hustling was granted at the first meeting of the incorporators and subscribers of the newly formed E. I. du Pont de Nemours Company. Held on 28 February 1902, one day after the papers of incorporation were filed with the Delaware Secretary of State. At this meeting, the seven stockholders of the new company—T. Coleman, Alfred I., Pierre S., Francis G., Henry A., Alexis I., and Charles I. du Pont— proceeded to elect themselves to the board of directors. The directors then chose executive officers for the company. In accordance with the earlier understanding reached between Alfred and Coleman, the latter was elected president. Alfred was elected vice-president, and was to be in charge of plant management and production; Pierre, as treasurer, would supervise all financial operations; and Alexis accepted the position of company secretary, with the understanding that the routine duties of that office would largely be assumed by the treasurer.[3]

The three cousins now had the authority to demonstrate what they could do to revitalize a century-old company. They were eager to accept the responsibility and the challenge. As Pierre later wrote: "The ambition to make 'two blades of grass grow where only one grew before' and to carry 'The Company' into a second century of success was strong within the hearts of these purchasers and there was never thought or suggestion that they were paying too high a price."[4]

Alfred was completely satisfied with the results of the first organizational meeting of the new company. He had obtained the position he had always wanted, and, more important, the full authority to operate effectively in that office. As Coleman knew nothing about the explosives industry and apparently had no desire to become involved in product operations, Alfred anticipated no interference from on high. For the first time in his life, Alfred was free to do what he did best and loved to do most—turn out the best possible explosive powder, be it black, brown, or smokeless, in the most efficient, economical, and safe manner. There was now no Boss Henry against whose strong will he had once had to contend, or no Cousin Eugene against whose uncaring inertia Alfred had been unable to prevail.

The old guard, however, still remained a potential threat to a new order. Although they held a minority of the stock, as directors, Frank, Henry A., Alexis, and Charles could, by a margin of 4 to 3, outvote the new majority owners in any contest over basic company policy. That was, however, only a remote possibility, for in accepting Coleman's offer these four representatives of the past had made it clear that they were ready to give the new generation its opportunity to lead. Even this highly unlikely threat to the three cousins' authority was soon to be erased. Charles I. du Pont died ten months after he became a director of the new company, and the two brothers, Francis Gurney and Alexis I. du Pont, died in November 1904. Henry A. du Pont remained as the only relic of the third generation to hold an official position within the company, but his interest in its affairs was limited to a concern for the prompt payment of interest on the outstanding notes and a reassuring dividend check on his stock.

Alfred du Pont's first task as general manager of production was to get an overall picture of how each of the five plants operated—their areas of strength, and, more important, their deficiencies. Alfred hoped to do what Eugene had promised when he had taken over the company in 1889 but had never implemented: the integration of the five separate units into a smoothly operating whole. Alfred would be the final authority in all matters relating to production, but unlike his predecessors, Boss Henry and Cousin Eugene, he was willing to delegate authority to the able supervisors he intended to find for each unit.

Alfred also had to acquaint himself with the operations of the eighteen other explosive companies in which the Du Pont Company held substantial interest. He recognized his own ignorance of the whole science and technology of dynamite and smokeless powder production. There would have to be many short but intensive field trips to Carney's Point, Repauno, and the Eastern Dynamite Company for purposes of inspection, production analysis, and, above all, education. A monumental task confronted him.

Alfred's primary interest, however, would continue to be black powder production at the Brandywine mills. The two or three miles lying along the right bank of the Brandywine encompassed all that Alfred had claimed as his birthright when he had responded to Frank's angry rejection of his offer to buy the company. The Brandywine mills had been Alfred's nursery, his childhood playground, his school of apprenticeship, his life's work.

No one knew better than Alfred the deficiencies that existed within that plant. The land the mills occupied was too limited for needed expansion, the buildings were too small and scattered, the equipment was obsolete, and the traffic pattern was hopelessly inefficient. Alfred had repeatedly criticized and sought to remedy these defects over the preceding sixteen years, only to be blocked by his elders' inertia and indifference.

It could hardly have come as a surprise to Alfred, therefore, that Coleman found much to criticize in the Brandywine operations when, a few weeks after the three cousins had assumed command, Coleman unexpectedly showed up in the yards one morning to inspect the mills. With characteristic breeziness, Coleman announced that, as president of America's largest explosive industry, he supposed he should be given a tour of the original Du Pont plant—just so he could later say that he had at least seen one powder plant. Alfred asked his personal secretary, James L. Dashiell, to give the new emperor a first view of his recently acquired domain. Coleman was shocked by what he saw. Although this tour was his introduction into the arcane world of powdermaking, as a former steelmaker Coleman experienced enough in industrial production to assess very quickly how inefficient and obsolete the entire plant was. His tour guide readily agreed with Coleman's judgment. "In 1902," Dashiell later wrote, "the du Pont Company was being managed just like *it was in 1802.* Horses pulled the dump cars about the plant. Up to date plant management was unknown."[5]

Coleman was outraged by what appeared to be an early nineteenth-century plant trying to compete in a twentieth-century industrial world. Coleman stomped back into Alfred's office and in a pompously authoritarian voice demanded to know how Alfred had tolerated working under such conditions for as long as he had. Fortunately, Coleman said, there was an easy remedy for this sorry state of affairs. The first item for action on the new company's agenda would be the closing of the Brandywine mills.

It was now Alfred's turn to be shocked. Although he totally agreed with Coleman's critical evaluation of the plant, he could not accept the proposed remedy. Abandoning the Brandywine mills would be for Alfred as gross an act of desecration as would be the melting down of the Liberty Bell just because it was cracked. Hagley and the Upper and Lower Yards were the very essence of the E. I. du Pont de Nemours Company. For the new owners to inaugurate their accession to the power by vandalizing their heritage was unthinkable. Alfred was a tough-minded pragmatist and an imaginative innovator. He had long railed against Eugene's philosophy of continuing to do what had always been done. But there was another side of Alfred's nature which was less apparent but equally important. Alfred had a deep, sentimental attachment to the past. For this reason, he had clung to Swamp Hall and called it his home long after he had the financial resources to build a more modern and elegant residence. For Alfred, modernization would always have to be reconciled with an equally strong desire for preservation. To defend the Brandywine mills, Alfred was willing to take a stand even if it meant challenging the authority of the company's new president at the very onset of their association.

Alfred told his cousin that he had been trying in vain for nearly twenty years to improve plant efficiency. Alfred knew what was wrong and what needed to be done, but only now had he won the authority to make the necessary changes. Change, yes—but not the draconian solution that Coleman advocated.

Alfred knew that it would be useless to appeal to any sentimental attachment to the Brandywine, for Coleman had none. Alfred could point out, however, that antiquated as the plant was, only four years ago it had been able to produce more black powder for the military in a shorter period of time than any other plant in the country. This miracle of production during the Spanish American War had been possible because the Brandywine had the best labor force in the powder industry. What would become of this highly trained and loyal group of workers if the Brandywine mills were closed? Many of the powdermen, like Alfred himself, represented the fourth generation to be employed in the mills. They were doing the work that their great-grandfathers had done for the first E. I. du Pont. One could not expect men like Pierre Gentieu or Henry Miller to uproot themselves from the only environment they had known and move to Mooar, Iowa, or Sycamore, Tennessee. Nor could one toss such people on a trash heap like a broken camshaft. Never, Alfred shouted, would he as general production manager consent to the closing of the Brandywine yards.

Hearing the loud exchange of words, Pierre stuck his head into Alfred's office to inquire as to the cause of the commotion. Quickly apprised of the issue, Pierre sided with Alfred, for he shared to some degree Alfred's attachment to the Brandywine. Alfred assured Coleman that within two years there would be changes made within the Brandywine mills of such an extent that Coleman would not recognize the plant should he again inspect it. Furthermore, the renovation would be accomplished at a cost that would be compensated for many times over in the increased efficiency and production of the mills.

Confronted with the united opposition of his two executive officers and co-owners, Coleman was forced to yield. In this, their first encounter, Alfred emerged victorious. For Alfred it was a useful contest. Not only had he saved the Brandywine mills, but also he had firmly established his territorial rights within the new organization. Coleman might be the president, with his head swollen by grandiose dreams of corporate and industrial expansion, but Alfred was the vice-president in charge of plant production, and in the world of grimy-faced powdermen and grinding wheels, he, not Coleman, was the master.[6]

The second issue of contention to arise between the two cousins would not result in as complete a victory for Alfred. Although this dispute was not as serious a matter as had been that of closing the Brandywine mills, it also arose out of the relative importance the two men gave to modernization versus preservation. Having grown up in Louisville, gone to college in Boston, and worked in Cleveland, Chicago, Memphis, and Dallas, Coleman found the small town of Wilmington and the still smaller du Pont enclave on the Brandywine to be so provincial and limited in scope as to be stifling to his expansive, cosmopolitan spirit. He had his eye on the only truly metropolitan center in America—New York City. Coleman believed that the company he proposed to build needed to have its headquarters where the real corporate action took place. He abruptly announced that as soon as possible he intended to move the main office to New York.

Alfred again reacted with shock and dismay. It made no sense, when there were so many other pressing matters to waste time, energy, and money in establishing a new head office for the company. When Alfred had begun work at the mills in 1884 the main office had consisted of a small one-room frame structure. He regarded the office building Eugene had erected only ten years earlier as a spacious edifice whose architectural style harmonized with the gracious architectural style that prevailed in the Brandywine valley. Alfred was troubled by the dismissal of the office as being hardly large enough to serve as a cloakroom for the corporation Coleman had in mind to create. Far more disturbing was the idea of moving the Du Pont headquarters out of the Wilmington area. This small town and the du Pont family had over the past century fused their identities. The du Ponts were Wilmington, and Wilmington was the du Ponts, to have and to hold, to nurture and to cherish. To abandon Wilmington for New York, of all places, was unacceptable, for if Alfred detested one city above all others, it was New York. The du Ponts had always prided themselves on remaining aloof from all that that city represented—the financial chicanery of Wall Street, the ostentatious plutocracy of Fifth Avenue, the ugly squalor of the Bowery.

So again, Alfred vigorously protested, and once again, Pierre quietly concurred, although admitting that Coleman was probably correct in his assertion that the company needed more office space than Eugene's charming country edifice provided. A compromise was reached. The Du Pont headquarters would remain in the Delaware valley, but Coleman would have the *Lebensraum* he demanded. The main office would be moved to the Equitable building in downtown Wilmington and would occupy three floors of that eight-story structure, the city's one feeble claim to a skyscraper. Alfred was satisfied. He seldom was at his desk in the main office building anyway, and he did not much care where that desk was located as long as the letterhead on his stationery gave the company's address as Wilmington, Delaware.

With these two bothersome and, in Alfred's opinion, unnecessary issues settled in his favor, Alfred turned his attention to the task that he regarded as being the most urgent and one that he thoroughly enjoyed doing—modernizing the Brandywine mills. Of course, he could do nothing about the space limitations nature had imposed on the location. His faith in the future of the Brandywine mills may have been great, but it could not move hills. It would be a test of his ingenuity, not his faith, to provide greater utility to the available valley floor.

If Alfred could not change the width of the west bank of the Brandywine, he could at least interfere with the natural order of things in respect to the river itself. Alfred's first undertaking to modernize the mills was to build a new and higher dam above the power plant which increased the available power by providing a greater headway of water. An even more important result of the dam was the assurance it provided of continuing power during periods of drought. Never again would the mills have to close down, as they occasionally had had to do in the past, because of an inadequate water flow.[8]

Alfred next turned his attention to the question of machinery maintenance. The sorry excuse for a machine shop, which was little more than a tool shed containing a forge and employing only two mechanics, was replaced with a machine shop worthy of the name, employing a score of skilled mechanics who could not only repair broken parts but also build new machinery according to Alfred's specifications. It would no longer be necessary for the Du Pont Company to send a broken part into Wilmington to be repaired. The machinists under the able supervision of Edward Bader and the chief patternmaker, George Sykes, could service not only the Brandywine plant but all of the other Du Pont powder works as well.[9] Alfred's happiest hours were spent in the machine shop, sketching out designs for new machines or offering suggestions on how to adjust and alter existing equipment. No wonder he was indifferent to the size and furnishings of the vice-presidential office. The machine shop was his true office—one might even say his true home—during these early years of the new regime.

From the machine shop emerged two of the inventions that were Alfred's greatest contribution to the gunpowder industry: his tumbling, double-walled cylinder for glazing powder grains, and a new graining mill that would automatically stop operating if the wheels came into contact with any foreign particle. Never again would a graining mill suddenly explode because of a carelessly dropped nail.

The limited Brandywine area did not permit Alfred to institute the efficient traffic pattern that he would have liked, but by effecting a more rational rearrangement of the mill buildings he achieved a processional order in production that at least approximated assembly-line efficiency. His most signficiant innovation in intramural transportation was the gasoline-powered locomotive that he designed and Ed Bader built in 1903 to replace the horse-drawn and man-pushed carts within the yards. It was the first gasoline locomotive to be used in the United States and was quickly adopted throughout the industry.[10]

Alfred had prevailed over Coleman in preserving the Brandywine mills and, as he had promised, within five years the mills were no longer the nineteenth-century relics that had so greatly shocked his cousin. Modernization and efficiency were achieved, but they exacted a cost in the loss of simplicity and informality. There was now far greater plant security as well as more bureaucratic regulation. Guards demanded passes of everyone entering the gates, even of Alfred's children, who came to the mills to bring their father his lunch, but who could not stay and play in the yards as their father and grandfather had done when they were children. Scrupulously observant plant inspectors carrying triplicate forms upon which to record infractions of safety and cleanliness standards made frequent and unannounced visits that put everyone on edge. While these were necessary and laudable innovations for a modern industry, they did change the easy familial ambience that had prevailed for the past century.[11]

Alfred was aware of the effect these changes were having upon those most intimately connected with the mills—the workers. Certainly the employees appreciated the greater degree of safety which now prevailed. Alfred's new graining mill alone would save many lives in the future. But with the new technology came unemployment as new machine power replaced man power—the inevitable cost for achieving greater safety and efficiency.

For those laborers who remained, however, Alfred was determined to provide the blessings of modernization that would extend beyond the plants' gates. Despite the pressing tasks of reorganizing and modernizing the Upper and Lower and Hagley Yards, Alfred found time to attend to the workers' living, as well as their laboring, conditions. The small stone houses along Breck's Lane and "Flea Row," which had remained unimproved since the time they were built nearly a century earlier, were now repaired and modernized. Old roofs were replaced, septic tanks were dug, and running water replaced the old, often polluted household wells. The main road from Henry Clay village, where many of the workers lived, was graded and raised three feet. No longer would it be impassable because of mud in the Spring or completely submerged when the creek flooded. Alfred also persuaded the local transit company to extend its city trolley line several miles farther out to those small communities where most of the Du Pont employees lived. For a few cents, these families could now have quick and easy transportation into Wilmington. They would no longer be dependent upon either shank's mare or hitching rides from passing wagons in order to obtain those urban services unavailable to them in their home villages.[12]

The numerous changes that Alfred effected during the first few years in which he was in direct charge of black powder production are illustrative of his continuing attempt to reconcile modernization with preservation. He was determined to save the Brandywine mills and the small communities that surrounded the plant, but he was equally determined to modernize and make more efficient and healthy the environment in which his employees lived and worked. Alfred was eager to push today into tomorrow, but at the same time he insisted upon preserving yesterday. It was a paradox that did not trouble him in the least, but to Coleman it was utter nonsense and to him, Alfred would always be an enigma. Coleman believed that bigger and newer always meant better. Alfred believed the older was always good but could be made even better. Herein lay the seeds of discord between the two men that over the next decade would sprout and eventually produce a bitter harvest.

So preoccupied was Alfred with his tasks on the Brandywine and with the supervisory role he had assumed for all the company's other production units that he could give little attention to what his two cousins were up to during the first critical months of their co-ownership. He was soon to discover, however, in somewhat piece-meal fashion, that Coleman and Pierre were as busy as he.

Only six months after the memorable afternoon in February when Alfred had dared to rise up at the board of directors meeting to protest the sale of E. I. du Pont de Nemours and Company to Laflin & Rand, Coleman, with an equally dramatic flourish, announced that there had been a complete reversal in the fortunes of America's two largest powdermaking industries. The would-be seller had become the buyer. As of late August, the Du Pont Company now owned Laflin & Rand!

Coleman gave Alfred the impression that he had singlehandedly accomplished this miracle with an easy wave of his magic wand and, furthermore, that it had cost no more in the outlay of hard cash than the purchase of the Du Pont Company had cost the three cousins six months earlier. The actual transaction, however, was somewhat more complicated than Coleman implied, and the instigator and prime mover in the purchase of Laflin & Rand had been not Coleman but the quiet and unassuming Pierre du Pont.

As he began his inventory analysis of the Du Pont Company's numerous investments in rival companies, Pierre unearthed some interesting information. Beginning with the acquisition of the Hazard Powder Company and the Sycamore Mills in 1876, Boss Henry du Pont's major strategy in bringing stability to the gunpowder industry through the agency of the Gunpowder Trade Association had been to buy into competing companies so that he could force them to abide by the trade association's agreements with respect to price, production quotas, and areas of marketing. Du Pont had had sufficient resources to buy Hazard and Sycamore in toto, but with each period of prosperity within the industry, new companies would appear. The Du Pont Company was unable to continue its policy of swallowing each major competitor in one gulp. Thus began the policy of nibbling away at potentially serious rivals by purchasing enough stock to give Du Pont a controlling interest in the company. Even this policy eventually proved to be too great a strain on the company's resources, so it had entered into an arrangement with its one truly major competitor, Laflin & Rand, to work in concert to force other competitors to respect the Gunpowder Trade Association's governance of the industry. Together, Du Pont and Laflin & Rand had bought a majority control of the Lake Superior Powder Company and the Oriental Powder Mills. Later, the two companies purchased controlling blocks of stock in the Marcellus and Ohio Powder Companies; under Eugene du Pont's direction in the 1890s, they acquired 55 percent of the Chattanooga Company, 49 percent of Equitable, and a majority control of the Southern Powder Company, even though the Sherman Antitrust Act of 1890 had made the legality of these transactions highly questionable.

The two friendly rivals had also cooperated in dynamite production. They had built the Repauno Chemical Company in 1880 and two years later the Hercules Powder Company in Cleveland. They purchased controlling stock in the Atlantic Giant Company, and finally created the Eastern Dynamite Company, a holding company for their jointly owned dynamite firms.

Pierre found that the Du Pont Company was now so inextricably linked with Laflin & Rand, as to make impossible, the streamlined, centralized administration of Du Pont's widely scattered interests that he, Coleman, and Alfred were attempting to create. The only solution was for Du Pont to buy out Laflin & Rand, lock, stock, and barrel. Only then could Du Pont create the modern structure of production and marketing that would give it the same monopolistic control over the explosives industry that Standard Oil and United States Steel had achieved within their respective industries. With Laflin & Rand in its pocket, Du Pont could dispense with maintaining the fiction of competition. The increasingly inconsequential Gunpowder Trade Association could finally be openly recognized as being obsolete and unnecessary. The most important result of the acquisition of Laflin & Rand would be that Du Pont could completely dominate the dynamite field, which both Coleman and Pierre believed represented the real future of the explosives industry. The two cousins were convinced that Laflin & Rand had made far greater strides toward control of this area than had Du Pont, although they had no actual data to substantiate their belief.

Pierre lost no time in bringing his preliminary findings to Coleman's attention. He thought he would find a receptive audience, and he was not disappointed. Pierre' recommendation was precisely what Coleman, with his dreams of empire, enthusiastically welcomed.

But would Laflin & Rand sell? Could this venerable company possibly be in the same position that Du Pont had been in at the moment of Eugene du Pont's death? Was the old generation willing to step aside in exchange for a quick profit?

To find answers to these vital questions, Pierre turned to his uncle,

Henry Belin, who for the past several years had served as treasurer of one of Laflin & Rand's subsidiaries in the Philadelphia area, the Laflin Powder Manufacturing Company. Belin was well acquainted with the four men who now controlled Laflin & Rand, and he reported that the situation within the company was not unlike that within the Du Pont Company at the time of Eugene's death. The three major stockholders were: Henry M. Boies, whose mother was a Laflin and a direct descendant of one of the founders of the company, and whose father had for many years been manager of one of the separate corporations owned by Laflin & Rand, the Moosic Powder Company; John L. Riker, who since 1864 had been a director and vice-president of the company; and Schuyler Parsons, whose father had been one of the company's original trustees. These three elders, Belin believed, had all been eager to get out of the explosives industry ever since Solomon Turck had retired as president of Laflin & Rand in 1895.

The only young man in the top echelon was Riker's son-in-law, J. Amory Haskell, who had succeeded Turck as president. Haskell was an old friend and associate of the du Ponts. It had been he who had succeeded William du Pont as president of Repauno in 1892. He had kept his close ties to Hamilton Barksdale, who had succeeded him at Repauno, and would not assume the role that Alfred had played in opposing the sale of Du Pont to Laflin & Rand, now that the situation was reversed, providing that arrangements could be made that would guarantee him a comparable salary and position of authority within the Du Pont Company after the merger.[13]

Henry Belin, serving as intermediary, opened negotiations with the ruling elders of Laflin & Rand on behalf of the du Ponts. On 13 May, Belin wrote to Pierre, "Parsons reports having had a talk with Riker and states that everything looks favorable. Will do nothing, however, until Boies' return about July 1."[14] When Boies returned home in early July, Belin met with him and gave Pierre an encouraging report on their meeting. Boies was also willing to sell if Du Pont would meet Laflin & Rand's price.[15]

The price quoted by Parsons on 22 July was far steeper than either Pierre or Coleman had anticipated. The Big Four of Laflin & Rand would sell their 5,400 shares, which represented 54 percent of the company's stock for $700 a share. This was exactly twice the prevailing price of the few shares of Laflin & Rand stock that were available on the open market. In addition, Du Pont would have to purchase the 1,100 shares of the Moosic Powder Company, whcih Boies in particular was eager to unload, at the same price of $700 a share. The total cost of the combined sale would be $4,550,000, and the four major stockholders expected to be paid in cash.[16]

Both the price and the form of payment demanded presented serious problems. Clearly, $700 a share was a grossly inflated price for the Moosic Powder Company, but this purchase had to be considered as a tie-in arrangement. It was a situation comparable to Alfred's unfortunate contract with the Belgian firm thirteen years earlier in which Du Pont had had to buy the brown prismatic machinery in order to obtain Coopal's formula for smokeless powder. If Pierre and Coleman wanted Laflin & Rand, which they desperately did, they would have to take Moosic. The price demanded for the Laflin & Rand stock also seemed inflated, in view of prevailing market price, but Pierre had reason to suspect that the current evaluation of Laflin & Rand's assets, particularly in respect to plant facilities and its investments in other powder companies, was at least as undervalued as the Du Pont Company's assets had been. Pierre, therefore, was not ready to reject or even to haggle over the stated price. But, inasmuch as the du Pont cousins were determined to take over Laflin & Rand completely, which meant the purchase of as many as possible of the remaining 46 percent of the shares, held by numerous other stockholders, it was important that the price of $700 a share being paid to the four major stockholders be kept quiet. Otherwise, the price of the other outstanding shares would also rise to $700 a share. Messrs. Parsons, Boies, Riker, and Haskell, who had already promised to use their influence to persuade the minority stockholders to sell their shares to Du Pont, certainly appreciated the du Pont cousins' concern about this issue. A plan would have to be designed which would conceal the true price that Du Pont was paying for their stock. Pierre insisted that as a matter of record it must appear that Du Pont purchased the controlling interest in Laflin & Rand at only $400 a share.

The agreement reached between buyer and seller on this point was to aid Pierre in dealing with the remaining question of making payment in cash. After several lengthy negotiations, Pierre came up with a proposal for payment that the Big Four would accept in lieu of cash. The Du Pont Company would create two holding companies, the Delaware Securities Company, which would purchase and hold the Laflin & Rand stock, and the Delaware Investment Company, which would purchase and hold the Moosic Powder stock. The four majority holders of Laflin & Rand stock would each receive $400 worth of Delaware Securities bonds for each share of Laflin & Rand they turned in to the holding company. This then would be the stated sale price of Laflin & Rand shares and all other holders of that stock could obtain the same price for their shares. But in addition, Parsons, Boies, Riker, and Haskell would secretly receive $300 worth of Delaware Investment bonds for each share of Laflin & Rand stock. This bonus would not be offered to any of the other shareholders. The four major stockholders would also receive $700 worth of bonds for each share of the Moosic Powder Company stock.

The two holding companies would issue a total of $6,500,000 worth of 5 percent, twenty-year bonds and the same amount of preferred stock. The bonds issued by Delaware Securities and Delaware Investment would be secured by Du Pont Company stock that would be held by the two holding companies. As a special compensation for having agreed to accept the bonds in lieu of cash, the four majority stockholders of Laflin & Rand were also to receive an additional bonus of $1,00,000 of stock in the two holding companies. The remaining $5,500,000 of stock would be held by the Du Pont Company to protect its control of the two holding companies.[17]

It was a neat and tidy little arrangement that Pierre had concocted for it not only secured the purchase of Laflin & Rand with no outlay of cash but it also cleverly disguised the true price that the du Ponts paid for the controlling interest. Moreover, it pleased the sellers, for it gave the Big Four a continuing interest in the fortunes of the Du Pont Company through the stock bonus that the majority stockholders received. Pierre had made good use of the model Coleman had provided in his plan for the purchase of the Du Pont Company. Once again, the cousins had been able to purchase a major industry for an outlay of only $2,000 in cash for legal services. They were also, once again, putting another mortgage on the Du Pont Company's and their own futures.

This venture, however, was in one important way different from the previous purchase. The three cousins had purchased the Du Pont Company without having, as Pierre himself said, "the slightest idea of what we are buying." This time Pierre could not permit himself the luxury of relying with blind faith on "the character of the sellers." He insisted that before the contract was finally consummated he be given an opportunity to inspect Laflin & Rand's books. After months of studying the Du Pont inventory, Pierre was a seasoned appraiser of powder manufacturing assets. What he saw in Laflin & Rand's private ledgers astounded him. His earlier suspicion that the company's assets were undervalued was more than confirmed by even a cursory glance at the books. Its plant facilities carried a book value of only $111,100, but Pierre's own very conservative appraisal of the company's seven production plants came to well over $1,000,000. What truly astounded Pierre, however, was the extent of Laflin & Rand's investments in other explosives industries, particularly in the dynamite field. Laflin & Rand held stock in twenty-five other companies, the most important holding being 5,761 shares in the Eastern Dynamite Company, the major producer of dynamite on the East Coast, and these shares were listed as having a value of only $31 a share. Du Pont held only one-fourth as many shares in Eastern Dynamite, and Pierre had listed the value of each of those shares at $140. Laflin & Rand's total investment in rival companies was listed on its books as being $633,381. Pierre figured that these investments were worth, at a minimum, $1,973,705—three times the book value that Laflin & Rand had assigned to them.[18]

After spending only an hour looking over the ledger, Pierre dashed out of Laflin & Rand's main office on Cedar Street in New York and caught the next train back to Wilmington. He lost no time in urging Coleman to complete the transaction at once "lest our inspection," he later wrote, "might lead the owners to resurvey their property and retire from the commitment."[19]

Having received Pierre's enthusiastic report, Coleman needed no further prodding. Although the next day was a Saturday, Coleman left early that morning for New York to work with Parsons and Haskell over the weekend in drawing up the final papers. On Monday 26 August 1902, the contract was signed. E. I. du Pont de Nemours Company, through the agency of its two newly created holding companies, now owned Laflin & Rand. Thus did Francis Gurney du Pont's plan for the disposition of the Du Pont Company reach its ironic conclusion. Frank had tried very hard to get Jonah to swallow the whale, but the natural order of things had prevailed after all.

Only after Coleman had signed the papers in New York was the man who had been the initiator of the entire chain of events informed of its surprising conclusion.[20] Alfred openly expressed pleasure at what his two cousins had accomplished without his knowledge, but he must have been inwardly hurt at having been bypassed in the entire transaction. He saw no clear reason for his having been kept in the dark during the months of negotiation. Alfred could vaguely sense, however, that a wall was being deliberately erected around him that would eventually isolate him from the company.

Alfred had no time, however, to nurture his hurt feelings with sulking in his tent. The immediate problem for him, as general manager of production, was to assess just what the whale had swallowed by gulping down Laflin & Rand. Jonah proved to be large indeed—much larger than Pierre or Coleman had anticipated when they first began negotiations for the purchase. The acquisition of Laflin & Rand meant outright ownership of eight additional black powder mills plus the two Moosic black powder plants. More important for the future was the full control of four large and four small dynamite plants. After this purchase the Du Pont production units totaled eight dynamite plants, twenty-one black powder works, and two large smokeless powder works in New Jersey. Small wonder that when the news of the merger became public knowledge, the entire gunpowder industry was as shaken as had been the steel industry the previous year when Carnegie sold out his empire to the J. P. Morgan syndicate. Historians have called the merger "the most significant negotiation in the history of the American explosives industry."[21]

One would think that, after this transaction, Coleman would have been content to spend some time digesting his gargantuan meal, but his appetite was insatiable. Shortly after the final papers had been signed and the transfer of Laflin & Rand's properties was completed in October 1902, Pierre was authorized to begin planning for the creation of a consolidated company that would fully incorporate and integrate all of the many powder plants that Du Pont either controlled outright or in which it held an important but not a controlling interest, such as the Equitable, the Austin, and the Ohio powder company. Ably assisted in this task by Coleman's old mentor in the Johnson Steel Company, Arthur J. Moxham, Pierre worked throughout the winter of 1902–3 to create "one big company," which would in effect have a monopoly of the explosives industry. On 19 May 1903, the E. I. du Pont de Nemours Powder Company was formally incorporated in the state of New Jersey, with a capitalization of $50,000,000. To this new company, the E. I. du Pont de Nemours Company transferred all of its recently enlarged plant facilities and other investment assets. Henceforth, the old parent company would remain as a holding, not a producing, company. The companies that Du Pont did not control could join the consolidation by exchanging their outstanding stock for common shares in the E. I. du Pont de Nemours Powder Company. All of the major companies in the East, with the exception of those companies controlled by the Olin, the Lent, the Fay, and the King families, accepted the invitation to join the consolidation by June 1904.[22]

In the meantime, Coleman, leaving the details of consolidation to Pierre and Moxham, had gone to California to try to bring some order to the chaos that prevailed in the explosives industry on the West Coast. This region had never achieved the order and self-regulation that Boss Henry and Laflin & Rand had imposed upon the powder companies east of the Rockies by means of the Gun Powder Trade Association. Coleman was gone for six months, and during that time he accomplished a great deal. He drastically reduced the number of independent, competing companies, but most important for Du Pont's own interest, Coleman brought into the fold the Judson and the Vigorite dynamite companies and the powerful California Powder Works. Of the major explosives companies on the West Coast, only the Giant Powder Company remained an important independent competitor after Coleman returned to the Brandywine in August 1903.[23]

By the summer of 1904, the great barbecue was nearly over. All the companies that could be eaten up were now being digested in the capacious maw of the newly created E. I. du Pont de Nemours Powder Company. It now produced 64.6 percent of all soda blasting powder, 80 percent of all saltpeter blasting powder, 72.5 percent of all dynamite, 75 percent of all black sporting powder, 70.5 percent of all smokeless sporting powder, and

100 percent of all military smokeless powder produced in the United States.[24] The Gun Powder Trade Association had become superfluous. The Du Pont Company withdrew from the association in March 1904, and with its withdrawal, the old confederation of independent companies created by Boss Henry quietly expired. Confederation had given way to consolidation. For the word "association," one had now to read the name Du Pont.

Along with expansion and integration came the basic policy question of whether the new Du Pont Powder Company should adopt a federal or a centralized system. Should it resemble the political structure of the United States, in which the various companies would, like the individual states of the Union, keep their corporate identity in name, logo, and local administrative authority, or should it adopt a structure similar to that prevailing in France, in which departments were simply units having no individual significance other than a geographical identity? The top leadership was sharply divided on this question. Pierre du Pont and Arthur Moxham argued forcefully for the Gallic centralized state model. Not surprisingly, Amory Haskell, the former president of Laflin & Rand who had come to Du Pont as part of the purchase, and Hamilton Barksdale, the former president of Repauno, wanted some companies, such as Laflin & Rand, to keep their individual identities, even though they would be subordinate to the parent company. Haskell and Barksdale pointed out the advantages the federal system had over the centralized structure. Keeping the old names would be advantageous for marketing. Loyal customers of Laflin & Rand, Sycamore, and Oriental wanted to buy products bearing those names, not that of Du Pont. Moreover, the federal system might prevent the total unionization of the company. If the Oriental Powder Company existed in name and it was forced to adopt a closed shop, pressure for unionization might not spread throughout the entire organization, as it would if a single unit of the Du Pont Company should unionize.

Legal counsel was sought, and the two lawyers who were consulted, James Townsend and Edward Walker, sided with Pierre and Moxham. They pointed out that a single centralized company was less subject to legal action by the United States government under the terms of the Sherman Anti-Trust Act of 1890 than was a company consisting of many semi-autonomous subsidiaries, each maintaining a separate identity to keep alive the fiction of intra-organizational competition.

The debate that followed the lawyers' reports was enlightened and rational. It was particularly refreshing to have Moxham, a seasoned veteran of many a fierce competitive business struggle, argue for morality in the marketplace. Fictional competition was immoral as well as illegal, he argued. We cannot fool the government, and in the long run we cannot fool the customers either. Why pretend that Laflin & Rand still exists and is competing with the Du Pont Brandywine mills when everyone knows that it is simply a subunit of Du Pont?

The issue was referred to the Executive Committee of the Du Pont Company. By this time, Haskell and Barksdale were willing to concede that most of the recently acquired companies, with their distinct product lines and individual marketing agreements, should be dissolved, but as former presidents of Laflin & Rand and Repauno, they continued to hold out for allowing a few exceptions to the general policy of consolidation. The Executive Committee vote proved interesting. Pierre and Moxham, of course, voted for consolidation with no exceptions. They were joined by Francis I. du Pont, Frank Gurney's eldest son, who now held a seat on the governing board. Surprisingly, Coleman voted with Haskell and Barksdale to allow exceptions, and even more surprisingly, Alfred cast the deciding vote on the same side. Alfred was still torn between modernization and preservation; he was still fighting to keep the ancient name and integrity of the Brandywine mills. Exceptions to total consolidation would be allowed, but Moxham and Pierre had the last word when they obtained an agreement from the majority that exceptions would be allowed only after a careful consideration of each individual case by the Executive Committee. Ultimately, Pierre and Moxham had the final victory. Exceptions were consistently disallowed and consolidation eventually became the totality that they had sought.[25]

The resolution of the issue of consolidation in their favor was critical to the plans that Pierre and Moxham had already drawn up for the organizational structure of the new E. I. du Pont de Nemours Powder Company. Pierre could now put into effect the rational departmental structure he had already sketched out on paper, establishing administrative and operating departments comparable to the ministries in the French centralized state. The administrative departments consisted of Treasury, headed by Pierre; Sales, directed by Haskell; and Development, with Moxham as its head. The operating departments were Black Powder, directed by Alfred; Smokeless Powder under Francis I., who in 1904 became head of the new experimental station and was succeeded by Henry Baldwin, another brother-in-law of Coleman; and High Explosives (dynamite), headed by Hamilton Barksdale. Each of the department heads was responsible for the operation of his specified activity. He was to organize and set one departmental policy that would prevail throughout the many separate units of the company. He would delegate to the individual plant managers and administrative agents the responsibility for carrying out the policies and operational methods decided upon, ensuring uniformity throughout the organization. There would be, for example, one operational policy and method for making Du Pont black powder, whether that powder was made in the Hagley Yards on the Brandywine or at the Sycamore plant in Tennessee. Gone were local, autonomous sales agents like Elliot S. Rice, who had for decades run his sales office in Chicago like a private fiefdom. Now a single sales force operated under directives issued from Haskell's office in Wilmington.

Not everyone adjusted easily to tightly centralized control. Moxham, reporting to Coleman, who was in California during the first few months that the new organizational structure was in operation, wrote: "In the main Haskell has been the greatest disturbing element, his past method having been to attend to very minute matters to such an extent that practically no one under him was a free agent." Barksdale, too, wanted to interfere in the policies of departments not under his direction. Apparently it was not easy for a former president to accept the lesser position of cabinet minister. On the other hand, Alfred, although he was vice-president in charge of all production, seemed willing to allow the heads of the departments of high explosives and smokeless powder to run their own shows. Moxham was pleased to report that "as far as organization is concerned, [Alfred] is in no way a disturbing element, but in every way an aid."[26] Moxham should not have been surprised. Alfred had argued for this kind of rationality in company organization for twenty years.

Each department head was a member of the Executive Committee, which in effect served as President Thomas Coleman du Pont's cabinet. The Executive Committee was to deal with "only the big questions. . . . We have to learn to leave details to Heads of Departments instead of trying to make them subjects of executive action," Moxham repeatedly counseled.[27] It was precisely this advice that Haskell and Barksdale found difficult to accept.

Over the president and the Executive Committee was the Board of Directors. Pierre had insisted on preserving this organizational remnant of the old corporate structure as a means of keeping the du Pont family's connection with the new company.[28] On the Board sat the family elders; Henry A., and, until their deaths in 1904, Cousins Frank and Alexis I., along with Coleman, Alfred, and Pierre. Although many able outsiders—men like Moxham, Barksdale, Frank Connable of the Chattanooga Powder Company, and Haskell—had been brought into the organization, it remained a Du Pont Company, owned and managed by the du Pont family. Pierre, like Alfred, believed in preservation, if only in form. But the du Pont elders knew very well that it was largely form, not function. The real executive power lay not with the Board of Directors but with President T. C. du Pont and the Executive Committee. The du Ponts still owned their company, but only because Alfred, Coleman, and Pierre had been brash and able enough to take over.

The family's third generation could only watch in bewilderment as the triumphant triumvirate of the fourth generation built a vast industrial empire that in two years exceeded in both power and wealth the wildest dreams of that old trailblazer of the second generation, Boss Henry du Pont. The union of the three cousins proved to be a fortunate concatenation of talent. Coleman had the boldness and imagination to conceive of an empire, Pierre had the sharp mind needed to construct the imperial administrative and financial structure, and Alfred had the operational know-how and the mechanical genius to produce the actual profits to pay for this empire. All that was necessary now for the future security of the du Pont family and its company was for the three cousins to continue to work in concert. In 1904 none of the three had any doubt that this harmonious relationship would continue. Alfred's fear of conflict lay in Swamp Hall—not in the mills.

VII

Exit Bessie, Enter Alicia
1902–1907

D URING THE FIRST YEARS of the twentieth century, Alfred I. du
Pont endured the painful experience of living in two conflicting
worlds. There was, first, the world of the Brandywine mills, where he
reigned as king. It was a world of action and accomplishment. There he
could now implement the radical changes in organization, technology, and
process he had been futilely advocating for the previous twenty years. There
he produced the profits necessary to pay off the mortgages on the imperial
corporate structures that his two cousins were feverishly constructing and
provide the rich dividends that they and the other stockholders demanded.
There he was alive with the happiness that comes from creativity and
fulfillment.

The second world was that of Swamp Hall, where he was obliged to
play the role of court fool. It was a world of inefficacy and failure. There
he was ignored, or worse, ridiculed by his wife, embarrassed in front of
guests, and isolated by his own increasing deafness and by her increasingly
sharp denigration. There he was deadened by the misery that comes from
ineptitude and frustration. Only the unquestioning love given him by his
children, fifteen-year-old Madie, thirteen-year-old Bep, and his infant son,
Alfred Victor, provided him with a cloak of self-respect.

Alfred must have frequently asked himself what had gone wrong. How
had he failed both his own and his wife's expectations in this marriage?
The first years had been happy. Except for the tragic consequence of his
young brother Louis's unrequited love for Bessie, there had been no suggestion of miasmic gloom in the house that they called The Swamp. Indeed,
it had been to Swamp Hall that Alfred had hurried each evening to find
solace from the frustrations he endured daily in the mills. Within less
than three years of marriage, he and Bessie had two beautiful daughters,
and the old home on Breck's Lane, which twenty-five years earlier had

contained so much darkness and death, was now filled with light and life.

Bessie had made many friends among scholars in New Haven and among writers, artists, and actors in New York and Philadelphia. These intellectuals and creative artists were frequent visitors to Swamp Hall, and Alfred, still as stage-struck as he had been when he neglected his studies at M.I.T. to take walk-on roles at the Boston Opera Company, was thrilled to offer hospitality to young Shakespearean actors like James Hackett. Bessie, in turn, had professed great pleasure at the beginning of their marriage in Alfred's interest in music. She welcomed his Tankopanicum Musical Club into Swamp Hall for their weekly practices in the billiard room. She would sit perched on the game table, keeping time with a swinging foot and applauding each number, no matter how ragged the performance.

But then, at first almost imperceptibly, her attitude began to change. Alfred's attempted sallies of sophomoric humor at the dinner table were no longer met with appreciative laughter from Bessie. Instead an awkward silence followed by a cold insult would reduce Alfred to confused embarrassment. Gradually, Alfred sensed that his presence at Bessie's salons of intellectuals and artists was barely tolerated if not wholly unwelcome. He soon found it more comfortable to stay away from these gatherings, although some of his wife's protégés, such as Hackett, remained his lifelong friends.

Bessie also began to stay away from Alfred's music sessions in the billiard room. First by her absence, and later by her frank criticism, she let husband know that she no longer appreciated the Tankopanicum Musical Club. She considered the program selections crass, patriotic marching rousers, more appropriate for a philistine audience at a village square gazebo than for her cultured ears. Worse yet were the so-called musicians themselves. With a few exceptions, such as Cousin Pierre, who played the piano—rather badly, according to Alfred—they were rough day laborers from the mills who were decidedly out of place in a du Pont mansion. Alfred's band retreated for their practice sessions from the Swamp Hall billiard room to the nearby Rising Sun Tavern, where they were more appreciated.[1]

Month by month, the gulf widened between Alfred and Bessie. In his perplexity over their growing incompatibility, Alfred tried to be fair to Bessie, for he was still very much in love with her. He sensed that she was as lonely and unhappy as he. Bessie, much like his mother, had never been fully accepted into the du Pont family. Like Charlotte, she was an outsider, a distracting foreign intruder into xenophobic Brandywine society. If Charlotte had been too much of a Southerner, Bessie was too much of a New England Yankee in this border region which looked neither North nor

South, but gazed with appreciation only upon its own navel. When Bessie had first arrived at The Swamp in 1886, Granny Meta and Alfred's aunts, Sophie, Louisa, and Polly, had quickly sized her up as an over-educated Brahmin snob. They resented her highfalutin airs, such as her calling the noon meal "lunch" and the evening supper "dinner." Every good Brandy-winean knew that the proper time for the main meal was at midday. Even Lammot's widow, Mary Belin, was a more integral part of the du Pont family, in spite of her Jewish blood, than was Bessie, for Mary Belin had grown up in the Brandywine valley. Bessie did have ancestral ties to the family, but they were the wrong ties, and all the old aunts and the younger cousins knew full well what Charlotte Henderson had given to the du Ponts as a dowry. Bessie could not look to the family for companionship.

Nor, as Alfred was only too painfully aware, could Bessie find easy companionship with her own husband. Alfred blamed much of the growing estrangement between them on his deafness. Intimate tête-á-têtes had become impossible. Small wonder that Bessie had to import her friends from New York and New Haven.

On those ordinary evenings when no guests were present, she generally retired early to her own bedchamber, and Alfred discovered that her bedroom door was almost always locked. Alfred later learned from the butler, Eben Parker, that his last duty after Mrs. du Pont had retired was "to carry a full quart decanter of Sp. Frumentia to her room," and his first duty in the morning was "to carry a pot of coffee and bring back the empty decanter . . . I could very clearly observe that there were strained relationships between her & her husband,—thus I accounted for this great consumption from the decanter."[2] Alfred, although at the time not privy to his butler's nocturnal and matutinal assignments, must have realized that Bessie had begun to drink more than her social obligations warranted. This also worried him and further provoked his feeling of guilt for having somehow failed her as a husband and companion.

One consequence of Bessie's increasingly cold and critical attitude that Alfred found particularly difficult to bear was that his own siblings would no longer visit his home. Since the death of their parents in 1877, when the five orphans had made a united stand against the family elders in order to preserve their home intact, a peculiarly strong bond had existed among Alfred and his two sisters, Anne and Marguerite, and his two brothers, Maurice and Louis. Although Alfred had been the only one to stay on at Swamp Hall, he had always been emphatic in his insistence that Swamp Hall should continue to be his brothers' and sisters' home as well. The door was always open to them, and no special invitations were ever needed. Bessie had accepted this arrangement as an integral part of their marriage contract. When Maurice had shocked the entire du Pont family by his much-publicized marriage to an Irish "barmaid," and had brought his bride back to the Brandywine, the only door that was at first open to him was that of Swamp Hall. It was to Swamp Hall that the distraught Louis had returned in 1892 just before he ended his tormented life. It was a place for celebration and for lamentation, for loud and public Fourth of July fireworks, and for quiet and private sibling consultations. It had been, in short, a communal home for Irénée's and Charlotte's orphaned children, and Alfred was determined that it should remain so for as long as any of them were alive.

Alfred was unable, however, to preserve the Swamp Hall of the past even though its preservation meant as much to him as did that of the Brandywine mills. During the 1890s he was to learn that Swamp Hall, with Bessie as its mistress, could not continue to be the same place for visiting adults as it had once been for orphans in residence. Alfred wanted to keep familial unity alive, but Annie, Marguerite, and Maurice knew that when they now returned to Swamp Hall, it was as guests, not as occupants.

Alfred might have been able to adjust to this changed relationship of his siblings to their childhood home if they had continued to feel free to return to Swamp Hall as welcome visitors, but Bessie's increasingly overt and public debasement of her husband made even visiting impossible. Marguerite was the first to sever ties with the old place. Some time after the birth of Bep, Marguerite came to Swamp Hall for a week's stay to become better acquainted with her new niece and to see relatives and old friends in the Brandywine valley. The first day of her visit was enough to reveal to Marguerite the low esteem in which her brother was held by his wife, and the cold, even rude, indifference with which Bessie regarded her in-laws.

A more timid person would have retreated before Bessie's scarcely veiled insults and, mumbling some weak excuse, would have departed the next day, but Marguerite came prepared to stay a week and remain a week she would. If Bessie thought she could drive Marguerite away in tears, she was mistaken. Marguerite stayed the allotted period, using her time to good advantage by paying calls on Granny Meta, Aunt Polly, and other assorted du Pont relatives, giving her fascinated auditors all of the concrete evidence necessary to confirm their worst suspicions of that New England snob. On the seventh day, Marguerite departed promptly on schedule, but before leaving, she told both Bessie and the unhappy Alfred that never before had she been so rudely insulted as she had now been by her hostess. If Alfred wished to see her in the future, he would have to come to Washington or meet her on some neutral ground, for this would be her last visit ever to Swamp Hall.

Not content with her own renunciation of her ancestral home, Marguerite was determined that her sister should join her in the boycott. She wrote Annie a long, detailed account of her visit, replete with "Bessie said this to Alfred" and "she had the effrontery to do this to me." Annie knew that Marguerite had a quick temper and an easily bruised *amour propre,* but she also knew that her sister was not one to manufacture false gossip. The specifics Marguerite provided in her letter must have been serious and convincing enough to persuade the level-headed and usually forgiving Annie also to sever relations with her sister-in-law. She announced that she would never again visit The Swamp, a promise she was to keep for the few remaining years of her life.

Soon afterward Maurice and Margery came up from Asheville for a visit, and they received the same kind of reception from Bessie. They cut short their stay. Margery, who had a temper at least as volatile as Marguerite's, also wanted to declare Swamp Hall out of bounds, but Maurice felt this would be cruelly unfair to Alfred. He insisted that whenever they came to Wilmington, where, he conceded, they would hereafter stay at a hotel, they must pay a short courtesy call on Alfred and his family.[3] In the future, however, Maurice depended mainly on Chesapeake hunting trips, which were exclusively male ventures, for companionship with his brother. Bessie had bolted shut more than just her bedroom door at Swamp Hall, in excluding Alfred from the personal contacts he desired and needed.

As inexplicable to Alfred as Bessie's growing hostility toward him and his siblings were those occasional instances when she would suddenly display the same warmth of feeling and friendly interest that he had known in the first years of their marriage. There would be Bessie's smiling attentiveness at the dinner table to encourage Alfred's participation in conversation and a bedroom door left ajar upon her retiring at night.

In the last year of the old century these happy occasions came more frequently. Perhaps during the feverishly active months of the Spanish American War when Alfred had been obliged to spend twenty-four hours a day at the mills, Bessie had come to realize that she did miss his companionship after all. To Alfred's surprise, she became the faithful Penelope, welcoming a weary Ulysses back to his home. In the summer of 1899, Bessie, after eleven years, became pregnant again. With the birth of their son the following March, Alfred became convinced that the long misery of alienation had ended. Home once again became the place he could escape the frustrations he experienced at the mills.

This surcease of pain, however, proved to be of short duration. The bad times of cold indifference and closed doors returned. The Swamp once more bore an appropriate name, a place of murky gloom where one wandered lost in a bewildering fog without a signpost for guidance to solid ground.

In the spring of 1902 a brief respite from domestic conflict once again aroused Alfred's hopes that his marriage could be saved. Bessie looked with new respect on Alfred for his having courageously stepped forward to save the Du Pont Company for the family. She shared his pride in his new status in the company and anticipated along with him the family prestige, power, and eventual wealth the reorganization of the company would bring. Alfred treasured the letters she wrote to him while he was away on an inspection tour of plants soon after the three cousins had assumed the company directorship. Alfred responded with a barely legible letter written in the parlor car as his train bumped along toward its next stop:

> My dear little Woman:
>
> I am on the train. . . . The road bed is rather rough & there are so many curves that it makes writing almost an impossibility. I had a fairly successful visit and was most royally entertained by Mr. & Mrs. Scott Sr. & Mr. and Mrs. Scott Jr. who live in adjoining palaces. . . . I have rec'd so many letters from you, one or two every day, that I feel overwhelmed. The children have written very often. . . . I can make but a poor attempt at returning all this thought & loving kindness from you all as my replies will be few and uninteresting. . . . Your last letter the one I read at Wap [Wapwallopen] has made me quite miserable. I have never known you to write so blue a letter in all our married life, and I realize fully that you must be in low spirits indeed to crave the companionship of your dreary old husband. I am afraid that you and Elly [Ellen La Motte] have been imbibing too much ginger ale. You must know that my loneliness is greater than yours. You have but me to miss and wish for even if it is so. While I have you all. I am going to try & take a week off in August. I am sure by that time I will feel the need of a rest, and think by that time I can get things in shape. . . . I do wish I was [sic] with you all this evening it would do me world of good & give me no end of happiness. Love & kisses to all.
>
> Your worthless husband.[4]

This is a tender letter, written by a man obviously very much in love with his wife, responding to numerous letters from her apparently manifesting the same sentiments. The note is significant not only for its expression of Alfred's continuing affection for Bessie but also for the clumsy words of self-denigration with which he presented himself to his wife. Such phrases as "your dreary old husband" and "your worthless husband" went beyond modesty to masochistic debasement, hardly conducive to gaining from Bessie the denial of his dreariness and worthlessness that he longed to hear. Had he been more assertive of his own merits he might have been more successful in securing his wife's respect. Too often, Bessie had simply to elaborate upon Alfred's own descriptions of himself.

Still, there were occasional good times that Alfred's children would always remember: Christmas parties at Swamp Hall for the millworkers' children, with their father, properly garbed in a red velvet suit and with nose and cheeks painted red, playing Santa Claus and handing out presents and boxes of candy to two hundred pairs of eagerly outstretched hands; sleigh rides down the narrow country roads of the Brandywine valley; and impromptu family concerts, with Madeleine playing the piano, Bep the violin, and their father his beloved flute.[5]

And although ever more rarely, Bessie sometimes still offered love to her husband and welcomed him into her bed. Unpredictability became the only certainty in their relationship. Fortunately for Alfred's sanity, there were the mills. There he found welcome refuge, and in his newly constructed machine shop he had the security of acceptance that he could not find at home.

It was into this crazy, see-saw world of vagary which had become Alfred's life that Alicia Heyward Bradford made her sudden and fateful appearance in 1900. Alicia was the second child and second daughter of Eleuthera du Pont and Judge Edward Green Bradford. Eleuthera, the daughter of Alexis Irénée du Pont I, was a first cousin of Alfred's father, Irénée. Although nearly twenty years younger than her cousin, Eleuthera had been a close friend of Irénée and Charlotte, and their children, in turn, had looked upon her not as a remote second cousin but rather as a young and doting aunt. Alfred, in particular, was always Eleuthera's favorite du Pont relative. His mechanical skills, his love of the black powder mills, and his camaraderie with the workers reminded her of her father, whom of all his du Pont forebears, Alfred most closely resembled in both personality and talent.

Although Alfred maintained affectionate family ties with Eleuthera, the age differential was such that he had only a remote association with her four children, although they were of the same du Pont generation as he. Alicia was just a pretty little schoolgirl of eleven when Alfred and Bessie married in 1887—someone whom Alfred was vaguely aware of when the family gathered at weddings and funerals but whom he knew only as one of Eleuthera's and the Judge's three daughters.

Edward Bradford was a pompous, self-righteous martinet. He was known on the bench as a "hanging judge," one who could never forgive or forget the slightest infraction of society's rules. The phrase "extenuating circumstances" could not be found in his legal lexicon, and no mercy dropped as gentle rain from his stern Calvinistic heaven. The Judge ruled over his family as he presided over his courtroom. The slightest act of disobedience became the major crime of lese majesty against his patriarchal authority and as such was appropriately punished. Both his children and his wife lived in terror of his judgment and were cowed into submissive acceptance of his rules.

Judge Bradford's closest friend and ally within the du Pont family was his brother-in-law Leighton Coleman, who had married Eleuthera's older sister Frances—the same Fanny who as a young girl in the 1850s had been the star pupil in the classroom of the family's young tutor, Page Williams. Fifty years later, Fanny's and Eleuthera's husbands assumed for themselves the roles once held by Aunt Victorine and Sophie and Uncle Henry. Leighton Coleman was the Episcopalian Bishop of Delaware, and from their exalted positions he and Judge Bradford joined together the canons of law and church to impose a theocratic standard of conduct on the entire du Pont family. As self-appointed moral censors they drove into exile the young clergyman who had dared to perform the marriage ceremony for the recently divorced May du Pont and her true love Willard Saulsbury. They had lent authority to the decree of family ostracism against all the errant du Ponts—William du Pont and the Saulsburys and, for a brief time, Maurice du Pont and his Irish Catholic bride.

By the last decade of the nineteenth century, however, the pious Bishop and the stern Judge had become anachronisms within the du Pont family. There were too many errant young du Ponts who ignored their commandments, too many du Pont scandals which they could not suppress, too many du Pont sinners who could not be properly punished. Not even in their gloomiest jeremiads, however, did the Bishop or the Judge imagine that the worst scandal ever to rock the du Pont family would be perpetrated by the Judge's own daughter in consort with the Vice-President of the Du Pont Company.

After his first child was not the boy he had hoped for, Judge Bradford even more eagerly awaited the birth of his second child, only to be disappointed again when Alicia was born in 1875. The third child was the hoped for son and was given the name of his proud father, Edward Green Bradford. Little Eddie, however, very early proved a disappointment to his father. Not that he was an unruly child—quite the contrary. He was as submissive to his father's will as was his mother, whom the boy closely resembled in both personality and temperament, and Edward Jr. did not have the intelligence, the strength, or the ambition that the Judge expected in a son. Instead, Alicia evinced the character and intellect that any Victorian father hoped to find in a son.

Judge Bradford realized very early on, that Alicia was a truly remarkable child. Not only was she the most beautiful of his three daughters (and quite possibly the most beautiful female the du Pont family had ever produced), but also she was the most intelligent, the most adventuresome, and the most strong-willed of his four children.

With each passing year, however, Edward Bradford's resentment against Alicia for having had the bad grace to be born a girl seemed to grow.

Although it could not be said that any of his children had been spoiled by his having spared the rod, the Judge seemed to take a particular, almost sadistic, pleasure in punishing Alicia. He was determined to tame this proud, strong-willed child, and had he been able, he would gladly have emulated his London counterpart, Edward Barrett of Wimpole Street, in reducing his daughter to a sickly, timorous prisoner in her own home. But although she also wrote poetry in secret, Alicia was no Elizabeth Barrett, languishing on a couch and waiting passively for a poet lover to come to her rescue. Alicia would eventually manage her own escape from parental tyranny and would actively pursue her own choice of lovers.

As a child and a young girl at boarding school, Alicia did her best to obey the family rules and to do well in school in order to please her father. She tried to meet the high standards he set not only because she feared punishment but also because she wished to save her mother, whom she adored, additional stress. Alicia could never suppress, however, the attributes nature had given her: a keen intelligence, which made school learning so easy that it became boring; a restless energy, which made passivity intolerable; a sharp, cutting tongue which could not be muted; a free-ranging spirit which could not be confined by rules; and a classic beauty, which, even when she was a young girl, aroused envy in her female companions and an uncomfortable edginess among boys of her own age.

Upon reaching adolescence, Alicia was sent to the Misses Hebbs's boarding school on Pennsylvania Avenue in Wilmington. Her weekly letters to Colonel Henry A. du Pont's daughter, Louise, who, although two years younger than Alicia, was her closest friend and confidante, reveal a great deal about Alicia's character and lively spirit, which her father's discipline had not been able to crush.[6] The letters also show her propensity for introspection. She candidly speaks of both her strengths and weaknesses:

> Last summer was lots of fun and I was a dreadfully bad "young one" I know but I tried to be as good as the *fun* would allow.... We had lots of fun in the holidays oh dear and now I am back here! still school is one of those evils which have to be endured and quietly no one here knows how I would love to, oh long to "flee away like a bird to the mountains and be at peace or rest!!" but it goes grinding on and no one is the wiser of my *affairs*. I do not make "bosom friends" as some do in a few days. I have very good ones as I told you and have never had the slightest rub with any one but still Rats!!!!!![7]

The stern Calvinistic piety of her uncle and the harsh Old Testament judgment of her father had left their marks on Alicia. On Valentine's Day, 1893, she wrote to Louise:

> Just think tomorrow is the beginning of Lent, how hard it is to be good anyhow! It seems to me I get worse in [every way] every year

instead of better, the harder I push foreward [sic] the further back I slide! How are you going to do this Lent? I do not give up things at the table for it is just blowing one's trumpet on the street corners in a place like this. It is a good plan to set apart a few minutes in the middle of the day for prayer. I think even if it is a "few minutes" indeed it is a help. You do it too.[8]

Alicia also worried about the state of Louise's soul when her cousin once again postponed her confirmation in the church:

It certainly is rather perplexing about your confirmation—and oh! Louise I had set my heart on having you give yourself to God this year— perhaps that is rather a strong way of putting it—but it is the way I feel. Of course it would not do to desert your own church—but why don't you just pretend you are confirmed till next I mean—in acts—take a Sunday School class and start to work anyhow. When I come home I will take another too and we could start in fresh together.[9]

Try as hard as she could to be good, however, Alicia always encountered that old devil, Temptation, always ready to lure her away from the path of righteousness, as Augustine, the Bishop of Hippo, might have told her, and as her uncle, the Bishop of Delaware, most certainly did. In one of her weekly reports to Louise, Alicia told how, one night, after lights out, when all the girls were expected to be in their own beds asleep, several of them gathered in one room to tell ghost stories. They suddenly heard Miss Ruth Hebbs coming up the stairs. Alicia and a couple of other miscreants had the wit to hide under the bed. The headmistress severely reprimanded those she found in the room, but "never knew we were there for we are considered pretty much saints and besides we had left dummies in our beds. Well, we had a little talk and L [Lavinia] and I decided that it would be horribly mean to have the others caught and us thought good so out we bravely walked there stood the Ruthless thing with the candle. She just glared at us in astonishment and we walked down to our room without a remark. . . . I have to be angelic for awhile now to retrieve my name for its the first time I was caught but I think it was very nice of us to come out, I must say!"[10]

Although quickly assuming a role of leadership in the classroom as a prize-winning scholar and editor of the school's literary magazine as well as in extracurricular midnight pranks, Alicia did not regard herself at this stage in her life as a rebel. In one letter she told of going to a lecture on Woman's Rights. "My I was disgusted. I heard two or three strongminded creatures speak and they wanted women to be lawyers and doctors and ministers just like men (instead of minding their own affairs). Lavinia liked it though and she believes greatly in the elevation of women. I think it is all 'funnie.' A woman's business is to take care of their husbands (if they have any) and sew and be useful in general instead of all that!!!"[11]

Although Alicia in her many letters to Louise never mentioned any interest or special skill in sewing, she did clearly indicate an interest in boys. She fully intended to get herself a husband "to take care of and be useful in general" instead of all that "funnie" business of women's rights. Apparently it was the custom among young ladies attending exclusive female finishing schools to wear the colors of one of the men's Ivy League colleges, as a symbol of their undying allegiance to those hallowed institutions and to all the inhabitants thereof. Alicia was upset when Louise chose Princeton, "for I want us to think alike and we can't as long as you sport the orange and black!" As for Alicia it would always be the Yale blue. "The summer season most always changes a girl's colors, but I have kept mine always Yale! Am I not to be congratulated?"[12]

Although Alicia frequently complained to Louise of the drudgery and tedium of life at the Misses Hebbs's school, she also realized that, compared with living at home under her father's rule, she was almost as liberated from her "cage" as the high-flying bird she longed to be. Indeed, she sometimes worried about yielding too completely to the temptations that the relative freedom of school offered her. "This is study hour," she wrote Louise, "you must wonder how I have time to write so much but I tell you I'm very bad to do it now. School certainly makes people sin and I think it is a bad thing to put such temptations in a person's way! Yes it is very wrong?!!"[13] But inevitably, temptation would win out over the stern dictates of conscience, and she would confess that "tonight I have one of my hilarious humors on you know the kind when I behave like a savage and have a tendency to use slang!!!"[14]

As her numerous letters to Louise illustrate, Alicia had a disdain for the established rules of punctuation. Commas and periods were generally dispensed with as being troublesome breaks in the flow of her stream of thought. It was the exclamation point that Alicia truly loved. She sprinkled these little dagger points of excitement across the pages of her letters with reckless liberality.

In the quotation cited in which she philosophized over schools' leading people into sin, Alicia outdid herself in originality of punctuation. She concluded the sentence, "Yes it is very wrong" not only with the expected two exclamation points but also with an unexpected question mark. Although this use of the interrogatory sign may have been accidental, it was not inappropriate to that particular sentiment at that particular time. Question marks had begun to appear unexpectedly in her mind as well as in her writing. For the first time in her life she was starting to question not her own actions but the authority of those who told her what her actions should be. Who were these infallible standard setters, and who had given them the authority to say this action is right and that action is wrong? Soon Alicia began asking herself why there was one standard for boys and a different standard for girls. These were troublesome questions for a young girl who still devoutly believed in God and who so recently had disdainfully rejected the feminist demand for equality expressed by those "strongminded creatures" whom she had regarded as being "funnie."

As she pondered these questions, her search for answers always led her in one direction—back to her father. He had set the standards of proper conduct for his family, and while he had sometimes said Yes to her brother, he had always answered No to her and her sisters. His was the final authority from which there was no appeal, and his was the stern punishment from which there was no escape. No matter how hard Alicia tried to please him, she had never won from him a congratulatory word for her high marks in school, nor a smile in praise for her teaching a Sunday School class. Much later in life she confessed to Francis I. du Pont's wife Marianna, "As a child I was frightened all the time—terrified of everything. Suddenly it came to me that my father was the cause of this. He had wanted me to be a boy. I saw it all and made up my mind to get even."[5]

With this revelation of the source of all her fears and self-doubts, Alicia's course of action became clear to her. Her brother Eddie and her two sisters, Eleuthera and Joanna, could continue their unquestioning acceptance of their father's authority if mere survival by means of cowed submission to his authority was all that they wanted out of life, but Alicia, as soon as she had attained her majority and had graduated from the Hebbs School, would deny the legitimacy of her father's authority. She would "get even" by living her own life in her own way.

Alicia obtained both her diploma and legal adult status in the summer of 1893. She was now independent of her father's control, and she set out to explore the newly opened territory of her liberation. She discovered that there were almost no boundaries to her freedom, if she did not violate the civil law and did not care what old gossips in the du Pont family said about her. She found out that her father's No did not keep her from going unchaperoned to weekend parties in Annapolis, and no bolt of lightning from on high struck her if she smoked a cigarette or even said "God damn."

In the fall of 1893 Alicia celebrated her newly found freedom by going to Annapolis for an extended stay, ostensibly to visit a cousin, but mainly to attend Naval Academy football games and post-game parties. She continued to report on her activities to Louise, but now, instead of advising her younger cousin to set aside moments in each day for meditation and prayer, she offered a different counsel: "Do not study too hard. I advise you

to 'scrap' with your governess it is the best path to notoriety in the household. I speak from experience as you know."

Alicia then described in some detail the delights of freedom:

> Yes, I am having a fine time. You will still have time to answer this
> letter as Cousin A says I shan't go yet. Yesterday I went to another
> football match between the Cadets and—I'll tell you the other team
> when I can think of it—Marshall something. It began at 1^{30} which left
> a good afternoon and the *almost* nicest cadet who plays the mandolin
> came I think I spoke of him before and stayed till tea. He certainly is
> fine. I couldn't help thinking so with his brass buttons flashing on the
> sofa beside me. There is a tremendous chance at vicars for me—a Pres-
> byterian minister came to see me the other night. Oh! such a character.
> I will take him off for you when I come home. I have an invite to dinner
> for tomorrow night. It is given for me and Mr. Striker (the young cler-
> gyman I mentioned as cutting a pigeon wing on the hardwood floor)
> and lots of St. Johnners are to be there. I am especially requested to
> bring my banjo.[16]

In classic Victorian melodrama, Alicia's declaration of independence
from patriarchal authority would have meant her expulsion from home had
Judge Bradford felt free to act upon his angry inclination. But for once in
her life, his wife dared to contradict her husband. Eleuthera implored, even
demanded, that Alicia be allowed to stay in her own home.

Surprisingly, the Judge reluctantly acquiesced, not because of Eleuthera's entreaties, but because of the realization that Alicia's banishment
would but confirm what were still only whispered rumors of her carryings-
on at New Haven and Annapolis. All civil communication between him
and his daughter ended, however, and it was only his wife's gentle patience
and continuing support of her daughter that preserved any semblance of
normal family life within the household. Alicia took advantage of every
opportunity offered to be absent from home.

Although Alicia continued to profess to Louise her undying loyalty to
the Yale blue, and had asked one of her many friends at the university to
send her an ample supply of Yale blue garter elastic so that she could
proudly if somewhat shockingly display her fealty to the Elis,[17] the flashing
brass buttons of the naval cadets had clearly made their impression. She
spent more and more time in Annapolis. It was not at the Naval Academy,
however, that she was to find her real love, but at the much smaller liberal
arts college, St. John's in Annapolis. In one of her last letters to Louise that
is still extant, she mentioned that on the following Wednesday she was
going "to the football match between St. Johns and the Cadets. My situation will be very divided as of course I want the cadets to win and a St.

Johns man is going to take me!"[18] Alicia did not identify her escort by name, but quite possibly it was George Amory Maddox, for soon thereafter Alicia was dating and had fallen very much in love with this handsome St. John's student. The Yale blue had been replaced by the St. John's orange and black.

Maddox came from an old, long-established Charles County, Maryland, family. His ancestors had emigrated from Wales in the seventeenth century to seek a haven in Catholic Maryland from the religious persecution they had suffered under the Protestants. The Maddoxes had married into some of the distinguished families of Maryland and somewhere in the process, they, or at least Amory Maddox's branch, had exchanged their Catholicism for the Episcopalian faith.[19]

Alicia's and Amory's pursuit of each other prospered, and one year after his graduation from St. John's, while Amory was a first-year student in law school, they became engaged. Alicia's mother announced the engagement in the summer of 1897, just prior to the marriage of her eldest daughter, Eleuthera, to her cousin Henry Belin du Pont, Pierre's younger brother. In a letter to Colonel Henry A.'s wife, Alicia's mother wrote:

> She [Alicia] met him [Maddox] several years ago at Annapolis while he was at St. John's College there. He is studying law, and this will be a long engagement, so I will not have to part with another daughter soon. We have known him for some time, and like him very much. He is of a good old family and a member of our church.[20]

To his wife's announcement of Alicia's engagement, the Judge would have added, had he been interested in discussing Alicia's affairs, that the Maddoxes may have been a good old family but they had no great wealth. As for their son, he seemed to lack both ambition and talent, and the Judge did not like him at all.

Eleuthera du Pont had been correct in saying that it would be a "long engagement." Amory, in particular, seemed in no great rush to conclude the engagement with marriage. He proved to be even less successful in his law studies than in his undergraduate courses at St. John's. He kept postponing the marriage date until such time, as he said, he could find himself. He didn't seem particularly eager to find himself in any occupation, however, and in the meantime he thoroughly enjoyed the freedom of bachelorhood, while Alicia in turn flaunted her freedom from paternal authority before shocked relatives and friends along the Brandywine.

It was during this five-year period between her engagement to Maddox in 1897 and her marriage in 1902 that Alicia first became a significant force in Alfred du Pont's life. When Marquis James reached this point in writing Alfred's biography, he told Mrs. du Pont that he did not look forward to the next chapter.[21] Subsequent chroniclers of Alfred du Pont's life must also share James's sense of inadequacy in attempting to distill, out of the welter of innuendo, gossip, and self-serving asseveration, the actual facts of the developing relationship between Alfred and Alicia. For anyone more interested in purveying sensationalism than truth, the task is simple: One accepts the version of events as given by Judge Bradford and as later elaborately embroidered by succeeding generations of du Ponts, and one has a tale that is certain to titillate the public's fascination for scandal in high places.[22] All the essential actors to provide an absorbing drama for the readers of the *Police Gazette,* and W. R. Heart's *New York Journal* are here: the spoiled and unruly daughter of a distinguished United States judge, herself an heiress to a small portion of the du Pont wealth; the co-owner, co-manager of the Du Pont industrial empire, unhappy in his marriage to a wife who has spurned his love; and a feckless youth who, in return for a high-paying job at the Du Pont Company and an elegant, old du Pont family mansion, is willing to play the role of husband to the beautiful heiress in order to provide a cover for her adulterous liaison with the unhappy Vice-President of the Du Pont Company. These are such stuff as scandal sheet circulation managers dreams are made on.

To weave this sensational tale, however, one must ignore certain facts, such as Alicia's enduring love for and long engagement to Maddox, as well as the testimony provided by one of Alicia's most intimate friends, which contradicts the account given by Alicia's vindictive father. To arrive at an understanding of what actually transpired in the changing relationship between Alfred and Alicia, one must have some insight into the characters of the actors in the drama.

It was not long after the first whispered gossip about Judge Bradford's wayward daughter began to circulate within the du Pont family that Alicia attracted the attention of Alfred du Pont. Prior to that time he had viewed her as Cousin Eleuthera's pretty little schoolgirl daughter, hardly distinguishable from his many other younger second and third cousins. As the whispers of Alicia's unconventional behavior grew louder and more detailed, however, Alfred began to take notice of the beautiful young maverick. Given his propensity for providing a sanctuary for other violators of the du Pont social conventions, the family was not too surprised when the door of Swamp Hall was opened wide to Alicia at the very moment that other du Pont doors were being quietly but firmly closed. Nothing could engage Alfred's protective instinct as completely as malicious gossip directed against one member of his family by the other members of the tribe.

By 1900, Alfred had become Alicia's most powerful friend and sympathizer within the du Pont family. As he came to know her, he also became her chief admirer. Keen intelligence, quick wit and toughness of character were all qualities that Alfred admired in a woman as well as in a man. Unlike many men, Alfred was not intimidated by either brains or a strong will in a woman. What many males found attractive—soft, simpering females who deliberately feigned weakness and stupidity—were repellant to Alfred. Alicia could meet him on equal footing, and he liked that. She was also a fighter who refused to be driven out of the Brandywine valley when she insisted upon living her own life in her own way, and he liked that even better. She could even outmatch him in sharpness of wit, and being a talented mimic, she would send him into paroxysms of laughter with imitations of her father, Bishop Coleman, and Aunt Elizabeth Bradford du Pont. The qualities that made her an *enfant terrible* to most of the family were precisely those that endeared her to Alfred.

But had the relationship between these two congenial cousins gone beyond mutual admiration to a more passionate attachment prior to Alicia's marriage to Maddox in 1902? One can safely assume that in Alicia's case it had not. She was, of course, tremendously pleased and flattered to have such a powerful and so generally popular a man as her champion, and she would always be grateful for Alfred's sympathy and support. Even more than her own mother, Alfred was the one person who truly understood and appreciated her for what she was. But with the limited vision that youth has in respect to age differentials, Alicia at twenty-five saw the thirty-six-year-old Alfred as an elder who belonged to the preceding generation. She needed and wanted Alfred as a counsellor and friend, but it was Amory she wanted as a lover and husband, as she told her cousin and closest confidante, Edith Goldsborough.[23] At this time, Alicia apparently had not considered the possibility of a sexual liaison with Alfred.

In contrast, Alfred's feelings for Alicia had undoubtedly progressed from avuncular affection to something like those of an ardent suitor. Deprived of the love and companionship he wanted from his wife, he must have been flattered to have a beautiful young woman dependent upon him for counsel and consolation. Alicia had taken full advantage of his initial friendship and apparently felt free to impose upon him at any time for his support. She was a frequent visitor to Swamp Hall, where Bessie, also somewhat isolated from the du Pont family, initially welcomed her. Later, as relations between the two women cooled, Alicia would call Alfred to come to her during the day when her father was away, and her mother would be as happy as she to welcome their favorite cousin into their home. No matter how busy Alfred was, he always came when called. The curious bond of being aliens in their own homes tied mother, daughter, and cousin together during these years.

Alfred's solicitous attentiveness to Alicia did not escape the sharp eye
of her father. If Alicia chose to see only Alfred's familial affection, the Judge
saw something else more intense. The Judge called it "blind infatuation,"
and he suspected that Alfred's congeniality toward his daughter had as its
motivation more than a simple desire to be helpful to a cousin.

Alicia made frequent trips to Maryland to see Amory, since her father's
attitude toward her suitor made visits by him to the Brandywine impossible,
and as the months and years of their engagement dragged on, she grew
increasingly impatient for Amory "to find himself." He was now studying
engineering, having failed in his law studies. In the early spring of 1900,
the young couple actually began to make definite plans for their wedding.
Alicia's cousin Louisa du Pont wrote to her brother Pierre:

> Today I lunched at Alfred's. . . . While I was there Alicia came in.
> She asked me most particularly just what dates in June we were going
> abroad, and if we would be here until just before we sailed. She asked
> so many questions I really think she must be thinking of getting
> married.[24]

Two months later, Louisa reported to her brother that Alicia had just
announced "her nuptials for June," but then in the same letter she added
a postscript: "We heard at Bess's that Alicia is not to be married in June
after all, Mr. and Mrs. Maddox having met with some pecuniary loss. They
say Alicia is broken hearted. It must be discouraging to be engaged in an
indefinite manner."[25]

The pecuniary failure, unfortunately, was Amory's, not his parents'.
Without a professional degree, he was unable to find the kind of job he
wanted, and Alicia was not prepared to bring before the family a penniless,
unemployed groom. By using Alicia's family connection, Maddox was
finally able to get a job at the Johnstown steel mills in western Pennsylvania,
but this hardly advanced the prospect for his marriage. Alicia was not willing to exile herself to a dreary spot like Johnstown. This move would be a
final humiliating submission to her father, who was so eager to have her
leave the Brandywine. She was determined to stay, in defiance of her father,
and she wanted Amory there with her in a place of importance of which
she could be proud.

Alfred's sudden elevation in February 1902 to co-owner and vice-president in charge of all production of the Du Pont Company at last gave
Alicia the opportunity she had been hoping for. She asked her patron to
grant her the greatest favor he had yet bestowed upon her. She asked Alfred
to give Amory a supervisory job in the mills. It could not have been easy
for Alfred to provide the means by which Alicia could marry another man,

but he was unable to deny her anything. He not only promised Amory Maddox a job directly under him in the plant, but now, as the dispenser of the company-owned homes in the valley, he assigned to Alicia and her husband-to-be the old family mansion, Louviers, the former home of Admiral du Pont and Great Aunt Sophie.

An exultant Alicia proudly announced the glad news to her parents, and, predictably, the Judge exploded in rage. He was unalterably opposed to his daughter's marriage to that "worthless puppy," as he called Amory, and he directed his anger against Alfred. Had this intruder into the Judge's family affairs not promised Maddox a job—and of all things, a du Pont house—there would be no marriage. The Judge had been counting on the wearying effect of an interminably long engagement. He hoped Alicia would eventually find someone else more suitable to her class. She might then settle down (preferably anywhere but in the Wilmington area), to become if not a respectable, at least a distant member of the du-Pont-Bradford families.

Why would Alfred, who, the Judge was convinced, was himself madly in love with Alicia, provide the means by which she could marry another man? This act of unselfish service to another was beyond Bradford's comprehension. Suddenly, he thought he understood. This was not an act of pure altruism upon Alfred's part, but rather a diabolical scheme that he and Alicia had worked out to provide a cover for their adulterous love. Marry off Alicia to an obliging dupe, provide the couple with a home that could serve as a trysting place, and supply a convenient husband of record to accept the responsibility for paternity if Alicia became pregnant. All very neat—but his wicked daughter and her distinguished paramour hadn't been able to fool the Judge. Not at all.

Thus, inspired by rage, Judge Bradford concocted a malicious tale that seemed to him to be entirely plausible and that helped to alleviate his frustration in having to face the prospect of Alicia's marriage. He lost no time in carrying his story to the one person who could be most helpful in blocking Alfred's nefarious scheme—Coleman du Pont, president of the Du Pont Company and Alfred's superior. Because Coleman was a newcomer to the Brandywine the Judge knew him only by name and reputation, Bradford went first to Coleman's and Alfred's Aunt Pauline to ask her to serve as intermediary in presenting the facts of the case, as the Judge saw them, to Coleman. He poured out his sordid tale to Aunt Polly, knowing that if nothing else she could be depended upon to disseminate the story throughout the family.

Aunt Polly was properly shocked and at once called Coleman and asked if she could see him that evening. As the Judge related the entire incident

to his son-in-law, Henry Belin du Pont, who was then in Phoenix mortally
ill of tuberculosis:

> Coleman called at my office on the following Monday, and we had an
> interview of about an hour and a half. Pierre had spoken to him as well
> as his aunt, and Coleman has now no use for Alicia. He thinks it
> extremely desirable that she and Maddox should not be there and dis-
> played such kindness, sympathy and generosity as I shall never for-
> get.... He now fully understands the situation.... He said, however,
> that Alfred was in charge of the manufacturing department and had a
> right to employ whom he would.... I asked Coleman whether Alfred
> had the right to select any house he chose and install Alicia in it. He
> said that he had not, as against the wish of a majority of the Board of
> Directors. The older members of the board evidently do not want to do
> anything that would anger Alfred.... Alfred ... has gone straight
> ahead, given all assurances to Alicia ... and apparently will carry his
> insane project through.[26]

Coleman, in spite of the sympathy that the Judge said he displayed
toward him, was apparently not willing to make an issue out of the whole
affair. He said he would consult with Pierre and then write Bradford.
Shortly thereafter, Bradford received the following note from Coleman, a
copy of which Bradford enclosed in his letter to his son-in-law:

> Dear Coz Ned:
> I talked with Pierre as promised. Under existing conditions of which
> you told me I am inclined to think that any radical action would not
> help matters. And I feel too that time will change and improve condi-
> tions materially.
>
> > Yours sincerely
> > Coleman du Pont[27]

Apparently Coleman and Pierre, in the midst of building an industrial
empire, were not willing to be side-tracked by having a showdown with
their vice-president, upon whom they were dependent for production. Even
if the story told by the hysterical Bradford could be accepted at face value,
about which Coleman may have had some doubts, it was not worth having
a family row over, providing Alfred and Alicia were discreet. Coleman
moreover, had a somewhat more relaxed moral standard than did Judge
Bradford.

So Amory got a job in Alfred's office, Alicia got Louviers, and the 30th
of April 1902 was set as their wedding date. Utterly frustrated in his efforts
to block the arrangements, Bradford could only write Coleman:

It seems like the irony of fate that, after waiting and hoping for years for her marriage and removal from the Brandywine, her marriage should now serve as the very means of retaining her here and perpetuating the evil sought to be averted. While I have no right to complain, it is to me a matter of intense regret that the organization of the company is of such a character that the just claims of flesh and blood apparently can receive no practical recognition, owing to the wild infatuation of an unbalanced mind for an unworthy and cruel woman—an infatuation which, if it lasts, I fear is only too likely to produce scandal in our midst.[28]

If Judge Bradford was no longer able to direct Alicia's destiny, he was at least still master in his own home. He gave strict orders to his wife, son, and youngest daughter. None of them was permitted to attend Alicia's wedding, and he refused to allow his wife to send a wedding gift. Bradford also had the satisfaction of knowing that his tale of the illicit love affair and the "sham" marriage had spread like an ugly spill throughout the valley. The stain could never be removed and became an indelible part of the du Pont legend, accepted by most of the family and by many future chroniclers long after all of the actors had departed from the scene. With her own immediate family's boycott of the wedding, it must have come as no surprise to Alicia that few other du Ponts showed up at Christ Church for the ceremony. Alfred accompanied the beautiful bride down the aisle and, in lieu of her absent father, gave her in marriage to the handsome groom. It would not be the last time that Alfred and Alicia would stand together in proud defiance of the whole "goddamned tribe."

Alicia would always remember the first two years following her marriage to Amory as the happiest period of her life. At last she was mistress of her own home, not a barely tolerated occupant of her father's house. Stately old Louviers, with its beautiful gardens, large, airy rooms, and a commanding view of the Brandywine valley, was indeed a house of pride. Alicia had been placed upon a very elegant pedestal.

She was also very much in love with her handsome young husband. Amory was the ardent lover she had imagined he would be. He also seemed at last to have "found himself" in the business world. Although Amory started out at the usual Du Pont beginning salary of $1,000 a year, under Alfred's auspices he made rapid advancement in both salary and responsibility. His progress was not due solely to the favoritism he enjoyed from his boss. To Alfred's initial surprise, Amory proved to be more capable than many of the salaried executives who had come as refugees from Laflin & Rand and other powder companies that Du Pont had absorbed and for whom Alfred had been obliged to find sinecures.[29]

It was a proud Alicia who drove the small two-wheeled horse carriage to the Hagley Mills gate every afternoon to pick up her husband. The couple, with heads held high, would ride past other du Pont homes, many of whose doors were closed to them, rattle over the old iron bridge that crossed the Brandywine, and drive on up the hill to Louviers. For the first time ever Alicia was supremely happy.[30]

She cared little about those relatives who no longer acknowledged her existence. She still had some close friends among the du Ponts: Francis I. du Pont and his wife Marianna, Pierre and his sister Louisa, and above any other, Alfred, although she saw much less of him now than she would have liked. Not even the restriction imposed by her father, which barred her from ever entering the old family home, Hagley, bothered her too much. She had no desire to see her father or her younger sister, Joanna, who was still at home. She regretted only that it made contacts with her mother more difficult, but even that barrier was breached by daily notes and little gifts of flowers and garden produce, which mother and daughter exchanged by means of the Bradfords' obliging and discreet yardman.[31] At least twice a week, weather and her health permitting, Eleuthera Bradford would make a surreptitious visit to Louviers to see her daughter during the day, when the Judge was absent from home. They would have tea and exchange gossip. Alicia was always particularly interested in news of Alfred and any information regarding his marital relations with Bessie. In the spring of 1903, Alicia was interested to learn from her mother that Bessie was pregnant. Apparently, Alfred's marriage was once again on a more secure footing.

Alicia was also pregnant, and in mid-summer of 1903 Eleuthera made plans for being with her daughter when she gave birth. She sent a hurried note to Alicia in late July

> Please tell your messenger, if you send him for me in the night, and ask Miss R. [Alicia housekeeper] to tell him, to call, and throw stones (gravel!) at the windows in the *front* 2nd story, where a light will be burning. I have decided to spend the latter part of every night at present in the billiard room on a cot. . . . *Look out* for being waked up from a *comfortable sleep,* in the night, or towards morning! This is the *orthodox* way, followed in our family! Though of course one can never tell!

Eleuthera did not want Alicia to send a carriage for her lest the Judge hear the sound of its wheels in the drive. "I will *walk* with your man, if he comes for me, 'rain or shine' as that is the best way to come, & I *prefer* it," she added as a postscript. "Instruct him *which* house, so as not to summon your Uncle Frank!"[32]

The baby was born on the first day of August 1903—a little girl whom they named Alicia Amory du Pont Maddox. Alicia's happiness was now complete, and so was the proud grandmother's. "How lovely Friday was! I enjoyed every minute at Louviers, and Honey is too perfect for words! And her new things, so sweet, & pretty." Eleuthera wrote after one of her visits to see her new granddaughter.[33]

The following month Alfred's and Bessie's fourth and last child was born—also a little girl, christened Victorine, a proud old du Pont name that Alfred must have hoped would also be symbolic of a new era of conciliation and peace within his marriage. Alfred's happiness at this moment was at least the equal of Alicia's. With the same optimism he had known after the birth of Alfred Victor three years earlier, Alfred was now convinced he could be as successful in rebuilding his marriage with Bessie as he was in rebuilding the obsolete Brandywine mills and the run-down workers' cottages in the surrounding villages.

Alicia's and Alfred's marital happiness proved short-lived, however. Within a year after the birth of their daughters, both of their marriages began to unravel. When Alfred had accompanied Alicia down the aisle in April 1902 to give her in marriage to Amory Maddox, he must have thought that this event marked the conclusion of their close association. But Alicia could not be so easily dismissed from his life. Once again family hostility brought them together, but this time they would find themselves bound into an unexpected and lasting union.

Alfred had no more adequate an explanation for Bessie's sudden changed attitude toward him shortly after the birth of Victorine than he had had on previous occasions. All that he knew was that this time her hostility was more intense. In the summer of 1904 she suddenly announced to him that she thought it better for both of them if she left for an extended stay in Europe. She had made arrangements to go to Brussels and to take the four children with her. She would place the two older girls, Madie and Bep, in a boarding school where they could get a proper education in language, music, and art, while she and the two younger children would live in an apartment that she had contracted to rent.

The news of this estrangement quickly became common knowledge in the family, and in a reversal of roles it was now Alfred who came to Alicia for sympathy. Frequently, Eleuthera would be at Louviers, too, to offer Alfred her support. In one note to Alicia she expressed her regret that she would be unable to see Alfred during the next several days.

> I would greatly like to meet him at Louviers. . . . [and] will try to do so. But at present my time is not my own, except sometimes in the morning, and I know he is busy then. "The Judge" is taking his holiday, coming home at 1 o'clock each day and sometimes staying at home all day. Later he is going away. I, of course, am always at home to drive,

or read, or play cards as he wishes. So you see it will be difficult to fix a time for Louviers.

Eleuthera wanted Alfred to know that he also had the support of others in the family.

> I think that "The family" will *all side* with him, except possibly the Eugene's and Alexis'. No influence exerted could *make me* anything but a loving cousin to him. I have been fond of him all his life, and seeing him and knowing him as I have the last two years makes me love and admire him very much. I only hope that soon he may have the peace and comfort in life which he so deserves! It is all *too bad,* but what can we expect otherwise? Things cannot go on this way indefinitely. And how foolish *she* is.[34]

Eleuthera greatly regretted the malicious gossip that was being circulated throughout the valley regarding Bessie's departure for Europe. "It has been a great mistake, among many, that our immediate family has 'talked abroad,'" she wrote Alicia soon afterward. Surprisingly, however, the one person who had heard nothing about the matter was Eleuthera's brother-in-law, Bishop Coleman. He had written to Bessie to ask her to be "one of the Babies' hospital managers—(not knowing anything,) and she wrote that 'she could not, for she had new *responsibilities* now, which would prevent.' He asked me about this and I 'said a few remarks' as to the *truth* of the matter, and Alfred's goodness."[35]

By late August, Bessie and the four children had left for Brussels, with no indication as to when they might return. Alicia had taken little Alicia to Nantucket to escape the August heat of Wilmington. Alone at Swamp Hall, Alfred devoted all of his time and most of his thoughts to the reorganization of the company and to remodeling the Brandywine mills. Even in Alicia's absence, he continued to meet Eleuthera at Louviers, when she could get away from the Judge, to receive news of Alicia and to be brought up to date on other family news, for Eleuthera was amazingly *au courant* for one who led so sheltered a life.[36] In this secluded retreat his cousin could shout out her news loudly enough so that Alfred could hear, and in this unsubtle fashion she let him know that not all was well with Alicia's marriage. It was not difficult for Alfred to infer from what little she said about Amory and by her obvious affection for and high esteem of Alfred himself that nothing would please Eleuthera more than to see Alicia and him together if only they were not both shackled to unworthy mates. All in all, Alfred found his cousin to be a most congenial conversationalist.[37]

The fall of 1904 was a lonely time for Alfred. He had little to occupy his time outside his work. His deafness had become so acute that he had to abandon music altogether and had turned the directorship of his Tankopanicum Musical Club over to his secretary and close friend, Jimmy Dashiell. For a brief moment in 1902, he had had hopes that he could recover his hearing. His actor friend, James Hackett, told Alfred of an inventor living in New York who had developed a remarkable device, called an akouphone. The inventor, Miller Reese Hutchison, after two years of medical study, had completed his professional training in electrical engineering and had earned a Ph.D. in that field. The akouphone had been so successful in restoring Queen Alexandra's hearing that King Edward VII had awarded Dr. Hutchison a gold medal. Alfred, desperate to try anything that might help, went to New York for a consultation. Hutchison designed an instrument for him, and Alfred tried it out by taking Hutchison with him to a concert. Alfred was delighted. For the first time in years he could hear every single note played, even the soft, high tones of the violin.

Alfred had once told Jimmy Dashiell that if he could only have his "ears back for three hours a day" he would spend two of those hours playing the violin. Now, he hoped that Dr. Hutchison had given him back his ears. But Hutchison was an honest man. He told Alfred that his condition was a serious degenerative condition, and that the akouphone would not help him for much longer than a year or two. By 1904, Hutchison's prognosis had proved accurate, and as far as music was concerned, for Alfred the rest was to be silence. His good friend, Thomas Edison, who suffered from the same affliction, tried to console him. "It's a boon, deafness," Edison told Alfred. "Like me, you'll miss a lot of things not worth hearing. Just think of not having to go to church again."[38] Edison's wry humor helped to ease the pain, but it could not fill the void in Alfred's life that the loss of music had created. Even as a small child, Alfred had always been able to use music as an escape from frustration and depression, and his dependency upon music had grown along with the increased pressures of adulthood. Whether conducting the Tankopanicum band at the Rising Sun Tavern or sitting alone in his study composing a march or playing his violin, Alfred had been able to push aside business weariness or domestic unhappiness and find euphoric release in the orderly, mathematical beauty of melody and harmony. Now that blessed opiate was lost, and there would never be another analgesic that could serve as an adequate substitute.

Alfred's deafness also deprived him of the pleasure of the theatre, and the silent movies of that day were hardly a satisfactory replacement. The only diversions left to him for which the loss of hearing was not a handicap were hunting and fishing. Trolling down the Chesapeake in a boat or stalking through underbrush with a few good companions still offered him the relaxation he sometimes desperately needed. Whenever possible he made his escape to his cabin on Cherry Island in the Bay, and at least once a year he managed to get back to Harding's Landing on Dividing Creek, Virginia. The Harding and Ball households offered him all of the warmth and respect that he so greatly missed at home. On the last night of his short stays with the Hardings, a big dance was always held in his honor at the community hall on Ball's Neck. Although he could not hear the music, Alfred could feel the rhythm by the pounding of the dancers' feet on the wooden floor, and he would dance every Virginia reel and two-step with the local belles. His favorite partner was Captain Ball's second daughter, Jessie, who had graduated from high school and was teaching in a local country school. After one of these visits, Alfred wrote to Miss Jessie:

> You see I can keep my promise about writing to you although I am quite sure you have forgotten all about it (the promise) and its author. . . . My little visit was most delightful, and I was very blue at having to leave. . . . Did we not have a fine time the night of the dance? I'll never forgive you for preferring the orchestra to me. I always knew I danced poorly, but never had it rubbed in like that. However, my revenge will come some day.[39]

Now in 1904, in the autumn of his loneliness, he thought ever more often of Harding's Landing. Early in November, when the hunting would be at its best, Alfred headed for Harding's Landing, taking with him his childhood friend, Frank Mathewson, along with Bill Scott, whom Alfred had recently brought into the company and placed in charge of the Wapwallopen mill, and Joe Weldy, a Wilmington fireman.

The first morning at Ball's Neck was perfect for hunting—a cold, crisp November day with a slight dusting of frost and a thin mist hanging low on the barren fields. The four men split up into two parties, with Alfred and Mathewson moving slowly along a hedge that divided the two fields and Weldy and Scott taking the other side of the barrier. The scouting party had gone only a short distance when Scott, who was something of a novice to hunting, thought he heard a game bird rising up from the brush. Without thinking of the possible location of his two companions, who could not be seen on the other side of the hedge some thirty yards away, Scott whirled and fired into the thicket. Mathewson, who may have heard the small sound of Scott's shotgun being cocked and had responded with an automatic defensive action, instantly dropped down. Alfred, who could hear nothing, of course, remained standing. Mathewson saw Alfred's hat fly into the air and Alfred himself fall flat on the ground. Mathewson shouted and dashed to his fallen comrade, who lay supine with blood pouring down his face. Mathewson was sure his old friend had been killed, but fortunately Alfred's high-crowned hat had taken the main force of the blast. Only a portion of the shot had struck him, in his left eye. He was able to stand, and with

Mathewson's help he walked back to the farmhouse. Scott, who had collapsed when he heard Mathewson's angry shout, needed as much assistance as Alfred in making it back to the Hardings'. Fannie Harding applied hot compresses to Alfred's face while one of her boys hurriedly rode to summon the nearest doctor.

The local physician could do little more than bandage Alfred's head and recommend that he be transferred to a major hospital as quickly and as carefully as possible. Carried first by improvised stretcher to Harding's Landing and then by a boat to the nearest train station, Alfred, accompanied by Mathewson, made the long, arduous trip to the University of Pennsylvania Hospital in Philadelphia, where one of the country's leading experts in the care of eye injuries, Dr. George E. de Schweinitz, took charge.

By the time that the initial examination was finished, a small group of Alfred's friends and relatives, whom he asked be informed of the accident, were gathered in the reception room: Alfred's sister, Marguerite; his two cousins who were closest to him, Alicia and Ellen La Motte; his old friend Wallis Huidekoper, who was a rancher in South Dakota but happened to be back in Philadelphia for a visit; and Father W. J. Scott, pastor of St. Joseph's Church in Wilmington. Dr. Schweinitz was able to report hopeful news to them. Alfred would recover, his life was not in danger, and with luck, providing he kept absolutely quiet and was spared any excitement or emotional stress, Alfred might even regain the sight in his injured eye. The next few weeks would be critical. Only one visitor of Mr. du Pont's choosing would be permitted at a time and then only for a vew few minutes.

Alfred requested that Huidekoper write to Bessie in Brussels to inform her of the accident and to assure her that all was well. Huidekoper promised to keep her informed about Alfred's progress, and at the patient's instructions he asked her not to return from Europe. Her presence might unduly excite and agitate him.[40]

Ten days after receiving the letter from Huidekoper, Bessie, accompanied by their eldest daughter, Madeleine, who had insisted upon coming with her mother, appeared at the reception desk at the University Hospital and asked to be taken to Alfred's room. At first Alfred refused to see them, but after Madeleine had pleaded that she at least be allowed to go to her father, Huidekoper, who had been in constant attendance to his friend, said that Mr. du Pont would see both of them the following morning. Alfred's special nurse, Rebecca Abbott, warned the two visitors when they arrived the next day that they must be as quiet and unobtrusive as possible in approaching Mr. du Pont. Nurse Abbott, however, would have been as successful in ordering a six-month-old St. Bernard puppy not to leap in loving greeting of its master as she was in cautioning Madeleine to observe proper sickroom decorum. Madie, whose natural, abounding affection for everyone had always been particularly directed toward her father, took one look at Alfred, lying flat on his back, both eyes heavily bandaged, and, with a cry of anguish, leaped to the side of the bed and gathered him into her arms. Bessie, other than quietly announcing, rather unnecessarily, that she and Madeleine had arrived, made no gesture of affection toward her stricken husband. Perhaps she was observing the nurse's strictures against exciting the patient. If so, her daughter felt she was being overly cautious, for as Madeleine later bitterly wrote of her mother, "But not one word of sympathy escaped Mrs. B. G.'s lips—not a kiss—no demonstration of pity or sympathy. From that moment on I had decided where my natural affections & sympathy belonged."[41]

It would be difficult to say which did more damage to Alfred's state of health—Madie's exuberance or Bessie's aloofness. Nurse Abbott would later testify, however, that for whatever reason, Mr. du Pont, who had been making excellent progress in his recovery, suffered a relapse after this reunion.[42]

Bessie and Madeleine stayed a few more days, but each of their subsequent daily visits to Alfred proved to be as disturbing as the first. Finally, Bessie announced, perhaps at the suggestion of the doctor, that she was returning to Brussels. She told Alfred that she had to get back to the other children before Christmas. Madeleine asked to stay with her father, but he insisted that she return and finish her studies. Bessie's visit had been as disastrous as Alfred had feared it would be.[43]

In contrast, Alfred greatly welcomed Alicia's frequent calls, and she in turn was able to keep her mother informed of Alfred's condition. Eleuthera sent him little gifts of her wine jelly and home made candy and wrote Alicia that it was "so dreadful that all this should be caused by stupid carelessness. It would have been bad enough if it had happened through his scientific inventions, but this is *worse!*"[44]

In late December Alfred persuaded his doctor to permit him to return home. He gave his usual Christmas party at Swamp Hall for the millworkers' children, and then spent a happy Christmas day with Alicia, Amory, and little Alicia. Two days after Christmas, he was back in the University Hospital, having suffered a serious relapse. Dr. Schweinitz told him that he had pushed his recovery too hard and too fast. There was now no hope of saving his left eye. It had to be removed immediately or he would lose the sight in the other eye. After the operation and another three weeks in the hospital, Alfred went south to Florida to recuperate.

In March, after a visit by Pierre du Pont, who was in Europe on a business trip, Bessie suddenly informed her children that they would not be finishing their school terms. They were all going home as quickly as possible. "Cousin Pierre and the family wish it," she told her surprised children. Madeleine thought that this was unwise. "I tried to convince Mother not to do it without absolute permission from Father. I told her how he had told me in the Hospital that all he needed would be rest & quiet & then all would go well—But no, she had the tickets—the boat was leaving in a few days. All I could do was obey."[45]

Alfred was at the dock in New York to meet his family upon their return. His greeting was not warm. Without speaking a word to his wife, Alfred told Madeleine to take all of them to the Waldorf, where he had reserved rooms for them, and he would join them shortly. After they were settled in their rooms, Alfred came first to Madeleine's room. Thirty-two years later, Madeleine wrote an account of this meeting and of the subsequent events that led to the dissolution of her parents' marriage as she, a bewildered and frightened seventeen-year-old schoolgirl, saw them:

> I told him of all that had gone on before. He told me he heard of our arrival only 24 hours before we docked. . . . That he was in no physical or mental condition to put up with social functions and that he must do all he can to get on his feet again. I naturally felt that at such a time anyone must be willing to help in every way & suggested his giving Mrs. B.G. a chance.
>
> We called Bep in & together we decided to put it up to Mrs. B.G. We told her it was up to her—that it was a case of either going back to Europe or Father's being able to lead a quiet peaceful home life, our only interests being to get Father well—Between sobs & sighs she agreed so we took her into the room where father was and on her knees, the tears rolling over her cheeks, kissing his hand she promised—
>
> And inside of two weeks Andrew Grey [a prominent Wilmington lawyer and a close friend of Bessie] was at Breck's Lane for dinner, drunk.—Shortly after on one occasion Bessie [Bep] and I were sitting on a bench in the old billiard room, Mrs. B.G. & Andrew Grey on the opposite bench. Both had had a good bit of wine. I remember how scared I was when Dad suddenly appeared & ordered us children to bed & Andrew Grey out of the house.
>
> A few days later Dad told us that he wished us in his den the following morning. We realized something was decidedly wrong. But we were terribly upset when Dad informed us before our mother that he was going to leave the house. "You are making it impossible for me to live here he said to her and I have decided to go inasmuch as you refuse to take the children back to Europe and let them finish their music. But before I leave there is something I wish you to answer me before the children—Have I ever given you any cause to be dissatisfied with me?"
>
> "No, Alfred," she answered. "You have always been good & kind to me."

"Have I ever paid any attention to any woman or done any thing of this kind which you found reason to disapprove of?"

"No, Alfred, never"—she said.

So he then told us what plans he had made financially for us all— that he felt it best for us to stay with Mrs. B.G. but that if ever any of us wanted to come to live with him later we could. . . .

Manners [sic] in Breck's Lane became most uncongenial. Mrs. B.G. was informed that he was going to divorce her—she was evidently uncertain on what grounds. She began to insinuate vile things about him, flirtations with french governesses, etc. When I reminded her of what answer she had given him in our presence, I received the answer to mind my business. . . .

This is the true story of why my father was divorced—why he left his home—and I am at every time willing to swear to the truth of this statement.[46]

Another detailed statement, recording the events that led to the dissolution of Alicia's marriage to Amory Maddox, was made long after the event by Edith Goldsborough, Alicia's cousin and apparently second only to Alicia's mother in being her closest confidante. Mrs. Goldsborough wrote this account to Alicia's daughter in 1921, after the girl had specifically asked her mother's friend to tell her everything she knew about her mother's relationship with her father and with her cousin Alfred. Young Alicia had heard disturbing gossip about her mother's divorce and even uglier stories about her own true parentage, and now, one year after her mother's death, she wanted to know the truth about her mother, her father, and herself.

With considerable pain, Cousin Edith complied:

Somehow, stupidly, it had never occurred to me, the possibility of your wanting me to *write* you about this, or I doubt if I should have had the courage to mention it to you. This will be a terribly difficult distressing letter to write, & for you to read. . . . It makes me *sick* to think of what shame to write you, and I could never do it if I didn't believe it was your *right* to know, and because Alicia wanted you to know. I can promise one thing. On my honor, I will do the best of my memory exaggerate nothing, and tell you as simply as I can only what Alicia told me. I cannot make a better start than by telling you the reasons she gave me for deciding to tell you. Besides the certain fact there were always people to tell you, under the guise of "friendship" the cruel things said about you, she felt as you got older or if anything happened to her, Mr. Maddox might make an effort to gain you for himself and "while your heart would, she knew, hold her justified, she wanted the approval of your mind." I remember her exact words. She also felt you had his blood in your veins too, and if you knew—what it stood for you could guard against it. The final decisive fact making her want you to know—was that one of the above-mentioned "friends" had finally told her it was openly said she & Mr. du Pont had agreed to marry, and that Mr. du Pont had paid Mr. M. one half million to allow the divorce to go through. Oh, Infant dear, it's hard to tell you all this—but I, too, through my known friendship to your Mother, have had that story hurled at me often!

Cousin Edith told the story as it had been given to her by Alicia fifteen years earlier. It must have been with shocked fascination that the eighteen-year-old daughter read this letter recounting events long past but which for the girl still had a painful immediacy. Alicia did not comply with Edith's request that she burn the letter as soon as she read it. She gave the letter to her stepfather, Alfred, and fortunately, it was preserved.

Because Alicia Bradford Maddox's version of the events related to her two marriages (as she told it to Edith Goldsborough) is so at variance with those that the family, her contemporaries, and subsequently most biographers of the du Ponts have generally accepted as being true, it deserves to be told in full:

He [Amory Maddox] was a handsome, dashing youngster, just the type to take a young girl's fancy. Even in those days, though, he carried on with women. . . . Your mother was beautiful and, as she acknowledged, in those days "as proud as Lucifer." Judge Bradford disapproved of the match . . . [but] Alicia, being madly in love, naturally went ahead and married him. . . . Alicia said those first two or three years were the happiest of her life: knowing he was being rapidly promoted she felt her choice was being publicly justified. She had you to add further to her pride & happiness for she said you were a "perfect" baby. In her high youth and pride she made a point of always driving in for him—driving a dog cart herself—to bring him from work—"flaunting her pride in him" as she said, when she met him then & on his return from many business trips with the other officers of the Co. Her only unhappiness was she used to think the men were cool to him.

A day or two after his return from a trip to Pittsburgh, a letter came addressed so illegibly she thought it read "Mrs." Maddox & opened it. It began "Darling Sweetheart:" & the first page showed it was from a woman who had met Mr. M. in Pittsburgh & spent the night with him there. If I remember correctly Mr. M. was again away from home. She took it instantly to a lawyer saying she did not believe it, and thought it a scheme to get money. Later the lawyer reported he had interviewed many of these men who made the trips with him, & it was true. They said he met this same woman every time he went to Pittsburgh. (She was a school teacher) and even worse, whatever place they went he openly picked up some woman each night & disappeared with her—

rejoining them the next morning. He also had proof from Pittsburgh. Alicia then told Mr. Maddox—told him utterly crushed and heartbroken—and he *laughed*—just howled, saying all men did that. Later he claimed to be bitterly ashamed and repentant, and acted so well your mother forgave him. Things went on Alicia trying to believe him, finding things out and forgiving him again—largely for you. She found she was pregnant, and, still loving him, she made herself believe he was true. One night, not expecting him till later, she was in the pantry at Louviers fixing some grapefruit for breakfast as he always said she did it better than anyone else. An exceedingly pretty young maid was in the kitchen, when the back door opened and in slid your father who, without seeing Alicia & before she could speak threw both arms around the maid & kissed her. Alicia let out a gasp & then he saw her, and right before the maid swore at her, telling her she was a "damned sneaking spy." Youngster, can you conceive your mother's feelings—finding him capable of carrying on an affair with a servant, in her own home, and swearing at her?

Soon the baby was born dead—from shock they said. When she was able to move she went North with Mrs. Bradford, to regain her health. When this scene with the maid occurred, she again told the lawyer she *must* know the truth, how he was acting, and to have him watched.

Before leaving for the North he so impressed her with his repentance and love, she agreed to put him on probation for a year, and if he stayed true to her for that time, she would forgive him & return to him as a wife—and she called off the lawyer. To quite appreciate the horror of this for her, dear, you must realize he was the one big love of her life & she *still* loved him dearly. About four months later he turned up one morning at York (I think it was there she said they were) begging and pleading to be forgiven. He said everything a man could say to move a woman. These four months had shown him life was worthless without her and her love. About 4 p.m. he had to leave for Boston to keep a business appointment and go on another trip from there. The upshot was Alicia forgave him, and being generous, forgave him so utterly, she found a few weeks later, she had conceived that afternoon and was again to be a mother. Alicia, that very same day she knew her condition she heard from her lawyer, saying he had been so suspicious he had disobeyed her and sending her a mass of papers among them a certified copy of a telegram Mr. Maddox sent from York, from the station on his way to Boston, to a professional prostitute to meet him. He also had a photo of the hotel register where he had registered with her and spent the night—the very night he left your mother. When confronted with this he confessed it true.

For appearance he was allowed to stay at Louviers until the child was born—dead. For months before Alicia was desperately ill, at death's door, & when the child was born, went temporarily blind with all the shock. When she was finally able to take up her life again her own mother (a strict churchwoman) her Drs. & Dr. John Clarke of Phila. who was there in consultation, and her lawyers, all told her she must divorce him. They read her (she was still blind) letters he had written from there, during these months, to this woman in Pittsburgh and others, bewailing being forced to stay around with "a sick, weeping woman" & promising to meet when it was over. The Drs. said she was jeopardizing her life and yours living with him as he was utterly careless whom he took as his woman—any common street walker even, he had taken. Alicia dear its terrible to tell you all this but I must for you to understand.

Then and not till then did she divorce him. I can't say how often she forgave him & gave him another chance, but his first infidelity she knew of before you were a year old. Does that sound like the divorce being a put up job and can you imagine a woman having a greater *right* to one?

Your grandmother died soon after, & Alicia took you abroad where she regained her sight. I don't know just how long she was there. She finally came home, utterly at sea what to do with her life, and with not much money. Mr. du Pont came to meet her and asked her to marry him. She had so certainly never thought of him that way, though she was devoted to him, that she burst out laughing & refused to consider it. She has often told me that was the only time in her life she saw Mr. du Pont forget his manners—he was so angry he went out and *slammed* the door in her face. Finally, however, realizing how deeply he loved her and you too, she consented, but before the marriage he settled outright on her enough money to make you both independent for life, in case she found she was not happy with him. Though he knew it would have almost instantly decided her (for she idolized Mrs. Bradford) he never told her until after they were married, that her mother, before her death, had told him she prayed Alicia would marry him. Alicia was—and is, I believe—the one big, true, love of his life, but he always knew, and never let her be hurt by his knowledge, that he was not the same to her. Alicia had given all of that love she had to give, to Maddox. She was deeply devoted to Mr. du Pont though and said, again and again, that *no one else* could have done *just* what he did for her at that time. Her pride had been literally dragged in the mud, and she said she felt she had been stripped naked and covered with filth. Mr. du Pont's one desire, from the beginning, was to sooth [sic] it for her—to put her, by his love, his position and his wealth, on such a pedastel [sic] she could be above everything that could touch her. Infant, that was one thing she so particularly wanted you to understand—his wonderful gentleness, patience & consideration to her. She said that loving her as she knew he did, the way he never tried to force from her more than she could freely give, the independence he gave her, and the constant, never-failing devotion, all this made his love for her. Then as if, after a "frightful storm, she, and you, had come into a pleasant peaceful harbor"—I remember her words . . .

Dear old kiddie, I know this has been hard reading and hurt you badly. . . . For no one else but you two dear Alicias *could* I have written this letter—it has taken me four days—but your mother was too splendid and had gone through so much so bravely, you must be in a position to judge her as she wanted you to—"with your brain" as well as your heart. You were the final inspiration left in her life. I never saw such delight and pride as she took in your *dignity* (that meant so much to her, dreading your inheritance) & your talent, and she said you made everything she had gone through worth it. Always, though, she rejoiced in the love Mr. du Pont had for you & your affection to him.

Now dear I am done. I may have mixed some of the details & sequences of things, but the facts are as Alicia told them to me. . . . If you are ever puzzled over anything do write me, for I hope you know you can count on my love & my loyalty.[47]

The two letters written in defense of Alfred and Alicia, the one by Madeleine du Pont Ruoff and the other by Edith Goldsborough, present an account of the couple's relationship different from the commonly accepted version. Neither Madeleine nor Edith Goldsborough can be considered an unbiased source, but then neither can the originator of the other version, Judge Bradford, nor can those who so eagerly disseminated his tale be regarded as disinterested observers. The historian who, in chronicling these events, seeks objectivity must depend upon an almost intuitive understanding of the people involved to evaluate the available sources. Although Edith Goldsborough's letter was written long after the events that she described and she clearly, as she herself admits, "may have mixed some of the details & sequences of things," her account has the ring of truth and can be in part verified by other sources. Particularly convincing are such details as: Alicia's evaluation of herself as being as "proud as Lucifer"; her continuing physical infatuation with her husband in spite of his many acts of unfaithfulness; and her especially revealing admission that she laughed in Alfred's face when he first proposed marriage to her. These self-deprecating revelations provide a note of verity to her account, which, taken in conjunction with the known characteristics of the personalities involved, offers—at least to this biographer—a convincing refutation of the more widely publicized and generally accepted version of the Alfred/Bessie–Alicia/Amory relationships.

The events that occurred subsequent to the dissolution of both marriages are not subject to the same divergent interpretations. They are a matter of record. On 26 September 1905, some months after Alfred had moved out of Swamp Hall and settled in a comfortable old farmhouse called Rock Farms, which he owned, he and Bessie met at a lawyer's office to sign a document that Alfred had had drawn up to establish a trust fund for Bessie and the four children.

By the terms of this instrument, a fund consisting of $600,000 worth of Du Pont bonds, with an annual return of 4 percent, was established. An income of $24,000 a year was guaranteed to Bessie for as long as she remained unmarried. She was to use this income to support herself and the four children, and provide for the children's education. If she remarried or died, the income would revert to Alfred. Should Alfred also die, the income was to be divided equally among the four children until the youngest had reached the age of twenty-five. Alfred guaranteed an annual income of $24,000, no more and no less, regardless of what the subsequent value of the bonds and the annual return from those bonds might be.

Under the terms of the agreement Alfred and Bessie would each select one trustee to administer the trust fund and arbitrate any parental disputes that might arise. Alfred chose as his trustee George Quintard Horwitz, a Philadelphia attorney, and he fully expected Bessie also to name a lawyer as her trustee. It would have pleased Alfred and further strengthened his hand if Bessie had chosen her good friend and frequent drinking companion, the Wilmington lawyer Andrew Grey. Bessie, however, was much too clever to do that. In a brilliant tactical move, she asked Pierre S. du Pont to serve as her trustee. To Alfred's great surprise, Pierre accepted the assignment.[48] By choosing Pierre, Bessie had greatly strengthened her hitherto tenuous relationship with the family, and for the first time Alfred had a clear indication of the serious rift developing between him and his cousin Pierre that would eventually affect their business relationship.

The estranged parents were to have joint custody of the children and would decide between themselves the amount of time the children would spend with each parent. In case they failed to reach an agreement, the two trustees of the trust fund would then decide "with whom or for how long a period each child shall live or dwell" during each year.

It did not take long for the terms of the financial settlement between Alfred and Bessie to become common knowledge within the family. Pierre, who knew precisely Alfred's current financial status, was particularly shocked by the niggardly support offered to Bessie and the children. Alfred in 1905 owned $1,200,000 in company bonds and common stock with a par value of $2,460,000. His income from these holdings during that year was approximately $135,000. In addition, he received a salary of $25,000 as vice-president of the Du Pont Company.[49] His assets also included real estate and numerous other small investments outside the company. Yet, Bessie's share of Alfred's annual income was to be but a little over 10 percent, out of which she had to maintain Swamp Hall and support herself and their four children. This ungenerous settlement may not have given Bessie anything approaching an equitable division of the family income, but it did at least provide her with an abundance of family sympathy. Even among some of Alfred's strongest supporters within the family, who had previously seen Bessie as a frigid, shrewish wife who had driven a devoted husband from his home, there was a growing consensus that she was the injured party. It is doubtful that Alfred then, or ever, fully realized what this financial settlement with Bessie was to cost him in family support or how effectively it served to corroborate the ugly gossip concerning him and Alicia. Only a few loyal allies within the family, the Francis I. du Ponts and Alicia's mother Eleuthera, would continue to insist that Alfred never intended the trust to be his final bequest to his children. Alfred himself defended his action to those few persons (such as Pierre) who dared to question the fairness of the settlement with the statement that the $24,000 he had allotted Bessie was more than was currently being spent to maintain his family at Swamp Hall. He had no desire to have his children spoiled by a sudden increase in their annual living allowance at a time when he was no longer at home to supervise the expenditures. On this last point, at least, all of the du Ponts could agree—Alfred was certainly not spoiling his children with an overly generous allotment from his annual income.

With the signing of the trust agreement, the marriage of Alfred and Bessie du Pont was terminated de facto, even if not yet de jure. Alfred never again spent another day in his beloved old Swamp, and all further communication between him and his estranged wife was conducted through their respective trustees. It was a dismally bleak and abrupt ending to what had been a tiresomely prolonged eighteen-year marriage. There had not been even a fiercely contested legal fight to give some passionate heat to the finale. Bessie, who was now fully enjoying the attention she received in the role of the long-suffering martyred wife, did not contest the terms of the financial settlement. Instead, with quiet resignation, she accepted what was offered her. She informed her growing audience of sympathizers that it would be futile to protest. What could she, a mere woman, do against the power of Alfred I. du Pont? Moreover, she added, poor Alfred was not responsible for his actions. He was now, she feared, quite insane. Eleuthera Bradford, when she heard this comment, was outraged. She wrote to Alicia, "I met Bessie and her daughters on the turnpike this morning, and she certainly looks *sour,* or as we used to say 'poisonous.' How cruel of her to say *he* is insane, making the calamity of his family *useful to herself!* I only hope that she did 'burn her bridges.'"[50]

Eleuthera had used the wrong metaphor. The effect of Bessie's evaluation of her husband's behavior was not such as would "burn her bridges" to the family. Rather, her declaration of her husband's insanity was a cruel stiletto thrust against Alfred, a deliberate reminder to all the du Ponts of the "bad blood" that had been his mother's tragic inheritance.

Surprisingly, Alfred waited six months after reaching a financial settlement with Bessie before taking the necessary steps to obtain a divorce. Not until late May 1906, did Alfred finally decide that the time had come for action. He abruptly left for South Dakota, ostensibly on a business trip to investigate the possibility of opening a powder plant in that state. The family must have realized, however, the true import of this trip, for Alfred was now following the same route that his cousin Willie du Pont had traveled a decade earlier. At the turn of the century "going to Sioux Falls" had the same obvious connotation that "going to Reno" would have a generation later. In 1900, South Dakota had the most liberal residency requirement of any state, requiring only a six-month stay in order to obtain a divorce in its courts.

Alfred, accompanied by Ira Williamson, who had served as his nurse during the long convalescence from his eye injury, established his official residency in Sioux Falls with the purchase of a small home. It was not until late summer, however, when Alfred went out to the Black Hills on a fishing trip with his old friend, Wallis Huidekoper, that the real objective of his South Dakota visit became public knowledge. Checking into a hotel in Hot Springs, Alfred signed the hotel register, listing his home town as Sioux Falls instead of Wilmington. The hotel clerk passed on this interesting bit of news to a reporter, and when confronted by the local press, Alfred confirmed that he was indeed establishing residency in South Dakota for the purpose of obtaining a divorce.[51]

The divorce complaint was filed by Alfred's attorney, William G. Porter, on the grounds of "barbarous and inhuman treatment." Supporting testimony was offered by Alfred's sister, Marguerite du Pont Lee, who, in her written deposition, described in vivid detail Bessie's treatment of her husband as Marguerite had witnessed it on her one visit to their home. Additional depositions were given by Ellen La Motte, Wallis Huidekoper, and Rebecca Abbott regarding the deleterious effect Bessie's visit had had upon Alfred when he was a patient in the University of Pennsylvania Hospital.[52] Bessie was also represented by counsel, Charles W. Brown, but she did not contest the suit. The proceedings were quickly over. The divorce was granted on 5 December 1906 on the grounds of "grevious mental suffering." Alfred and Bessie were given joint custody of the four children.[53]

With the divorce decree in hand, Alfred hurried back to Wilmington. After a six-month absence, he was eager to assume once again full command of the company's production facilities. It was, however, a different and much more lonely Brandywine valley to which he was returning. The two people who meant the most to him, Alicia Maddox and her mother Eleuthera du Pont Bradford, were no longer there.

Just prior to Alfred's departure for South Dakota in May 1906, Alicia, after a very difficult pregnancy, had given birth to her third child, a boy. This infant also had lived only one day, and Alicia herself had nearly died. Left temporarily blind following the birth, Alicia had been sent to a hospital in Philadelphia for treatment. While Alicia was recuperating and slowly regaining her sight, her mother had suddenly died following a heart attack in June. The stress Eleuthera had endured over her concern for her daughter had proved to be too much for her frail constitution. At least, Eleuthera had the satisfaction of knowing before her death that Alicia intended to institute divorce proceedings against Maddox. There was now a chance that Eleuthera's dearest wish—the marriage of Alfred and Alicia—might at last be realized, but she kept that hope to herself and died before it could be fulfilled.

With her mother dead and Alfred far away in Sioux Falls, there was now no one to whom Alicia could turn for support. As soon as she was able to travel, she and her little daughter, accompanied by the child's nurse, left for an extended stay in France. In Paris, Alicia rented an apartment on the Avenue de Bois de Boulogne. Alicia loved Paris; there, far away from all the pain and sorrow she had endured on the Brandywine, she hoped she might regain her health and find a new direction for her disordered life.

Sometime during the spring of 1907, perhaps at Alfred's urging, Alicia returned to the United States. She realized that she had to keep her promise to her mother and could no longer delay divorce proceedings against her husband. After her marriage to Amory had been terminated, she could decide where to go next. Perhaps she and little Alicia would return to Paris or they might live in New York. Of one thing she was certain—she no longer had a home in the Brandywine valley.

Alfred was at the dock to meet her when she arrived in New York. It was then, according to Edith Goldsborough, after Alicia had informed Alfred that she intended to divorce Amory, that he proposed marriage. The outburst of surprised laughter with which she refused his proposal may have turned Alfred away in anger, but not for long. Alicia had no one else to whom to look for help, and Alfred could not leave the woman he loved stranded to fend for herself. Even if she would not be his wife, she and her child were now his responsibility. So Alfred quickly returned to her, and plans were made for the immediate future.

Alicia wanted to get her divorce in Pennsylvania. She could not bear to go back to the Brandywine. Her mother was dead, and there was no one else there she wanted to see or to be with. So Alfred found a lovely old mansion in Carlisle, Pennsylvania, which he rented for her. No one but

Alfred and her lawyer knew where Alicia was, and the family assumed she was still in Paris.

Living in the anonymity she desperately desired and cut off from all social contacts except for an occasional visit from Alfred, Alicia had ample time to consider her future. Her only income was the $6,000 she received annually as her share of her grandfather Alexis's estate plus the small inheritance she had received from her mother. This was hardly sufficient to maintain her and her daughter in the fashion that she had always known. She neither expected nor wanted support from Amory. To whom else could she turn? There was, she realized, only one answer to that question—Alfred. If Alfred was willing to take her as his wife, knowing that she could never give him the love that she had given to Amory, then she would be proud and honored to be his wife. Alfred was quite willing. They would be married as soon as her divorce was final.

Only then did Alicia give Alfred the full details of her life with Amory. Alfred was horrified, and with a characteristic outburst of anger, rushed back to Wilmington and summarily dispatched a letter, dated 18 May 1907, to Amory:

> You are hereby discharged from the employ of the E. I. du Pont de Nemours Powder Company. This discharge is to go into effect at once.
> You will immediately turn over all records, drawings, passes, etc. belonging to the Company to Mr. George Patterson. Yours truly,
>
> Alfred I. du Pont
> Vice President.[54]

Alfred also wrote to Coleman: "I regret to inform you that I have been compelled to peremptorily discharge Mr. G. A. Maddox from the Company's employ." Similar notes went to Pierre, George Patterson, the General Superintendent, and Frank Connable, General Manager.[55]

Shortly after Maddox's discharge from the company, Alicia's lawyer found a judge in Carlisle who would accept the petition of Alicia Bradford Maddox, now a resident of Carlisle, Pennsylvania, that she had in effect established residency in Pennsylvania the previous year while a patient in the Bryn Mawr hospital and that she had had no other residence in this country except in the state of Pennsylvania since that time. The way was now cleared for Alicia to institute divorce proceedings immediately. On 8 October 1907, Judge W. F. Sadler granted Alicia Maddox her uncontested divorce from George Amory Maddox on the grounds of abandonment. She was granted sole custody of their only child.

One week later, Alfred handed Jimmy Dashiell a sealed letter addressed to the *Wilmington News*. Alfred said that he was leaving that morning for New York and might be away for several weeks but that he would call

Dashiell that evening and give him instructions concerning the delivery of the letter. At nine-thirty that evening, Dashiell received a call from Alfred telling him to deliver the letter to the addressee.[56]

The following day, 16 October 1907, the *Wilmington News* carried on its front page the following story, written by Alfred:

> Alfred I. du Pont de Nemours, vice president of E. I. du Pont de Nemours and Company, and Mrs. Alicia Maddox, his cousin, were married at the Plaza Hotel in New York City yesterday [15 October 1907]. Only their immediate families were present. Immediately after the ceremony Mr. and Mrs. du Pont left on an extended motor trip in Mr. du Pont's 70 horsepower French car. On their return Mr. and Mrs. du Pont will make their home at Rock Farms, near Wilmington, one of the several estates owned by Mr. du Pont.[57]

Once again, a du Pont marriage was announced not on the society page, but on page one of the Wilmington and Philadelphia newspapers. The entire du Pont family, including Coleman, Pierre, and the bride's father, were astounded. Everyone thought Alicia was still in Paris, still married to her no-good husband. Judge Bradford's eye lingered long over two phrases in the announcement: "their immediate families were present" and "Mr. and Mrs. du Pont will make their home at Rock Farms, near Wilmington." He found it impossible to believe that any member of Alicia's family was present at the ceremony. He would quickly prove that to be false. Worst of all was the announcement that she would be making her home at Rock Farms. The Judge was not losing a daughter after all. She would soon be back in the Brandywine valley.

Alfred's announcement was erroneous in one detail even before it was published. As soon as the ceremony was over, Alfred's best man, his brother Maurice, whispered to Alfred that he had just learned from the evening New York newspapers that there had been a terrible explosion at the Du Pont powder plant in Fontanet, Indiana, which the company had only recently acquired. Alfred and Alicia did not leave in his "70 horsepower French automobile" for an extended honeymoon in the South. The evening of their wedding day, the bridal couple boarded a train for Indiana, to render what assistance they could at the scene of the disaster.[58] It was an inauspicious but not inappropriate beginning to their marriage.

VIII

Du Pont vs. Du Pont: The Family 1907–1915

ALFRED'S AND ALICIA'S honeymoon trip was not what Alfred had planned. No leisurely motor trip down the east coast from New York to Florida but, instead, a hurried train trip to the scene of devastation in Fontanet, Indiana, followed by calls on the injured survivors in the hospital and even more difficult visits to the homes of the twenty-seven men who had perished in the explosion, where they were confronted with the weeping hostility of the victims' families. Although the sad aftermath of a plant accident was familiar experience for Alfred, he had never become inured to it. He and Alicia did what they could to provide support, by words of condolence and by promises of monetary compensation. Alfred knew only too well that an employer's sympathy could do little to ease the victim's pain, and money could never adequately compensate for limbs and lives lost, but gestures were at least better than nothing, and it would have been nothing if he and Alicia had not forgone their honeymoon. Alfred was outraged that there was no other official response from the Du Pont headquarters to the disaster.

After viewing the totality of the destruction, Alfred ordered the permanent abandonment of the Fontanet plant. The bitter honeymoon trip at last being thankfully over, he and Alicia hurriedly returned to Wilmington, where they were confronted by a different scene of devastation—the aftermath of the explosion that their marriage had produced within the family.

Alfred had fully intended to upbraid Coleman for dereliction of duty in neither going himself nor in sending a representative to the stricken community, but upon checking into the company's main office Alfred had no opportunity to reprove his cousin, for Coleman immediately took the offensive. As soon as Alfred walked into the president's office, Coleman, red in the face, shouted out, "Al, now you've done it. The family will never stand for this." Then came the unforgivable follow-up: "Don't you think you'd better sell out and get away from here?"

Alfred must have been astounded by this brutally blunt greeting. Neither he nor Alicia was naive enough to believe that they would be welcomed home by the family with showers of rice and joyful shouts of congratulations. Alfred, however, did expect something different from Cousin Coley—if nothing else, at least some initial inquiry about his trip to Fontanet. He was quite unprepared for a bombastic ultimatum demanding repentance and surrender. Alfred had always known that his cousin was completely self-centered and inordinately covetous, but this time Coleman seemed to exceed all of his previous grasps for power and wealth. Could Coleman have possibly imagined that Alfred would meekly acquiesce to this offer to buy his share of the company, and then, like Cousin Willie, take his new bride by the hand and quietly disappear into the Virginia countryside? Alfred suddenly realized that even after all of their years of close association, Coleman did not really know him at all.

Alfred may have been completely astounded by his cousin's gall, but he was not struck dumb. With equal bluntness he shouted back that before he would sell out and run he would quite happily see Coleman and the entire family in hell.[1] With that rejoinder, Alfred turned and abruptly left the room. There had been no talk of Fontanet after all. One brief, angry exchange of words had blown away twenty-five years of collegial and familial friendship and had destroyed any prospects for future harmonious business relations between the company's two top officers. After his brief encounter with Coleman Alfred now knew who was his opponent in the company. He would more slowly learn who were his enemies within the family.

One thing that can be said in Coleman's favor is that he had at least been unabashedly open in his hostility toward Alfred and patently obvious in his desire to take personal advantage of the scandal that his cousin's marriage had precipitated. He had put his cards face up on the table, and Alfred could have no doubts as to the kind of game his cousin was playing. The same could not be said of the other members of the family who were most eager to destroy Alfred and Alicia. Coleman had been quite overt when he bluntly said, "Al, now you've done it. The family will never stand for this." The family—or at least most of its members—were indeed outraged by the marriage and determined that the principals be punished. Unlike Coleman, these du Ponts were not prepared to challenge Alfred directly and vent their feelings openly. They preferred the slow poison of gossip and social ostracism, which in the long run were to prove more effective because these attacks could not be as easily ignored as Coleman's push for voluntary exile. The du Ponts punished Alfred by the crueler torture of inflicting injury upon those closest to Alfred—upon his secretary, Jimmy Dashiell, his cousin, Francis I. du Pont, his newly acquired brother-in-law, Edward Bradford, Jr., and above all, upon his wife, Alicia.

The family's attack began even before the newlyweds had returned to the Brandywine. Alicia's father and her uncle, Bishop Coleman, the two most vigilant guardians of the family's morals, were even more outraged, if that was possible, by the publicity the wedding had received in the local press than by the marriage itself. Judge Bradford was particularly irate about the statement that had appeared in the *Wilmington News* that "only their [Alfred's and Alicia's] immediate family were present." Nothing would give the Judge and the Bishop greater satisfaction than to demand that the paper print a retraction of this gross falsehood, for surely no member of the Judge's immediate family could have dared to attend what Bradford regarded as an obscene ceremony. He knew for a fact that his two daughters, Eleuthera and Joanna, were not present, for they were at home with him on that day, and they were as disapproving of their sister's behavior as was their father. That left only Alicia's brother Edward as a possible participant. Judge Bradford, unfortunately, could not state with certainty his son's whereabouts on 15 October, since Eddie had spent two weeks cruising along the Altantic coast on a friend's yacht. It was possible, although highly improbable, that Eddie was present at the wedding. Before accusing the editor of the *News* of having printed a lie, he needed Edward Jr.'s assurance that he had not been present at his sister's wedding.

Judge Bradford asked his brother-in-law to communicate with Edward, for the Bishop could give the moral authority of the church to his request for an accounting of the youth's recent activities. The Bishop was only too happy to serve as inquisitor and hurriedly sent off a telegram to his nephew. "Please telegraph me whether or not you attended Alicia's so-called wedding. [Your] father also wishes to know reply prepaid. Leighton Coleman." He followed up his wire with a more detailed letter:

> Let me first congratulate you upon your safe return. You must have had a charming trip. This morning's paper contains an announcement of Alicia's *so-called* marriage yesterday. In it a statement was made that "only the immediate members of the family were present." I know that this must be a lie as far as your father's family is concerned. He wants you *to telegraph me* at No. 107 Nth 6th St. Richmond whether or not you were present. We both believe to the contrary, but we want your own authority to contradict this fabrication.[2]

Edward's quite unexpected reply was received the following day:

Dear Uncle Leighton:

Your telegram and letter received. Since you feel so strongly about the matter, I regret to inform you that I lent countenance to that which was displeasing to you. But in this, whatever may have been my feelings, & did what I myself believed, and still believe, to be right. I may have made a mistake. I may be criticized but I pray you not to make anything I may have to bear heavier by condemning the integrity of my motives instead of my judgment.

My trip was indeed a pleasure and I enjoyed every minute of it, barring the poignant period of acute seasickness which I suffered once.[3]

For Edward this response was the most courageous act of self-assertion in his timorous life. The nausea that his incredulous uncle and father felt upon reading Edward's letter was at least equal to that which the youth had experienced while cruising on the rough Atlantic. There could now not be any retraction printed in the *Wilmington News,* regarding the guests attending the Maddox-du Pont wedding. There would also be no merciful absolution granted to Edward by the stern Bishop of Delaware. His nephew had used much too mild a term when he wrote his uncle that he might be "criticized" for having attended the wedding. Guilt by association with Alicia was not a minor social faux pas to be lightly dismissed with mild criticism. To Edward's father and uncle it was a mortal sin for which there could never be expiation or forgiveness. Because of his transgression, Edward must be forever barred from his father's house and his uncle's church.

If the effectiveness of any punishment is determined in part by how well it serves as a deterrent to others who might be disposed to commit the same act, then it must be said that the penalty of exclusion from the family circle imposed upon Edward by the Bishop and the Judge was truly efficacious. There would be few du Ponts courageous or defiant enough to invite the same wrath of these powerful elders by maintaining a friendly relationship with their two wayward relatives. Alfred and Alicia could still count upon the friendship of Alfred's brother, Maurice, and Alicia's brother, Edward, and their cousin, Willie du Pont, living in exile in Virginia, but they were family mavericks who had already strayed or been driven from the du Pont circle. Other than these social outcasts, only Cousin Francis I. du Pont and his wife and children kept the door of their home open in warm welcome to Alfred and Alicia. For their continuing loyalty and friendship, they too would eventually pay the penalty of loss of position and wealth within the company and social prominence with the family.

As the interdiction decreed by the elders against Alfred and Alicia attracted ever wider support among the younger members of the family, the stand that Cousin Pierre would take in response to Judge Bradford's, Bishop Coleman's, and Colonel Henry A. du Pont's public denouncement became critical to Alfred's future relations with the company and the family. Only Pierre was in a position powerful and secure enough to offer meaningful support within the company in offsetting President Coleman du Pont's hostility toward Alfred, and only Pierre, as the recognized head of the very extensive Lammot branch of the du Pont family, could ensure that his many brothers, sisters, nephews, and nieces would not participate in the social ostracism that had been imposed upon his two cousins.

Pierre's attitude was at first ambivalent. Probably he had been as unpleasantly shocked by the news of Alfred's and Alicia's marriage as anyone else in the family. Straight-laced in his own observance of the moral code and never having himself experienced a strong sexual attraction toward any woman, he found it difficult to understand how any man could risk everything—career, family ties, and friendship—for the love of a woman, as Alfred had done for Alicia.

Pierre was not prepared, however, to condemn Alfred outright. Unlike his cousin, Pierre hated open confrontation and always sought to avoid controversy, particularly within the family. Moreover, his affection for Alfred lay much deeper and was far stronger than was true of Coleman. Pierre could not easily forget the days when he had first joined the company. In spite of his lack of talent as a piano player, Alfred had welcomed him into the Tankopanicum Band. And there had been those long talks as they walked home together after band practice in which they had shared their current frustrations and their hopes for the future. Alfred had always set an unattainable ideal for the younger Pierre—whether as the leader in their boyhood games or as powdermaker and mechanical genius within the company. Nor had Pierre forgotten, as Coleman so conveniently had done, that it had been Alfred who had saved the company for the du Pont family, and had asked him and Coleman to be a part of the triumvirate that would own and run the company. Coleman, much to Alfred's irritation, now always referred to the Du Pont Company, as Uncle Henry had done before him, as "my company," but Pierre, even after his break with Alfred had become total and irreparable, would always continue to give Alfred the credit for having first provided the opportunity for him to reach the throne of the Du Pont empire.

Pierre, therefore, was not willing to sever abruptly the ties of friendship with Alfred, although he was unable to understand or to approve of his cousin's marital ventures. The punishment decreed by the du Pont elders did not, in Pierre's opinion, fit the crime. Social ostracism was so excessively draconian as to be properly termed "cruel and unusual punishment." Pierre, more than most of his du Pont relatives, had reason to be sensitive on this topic. His own parents had been driven from the Brandywine because of Grandmother Meta's refusal to extend social recognition to his mother. The cause for the social ostracism inflicted upon Mary Belin was, to be sure, quite dissimilar from that in the case of Alfred and Alicia. The latter two by their own actions had brought the wrath of the family upon themselves, while poor Mary Belin had been judged a social pariah simply because of her one-quarter Jewish ancestry. Nevertheless, Pierre was not willing to join in the hue and cry raised against Alfred and Alicia. He and Alfred were to remain on friendly terms, for a number of years; they would still sign their letters to each other "Your affectionate cousin"; and Pierre, a bachelor, would even dare to accept an occasional invitation to visit his two cousins at Rock Farm.

Pierre would not, however, take a public stand on Alfred's behalf. Nor would he use his position of authority within his own immediate family to urge his brothers and sisters to maintain normal social relations with Alfred and Alicia. Pierre kept his own counsel, and he attempted no rescue mission as the flood waters of opprobrium swirled about his unfortunate cousins. Instead, he drifted with a current that eventually would separate him forever from the du Pont he had once most admired. Indeed, by having agreed to be Bessie Gardner du Pont's representative in the financial trust created by Alfred, he had in effect determined the direction in which he would allow himself to be carried toward the shore of total alienation.

In all fairness to Pierre, it must be said that there was little possibility that either he or any other member of the du Pont family could have protected Alfred and Alicia or have persuaded the rest of the family to welcome them back into the fold. Only the two principals in the case could finally define and determine their future relationship with the family. In responding to the shocked disapproval that their marriage had elicited, they had two courses open to them that could improve their unhappy situation. They could accept Coleman's advice to sell out and leave the Brandywine, or they could stay and by tact, patience, and accommodating good will toward all, slowly erase the mark of infamy with which they had been branded by Judge Bradford and his close associates.

There were precedents within the family for each course of action. Cousin Willie had taken the first option. He sold out his interest in the company and, with his beloved second wife Annie, had gone to Montpelier, Virginia, where the two lived happily ever after. Cousin Willie's first wife, May du Pont, and her true love, Willard Saulsbury, had taken the second option. They stayed, and although it had been difficult for several years, they had eventually regained a place of respect within the family. Pierre's mother, Mary Belin, had tried both possibilities. She and her husband had moved to Philadelphia, but after Lammot's death she brought their children back to the Brandywine, and with a remarkable degree of tolerant geniality toward all of her in-laws, she had slowly gained acceptance—even from that formidable grande dame, her mother-in-law, Margaretta du Pont. These several instances had clearly demonstrated that social ostracism need not be a permanent condition of life. One could escape it or with patient acceptance eventually outlast it.

The first alternative was never a viable option, for either Alfred or Alicia. Both would always prefer fight to flight, and neither for one moment thought of allowing Coleman to profit at their expense. The Brandywine valley had always been home to each of them, and they were determined it would continue to be now that they were together.

The second alternative carried a price tag that neither Alfred nor Alicia was willing to pay. It required a patient forbearance that Alfred would never give his enemies, and a tactfulness and deference to others that Alicia in her pride did not possess.

Alfred, had some very positive resources that might well have ensured his success in this family contest had he been willing to accommodate himself to the rules of the game—and had he been married to a different kind of woman. He had available to him for use in the fight against Judge Bradford, Bishop Coleman, Colonel Henry A. du Pont, and President Coleman du Pont a vast reservoir of love and genuine admiration within the family and the company which he could have drawn upon had he tapped it in the proper fashion. As a young boy, Alfred had had to assume the role of the male head of his family, and he had early been the favorite of the elders in the family who really counted—Grandmother Margaretta du Pont, Great Aunt Sophie, and Great Uncle "Boss" Henry du Pont. Later, as a young man employed by the company, he had quickly won recognition, however begrudgingly, from the older du Pont directors—Cousins Eugene and Frank and Alexis Jr.—as a great powdermaker, the true successor to the first Alexis and to Lammot du Pont, and no one, not even Great Uncle Alexis, had ever had such complete fidelity from the employees as did he. To the younger du Ponts—his many cousins both within and outside the company—he was the heroic savior who had dared to challenge the tired old leaders who had wanted to sell out. To most of the family, Alfred was the most important and admired of the triumvirate who had taken control of the company in 1902. Although Coleman was the president, he would always be an outsider in the Brandywine valley, the Kentucky cousin who had moved in to grab the power and a major portion of the company's shares but who knew nothing about powdermaking and whose braggadocio most du Ponts found very unattractive. Pierre's astute financial management was admired, but his accomplishments in this area were neither as visible to the outsider nor as easily understood by the family as were Alfred's achievements. Alfred was the man who had produced tangible results by modernizing the facilities, increasing the production, and accruing the profits that they all enjoyed. The young du Ponts in particular appreciated Alfred's sense of humor, his unpretentious openness, and his courage in defending those who dared to be different. Had Alfred continued to be the same Alfred the family knew and loved, he would have had little difficulty in contending with the forces aligned aginst him. Neither judge nor bishop, army colonel nor company president could have prevailed over him.

Alfred had been changed, however, by both the intensity of his hostility to his first wife and the intensity of his infatuation with his second wife. Driven by the blind passions of love and hate, he moved like one possessed from one irrational and self-destructive act to another. Each instance of aberrant behavior seemed designed to turn away those who would naturally have supported him and further depleted the treasury of grace that had been his to draw upon. In the final analysis, Alfred must be judged to have been his own worst enemy.

There was, first, his alienation from his children. By the terms of the trust he had established for his family and by the divorce decree granted by the South Dakota court, he and Bessie were given joint custody of the four children. During the short two-year period between the time of his separation from his first wife and his marriage to Alicia, Alfred took his custodial rights very seriously. The children spent the weekends with him, and during the week he would frequently pass by Swamp Hall in the late afternoon. He would never enter the house, but if the children were out playing in the front yard he would stop and talk with them for a few minutes.

There were frequent rides out into the surrounding countryside in Alfred's newest French Renault auto, and his son would have the thrill of sitting on his father's lap and being allowed to steer the car. In the late winter of 1907, Alfred took the two oldest girls, Madie and Bep, with him to Florida, and from there he wrote to his seven-year-old son:

> My dear little Man,
> When I got back from a little two day trip up a funny river with a funny name, called Ocklawaha, I found your nice letter awaiting me here at the Hotel.
> I want to tell you that I think your hand writing is very much improved, and I only wish I could write as well myself. I am especially taken with your question marks, which are nothing more of less than wonderful. . . .
> I sent you a card with a picture on it, which shows the river we went up in a funny looking steamboat, which had two smoke stacks and burned pine wood instead of coal. It made so much smoke that it got in my hair and made my head black. . . .

The river is very crooked, and winds like a snake in between two rows of tall trees for nearly 100 miles. The water is very clear; and the bottom of the river is white, so you can see the fishes, turtles and snakes when they are swimming around. There were some very funny looking fishes with long noses, and a good many snakes, but only a few alligators. I took some photographs, and I think they will be ready to show you when I come home.

I expect to get home on Sunday, and I want you to call me up on the telephone and talk to me. . . .

Kiss yourself in the glass for me and see how it tastes. With lots of love,

Your old Dad.[4]

Soon after Alfred's divorce was final and he had returned from South Dakota, his nineteen-year-old daughter Madeleine, who was no longer on speaking terms with her mother, moved out of Swamp Hall and came to live at Rock Farms with her father. Those months when she at last had her father entirely to herself were one of the happiest periods of her life. She later recalled an evening shortly before Christmas when they were sitting alone by the fire:

Dad reading, I sewing, when he suddenly dropped his paper and remarked, "How about having some real fun this Xmas. Suppose you go up to Philadelphia and blow in a lot of money. Buy bunches of pretty spring flowers and toys. Just have a good time and then you and I will go on Xmas morning and visit the hospital." No sooner said than I was on the telephone ascertaining the number of patients—men, women and children.

Bouquets of lovely spring flowers which we bound ourselves on Xmas Eve—that was fun, sitting and chatting alone together. What a happy 'boy' he was at such times. And then the next morning—a crisp, snowy day, full of sweetness—and the happy surprised faces when Dad would stop at a bed with a kind comforting word, the flowers, and in cases of poverty, a few dollars.[5]

Soon after his marriage to Alicia the close relationship that Alfred had maintained with his children came to an abrupt end. Alfred Victor would always remember the morning that he was taken by his father to Louviers, where Alicia was engaged in overseeing the packing of her personal possessions, which were to be moved to her new home at Rock Farms. Alicia greeted the child warmly and asked, "How would you like to have me for a mother?"

The seven-year-old boy answered with surprise, "Why, Cousin Alicia, I wouldn't like it at all."[6] The child had given an honest answer to one of those incomprehensibly silly questions with which adults frequently bewilder children. The boy didn't need Cousin Alicia for a mother. He already had one, and no one he knew ever had *two* mothers. The boy had available only one father, however, and unfortunately he was soon to lose him. Alicia may not have been unduly offended by the child's direct rebuff, but apparently Alfred was, for the incident seemed to mark the beginning of Alfred's estrangement. As far as Alfred Victor could remember, he never saw Cousin Alicia again. Far more disturbing to the child was the fact that the weekends spent with his father, the automobile rides, the long walks in the woods along the Brandywine, and the jolly and loving letters also ceased. Except for Madeleine, who for a brief time continued to live at Rock Farms with her father and his new wife, Alfred's children were to be as completely divorced from their father as was their mother. Little Victorine, who was only four years old at the time when all contact with her father ceased, soon had only the dimmest recollection of a tall man dressed in knickerbockers who used to come around to take her for auto rides, a man she called "Father." For years afterward when she said her prayers at night, "Our Father who are in heaven" was for her that mysterious person dressed in tweed knickers who had suddenly vanished from her life.[7]

Two months after her father's marriage to Alicia, Madeleine married John Bancroft, whose family had been manufacturers on the Brandywine for a longer period of time than the du Ponts and were second only to them in wealth and social prestige. With the departure of Madie to her new home in the center of town, Alfred was bereft of all his children. Alicia's little daughter, who was the same age as Victorine, became the sole object of Alfred's paternal affection. He lavished upon her all the love he had once shown to his own children, and he did it in so demonstrative and public a fashion as to strengthen the suspicious gossip that he was little Alicia's biological father. Alfred frequently addressed the child, both in conversation and in letters, as "My little Pechette." One biographer of the family, eager to exploit to the fullest the old du Pont scandals, translated "Pechette" as "little sin."[8] The French word *peche* has, however, two quite distinct meanings. It is most unlikely that Alfred would have used "little sin" as a term of endearment for a much loved child. "Little peach" would seem to be the meaning Alfred intended in this form of address.

Alfred's abrupt and complete alienation from his children was given various interpretations by the family. Those who were the closest to him blamed the severing of relations entirely upon Bessie. It was she, they said, who had poisoned the children's minds and turned them against their father. All three of the younger children later insisted that this was not true. It was not that their father was turned into an ogre by their mother. It was his absence and his never being mentioned by their mother or by them in her presence that transformed him into a mythical figure, a man who had simply disappeared from their lives without leaving a trace.

Others in the family, because of their dislike of Alicia, believed that she, out of jealousy, had kept her husband away from his children. Again there is no evidence to support this charge. On the contrary, Alicia always demonstrated a warmth and sweetness with children that she seldom displayed with adults. The loss of the two children she had borne after the birth of her daughter Alicia had been very traumatic for her, and she would have welcomed the opportunity to be a mother to Alfred Victor and little Victorine. It would have been to her distinct advantage, moreover, to have won their love and respect, for it would have improved her own standing within the family.

The principal responsibility for his alienation from his children was Alfred's. His abrupt dismissal of those with whom there had been such a mutuality of love cannot be excused, and it can only be explained by the growing intensity of his hatred for Bessie. As he continued to brood about the failure of his first marriage, Bessie became an increasingly despicable figure in his memories of their life together. He remembered only her coldness and her derision, and he came to believe that there had never been anything else—no love, warmth, or understanding—in their marital relationship. But his four beautiful and loving children were in themselves evidence that his marriage had been something more endearing and fruitful than the desolate wasteland that he now preferred to believe it had been. So the ties had to be brutally cut. For the next fifteen years he would never once see nor communicate directly with any of the three younger children. Only with Madeleine, whose hostility toward her mother was as strong as his, did Alfred continue to maintain a paternal relationship.

The termination of all personal contact with his three younger children did not mean that Alfred relinquished his supervisory custodial rights, particularly in regard to his son. Although he offered no advice or seemingly took little interest in the education or social activities of his two daughters who were still at home with their mother, he continued to express concern for Alfred Victor, who, he feared, would suffer the most by being brought up in a fatherless home.

An acrimonious dispute over the boy's education arose between him and Bessie. Alfred was furious when he learned that Bessie, without consulting him, had entered their son's name for possible later application for admission at the Groton School. At his orders, George Horwitz, who represented him as trustee, wrote to Pierre, Bessie's representative, that Mrs. B. G. du Pont had "no right to have entered her son at the Groton School or at any institution whatsoever. The proper education of the son, and his proper bringing up should rest with the Father, and not with the Mother. This is the law recognized in England, and generally recognized in the United States."[9]

Pierre, in his usual placating manner, answered that "this affair does not seem a matter of moment. The entering of a boy at a school implies no commitment as to his final education, and parents so frequently enter their sons at a number of good schools, that I cannot see why a decision should be reached immediately in this case . . . No application can be made now, as it is doubtful if many schools will be available in the course of several years. . . . I suggest that the matter be allowed to rest for the present."[10]

Horwitz, in response, said that while he could agree that "this particular affair 'does not seem a matter of moment,' nevertheless if it were allowed to go without notice from Mr. Alfred I. du Pont, I have no doubt that it would result in Mrs. Bessie G. du Pont's thinking that she had the right to determine the proper provision for Mr. du Pont's son . . . The boy is now rapidly approaching the age when he will be more benefited under the command and advice of a man rather than that of a woman. Mrs. Bessie G. du Pont should not gather the impression that . . . she alone has the sole right to decide as to where they [the children] shall live or the manner in which they shall be brought up."[11]

In April 1909, one month after Alfred Victor's ninth birthday, Alfred wrote Horwitz, "I have decided that it is best for the future of my son, Alfred Victor du Pont, that he be placed at school where he will have the advantage accruing from associating with boys of his own age and the benefit resulting from the guidance and supervision of men. I deem this necessary in order that such manly qualities as he may possess will have every chance for full development, and he has now arrived at an age when this development should begin. I will ask you to inform Mrs. B. G. du Pont of my decision, so that the boy may be prepared to leave. When my arrangements have been perfected, due notice of which will be given you."[12]

The school that Alfred had selected was the Hallock School in Great Barrington, Massachusetts, which had recently been founded by Gerard Hallock. Hallock's educational philosophy, which stressed physical over academic discipline, appealed to Alfred, who was convinced that it was just what the boy needed to develop those "manly qualities" that Alfred felt his son so sadly lacked, thanks to his mother's permissive and ineffectual tutelage. Bessie, who had plans to send the boy to Groton or to his father's old school, Phillips Andover, was outraged by Alfred's choice, a new school that no one had ever heard of and whose academic program, she was convinced, must surely be sub-standard. She had angry conferences with the two trustees, but to no avail. Alfred was the father, and he held the purse-strings.

At age nine, the boy, who didn't want to leave Swamp Hall and his mother—the only two constants in his otherwise disrupted childhood—was hustled off to Great Barrington. Alfred even made arrangements with the obliging Hallock, who was expecting and got more than just the boy's tuition in the way of financial support for his school, to keep the boy in his own home during the month-long interim between the closing of school in early June and the opening of summer camp, where the child would spend the remainder of the summer.[13] Thus the boys' time at home was reduced to the absolute minimum of the Christmas and Easter holidays. Alfred Victor spent the next five years of his life under this regimen, hating every minute of it. He never saw his father and only rarely saw his mother. It was not a healthy and happy childhood that Alfred had provided for his son.

Another rancorous custodial dispute arose over the question of who was to provide support for Madeleine after she left her mother and came to live with her father. Bessie vigorously protested Alfred's demand that she provide an allowance for her daughter's clothes, entertainment, and travel from the annual income she received from the trust fund. Once again, however, Bessie was forced to yield. Rather than face a court suit, she agreed to send Madeleine $2,000 a year, which was continued even after her daughter's marriage to Bancroft.[14]

Serving as trustee for two such contentious parties, who would communicate with each other only through their individual trustee representatives, was no enviable assignment. Alfred's actions in regard to his children and his former wife had the effect of increasingly alienating him from Pierre and driving his cousin more firmly into Bessie's camp. As for George Horwitz, who felt no strong familial obligation to serve, the constant bickering between Bessie and Alfred became, after four years, too much to endure. In May 1909 Horwitz wrote Bessie du Pont, "For various reasons, I have tendered my resignation as Trustee under the Trust Agreement of September 26, 1905, to take effect on June 2nd, 1909. . . . Mr. Alfred I. du Pont has accepted my resignation, and as my successor has chosen J. Harvey Whiteman of Wilmington, whose selection, I have no doubt, will be satisfactory to both you and Mr. Pierre S. du Pont, my present co-Trustee."[15]

Whiteman, however, as soon as he had been informed of some of the difficulties that the assignment encompassed, refused to serve. It took Alfred two years to find someone who would, even for a handsome fee, undertake so unpleasant a task. In the interim, Pierre was placed in the impossible position of serving as trustee to both parties, which certainly did little to better relations between him and Alfred.

Backed with the authority of law, which was essentially patriarchal, Alfred could enforce compliance to his demands: Alfred Victor was sent off to a school of Alfred's choosing; Bessie paid for the support of a married daughter whom she despised. Alfred seemed to find a sadistic pleasure in imposing his will upon Bessie, but it was not the unfinished business of his first marriage that caused him the most anxiety during these years.

It was the still undetermined character of his second marriage that was Alfred's major concern. He could demand and get obedience from his former wife, but he could not order love and honor from his present wife. He desperately sought from Alicia not just her gratitude for his having served as her champion but also the same total commitment of affection that she had once given so freely and so foolishly to Amory Maddox. To secure her love, he was ready to lavish his wealth upon her and to wage unremitting war against any and all who would malign her. But for Alfred, omnia vincit amor would not be a truism but a cruel delusion. Ironically, he would find it much easier to achieve success in resolving his matrimonial problems when motivated by hate than when motivated by love.

In accepting his proposal of marriage, Alicia with customary bluntness had let Alfred know that she could never love him in the same way that he loved her. She had spent her passion on Amory, and there could never again be the same ardor of affection for anyone else—not even for Alfred, to whom she would forever be indebted. If he wanted her, he would have to take her as she was and as she would always be. This did not mean he would be subjected to the same uxorial frigidity that had destroyed his first marriage. There would be no derisive scorn at the dinner table, no locked door to her bedroom. She would accept the totality of his love even if she could not fully reciprocate that love, for Alicia at thirty-two was as eager as Alfred to have another child. Only then could she hope to erase the pain of having lost her last two babies.

Alfred had been so happy to have Alicia's acceptance of his proposal that at first he was not bothered by the limitations she had imposed upon their marriage. He undoubtedly believed that with marriage would also come love. There would be nothing that she ever wanted that she could not have—his wealth would ensure that. He knew he already had her gratitude and her admiration. With the supreme self-confidence that their marriage vows had given him, he truly believed that admiration and gratitude were the first steps leading inevitably to complete love. It was some time before Alfred realized that the limitations of her affection were indeed absolute, and that a sense of indebtedness was not a sound foundation upon which to build a romantic marriage. The debtor seldom passionately embraces the creditor, and obligations owed to another are more often paid with other currency than love.

At first it was enough for Alfred to have Alicia as his wife, and he set out to buy her love: French Renault and German Mercedes automobiles; three Norfolk breed mares, "the most perfectly beautiful and possessed of the most stylish action of anything I have ever seen outside of a circus," were purchased for Alicia from Messrs. W. J. Smith of London at a cost of 4,000 pounds sterling;[16] and jewelry from Tiffany and Cartier, gowns and furs from B. Altman and Bergdorf Goodman. All of these gifts were purchased at the same time that Alfred was quarreling with Bessie over the $225.32 Madeleine had spent on her trousseau, which Alfred insisted upon Bessie's paying.[17]

Alfred's most spectacular gift to Alicia, which was offered as a fitting symbol of his extravagant love, was the new home he built on the four-hundred-acre estate he had recently acquired at the junction of Rockland Road and Murphy Road, just east of the Brandywine. Alfred engaged the New York firm of Carrère and Hastings, architects for such imposing edifices as the New York Public Library, the Senate Office Building, and the Carnegie Institution in Washington, D.C., to draw up plans for the mansion. It was to be built of limestone, three stories high, and because Alicia was enamoured of eighteenth-century French culture, the architects took as their basic design the classic French symmetry of Marie Antoinette's petite Trianon. But there was to be nothing petite about this mansion, with its seventy-seven rooms set in the center of an estate that would eventually encompass 1,500 acres of land, including 400 acres of virgin forest. Alfred called their new home Nemours, after the small French town that his great-great-grandfather, Pierre Samuel, had once represented in the Estates General of 1789, and whose name he had taken when he received his patent of nobility from Louis XVI. Alfred had an historic claim to the name Nemours, but a more appropriate appellation would have been Versailles.

Against the wishes of the New York architects, Alfred chose his own contractor, James Smyth, an old friend and employee of the Du Pont Company who had built the smokeless powder plant at Carney's Point and had remodelled Swamp Hall for Alfred and Bessie. Alfred also tinkered with the carefully drawn and historically authentic designs of Messrs. Carrère and Hastings, insisting that there be a recessed middle portion of the front facade, to permit a wide front verandah enclosed by six stone Corinthian columns. The result was an unusual hybrid of French chateau and Southern plantation architecture, which the architects found stylistically disturbing and which they undoubtedly must have denigrated with that most damning of all architectural phrases "too greatly client influenced," but Alfred liked it, and curiously enough, it worked both aesthetically and functionally. That bible of the wealthy suburban class, *Town and Country,* hailed it as "one of the purist examples of French architecture to be found in this country. The grounds are spacious, and the house is spacious, suggesting plenty of everything—of light, of room, of hospitality. The lines of the [front] hall are

The du Pont de Nemours coat of arms

ALFRED I. DU PONT'S FOREBEARS

Pierre Samuel du Pont de Nemours, 1739–1817. Founder of the family in America, A. I.'s great-great-grandfather.

Samuel du Pont, 1708–1776. Father of Pierre Samuel du Pont de Nemours.

Eleuthère Irénée du Pont, 1771–1834. Second son of Pierre Samuel, founder of the Du Pont Company, A. I.'s great-grandfather.

Alfred Victor du Pont, 1798–1856. First son of Eleuthère Irénée, A. I.'s grandfather.

Margaretta Lammot du Pont, 1807–1898. Wife of Alfred Victor, A. I.'s grandmother.

Henry du Pont, 1812–1889. Second son of Eleuthère Irénée, "Boss" of the company and the family from 1856 until his death.

Eleuthère Irénée du Pont II, 1829–1877. First son of Alfred Victor, A. I.'s father.

Charlotte Shepard Henderson du Pont, 1835–1877. Wife of Eleuthère Irénée II, A. I.'s mother.

Anne du Pont (Waller), 1860–1899. First daughter of Eleuthère Irénée and Charlotte du Pont.

Marguerite du Pont (Lee), 1862–1936. Second daughter of Eleuthère Irénée and Charlotte du Pont.

Maurice du Pont, 1866–1941.
Second son of Eleuthère Irénée
and Charlotte du Pont.

Louis Cazenove du Pont, 1868–
1892. Third son of Eleuthère
Irénée and Charlotte du Pont.

Bessie Gardner du Pont, 1864–1949. Married A. I. in 1886, divorced in 1906.

Alicia Bradford (Maddox) du Pont, 1875–1920. Married A. I. in 1907, died in 1920.

ALFRED I. DU PONT'S WIVES

Jessie Ball du Pont, 1884–1970. Married A. I. in 1921.

ALFRED I. DU PONT'S
CHILDREN

Madeleine du Pont (Bancroft,
Hiebler, Ruoff), 1887–1965.
Picture taken at age 15.

Bessie Cazenove du Pont
(Huidekoper, Fay), 1889–1973.
Picture taken at age 13.

Alfred Victor du Pont, 1900–1970. Picture taken at age 11.

Victorine Elise du Pont (Dent), 1903–1965. Picture taken at age 8.

Foster daughter Alicia Maddox (Glendening), 1903–1977. Daughter of Alicia Bradford and Amory Maddox. Picture taken at age 21 with her infant son Alan.

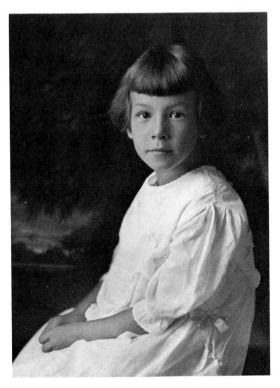

Foster daughter Denise du Pont (Zappfe), 1915– . French orphan, adopted by Alicia Bradford du Pont, raised by Alfred I. and Jessie du Pont. Picture taken at age 8.

Alfred I. du Pont

Pierre Samuel du Pont II, 1870–1954

THE TRIUMPHANT TRIUMVIRATE

These pictures taken circa 1903

Thomas Coleman du Pont,
1863–1930

Swamp Hall on Breck's Lane on the Brandywine *Left to right,* Louis, Maurice, and Alfred in mock battle on the roof.

ALFRED I. DU PONT'S HOMES

Nemours mansion, Wilmington.

Epping Forest, Jacksonville.

A meeting of the secret Holy Brotherhood Club. Alfred, standing center, and Maurice, seated left, meet with their "holy brothers" for poker and other forbidden pleasures. (Note the initials "H B" formed with the playing cards in the foreground.) Picture taken circa 1880 while A. I. was a student at Phillips Academy, Andover.

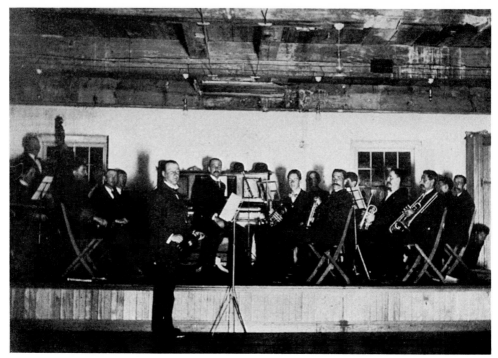

The Tankopanicum Musical Club meets at the Rising Sun Tavern for a Sunday morning rehearsal. A. I. in the foreground conducts, Pierre S. is at the piano. The other members of the orchestra are all mill workers.

"Commodore" Alfred I. du Pont seated aft on the *Nenemoosha*.

The black powder mills of the E. I. du Pont de Nemours Co. on the Brandy-wine circa 1806. A tempera painting by John W. Mccoy II, done from an original sketch by Charles Dalmas, E. I. du Pont's brother-in-law, in the Collections at the Hagley Museum.

Upper Yards, Hagley mills following the explosion of 1890.

Work begins on the Port St. Joe Paper Company shortly after Alfred I. du Pont's death in 1935.

Edward Ball, 1892–1981.

Alfred I. du Pont inspecting orange trees in Florida in the late 1920s at the time when he hoped to modernize the citrus fruit industry.

especially good and the stateliness without heaviness and restraint, without undue formality and coldness, give the keynote to the house.''[18]

Messrs. Carrère and Hastings would have disagreed that it was a "pure example of French architecture," and many du Ponts would question the suggestion of "plenty of hospitality," but no one could deny the spaciousness of either house or grounds.

Construction began in the summer of 1909, and the house was ready for occupancy in December 1910. During the intervening eighteen months, the entire du Pont family had ample time to express their curiosity, their criticism, and their ill-concealed envy of this extravaganza of Veblenian conspicuous consumption that was emerging within their enclave. To them, Nemours must have seemed absurdly out of place in this rural valley, inhabited by plain-living Quakers and thrift-minded du Ponts. Alfred, however, did not plan this structure to gratify the tastes of his neighbors. Alfred was building a palace of space and light, a home which he hoped would inspire a reciprocating love from his wife.

There was, at that time, no other mansion like Nemours in all of Delaware—or Pennsylvania for that matter. One would have to travel to Biltmore in Asheville, North Carolina, or to the Breakers in Newport, Rhode Island, to find a private residence that surpassed Nemours in space, grandeur—or expense. Only Alfred and his contractor, James Smyth, ever knew how much Nemours actually cost, for Alfred ordered Smyth to destroy all the account books as soon as the final bill had been paid. Conservative estimates placed the cost, in 1910 dollars, at $2,000,000—twice the cost of Andrew Carnegie's mansion in New York City built at approximately the same time.

A frequently asked question on the streets of Wilmington must have been, "Have you even seen the like of Nemours?" To which the correct answer would have been, "No—and I haven't seen Nemours either," for Alfred guarded the construction of his palace with as much zeal as the Du Pont Company guarded its high-explosive factories from the idle scrutiny of the public. Guards were posted at all entry points into the grounds, and the only persons other than the builders themselves who were allowed in were a few of Alfred's most trusted workers from the Brandywine mills, whose comments on the construction were welcomed by the owner.

To ensure his privacy even further, Alfred hired Tom Montgomery, one of the many stone masons whom Boss Henry had once employed in the 1870s, to build a nine-foot stone wall around the four hundred acres that comprised the main grounds of the Nemours estate. Small boys from all over the valley were paid a penny for each glass bottle they brought to Montgomery. The bottles were broken and the jagged pieces were set in concrete on top of the wall, forming a formidable crowning barrier that discouraged uninvited visitors who might consider scaling the wall to gain entry.[19] It was widely repeated throughout the community that when Alfred commissioned Montgomery to build the wall he had told the mason, "I want a wall high enough to keep out intruders, mainly those of the name of du Pont."[20]

Apochryphal stories like this alleged remark of Alfred did not in any way help to decrease the criticism of Nemours that was circulating among the family. Most of the blame for the building of Nemours was unfairly laid upon Alicia, for the du Ponts found it hard to believe that good old unpretentious Alfred, who always looked and acted as commonplace as any of his millworkers, was responsible for so pretentious a building. Surely, they said, it must have been Alicia, that stuck-up and selfish hussy, who had insisted upon flaunting this extravagance before the family and had goaded Alfred into building Nemours for her. The cousins, aunts, and other assorted in-laws made harsh comparisons between Alicia, being driven down Pennsylvania Avenue by a liveried chauffeur in her Mercedes, and Bessie, struggling to maintain a decent home for Alfred's children in old Swamp Hall. The cost of the atrociously forbidding wall around Nemours would in itself have kept Alfred's family in luxury for five years. So the talk circulated among the tea tables all over the valley. Bessie had now become "poor Bessie," an object for sympathy and solicitous concern among all the du Pont ladies who had once ignored her. And Bessie played her role to perfection, smiling her brave little smile as others clucked sympathetically and keeping a discreet and polite silence as the stories about Alicia and Alfred swirled and washed about her.

What Alfred had intended as a lovely and very personal gift for Alicia had instead become an ugly and very public scandal, coarsely bruited about the valley. Before the couple had even moved into the house, Nemours had been tarnished by malicious gossip, and it would never be for Alicia that shining token of love Alfred had wished to present to her.

If Alfred and Alicia could have emulated Bessie's conduct, if they had simply ignored the gossip, if they could have held their heads high and refused to respond to their critics, they might have been able to ride out the storm. Alfred should have opened the gates of Nemours to the du Pont family and invited one and all to inspect the marvels of construction and engineering that he and Smyth together were fashioning. He would not have dispelled the envy, but he might have turned the contempt into awe.

Neither Alfred nor Alicia could respond in that manner, however. Alicia would never play the role of "poor Bessie." Her tongue grew ever more caustic, her bitterness toward her father and sisters ever more intense. She sought neither support nor sympathy. She would stand apart and live her own life in her own way as she had done since she had first, at the age of eighteen, declared her indepdence from her father's tyranny. Let the family be damned if, indeed, they weren't already.

Alfred also behaved in characteristic fashion. Long years before, as a member of the Down-the-Creek Gang, he had been taught that the best defense is an offense. So with head lowered and fists raised, he came charging out of his corner with the ferocity that he had learned form John L. Sullivan. He would give those goddamned hell-cats something they could really gossip about. He entered a suit against Mary H. Bush, the mother-in-law of Alicia's younger sister, Joanna, whom Alicia's mother had long ago identified as the worst gossip in town. Four weeks later a second suit was filed in the Delaware Superior Court against Judge Bradford's sister, Elizabeth Bradford du Pont, the widow of Dr. Alexis I. du Pont.

Under Delaware law at that time, the filing of papers in a civil suit did not require the plaintiff to state the cause or particulars of the suit. The initial filing was simply a statement of intent to sue the named defendants, and the plaintiff was given a "reasonable time" in which to declare a bill of particulars. Alfred appeared to be in no hurry to let the defendants know the exact nature or details of the suit. Let the old girls stew awhile over why they were being sued. Alfred named as his lawyer, J. Harvey Whiteman, the man who had refused to serve as Horwitz's successor in the trust agreement with Bessie. Whiteman would have been well advised to have turned down this assignment as well.

The Mesdames Bush and du Pont lost no time in securing their counsel—the best legal talent available in Wilmington. Elizabeth Bradford du Pont engaged the services of her son-in-law, Thomas Francis Bayard, Jr., scion of the Bayard family, which for over a century had been the most distinguished political dynasty in Delaware. The legal counsel that Mary Bush secured was an even greater coup—no other than Willard Saulsbury, who once, along with his wife May du Pont Saulsbury, had been ostracized by all the du Ponts except Alfred. The Saulsburys, however, unlike Alfred and Alicia, had been willing to play the game of complaisance, and now in accepting Mrs. Bush's case, the man who had once called his in-laws "that Goddamned tribe" propelled himself into the very center of the Queen's row on the du Pont chessboard.

For months, the defendants' lawyers waited for Alfred to make his next move. The carefully muzzled Wilmington press kept a discreet silence about the whole affair. Finally, however, a disgruntled Wilmington reporter leaked the court report to a Philadelphia newspaper, and the story exploded across the nation. The Philadelphia *North American* headlined the story with the banner: "THE WOMEN'S WAR THAT CONVULSES DELAWARE." Even the conservative New York *Sun* gave it full play: "Alfred

I. du Pont, vice-president and recognized head of the $50,000,000 Du Pont Powder Company, [has accused] kinswomen of originating and circulating untrue and vulgar statements about his wife." The reporters found few du Ponts available for interview, but Alicia's brother, Edward, whose break with his father was now as complete as hers, did issue a statement. "The people who have circulated the scandalous stories about Mrs. du Pont do not even know her," he said. "She has not associated with them for years." For all of their diligent efforts at investigative reporting, however, the avid journalists could not come up with any concrete facts as to the nature of the suit. The New York *World* suggested that the slanderous stories that the defendants had allegedly circulated "are of such a nature that they cannot be published."[21] Such speculation was infininetly worse than the full truth could ever have been. The du Ponts who were still speaking to Alfred intensified their pressure on him to drop the suits.

Alfred stubbornly refused to drop the suits, but neither would he file a bill of particulars. The suits were put over from the November 1909 term to the January session of the court. Either Alfred feared he did not have a case that could stand in court or, what is more likely, he wanted Alicia's tormenters to be themselves tormented a little longer. In November, he wrote to his cousin Frank Cazenove Jones to thank him for a letter that Jones had written expressing sympathy and love for Alfred and Alicia:

> It is impossible for me to tell you of my appreciation for your kind letter. . . . Alicia has been subject to persecution from certain members of our own family for years. It began when she was but a girl and was caused by jealousy on account of her beauty and the great attention which she received as a child and a young girl. These people have been unremitting in their attempts to injure her at every turn, and especially so since she became my wife, as she was thereby placed in a much stronger position than any of them. . . .
>
> I intended two years ago to stop their malicious lies by legal action, but was dissuaded from so doing by Alicia who claimed that they were not worth attention or time that such action would involve. So long as they confined their maliciousness to this immediate vicinity, it made little difference. One's position in life is defined by the love of one's fellow beings, and it is not for any member or members of the du Pont family or convention of cats making up a self constituted society to determine one's position in life, but when they started their abuse on an international scale,[22] I determined that something would have to be done. . . .
>
> There is no arraignment for this sort of thing too severe, and anyone who is instrumental in preventing malicious hell cats operating under the guise of women from injuring other members of the human race who are forced to tolerate their existence would be working in a good cause. . . .

As you can readily understand, I do not care for what money they will have to pay if the suit comes to trial. My one aim is to forever stop their tongues, which I propose to do. I have no desire to parade many of my own relatives under the worst guise, namely that of scandal mongers, but unless I can get positive assurance that they will behave themselves, I most certainly intend to do so. When I am through, the atmosphere about here will be much cleaner, of that I am confident.[23]

The March 1910 term of Superior Court opened with Alfred's suit still on the docket but with no bill of particulars filed. On the tenth of March, Alicia gave birth to her and Alfred's first child, a son whom they named Samuel. The boy lived only twenty days, and Alicia was once again plunged into despondency over the third tragic death of a newborn child that she had had to endure.[24] Never having favored the court action in the first place, she now pleaded with Alfred to drop the suits. Hers was the only appeal that Alfred could not ignore. The day after the death of his son, Alfred asked his attorney to withdraw the suits from the court docket. In a terse one-sentence statement to the press, he offered the explanation that "Mrs. du Pont has never been fully in sympathy with the actions I was forced to bring to protect her good name."[25]

Thus the cases of *Du Pont* v. *Bush* and *Du Pont* v. *du Pont* were abruptly terminated without resolution. For Alfred the entire affair had been a fiasco. He had not "forever stopped the tongues of the malicious hell cats"—they now wagged with greater vigor. Nor did he and Alicia find "the atmosphere much cleaner" in the Brandywine valley—it was fetid with the pollution of innuendo and falsehoods that Alfred's abortive legal action had stirred up, not cleared away.

Frustrated in his efforts to attack his and Alicia's enemies in the courts, Alfred struck out in a blind fury against the one person who, he firmly believed, was the real source of all his personal torment and grief—Bessie Gardner du Pont. Alfred never went down Breck's Lane anymore, but even while building his new mansion for Alicia, he must have often thought of his old home—of what Swamp Hall had meant to him as a child, and of the bright hopes and great happiness he had known when this same contractor, James Smyth, had been engaged in remodeling the Swamp for him and Bessie. The hopes and happiness had long since turned into disillusionment and disdain, but Bessie was still firmly ensconced in his Swamp Hall, still mistress of a home that she had no legal claim to. He could not bear the image of her seated regally in his parlor, serving tea from his mother's silver tea service to the family hell-cats whom he had not been able to silence, and smiling indulgently at their latest stories about the "so-called second Mrs. Alfred I."

So, without a thought for the consequences, Alfred called James Smyth into his office one morning soon after the suits had been withdrawn and told the contractor that he was giving Mrs. B. G. du Pont one week's notice to vacate Swamp Hall. As soon as she was out, Alfred wanted the work on Nemours to be temporarily halted while Smyth and his crew razed Swamp Hall. Smyth at first could not believe Alfred was serious. The Swamp was a Brandywine treasure. It was as much a part of the history of the valley as the Hagley mills themselves. But Alfred was deadly serious. When Smyth tried to protest, Alfred cut him short. If Smyth would not do it, then there was certainly someone else who would. The contractor had no choice but to comply. Within a few days, the graceful old mansion was gone—totally obliterated. Alfred, like a latter-day Cato crying "Delenda est Carthago," had ordered that even the foundation stones be dug up and the basement filled in. The ground was levelled and grass planted, leaving not even a small mound to mark the grave of what had once been Alfred's ancestral home.[26]

There were a few du Ponts, of course, who remembered dark stories of Alfred's mother, mad Charlotte, who in her frenzy would sometimes beat her fists impotently against her bedroom walls, while cursing Swamp Hall and screaming for its destruction. Now her son, bedeviled by his own particular fury, had accomplished what she had been powerless to do.

Bessie and the three children moved into a rented house on Red Oak Road, just off Pennsylvania Avenue in Wilmington. Compared to Swamp Hall, it was a very modest dwelling indeed, and compared to the mansion Alfred was building, it was a hovel. The whole town noted how bravely Bessie accepted without protest the latest outrage inflicted upon her. Her only public comment was that she actually preferred living closer to town, for, after all, she had no fleet of automobiles at *her* disposal. Besides, she added mournfully, Swamp Hall held no happy memories for her either.[27] Still playing her role beautifully, she had now become "poor, poor, *dear* Bessie," and once again Alfred's attempt to humiliate her had only further enhanced her position within the family.

Several years later, when his break with Alfred was complete, Pierre built, at his own expense, a charming French chateau on Kennett Pike for Bessie. It was, of course, much smaller than Nemours, but architecturally it was more authentically French. Bessie called her new home Chevannes. The significance of this choice of name was not lost on Alfred or the rest of the family. Nemours was the name of the town in France that Pierre Samuel du Pont had briefly represented in the Estates General, but the du Ponts had never lived there. Chevannes, on the other hand, was the name of Pierre Samuel's actual home village and in its churchyard lay buried his first wife, Marie Le Dée, the mother of his children and the first lady of the du Pont dynasty. Alicia might live in regal isolation behind the walls at a place called Nemours, but Bessie had appropriated for herself the name of the historic spot that marked the true beginnings of the du Pont family. It would be at Bessie's Chevannes over the next several decades that Cousins Pierre, Coleman, and Colonel Henry A. du Pont would come to call and would provide Bessie with letters, records, and memories for the several histories she would write of the du Pont family and the Du Pont Company. These same cousins were never invited to and never came to Nemours. Bessie, not Alfred, now sat secure in the center of the family's affections and their history. Alfred's and Alicia's long-awaited move into Nemours in December 1910 proved to be something of an anticlimax to the high drama that Alfred had provided the family by his destruction of Swamp Hall.

Nemours was as beautiful in its spaciousness and light as Alfred had hoped it would be, and for many months Alicia took pleasure in decorating her mansion. Not restricted by any limit on expenditures, she was able to furnish the large rooms in impeccable taste with Louis *Seize* furniture. In the great dining room hung a larger than life portrait of Louis XVI, impressively garbed in his royal blue, gold fleur-de-lis embroidered state robe. Alicia might have made a happier, although certainly no more appropriate choice of French monarchs to display at Nemours than this unfortunate Bourbon king who, along with his ill-fated and hated Austrian wife, was destined to become a prisoner in his own palace.

There must have been times when Alfred and particularly Alicia felt that Nemours, like the Tuilleries for Louis and Marie Antoinette, had become their prison. Neither of them, when they returned to Wilmington after their wedding trip, had at first fully appreciated how cruel and permanent the family ostracism would be. Alicia had thought, in her pride, that it could be simply ignored. She had no affection for or desire to associate with most of her du Pont relatives anyway. Alfred, in his pride, believed that he was in a powerful enough position both within the family and within the company to destroy those who maligned him and Alicia. But as the months and years passed with no lowering of the social bars that had been raised against them and with no lessening of the malicious gossip that circulated about them, their isolation became ever more oppressive. A few people remained loyal friends: Alfred's brother Maurice and his wife; Alfred's sister Marguerite and her two sons; Alicia's cousin Edith Goldsborough; and a few old friends, like the actor James Hackett and the inventor Miller Hutchinson— but these people did not live in the Wilmington area and rarely came to Nemours. The few remaining friends who lived close by and who were always welcome visitors to Nemours—the Francis I. du Pont family and

Jimmy Dashiell—found Alicia's comments about her life in general and Wilmington society in particular so caustic as to make a visit to Nemours an uncomfortable and hence an ever more infrequent experience.

Alicia and Alfred were perforce increasingly dependent solely upon each other for companionship, and this close association was not good for either of them. Alicia's bitterness adversely affected Alfred's good judgment and fed his paranoid rage against Bessie. Unfortunately, Alfred's love for Alicia could neither alleviate her unhappiness nor compensate for her loneliness. In her frustration, Alicia frequently turned the anger she felt toward others against the one person who loved her, and the ensuing quarrels only increased their mutual misery. At times, Nemours became for them the confinement of Jean Paul Sartre's "no exit" hell.[28]

Seven months after the move into Nemours, Alicia was once again pregnant. Both she and Alfred rejoiced in having yet another chance for what they both desperately wanted—a healthy baby who would give meaning to their lives. On 4 April 1912, the child was born—a little girl whom they named Eleuthera Paulina du Pont after Alicia's mother, whose love they had once shared. The baby lived only six days. The same sad experience for Alicia had been repeated for now the fourth time.[29]

The record of Alicia's pregnancies—a first child who survived, followed by one stillbirth and three more births in which the infant survived only a few days—clearly fits the pattern of Rh incompatibility. If the mother is Rh-negative, the father Rh-positive, and the fetus is also Rh-positive (and the probability is very high that it will be), the mother's body will gradually produce antibodies, called anti-Rh agglutinins, which in succeeding pregnancies can destroy an Rh-positive fetus's red blood cells, resulting in either a stillbirth or a live infant who cannot survive more than a few hours or days. Undoubtedly both Amory Maddox and Alfred were Rh-positive, as is 85% of the population, and consequently, after the birth of her first child, Alicia had very little chance of producing viable offspring from a union with either of her two husbands.

The Rh factor in human blood group typing would not be discovered, however, for another thirty years, and for Alicia in 1912, there was no adequate medical explanation for her continual failure to give birth to a healthy child. There were those in Wilmington who where only too ready to offer a moral explanation for this persistent, tragic phenomenon. It must be the judgment of God upon the parents for their past sins. Neither Alfred nor Alicia could, of course, accept this wrath-of-God explanation, but the story was cruel enough and the cause for the deaths of their two babies mysterious enough to sink Alicia into deep despondency and to drive Alfred to the brink of madness.

Why, indeed, should Alicia, at the repeated risk of her own life, produce perfectly formed babies who for no apparent reason quickly wasted and died, and yet Bessie could so easily give birth to four babies who vigorously thrived? Neither doctor nor priest could provide a satisfactory answer. In his frustration Alfred angrily took another giant step in his march toward folly. Once again, his main target was Bessie, but this time he sought to injure her by attacking their only son, twelve-year-old Alfred Victor. It was the most irrational and inexcusable tactic he was to employ in his long war against his former wife and her allies within the du Pont family.

Early in January 1913, Alfred contacted a Wilmington lawyer, Thomas Bayard Heisler, asking him to have introduced in the current session of the Delaware state legislature a private bill that would authorize the change of the legal name of his son from Alfred Victor du Pont to Dorsey Canzenove du Pont.[30] Heisler had been Madeleine du Pont's lawyer and had handled her trust funds, but, more to the point for Alfred's purposes, Heisler was also the most powerful Democratic boss in the state, and in 1913 the Democrats were the majority party in both houses of the legislature.

An obliging party hack, Representative Charles Swan from Delaware City, agreed to introduce the bill into the House of Representatives. It was a strange bit of legislative business, but Swan was given to understand that it came at the request of both parents. Who was to question the why of what those du Ponts might want? So House Bill Number 97, "An Act to Change the Name of Alfred Victor Dupont to Dorsey Cazenove Dupont," was duly introduced by Representative Swan on 3 February 1913. The bill was unusual enough to cause some raised eyebrows in the House, but the word had been dutifully passed by Swan to all the Democrats that it was a noncontroversial matter that had the support of both of the boy's parents. The Republicans in the House looked for guidance from one of their members, none other than Edward G. Bradford, Jr., Alfred du Pont's brother-in-law. By his silence Representative Bradford appeared to be giving his consent and that was good enough for the Republicans. The bill was quickly referred to the Miscellaneous Committee of the House, where it received immediate committee approval, and was reported back to the full House within four hours. It passed the House that same afternoon with almost no opposition and was sent on to the Senate. Most of the representatives were far more concerned with another bill that Representative Swan had introduced that day dealing with the regulation of shad fishing off the Rehoboth beach area of the southern Delaware coast.

Prior to the actual introduction of the bill in the legislature, at Alfred's request, his lawyer Heisler had sent Bessie du Pont a copy of Alfred's letter of intent to file the bill. Alfred's letter was blunt and brutal:

Owing to the gross immorality of Mrs. B. G. du Pont . . . and knowing of indecent conduct on the part . . . of the children of this person, I
am determined that her son shall no longer bear my name or that of my
grand-father. I will, therefore, make application to the Legislature now
in session to change his name from Alfred Victor du Pont to Dorsey
Cazenove du Pont.

Should Mrs. B. G. du Pont have another name to suggest, kindly
inform me within forty-eight hours, and the same will be given
consideration.[31]

This attack upon Bessie was not one she was prepared to accept quietly
and bravely. She had yielded on the support of Madeleine, on her son's
schooling, and on vacating Swamp Hall, but this latest outrageous demand
was too much. Her lawyer, friend, and, as Alfred suspected, her paramour,
Andrew Gray, hurried down to Dover the next morning and very quickly
spread the word that the House had been gravely misinformed in thinking
that the bill had the support of both parents. A motion to recall the bill
from the Senate carried unanimously.

Alfred then felt obliged to make a public statement regarding the origin
of the bill. "I wish to say that this bill was introduced by me and not by
the boy's mother, Mrs. B. G. du Pont, as certain papers have erroneously
stated," he told the press. "This change was requested by me for reasons
to me sufficient and at the moment not of public concern."[32]

The Delaware legislature may have been more interested in shad fishing
than in du Pont name changes, but such was not the case with the large,
eastern metropolitan press. This curious piece of legislation raised titillating
speculations of the sort that news editors feast upon. What was Alfred I.
du Pont's objective in offering this bill? Was he denying his paternity of the
boy? And who was the Dorsey Cazenove whose name Alfred was now trying to impose upon his son? (This last question could have been easily
answered, but wasn't. Dorsey Cazenove were the given names of Bessie's
late father, Dorsey C. Gardner.) Again Alfred had raised a storm of journalistic gossip that appalled the entire family. The ill-fated bill was recon-
sidered in the Miscellaneous Committee of the House. At first declining to
give any reason for wishing a change in his son's name, Alfred finally said,
when pressured by the Committee, that he feared the boy would "bring
disgrace" upon the family name. As the New York *Times* correspondent
quite correctly reported, Alfred did not make a "happy impression" upon
the committee.[33]

Bessie's lawyer had little difficulty in refuting this vague, unsubstan-
tiated charge. Gray offered statements from the boy's teachers attesting to
his good character and transcripts of his academic record showing how well
he was doing in his studies. He also offered a letter from the twelve-year-

old Alfred Victor to his mother asking plaintively why father wanted him to change his name.[34]

At this point Alfred should have realized his folly and have allowed the bill to die in committee. But with foolish stubbornness, he insisted upon a vote in the full House. The bill was finally pried out of committee on 13 March and by a vote of 17 to 15 the House rejected the measure. Five Democrats had refused to follow the party line laid down by Heisler. The only Republican who voted for the bill was Alicia's brother, Edward Bradford, Jr.[35]

Alfred may have taken some comfort in the fact that the vote was as close as it was, but, if he did, he shouldn't have. The whole affair had been a totally unnecessary and inexcusable action upon his part that had caused irreparable damage to him and Alicia. Not even his staunchest defenders within the family and the community could understand or excuse his behavior. As one devoted cousin would say to Marquis James many years later, "It was the greatest mistake of Cousin Alfred's life."[36]

Alfred would be permanently scarred by this irrational attack upon his young son. There were few in the family who could ever forgive him. Years later Alfred built upon the grounds at Nemours a campanile that at the time was the highest structure in the state of Delaware. A red light was placed at the top of the tower as a warning to low-flying aircraft. One member of the du Pont family remembered fifty years later that, as a child looking out of her bedroom window one night, she saw the light and asked her mother, "What is that red light that I can see way over there?" Her mother replied, "Why, that is the evil eye of your Cousin Alfred."[37]

The poet William Blake believed that "Excess leads to wisdom." That sentiment is perhaps the most sanguine evaluation one can make of Alfred's behavior in the war he conducted against his own family. It was excessively wrong both to himself and to others, but out of the very excesses he committed he gained a certain wisdom. The family war was far from over in 1913. There lay ahead the battles for control of the company and for political control of the state. In these contests, shrewdness, restraint, and self-control upon Alfred's part would be conspicuous by their presence, not as previously, by their absence. Never again would Alfred pursue the irrational and self-destructive attack as he had in the first round of du Pont versus du Pont.[38]

IX

Du Pont vs. Du Pont: The Company 1906–1916

THE SECOND BATTLE in the long war that Alfred I. du Pont waged against various members of his family was for control of the Du Pont Company. Here the stakes were much higher and the historic significance of the struggle was far greater than they had been in the first battle, which had been centered on Alfred's and Alicia's social standing and their personal reputations within the family.

The second engagement, like the first, ultimately ended in defeat for Alfred. In this instance, however, fair play and justice would more clearly be Alfred's to claim. He would make some tactical errors that would prove costly, but they were errors of rational judgment, not errors of irrational passion engendered by infatuation for Alicia and hatred of Bessie. In the conduct of this campaign there would not be any foolish sorties like his attempt to change the name of his son, any purposeless acts of vandalism like the razing of Swamp Hall. Alfred would emerge from the second encounter with his honor intact, and that would be in itself no small victory when compared with the devastating results of his first defeat.

Alfred himself had precipitated the first familial battle, but the cause of the second battle could in no way be attributed to Alfred. Indeed, it took several years before Alfred, preoccupied as he was with the social bans that had been laid upon him and Alicia, even realized that powerful forces were assembling that could jeopardize his position in the company. Almost from the start, of course, there had been antagonism between him and his cousin Coleman, but in the preliminary brushes over such issues as the closing of the Brandywine mills and the moving of the company headquarters to New York, Alfred, with Pierre's support, had prevailed.

Coleman's self-aggrandizement and greed, to be sure, became ever more irksome, and Alfred frequently regretted that he had been unnecessarily generous in acceding to Coleman's demands for the presidency and for 50 percent of the stock allotted to the three cousins as his price for joining the triumvirate. Alfred didn't really care about the presidency except that Coleman regarded it as an opportunity to use the first-person possessive "my" in referring to *their* company. Titles never amounted to much in Alfred's opinion, but the number of voting shares held by an individual obviously did. Nor was it the additional wealth that his 50 percent share of the stock brought to Coleman that Alfred resented but the predominant power within the company that this percentage gave to its owner. In any showdown with Coleman over company policy, Alfred could prevail only if Pierre joined forces with him, and then only because of the small percentage of additional shares that belonged to Alfred because of his having been a partner in the company prior to the three cousins' takeover in 1902. In this unfairly weighted balance of power between Alfred and Coleman, Pierre, by joining Alfred, could provide the necessary shares to give to the latter a 52 percent control of the triumvirates' voting stock. But if Pierre voted with Coleman, Alfred would be crushed, by a margin of his cousins' combined 72.5 percentage of the stock to his 27.5. It was a situation that carried the possibility of great trouble for Alfred, but it would be some time before Pierre appreciated his potential advantage and it took even longer for Alfred to realize the peril that confronted him.

Pierre of course, could be defeated by an even greater margin if Alfred and Coleman ever combined forces in a confrontation with him, but that possibility never seemed to have occurred to Pierre. There was little likelihood that Pierre's two older cousins would ever form an alliance from which he would be excluded, and even more unlikely was the idea of Alfred's betraying the trust and friendship that had long existed between him and Pierre. It was Alfred who bore the major responsibility for maintaining a harmonious relationship, if for no other reason than to protect his own self-interest.

Initially, this proved to be not too onerous a task. Coleman and Pierre were busy building the corporate empire by the acquisition of Laflin & Rand and by achieving a virtual monopoly in the high explosives field. Alfred was content to stay out of their ventures into high finance and corporate restructuring if they left him free to manage production facilities. The division of power was satisfactory to all three cousins as long as Coleman and Pierre still had new worlds to conquer. Real trouble would come only when the empire was complete, and Pierre, in particular, turned his attention to internal structure and its operations.

In the five-year period after the three cousins had taken over the Du Pont Company in 1902, Alfred's achievements as general manager of production were truly remarkable, especially considering his personal problems which frequently interrupted his direct supervision of the many tasks that confronted him. There was first the tragic hunting accident in November 1904 followed by long months of convalescence during which time Alfred was unable to deal even indirectly with the many issues that needed his attention. He was not sufficiently recovered to be able to resume an active role in the company's affairs until the summer of 1905.

Only one year later, Alfred once again had to leave the Brandywine for the six-month required residency in South Dakota in order to obtain his divorce from Bessie. This time, however, he could and did keep in close touch with operations by means of letters, telephone calls, and occasional hurried visits to Chicago or Wilmington to meet with plant supervisors and regional managers. During this extended absence Alfred was fortunate to have as his deputy Frank Connable, former president of the Chattanooga Powder Company, who had joined the Du Pont Company in 1903 when his own firm was taken over by Du Pont. Connable's good judgment, tact, and technical expertise in powdermaking made him an entirely satisfactory stand-in during the half-year that Alfred was away, but nevertheless in the view of Alfred's associates and the powdermen in the mills there could be no adequate substitute for his own presence.

In spite of two prolonged breaks, totaling nearly fifteen months, Alfred was able to accomplish more from 1902 through 1907 than at any other period during his long career with the Du Pont Company. Not only did he completely reorganize and modernize the original mills on the Brandywine, but also, as vice-president and general supervisor in charge of production, he achieved the coordination of facilities throughout the entire and rapidly expanding Du Pont empire. Alfred was a black powder man, and he never professed mastery in any other field. In the reorganization of the company that followed the three cousins' takeover in 1902, Alfred had taken for himself direct responsibility for all black powder operations. Hamilton Barksdale, the former president of both Repauno and the Eastern Dynamite Company, headed the High Explosives (dynamite) Department, and Francis I. du Pont, Alfred's closest friend and ally within the du Pont clan, at Alfred's insistence, was placed in charge of the much smaller Smokeless Powder Department, with its two plants at Carney's Point and Pompton, New Jersey.

Technically, production in all three fields was under Alfred's supervision as vice-president of the company, but one measure of true executive talent is the ability to delegate as well as to assume managerial authority. In Barksdale and young Francis I. Alfred found the two best managers possible for their respective fields of expertise. The two men had his complete confidence. He was satisfied to let both Barksdale and Francis run their own departments without interference from him. Letters and memoranda still

extant that were sent between him and his two colleagues indicate, however, that Alfred was not as uninformed or as disinterested in dynamite and
smokeless powder production as he was reputed to be. He could communicate with both men about facilities, supplies, and production goals with
a breadth of knowledge that showed that he spoke their language and
understood their problems and needs.

Unfortunately, very few such pieces of written evidence have survived
to provide an insight into Alfred's managerial style and business philosophy.
The great bulk of Alfred's official correspondence was apparently consigned
to the trash bin in the general house-cleaning of his files that followed his
forced departure from the company. One notable and valuable exception is
the file that Frank Connable kept of his correspondence with Alfred during
the latter's six-month sojourn in South Dakota.' During this brief exile to
the West Alfred wrote frequently to Connable—sometimes as often as two
or three times a day. Although most immediately concerned with the specific problems of personnel, supplies, and labor relations that Connable had
presented to him, Alfred's responses would often place these issues within
a larger context. In so doing, he revealed his broad understanding of the
business and his overall attitudes toward capital, labor, competition, and
expansion. Somehow Connable's small collection of letters escaped destruction, and thus the six months that Alfred spent in South Dakota have
proved to be a boon to the historian. This one file of letters provides an
understanding of Alfred's role in the company at the time that he was at
the height of his power and effectiveness as a corporate executive.

During the summer and autumn of 1906 the major issue that Alfred
had to deal with long-distance, by means of letters and telegrams to Connable, was the first serious labor dispute in the one hundred year history of
the Brandywine mills. The mills had never been unionized, and indeed,
during most of that long period there had never been much discussion of,
or even interest in, the possibility of forming a union among the workers.
The first group of employees were largely French emigrés, like the du Ponts
themselves. The idea of a union would have seemed as absurd to both management and labor as would the proposal for a union of children organized
against their parents. From the time of the first E. I. du Pont through the
long reign of Boss Henry, the company had remained a paradigm of patriarchal management. Owners and workers lived in a close communal relationship, residing in company-owned houses, working side by side in the
mills, undergoing the same dangers and occasionally sharing the same
fate—being blown "across the crick." Eventually, the labor supply drawn
from the old French families died out, and new sources of immigrant labor
had to be found. Names like Pierre Gentieu and François Pavent were

replaced by names like Patrick Murphy and Michael Halloran on the company's rolls, but the Irish quickly proved to be as amenable to patriarchal rule as had the French.

The old familial order, which must have appeared eternal to the first three generations of du Ponts, did not survive into the twentieth century, however. It was doomed for a variety of reasons, the foremost being the company's growth. Bureaucracy inevitably replaced community, organizational flow charts supplanted family ties, and as both labor and management became capitalized proper nouns, they drew apart into separate entities. No longer would the president of the company make frequent and uninvited calls on his workers, either at the mills or in their homes as Boss Henry had so often done. Between the chief executive and the millworkers lay an intervening screen of managers, superintendents, clerks, and foremen. It is doubtful if, during the fifteen years of his presidency, T. Coleman du Pont ever met or spoke to even one of the company's multitude of day laborers. Only a Du Pont executive who had grown up within the old system really knew and appreciated the men who operated the mills at their most basic production level. By 1906, Vice-President Alfred I. du Pont was the only du Pont active in the company who had begun his career under the patriarchal rule of Henry du Pont. Father Henry had been replaced by the Big Company, and the workers were no longer individual members of the Du Pont Company family, but simply a resource in production, like Chilean nitrate or Louisiana sulphur. For the old-time employees and Alfred it had not been easy to make an adjustment to the new order, but for the new crop of workers, largely Italian immigrants, there were no familial ties to cast off. For the latter workers, bosses always had been and always would be the exploiters, workers would always be the exploited, and the only issues that mattered to either side were wages and hours. In this adversarial relationship, unionization became as natural and acceptable a weapon for labor as cost-accounting and cartel agreements were to management.

The rapid expansion of the Du Pont Company after 1902 contributed in another way to the growth of union settlement among the workers at the Brandywine mills. Several of the new companies that Du Pont acquired had already been unionized, and their union contracts could not be easily abrogated without considerable and costly labor disruption. The difficulty of trying to consolidate existing union plants within an essentially non-union company like Du Pont had provided Hamilton Barksdale, Amory Haskell, and Alfred du Pont with their strongest argument for allowing the newly acquired companies to retain their old corporate identity and for maintaining the pretense of their being autonomous units. Albert Moxham and Pierre du Pont, however, pointed to recent consolidations in the steel industry in which union and non-union plants existed side by side within the same corporation. They argued that eventually the minority group of unionized units would have to yield to the overall non-union policy of the parent company, just as was currently happening within United States Steel. If, however, the unionized companies were allowed to keep their separate identities, the existing unions would never be dislodged. This argument of Pierre and Moxham finally prevailed, but, just as Alfred had feared, for several years after consolidation, the push for unionization was strong in the non-union mills. It would not be easy to explain why Du Pont allowed a union in some of its newly acquired mills but denied the same rights to the workers in other mills.

Precisely that question was being asked in the Brandywine mills at the moment that Alfred was settling in to his temporary Sioux Falls home. By telegram and letter, Connable informed his absent chief that the workers at the home plant were threatening to strike unless the company met their two demands: the recognition of a union and the establishment of an eight-hour day.[2] Alfred's response to Connable was that he should close the mills and not attempt to bring in substitute labor. Business was slack anyway in the summer of 1906, and the company had an adequate supply of black powder to meet current demands. The company could well afford a lock-out, but it could not afford a strike called by an incipient union. Above all, the company could not afford to run the risk of operating so dangerous a business as powdermaking with inexperienced scab labor. If the scabs' appearance at the hiring gate didn't provoke an explosion on the part of the striking employees, then surely these new men, once they were on the job, would accidentally produce a far more devastating explosion within the graining and press mills, an explosion that could make the massacre at the Homestead steel mill of 1892 look like a Sunday School outing by comparison.

Alfred did not believe that these labor demands could have originated among his old and trusted Irish employees. He blamed the whole situation upon the new employees—"the undesirable dago element," as he crudely labeled recently hired Italian employees.[3] Italians had been substituted for Irish workers because they could be had for a few pennies less a day than the Irish—$1.50 as a daily wage instead of the standard $1.80. In Alfred's judgment this was one more instance of Coleman's misguided policies.

The lock-out could be a blessing in disguise, however, if Connable would take immediate advantage of it by serving notice on those identified as the leaders and troublemakers to vacate their company-owned homes within three weeks or be forcibly evicted.[4] Connable was happy to carry out this order. "Before starting the mills," he wrote Alfred, "we intend purging the Brandywine of the undesirable element. This is primarily the greatest work we have to accomplish, and while the opportunity offers we must avail ourselves of it. This is a chance you have waited for many years; we won't let it slip by."[5]

Alfred wanted "the undesirable element" not ony out of their mills and their homes but also out of the Wilmington area as well. He wrote Connable:

> It has occurred to me that it might be advisable to take up with Messrs. Bancroft the question as to whether they would not be willing to issue instructions to the effect that none of our discharged men were to be employed by their foremen. We have had no end of trouble in years gone by, due to their taking our men that we have had to lay off owing to their habits or general unworthiness. They are so near to us when working at Bancroft's, as to be continually a thorn in our sides. . . . Tomorrow is the last day [before eviction], and their [sic] should be some empty houses on the Brandywine soon. I would not be in any hurry to fill them, as we will need them for new men which we take on.[6]

Alfred intended to make sure that in future the empty houses would be filled with experienced Irish workers, not Italian troublemakers. Thus, in addition to the lock-out and house-eviction orders, Alfred sought to impose the penalty of the blacklist. He intended to make it clear that the union agitation would be crushed ruthlessly and permanently.

In response to the workers' demand for an eight-hour day, Alfred was as adamant in his opposition as he was to union recognition. He expressed his views forcefully to Frank Connable on 4 August:

> Keep a perfectly stiff upper lip and I feel sure everything will come around as we wish. We do not want any eight hour law at all. The understanding on the Brandy Wine [sic] is that the men would work until their daily task is accomplished, whatever that happens to be. If you make a concession of eight hours, it would only be an opening wedge for others. . . . They are making this demand in order that they may limit their mills in output; in other words, if they have a day of so many hours they will loaf and turn out very little material, if they have a day of so much work they will try to get it out as quickly as possible.[7]

Although the workers had not themselves raised the question of wages, higher pay was the carrot that Alfred was willing to hold out to those workers he was eager to keep, at the same time that he was wielding the stick against those whom he wished to eliminate. In his letters to Connable he repeatedly pushed for a revision of the existing wage scales. "I, myself, think that the men had some grounds for complaint and that they were underpaid," he wrote to Connable regarding a labor disturbance at the Chattanooga Powder Company. "It is always a wise policy and one I have always recognized as most important, namely, never underpay any person whom you wish to stay with you, but most decidedly vice versa. In other words there is no quicker way of getting rid of a person than to underpay him."[8]

In meeting the stand-off between labor and management that resulted from the lock-out in the Brandywine mills, Alfred asked for and got the Brandywine wage list from Connable. After giving it careful study, he wrote

> It seems to me the ordinary rate of labor should be advanced, that is to say, [to] the wages paid the ordinary men in the powder—which now seems to be about at the rate of $1.80 a day. It strikes me there is too much difference between the so-called first and second men in various operations, and, as a rule, they all have to do about the same amount of work. . . . I see in some instances where there is as much as 40¢ a day . . . difference between 'first' and 'second' men. It is the rank and file that are more apt to breed discord, and I should think there would be more apt to be general satisfaction if the rate of ordinary labor was raised a trifle.[9]

As in other issues, Alfred's chief lieutenant on the scene was in complete agreement. "So long as we can maintain an average wage scale at non-union mills, which equals or very nearly so the union pay, the incentive for organizing becomes less and less, which is one way of defeating it. In the new scale we shall keep in mind the great disparity which might exist between men in the same mill; for instance, first and second press."[10]

Alfred urged pay increases for both daily wage and monthly salaried employees. He was convinced that there was soon to be "a labor famine" in the United States, and he sent Connable an article he had found in the *New York Herald* that supported this view.

> We must weigh general conditions most carefully, so as to be prepared to take such steps as will prevent our men leaving us. There is no business that I know of in which unskilled labor is so important as ours, and where the question of proficiency is measured, to a very large degree, by the length of service in our employ. It is not only a question of actual dollars and cents paid which we have to weigh but the increase in cost of production and increase in the factor of danger due to constantly changing our men. To put it more concisely: I think a man running a corn mill, and who has done so for five or ten years, is actually worth, in dollars and cents, twice as much as a new man, for we could readily afford to pay the old hand a good deal higher and still in the end, come out to the good. . . . If the conditions, as stated in the New York 'Herald,' are maintained for any length of time, we can't hope to pay the same wages we are now, without losing a great many of our best men.[11]

Thus Alfred anticipated by some twenty years Henry Ford's policy of paying wages that kept trained men on the job and out of the union.

Alfred's new wage and salary scales did not escape the notice of President T. C. du Pont. Coleman wrote to his vice-president that he was surprised to note the rapid increase in labor costs within the black powder department. Alfred, in irritation, wrote back, "Of course, I knew that the salary list during the last year had been increased considerably. . . . Roughly speaking it seems that the increase amounts to about $2,500 per month. . . . I have in mind the necessity of keeping our salary roll down as low as possible, but I think it poor economy to shut our eyes to the future necessities of our organization. . . . The question in my mind is whether the present salary list is high enough. I wish I could think so."[12] The list of differences over company policy between the president and the vice-president continued to mount, along with their mutual irritation.

The union movement abruptly collapsed in late October, and the Brandywine mills promptly reopened. The outcome was a total victory for Alfred. There would be no union in the Brandywine mills and no change in hourly schedules, but there were the changes that Alfred wanted: the "undesirable elements" had been sent packing and those who returned to their old jobs, as well as the new men who were added to the labor force, were to receive higher compensation for their labor.[13]

Alfred's great regret was that he was away during this difficult period. "I regret the necessity of my continued absence from Wilmington, especially at this time [of reopening the mills], as the personal influence which I have over the Brandywine men would go a long way towards insuring satisfactory operation," he wrote Connable soon after the lock-out ended.[14] Indeed, Alfred believed that there would have been no labor disturbance at all if he had been on the scene. Both the company and Alfred, however, had been fortunate in Frank Connable's ability to put into effect the tactics and policies that Alfred had dictated from afar.

Although the labor situation was the major concern of Alfred's department during the time that he was away, a multitude of other issues also had to be dealt with through correspondence and by reliance upon Connable to carry out Alfred's policies. Expanding the production facilities, for example, was a major consideration, for expansion continued to be the identifying signature of company policy in this decade of its history. On this one theme song, at least, the three cousins believed they could join in harmony, but as each sang a quite different part their voices occasionally produced dissonance. To Pierre, expansion meant taking over competing concerns like Laflin & Rand and Eastern Dynamite, who were major producers and whose successful operations would contribute greatly to Du Pont's future monopolistic position. Coleman had no quarrel with this line of expansion, but in addition he wanted to buy out newly organized companies that might at a later date develop into real competitors. To this policy, Alfred was unalterably opposed. "I agree with you as to the inadvisability of purchasing competitors' plants," he wrote to Connable, "but I have so often stated my views on this subject, that reiteration seems superfluous."[5] The risk of being charged with redundancy did not, however, prevent Alfred from reiterating this view with great frequency to Connable:

> I am distinctly against the absorption of competitive plants for the purpose of operation. It may be advisable to buy out competitors sometimes at a very low price, but on this point I am also uncertain. But the practice of purchasing a competitive plant in order to augment our productive capacity should not be considered for one moment. The practice is wrong; the precedence which it establishes is most harmful and the plant acquisition of questionable value. You are right in stating our plants should be up-to-date and furnished only with such managers and employees as will permit our manufacturing at the lowest possible cost and with the greatest possible facture [sic] of safety.[16]

Unlike Coleman and Haskell, Alfred truly believed in the free-market philosophy that all American entrepreneurs loudly professed but few wished to maintain within their own industry. When a question arose as to the advisability of the company's selling a small, inefficient plant at Dorner, Alfred urged Connable to sell it:

> If we are to have more competition, now is the time to have it, and even if I knew it was to be bought for that purpose, I should even then advocate selling it, and perhaps at a lower figure to a proposed competitor than to anybody else. I am a full believer in the theory that you must have a certain amount of competition. There is no quicker way to determine this natural amount of competition than by the process of elimination, based on the survival of the fittest. Competition cannot be curtailed either, by refusing to sell land to a competitor any more than by buying up competitors' plants already in existance [sic]. . . . Unless there is a possibility of the land in question appreciating, I would most certainly advise this sale.[17]

Alfred was as committed to expansion as his two cousins, but he wanted to expand by building new facilities, not by purchasing the facilities of competitors. Pierre's program of buying up strong competitors and using their existing facilities, which was essentially Boss Henry's old policy, violated Alfred's concept of the free market and the desirability of maintaining competition. Even worse, however, was Coleman's variation of Pierre's program, for Coleman wanted to purchase not only well-established companies with good production facilities but also every small plant that suddenly appeared on the scene, out of fear that it might some day develop into a formidable opponent. Alfred was convinced that most of the new fly-by-night ventures had been launched not to produce powder but to produce quick profits for their owners by selling out to Coleman! They were "spite companies," and Alfred, having served under Cousin Eugene, knew from personal experience how expensive and ultimately disastrous purchasing them could be.

The only ancestral model for expansion that Alfred wished to follow was not that of either Boss Henry or Cousin Eugene, but that of his great-grandfather, E. I. du Pont de Nemours, the founder of the company. When Irénée saw the need for expanded facilities during the War of 1812, he didn't look for competitors' facilities to purchase in nearby Pennsylvania. He built the Hagley Mills. In that way, he got what *he* wanted, not what someone else had wanted and had built. Like E.I., Alfred knew that expansion was your own business; what your competitors did was not your business. In the long run, it was always more expensive to try to eliminate competition through purchase than it was to live with competition and outstrip it by superior performance and facilities.

Through his months of exile, therefore, Alfred constantly urged his lieutenant to push vigorously ahead with the construction of new black powder, smokeless powder, and dynamite plants in the Springfield district of Illinois, in the Indian Territory of Oklahoma, and farther west in Colorado and Washington.[18] He was greatly disturbed to read Connable's report to the Executive Committee, in which Connable first recommended that a new mill be built in the Springfield area but then added in a supplementary report that "our construction department is already overtaxed [and] . . . no new mill [should] be built for the present." This report should not go to the Executive Committee with two distinct recommendations," Alfred scolded Connable. "I agree absolutely with the original study [on the advisability of building the plant]. . . . If our construction department is not of sufficient caliber to construct mills as rapidly as we need them, then steps should be taken to augment it at once, as procrastination is not going to help us and a year from to-day will find us in a much worse plight unless radical steps for improvement are taken at once."[19] Build, not buy, was what expansion meant to Alfred, and this simple definition would only further separate him from his two cousins.

New construction was as paramount in Alfred's thoughts at this time as was cartelization in Coleman's and Pierre's, but the many mundane details of management—supplies, personnel, and production quotas—in all of the Du Pont plants also demanded his attention. He fussed about the shortage of kegs at many of the plants and the inexcusable negligence of allowing the supplies of plumbago to run short at the Rosendale mills, and he demanded to know "why our Columbus mills ran out of empty packages, causing a shut down of the plant of many days."[20]

Connable replied that these shortages were not due to the negligence of the plant superintendents but to the failure of the Purchasing Department to anticipate needs. He reminded Alfred that under Pierre's new organizational structure only the Purchasing Department could requisition material for distribution to the individual mills. Alfred angrily protested this new procedure:

> Some one person should be directly responsible for the continuous operation of each and every plant. That person should be the Division Superintendent.... In order to keep a plant in continuous operation, some essential materials are absolutely necessary.... If it should happen that any of the plants would be forced to close for lack of essential material, under our present regulations, the division superintendent naturally says, "I am sorry. I am not responsible for this shut down, as I have nothing to do with ordering those materials." ... I, therefore, for this reason think it was a bad move to withdraw from the division superintendent's department the making of requisitions ... and I feel quite sure that the future will prove these conclusions.... A certain amount of red tape is all very good and necessary to strict discipline. Too much red tape is bad, as it prevents free breathing and is apt to produce negative results.[21]

Although Alfred, as vice-president, was himself part of the bureaucracy, he would never allow himself to be so removed from the daily line operations as to look upon his district superintendents and plant managers as small squares on a corporate organizational chart. He not only knew the men in the lower echelons by name, but also knew them as distinct personalities, and he was remarkably perceptive about how they interacted with their counterparts on the chart. "I have given the question of the Anthracite re-organization some further thought," he wrote to Connable. "I am not altogether certain that it is a wise move to place Mr. Victor du Pont in the position of Division Superintendent.... As to the proper person to place in charge of that Division, I am not altogether clear. Mr. John S. Scott would be an excellent man were it not for the fact that I do not think that he and Mr. Belin would get along together. Besides that his own father would be under him at Wapwallopen. I think, therefore, it might be best to put Mr. Ferriday in charge. He has not had much experience with this particular kind of powder, but he would soon learn all that is necessary."[22] This letter is only one example of how thoroughly conversant Alfred was with personnel matters in the plants. He was knowledgeable about and could empathize with his subordinates' sensitivity regarding the prerogatives of their position. He found it necessary to reprimand Connable for having asked Lammot du Pont, Pierre's younger brother, to assist in the development of a new powder for the Reading Company without first consulting with George Patterson, the manager of the Black Powder Department, under whose auspices such research should have been conducted:

> I feel sure that Patterson is right on this subject and that you have invartantly [sic] trod on his toe. You must be most careful not to induce friction by what may be construed as evidence of want of consideration.... I have always been most careful in my dealings with all with whom I am associated in business, not to give them any cause to feel that their prerogatives have been overlooked.... While the ideal business man is nothing but a machine, we are unfortunately endowed to a greater or less extent with certain qualities which may be classified as senses and to which our mental attributes are strangely subservient, and it is, therefore, necessary in dealing with another man to recognize this axiom.... A man in as high a position as you are, must be loved personally by those who are under him.... and in order to win the affection of other men, constant consideration must be the watch-word. My success in gaining the love of my fellow beings has been largely due to the strict recognition of this principle and, while I know that I have overdone it, I feel that the error is better on the side of profligacy than on the other and that, if recognized, the error can be corrected.
>
> You will pardon this long dissertation on moral philosophy but, as I am away from Wilmington, I am anxious that my absence should not result in the slightest discord among my assistants.[23]

This degree of sensitivity to subordinates' feelings is a rare quality in executive officers. Certainly it was not a trait that either Coleman or Pierre ever demonstrated to such a degree as to give evidence of their having erred "on the side of profligacy," as Alfred was shortly to discover.

It was not easy for Alfred to conduct by remote control his duties as general supervisor of all production. To add to his difficulties, Alfred was plagued by a serious ear infection that necessitated a hurried trip to Chicago in early August to have abscesses in both ears lanced. The months of exile must have seemed excessively painful and tedious to Alfred. There were moments of relaxation, however. There were the occasional duck-hunting and fishing expeditions with his old friend Wallis Huidekoper. Three Du Pont sales agents located in the western mountain region took Alfred on the obligatory tour of the Black Hills. They visited America's largest gold mine in Lead, South Dakota, and Alfred spent several days in the old Wild West mining community of Deadwood. He reported to Connable:

> I am rather favorably impressed with this town. In the afternoon, I drove ... up on top of a mountain, where I had an interview with

Calamity Jane and Wild Bill [Hickok], both of which [sic] are safely planted in the bone orchard situated on the hill in question. There are other noted characters buried there, all of whom were intimately connected with the early days of this town, when everything ran wide open, and there was no law of either God or man. . . . Wild Bill also has his likeness [carved] in sandstone, but Calamity Jane, for some reason or other seems to have been neglected. This I considered neither proper nor fair, and I am going to try to find out before I go how much it would cost to have her properly recognized, in a white silk gown with an empire waist, all cut out of native rock. I understand that she was somewhat looked down upon, because she had the reputation of being more generously disposed toward her fellow beings than most of her sex. . . . The fact remains the same, that she had sufficient pluck and grit to help hold up the old Deadwood coach several times and make the passengers ante up, which shows a great deal more character than the average bunch of milktoast that we run across nowadays.[24]

Alfred found the Black Hills so attractive that he stayed in Deadwood for a couple of weeks and returned later in the fall for the hunting season. But if Huidekoper had any hope that his friend might be converted into a true South Dakotan and settle down on a ranch, as he had done, Alfred was to disappoint him. As soon as his divorce became final, Alfred hurried back to Delaware. There might not be any mountains towering over the banks of the Brandywine, but one could find in Wilmington plenty of action. Even Calamity Jane and Wild Bill Hickok might have taken a few lessons in the quick draw from such a wheeling and dealing artist as Cousin Coleman. Alfred would not lack for excitement in the so-called effete East.

The six months immediately after Alfred's return from South Dakota were the high point in his long years of service with the Du Pont Company. He was in control of that aspect of the business that he enjoyed most and that to him was obviously the most important—the actual production of gunpowder and high explosives. He had demonstrated, even in absentia, that no one else in the company had as complete an understanding and mastery over every aspect of operational production as he, whether it be in settling labor disputes, selecting key supervisory personnel, anticipating materiel needs, or in expanding plant facilities. There were fundamental points of difference between him and his two cousins, particularly between him and Coleman, over the proper management of the company, but so far, when these issues had emerged into open dispute, Alfred usually prevailed in matters concerned with production.

The allocation of managerial functions that had emerged since the cousins took over the company in 1902—giving Coleman control of sales and public relations, Pierre the internal corporate structure and the financial affairs, and Alfred line operations—was a reasonable and acceptable distribution of power. As long as Coleman did not try to exert his power to close the Brandywine mills and to stop Alfred from building new black powder plants, or Pierre did not bind operations too tightly with restrictive bureaucratic red tape, Alfred was content to work in harmony with them. As far as he was concerned, the troika that had been fashioned in 1902 was a serviceable vehicle, but it required each member of the team to stay within his own traces and pull his own share of the weight.

In his own personal life as well, the early months of 1907 brought Alfred a sense of relief and hope that he had not experienced in many years. His unhappy marriage had been legally terminated, and after Alicia's return from her extended stay in Paris in order to obtain a divorce from Amory, Alfred had reason to hope that he might finally find the marital happiness he had so long been denied. Alfred could not have been oblivious to the fact that his divorce had not been well received by the family and that it had given substance to the ugly rumors that Judge Bradford had so busily circulated concerning Alfred's relationship with Alicia, but all of this gossip mongering, Alfred believed, could be simply ignored. What did it matter if there was talk among the busybodies within the family? Alfred was in love and as never before, he felt himself to be in command of his own life.

This euphoria proved to be of short duration. His marriage to Alicia in October 1907 shattered all hope of a harmonious relationship with most of the du Pont family, and the social ostracism that ensued proved to be a far more oppressive and lasting punishment than either he or Alicia had anticipated. It produced an environment that was certain to be inimical to a happy marriage.

Alfred had naively believed, however, that somehow he could keep his personal life separate from his business career, and he had been shocked by the Coleman's reception upon his return from his wedding trip. Hence forward he knew that Coleman was his enemy and that he wanted him out of the company. He must also have realized that his treatment of Bessie and his children was adversely affecting his relationship with Pierre, upon whose counsel and support his former wife increasingly depended. The loss of the camaraderie he had once enjoyed with his two cousins, particularly with Pierre, was certainly regrettable, but Alfred continued to believe that it in no way should affect their formal business association. Letters might be signed "Sincerely yours," instead of "Your affectionate cousin," but the letter writing would continue, simply because the business demanded it— and more important, the business demanded Alfred's services. The Du Pont Powder Company needed Alfred far more than Alfred needed Coleman's and Pierre's affection and socializing. Coleman might know the art of selling and Pierre might know the arcane intricacies of high finance, but Alfred knew how to make gunpowder, and as long as that was their business, the business was essentially Alfred's to manage and produce—or so he thought. Alfred fully expected the troika to continue to pull the company; a two-horse coach or a one-horse shay just wouldn't work.

What Alfred never realized was that neither Coleman nor Pierre could separate, as did he, Alfred's contributory role within the company from his disruptive role within the family. He never understood that his lack of judgment in threatening to bring slander suits against his own kin and in attempting through a legislative act to change the name of his own son would inevitably give rise to doubts about his judgment in ordering a new black powder plant in Springfield or in dismissing Victor du Pont as district superintendent. As had been true for his father before him, there was for Alfred the rational and controllable world of the mills and there was the irrational and unpredictable world at home with no bridge connecting the two. One kept one's sanity and effectiveness as a man only by keeping these worlds separate. His business associates, who were fortunate enough never to have had to adapt to such a schizoid existence were to judge Alfred, however, by the totality of his actions. For Coleman and Pierre, at least, the failing grades Alfred received from the family as a husband and father far outweighed the A+ given him by his subordinates for his performance as general manager of the mills.

In the coming struggle to maintain his position within the company, Alfred was handicapped by more than the disapprobation of his family. There were also his physical disabilities. The loss of one eye made the reading of lengthy reports a difficult process. A far more serious handicap was his deafness. Subsequent to the ear infection he had suffered in South Dakota, he lost almost the last vestiges of his hearing. Not even Miller Hutchinson's improved acousticon could be of much help to Alfred now, and it was sheer torture for him to sit through long committee meetings without knowing what was being said unless the speaker sat close to him and remembered to shout directly into the machine.

At least in respect to his physical afflictions, Alfred was cognizant of the difficulties under which he had to operate. While in South Dakota, he frequently complained to Connable of his problem in finding secretarial help whose competency could compensate for his difficulties in reading and hearing. He wrote to Connable that he desperately needed a secretary who didn't "waste time asking foolish questions. The latter virtue is appreciated above all by one who has no ears [and] but one eye."[25] When he returned to Wilmington for a hurried visit in mid-September, he wrote Coleman a brief note saying that although he was in town, he would not be able to be present at a meeting of the Executive Committee to be held on the coming Thursday afternoon. He did not mention the recent ear operation he had undergone in Chicago but said that he found "my hearing is much worse since I returned East, due to change in climate I presume, and it is hardly possible that I would be of any use. This continued failure of my hearing apparatus would seem to indicate that my usefulness, as a member of the Executive Committee is about exhausted. As a matter of fact, I think this has been so for some time, although a man never likes to admit even though he is down and out. I want to discuss with you a plan which I have in mind whereby this Department may be more equitably represented."[26]

Alfred's plan was that Frank Connable should replace him on the Executive Committee. Coleman was only too pleased to have Alfred remove himself from the Executive Committee, and so before Alfred returned to South Dakota to serve out the remaining two months of his residency, the change had been effected. Since Coleman and Pierre usually conveniently forgot to inform him of Executive Committee business, Alfred became totally dependent upon Connable for information at the highest level of policy making within the Company. Even so, much behind-the-scenes discussion took place between Coleman and Pierre to which Connable was not privy. The doors to the inner sancta of power were beginning to close in front of Alfred.

Even Connable proved to be less than the dependable ears and eyes in the Executive Committee that Alfred had hoped he would be. Connable was young, bright, and above all, personally ambitious. Once he had Alfred's seat on the Committee, he quickly sized up the situation among the three cousins, and he was intelligent enough to figure out how he must conduct himself to further his own best interests. He knew that Coleman didn't like him, for in the president's view Connable would always be "Alfred's man." Pierre was a different matter. It did not take Connable long to realize who was in the lead trace of this troika. Pierre was clearly Coleman's superior in intelligence and Alfred's superior in power and influence within both the company and the family.

Jimmy Dashiell, Alfred's personal secretary and loyal confidant, had long made it his business to ferret out information that might prove useful to his esteemed employer. Dashiell felt it important to keep an eye on Alfred's representative on the Executive Committee. It quickly became apparent to him how very smoothly and subtly Connable was promoting his own, not Alfred's, interests. Dashiell would later say that Connable "had the knack of asking practical men their opinions on matters and then absorbing them and issuing orders as coming from him. He also had the knack of first finding the 'Iceberg' of a concern and then playing up to him."[27] In Pierre, Connable had found the real "iceberg" of the Du Pont Company, the bulk of whose weight within the organization lay three-quarters submerged below the surface. Connable was determined to melt him, and surprisingly Pierre proved to be quite approachable and receptive to

Connable's cautious but persistent advances. For Pierre, being sharper than Coleman, did not turn Connable away simply because he was "Alfred's man." Instead, he realized how useful it could be in the future to turn Connable into "Pierre's man."

Connable's wife, in a careless moment, had once remarked to Dashiell that poor Frank "did not know where he stood" in his relationship to Alfred.[28] In actuality, the situation was exactly the reverse. It was Alfred who did not know where he stood in relation to Frank. Dashiell felt it necessary to enlighten his boss. Soon Alfred's secretary was reporting on the frequent lunches Pierre and Connable were having at Hanne's Restaurant in downtown Wilmington. No one else was ever invited to accompany them, and the two men could be seen engaged in serious but apparently mutually fascinating conversation.[29] Similar reports, coupled with the Dashiell's persistent urging, finally persuaded Alfred that "no ears" on the Executive Committee were far better than "false tongues." Early in 1909, to Pierre's and Connable's surprise and no little annoyance, Alfred suddenly insisted upon resuming his seat on the Executive Committee. Thereafter, as the committee minutes confirm, for the first time since his appointment to the Committee Alfred became, according to Pierre, "a very regular attendant at the meetings."[30] It had now become disturbingly apparent to Alfred that he could no longer take for granted Pierre's friendship nor place his trust in his younger cousin's honesty in dealing with him. Pierre had crossed over and now stood side by side with Coleman on the other side of the chasm that separated Alfred from so many of his relatives.

The troika had ceased to exist, for it was no longer appropriate to use as a figure of speech, an image of the three cousins pulling together as a team. Coleman and Pierre had changed the metaphor. It was now a game of two on one, with Alfred as the lonely one, increasingly vulnerable to attack from either flank. The game still meant so much to Alfred, however, that he was willing to continue the play, even at these odds. All he wanted was to be allowed to go on doing what he did best, which was, as he once told Connable, to manufacture the highest quality of explosives in the world "at the lowest possible cost and the greatest possible factor of safety." Unfortunately, time was running out for Alfred. The whistle-blowers were poised, ready to sound the signal which would send Alfred out of the game in order to be replaced by a more tractable if less skilled substitute.

Coleman had wanted Alfred out as early as 1907 and had bluntly told his cousin so. But Alfred had refused to leave, and Coleman did not yet have the necessary support among the board of directors to force the issue. It is difficult to say precisely when Pierre finally decided that Alfred must go. Certainly, by the time that Alfred returned to his seat on the Executive Committee in 1909, Pierre was already watching closely for any error on

Alfred's part that would justify the full board in removing Alfred as general manager of production.

Alfred provided the opportunity for an attack early in 1910, when he authorized, as part of his overall expansion program, a new black powder facility at Wilpen, Minnesota, to be built at a cost of $350,000. Pierre made sure that his objections to the proposal were recorded. Eventually, Pierre was proved right and Alfred wrong. Black powder, which had once been the be-all and end-all of Du Pont production, was by 1910 running a distant second to dynamite in its share of company sales. A new black powder plant had not been needed, and the Wilpen facility never paid for the cost of its construction.

It had long been axiomatic within the Du Pont Company that Alfred was a black powder man and, like his Uncle Henry before him, regarded dynamite as being, at best, but a necessary evil in the explosives trade. Alfred had no quarrel with this statement. He always had been and always would be a black powder man. However, he would never accept the additional charge that he was totally ignorant of the high explosives field or that he was so set in his ways as to be unalterably opposed to new developments within the industry. The letters still extant, particularly his correspondence with Connable, show how knowledgeable Alfred was of all forms of high explosives manufacture, and how interested he was in chemical research. In 1904 he had been very instrumental in setting up the first experimental station exclusively devoted to the advancement of chemical research. It was at his urging that his cousin, Francis I. du Pont, was placed in charge of the facility, and the close bonds of friendship that had long existed between these two cousins were further strengthened by Alfred's keen interest in Francis's research. Alfred might not care much for dynamite or smokeless powder as explosives, but he was fascinated by the prospects that experimentation with cellulose provided for the possible future development of synthetic fabrics. Yet so assiduously has the theme of Alfred's obsolete antiquarian interests been pushed, that in time it has become accepted even by the historians and biographers who are the most sympathetic to Alfred's career.[31] Black powder became as much a part of the Alfred I. du Pont legend and as limiting to his true historical significance as stub candles and goose quill pens were to the Boss Henry du Pont story.

Even more than Alfred's eagerness to expand the black powder facilities, his response to the antitrust suit that the United States government brought against the Du Pont Company in 1907 provoked his two cousins' displeasure, bringing to a head their determination to remove Alfred from his position in charge of production. This protracted and momentous trust-busting suit had a somewhat trivial beginning in a civil suit brought by a former employee of the Du Pont Company.[32] Robert S. Waddell had begun his career in the explosives trade as a salesman for Du Pont powder in Ohio. With his flashy manners, glib talk, and grandiose promises, Waddell was exactly the type of salesman who would appeal to Coleman. To the disgust of old-time sales personnel like E. S. Rice, Coleman, soon after taking over the presidency of the company, made Waddell general manager of the Sales Department. In this influential position Waddell was able to obtain a great deal of information regarding the inner workings of the company, including trade agreements with its own subsidiaries and with supposedly active competitors, both domestic and foreign. Waddell learned how much it cost Du Pont to produce a pound of smokeless powder (47¢) and how much it cost the U.S. military forces to purchase a pound of this vital material (69¢), over the manufacture of which Du Pont had a virtual monopoly. A profit of 22¢ per pound was not a bad return. All of this information Waddell carefully filed away for future use.

In 1903, Waddell left Du Pont to become president of his own company, the Buckeye Powder Company. The move came as a surprise to Coleman, since Buckeye was a small, inefficient operation that had no possible chance of success in the powder industry, but the game its former general sales manager was playing soon became clear to the Du Pont Company. Waddell offered his company for sale to Du Pont at a price that would return to him a very handsome profit. His was one of the many "spite companies" created solely for the purpose of being bought out which Alfred had adamantly opposed buying. This was a spite company with an added punch, however, for Waddell had accurate information that could prove very damaging to Du Pont if made public. And Waddell had every intention of going public if Du Pont did not meet his price. It was a plain case of blackmail, and even Coleman agreed this time with Alfred that here was one small competitor that should not be bought. To yield to Waddell's blackmail attempt would be sheer folly.

At first, neither Coleman nor Pierre took Waddell's threat seriously. Both men underestimated the importance of the data that Waddell had been able to collect and to take with him when he left the company. Both, moreover, clearly failed to appreciate the climate of the times. The year 1906 marked a high point of antitrust sentiment in the United States. Teddy Roosevelt in the White House was swinging his very big stick and speaking not softly, but very loudly indeed, about the "malefactors of great wealth." The press, with the Hearst papers in the vanguard, was screaming for action by the federal government against the evil trusts that were extorting their ill-gotten pelf from a helpless consumer public. Roosevelt, ever sensitive to the political climate, was making a valiant attempt to oblige by seeking the scalps of such giants as the Northern Securities Company, Standard Oil, and the American Tobacco Company.

The Du Pont Company would undoubtedly have come under scrutiny even if Waddell had never brought a civil suit against Du Pont. Nevertheless, the inside and detailed information that Waddell as plaintiff could provide the courts certainly facilitated matters for the U.S. Justice Department. Waddell made sure that not only the courts and the Attorney General received copies of his brief, but also every member of Congress. It made fascinating reading for Congressmen, and Senators, and the general public as well, thanks to the publicity that the press gave to Waddell's charges. It is difficult to understand how the three cousins could have been surprised as they apparently were when on 31 July 1907, the Department of Justice itself brought formal suit against the Du Pont Company and all of its executive officers. It was no longer a minor civil suite of *Waddell* vs. *Du Pont,* for damages. It had now become a major case of *The People of the United States* vs. *Du Pont,* carrying with it the possibility of criminal charges against those who were in command of the company.

Each of the three cousins made his own distinct response to the suit in which he found himself a principal defendant. Coleman was the most confident that the suit represented no threat to the Du Pont empire. He regarded it as a cheap blackmail attempt on the part of Waddell and an obvious ploy on the part of the Democratic minority in Congress and a few grandstanding Republican Progressives like Roosevelt and Senator Robert LaFollette of Wisconsin to make political hay with the general public at the expense of American capital. Coleman's tactics were simple: expose Waddell for what he was—a dirty blackmailer, and play the game of politics with assurance of victory, since Coleman believed he held all the high cards. He could not believe the administration was really serious in wishing to prosecute the Du Pont Company. Hadn't Roosevelt himself said that not all trusts should be destroyed? Hadn't he differentiated between good trusts and bad trusts? And where would Teddy's beloved navy be without Du Pont's smokeless powder?

When, however, Roosevelt's Attorney General, William H. Moody, gave every indication that he would pursue this case as vigorously as he had the suit against Swift and the meat-packing trust, Coleman still saw no need for concern. Roosevelt would soon be out of office, and his likely successor, William Howard Taft, was, thank God, no rough-riding cowboy but a good and true Republican conservative from the good and true Republican state of Ohio. Taft's campaign manager was none other than Coleman's old political crony and business partner in New York real estate, Frank Hitchcock. With Taft in the White House and Frank Hitchcock the dominant political advisor in his administration, the suit would be quickly dismissed and forgotten.

Coleman was proved wrong on two counts. First, Taft was a stricter jurist than he was a good-old-boy conservative. He believed in law enforcement, and when the Sherman Act stated that no company engaged in interstate commerce could operate in such a way as to effect a restraint of trade, it meant precisely what it said. As chief executive officer, Taft had no option but to enforce that law. Taft was not prepared to make subjective judgments between "good" trusts and "bad" trusts. This large, unenergetic man who looked like the very embodiment of the cartoonist's bloated capitalist, proved to be a more tireless trust-buster than the pugnacious Teddy ever was.

Coleman also erred in placing his hope in Hitchcock's having influence within the Taft inner circle. It was soon apparent that his old pal, who had been shunted off to the Postmaster Generalship as his reward for his campaign efforts, was all but ignored in the Taft cabinet. The man whose influence was pervasive was Attorney General George Wickersham, who was equally determined to enforce the Sherman Act. Coleman found few friendly faces in the Taft administration.

Coleman, however, never gave up hope that his political connections would pay off. He continued to work on his many friends in the United States Senate, who represented a broad range of the political spectrum from Senator William E. Borah of Idaho on the progressive left to John Spooner of Wisconsin on the conservative right. Even after a unanimous decision of the United States Circuit Court found that the Du Pont Company had violated the Sherman Antitrust Act, Coleman kept on trying to influence Wickersham to reopen the case for a second hearing. As Pierre du Pont's biographers have said, "Hatred and anger so obsessed Coleman that he began to lose touch with reality. . . . Nothing that had happened convinced Coleman that the government was serious."[33] He continued to badger Wickersham to grant him a private audience to discuss the case, which the Attorney General, in recognition of Coleman's standing within the party, reluctantly agreed to on several occasions. But Wickersham could do nothing except to reiterate that he had no alternative but to enforce the court decree; unless, of course, on appeal, the U.S. Supreme Court reversed the decision of the Circuit Court.

Throughout the four years that the case was in litigation, Coleman accomplished nothing constructive in his company's defense. His political lobbying was, if anything, counterproductive. Even worse in its ultimate effect upon the courts was his refusal to appear in person to testify. As chief executive officer, he had a responsibility to direct and participate in the defense, but Coleman, pleading ill health resulting from complications of gallbladder surgery, successfully avoided a subpoena and left to his two cousins the unpleasant task of spending many hours on the witness stand. It could not have entirely escaped notice of the court, however, that Coleman's health apparently was not so seriously impaired as to prevent him from meeting with Senators and administrative officials throughout this period in an attempt to win their political support.

Pierre's initial response to the suit was very similar to Coleman's. He could not believe that the government was serious in seeking to break up the Du Pont Company upon which it was dependent for its most advanced ordnance. Pierre was hopeful that Coleman's political influence would pay off, and he was at first willing to allow his senior cousin to take the lead in designing their defense. In the meantime, however, it was necessary to observe the proper legal procedures, to present a brief answering the charges brought, and to appear as a witness in defense of current company policies. Since Pierre knew more about the restructuring and expansion of the Du Pont empire than anyone else, the responsbility for preparing the case and working with the defense lawyers fell largely upon him. Pierre's optimism matched Coleman's but for a somewhat different reason. He was confident that the dissolution of the Gun Powder Trade Association in 1904 and the reorganization that had followed the acquisition of Laflin & Rand, in which all of the competing companies recently acquired had been, with only a couple of exceptions, consolidated into one large company, exonerated the Du Pont Company from having entered into illegal trade agreements with competing companies in order to effect a restraint of free trade. He knew now that he and Moxham had been right in insisting upon this above-board consolidation rather than continuing the fiction of supposed competition by allowing these newly acquired companies to keep their old corporate identity. The few exceptions that remained, such as the King Powder Company, were now also belatedly and hastily brought into the Du Pont tent, but not soon or fast enough to escape the notice or concern of the court, as the court's opinion later stressed.

When the court's decision was rendered in June 1911, Pierre was as angry and as surprised as Coleman. It was finally apparent to Pierre that he could expect no salvation emanating from Coleman's highly touted but totally disastrous political maneuvering. Two options were now open to Pierre: either accept the court's decree and attempt to work out the best possible proposal for the dissolution of the company that the government would accept or appeal the decision to the United States Supreme Court. Pierre was at first inclined to accept the second alternative, which, as might be expected, the defense lawyers also favored. There was, to be sure, the third alternative which Coleman was vigorously pursuing, that of attempting to persuade Wickersham to agree to a rehearing of the case, at which

Coleman promised this time he would testify. Pierre, however, was now convinced that Coleman's pressure upon the Attorney General was accomplishing nothing.

From the beginning it was Alfred who was most realistic in his appraisal of the situation. He had never put any faith in Coleman's political tactics. The government had brought serious charges, and they must be met with powerful and straightforward defense. Alfred did not hesitate to take the witness stand, but his forthright answers to the prosecuting attorney's questions quickly revealed that he knew nothing about the various trade agreements that had existed prior to 1902, when he had been a very junior partner under Cousin Eugene, or later under the presidency of T. Coleman du Pont. Although he was vice-president and nominally second in command, he had not even been informed about the purchase of Laflin & Rand until it was an accomplished fact. He could state unequivocally without fear of perjuring himself that he had known nothing about the contract with the King Powder Company or the many other deals that Pierre had made in expanding and consolidating the Du Pont empire. He had read the charges presented by Waddell and the Justice Department with much the same surprise as had the general public. All he could provide the court was detailed information on the plant facilities and their production. Much of Alfred's testimony would be of great interest to historians of the company, but it did not produce evidence helpful to the defense. The fact that Alfred, a senior officer, had not been informed by Coleman and Pierre about their methods in building the Big Company made their activities seem clandestine and especially nefarious to the public and to the court. Alfred's testimony did not endear him to his two cousins. They could only have wished that Alfred had followed Coleman's example and had not testified at all.[34]

After the Circuit Court returned its decision, finding the Du Pont Company guilty of having violated the Sherman Act, it was again Alfred who was prepared to accept the harsh reality of the decision. He insisted that there was no course open but to work out a plan for dissolution that would do the least damage to the company and at the same time be acceptable to the government. He argued against taking an appeal to the U.S. Supreme Court, which he thought would only further delay the day of reckoning and could well produce even more stringent demands upon the part of the Justice Department for the reorganization of the company.

At the conference arranged by the Circuit Court in March 1912 to discuss a possible settlement, Pierre and Alfred attended as representatives of the company. As some indication of the importance of the case, not only was Attorney General Wickersham in attendance to represent the government but he was accompanied by President Taft. During a recess in negotiations, Alfred accidentally encountered the attorney general in the hallway. Wickersham asked Alfred to join him, the assistant attorney general, William Glasgow, and President Taft for a private talk in an adjoining conference room. It was made clear to Alfred that Taft did not trust Coleman, whom, it was later reported to Alfred, the President had once described as being "as slippery as an eel and as crooked as a ram's horn."[35] Both Taft and Wickersham greatly resented Coleman's repeated efforts to get the attorney general to reopen the case. They appreciated Alfred's willingness to accept the court decision as being final, and they now wanted to inform him privately that he should urge his associates to come up quickly with an acceptable and meaningful plan for reorganization that would put the company in compliance with the Sherman Act. If this was not done in the immediate future, the government was prepared to ask the court to put the company into receivership. A court-appointed agent would then most certainly do the job, and it would probably not be one that the Du Pont executives would like.[36]

Alfred at once relayed this information to Pierre. President Taft's alarming bit of confidential advice to Alfred was enough to convince Pierre that there should be no appeal to a higher court. The time for further court hearings had passed. Receivership could mean the end of the du Pont family's control of the company and its total dissolution. Negotiations now began in earnest, and two months later a plan had been devised that the court accepted. Under the terms of the Final Decree issued on 13 June 1912, two new companies, the Hercules and the Atlas Powder Companies, were to be created. Du Pont was required to turn over to these two companies plants with the capacity to produce 42 percent of its dynamite production and 50 percent of its black powder production. The two companies were to be truly competitive with Du Pont. No secret understandings in regard to market and price would be countenanced. The facilities that Du Pont would be allowed to keep and those that the company would have to give up were specifically named. The plants that Du Pont could continue to hold included the two that meant the most to Alfred—the Brandywine mills and the black powder blasting plant at Mooar, Iowa.[37] Coleman would hardly share Alfred's pleasure in saving the Brandywine mills, but both he and Pierre were delighted to learn that the court would allow them to keep all of their smokeless powder facilities. This was particularly surprising in view of the fact that it was in this one area of smokeless powder that Du Pont had achieved an absolute monopoly, and it was this control that Waddell had especially stressed in presenting his initial charge of monopoly against Du Pont. Clearly, the Taft administration, although dedicated to a strict enforcement of the Sherman Anti-Trust Act, was in this instance willing to shut its eyes to a glaring violation of the anti-monopoly legislation in order to safeguard the national security. As had been true in 1812, 1846,

1861, and 1898, the military had once again proved to be the Dupont Company's best friend.

As for Robert Waddell, the man who had been the instigator of the entire judicial action against Du Pont, he found his own suit had been completely overshadowed by the case of United States vs. Du Pont, and he was to emerge from the affair as the big loser. In February 1914 a jury accepted the defendants' argument that Waddell had taken over the Buckeye Powder Company for the sole purpose of blackmailing Du Pont. Not only did Waddell not receive one cent of the $5,000,000 damages he sought, but he was left bankrupt by the staggering legal fees that his unsuccessful suit had cost him. A vindictive Coleman had wanted to heap further injury upon him by having him prosecuted for perjury, but Pierre and Alfred could see little point in pursuing the matter further, and their counsel prevailed.[38] It was enough satisfaction to them to know that the Du Pont Company had survived the judicial attacks by both Waddell and the people of the United States as well as it had. The trust-busting era had exacted several pounds of flesh from the body of the old company, but E. I. du Pont de Nemours Company still survived, and for this the three cousins could be grateful.

Pierre should also have been grateful to Alfred for having taken a realistic approach to the antitrust affair. It had been Alfred whom Wickersham and Taft had taken into their confidence, and the information thus gained proved decisive in forcing Pierre to devise a plan for dissolution that the government would accept. Alfred's insistence upon reaching a settlement may have saved the company for the du Pont family once again, just as he had saved it in 1902, but he received even less credit for his action this time from his two cousins. No one loves the messenger who brings bad news, and certainly neither Coleman nor Pierre loved Alfred for the stand he had taken throughout this entire difficult period. Although they themselves were responsible for his having been kept in ignorance of the expansion activities that had lead to the government's action against the company, they deeply resented his having informed the court of this fact. Coleman and Pierre saw Alfred as the man who had publicly washed his hands of responsibility for the company's violation of the Sherman Anti-Trust Act, and no one has ever had a high regard for a Pontius Pilate. Long before the court's Final Decree went into effect in 1913, Alfred would receive his own personal decree of punishment from his two cousins.

The government's suit against the Du Pont Company not only widened the gulf between Alfred and his two cousins but it also provided the opportunity for Pierre and Coleman to reorganize the top managerial structure in such a way as to force Alfred out of control of production.

In should be noted, however, that the reorganization, pushed through in 1911, was not done solely as a means of punishing Alfred. That was simply an added bonus that Coleman and Pierre found particularly attractive.

Pierre had long wanted a different kind of corporate structure than the one the three cousins had created in 1902. Along with his close ally, Arthur Moxham, Pierre had consistently argued against the organizational structure that Hamilton Barksdale, Amory Haskell, and, to a lesser degree, Coleman and Alfred favored. Pierre and Moxham had wanted only one Big Company, bearing the name E. I. du Pont de Nemours, which would absorb and totally integrate within itself all other companies that Du Pont might acquire. Formerly independent companies should not be allowed to maintain their old corporate identity in order to foster the fiction of their competition with Du Pont, as had been the case with the Hercules and Sycamore Powder Companies during the regimes of Henry and Eugene du Pont. There would, of course, be clearly defined divisions within this closely integrated empire, but these separate units would not consist of corporate satellites revolving around the Du Pont sun. They would instead be departments rationally designed in order to represent service functions and manufactured products. The heads of these divisions would have considerable autonomy and would be responsible only to the president who in turn was responsible to the Board of Directors.

This was the ideal corporate organization which Pierre and Moxham were only partially successful in realizing during the early years of 1903–4. They did integrate most of the companies that they acquired within the Big Company, but as a concession to Barksdale and Haskell a few exceptions had been allowed, and just as Pierre feared, these exceptions proved to be highly embarrassing to Du Pont in 1907 in its defense against the charges brought under the Sherman Anti-Trust Act.

Pierre had been even less successful in getting the internal divisional structure he wanted. Over Alfred's and Frank Connable's strong objections, particularly with respect to a company-wide Purchasing Department, Pierre was able to establish the functional service departments organized into a chain of command along the lines he advocated. But there remained the major areas of explosives production—black powder, smokeless powder, and dynamite—which were not organized into autonomous departments under department heads who reported only to the president. Instead, these three production units were under the control of Vice-President Alfred I. du Pont. This arrangement was particularly irksome to Pierre and Coleman, not only for the independent authority it gave to their cousin in the all-important field of product manufacturing, but also because they believed that Alfred was really interested only in black powder, and that in his plans for plant expansion, he unfairly favored that area of production to the detriment of smokeless powder and dynamite.

In 1910, under the pressure of the government suit, Pierre at last had the opportunity to push through the kind of reorganization he had long advocated both in respect to the affiliated companies and to the internal divisional structure. There would no longer be any exceptions to a totally integrated company. King and International Smokeless quickly lost their separate corporate identity and their stockholders were required to exchange those companies' stock for Du Pont Powder stock.

In effecting the reorganization of the internal structure, Pierre at last had the complete cooperation of Coleman. Indeed, Coleman himself worked out the details of the new organizational structure while in residence at his estate on Maryland's Eastern Shore, where he was recuperating from yet another serious bout of illness. The basic premise underlying Coleman's plan was a more explicit separation of routine operational administration from long range policy-making. The Executive Committee would remain in its dominant position as the special center of power of the old guard. Its seven members consisted of Coleman, Pierre (acting president during Coleman's illness), Alfred, Barksdale, Moxham, Charles Patterson (the vice-president in charge of sales), and Amory Haskell. All members of the Executive Committee were required to give up whatever operational titles and duties they had previously held in order to concern themselves with overall policy. This of course included Alfred, who would have to surrender his title of vice-president in charge of production. But the separation of operations from policy-making was blurred by Pierre's insistence that Hamilton Barksdale, even though a member of the Executive Committee, be made general manager of production, essentially replacing Alfred in that key position. Coleman was certainly not happy with this suggestion, but Pierre was insistent. Unless Barksdale could be won over to the plan, he would join Alfred in opposition and could possibly convince Haskell to join them. The entire plan would then go down in defeat. So Coleman yielded, and Barksdale was won over by being offered the one position in the company that he most coveted.

Neither Pierre nor Coleman apparently worried about winning Alfred's support prior to presenting the plan to the Executive Committee. They did not even bother to inform him that a reorganizational proposal was being considered. Pierre's biographers, Alfred Chandler and Stephen Salsbury, have offered the following explanations for this unpardonable failure to communicate with the company's second officer, second largest stockholder, and the man who would be most adversely affected by this proposal:

Coleman's physical absence from Wilmington, the coolness between
the two cousins, together with preoccupation with the anti-trust suit,
helped explain why Alfred was not consulted about the new organization. Furthermore, administrative details bored Alfred, and it was natural that Pierre would not seek Alfred's advice.[39]

This lame statement does not "help explain" or provide justification for
Pierre's and Coleman's callous treatment of their cousin. Coleman may
have been absent from Wilmington, but it did not prevent his being in daily
communication, by phone and letter, with Pierre. And even if Coleman was
absent from the scene, Pierre was in Wilmington. He had both the time
and the desire to consult with Barksdale and to win his approval of the
scheme by promising him Alfred's position as the supervisor of production.
To say that "administrative details bored Alfred, and it was natural that
Pierre would not seek Alfred's advice" on this particular matter is to indulge
in a sophistry that would be comic if it were not so insensitive. Surely this
was one "administrative detail" that Alfred would not have found boring,
since it removed him from his major position of authority within the company. There was nothing "natural" in Pierre's failure to communicate with
Alfred on this matter. It can only be seen as a deliberate effort to keep
Alfred in ignorance of the entire plan until he could be presented with a
fait accompli.

On 18 January 1911, Coleman made one of his rare visits to Wilmington to present his proposal to the Executive Committee. The meeting was
held when Alfred was conveniently absent from Wilmington on an inspection tour of the black powder plants. The plan was briefly discussed and
quickly approved by the committee. It was agreed that the new organization would go into effect on 1 February 1911.[40]

It was only after returning to Wilmington a week later that Alfred first
heard of this reorganization which so drastically altered his position within
the company. Without warning he had been summarily removed from his
supervisory role of all explosives production. Even harder to accept was his
being forced out as the department head of black powder manufacturing.
His beloved Brandywine mills, which he had over many years nurtured,
preserved, and rebuilt, were no longer his to protect and to govern as his
personal domain. As might be expected, Alfred was stunned and outraged
by the news.

As might not have been expected, however, in view of his many intem-
perate responses to the slander circulated within the family regarding his
personal life, Alfred kept his balance in reacting to this startling development. If his cousins had anticipated and perhaps even hoped that Alfred
would display behavior that could be branded as violent instability, they

were disappointed. Alfred did not even bother to communicate with Pierre, who sat in an adjoining office. Instead, he wrote directly to Coleman, pointing out the obvious fact that the change had been made without providing him the opportunity "to express my views for and against it."[41]

Immediately after writing to Coleman Alfred did express his views on the matter in a letter addressed to the entire membership of the Executive Committee. In light of what had been done to him, it was a remarkably reasonable and dispassionate letter. While expressing regret that the change had not been given "the careful thought which it should have had before being finally adopted," Alfred did not ask that the action be rescinded or even brought up for reconsideration in his presence. He did point out that during the past decade the company had "thrived under the old plan of organization to a remarkable extent" and that "this success had been due largely to past methods and not in spite of them." It was not his purpose, however, to present a brief for the status quo. Alfred had considerable sympathy for Coleman's expressed desire to advance younger men like Frank Connable and Pierre's two younger brothers, Irénée and Lammot, as quickly as possible into positions of authority. He remembered only too well how Eugene and Cousin Frank had kept him in a subservient position long after his experience and talent merited a position of greater authority. Change was inevitable and desirable in organizational structure as in everything else, but changes should be made after careful consideration and with due deliberation, not unseemly dispatch. "Certain changes could be made at once," he conceded, "and others follow from time to time as the wisdom of each step had been made clear. If the plan as a whole is the right one, by this method it will finally be put into effect as it stands approved. If wrong as to detail opportunity will be given to detect errors." He asked for at least six months before the entire organizational plan was implemented. "The present Heads of the Operations and Sales Departments" should "retain advisory roles" which would allow "the new Department Heads to consult them."[42]

Coleman was apparently so impressed with the surprising reasonableness of Alfred's position that he answered with a conciliatory note. In response to Alfred's recommendations for implementing the changes over a period of time, Coleman wrote: "This is a matter that should have consideration by the Committee and I have asked Mr. Pierre du Pont to bring it up at the next meeting . . . so that you can have a full opportunity of presenting your views to the members of the Committee."[43]

Pierre, however, would have no part of a proposal which would delay Alfred's removal from his directorship in manufacturing. He flatly refused to bring the matter up for further discussion. As his biographers have correctly stated, "He [Pierre] realized that the damage had been done and that further discussion would only deepen the wound. 'The less said the sooner mended' was his motto."[44]

Given Pierre's attitude, Alfred also could see little point in further discussion. He pointed out to Coleman in a follow-up letter, "I beg to call your attention to that fact that my views are clearly expressed in my letter to the Executive Committe. . . . I am confident that my logic is sound and it may have the effect of making the Executive Committee's action somewhat conservative. But to bring up the letter for discussion . . . after the scheme of reorganization has been fully approved by every member save myself, it will only result in a waste of valuable time and nothing will be accomplished."[45] The only effect of Alfred's letter to the Executive Committee that can possibly be attributed to "making the Executive Committee's action somewhat more conservative," was to delay the plan's being put into effect until 1 March instead of 1 February as originally scheduled.

At least part of Pierre's motto was realized. He got his wish for "the less said," but he was to be sadly disappointed in achieving "the sooner mended" objective. Relations between him and his elder cousin, who had once served him as mentor, patron, and close friend, could never be mended, either sooner or later.

If nothing more was said in the Executive Committee, a great deal was said in the production plants throughout the Du Pont realm, particularly in the Brandywine mills, where Alfred was regarded by the workers to be as much a necessary and eternal part of the natural order as the Brandywine river itself. The news that Alfred I. du Pont would no longer be department head of all black powder operations was given to the workers on 11 February just as the six o'clock morning shift checked in. The news was presented to them not as Alfred's dismissal but as his promotion. No longer would their boss have to devote his time and energy to mundane operations, but as executive vice-president he could give his undivided attention to long-range planning for the company. The announcement seemed to imply that the change was something Alfred himself had wanted and worked for—a nice piece of fiction that the plant superintendents felt obliged to circulate. But the workers were not fooled. They knew Mister Alfred would not willingly leave the one place where he felt at home to spend all of his time at a desk and conference tables in that big, downtown office building.

Alfred's skilled machinist and close friend, Ed Bader, had over the years kept a diary of sorts in the small notebook he carried in his jacket pocket. The entries were irregular, since Bader recorded only unusual happenings, such as: "Aug 7 [1909] S[aturday] 10 [hours]. Graining mill blew. U. G. Mott killed," or "Jan 11 [1911] W[ednesday] 15 [hours] L[ower] Y[ard] water wheel started." On 11 February he wrote in his usual terse fashion: "Feb. 11 S[aturday] 6[hours] A. I. du Pont out!"[46] During the years he spent in the machine shop, Bader had had to record many explosions within the mills, but never before had an entry merited an exclamation point. For him and his co-workers, Alfred's ouster was truly an explosion that rocked the entire community. Alfred had "gone across the crick" in a somewhat different fashion than those who were blown across by exploding gun powder, but the result was the same. Alfred was lost to the Brandywine, and nothing would ever be the same again for his men and their families.

Alfred spent the month that was left to him before the new organization took effect in getting his department in order for the incoming regime. He persuaded Coleman to appoint Frank Connable as his successor in heading the Black Powder Department, rather than Pierre's youngest brother, Lammot. Alfred was confident that the department would be in good hand with Connable. The workers knew and respected Connable and would work well with him although there could never be the same kind of relationship that they had known with Alfred. Connable, moreover, was the one man Pierre would be willing to take in lieu of his brother.

There were a few employees who had special ties with Alfred and might find themselves in jeopardy with the company upon Alfred's departure. His personal secretary, Jimmy Dashiell, had for many years served as Alfred's private ears and eyes within the company and the community. For these special services to his boss, Dashiell had earned Pierre's enmity. To prevent Dashiell's being summarily fired as soon as Alfred left his office on the Brandywine, Alfred made use of his many contacts in Philadelphia and secured for his secretary a good position with the Wanamaker department store.

Alfred was also determined to protect his old boyhood friend and fellow pioneer electrician, Frank Mathewson, now employed as chief inspector of black powder machinery. When Mathewson came to his employer to express regrets over his departure from the Brandywine, Alfred asked him what his own plans were. Mathewson replied that he would stay on only until such time as he would be eligible for his pension and would then retire. Alfred urged him not to remain until retirement but to resign immediately. "You won't need any pension if you go with me," Alfred told his old friend. "You and yours will be taken care of. I mention this, Francis, because if you stay with the Company some of those who have it in for me may take it out on you." Frank immediately sent in his resignation and became Alfred's supervisor of various properties, with special attention being given to the maintenance of Alfred's hunting and fishing retreat on Cherry Island.[47]

There remained one other person whose future in respect to the Du Pont Company required careful consideration—Alfred himself. Stripped of all operational responsibilities and authority, which he truly enjoyed, Alfred saw before him only the bleak prospect of endless committee meetings, which he thoroughly detested—long sessions of discussion in which he could hear only a fraction of what was being said and crucial votes in which he would usually find himself in a lonely minority.

Several years before, when he had gone to Dr. Miller Hutchison to get his first akouphone, the two men struck up a friendship that had persisted over the years. They had on occasion talked of the possibility of joining forces in establishing a research laboratory in electricity. With Hutchison's training in medicine and his remarkable inventive mind and Alfred's mechanical skills and wealth to finance such a venture, they could produce technological changes that would help advance medical science.

Alfred took a train up to Llewellyn Park, New Jersey, where Hutchison lived, close by the Edison laboratories, where he served as chief engineer for the Wizard of Menlo Park. Alfred wasted no time in presenting his proposition.

"Hutch," he said, "the time has come. Let's strike out together. I'll put a quarter of a million dollars in a laboratory and furnish all the capital needed to exploit the things we work out. What do you say?"

Hutchison shook his head. He told Alfred that Edison was paying him $300,000 a year, but money wasn't the issue. He had been offered more by United States Steel to go with them, but he felt he couldn't leave Edison who really needed him. "I just can't leave the Old Man," he said.[48]

Alfred could understand the loyalty that tied Hutchison to his employer. It was something he himself once felt toward the Du Pont Company. There was nothing more to be said, and Alfred took the next train back to Wilmington.

Soon after the first of March, he and Alicia sailed for Europe for an extended stay in France. Alfred could see little point in standing by as witness to the birth of Coleman's and Pierre's new order.

When Alfred returned home in June, his old employees in the Brandywine mills at last had the opportunity to present him with a silver cup and to express formally their sense of loss. Frank Pyle, as spokesman for them all, read the resolution that they had passed,

> Whereas, our Vice-President, Mr. Alfred I. du Pont, has left as executive in charge of the black powder operating department for some higher office . . . we have hereby resolved that we feel keenly the loss of our leader and chief . . . [whom] we have always found a friend.

Alfred found it difficult to deliver the remarks he had prepared for the occasion:

I presume that the time comes in the life of every man when he asks himself, "Have I made a success of life, or has it been a failure?" In order to answer that question, he must have clearly defined in his mind what in his opinion indicates success in life. . . .

Some men measure their success with the wealth they have attained; others have political ambitions; and so on. I, myself, have always believed that the man most to be envied was he who, through life, had won the love and esteem of the greatest number of his fellow men. . . .

Men in the Brandywine have always held the highest place in my affection. With them I began my life work and for years we have toiled as brothers, through days of sunshine and days of shadow. You may therefore feel assured that no gift of the past or of the future will be greatly cherished as this exquisite token that you give to my keeping.[49]

With tears in his eyes and his voice choked with emotion, Alfred could say no more. Looking out at the crowd of day laborers assembled to pay him tribute, Alfred could take satisfaction in knowing that his life had been a success, as he himself had defined success. It was not Coleman's and Pierre's definition of success, but then they would never know a day when they would be presented with a silver cup by the workers, paid for out of their two-dollar-a-day wages.

As Alfred himself had implied, success is not an absolute, but a relative quality of status. It is measured against a yardstick of one's own choosing, and by Alfred's scale of values, he left the Brandywine not in defeat but in victory.

On that warm June afternoon in 1911, upon the conclusion of the farewell ceremonies in front of the machinist shop at the Brandywine mills, Alfred had a sudden impulse to walk again over the terrain he knew so well. Accompanied by Mike Maloney from the blacksmith shop, Alfred walked through Hagley Yards and out the wrought-iron gates that, at his request, Ed Bader had fashioned in 1902 to commemorate the company's centennial. Then by almost reflex action, Alfred turned up Breck's Lane, passing with hardly a glance the empty lot where Swamp Hall had once stood, and continued on until he reached the door of the Rising Sun Tavern, where he had known some of his happiest moments as conductor of the Tankopanicum orchestra. Strange that a short fifteen-minute walk could encompass a man's entire life. But there it all was—ancestral history, birth, childhood, marriage, parenthood, life's work—all of that which was his creation and recreation neatly packaged within a tight little geographic circle.

Nostalgia always made Alfred uncomfortable, and he must have wondered why he had bothered to take this walk. It was the last time that he would indulge himself in this particular sentimental journey. Turning to Maloney he said, "I won't be here much any more. I won't feel like coming. But I intend to keep track of the Brandywine people. Maloney, if I ever hear of one of them being in need and neglected, I'll hold you responsible. When anyone needs help come and see me."[50]

With that final instruction, Alfred turned and hurried back to the Hagley gates, where a car was waiting to take him into Wilmington. Thus ended his farewell to the Brandywine.

It must also have seemed to him that the day marked the end of the long power struggle of one du Pont against two du Ponts. Alfred had no illusions that his change of position within the company meant a promotion for him. Coleman and Pierre had wanted him out of production, where his authority was real and meaningful, and they had won out. As vice-president without portfolio Alfred did not expect to carry much weight on either the Executive Committee or the Finance Committee, and he wasn't at all sure how he would manage to fill up a normal work day.

Much to Alfred's surprise, however, he quickly discovered that he was not as bored or as isolated in some inane bureaucratic limbo as he thought he would be. Frank Connable was considerate enough to keep Alfred personally informed of developments within the Black Powder Department and was not too proud to consult occasionally with his former boss on issues where his advice could prove valuable.

Alfred now had the time to make an intensive study of high explosives. He quickly learned enough about the technology of dynamite and smokeless powder manufacturing to be able to evaluate and pass judgment on various mechanical inventions that were submitted to the company for consideration. Most of these so-called improvements proved to be worthless, but Alfred kept waiting for a genius to come along who could translate some of Alfred's own imaginative dreams of chemical breakthroughs into patentable realities.

Alfred fully expected that that genius would be his younger cousin, Francis I. du Pont, now in charge of the experimental station Alfred had built in 1904. The two men had along ago discovered themselves to be kindred spirits within both the company and the family, but never had they been as close as they became in these years immediately following Alfred's removal from the Black Powder Department.

Alfred and Francis frequently talked over some of the visionary schemes they shared, and one that stood foremost in their minds was the possibility of extracting nitrates from the air. This idea was not unique to them, of course. Other powdermen had the same tantalizing dream. What a boon it would be to the industry if it no longer had to be dependent on the nitrate fields of such distant places as India and Chile for potassium or sodium nitrate. Since 78 percent of the earth's atmosphere consists of nitrogen, an inexhaustible source of this indispensible product would be available if only a way could be found to "fix" the atmospheric nitrogen gas, that is, to combine it with either hydrogen or oxygen to produce nitric acid and ammonia. Some of the lowest and simplest forms of life on the planet, certain soil bacteria, either free-living within the soil or attached to the roots of legume plants, such as alfalfa and peas, have this capacity to fix nitrogen and convert it into amino acids essential for all life, but no laboratory had been able to duplicate this remarkable feat.

Francis turned his attention to the task of fixing nitrogen, and eventually he actually succeeded, in a very primitive fashion, in extracting nitrates from the air—but not in sufficient quantity to make it seem commercially practical. In would take World War I and the British blockade of Germany to provide a necessity compelling enough to mother the invention of the process that would fix nitrogen in order to provide the nitrates essential for Germany to manufacture explosives and wage war without benefit of the Chilean nitrate fields.

Other fascinating possibilities for chemical research kept occurring to Alfred during these years. There was the problem of the waste products that the manufacture of high explosives left as noxious residue to pollute the land and streams of New Jersey where the plants were located. Alfred kept urging others to find methods for recycling these chemical wastes which would be of immense benefit to both the company and the environment, but he elicited little interest or cooperation from the plant personnel. Production was industry's only concern. A concern for conservation and environmental protection lay far in the future.

Alfred also became increasingly interested in the question of diversification. There was a limit to the market for explosives in industry and sport—only so many tunnels to blast out of solid rock, only so many hunters to take aim and fire on only so many ducks. There was, to be sure, that other market that had in the past so often proved to be a source of rich profits for the Du Pont Company—the military. The major powers of the world in this first decade of the twentieth century were feverishly building up great navies and stockpiling vast quantities of ordnances for possible future wartime use. Alfred's concern was that Du Pont and other powder-makers were too dependent upon the military market, which might become even more limited in its demands for explosives than the commercial industrial market. For if the first decade of the new century was a time of battleship building and inflammatory jingoism, it was also paradoxically a moment when the ancient dream of international peace seemed to be not an utopian illusion but a practical and attainable reality. No one who was as intelligent and as careful a reader of newspapers and journals as Alfred was in the years between 1900 and 1914 could fail to be impressed by such developments as the establishment of a World Court at the Hague in 1902, by President Taft's arbitration treaties of 1911, or by the millions Andrew Carnegie spent in the hope of purchasing world peace. Crazy as it might seem, perhaps the pacificists who were now confidently asserting that the last Great War had already been fought were right.

These were airy and heady speculations, but they led to a very down-to-earth conclusion. Du Pont needed to explore the possibility of diversification of its products. The nineteenth-century amateur pioneers in chemical research, men like John Hyatt and James Brown in England and Count Henri de Chardonnet in France, had already shown that cellulose nitrate could be used to produce something other than a big bang—celluloid billiard balls and a new synthetic filament that Alfred's friend, Edison, found to be exactly right for his incandescent lamps.[51] A whole brave new synthetic world was waiting to be spun out of cellulose nitrate, and the Du Pont Company would be well advised to work on developing products that would attract customers other than military officers, construction engineers, and hunters.

Unfortunately, Frances I. du Pont could not offer either his scientific facilities or his laboratory to further Alfred's interest in diversification. Francis was required by the terms of his employment with the company to utilize his time and his facilities only for experiments that were directly applicable to the development of explosives. Thus being deprived of both Miller Hutchison's medical expertise and Cousin Francis's chemical research, Alfred had to strike out on his own to explore the entirely new field of scientific research that had suddenly attracted his attention.

In his autobiography Henry Adams wrote of the effect that this new science, or "vis nova" as he called it, had on his own thinking. Any "man of science," he wrote, "must have been sleepy indeed who did not jump from his chair like a scared dog when in 1898, Mme. Curie threw on his desk the metaphysical bomb she called radium. There remained no hole to hide in."[52] Alfred was no "sleepy scientist." He did indeed jump from his chair when he first read of Madame Curie's discovery. The strange new force that she had extracted from pitchblende seemed to defy all the well-established laws of an orderly Newtonian universe. Alfred jumped, not like "a scared dog," but like an eager pointer who has just got the scent of an exotic new game. He was not seeking a hole to hide but rather to dig in.

In 1911, Alfred entered into a partnership with Dr. Howard Kelley, a Baltimore surgeon and member of the Johns Hopkins medical faculty, who promised to do research on cancer treatment if Alfred would furnish the radium. Alfred proceeded to do just that. He hurried out to Colorado and bought, for the bargain price of $15,000, four uranium mines near Central City that his geologist friends believed to contain "the highest grade of pitchblende yet discovered." The Wilmington *Morning News,* belatedly reporting on this new venture, stated on 11 October 1913 in an editorial titled "Important Work for Humanity," that "medical and other scientific men will be greatly interested in the hope expressed by Mr. Alfred I. du Pont of this city that radium will be produced in sufficient quantities so that it can be utilized in the ordinary treatment of disease. . . . Mr. du Pont's plan for the development of the mines is of world-wide importance and significance, and the spirit that has prompted him in the undertaking deserves and will receive the highest commendation."[53]

Over the next few years Alfred invested $80,000 in the four uranium mines, and 140 tons of pitchblende were extracted, of which about 15 tons were of a sufficiently high quality to produce radium. He was able to present Dr. Kelley with one gram of radium for his medical experiments.

Other interests and concerns crowded in to demand Alfred's attention, however, and in the summer of 1914, just prior to the outbreak of the Great War, Alfred sold his mines to a German company that, interestingly enough, was sponsored and supported by the German government.[54] Whatever may have been the motivation for the Germans' taking over the American uranium mines in 1914, Alfred's interest seems to have been entirely humanitarian. He had proved to his own and to the medical profession's satisfaction that America had the natural resources and the technical capability to produce a gram of Madame Curie's strange new force, which, he hoped might prove to be the "magic bullet" with which to shoot down cancer.

Henry Adams had been more correct when he called the newly discovered element "a metaphysical bomb." It took thirty more years of highly theoretical speculation and intensely technical experimentation, however, before Einstein, Fermi, Urey, and Oppenheimer could convert the abstract metaphysical into the terrible physical bomb with which to destory Hiroshima. Certainly, Alfred had no idea, on the proud day when he presented Dr. Kelley with the gift of radium, that out of the same ore which they hoped would prove to be an elixir of life there would also come another element that had the potential to destroy all life. Alfred would not live long enough to appreciate the grim irony that lay in the fact that he, a du Pont, had been one of the first men in America to promote the development of uranium mining that would produce the substance to make even the high explosives of the Du Pont Company seem, by comparison, like the puny pops of a child's cap gun.

With this catholicity of interests that led him from the waste dumps at

Repauno to the uranium mines of Central City, Alfred had little time to mope in boredom in his vice-presidential office in Wilmington. He had never been more intellectually stimulated than he was during the years that he had originally feared would prove to be nothing but a meaningless cul-de-sac to his association with the Du Pont Company. Instead of a dead end, he had found to his delight that he was at the hub from which radiated a hundred different avenues of inquiry. He now had the time to ask questions, and, even more important, to seek answers to those questions.

There is every indication that Alfred would have been content, indeed even eager, to have continued in the new role that had been assigned to him until terminated by either the infirmities of old age or death. Other circumstances soon arose, however, that preempted such a natural conclusion to his life career. Alred had been wrong in thinking that the issue of control of the company had been settled on the day in January 1911 when Coleman and Pierre had forced through their reorganization plan. One final battle remained to be fought before the issue of company control was settled.

Soon after forming their original triumvirate, the three cousins realized that divided authority creates problems. A troika will not move at all if there are simultaneous efforts to pull it in three different directions. With Alfred's ouster from the management of production in 1911, the triarchy had been reduced to a dyarchy, but the old question "Who's in charge here?" still had no clear-cut answer. Alfred, if he had been so inclined, could have found a certain sadistic pleasure in observing Cousins Coly and Pierre, who had so effectively ganged up against him, now turning their big guns against each other, for relations between the two remaining cousins were anything but harmonious after Alfred left the center of power. Against his better judgment Coleman had had to yield to Pierre's demand that Barksdale be made general manager of production. Pierre, in turn, was vehemently opposed to Coleman's suggestion that William du Pont, who had long been in exile from the company after his scandalous divorce and second marriage, be allowed to return to the board of directors.

The argument between Coleman and Pierre over William's reinstatement produced a rupture in their former alliance that would never be satisfactorily healed. Alfred, for once, found himself in complete accord with Coleman and strongly supported Willie's election to the board. During the course of this particular dispute Coleman wrote a long letter to Pierre suggesting several alternatives to resolve the growing difficulties arising out of a two-man rule: 1) Pierre and his brothers could buy out Coleman's interest and take over the company; or 2) Coleman could buy out Pierre's interest; or 3) the two men could agree "to leave to arbitration any differences between us by the three best men we can get not personally interested, paying them well for their work and agree that we will both be bound by their decision and on any point that we may differ carry out their decisions."[55]

Pierre quickly rejected all three suggestions. He certainly was not willing to sell out to Coleman nor was he at this time prepared to buy Coleman out. The third alternative, an arbitration board, was, of course, so utterly absurd as not to merit consideration. If they had to resort to that to settle their differences, they both might just as well get out.

A truce was reached a few days later when Coleman unexpectedly wrote Pierre a short note on 27 August 1914 asking him to serve once again as acting president in order that Coleman could devote his time to his many pressing business affairs in New York. Coleman promised "to keep hands off any future questions unless you want to talk to me." Pierre was happy to accept this arrangement, and in turn promised not to block Cousin Willie's election to the board of directors.[56]

It was in this spirit of detente that Coleman and Pierre pushed through one more reorganization which more fully complied with Coleman's original concept for reorganization. Under the terms of this latest reshuffle, Barksdale was removed as general manager of production and the position was abolished. The several operational departments were now to be autonomous units, reporting directly to the Executive Committee with no intermediary supervisor or committees over them. Harry Haskell was placed in charge of high explosives (dynamite), H. F. Borwin in charge of smokeless powder, and Pierre's brother, Lammot, got Alfred's old job as director of black powder. Frank Connable, Alfred's immediate successor in black powder, was made vice-president in charge of purchasing, the position that both Connable and Alfred had opposed when Pierre had first created it in 1905.

The major change that the reorganization of 1914 effected, however, was in the composition and assigned duties of the Executive Committee. This former bastion of the old guard was now to become the center of administration under the control of the younger men in the company. Coleman, Alfred, Barksdale, Moxham, Amory Haskell, and even Pierre were required to surrender their seats on the Executive Committee and were replaced by the three product directors—Harry Haskell, Brown, and Lammot du Pont— along with Frank Connable; Pierre's brother-in-law, R. R. M. Carpenter, the director of Development; William Coyne, director of Sales; and John Raskob, Pierre's long-time secretary who had replaced him as Treasurer. Pierre's brother, Irénée, would serve as chairman of the executive committee. As Pierre proudly announced, the new generation was at last moving into its proper seat of power. It was also, as Pierre had carefully made sure but did not find the need to proclaim loudly, an Executive Committee filled with his kin and allies whom he could control.

The old troika—Coleman, Pierre, and Alfred, along with William du Pont—maintained their very real presence in the company by keeping their seats on the Finance Committee. Significantly, this committee would no longer be responsible to the Executive Committee, but only to the Board of Directors, which meant in reality being responsible only to themselves, since the four men held the preponderance of the company's stock. In the words of Tennyson, "the old order changeth," but neither Coleman nor especially Pierre was ready to "yield place to new," for the Finance Committee would determine the annual dividends, approve of all appropriations, and authorize all expenditures of $150,000 or more.

The new organizational plan took effect on 1 September 1914. Both Pierre and Coleman hoped that they had at last achieved a truce that would endure, and since each seemed satisfied that victory was his, peace appeared to be a real possibility. Coleman believed that he had achieved the organizational structure he had wanted in 1911 but had failed to get because of Pierre's holding on to Barksdale and Haskell. Above all, Coleman was now free to give his full attention to his New York business interests. Pierre was pleased because he had won the real power in the company, even if his title as president was qualified by the word "acting." He and his two brothers occupied the key positions, and he had Coleman's promise to "keep hands off any future questions." Cousin Willie's seat on the board was a small price to pay for that.

Both Coleman and Pierre thought it particularly fortunate for the Du Pont Company that an armistice had been reached in Wilmington, for in the final days of the fateful month of August 1914 the "guns of August" had already begun to boom from the Belgian coast to the Baltic Sea. By 1 September 1914, Pierre like Alfred in January 1911, believed that the struggle of du Pont vs. du Pont for the control of the company was over. Unlike Alfred earlier, however, Pierre believed that he had emerged the victor. Alfred appeared to be resigned to his subordinate role, and Coleman had discreetly withdrawn from the field of combat. The Du Pont Powder Company, as restructured in accordance with Coleman's wishes and with Pierre's steady hand at the helm, was now prepared to give full attention to and profit from the Great War raging in Europe.

When the new organizational plan became operative on 1 September 1914, the condominium of shared power that the three du Pont cousins had created twelve years earlier was essentially terminated, although Coleman still held titular claim to the presidency. Each of the three cousins responded to this reality in his own way. Alfred had accepted the ending of the triumvirate with what for him was an atypically stoical resignation. Coleman welcomed with relief his being at last disencumbered of the burdens that had prevented him from fully participating in the role of the big time New York operator that he had longed to assume. As for Pierre, his response must have been one of joy similar to that experienced by Octavian when the triumvirate he had formed had finally been terminated at Actium. The laurel crown of the victor was on his head.

None of the three expected a resumption of the power struggle which for over a decade had confused the management of the company. Alfred had no hope, Coleman had no desire, and Pierre had no reason to alter the status quo. Yet, within six months after the seemingly permanent resolution of the conflict, a new crisis arose that would lead to the most fiercely waged battle of all and would further divide the du Pont family for the next half century.

It was Coleman du Pont who, without intent or malice aforethought, precipitated the final contest between Alfred and Pierre for control of the company. Coleman himself was not an active participant in the ensuing struggle, but he served as the catalyst.

Coleman was a big man in every way—in physique, in spirit, and in behavior. His appetite was gargantuan—for food, for drink, for sex, for money, and, above all, for power. Even as a boy in Louisville he had always insisted upon being the leader in every game. If he couldn't be captain of the sandlot baseball team, he would take his ball and bat and go home. He always got to be captain. When Alfred came to him thirty years later to ask him to become a partner in taking over the family's powder company, Coleman would join the game only if he could be president. He got the presidency.

But long as his arms were, Coleman's ambition always exceeded his reach. He had yet to find a game that he considered big-league enough to match his talent or his appetite. From the moment he arrived in Wilmington, he had felt cramped by the smallness of the arena. One of his first suggestions as president was to move company headquarters from the banks of the Brandywine to the towers of Manhattan, but his two conservative cousins, unduly obsessed, in his opinion, with ancestor worship, had blocked the move.

Over the next few years, Coleman did his best with what he had— which was considerable—to build an industrial empire. He was eminently successful. Within five years after he took command the Du Pont Powder Company had swallowed up its single largest competitor, Laflin & Rand, along with scores of smaller companies from coast to coast, and had achieved a virtual American monopoly in black powder and dynamite and an actual monopoly in smokeless powder. Yet all of this was not enough for him. In spite of its remarkable growth, the Du Pont Company in 1910 was still only the twenty-seventh largest corporation in America, and that was not big enough. By almost anyone's standard of measurement, Coleman was a rich man, but in the race for wealth such fleet-footed plutocrats as Rockefeller and Carnegie were far ahead of him. Coleman was still not running fast enough. Or running on the right track, Coleman thought. It would be another fifty years before some New York public relations genius would coin the term "Big Apple," but Coleman already intuitively knew that that was the prize he was seeking. Only those who made it big in New York could claim they had made it big in the world of high finance. Neither Pierre nor especially Alfred had enough vision to see that. They seemed to fear the big city and were content to stay in little Wilmington, but for Coleman no place was out of bounds, and there was no such thing as forbidden fruit. The old serpent Ambition continued to hiss in his ear, and in spite of the demands of Du Pont empire building and recurrent bouts of illness, Coleman still found the time and energy to make quick raids on the big apple orchard.

His New York action began with the purchase of real estate in Manhattan. Then, in 1910, he formed a syndicate with Charles P. Taft, the President's brother, to build the McAlpin Hotel at the corner of 34th Street and Broadway. Coleman put half a million dollars into the venture, which he envisioned as only the first of a chain of luxury hotels in New York, San Francisco, Chicago, and Philadelphia. Coleman took a penthouse suite in the McAlpin, which became his most frequented residence outside of Wilmington. There he could entertain a whole chorus line of girls from a current Broadway show at lavish, after-the-show, all-night parties. To Coleman this was heady stuff. It was a long way from the newly opened Hotel Du Pont on the Brandywine to the newly opened Hotel McAlpin on Broadway, and Coleman had had his fill of Brandywine livng. He wanted more, a great deal more, of Broadway lights.

For Coleman, his first few years as president of the Du Pont Company had been great fun. Coleman was an empire builder, but he took little pleasure in empire management, and, like Winston Churchill, he had no interest whatsoever in empire dissolution. After the government's antitrust suit went against the company, and Du Pont was ordered to divide itself, like Gaul, into three parts, Coleman lost what little interest he still had in remaining as its captain. Happily for him, a new opportunity presented itself in New York at that time. In January 1912, the home office building of the New York Equitable Life Assurance Society, America's largest insurance company, burned to the ground. The ashes were barely cool before Coleman had formed a corporation to build a new office building on the same site. It cost him $13.5 million to purchase the land where the old building had stood. On this plot of precious Manhattan rock he proposed to build the world's tallest office building—thirty-six stories high—at an estimated cost of $14.9 million.

At last Coleman found himself running on the big track, but it quickly proved to be a much faster track than he was prepared for. The cost of the Equitable land alone had far exceeded his own personal wealth, but Coleman still had the magical touch. He could still pull fat rabbits out of empty hats. With his fast talk, he persuaded Equitable to take a long-term mortgage in lieu of the cash the company had wanted for the purchase of the land. The cost estimate he had been given for building the structure quickly proved to be much too low. The overrun would double the original cost estimate. With another wave of his magic wand, Coleman got a loan of $20.5 million from Equitable, for which he put up 10,500 shares of his Du Pont stock as collateral—not nearly enough, of course, to cover the loan, but apparently enough to satisfy Equitable of his good faith and credit standing. Even after these remarkable feats of legerdemain, however, Coleman was still some $3,000,000 short of meeting the cost of his spectacular skyscraper.

At this moment of crisis in his personal financial affairs, Coleman precipitated the great crisis within the Du Pont Company and the du Pont family. He offered 20,000 shares of his stock, representing nearly one-third of his total holdings, for sale to the company. He made this unexpected offer orally to Pierre shortly before departing for Rochester, Minnesota, on 14 December 1914 for yet another abdominal operation. Coleman set a price of $160 per share for his stock, which would provide him with the $3.2 million he desperately needed for his Equitable venture.

Pierre was surprised but delighted to have Coleman's offer. The details were quickly worked out. Pierre did not object to Coleman's price, but of course the sale would have to be approved by the Finance Committee. Once the company had purchased the stock, it would make the shares available to the salaried employees who held managerial positions and whose monthly salary was at least $500. Coleman sketched out the plan in writing just before leaving for Rochester and advised Pierre to talk the matter over "with Alfred to see whether he approved of such a procedure."[57]

Pierre brought the plan to Alfred, who also responded favorably to the general idea of Coleman's proposal. Alfred had long been an advocate of bringing younger men of ability into the partnership. He had been instrumental in 1906 in instituting one of the first bonus plans to be adopted by any major American industry, and his interest in employee participation extended beyond management. He also advocated a profit-sharing plan that would permit hourly wage-earners to share in company profits. The most immediately pressing issue in 1914, however, was to give an appropriate share of company ownership to the new generation of management that was now coming to the fore. Heretofore, no member of either the Executive or Finance Committee had been permitted to share in the 1906 stock bonus plan, for the very justifiable reason that the membership of the two committees consisted of the company's large stockholders. With the reorganization of 1914, however, the new Executive Committee had younger members who owned few if any shares of Du Pont stock. Because in 1914 the Du Pont Company was not a public corporation and its stock was not listed on the New York Stock Exchange, the only way these younger men could get stock was to buy shares over the counter from present owners. Remembering his own youthful struggles to wrest a few shares of stock from Cousins Eugene and Henry in 1889, Alfred was so sensitive to the present anomaly of having in top management men who had no stake in company ownership that he sold 1,000 shares out of his own holdings to Pierre's secretary and alter-ego, John Raskob. It is hardly surprising, then, that Coleman's sudden offer to sell a large block of his stock to the company was welcome news to both Pierre and Alfred. Here was a new reservoir of stock that would enable all of the young executives to become significant partners within the Du Pont Company and still leave a reserve pool ample enough to continue the stock bonus plan for the foreseeable future.

Only one feature of the proposal sketched out by Coleman gave Alfred pause—the price that Coleman was asking for his stock. In the late summer of 1914, at the moment that the great powers of Europe went to war with each other, the Du Pont stock available for purchase over the counter in New York was selling at $118 a share. The first few months of the Great War brought a rise in the value of Du Pont shares as orders for explosives from the British and French governments began to trickle into Wilmington—not a flood of orders as yet, but enough to raise the price of Du Pont stock to $160 a share by the end of November 1914. Coleman was therefore asking the company to purchase his stock only at the price currently being asked for over-the-counter sales in New York. Coleman quite rightly believed that he was setting a fair price that would within a few months appear to be a bargain if the war in Europe continued to make its demands on Du Pont production.

Alfred, however, in considering the price Coleman had set for his stock used a different standard than that of present and future market price. If the underlying purpose of the company's purchase of this stock was to make it available to its junior executives, then the price had to be one that the young men could manage out of their own resources. It also had to be low enough to give them a fair return on their investment. In 1914 the dividend on Du Pont stock was eight dollars per share. Alfred made an estimate as to what the price of the shares would have to be in order to give to these potential purchasers a rate of return on their investment that would be attractive enough to induce them to buy the stock. If they paid $160 a share for Coleman's stock, they would be realizing less than a 5 percent return on their investment. Alfred believed that to be much too low to make the stock attractive.

On 14 December, Coleman left for Minnesota thinking that the deal was settled. On the same day, Alfred wrote a short note to Pierre in which he expressed the opinion that Coleman's asking price was too high. Alfred also wrote to William du Pont stating that he favored the sale but opposed paying more than $150 a share. "My opinion is that 160 is too high as 150 would be purchasing it on a 6% basis, and 160 would be a 5% basis which, in my opinion, is too low a rate for the Company to invest its money and too low a basis on which to offer it to any of the Company's employees. I am simply offering these figures as subjects for thought." Alfred also expressed the opinion to William that he considered the distribution of the stock to be "a matter of grave importance, as you must recognize, and should only be done under a plan which will bring the best results to the organization." In a follow-up letter to his earlier note to Pierre, Alfred went into further detail as to why he thought $160 a share was too high. Alfred suggested a price of $125.[58] Clearly what Alfred implied in his letters to Willie and Pierre was that there should be further negotiations with Coleman about the price. He favored starting the bargaining at $125 a share and setting $150 as the top price.

On 23 December, the Finance Committee met to give formal consideration to Coleman's offer. Present at the meeting were Pierre, Alfred, and William. Pierre spoke briefly in support of the proposal and said that he considered $160 a share to be a fair price. Alfred then spoke up, saying that he also favored the proposal in principle, but outlining in detail his reasons for objecting to the price Coleman had set. He wanted further discussion with Coleman on that point. As he had told Willie, the matter "is an important one, and should not be decided hurriedly." William, as was his wont, did not waste words. He said only that he concurred with Alfred in everything he had said.

The minutes of the Finance Committee's meeting of 23 December 1914 contain the following notation of the action taken:

> Mr. P. S. du Pont presented a letter from Mr. T. C. du Pont offering to sell 20,700 shares of Common Stock of this company at $160 per share. After discussion it was voted and carried (Mr. P. S. du Pont voting in the negative) that Mr. P. S. du Pont be instructed to advise Mr. L. L. Dunham, attorney for Mr. T. C. du Pont that we do not feel justified in paying more than $125 per share for this stock.[59]

This single entry in the committee minutes later became the major piece of evidence to support Pierre's contention that he considered Coleman's offer had been, by a vote of two to one, unequivocally and finally rejected by the Finance Committee over his objection, and that he therefore was free to negotiate on his own with Coleman. In rebuttal, Alfred and William maintained that their understanding of the action taken by the committee on that fateful day was to raise an objection to the price of $160 per share, but that Pierre was to enter into further negotiations with Coleman. Both Alfred and William insisted that the minutes as recorded had, perhaps inadvertently but certainly conveniently for Pierre, failed to carry an important qualifying phrase "at the present time" after the clause "we do not feel justified in paying more than $125 per share for this stock." Nevertheless, both Alfred and William could not deny that they had, after the meeting, initialed the minutes as being correct as stated. It was a minor slip-up on Alfred's and William's part, but Pierre was to make the most of their negligence. Pierre, however, made a similar mistake when he initialed as being correct the report of the action taken by the Finance Committee that was sent to the Board of Directors. This report stated:

> The [Finance] Committee expressed that we were not justified in paying more than $125 per share and asked Mr. P. S. du Pont to take the matter up with Mr. T. C. du Pont further.[60]

This communication from the Finance Committee to the Board of Directors clearly indicated that further negotiations were to take place. In initialing this report Pierre had done as much damage to his case as Alfred and William had done in approving the committee minutes.

A far more serious and fundamental mistake committed by Alfred at this time was his objection to the price Coleman had set for his stock. At the very moment that Alfred was protesting Coleman's asking price of $160 a share, Du Pont stock had already passed the $180 mark on the open market and four months later would soar to $300 a share—with no ceiling in sight. It is easy, with the wisdom of hindsight, to make the judgment that Alfred, who had been kept fully informed as to the volume of the wartime orders coming into Du Pont from Europe and the plans the company had for the expansion of its facilities to meet those orders, should have seen that Coleman was offering his stock at a bargain price and should have leaped at the opportunity to buy twenty thousand shares at $160. Events quickly proved how wrong Alfred had been to question this price. Had he and William joined Pierre in voting to accept the offer, his own and the du Pont family's history would have been very different. Relations between him and his two cousins, to be sure, could never return to what they had been prior to 1906, but Alfred would undoubtedly have remained with the company to which he had dedicated his life and would have been a part of the exciting postwar developments in chemical research and product diversification that he had long advocated and encouraged. Above all, there would have been no great schism within the du Pont family which would condemn his allies within the family to the ostracism which he himself had suffered.

Alfred, however, had to evaluate the situation as he saw it in December 1914, and his evaluation, although diametrically opposed to Pierre's, seemed valid and realistic at the time. The great German war machine, after its initial blitzkrieg drive through Belgium to the environs of Paris, now seemed to have stalled. Neither the Central Powers nor the Allied Powers seemed to have the means to destroy the other side. Except on the eastern front, the war had reached a stalemate, and there were few who could foresee that this conflict of trench warfare attrition could last another four years. Most military and political analysts believed that as an absolute stalemate now existed between the two opposing forces, a settlement similar to the Treaty of Berlin in 1878 or the Algeciras Agreement of 1906 would have to be concluded within the next few months. President Woodrow Wilson, after proclaiming that America would be "neutral in thought as well as in action," stood waiting in the wings to step on stage when called to play Count Bismarck's former role of "honest broker" in settling all outstanding differences between the great powers.

Even if the war continued another year, which Alfred seriously doubted, there was a real possibility that Congress would be persuaded by the pacifist sentiment prevalent in the country and openly expressed by the President and his Secretary of State, William Jennings Bryan, to place an embargo upon the shipment of munitions to the belligerent powers.

In December 1914 Alfred could argue very plausibly that Coleman's stock, which had risen dramatically in the past six weeks and might go still higher in the next month or two, represented a high risk involvement at $160 a share that the company should not make. What would happen to Du Pont stock prices if the war ended in March 1915, or if Congress passed a neutrality act forbidding the shipment of munitions, as it was threatening to do? Of course, it was to Coleman's advantage to talk glibly of the great sacrifice he was making for his colleagues' benefit and to make wild predictions that the shares would soon be worth $200 or even $300. Alfred had heard Coleman's sales talk many times before this. Coleman might be simply trying to unload while the market was at its peak. Alfred's judgment of the situation was clearly affected by his basic distrust of Coleman and his method of operation. Alfred could not say it openly in committee meeting, but what he thoroughly subscribed to was the ancient Laoccoön admonition to beware of Greeks bearing gifts.

The facts regarding both the international and domestic situation were, of course, as well known to Pierre, as they were to Alfred, and yet he had come to the opposite conclusion. Pierre saw the stalemate in Europe not as an argument for a quick end to the conflict but as evidence that this would be a prolonged war of attrition in which both sides would expend a fantastic amount of munitions in futile efforts to rout the opposition. There was too much at stake to expect an immediate compromise settlement. This was no minor Balkan quarrel to be settled quickly at a Congress of great powers nor was it a peripheral African imperialist dispute to be resolved at some minor conference of diplomats. The issue at hand was nothing less then the future control of Europe, and both sides would fight on, not until one side had won, but until one side dropped in exhaustion. Pierre, who, in spite of his quiet manner, was always more of a gambler than Alfred, accepted Coleman's hard sell as being true.

Pierre also knew more about Coleman's true state of affairs than did Alfred. He knew his cousin was in over his head in the swirling waters of New York finance and desperately needed someone to throw him a life preserver. Pierre's great fear was that if the Du Pont Company did not buy the stock, Coleman would sell to an outside agency. This fear was greatly strengthened when a representative from the British Nobel Dynamite Trust, Edward Kraftmeier, came to the United States in January 1915 to make purchases for the British army. Kraftmeier secretly informed Pierre that his government had had a report that "Kuhn, Loeb & Company of New York (who are pro-German) had gained control of our Company through the embarrassment of one of our largest stockholders and that they on that account had fears concerning the orders placed with us."[61] Fortunately Pierre could at that time assure Kraftmeier and the British government that there was no truth in such rumors, but the haunting fear remained that Pierre would not be able to give such an assurance a month later if the Du Pont Company failed to take advantage of Coleman's offer. T. C. needed cash right now, and he was determined to get it from one source or another. Pierre even tried to persuade Coleman to agree to a pooling of all three cousins' stock, which would so entail their holdings that no one of them could sell his stock without the consent of the other two. Coleman, of course, would not even consider such a restriction on his right to sell.

In the six weeks following the Finance Committee meeting of 23 December, Alfred apparently gave very little further thought to Coleman's offer to sell. He assumed that Pierre was negotiating with either Coleman or his designated representative, Lewis Dunham, about the price. No definitive action could be taken anyway until Coleman had undergone his very serious operation and the results of the surgery were known. Alfred was not averse to having the matter lie dormant for a while. Another month or two should bring clearer indications as to the direction the war would take and also what the American government's response to the war would be. Alfred busied himself with other matters and did not press the issue.

Pierre, however, pressed the issue vigorously. On 4 January 1915, as soon as he had word from Dunham that Coleman had not only survived the operation but was making a good recovery, Pierre wrote to his cousin to report on the action of the Finance Committee: "Unfortunately, Alfred, who had approved the plan before you went away, got somewhat crosswise in the meeting and I think it wise to let the matter rest for the moment, preferably until I can see you, before taking any other step. I am sorry and provoked that the proposition did not go through . . . but like many other things the final result cannot be obtained quickly."[62] He gave no reason as to why Alfred "got crosswise" in the meeting, and he made no mention of the committee's wish to negotiate with Coleman on the price. Pierre simply left the impression that Alfred was being an obstructionist, perhaps out of spite toward his cousins. Coleman replied that the refusal might necessitate his opening negotiations with New York bankers—the very thing Pierre feared. Pierre's letter and a second follow-up letter to Coleman on 9 January did not, however, substantiate his later argument that the committee had totally rejected the offer and that all negotiations were at an end. Indeed, in his letter of 9 January, Pierre added the sentence, "As it remains now, I feel that the propostion is open for reconsideration at your option of course. . . . My judgment is that the deal will go through this time."[63] Pierre would later have a very difficult time squaring this correspondence with his claims that the Finance Committee had closed the door on Coleman's offer.

It was not until 10 February 1915 that Alfred brought up the question of negotiations with Coleman over the purchase of his stock. At the conclusion of a meeting of the Finance Committee on that date, Alfred, as he later remembered the conversation, casually asked Pierre, "How are the negotiations for the Coleman du Pont stock progressing?"

He was stunned by Pierre's answer, "Why they are all off."

"Since when?" Alfred demanded to know.

"They were called off shortly after you and Willie turned down the offer in December."

"But the offer was not turned down," Alfred protested. "There was merely a difference of opinion as to price, and it was my understanding that we believe the price of $160 a share excessive, and we suggested $125 as a proper price at that time."

"That was not my understanding," Pierre replied. "My understanding was that you turned down T. C.'s offer definitely."

Alfred then turned to William. Although the phlegmatic Willie never allowed his face to reveal his inner feelings, he must have been equally astounded by Pierre's announcement. He simply said that Alfred's understanding that negotiations would be continued was consistent with his own recollection of the discussion.

Alfred, still willing to give Pierre the benefit of the doubt, then said, "There seems to have been some misunderstanding as to the position taken by Willie and myself at the meeting in December, and I desire to have this matter cleared up. You have unintentionally misinformed T. C., and I suggest that Willie and I write T. C. setting forth our views."

Pierre politely said that that was an excellent idea. Alfred then asked Pierre to send him copies of his correspondence with Coleman regarding the Finance Committee's action in December and subsequent correspondence, if any, concerning this matter. Pierre agreed to do so, and with that the meeting was adjourned.[64]

Pierre did not send the correspondence Alfred had requested; instead, he sent only a single letter relating to the Kraftmeier incident and the suggestion that the four cousins pool their stock. Alfred notified Pierre that the wrong letter had been sent and once again requested copies of the correspondence he had asked for and needed to have before writing to Coleman. He received no reply from Pierre, so without benefit of knowing what Pierre had told Coleman, Alfred on 16 February wrote to his ailing cousin, explaining in some detail why he had objected to the price, which had been his only objection to the plan, and why he thought at that time that $125 a share was a fair price. Much had happened, of course, since the December meeting, both to the value of the Du Pont stock (very bullish) and to the prospects for an early end to the European war (very bearish). Once again, Alfred had an opportunity to offer a higher price for Coleman's stock, but he did not give a specific figure which he would find acceptable and left the impression that he was still opposed to any price above $125.[65]

Alfred did not know that, on the day prior to his writing to Coleman to explain belatedly his position in regard to the purchase, Pierre had received a letter from Coleman offering to sell an even larger share of his holdings. Once again Coleman stated his firm belief that the price would soon rise to $300, but he was convinced that the executives "now at the helm and actually doing things should make the profit. . . . So clear am I that this is the right time to get them interested that I am willing to let go . . . 20,000, 30,000 or even 40,000 shares at today's market, which I assume is about $200."[66]

As might be expected, Pierre did not let either Alfred or William know of Coleman's latest offer, but his reaction to this surprising development must have been similar to that of his ancestor and namesake Pierre Samuel du Pont I upon discovering that the Emperor Napoleon was willing to sell the whole territory of Louisiana. In cannot be accurately determined, of course, just when Pierre may first have conceived of the idea of getting Coleman's 20,000 shares for himself and a few close associates whom he wished to reward. His failure to communicate accurately to Coleman the reason for Alfred's objections to the proposal, even though both Alfred and William wished to continue negotiations, seems to indicate that the idea had occurred to him very soon after the committee meeting in December. With this new offer from Coleman, which spoke only of making the stock available to the men "now at the helm," there was no longer any question as to what Pierre's real objective was. With his head reeling from the excitement that this offer aroused, Pierre hurried off to confer with John Raskob.

Alfred's letter of 16 February to Coleman, written in ignorance of the communications that had taken place between Pierre and Coleman since December, was read by his cousin with great puzzlement. Why on earth would Alfred still be talking about the offer of last December and trying to justify his position of offering only $125 a share for Coleman's stock?

Immediately upon receiving Alfred's letter of 19 February, Coleman wrote:

> I am in receipt of your letter of February 16th and have read it several times. I cannot, however, make out why you wrote it. . . . You say, "The one point on which there seemed to be a difference of opinion was the question of price."
>
> If you will have the date of my proposition to the Finance Committee looked up, you will find that the then accepted orders, and the money in our hands, was sufficient to make the stock have a book value beyond the price which I offered it, and the earnings by reason of the accepted orders, would make the 6½% you mention . . . look like a drink of water. . . .
>
> To guide me, won't you please advise how much of your common stock you are willing to let go at this time to important employees, at price suggested by you, $125 per share. Probably I can join you in [such] an offer.[67]

Coleman's letter, especially the final sarcastic paragraph, was as puzzling to Alfred as his letter had been to Coleman, for Alfred still knew nothing about Coleman's new offer to sell most, perhaps all, of his stock at $200 a share, and Pierre's enthusiastic response to that proposal.

Pierre had been very busy during the few days that Coleman and Alfred had been hashing over what was now ancient history. On 18 February, after having been assured by Dunham that Coleman was indeed prepared to sell all his stock, Pierre met with his brothers Irénée and Lammot, his brother-in-law R. R. M. Carpenter, and, of course, the ever-attentive John Raskob to discuss how they could possibly finance this spectacular undertaking. It was agreed that Raskob should go to New York that afternoon to confer with William Porter, a partner in the House of Morgan, for it was clear that Pierre and his four associates would have to arrange a sizeable loan if they were to take advantage of Coleman's offer. His 63,214 shares of common stock, at $200 a share, would cost $12.6 million. In addition, Coleman wanted to sell his 13,989 shares of preferred Du Pont stock for $85 a share. The grand total for all of Coleman's holdings in the Du Pont Company would amount to $13,831,865. The total assets of all four prospective purchasers amounted to only $11,000,000, even if they pooled everything—stocks, real estate, and personal property.[68] A loan of at least ten million would be necessary to swing the deal.

Raskob returned from New York on 19 February with good news. Porter had assured him that J. P. Morgan and Company would be able to arrange a loan sufficient to meet Pierre's needs. With this assurance from the country's most powerful banking firm Pierre on 20 February wired Coleman and then, in a lengthy telephone conversation, worked out the final details satisfactory to both parties. Pierre also sent a follow-up telegram:

> I understand I have purchased from you 63,214 shares of the common stock of the E. I. du Pont de Nemours Powder Company at $200 per share and 13,989 shares of the preferred stock . . . at $85 per share. Also you are to be paid $8,000,000 in cash and $5,831,865 in 6 percent 7 year notes . . . the collateral on the notes being 36,450 shares of common stock.[69]

Coleman confirmed the transaction by return wire.

Thus while Alfred was reading Coleman's letter of 19 February and flushing with anger over his cousin's sarcastic offer to sell as many shares of his stock at $125 as Alfred would be willing to sell at that price, Coleman was in the process of selling everything he owned in the Du Pont Company to Pierre. It had taken only five days after receiving Coleman's offer for Pierre to consummate the deal that made him the single largest stockholder in the history of the company. Providing he could hold fast to the laurel crown, Pierre's victory was now complete.

Coleman had made his offer to sell to "the men now at the helm and actually getting things done." No one of course was "more at the helm" or more active "in getting things done" than Pierre himself. He planned to take the lion's share—50 percent—of Coleman's stock for himself. The remaining half would be distributed in part to the four associates who had joined him in the secret transaction—his brothers Irénée and Lammot, his brother-in-law Ruly Carpenter, and John Raskob. Into this elite company Pierre also admitted a sixth man, A. Felix du Pont, brother of Francis I.

du Pont and the one son of Francis Gurney du Pont who had over the years consistently sided with Pierre in every conflict with Alfred.

Only one more step remained before Pierre could take control of the Du Pont Company. Although he could now claim as his own one-half of Coleman's shares in the company, this holding did not in itself ensure Pierre's absolute control. Having finally won out over both Alfred and Coleman, Pierre never again wanted to contend with a division of authority within the company. If Coleman's shares were sold directly to individual executives, as was Coleman's original intention, there was always the possibility, however, remote, that these shareholders might vote their shares contrary to Pierre's wishes. Some intermediary agency had to be devised that would give absolute voting control of all of Coleman's former holdings to Pierre, even though he owned outright only 50 percent of those shares.

For a master financial manipulator like Pierre, this control was easily arranged. He hastily created a new corporation, the Du Pont Securities Company, incorporated under the laws of Delaware and having a capital issue of 75,000 shares. It was this holding company which became the formal purchaser of all of Coleman's stock. Its total assets consisted of the stock purchased from Coleman plus enough Du Pont shares from Pierre's own holdings to give to Pierre 60 percent of the Du Pont Securities Company stock, amounting to 45,000 shares. Each of the other five members of the syndicate received 1,250 shares, and a healthy reserve of 23,750 shares remained for future dispersal to those whom Pierre might wish to influence or reward.[70] In any issue that might arise in the future, the Du Pont Securities Company would vote its holdings of Du Pont Powder Company as a single block—a block tightly held in the hands of Pierre. It was a neat and tidy package with no loose strings dangling that might be pulled the wrong way. At last Pierre had the kind of orderly arrangement that he had always sought for the Du Pont Company.

It is difficult to understand how all of the feverish activity occurring during the week of 18–25 February in New York, Rochester, Minnesota, and Wilmington, Delaware, could have gone unnoticed and unreported. All six men in the syndicate were sworn to secrecy, to be sure, as was Coleman, who, given his propensity for talk, was fortunately in a hospital a thousand miles away from the Du Pont headquarters.

Alfred was in Wilmington, sitting in an office adjacent to Pierre's, yet he noticed nothing to make him suspect that something momentous was afoot. He was still pondering the pros and cons of answering Coleman's insulting letter of 19 February and perhaps even of suggesting a specific price for Coleman's stock that his cousin might find acceptable.

Although Pierre and his syndicate were entirely successful in imposing a cover of secrecy over their activities in Wilmington, they did not have the power to extend that same protective cloak to other places particularly to that notorious passageway for rumor, Wall Street. On Friday, 26 February, Seward Prosser of the Bankers Trust Company, one of the several New York banks that had joined with Morgan in providing the ten million dollar loan to Pierre, called Raskob to tell him that information of the deal had unfortunately leaked out, and Mr. du Pont might prefer to make his own public statement before the press began to publish the rumors that were now circulating. Mr. du Pont did so prefer, and quickly provided the *Wilmington Star* with as brief an account of the transaction as possible. He refused, when questioned, to name the other participants in the purchase but did say that all were actually engaged in the management of the company. Negotiations, Pierre admitted, had been in progress for several weeks. The acting president wished to emphasize that there would be no change in the present management.

The *Wilmington Star* carried the story on its front page on Sunday 28 February under the heading: GENERAL DU PONT SELLS HIS ENTIRE POWDER HOLDINGS. Considering the magnitude of the news for the Wilmington public, it was a modest little account, for Pierre had requested that the story not be blown up sensationally, and the *Star* always tried to please the du Ponts.[71] For Alfred, however, the story leaped out from the page with as great a force as a two-inch banner headline announcing the United States had entered the war would have had. His two cousins had surprised him several times before in their long years of association, but never so completely and so devastatingly.

Alfred was in his office early the next morning expecting a call from Pierre, but Pierre, apparently still holding to his motto of "the less said the sooner mended," sent no word. Finally, at four in the afternoon, Alfred could wait no longer for the confrontation that had to be. He called Pierre and asked him if he would kindly come over to see him. Pierre entered Alfred's office as pleasantly unperturbed as if this were a normal visit. Alfred wasted no time in coming to the point. As he later would recall their conversation, Alfred blurted out:

> "Pierre du Pont, don't do this thing! . . . It is wrong."
> Pierre in surprise asked why it was wrong.
> "Because," Alfred responded, "you have accomplished something by virtue of the power and influence vested in you as an officer of the company. . . . For that reason the stock which you have acquired does not belong to you but to the company which you represent. I therefore ask you to turn the stock over to the company."
> Pierre shook his head and said that he was very sorry to have to disagree with Alfred, but what he had done he had done on his own.

He assured Alfred that he had not "used the Company's credit in any way."

"Pierre," Alfred said, "your father and my father were brothers. Neither of those men would have approved, I am confident, of what you have done. For their sake, as well as for your own, put that stock in the company's treasury. You can't afford to . . . injure your business reputation. Pierre, I ask you."

Pierre again shook his head.

"Then you refuse to make this concession I ask?"

"I do," Pierre replied and left the room.[72]

It would be the last time Alfred would ever make an appeal to Pierre for the sake of kinship and friendship to do the right thing. This line of personal communication between them, which had once meant a great deal to both men, had been forever cut.

The *Wilmington Star* had had a major scoop for its Sunday edition, but unfortunately because of the stricture laid upon it by Pierre it had not been free to give the story the full play that its editors would have liked. The press outside Wilmington was under no such restriction, however, and every major paper in the country gave it full play the following morning. Pierre's terse announcement, which had omitted such important details as the membership of the syndicate that had purchased Coleman's stock, the price paid, and how the transaction had been financed, simply invited the very journalistic speculation that Pierre had sought to avoid. The most commonly reported figure on the price tag was twenty million dollars. Coleman must have found little humor in that bit of journalistic exaggeration, nor would Alfred appreciate the Philadelphia *Public Inquirer's* assertion that, in addition to Pierre, the purchasers must surely have included Alfred and William de Pont.[73]

Cousin Willie, at his winter home in Sea Island, Georgia, did not get the news until Monday when he read it in the local paper. He at once sent Pierre a telegram, "Paper states you have purchased Coleman's stock. I presume for the company. Any other action I should consider a breach of faith."[74] After receiving a telegram from Alfred, however, William realized that his presumption was wrong, and he hurried back to Wilmington for consultation.

Francis I. du Pont, upon learning from his brother A. Felix just who the members of the syndicate were, called for a meeting of all the major stockholders who had not participated in the transaction, to be held on the evening of 4 March in Alfred's office. In addition to Francis I. and Alfred there were present: the two sons of the late Dr. Alexis I. du Pont, Philip F. and Eugene E. du Pont; the son of the late Eugene du Pont, Alexis I.

du Pont III; the son of Colonel Henry A. du Pont, Harry F. du Pont; and of course, the Colonel's brother, William du Pont. These du Pont cousins had, like Alfred, been kept in total ignorance of the stock transaction. Each of the "outsiders" had responded to the startling news of the purchase in varying degrees of amazement and anger. Harry F. du Pont, who had taken over his father's very large block of stock when the Colonel was elected to the United States Senate, was more impressed than upset by Pierre's action. Two of the younger cousins, Eugene E. and Alexis I. du Pont III, were simply puzzled as to how the development would affect their own financial status within the company, but the elders of the group—Alfred, William, and Francis I. du Pont—were obviously outraged and were prepared to take the leadership in opposing the *fait accompli* that Pierre had presented to them.

Before any of the group could express his views or suggest any course of action, the door to Alfred's office opened and Pierre and his brother Irénée entered the room. To the amazement of the group assembled, Pierre smiled genially to all of his cousins and politely asked Alfred if he and Irénée might join them. There followed a half-hour of family chit-chat in which Pierre was at his most charming, inquiring about the health and current social activities of his cousins, while Alfred sat at the head of the table and stared at the ceiling. Finally, when not even he could maintain this farce any longer, Pierre broached the subject that was on all their minds. He said that he understood that some of his cousins present had taken exception to a transaction that he, Irénée, and a few others not present had recently completed. Cousin Willie had even suggested to him that the act had been a breach of faith. Pierre now politely asked Willie to explain his statement. William did so with his usual terseness. Pierre could not possibly have made this large purchase without having used the company's credit. The company had made the purchase possible. Therefore, the stock belonged to the company. It was as simple as that. Alfred concurred in William's brief statement. Francis I. went even further, pointing out that the Du Pont Securities Company now controlled so much of the Powder Company's stock that it could determine all of Du Pont Company policies and investments. It had been this control of the company's treasury that the Coleman stock purchase had given him that had provided Pierre with the collateral he needed to back his notes of indebtedness.

Confronted by these charges, Pierre lost his usual composure and angrily replied that he had in no way used the company's credit to get his loan. He deeply resented the accusations. Alfred told him there was one way in which he could absolve himself from having acted in bad faith. He could now offer to sell the shares to the company. Pierre said he would not, and with that blunt refusal he and Irénée left the meeting.[75]

If the purpose of Pierre's surprise appearance at a meeting to which he had not been invited was to disconcert the gathering, then that goal was fully realized. The meeting in Alfred's office broke up shortly after Pierre's departure, without the full discussion Alfred and Francis had hoped for. No common response to Pierre's purchase had been agreed upon.

Pierre himself was more than a little disconcerted by the confrontation. He had been struck by the imposing group of du Ponts who had assembled at Francis's invitation; representatives of nearly every branch of the family except Coleman's and his own had gathered to discuss what he had done. It looked like the family councils of old in the days of Boss Henry, and this made Pierre uneasy. The possibility of Alfred's forming an united opposition to him within the company and the family could not be lightly dismissed. Later that night, as he mulled over the whole affair, Pierre realized he had made a tactical error in giving a quick and unequivocal no to Alfred's request that he make the stock available to the company. The next morning Pierre hurriedly sent out a letter to all of the directors of the company stating, "I withdraw a statement made on . . . March 4th that I will not sell the stock. I am now open to consider a proposition. I have not formed a final opinion, nor have I and my associates discussed the acceptance of a proposition. The question is absolutely open as far as we are concerned."[76]

At a special meeting of the board of directors on the following day, 6 March 1915, Pierre quickly revealed how open the question really was as far as he and his associates were concerned. After first demanding and getting a vote of confidence in his position as acting president of the company, Pierre then formally offered to sell to the Du Pont Powder Company all of the stock that the Du Pont Securities Company had purchased from Coleman. This was the cue for J. P. Laffey, the company's general legal counsel, to enter the board room, bearing in his hand an impressive legal tome. Pierre asked counsel to advise the board on the legality of the company's purchasing the stock from him and his associates. Laffey did his job well, for which later that day he would be rewarded with a gift of $50,000 worth of stock in the Du Pont Securities Company. Laffey gave his solemn legal opinion that the Du Pont Company could only use its surplus to buy stock. Since at the moment there were only six million dollars in the surplus funds, the company could not legally pay the thirteen million dollars required to obtain Coleman's stock. Taken aback by this judgment, the board then postponed action on Pierre's offer to sell until it could have a recommendation from the Finance Committee, which it asked for within four days.

Pierre next read a letter from T. Coleman du Pont in which he formally resigned as president of the Du Pont Company, as chairman of the Finance Committee, and as a member of the board of directors. Promptly and unanimously, the board then elected Pierre du Pont president and appointed him chairman of the Finance Committee. Pierre's brother, Irénée, was elected to membership on the Finance Committee to replace Coleman, and Francis's brother A. Felix du Pont was made a director. The pieces were all falling neatly into place. Pierre at last had the title as well as the substance of power, and his control of both the Finance Committee and the Board of Directors was assured.

On 8 March 1915, the Finance Committee met to consider Pierre's offer to sell the Coleman stock. A motion to recommend the purchase would require the approval of a majority of the committee. With the four-man committee evenly divided between Pierre and his brother on one side and Alfred and William on the other, there was no possibility of the committee's recommending the purchase. Nevertheless, the four went through the meaningless process. Alfred moved and William seconded a motion to recommend the purchase. Irénée cast his vote against the motion and that killed it, since Alfred's and William's two favorable votes did not constitute a majority of the Committee.[77]

Two days later, the Finance Committee reported to the board of directors that it was unable to make a recommendation either for or against the purchase. Since the Finance Committee was deadlocked and a sharp division existed among the directors, Alfred then proposed that the board ask a group of outside economic and legal advisers "not in any way associated with the Company" to consider the matter in all of its legal and economic ramifications and to make its recommendation to the board. Francis enthusiastically seconded the motion. Irénée argued vigorously against this proposal, saying, "I do not see why we should delegate our duties to others or seek advice in a matter we know more about than others." From the chair Pierre warned the board to consider carefully Mr. Laffey's opinion that any such purchase would be illegal. The vote on Alfred's motion proved to be a test of Pierre's control of the board of directors. Alfred's motion lost by a vote of 14 to 4, with only Alfred, William, Francis, and Frank Connable voting in favor.

John Raskob then moved that the company accept Pierre's offer to sell the stock at cost plus $250,000 for expenses. The fateful moment of decision had arrived. Raskob's motion was defeated by a vote of 14 to 3. Only Alfred, William, and Francis du Pont voted in favor of the motion. Frank Connable, caught in a vise of conflicting loyalties, abstained. Raskob, of course, had voted against his own motion.[78] The directors' meeting adjourned with Pierre confident that the issue was now settled. He had made his offer to sell, which should absolve him from any charge of having acted in bad faith. After full discussion the board of directors had overwhelmingly voted to reject the offer and had thereby legitimized the action he and his five associates had taken. And that should be that. Period.

To minimize any hard feelings that might still remain as an unpleasant residue of this transaction, Pierre offered shares of Du Pont Securities Company stock to various members of the family, particularly to those cousins who had gathered in Alfred's office on the evening of 4 March. Pierre invited them to exchange 223 shares of their Du Pont Powder Company stock, worth $66,825 on the current market, for 500 shares on Du Pont Securities Company stock, worth $99,100. The only two cousins who were not given this opportunity to enrich themselves were Alfred and William. Pierre knew that they would reject the offer out of hand, but if they could not be won over, they could be isolated by this tactic. Even Francis I. du Pont was invited to participate.

It was a tempting offer, and it is hardly surprising that most of the du Ponts grabbed at the bait. What is surprising, however, is the number of du Ponts who refused the lure. Pierre made a special effort to persuade Francis but Francis told him that the offer could not be accepted because to do so would be "against my convictions." Three of Francis's brothers—Ernest, Archibald, and E. Paul du Pont—also refused to participate, as did three of their cousins, Philip F., Eugene, and Alexis I. du Pont III.[79] If Pierre had found this distribution of Du Pont Securities stock to be useful in winning friends, Alfred also found it to be useful in determining who within the family he could count upon to be his allies, for as far as Alfred was concerned, the issue was not yet settled. A counterattack would be made, but he needed time to work out the best tactics to undo what Pierre now believed could not be undone.

Confident that victory was his, Pierre turned his attention to matters that needed immediate attention. During the 115 years of its existence the Du Pont Company had been a major supplier of munitions for four American wars, and extraordinary demands had been made on its facilities, especially during the prolonged Civil War and the brief Spanish American War. But never had the company known such pressure as it now experienced during the years 1915 and 1916. All of the existing Du Pont plants were running twenty-four hours a day at full capacity to supply ordnance for a world at war and still the demands of the Allied Powers for high explosives and smokeless powder could not be met. Plant expansion was proceeding at a feverish pace, with ample funds to finance construction, and still the unspent profits, like the demand, continued to mount. In 1912, the Du Pont Powder Company's net earnings were $6,871,000. In 1916, its net earnings had climbed to $82,107,000. In the one year of 1916 alone, the Du Pont Company's profits exceeded the combined totals of all the years from 1902, when the three cousins had taken over the company, through 1915.[80]

As the profits increased, so did the value of the Du Pont stock, as well as the dividends paid on the stock. Coleman's shares, which Alfred had wanted the company to buy in December 1914 for $125 a share and for which Coleman received $200 in Feburary 1915, were worth $900 a share by October. The dividend paid on each share rose from $8 in 1914 to $82 in 1915. Coleman's prediction that Alfred's projected return on Du Pont stock would soon seem "like a drink of water" had been realized beyond both cousins' wildest imagining. Coleman's stock, which he had eagerly sold for $13.8 million, was now worth $58.1 million six months later. Poor Coleman. In dashing off to New York to find a faster track, he had foolishly abandoned what had become America's real speedway, right where he had been in little old, provincial Wilmington. The prize that Pierre had seized by staying put was becoming a richer purse with each passing week.

Alfred, still the second largest stockholder (albeit now a distant second to Pierre), also shared handsomely in the wartime bonanza. His monthly income was in six figures, but he felt he was doing little to earn it. He dutifully attended board and committee meetings, although he must have found it painful to sit in the same room where Pierre was the presiding officer. He gave no indication to the new powers controlling the company that he intended any further opposition to their regime. He appeared to be simply a tired and defeated old man whom time and events had made superfluous. He was, in short, someone whom Pierre and his brothers believed they could safely ignore as they devoted all their thoughts and energy to supplying the Allied Powers with the matériel of war.

Alfred, however, was not sitting quietly with hands folded in resignation during the summer of 1915. He met frequently in secret conference with his closest allies, Francis and William. Alfred also discovered to his surprise a third and most determined cohort—Philip F. du Pont, the eldest son of Dr. Alexis I. du Pont. This particular cousin had always been considered a bit strange by the du Pont family. As a boy he had showed an interest in only two things: writing poetry and dashing off to witness fires whenever he heard the Wilmington sirens blow. This latter avocation had earned him the nickname of Fireman Phil. As a young man he had shown no interest whatsoever in the family business, and upon the death of his father in 1904, he had taken his share of the estate and had gone off with his wife to Merion, Pennsylvania, to become something of a recluse, writing poetry that no one read, and making shrewd investments in Wall Street to increase his fortune. Since he shunned all family gatherings, Alfred hardly knew anything about him when he suddenly made his appearance to offer his support in opposing Pierre. That he should be Alfred's ally was particularly surprising, inasmuch as it had been against Philip's own mother, Elizabeth Bradford du Pont, that Alfred had brought his ill-advised suit for slander only a few years earlier. Perhaps, however, it was this very suit that helped to endear Alfred to Philip, who may have disliked his mother as much as he disliked most of the other du Ponts.

Whatever his motivation, Fireman Phil was ready to dash into battle at Alfred's side with the same eagerness that he had once shown in racing to fires in Wilmington. His support quickly proved to be more valuable than Alfred had anticipated. During his years of activity in the stock market, Philip had made some valuable contacts, and high on his list of important acquaintances in the business world was John G. Johnson, America's most noted corporation lawyer. Johnson had become rich and famous defending such giants as Standard Oil, American Tobacco, and the Northern Securities Company. Although the government had prevailed in many of their most important suits, nevertheless it had been Johnson's doctrine of "the rule of reason" that the U.S. Supreme Court had accepted as its guide to interpreting the Sherman Anti-Trust Act, and it had been that doctrine that had proved so effective in the emasculation of that act.

It was to this giant in corporation law that Philip du Pont now turned for legal counsel in preparing the attack on Pierre. Living in virtual retirement on his estate near Philadelphia, Johnson was now devoting most of his time to his art collection, which was one of the finest private collections in the country. Philip was not at all sure that Johnson would take the case, but was delighted to find that the old man was intrigued by the details as outlined to him. Later Johnson told Alfred that he had spent his life defending the majority interests of corporation, and now at the conclusion of his career, he decided it was time to give some attention to the interest of the minority stockholders.[81]

In addition to Johnson's services, Alfred acquired the services of William A. Glasgow, the former assistant attorney general under Taft who, along with Attorney General Wickersham, had successfully prosecuted the government's case against the Du Pont Company in 1911. Alfred had been so impressed with Glasgow's performance in that suit that he had vowed that if ever again he was involved in a court case, he would like to have Glasgow on his side.

The major burden of pushing the case against Pierre and his associates in the Du Pont Securities Company fell upon Alfred, who all through the autumn of 1915 worked closely with his distinguished panel of lawyers in collecting data, conferring with friendly witnesses, and preparing a bill of complaint that would accuse Pierre du Pont of having acted for himself and a few colleagues in securing T. Coleman du Pont's stock "in violation of his trust as an officer director and confidential representative of the E. I. du Pont de Nemours Powder Company."

Alfred had assumed that it would be he who would bring the suit against Pierre and that he would be joined in that suit by his three cousins as well as any other persons who wished to be named as co-plaintiffs. Philip, however, insisted that he be allowed to file the suit. Alfred and his lawyers quickly agreed. Having Philip F. du Pont file the bill of complaint would come as a complete surprise to Pierre and his associates and would catch them off guard. It might even give rise to apprehension among the defendants as to how broadly based the opposition was if this little-known cousin was listed as first plaintiff. Thus did Fireman Phil win for himself an enduring place in the history of the du Ponts, by having the family's most notorious law case bear the official title *Philip F. du Pont, et al.* vs. *Pierre S. du Pont, et al.*

All of these preparations were made during the summer and autumn of 1915, with the closely guarded secrecy that even that master of secret operations, Pierre S. du Pont, might have envied. Neither he nor any of the other directors and top executives of the company had any inkling that charges were being prepared against them until on 9 December 1915 a court officer for the U.S. Court for the District of Delaware appeared in Pierre's office to serve the papers stating the bill of complaint and the intention to bring suit in equity.

It was now Pierre's turn to be astounded by what he read in the papers—in this case, court papers—and it was now Pierre who called Alfred to demand an interview. Alfred refused to discuss the matter with Pierre, claiming that he was not the plaintiff. Pierre never doubted for a minute, however, that Alfred was the real instigator. Pierre struck back, hard and fast. At a meeting of the Board of Directors on 10 January 1916, he again demanded a vote of confidence by asking for the removal of Alfred du Pont as vice-president of the company and as a member of the Finance Committee. The board approved of this action, with only Francis I. du Pont voting in the negative and Alexis I. du Pont III abstaining.[82] Alfred, who along with William and Frank Connable had not attended the meeting, countered this action of the board by announcing his intention to join Philip in his suit. In addition to Alfred, Francis I. and three of Francis's brothers and one of his sisters, Eleanor du Pont Perot, also became part of the et al. in Philip du Pont's suit.

Pierre was not yet finished with his punitive action. He would not be satisfied until he had severed the last tie connecting Alfred with the company. The announcement of the annual meeting of the Du Pont stockholders to be held on 13 March 1916 carried the notice that the board recommended the election of three new directors: Hamilton Barksdale, William Ramsay of the Engineering Department, and Frank Tallman, director of purchasing, to replace Alfred, William, and Francis I. du Pont. Pierre began the collection of proxy votes for this election. The vote was a lop-sided victory for Pierre. His slate of directors received, with the aid of the proxies from Coleman's stock, 411,053 votes, against a token 3,621 write-in votes for Alfred, William, and Francis, since Alfred had urged his supporters not to send in their proxies for what he called an illegal election.

On the morning after the annual meeting the Philadelphia *Public Ledger* carried what it regarded as the obituary notice of Alfred I. du Pont's long career with the du Pont Company:

> With the smoothness of a well-oiled machine, the annual meeting of the E. I. du Pont de Nemours & Co., the big powder company was held at noon ... and at the close of the meeting Alfred I. du Pont, who has been connected with the powder manufacturing business in this country as an active official all of his life; William du Pont also interested actively in the powder business for many years; and Francis I. du Pont, chemist of note and inventor of a number of smokeless powder processes, had ceased to be directors of the biggest powder company in the world.[83]

By mid-March 1916, the stage had largely been cleared of all minor actors and the spotlight was now focused on the two protagonists who had emerged as contenders for the empire that four generations of du Ponts had built over the past one hundred years. The du Ponts had long been a disputatious lot, but all previous quarrels had been family spats compared to this confrontation, for never before had there been so much at stake. Never before had the whole family been forced to take sides, for in this conflict neutrality was not a possible option. One must cast one's lot with either Alfred or Pierre, and the unhappy choice that had to be made would separate brother Francis from brother A. Felix, sister Eleanor from sister Irene, mother Elizabeth from son Philip, uncle William from nephew Harry. It was a true civil war within the family, but there was nothing civil in the manner in which it was conducted, and after the war, there would be no reconstruction of former relationships within the lifetime of any of them. Small wonder that this domestic conflict captured the attention of the country as no other intramural struggle for corporate and family power has ever done.

Each of the two protagonists was armed with his own formidable weapons. Alfred, for his part, was confident that the evidence at hand would show conclusively that he was correct on the two major points at issue: 1) that Pierre had deliberately violated his trust as acting president of the company in his negotiations with Coleman in order to advance his own and not the company's interest, and 2) that Pierre had used the company's credit to obtain the loan that enabled him to buy Coleman's stock. With the evidence available, a good lawyer should be able to make his arguments so effectively and so conclusively on these points as to convince any unbiased judge in the land. Alfred, fortunately, had secured the services of the best corporate lawyer in the country. As one historian has evaluated John G. Johnson's courtroom effectiveness: "It had become almost proverbial among financiers and others that his opinions were equivalent in value to judicial decisions. His strength before the courts was due not only to the vigorous power of his accurate reasoning, but still more to the fact that the courts felt absolute trust in the fidelity of his presentation of his cases."[84] It would be a most unusual judge—either unusually stupid or unusually prejudiced—who would not be favorably impressed with the thrust of Johnson's arguments. Even Pierre had to admit that Alfred had the better legal counsel, for although Pierre had secured the services of George Graham, a very able Philadelphia trial lawyer, Graham was no equal to Johnson in either knowledge or experience in corporate law. Pierre was also incensed by Alfred's employing as his second legal counsel, William Glasgow, the man who had effectively prosecuted the antitrust case against the Du Pont Company. The fact that Glasglow and Alfred had become good friends after that suit simply confirmed Pierre's suspicion that Alfred had not really wanted the company to win its case against the government but instead had hoped for a dissolution of the powder trust Coleman had built.

Alfred realized, however, that superiority in legal counsel and the winning of a favorable verdict from the courts would not in itself ensure his victory over Pierre. A judge could, and Alfred firmly believed any judge would, order Pierre and his associates to make available to the company the stock they had purchased from Coleman, but no judge could force the company's stockholders to purchase that released stock. It would be necessary for Alfred to persuade a majority of the shareholders to approve of the purchase. To get this majority Alfred would have to win over a great many of the smaller stockholders, both those from the general public and those within the family, to agree to the purchase. Alfred was far less certain of victory in this arena than he was of victory in the courtroom. Although it might appear on the surface that it would be of such a distinct financial advantage to all of those stockholders who did not possess any of Coleman's former stock to have that stock revert to the company, nevertheless Alfred was enough of a realist to appreciate that the task of persuasion would not be easy. Pierre had cleverly distributed the Coleman stock to many of those within the family who already held large blocks of Du Pont voting stock, men like Harry F. du Pont and Eugene E. du Pont, and this would ensure their votes if nothing else did.

A few simple calculations revealed the kind of odds that confronted Alfred. On his side, the only two large blocks of votes were his own 75,534 shares and William's 27,994 shares. His other allies' holdings were relatively small: Francis had 1,634 shares; Philip, 1,732; Ernest, 1,730, Eleanor Perot, 108, Archibald, 68; and E. Paul du Pont just 8 shares. Alfred thus had only 108,808 shares that he could safely count on.

Excluding the Coleman shares, Pierre and his fellow syndicate members controlled outright 100,000 shares. Pierre's closest family relatives could contribute another 58,000. The directors of the company who were not members of the syndicate but who were indebted to Pierre for their places on the board had another 14,000 votes. An additional 50,000 shares were held by company employees and by pensioners who could also be pressured to vote Pierre's way. This meant that Pierre was fairly certain of having 212,000 votes, even if the Coleman shares were not voted. There remained 139,000 scatter shares, which were up for the taking by either side. Pierre had to win only 18,000 of these votes to ensure his victory, but Alfred needed to get at least 122,000.[85]

To win the uncommitted votes, Alfred could appeal to the shareholders' sense of justice, to their loyalty to him, and to their appreciation for his long years of service as a great powderman. (If only the millworkers could have voted, Alfred would have won hands down.) Although the scandals associated with his personal life and the social ostracism that had resulted had largely dissipated the deep reservoir of family affection and admiration that had once been Alfred's, he still hoped he could appeal to the family's strong regard for history. He had saved the company for the family when the ruling elders had wanted to sell out in 1902, and he was, after all, the eldest son of the eldest son of the eldest son of the founder. His rightful place was in the company. Above all, he could appeal to the small shareholders' mercenary instincts. That would be a powerful appeal, but would it be enough?

Pierre's greatest strength lay in the fact that he was the incumbent in the seat of power. During his administration the company had grown rich beyond the wildest dreams of avarice. With a war in Europe it could hardly have failed to prosper, but nevertheless this prosperity had occurred under Pierre's regime. If Alfred could claim the title of the great powdermaker, Pierre could claim the even more appealing title of the great moneymaker. No one, not even Alfred, could deny Pierre's proven skills as financier and manager. Who within Alfred's camp could replace Pierre if he were ousted, as he had ousted Alfred? Could Alfred himself, or Francis, or William? Who, in short, was the more dispensable—Alfred or Pierre? These were the hard questions that Pierre would raise and that Alfred would have to answer if he could.

First, however, there was the trial itself. All other speculations were largely dependent upon the opinion that the court would render after reviewing the evidence and considering the arguments presented by the legal counsel on both sides.

The trial opened on 28 June 1916 in the Federal District Court in Wilmington, with Judge J. Whitaker Thompson of Philadelphia presiding. This was the court in which Alicia's father, Judge Bradford, normally presided, but it was, of course, unthinkable that Alfred's avowed enemy would be permitted to hear the case. Reluctantly, Bradford had been obliged to ask to be excused because, as he said, he was "related to both the plaintiff and the defendant."

All through the hot weeks of July 1916, the case was argued. The courtroom was stifling in its oppressive heat, but neither the press nor the public was deterred from packing all available space. Philadelphia and New York newspapers ran pictures of the du Pont ladies trooping up the stairs of the courthouse dressed in their finest summer clothes, as if they were attending an afternoon leveé at the Wilmington Country Club. It was the biggest show on the whole East Coast.

Both the heat and old age considerably slowed down the physical movements of the grand old man of the American court room, John G. Johnson, but neither had affected his mind or his tongue. He was relentless in his cross-examination of Pierre's testimony and coldly logical in his arguments for the plaintiffs. Graham was also at his oratorical best in portraying Alfred as the spiteful third man out, who "thought T. Coleman du Pont was in trouble and was waiting to get his grip on that which would put him in control of this Company."[86] Graham did not suggest that this statement applied equally well to Pierre.

The hundreds of exhibits and the weeks of testimony and arguments presented to the court would, when printed, fill many volumes, but the entire case could be condensed to the two simple issues that Alfred had initially raised: One, had Pierre fairly and honestly represented the Finance Committee's position in regard to the purchase when he negotiated with Coleman in January 1915? Two, had Pierre and his associates either openly or by inference made use of the Du Pont Company's credit and wealth in order to obtain their ten million dollar loan from the New York banks? It was now up to one man to provide the answers to those two questions.

In his closing argument to the court (and what proved to be the closing argument of his career), John G. Johnson urged Judge Thompson to "teach these people a lesson that, higher than the mere acquisition of wealth . . . it is better to do right, and that when they attempt to misuse their position and to destroy the confidence which has been put in them and which they have agreed to maintain, that such action can never be attended by any results favorable to the man who undertakes it."[87]

The long, hot ordeal was over. The judge indicated that it would be some time before he would be able to render a verdict, and so it proved to be. During the months of waiting, Pierre, seemingly confident that victory was his, was preoccupied with ever-mounting demands for munitions, not only from the Allied Powers, but also from his own government, as the United States prepared its defenses and moved closer to war.

Alfred, equally confident of victory, was also busy adjusting to a new life outside of the Du Pont Company. He moved his office from the Du Pont Building into the Odd Fellows Building (an appropriate location for him, in the opinion of some of his more hostile cousins). He also withdrew all of his funds from the Wilmington Trust bank, which the Du Pont Company controlled, and placed his accumulated wealth in the Delaware Trust Company, a bank that he and William purchased—Alfred's first venture into banking. He was delighted to learn that most of the Brandywine powdermen followed suit and transferred their small deposits into his bank.[88]

During these months Alfred also became involved for the first time in his life in Delaware politics, not for the purpose of seeking office himself but in order to break up the Republican machine that his cousins Colonel Henry du Pont and Coleman had constructed over the years. Alfred did not lack activities to keep him occupied during the last months of 1916 while he waited for Judge Thompson to announce his decision.

In spite of his many pursuits, however, the suit was never out of Alfred's thoughts, and as the months passed with no word from the court, the wait became increasingly more difficult to bear. On 2 April 1917, President Wilson appeared before a joint session of Congress and asked for a declaration of war against Germany. Four days later, Congress acceded to the President's request. The United States had joined the great crusade "to make the world safe for democracy."

Alfred and his friends had a new cause for anxiety. Now not only Britain and France were dependent upon the Du Pont Powder Company, but also their own nation. Since Du Pont had a monopoly on the production of the absolutely essential smokeless powder, would Judge Thompson dare to make a ruling that might in even the slightest and most temporary way disrupt the most vital industry in America at that moment? Might he not regard it as a patriotic duty to sustain the present management, even if he was convinced that justice would be served by a ruling that was adverse to its interests?

These very real fears which Alfred held were soon to be put to rest. On 12 April 1917, six days after the United States had entered the war and patriotic hysteria was at its height, Judge Thompson handed down his long-awaited decision. It was a bold assertion of judicial independence, even in time of war. The judge neither minced nor economized on words. In a lengthy text of over one hundred pages, Thompson had not only dared to rule against the present Du Pont management but also had done so with such force as to indicate unequivocally that a judge could best express his patriotism not by laying aside the principle of justice in the name of expediency but in asserting that principle with great vigor no matter what the practical consequences might be.

Judge Thompson first took up what he termed the "fundamental fact to be determined." He found that Pierre had knowingly deceived Coleman by not providing him with the full details as to why on 23 December 1914 the Finance Committee had not accepted his offer of 20,000 shares at $160 a share. Instead, Pierre had given Coleman to understand that Alfred and William had rejected the offer out of hand and had made no mention of their wish to negotiate over price. Pierre had also deceived the Finance Committee by telling them on 10 February that the deal was completely closed when in actuality he was still in secret negotiations with Coleman for the purchase of the stock for himself. The judge sharply criticized Pierre for not making available the correspondence that had been exchanged between him and Coleman, even when Alfred specifically requested copies of those letters.[89] In a strong indictment of Pierre, Thompson wrote, "He [Pierre] in whom confidence was reposed of faithfully representing the Company in the transaction, cannot, after betraying his trust through the suppression of facts and through the abandonment of his principal [the Finance Committee] . . . now take advantage of his own wrong and successfully claim . . . [that this] relieves him of any further responsibility and puts him in a position to thereafter deal for himself."[90]

Although the court failed to find any clear evidence that Pierre and his associates had directly used the company's credit to obtain the necessary loans from the New York banks to purchase the stock, Judge Thompson found it hard to escape the conclusion that the securing of the ten million dollar loan had been greatly facilitated by the fact that the borrowers were executives of a firm that had millions on hand for deposit in the banks that would accede to their request for a loan. The fact that in eleven of the fifteen banks who provided the loan, the Du Pont Company increased its deposits by 300 percent the next day seemed to be more than the mere coincidence that Pierre and Raskob claimed it to be.[91]

Judge Thompson gave short shrift to John Laffey's legal judgment (a judgment for which he had been handsomely rewarded with a gift of stock from Pierre and a seat on the board of directors) that it would be illegal for the company to purchase the Coleman stock even if it wished to do so, inasmuch as it did not have $13 million in surplus funds. The Du Pont Company's credit was certainly good enough to enable it to borrow funds if it had been necessary, but the profits that were accumulating in the spring of 1915 were so great that in a matter of a few months the company had a surplus large enough to buy Coleman's stock four times over.[92]

Nor did the court consider the Board of Directors' rejection of Pierre's belated offer to sell the stock he had acquired from Coleman as being either a legitimate or a conclusive action. "That action was obtained by the votes of nine directors who . . . were interested in preventing the Company from acquiring the stock, and of six others who were kept in ignorance through willful misrepresentation and suppression of facts. . . . To uphold the action of the Board meeting of March 6th and 10th would be contrary to conscience and good morals."[93]

The court had ruled for the plaintiffs on every issue raised, and it had rejected every argument offered by Pierre et al. in defense of their actions. Pierre was stunned by the verdict, and his enraged brother, Irénée, wrote to their co-defendant, A. Felix du Pont, that Judge Thompson had given an opinion that might as well have been written by John G. Johnson himself "as far as its one-sidedness is concerned." Thompson had even gone "to the ridiculous extent of claiming that Pierre willfully prepared a fraud just as Johnson claimed."[94] Indeed, Johnson's arguments had been virtually synonymous with the court's decision. It was Johnson's final courtroom victory. Two days after Thompson handed down his opinion, Johnson dropped dead of a heart attack. Irénée told Felix that he saw the hand of Providence at work in that,[95] but unfortunately for Pierre and his brothers, Providence had been too slow on the draw.

The plaintiffs had won a notable victory in court. Their bill of complaint against Pierre du Pont and his associates had been accepted in toto and almost verbatim by Judge Thompson. Pierre had been branded deceitful, one who out of avarice had wrongfully seized for himself that which belonged to the company he was entrusted to represent. The night of 12 April 1917 was a night for celebration, and the champagne bottles popped at the great limestone chateau of Nemours, and the beer mugs foamed in the stone cottages on Breck's Lane. There was even a wild rumor that quickly circulated among the older powdermen in Henry Clay village that Pierre and his brothers were already in jail and that Mr. Alfred would be returning to his old office on the Brandywine the following morning.[96]

Alfred's own elation, however, was tempered by the cold realization that the victory over Pierre had only been partially achieved. The battle had reached precisely the turning point that Alfred had anticipated when he and his friends had carefully tallied the votes that each of the two contending parties could count upon if the issue should be taken to the stockholders for a final decision. Now it would go to the stockholders, for Judge Thompson had concluded his opinion with the judgment that there must now be a special election among the holders of Du Pont common shares to determine if they wished the Du Pont Company to purchase at cost, plus expenses, the stock that the Du Pont Securities Company had obtained through deceit from T. Coleman du Pont. The 63,000 shares formerly owned by Coleman could not, of course, be voted in this election since the ownership of these shares was itself the issue at stake. Judge Thompson appointed a special Master of the Court to supervise the election, which was set for 10 October 1917.[97]

Alfred knew that the odds were heavily weighted on Pierre's side in such an election, and he made a futile plea to the court to have the stock turned over to the company first, and then hold an election to determine if the company wished to retain the stock by paying Pierre for its cost. In other words, Alfred sought to reverse the procedure, by asking the shareholders if they wished to retain what had been given to them by the court rather than to vote on accepting what they did not as yet have.[98]

Judge Thompson, as might be expected, would not agree to Alfred's proposal. He could castigate Pierre for what he had done to get the stock but he could not force the stockholders to accept what they might not wish to have. After all, the plaintiffs themselves in their initial suit had asked the court to order precisely the procedure that Judge Thompson had now decreed must take place. The election would be held as scheduled. The great scramble for proxies was on, and both sides used extraordinarily high pressure to gather in the uncommitted votes.

Alfred very effectively used the court's indictment of Pierre to further his campaign. On the morning after the decision was given, the key sentences in Judge Thompson's opinion were printed in heavy bold type on the front page of the Wilmington *Morning News,* which Alfred now owned, and this issue of the paper was sent to every Du Pont stockholder in the country.[99] Pierre, in turn, had Barksdale send out a circular to all stockholders warning them that what was at stake was not the ownership of a few thousand shares of common stock but the very existence of the company. If a majority of the shareholders should support Alfred, then Pierre du Pont and all of the other defendants in this case would immediately resign their managerial positions and leave the company.[100] Did the shareholders really wish to kill the goose that was presently laying and would continue to lay such very rich golden eggs for them all? The few eggs that they might gather in by this foolish ploy were hardly worth that.

For the more sophisticated shareholders, Pierre commissioned the highly respected Charles Evans Hughes, former Associate Justice of the U.S. Supreme Court and recently defeated Republican presidential candidate, to write an informal opinion that would refute Judge Thompson's ruling. In his statement Hughes very cleverly differentiated between Coleman's original offer to sell 20,700 shares to the company, which had never been accepted, and his later offer to sell all of his holdings, which had never been made to the company but only to Pierre and the other men at the helm. It was an effective argument. Pierre had the statement printed in a booklet for the limited distribution that Hughes, for political reasons, insisted upon.[101]

The campaign for the votes of 461,432 shares of common stock was bitterly waged by both sides for six months. Even Coleman got in the act by issuing a statement in which he claimed Pierre had always kept him informed of every detail of the transaction. Still grieving over his loss of $45 million by having sold everything too quickly, Coleman was more interested in promoting his own interests than those of Pierre, however. He made the extraordinary claim that if the shareholders voted to take the stock away from the present owners, then it must revert to him. He would be only too happy to take all 63,000 shares back![102]

It was clear to Alfred long before the proxies were counted on 10 October that he was not going to win. With the two to one advantage that Pierre had among those already committed, Alfred could not hope to obtain the 122,000 of the 140,000 uncommitted shares that he needed for a majority. Even so, he did not expect the outcome to be the disastrous defeat it was. The final tally gave 140,842 votes in favor of the company's buying the stock from Pierre and his associates, but the negative votes amounted to 312,587. By a more than two to one margin, Alfred had lost the contest. Pierre finally sat secure in the seat of power, the golden goose sat secure on her nest.

In March 1915, soon after Coleman had made his surprising offer to sell all of his stock and the proposal had been quickly accepted, Pierre, somewhat bemusedly, had written to his cousin, "I still wonder, Coleman, why you did it."[103]

Coleman had answered, "It became evident that one of us had to take a back seat . . . and with your brothers to help and brothers in law with you, it was best for me to take a back seat, and you to take the lines."[104]

The same question, "Why did you do it?" would also be asked of Pierre, but unlike Coleman, he had no quick and easy response to that troublesome query. He had tried to provide an answer that would be acceptable to the Court of Equity in 1916, but Judge Thompson had rejected his explanation and had found his testimony to be false. Pierre had, to be sure, won out over Alfred in spite of the court's indictment. The overwhelming vote of confidence he had received from the shareholders should have forever stifled any further inquiry. The same old question persisted, however, and even as an old man in the 1940s—long after Alfred and his other foes were dead—Pierre continued to seek an answer that would satisfy an even more demanding judge than J. Whitaker Thompson—his own conscience. He wrote and rewrote the chapter on "The Schism of 1914–15" for the history of the Du Pont Company that he was preparing, but no version ever gained the complete approval of his own critical judgment. Every explanation he could offer for his action—that it was necessary to take the stock before Coleman sold it to others, or that it was essential that he "take the lines" for the good of the company, as Coleman had said, or that the deserving younger executives needed to be rewarded—could be refuted with the single statement that exactly the same thing could have been accomplished if the stock had gone to the company instead of to him.

Only one explanation would ring true, and Pierre's two friendly biographers gave it in their careful analysis of Pierre's motives. "Yet clearly," Alfred Chandler and Stephen Salsbury wrote, "the reason Pierre moved so fast and secretly when the offer from Coleman came was that he wanted to obtain control of the family company. He preferred to face Alfred and William with a *fait accompli*. These two cousins . . . would not have permitted a transaction to go through that gave Pierre voting control of the company any more than Pierre would have quietly accepted a similar move by Alfred and William."[105]

This is the one sound explanation that can be offered to the query, "Pierre, why did you do it?" It is the one honest explanation, however, that Pierre himself could never offer—even to himself. He could never admit that he had made the move that he himself would never have countenanced from any other person—a move that would never have occurred to Alfred and William to make and one that had caught them both in utter surprise when Pierre did it.

On 11 October 1917 the press hailed Pierre as the victor in the long and bitter dispute just as it had earlier and, as it turned out, prematurely given the laurel crown to Alfred in April. Yet the press and the public in general erred in proclaiming any one of the cousins who had once formed a triumphant triumvirate as the victor. All three proved to be losers. Certainly there was no doubt that Coleman was a big loser. In his rush to get his hands on thirteen million dollars he had forfeited his claim to an additional forty-five million dollars just one year later. Alfred, to be sure, still had his shares and the great wealth that these shares brought to him, but he had lost all other ties to the company to which he had dedicated himself, and for the moment he had no other direction or purpose to his life.

Pierre seemingly had everything: the presidency, the largest block of stock, and an absolute control that the company had not known since the days of Boss Henry. But Pierre had been the loser, too. He had been publicly branded by a federal court as being deceitful and false to the trust which had been placed in him. Neither wealth nor power nor even a later court decision could ever totally disguise the scar, for Pierre's hide was not as tough as that of the old railroad buccaneer, Jim Fisk, who had once proudly boasted, "Nothing is lost, save honor."

Perhaps the only winner was the Du Pont Company itself, for under Pierre's able management it would prosper in both war and peace as never before in its long history. But that too would have been realized even if Pierre had turned Coleman's stock over the company. As in most wars, so also in the du Pont civil war, nothing was gained that could justify the losses suffered.

X

Du Pont vs. Du Pont: The State
1916–1920

THE WAR that was waged among the du Pont cousins for fifteen years had begun as a sordid intramural scandal, precipitated by Alfred's divorce and subsequent marriage to his beautiful and generally unpopular cousin, Alicia, herself a recent divorcée. This first phase of the du Pont civil war consisted largely of guerrilla action conducted by du Pont in-laws—Judge Bradford, Bishop Coleman, and a bevy of aunts by marriage. Their campaign deployed the kind of stealthy tactics against which Alfred was least prepared by temperament to defend himself and his wife. He had struck back blindly and shamefully in a series of counterattacks, which occasionally were directed against innocent bystanders, such as his own son, but Alfred succeeded only in further weakening his own position within the family. His ignominious defeat resulted in social ostracism from a major portion of the family.

The second phase of the war extended far beyond the narrow confines of the club and drawing-room society of the Brandywine valley, for what was at stake in the second engagement was not simply social position within the family, but control of an industrial empire that reached from the nitrate fields of Chile for its raw materials to the bloody battlefields of France for its market. Only the family and a few gossip tattlers in the yellow press really cared to know whether Alicia had slept with Alfred prior to their marriage in 1907, but in 1915 the entire world had a vested interest in the fortunes and future of the Du Pont Company.

The second campaign, like the first, was neither anticipated nor desired by Alfred. He was perfectly content with the arrangement agreed to by all three cousins when they took over the company in 1902, and as far as he was concerned, the triumvirate could and should continue until voluntary retirement or death removed any one of the three cousins from his assigned roles. Coleman and Pierre, however, had had a different view of the triumvirate, both as to its desirability and its longevity, and they had struck without warning in 1911 with a reorganization plan that removed Alfred from his key role as general overseer of production. Pierre had struck a second time in 1915, when he grabbed for himself Coleman's stock which gave him complete control of the company. In both assaults, Alfred had been caught completely by surprise and had to respond as best he could to attacks that had achieved their objectives before he had even become aware that they had been launched.

In the second phase of the war, Alfred could at least understand the motivation that lay behind the attacks. In the previous raids on his and Alicia's privacy and honor Alfred could never understand what hope for personal satisfaction or material gain could possibly motivate the attackers. They had apparently attacked for the purely sadistic pleasure they found in causing pain, and unfortunately, in his response, Alfred had fallen to their level.

The reasons behind Coleman's and Pierre's attacks in the second phase, however, were no mystery to Alfred. He did know what power and wealth meant, and he knew what some men would do to secure these objectives. Much as he might decry the methods used by his two cousins, he was not at a loss in understanding their motivation, and consequently he could respond with rationality and not with blind passion. The result was that Alfred made a more effective and certainly a far more honorable counter-attack. For a brief moment in April 1917, he even appeared to have won the second round, only to have victory taken from him by the vote of the company stockholders.

Alfred learned a great deal from his first two battles with the family. He had learned the advantage of taking the offensive—of striking first without warning. Above all, he had learned that what counted was might. Might did not necessarily make for right, but it certainly did make for victory. And, he discovered, might could not be found in the courtroom. He had tried that approach both in his ill-conceived and poorly executed slander suit in 1909 and again in his well-conceived and brilliantly executed equity suite in 1916, but in each instance to the spoilers had belonged the victory. To win one needed not a court decree from a single judge but a popular mandate from an army of supporters.

Having lost the previous battles within the family and the company, there remained another sector in which he could win prestige and power and that was within the political arena of Delaware. It was here that Alfred decided to launch an attack that would strike at least two of his hostile cousins, Coleman and Henry A. du Pont, where their immense egos were most exposed and where, consequently, they might prove to be the most vulnerable.

This time Alfred planned to be the aggressor. Having graduated cum laude from that dear school which experience keeps, he could now apply the lessons he had learned from his masters, Pierre and Coleman. He would make his preparations in secret, he would not attack until ready, and he would then strike without warning in order to present his surprised opponents with a *fait accompli*.

Alfred was to be greatly aided in this strategy by the fact that the attack would come from a direction least expected by his opponents. For Alfred had never shown any real interest in politics at any level, local, state, or national. Like the rest of his family, he had always made his expected contribution to the GOP coffers and had usually voted the straight Republican ticket, but that was the extent of his political participation. He had very little respect for most of those Delawareans who sought office and certainly he had no such ambition for himself. Elections were, at best, a necessary ritual for democracy, like baptism or confirmation for Christianity, something that one felt obliged to observe at specific times in one's life.

Active participation in politics as either an office seeker or as a party manipulator, however, was a dirty little game that most gentlemen disdained as being akin to poker games run by shifty-eyed men named Doc or Slick in the back rooms of saloons. This was particularly true in Delaware, a state which, since the first days of its independence, had been notorious as the bargain basement of American politics, where cheap votes could be readily purchased over the counter—no credit standing needed, only cash accepted.

There were, of course, a few distinguished old families in Delaware, like the Bayards, the Saulsburys, and the Grays, who in return for seats in the United States Senate or in presidential cabinets were willing to lend their names to the game in order to provide a veneer of respectability. Throughout most of the nineteenth century, however, very few du Ponts evinced any interest in allowing their illustrious name to be used in exchange for an office—perhaps for the very good reason that there wasn't much of a marketplace in which they could barter. The du Ponts were Republicans in a state that was traditionally predominantly Democratic.

Delaware, like many other states, had had its political lines sharply drawn by the Civil War. Although it was only slightly larger in territory than Rhode Island, its smallness had not provided it with political homogeneity. Delaware was a border state, and like Tennessee it was divided into three distinct parts. Each of its three counties had political character that differentiated it from its neighbors. The northernmost county, New Castle, was dominated by Delaware's only large town, Wilmington, which in turn was a satellite of Philadelphia Republican politics. Fiercely loyal to the Union cause in the war, Wilmington's Republicanism had been further strengthened in the postwar years by the favors the national Republican administrations had provided its industrialists in the form of high protective tariffs, a sound currency, and a national banking system.

The middle county, Kent, although largely rural and Democratic, did contain the state capital at Dover and some pockets of industry. It had the potential of being a swing county that, given enough inducement in the form of hard cash, could be tilted sufficiently to elect a few Republican candidates.

Finally, Sussex, the southernmost county, was almost entirely rural and, for all practical political purposes, was a part of the solid Democratic south. Although only ninety miles separated Wilmington from Delmar on the southern border of the state, politically the distance between the two towns was as great as that between Boston and Atlanta. The only Republicans to be found in Delaware's "deep south" were the blacks, but they were as effectively barred from the polls as they would have been had they lived in Mississippi or Alabama.

Aided by a high poll tax that no black man, at least in the southern two counties, could ever pay even if he had the money, which was highly improbable, and every white man no matter how poor could pay providing he voted right, it was not difficult for the old corrupt Democratic machine to keep its hold in the state. There was also the very convenient Voters Assistant Act, first passed in 1891, which allowed a party pollwatcher to accompany a voter into the voting booth to "assist him in reading the ballot." As one party leader candidly said, "When I buy a horse, I want to see my horse safely locked up in my own stable."' Wilmington might continue to elect Republican mayors and city councilmen, and occasionally a state official if the Democratic incumbent proved to be too incompetent even for Delaware's relaxed standards of merit, but under ordinary circumstances, the statehouse in Dover and the courthouses in Kent and Sussex were securely within the Democratic paddock.

Given the state's limited area and small population, as well as its venerable tradition of vote buying, it was inevitable that an ambitious politician with a great deal of money in his carpetbag would arrive on the scene to buy the state for himself. Delaware was such a neat little package. It had fewer voters in the entire state than there were in the single city of Des Moines, Iowa, and it contained less land than the King cattle ranch in Texas. Yet Delaware had the same number of U.S. Senators as New York or Pennsylvania. What is surprising is that it took as long as it did for an outsider to take advantage of the political bargain that the little Blue Hen state offered.

In 1889, a quarter of a century before Alfred became involved in politics, a political opportunist arrived in Dover prepared to avail himself of this

tempting piece of merchandise. John Edward O'Sullivan Addicks, a Phila-
delphian by birth, had had the good fortune to marry the heiress of a
wealthy pork packer. He had also had the financial skill necessary to augment his wife's dowry many times over, first by speculation in grain futures
and then by organizing and controlling gas utility companies in Boston,
Brooklyn, and Philadelphia. By the late 1880s, "Gas" Addicks, as he was
not so affectionately called by those who knew him best, had almost everything that money could buy—everything except that which he wanted
most—a seat in the United States Senate. He was certain that money could
buy that for him as well, especially in Delaware, which was so temptingly
close to his business headquarters in Philadelphia.

Addicks built a palatial mansion in Clayton, a small town in Kent
County, in order to establish his residency in Delaware. He then lost no
time in hurrying to nearby Dover with his carpetbag full of money. He
quickly sized up the political situation and saw that it would be impossible
to buy a senatorial seat from the Democrats, even if his political philosophy
had moved him in that direction, which it didn't. The Democratic claim
to senatorial seats had long been a feudal right of the Bayards and the Saulsburys and their immediate allies. That left the Republican party, to which
Addicks now turned. The fact that only a rudimentary framework of a
Republican machine existed south of the Chesapeake and Delaware canal
was for Addicks not an obstacle but an opportunity. With his money, he
could build his own machine, and he immediately set out to do so. With
the arrival of "Gas" Addicks in Dover, Delaware politics would never again
be the same.[2]

Addicks was not an ordinary American political boss like Tweed of New
York or Hanna of Ohio. Party ideology meant little to him and he was not
in this game to line his own pockets—quite the contrary. He was there to
spend his wealth in order to satisfy his one all-consuming ambition. Since
U.S. senators in the 1890s were still elected by the state legislatures (and
would continue to be so elected until the 17th Amendment was added to
the Constitution in 1913), Addicks's task was clear cut. He would have to
buy enough seats in the Delaware state legislature to ensure his election
over any candidate the long-entrenched Democratic machine might present
to the state legislature.

Word quickly spread throughout the lower two counties of Delaware
that a new customer was shopping in the political market and was paying
the highest price ever offered for votes. Those who had votes for sale in
Dover, Milford, Georgetown, and on the small farms that surrounded these
towns came forward with their merchandise in a grand statewide trade fair.
There was something almost refreshingly honest in the very openness of
Addicks's bribery. He presented to the voters of Delaware a test of their

character and principles that would have fascinated Mark Twain, for here was a man eager to corrupt not a single town of Hadleyburg but an entire state of Delaware. The sweet potato farmer from Laurel and the oysterman from Lewes, who had imbibed their father's politics along with their mother's milk, could now choose between the faith of their fathers and the tangible reality of a crisp new fifty-dollar bill. Increasingly, cash won out over custom, and quite suddenly, to the complete surprise of the Republicans and the consternation of the Democrats, Republicans began to win seats in the Delaware legislature from districts in Kent and Sussex, where previously they had often not bothered to run a candidate. Even some of the blacks, with poll-tax receipts clutched in their hands, were being led by Addicks's men to the polls to vote a straight Republican ticket. The situation looked very good for Addicks. Almost single-handedly (a hand filled with money, to be sure), he had built a Republican party in Delaware that gave promise of becoming, for the first time, the major party of the state. There was to be a senatorial election in 1895, and by then, Addicks was convinced that he would have purchased enough votes in the Delaware state legislature to achieve his long-cherished goal.[3]

It was Addicks's misfortune, however, that his very success in giving life to the Republican party in Delaware also aroused an opposition that would ultimately defeat him. Suddenly, the Grand Old Party had become sufficiently attractive to be actively courted by others who had long wanted just what Addicks wanted. Now, thanks to Addicks, they saw a chance to realize what had heretofore seemed an impossible dream. Foremost among these would-be Republican candidates for the U.S. Senate was none other than the current reigning patriarch of the du Pont family, Colonel Henry A. du Pont. The Colonel had everything to make him a viable candidate for office—the du Pont name, the du Pont wealth (in 1895 he was the largest stockholder in the company), a distinguished war record that included the Congressional Medal of Honor, the highest social standing in the state, and, since he took his administrative duties as vice-president of the Du Pont Company very lightly, ample time to devote to political activities. Above all, no one could accuse the Colonel of being a johnny-come-lately carpetbagger, as he was so quick to accuse Addicks of being. Henry du Pont could even outspend Addicks, by a margin of two to one in buying votes, if it had to come to that.

The year 1895 was an exhilarating time to be a Republican in the state of Delaware. Addicks and du Pont money flowed across the state in a golden stream, and those who could not be bought could be terrified by horror stories of what would happen to the country if such wild Populist Democrats as Governor Peter Altgeld of Illinois or William Jennings Bryan of Nebraska should ever get control of the country and crucify American business on a cross of silver. The country was also suffering from the worst depression it had yet known, an economic catastrophe that could be conveniently blamed on the conservative Democratic administration in Washington. Not surprisingly, in the November election of 1894, the Republicans in Delaware gained both the governorship and control of the state legislature. When the newly elected legislature convened in January 1895, it was apparent that a Republican would be chosen as U.S. Senator. But which Republican? The state legislature was obliged to provide an answer. It began balloting on 15 January 1895. After 177 attempts to reach a majority vote of 27, the legislature on 9 May finally, by a one-vote margin, declared that Henry A. du Pont was the duly elected senator from the sovereign Commonwealth of Delaware. "Gas" Addicks had learned the hard lesson that in Delaware you could quite easily buy a horse. You could even lead him home to your stable, but you had no assurance that he would stay put if someone came along swinging a bigger bucket of oats.

It was a proud Colonel du Pont who left for Washington to claim the seat in the Senate that had been waiting for an occupant since the 4th of March. His arrival did not prove to be the triumphal entry he had anticipated, however, for the enraged Addicks faction of the party had sent its own lobbyists to Washington. They were armed with an impressively damning bundle of affidavits attesting to the methods used to obtain the election of Colonel du Pont. There was also a petition containing hundreds of Delaware Republicans' signatures urging the United States Senate to deny du Pont his ill-gotten spoils. The fight had simply shifted to another arena. The United States Senate proved to be as badly divided as the Delaware legislature had been, and it was not until May of the following year that the Senate, by a vote of 31 to 30, rejected Henry du Pont's credentials as having been properly elected to the Senate. A few Republican senators had been sufficiently appalled by the evidence of fraud to join with the Democratic minority in turning du Pont away. Finally, in February 1897, the Delaware legislature in desperation elected an obscure and quite innocuous Democrat, Richard Kenney, to take the Senate seat that had been vacant for two years.[4]

The battle between the du Pont and Addicks factions had certainly not been resolved by this weak compromise, however. For the next six years, the state was to be torn apart by what is known in Delaware as "the years of the Addicks frenzy." It might just as well have been called "the years of the du Pont frenzy." The Republicans continued to be in the majority in the state legislature, but they were in reality two parties, nearly equal in respect to both numbers and degree of corruption. There was only one political issue that received any attention during these years of turmoil.

Would it be Colonel du Pont or "Gas" Addicks who would finally get a Senate seat and thus bring to an end Delaware's civil war?

In 1899, when the term of Senator George Gray expired, the battle between the Addicks and du Pont factions flared up with a greater intensity than before. Gray had been senator for fourteen years. He represented the old order of Delaware Democratic machine politics, which in contrast to the existing Republican chaos looked like the Olympian respectability of George Washington and the Era of Good Feelings harmony of James Monroe rolled into one. There was, of course, no chance that the Republican legislature would re-elect Gray. It had been enough of an embarrassment for a Republican legislature to have elected one Democrat to the Senate. A second Democratic senator was unthinkable. There was, however, no chance that the legislature could elect anyone else. So Gray's seat remained vacant, and there was in the U.S. Senate from Delaware only one senator.

And then, two years later, there was none. The almost-forgotten Richard Kenney's term expired in 1901. Ballot after ballot failed to provide an occupant for either seat. Addicks by now had spent over three million dollars—an astronomical sum at that time—to buy a seat. He had come within three or four votes of achieving a majority, but he never could reach the magic number of 27. For two years, from 1901 until 1903, Delaware remained without representation in the United States Senate, which surely must have set a record in American politics. The state had become the laughingstock of the nation.[5]

An obvious solution, of course, would have been for the two Republican factions to agree to elect both du Pont and Addicks to the Senate. This was a resolution that Addicks would have been happy to accept, but the Colonel, still smarting from his forced retreat from Washington, let it be known he would never enter any room in which the despicable carpetbagger was present, let alone the sacred chamber of the United States Senate.

This deadlock, however, could not continue for much longer. The business community of Wilmington, in particular, pressured its favorite party to find a solution. Theodore Roosevelt was in the White House. His aggressively progressive Square Deal was a matter of great concern to entrepreneurs everywhere, and Delaware simply could not afford to be voiceless in the United States Senate. So in February 1903, a great compromise was reached between the two factions. Senator Gray's former seat, which had been vacant for four years, would go to Delaware's current Republican Congressman, L. Heisler Ball, who for two years had been the state's sole representative in the U.S. Congress. Ball could not be labeled a du Pont follower, but he was known to be anti-Addicks and was therefore acceptable to the Colonel. Kenney's seat, in return, would be given to anyone Addicks

cared to name, except himself, of course. Addicks, like Moses on Mt. Pisgah, was permitted to gaze upon the Promised Land, but could not enter
it himself. Addicks named a faithful henchman, J. Frank Allee, to occupy
the seat that had represented his last hope of glory. Addicks had got the
best of the deal (he at least had one of his own men in the Senate), but he
had lost the war.[6]

Two years later, when Ball's abbreviated term was due to expire,
Addicks, with his usual punctuality, reappeared in Dover, but by now he
was a pathetic figure. He had lost control of his gas utilities and had
declared bankruptcy. His clothes were shabby, and his once-stuffed carpetbag was now as empty as Mother Hubbard's cupboard. This time he got
no votes for Senator in the Delaware legislature. Even so, old animosities
died hard. It took another year before the legislature could elect a successor
to Ball, but ultimately Colonel du Pont could not be denied his place in the
sun. In June 1906, we was duly elected to the United States Senate, and
this time that august body opened its doors to him.[7]

Like hundreds of other Delawareans, Alfred du Pont had viewed these
past ten years of Delaware politics at its worst with mixed emotions of
disgust, amusement, and indifference. It was a spectacle that hardly inspired
confidence in the political process, nor one that would attract honest men
to seek office. Alfred, out of loyalty to the family, gave his personal support
to his Cousin Henry in the Colonel's long struggle to gain a Senate seat.
Not until after his divorce in 1906 did Alfred openly break with that particular branch of the family, and by that time Henry was safely seated in
the Senate.

By the following year, Colonel du Pont's machine was so completely in
control of the Republican party and the state legislature that the Colonel
had no difficulty in blocking Allee's re-election to a second term and in
replacing him with one of his own men, Henry A. Richardson of Dover.
The Addicks frenzy had now completely subsided, and from 1907 on the
du Pont machine ran with the same smooth, quiet efficiency that had been
the distinguishing characteristic of the Democratic machine in the time of
the Bayards and the Grays. There was plenty of money available to keep
the troops happy and the machine oiled, and no rising breeze of political
discontent appeared to ruffle the feathers of the little Blue Hen state.

Even if Alfred had had any interest—which he didn't—in participating
in the game of politics when it had reached a fever pitch of excitement
among the Addicks and du Pont players, he had been much too busy tending to Du Pont powder production to give more than a passing glance at
the political explosions that were rocking the lovely old Georgian State
House in Dover. But if Alfred had no time for the comic relief that Delaware politics could offer, Cousin Coleman, whose ambition and energy

seemed to have no limits, somehow found the time in addition to presiding over the Du Pont Company and becoming involved in New York financial ventures to take a hand in this little game as well. Although Coleman frequently complained about the parochial smallness of Wilmington and longed for the cosmopolitan bigness of New York, he like Addicks, could appreciate the advantages of playing political poker in a little state where the pot was as big as that of any other state.

In Cousin Henry's long, drawn-out bluffing game with Addicks, Coleman was such an active and helpful kibitzer that when Henry in 1906 finally raked in the pot, Cousin Coley was right at his elbow to help count the chips. So indispensible had Coleman become to the old man that he became the newly elected senator's unofficial political manager. From there, it was an easy step for Coleman to take in becoming national committeeman of the Republican Party of Delaware. Today, Dover; tomorrow, his own seat in the United States Senate; and who knew what the day after might bring—perhaps the Oval Office itself. So ran the dreams and the ambitions of T. Coleman du Pont.

In 1911, Coleman startled even the boodle-habituated voters of Delaware with the announcement that he planned to give the people of his adopted state a munificent gift as a token of his love and devotion. He would build "a monument one hundred miles high and lay it on the ground." This was his way of telling the voters, in his customary hyperbolic manner, that he planned to build a highway that would extend from Claymont on the Pennsylvania line southward to Delmar on the Maryland border. This would be no ordinary highway. In an age when even the major national highways were no more than gravelled, potholed roads and most state roads were nothing but one-lane mudpaths, Coleman proposed to build a paved, three-lane highway from one end of the state to the other. It was to be America's first great superhighway, although it would be intra-, not inter-, state. Coleman had no interest in extending his largesse to nonvoters.[8]

The financial difficulties he experienced in his Equitable venture two years later, which led to his selling all of his Du Pont stock, forced Coleman to abandon his highway construction scheme in 1913, before it even got started. This was something of a political embarrassment, to be sure, but the irrepressible Coleman assured the voters that it was only a temporary postponement of the project. Someday soon, the people of Delaware would have their glorious monument, and as they breezed along the T. Coleman du Pont Superhighway in their Model T Fords, they could bless the name of their patron saint at each milepost they passed. They could also give him their votes at the proper time.

In March 1917, Henry du Pont's second term would expire. Neither he nor his active campaign manager, Coleman du Pont, anticipated any difficulty in securing his re-election, even though they would now have to campaign for votes among the general public rather than within the state legislature. The du Pont machine, to be sure, had suffered a minor setback in 1912, when Teddy Roosevelt and his ill-fated Bull Moose party had split the Republican party wide open. For the first time in twenty years, Delaware had given its three electoral votes to a Democrat, Woodrow Wilson. The Democrats had also won the state elections, and, for the last time, the Delaware legislature had the dubious honor of electing a United States Senator. With the Democrats in control, the legislature chose another representative of the old Democratic aristocracy, Alfred's friend and cousin by marriage, Willard Saulsbury.

The 1912 election had been an aberration, however, and by 1916, the du Pont Republican machine seemed once again to be in complete control of the state's politics. Harmony was the keyword for 1916. Never again would the Republicans be foolish enough to hand an election to the Democrats on a silver platter by indulging in self-flagellation. Poor old Addicks, whose money had initially transformed the small Republican party in Wilmington into a statewide machine, was now living as a pauper in a single room in Brooklyn, New York, and he should provide a lesson to any other millionaire carpetbagger who might try to come into Delaware to buy the state. The state had now been purchased by the du Ponts and was not for sale to any others.

Coleman du Pont, in particular, regarded the political prospects for 1916 as being auspicious for him. The sale of his stock to Pierre had solved all of his financial difficulties in New York and had given him enough reserve capital so that he could not only construct the Equitable Building but could also buy the entire Equitable Life Assurance Society. He could even resume construction of his one hundred mile long monument for the people of Delaware. That surely should give him the name recognition throughout the entire state that he needed to satisfy his political ambitions. He would see to it that Cousin Henry got re-elected in 1916, and then perhaps the old Colonel, who was now 78 years old, would resign after a year or two and a compliant Republican governor would appoint Coleman to fill out his term. If not that, then Coleman would be a candidate in 1918 for Saulsbury's seat. It would be only fitting that both senators from Delaware be du Ponts.

Perhaps, however, Coleman might not even have to avail himself of this senatorial landing on his climb to the top of the political staircase. Just as construction on his highway got under way again in the early spring of 1916, a little item appeared in several metropolitan political news columns to the effect that friends of T. Coleman du Pont were urging him to consider the possibility of becoming a candidate for the presidency in 1916. It apparently took only a few friends and very little urging, for in March 1916 a Business Men's Presidential League was formed, and with much fanfare, it opened impressive headquarters in the Waldorf-Astoria Hotel in New York. A former political organizer for both Presidents Roosevelt and Taft, Ormsby McHarg, was in charge of operations. Ostensibly organized to ensure the nomination of any major business leader for the presidency in 1916, the League was in actuality a campaign headquarters for one man. When pressed by reporters for the names of the men whom the League would support, McHarg replied that there were of course many able business executives who would be excellent candidates, but one name topped all others, and that was the former president of America's largest gunpowder company and now president of America's largest insurance company, Thomas Coleman du Pont.[9] Coleman had begun to beat his own drum with vigor. Of the many Delawareans who found the din objectionable, one man in particular was determined to muffle the sound, even if it meant entering an arena that he had hitherto always carefully avoided.

Quite unwittingly, however, Alfred du Pont had already taken his first step into Delaware politics five years earlier when he purchased the Wilmington *Morning News* for $111,000.[10] Owning a Wilmington newspaper had never been a very profitable investment. The press of that town had always had difficulty in competing with the newspapers of Philadelphia. Over the past century and a half many journals had made their brief appearance on the streets of Wilmington, only to succumb to starvation for lack of readers and advertisers. At the time that Alfred entered the journalistic field, there were four newspapers in Wilmington, each struggling desperately to survive. The *Every Evening* and the *Journal* were afternoon papers and the *Star* published only a Sunday edition. Alfred's newly acquired *Morning News* was the city's only daily morning paper.[11] Three of the papers were Republican organs. It was generally believed that the *Journal's* most important financial backer was Colonel Henry du Pont, for it served as a faithful mouthpiece for him and his political machine. Only the *Every Evening,* owned by the Bancroft family, carried the banner for the Democratic party. All four papers, however, regardless of political persuasion, were generally respectful of the sensitive feelings of the du Pont family. If the people of Wilmington wished to learn the details of the latest du Pont scandal, they usually found it necessary to turn elsewhere for the news, which is one reason that the Philadelphia papers sold well in Wilmington.

Of the four papers, the Wilmington *Morning News* was the poorest in terms of equipment, circulation, and advertising revenue, even though it had the morning field all to itself. About all the *News* had in its asset column were its editor, Roy Jones, and its small corps of able but underpaid report-

ers, who stayed with the *News* only out of loyalty to the editor. Jones was
a relic of the old school of personal editors, which had once included such
journalistic luminaries as Horace Greeley, Charles Dana, and Henry Watterson, and his was the only paper in Wilmington that occasionally dared
to show a bit of spunky independence from the ruling patriarchs of the du
Pont clan.[12] For that reason, it had been to the *News* that Alfred had given
the prized journalistic scoop of his marriage to Alicia in 1907.

It was also because of its editor that Alfred had purchased the *Morning
News* in 1911, thus saving the paper from bankruptcy. Alfred had not
bought the *News* to launch himself into the political arena, however. He
had been persuaded by Alicia to do so in order that they might have at least
one friendly forum in Wilmington to present their side of the story in their
ongoing feud with many of their du Pont cousins.

Although Alfred had put up the money for the purchase, it was in the
beginning Alicia's paper. There were frequent secret editorial sessions at
Nemours in which Alicia would provide inside information on the du Pont
family and offer general guidelines for the direction that she wished the
editorial page to take. She and Jones got along famously. She thoroughly
appreciated his acerbic pen and his irreverence toward the sacred cows of
the Brandywine valley, and Jones in turn reveled in the freedom to "whack-
doodle" most of the du Ponts to his heart's content. It was Alicia who first
suspected that Coleman had secret ambitions for high political office, and
it was Jones who refused to swoon in delightful appreciation over Coleman's "monument," unlike most of his fellow Wilmington journalists. With
merciless ridicule, Jones exposed the gift for what it was—a giant billboard
upon which to emblazon the name of T. Coleman du Pont.[13]

Alfred initially paid little attention to his journalistic acquisition. After
putting up the purchase money, he provided no additional funds for operational costs. The paper continued to limp along with the same antiquated
presses and the same meager pay checks for its editor and his staff. Not
until he brought suit against Pierre in 1915 over the purchase of Coleman's
stock did Alfred began to appreciate the full value of having one's own
newspaper. The *Morning News* was the only periodical in Wilmington and
virtually the only paper in the state that presented his side of the controversy. Its circulation figures showed an impressive increase, for even Alfred's
enemies were finding it expedient to see what the *News* was saying about
the case. Alfred, in appreciation, began to pour some much needed monetary sustenance into the journal's famished body. New presses were purchased and additional staff hired. When Judge Thompson gave his decision
in favor of the plaintiffs, Alfred was so pleased and impressed with the
Morning News coverage that he used its issue of 13 April 1917 as his chief
instrument of appeal to the stockholders to persuade them to vote in favor

of the company's acceptance of the disputed stock. For this service alone, Alfred felt that he had been amply repaid for his relatively small investment in journalism.

The *Morning News* proved equally valuable to Alfred when in the spring of 1916, he decided to enter the political arena. Here was a vehicle all fueled and ready to transport Alfred's political views throughout the city and the state. Alicia had been quite prescient in her early suspicions that Cousin Coly had political ambitions that extended far beyond providing support to Cousin Henry in his bid for re-election. The ridiculous T. Coleman du Pont-for-President boom provided ample confirmation of Alicia's speculations. Another senatorial term for Henry du Pont was bad enough to comtemplate. A second du Pont in the United States Senate was unthinkable. Alfred could no longer sit quietly on the sidelines, muttering "A plague on both your houses."

It was former U.S. Senator Allee who would serve as Alfred's Virgil to escort him down into the Inferno that was Delaware politics. Alfred could not have found a better tour guide for no one knew those underground circles better than did J. Frank Allee. A former jeweler in Dover, Allee had attached himself to the coattails of "Gas" Addicks almost from the moment that that ambitious utilities magnate had arrived in Delaware's capital. Allee had quickly become Addicks's most useful and faithful henchman, and when it had become apparent in 1903 that neither Addicks nor Colonel du Pont could achieve the necessary majority to win a trip to Washington, it was Allee who went to represent the Addicks faction in the United States Senate.

Allee had enjoyed his brief tenure of three years in the upper chamber of Congress and had tried to ensure its continuation by deserting his former leader and throwing his support to du Pont in 1906 when the irrepressible Colonel tried once again, and this time successfully, to gain entry into the Senate. Allee quickly discovered, however, that in Delaware politics there was no such thing as honor among thieves. Far from appreciating the service Allee had rendered him, the Colonel succeeded in replacing the now senior Senator from Delaware with one of his own henchmen, Henry Richardson.

His brief moment of glory having been so quickly terminated, Allee since 1907 had nursed his hatred of Colonel du Pont and had lived for the moment when he might have his revenge. He might be a machine politician without a machine, but he was still one of the sharpest political operators in the state. He maintained many of his old contacts, and he had an encyclopedic knowledge of the foibles, strengths, and weaknesses of every ward heeler in Delaware.

Allee found it necessary to earn a living as well as play the game of politics. He owned and operated a vegetable canning factory near Dover, and he even for a time attempted to emulate the financial career of his former mentor, "Gas" Addicks, by organizing a company, the Delaware Gas Light Company, for the manufacture and distribution of gas streetlamps. It was that venture that first brought Allee to the personal attention of Alfred du Pont. Always a likely prospect for anyone offering a new technique in the field of illumination, be it gas or electricity, Alfred in 1912 invested over $100,000 in Allee's highly touted gas lamps. Alfred lost his investment, but without realizing it at the time, he gained an ally who would prove invaluable in future political battles with his cousins.[14]

Following the departure of Addicks from the scene in 1906, there was no longer a rival machine in Delaware to challenge the du Pont machine for political power within the Republican party. If Allee was to have his revenge on Henry du Pont, he would have to find some vehicle other than stalwart, stand-pat Republicanism in which to ride to victory. So willy-nilly, Allee became a progressive political reformer. This former wheeler-dealer *par excellence* of the Addicks brand of political corruption was still a wheeler-dealer, but now he was mouthing the slogans of clean politics and electoral reform with such conviction as to win for himself the seal of approval of political idealists like George Norris and Robert La Follette. Allee was slowly working his way out of the Inferno of Delaware politics into the Purgatorio. Like Dante's Virgil, he would never make it into Paradiso, but he was now eager to point the way.

In 1912 Allee took the leadership in promoting Roosevelt's Progressive party in Delaware. He did so with such effectiveness as to split the Republican party wide open and give the state to the Democrats. This was at the same time that he got Alfred to invest in his gas light company. He also tried to interest Alfred in playing an active role in Progressive party politics. He got from Alfred his vote and following the election of 1912, a note of appreciation. "I go with you in your appreciation as to the result of the Delaware election. The work done by the progressive party was most excellent, and deserves the thanks of all decent people," Alfred wrote the week after the election.[15] Although supportive of the goals of the Progressives, Alfred was not yet ready to accept Allee's offer to take the leadership that that wing of the Republican party desparately needed if it was to do more than simply ensure the victory of the Democratic machine.

Four years later, however, the situation had changed drastically for Alfred. Having been ousted from the company, thanks to the power that Coleman had handed to Pierre, Alfred was in the proper mood to listen when, in early March 1916, Allee came to Nemours, accompanied by Alicia's brother, Edward, to offer Alfred once again the command of the insurgent reform faction of the Republican party, the only political organization that had a chance of breaking the Henry/Coleman du Pont machine.

Allee came directly to the point. As Alfred well knew, his cousin Henry was a declared candidate for re-election to the U.S. Senate for a third term. That in itself was something that should be blocked if at all possible. What was even worse was the real possibility that the seventy-eight year-old senator, who had the poorest attendance record of any man in the Senate, was simply being used as a stalking horse for the even more noxious T. Coleman du Pont. Allee had heard from several of his many very good authorities that the plan was to get Henry re-elected and then have the old man resign. Governor Charles Miller, a loyal supporter of the du Pont machine, would immediately appoint Coleman to fill the vacancy until the next general election two years hence. By 1918, Coleman would be firmly entrenched in office and could win a full term in his own right. So ran the plan. Alicia, who was present at this meeting, could now say "I told you so," for indeed she had been telling Alfred that this might happen for the past several years.

The only way to defeat this nefarious scheme, Allee continued, was to meet it head on. The first test of strength between the machine and Allee's reform movement would occur the following month, when the Republican state convention was scheduled to meet in Dover to choose delegates to the national presidential convention. Delaware had been allotted six delegates to go to Chicago in June to nominate the next Republican candidate for president. Rural New Castle county, outside the city of Wilmington, was allowed to select one of these delegates. It was, of course, simply assumed by the party that Colonel Henry du Pont would be that delegate. Not to select the party's highest-ranking officeholder would be an unthinkable act of impudent independence. But, Allee said—and now his face was flushed with excitement as he leaned over to shout in Alfred's ear trumpet—he, Allee, had the votes already lined up to defeat the Colonel in his home base of New Castle county, providing of course that the reformers had an opposition candidate who would command the respect of the several disgruntled Republican delegates to the county convention whom Allee had contacted.

Alfred was definitely interested in any plan that might thwart the machine. But whom could they select as a candidate? Alfred thought a moment and then suggested Alicia's brother. Edward, however, demurred, and Allee, with all due respect to present company, said that the state convention would support no one to replace Senator du Pont as national delegate other than Alfred du Pont himself. "They will never desert a fully equipped man o'war to trust their chances to a leaky rowboat. They want a craft as substantial and reliable as the one they leave."[16]

Alfred still hesitated. It was a big step to pick up the political banner and lead the charge—something that he had said he would never do. He protested that there were already too many du Pont politicians trying to run the state, and the very name du Pont had become a stench in the nostrils of all decent people. But Allee told Alfred to give the voters of Delaware some credit for intelligence. They were smart enough to be able to tell one du Pont from another.

Then Alicia spoke up. She told Alfred that if he was the man she thought him to be, he could not refuse to meet Cousin Henry and Coleman on the field of battle. After that little speech, Alfred had no option but to accept. It was quickly settled. Alfred would allow Allee to present his name to the appropriate people and would himself come to Dover for the state convention on 11 April. In the meantime, all of this was to be kept secret. No mention of this plan in the *Morning News,* no discussion of it, even with their most intimate friends. The Old Guard must be caught completely off guard. This would be a surprise attack that even Pierre might envy.[17]

Allee made good on the promise he had given Alfred in the Nemours library. As the county delegates gathered in Dover on 11 April, Allee had the necessary twenty votes for Alfred safely secured in his pocket. When the roll was called for New Castle County, twenty votes were duly cast for Alfred I. du Pont to be New Castle's delegate to the national Republican presidential convention in Chicago, and fifteen votes were cast for Henry A. Du Pont. The impossible had happened, Senator du Pont would not head, or even be a member of, the Delaware delegation to Chicago.

Henry du Pont had been so sure of victory that he had not even bothered to attend the convention at Dover. Coleman was there, of course, but busy promoting himself more than he was lobbying for Cousin Henry, and he, too, suffered a slap in the face. He needed to have the convention instruct the Delaware delegation to vote for him at Chicago if his presidential bid was to be taken at all seriously by the national convention, but instead the state convention passed a vague resolution to the effect that it "endorsed" T. Coleman du Pont for president. Did this curiously worded motion mean that the Delaware delegates were committed to voting for their state's "favorite son" as long as he was in the running? When asked about this immediately after his surprising victory, Alfred left no doubt as to how he, at least, would interpret the convention's action. He said he intended to vote for "the man who will bring honor and dignity to this country and who will best serve the interests of the Republican party.[18] No one was so naive as to think that Alfred regarded his cousin as meeting those qualifications. Still committed to the principles of the Bull Moose party of 1912, Alfred announced that he would vote for Theodore Roosevelt.

The Wilmington *News* gleefully ran a banner headline, "BANG! T. C. DU PONT'S BOOM BLOWS UP."[9] Two weeks before the national convention was to assemble in Chicago, the Business Men's Presidential League was out of business. The fancy headquarters in the Waldorf-Astoria were hastily closed, and Ormsby McHarg quietly departed for his home in Waco, Texas, his dream of being a member of President du Pont's cabinet having proved to be as ephemeral as Coleman's own dream of occupying the Oval Office.

Coleman might have suffered further humiliation if Alfred had been willing to engage in the game of underhanded political betrayal that the Republican machine of Delaware delighted in playing. Shortly before Alfred was to leave for the national convention in Chicago, he was surprised to receive a visit from two politicians from Kent County, Alden Benson and James Hirons. These supposedly devoted lieutenants of the du Pont machine offered Alfred Coleman's job as Republican National Committeeman from Delaware. The price tag for this betrayal of their boss was that Alfred must drop Allee and make them his chief political advisors in the state. Impressed by Alfred's victory at Dover and seeing his star in the ascendancy, the Kent County men were in short offering to turn the old machine over to Alfred, providing they could be his chief officers. It is possible, of course, that this offer was a trap in order to reveal that Alfred's feet were made of the same clay as theirs, but whatever their motivation may have been, they were unsuccessful in the attempt. Alfred was not interested in their offer and curtly dismissed them. To underline his position, he made it a point to take J. Frank Allee along with him to Chicago as his guest.[20]

Alfred thoroughly enjoyed his week in Chicago. Alicia's brother had been selected as the alternate delegate from New Castle County, and the two men, with Allee serving as an experienced guide, found the climb up to the high level of American politics an exhilarating experience. It was also, for Alfred, a welcome respite from the tension he had experienced in preparing his suit against Pierre for the disposition of Coleman's stock. The trial was scheduled to begin a week after Alfred's return home, but for five days he could find relaxation in the relatively civilized contest over President making and forget the much more savage combat involved in president breaking that awaited him at home.

The Delaware delegation was sharply divided between the three du Pont machine men and the three Republican mavericks. Although the Coleman-for-President organization had opened an office in Chicago, it was clear from the start that Cousin Coley's candidacy was not going to go very far. Alfred, as he had earlier announced, voted for Theodore Roosevelt on the first two ballots. Associate Justice Charles Evans Hughes of the United States Supreme Court was, however, far in the lead on the first ballot, with

Senator John Weeks of Massachusetts and former Secretary of State Elihu
Root the distant second- and third-place runners. Among the fourteen
other hopefuls in the race, Roosevelt was fifth, with 65 votes, and Coleman
du Pont was eleventh, with only twelve supporters. After a second ballot in
which Hughes picked up additional strength, the convention adjourned for
the night. The next morning Roosevelt announced that he was not a candidate for the presidency on either the Republican or the Progressive party
ticket. (The remnant of the old Bull Moose party was meeting in Chicago
at the same time as the Republicans, and these progressive idealists were
determined to nominate their hero even if the G.O.P. did not.) Even after
withdrawing his name, however, Roosevelt still received 18½ votes on the
third ballot, and Coleman doggedly hung on with five votes. But by now
the bandwagon for Hughes could no longer be halted. On the fourth ballot,
Alfred joined the rush as the Associate Justice received 949½ out of the 987
possible votes and was quickly declared to be the unanimous choice of the
convention.[21] A year later, when Hughes at Pierre's request wrote his pamphlet to influence the company stockholders to vote in favor of Pierre's
keeping the stock he had acquired from Coleman, Alfred would sorely
regret that he had cast his vote in Chicago for Hughes's nomination.

Alfred returned home in mid-June refreshed and buoyed up by his experience in Chicago and eager to face the ordeal of the trial that awaited him.
He was also ready to continue his role as politician. He and Allee had won
an amazing victory in blocking Henry du Pont's election as delegate to the
national convention, but that was after all a minor and very temporary position of political preferment. The real test of the reform movement's
strength lay just ahead, when each county selected delegates to the state
convention, which would meet in Dover on 22 August to select the Republican candidates for the general election in November.

The prospects of a statewide victory for Allee's and Alfred's reformers
were not bright. First, they no longer could count on the advantage of a
surprise attack. The du Pont machine now knew the reformers were in the
field and just who their leaders were. Furthermore, the Old Guard had
control of the actual electoral machinery. The officials at the polls were all
loyal stalwarts; they controlled the election books; they determined which
ballots were valid and would be counted and which were "spoiled" and
would be tossed out; and they were the ones who would prepare and file
the official election returns. The Republican machine also had the support
of most of the newspapers in the state, and the columns of those papers
would hurl invectives at the reformers and label them "party wreckers."

The reformers' most serious weakness, however, lay in the fact that they
were unable to present the primary voters with an alternate slate to oppose

the machine-designated candidates. Allee was prepared to give names of men who might be interested in holding office and who were acceptable to the reformers, but it is hard to sell mere suggestions as viable candidates. Tremendous pressure was put on Alfred to declare himself a candidate for his cousin's seat in the U.S. Senate. Had he yielded to this pressure, it would have made a tremendous difference, for with Alfred heading the slate, other men of integrity and ability might also have declared their candidacy and the reformers would have had a slate that could compete with the machine's. Alfred, however, was adamant in his refusal to run. Not even Alicia could persuade him on this issue. He had entered politics to drive other du Ponts out of politics, and his candidacy for the Senate would, he believed, make a mockery out of his entire campaign for reform.

Even with all of these difficulties confronting them, the reformers did surprisingly well in the county primary elections. Although they did not have a specific list of candidates to back for state and national office, they did have able and loyal candidates to be delegates to the state convention where the Republican ticket for the general election would be determined. In each county, there was no question in the minds of the voters as to who were the du Pont machine candidates and who were the reform candidates. In Wilmington itself, the traditional center of Delaware Republicanism, the machine was able to elect only five more delegates to the state convention than the anti-machine faction. In rural New Castle County, the number of machine and anti-machine delegates chosen were exactly the same. In Sussex County, the reformers scored their greatest victory. They elected eight more delegates than did the machine. This gave the reform faction in two of Delaware's three counties a margin of three delegates over its opponent.[22]

Kent County, however, more than wiped out the surprising edge which had been given to the reformers. In this central county, the machine was the most firmly entrenched, and the county chairman, James Hirons, had given his guarantee that not a single reform candidate would be elected in his county. In spite of his most corrupt manipulations, however, fourteen reformers out of the county's total delegation of forty were elected to go to the state convention. This meant that the du Pont machine would control the state convention by a margin of thirty-seven votes. The Old Guard should have been satisfied with this majority, but it went too far in its eagerness to show its power. When the elected delegates assembled at Dover on 22 August, the credentials committee, controlled by the machine, brazenly denied recognition to the fourteen reformers from Kent County and declared that all forty machine delegates had been duly elected from that county.

This audacious act of robbery in broad daylight outraged even some of the Old Guard delegates elected from Sussex and New Castle, and lost the machine some of the votes it had counted on. As the polling began for the selection of the Republican ticket for the November general election, it quickly became apparent that the machine would not have as easy a time in ramrodding its entire slate through the convention as its numerical superiority of delegates should have ensured.

The reformers were able to prevail over the Old Guard's choice of William D. Denney for governor, and with the support of a sufficient number of machine delegates, they succeeded in nominating John G. Townsend, whom they found far less objectionable than Denney. The convention also nominated Thomas W. Miller for Delaware's one Congressional seat. The reformers were not wildly enthusiastic about Miller, but he was the best of the several machine possibilities. As for the rest of the Old Guard's slate for state offices, only two candidates were acceptable to the reformers— David Reinhardt, nominated for attorney general, and William J. Swain, for state treasurer.

The reformers failed, however, to block the one nomination that they found the most repugnant. Colonel Henry du Pont got his nomination for a third term in the Senate. Any enthusiasm the reformers had felt for their minor victories over the machine had been so dampened by this one major failure as to give little cause for cheering as Alfred and his supporters headed for their homes.

The most immediate issue that faced the reform Republicans after the state convention was the question of accepting the Republican slate as it was. Should they meekly submit to the Old Guard in the interest of party harmony or should they put up a third ticket for the general election? Alfred himself had little difficulty in resolving this issue. He had entered politics primarily to rid the party of the du Pont machine. He had given it a few hard blows but he had not dislodged Cousin Henry from the driver's seat nor toppled Henry's co-pilot, Cousin Coleman. There had to be a third ticket. Alfred made this clear in the announcement that he issued a week after the convention adjourned:

> The methods which obtained at the convention in order to railroad through the nomination of Senator du Pont . . . is [sic] a matter of record, and leaves [sic] no room for argument. . . . If the Delaware electoral vote is to be saved for Mr. Hughes and Mr. Fairbanks [the Republican vice-presidential candidate] at the election in November, no other way than by putting a third ticket in the field presents itself to me at the present time.[23]

Alfred du Pont had issued a clear call for rebellion, and most of the reformers within the state Republican party answered his appeal. Their first step was to make contact with the scattered remnants of the Bull Moose party of 1912, which had never been reconciled with the regular Republican party, even though their leader, Theodore Roosevelt, had now deserted them.

The Alfred du Pont insurgents and the Bull Moose Progressives agreed to meet in Dover on 27 September.[24] It was 1912 all over again, but this time the rebellious progressive faction would be selective in its rebellion. The leaders quickly decided against putting up an entirely new slate, which would only guarantee the Democrats a clean sweep of the state, as had happened four years before. Instead they would nominate Progressive Republicans only for those offices for which the machine nominees were the most clearly deficient in meeting the standards for honesty and enlightened Republicanism.

The first objective of the September Dover convention was to nominate a candidate to run against Henry du Pont for U.S. Senate. This would, of course, ensure the election of the Democratic candidate, Josiah Wolcott. Alfred and many of the other reformers were not particularly enthusiastic about Wolcott, who came from Alfred's own district in rural New Castle County. Alfred would have been far happier if the Democrats had nominated Henry Ridgely of Kent County, but anyone was better than Henry du Pont and his possible successor, Coleman du Pont. After some debate and behind-the-scenes maneuvering, the reformers finally agreed upon Dr. Hiram Burton of Lewes as the man who could attract the most votes away from Colonel du Pont.

The reformers and their Bull Moose allies were in agreement that they should select for their ticket the same three presidential electors whom the regular Republicans had chosen. In contrast to the election of 1912, the Progressives had no presidential candidate of their own and they were now supporting the national Republican ticket of Hughes and Fairbanks. They had no wish to divide the Republican vote for presidency and give the state to President Wilson, as had happened in 1912.

The insurgents then took a hard look at the rest of the regular Republican ticket. The reform delegates had earlier given lukewarm support to the nomination of Thomas Miller for U.S. Representative to Congress, but at this convention, the Bull Moosers, led by Robert Houston, editor of the *Sussex Republican,* refused to take Miller on the third ticket and demanded a replacement. Alicia's brother, Edward Bradford, was chosen as the reformer's nominee for Congress. The convention did accept three of the regular Republican nominees for state office: Townsend for governor, Reinhardt for attorney general, and Swain for state treasurer, mainly because the three were less clearly identified with the Republican machine and were preferable to their Democratic counterparts. For the other state offices, however, the convention of reformers chose its own candidates.

Their selectivity in accepting some of the regular Republican nominees and rejecting others gave to this band of reformers the opportunity of determining who would be elected in November. The Democrats were certain to win only if the Progressives put up a candidate in opposition to the machine Republicans. The election would, in effect, demonstrate to the machine just how powerful the insurgent movement had become.

The campaign that ensued was as lively and as vicious as any that the Blue Hen state had ever known. The regular Republican machine was outraged by the action of the September reformers' convention. Many of the leaders, including James Hirons and Coleman du Pont, would have preferred to have the whole Republican party go down to defeat, as it had in 1912, rather than have certain of its nominees owe their election to Alfred's band of reformers. Such stalwart machine papers as the Wilmington *Journal* demanded that Townsend, Swain, and Reinhardt either remove their names from the Progressive ticket or withdraw as Republican candidates and let some "real" Republicans run. But the three men, with the scent of victory sweet in their nostrils, refused to accept either option. They were quite happy to be the choice of both factions within the party.

The Progressive ticket was received with considerable enthusiasm among the many Republicans throughout the state, and particularly within Wilmington, which had grown increasingly irritated by the arrogance of the two du Pont cousins and the corruption of the Dover statehouse gang. New allies flocked to Alfred's side, including a leader of the railroad workers' union, E. M. Davis of Milford. In time, Davis would be second only to Allee as Alfred's political advisor and confidante. For the purposes of this campaign, Davis formed a Committee of One Hundred, with headquarters on Market Street in Wilmington, and this organization proved to be highly effective in organizing and turning out not only the labor vote but also those more traditional Republicans who were eager for a change but not willing to leave the Grand Old Party.[25]

Election day was every bit as good as Alfred, Allee, Davis, and Houston had hoped for. Hughes and Fairbanks carried the state, but only by a margin of 1,000 votes. If the reformers had not selected the same electors as the machine Republicans, Delaware would once again have been in the Democratic column, as it had been in 1912, and Wilson's re-election would not have been quite as close a call as it was.

What was even more gratifying to Alfred was the fact that all three of the state Republican nominees who also ran on the Progressive ticket were elected by comfortable margins, but not a single machine candidate who did not have the blessing of the reformers was victorious. The Democrats won the Congressional seat, the lieutenant governorship, and the offices of state auditor and insurance commissioner. Particularly sweet were the results in the race for the United States Senate. Henry du Pont lost his bid for a third term by the widest margin of any candidate seeking office. The little-known Democrat, Josiah Wolcott, had defeated the Colonel by 2,600 votes. Henry could now spend all of his time at Winterthur and not feel obliged to make even an occasional appearance in the Senate chamber. Best of all, his seat would now not be available to Cousin Coley. The du Pont political dyarchy had been dethroned.

A jubilant Alfred du Pont issued a victory statement:

> The people of Delaware ARE able to run their own state. . . . Senator du Pont, as is well known, listened to bad advisors, one of whom had his own reasons and ambitions for advising as he did. The Senate, however, is a place for serving the people, in which the Colonel signally failed. . . . The element represented by the Progressive Republicans—the independent vote—is in Delaware to stay. Nothing can eradicate it, no force can crush it. It is everywhere triumphant throughout the State. Machine managers, though artful, no longer can put undesirable men on the ticket and expect the endorsement of the voters. That is not Republicanism. We want Republicanism that will answer to Lincoln's. The old T. Coleman du Pont and Henry A. du Pont machine . . . is gone—dead at the age of fifteen.[26]

On election night Alfred and Alicia invited their political allies, including Allee, Davis, and brother Edward, to Nemours to hear the election returns as they were phoned in from the news desk at the Wilmington *Morning News,* and, as it turned out, to celebrate their great victory. In the course of the evening, Alfred stubbed out the cigar he was smoking and announced to the company assembled that he had just finished the last smoke of his life. If he was to be a politician, he at least would not make any further contributions to smoke-filled rooms. Alfred then asked why the one du Pont cousin whom he was always glad to see, Francis I., was not here to celebrate. Alicia said he was probably in his laboratory as usual. The phone rang. It was Francis, announcing that he had just become the father of twin girls. It was, all in all, a festive night—perhaps not for cigar smoking but certainly for toast making. Alfred needed a victory over Coleman, and that night, for a change, he had one.[27]

Victory over Henry and Coleman du Pont was succulent fruit to savor, but there was a hard stone inside the sweet meat. There was the sobering realization that something more than mere gloating over one's defeated enemies was now expected from the leader of the reformers. Campaign promises had been given and hopes for better government had been raised. The people of Delaware, both friends and foes alike, looked to Alfred to see what he would do with his victory.

They did not have long to wait. On 24 November 1916, only two weeks after the election, Alfred called a meeting of the independent Republican leaders in his office to discuss the program that the new leadership would present to the state legislature when it assembled in January. To the delight of many and the consternation of others, Alfred proposed that the one issue that politicians fear most—tax reform—be boldly confronted and resolutely dealt with by new legislation. After the meeting the leaders of the independent Republicans issued a statement in which they charged that Delaware

> is actually living under the most obsolete laws of any state in the Federal Union. Some of these laws are not only obsolete, but highly iniquitous, and wholly opposed to public progress. The most marked of these iniquitous statutes are those relative to revenue and taxation. In the main they are a century and a half old—fair and equitable enough in the day of their creation, but in view of the changed forms of wealth that exist in the present day, wholly inadequate, and actually inimical to all public progress. The largest part of the wealth of the state bears no part of the state's expense—the old system of taxation including only land and livestock, and exempting all that vast wealth of an interest-bearing character which so markedly characterizes the wealth of the present time. The consequence is that real estate and livestock are taxed to the limit, and the revenues therefrom being wholly inadequate for the purposes of the present day, the state has fallen behind in all that should make it worthy of a high place among the other states of the Union.[28]

Two days later, Alfred issued his own statement, in which he was even more condemnatory of Delaware's taxation policies:

> Delaware, per capita, is nearly as wealthy as any State in the Union. But the largest part of the wealth does not contribute its proper share of the State's expense in running its government, its schools and its various public institutions. . . . Once in the front rank as a State, it has now fallen behind in all things, and largely to the fact that such laws of recent enactment have been framed not in the interest of the people, but for some political party or individual. . . . According to the recent statement made for publication by Independent Republicans, and with which I fully agree, steps should be taken by the next Legislature providing a complete revision of our system of taxation, both state and county. . . . Property, real and personal, should be taxed alike and taxed equitably. If a man has a million to be levied upon, he should object no more than the man who has only a few hundred dollars in property. Surely it is

right that the person who can afford to pay taxes should pay them and consider it a distinct duty as well—an honor to himself and his State.

Education in Delaware needs greater encouragement. Archaic and lifeless from lack of funds, [it] is a sad commentary upon the method [of taxation]. . . . Had sufficient means been in hand to properly educate its children, Delaware today would not be the twelfth or thirteenth state in the point of illiteracy. . . . Delaware should have schools, roads and public institutions second to none in the United States. It is merely a question of securing needed funds in a manner fair to all.[29]

In his statement to the public, Alfred said that his "whole aim now is to endeavor to arouse the people of the State to the necessity of the enactment of new and more progressive laws."[30] In achieving this objective, Alfred was highly successful. The people of the state were aroused. The dirt farmer in Sussex County, the property owner in Milford, the renter of cheap housing in Wilmington, and school teachers and parents throughout the state who had never imagined they would hear a call for this particular reform from any politician, let alone from any du Pont in politics, were both amazed and wildly enthusiastic in their response. State legislators and newspaper editors received hundreds of letters, many of them scrawled upon bits of scrap paper and providing in their very context sad evidence of the state of literacy in Delaware. The proposal was hailed by labor union locals, and the Taxpayers League of Delaware hurriedly passed a resolution in support of tax reform.

Alfred had discovered his ownership of the Wilmington *Morning News* to be such a useful vehicle in carrying his equity suit case against Pierre and his political campaign against Coleman and Henry du Pont that he readily concurred with Allee's and Davis's suggestion that he should extend his journalistic influence throughout the state. During the next two years, Alfred, either through purchase or by heavy subsidies, took control of nine papers outside of Wilmington: the Dover *Sentinel,* the Georgetown *Union Republican,* the Harrington *Journal,* the Laurel *Leader,* the Lewes *Delaware Pilot,* the Middletown *Chronicle,* the Newark *Ledger,* and the *Seaford News.* Davis wrote to Alfred, "With these we will have practically all the newspapers in the state that are of any use to us and will keep TC from having any way to publish his hot air."[31]

On the issue of tax reform, however, Alfred was not solely dependent upon his own newspapers to promote his views. The Democratic press, including the influential Wilmington *Every Evening,* wrote strong editorials in support of the Independent Republican's position. Even the hostile Wilmington *Journal* and Dover *Republican* did not dare to oppose openly the proposal for tax reform. Instead, they asked for a two-year study of the problem—a grace period that would conveniently extend beyond the election of 1918, when they hoped that Coleman du Pont's machine would again be back in control.[32]

Once the issue had been brought forth, however, it could not be put aside for study. Over night, it had become the hottest political issue Delaware had known since the Civil War. Even such emotionally charged issues as prohibition and women's suffrage were pushed into the background. On 8 December, a second meeting of Independent Republican leaders was held in Alfred's office, and this group hammered out two concrete proposals for the legislature to take up in January: a bill calling for a graduated inheritance tax as a source of revenue for the state, and a second measure that would tax interest-bearing investments, which were currently tax exempt, at the same rate as real property. Revenue from the latter source would go to the counties for education and roads.

Coupon clippers and money lenders, including most of Alfred's own family, were horrified by these tax proposals. Alfred had proved to be a traitor not only to his political party but to his social class as well. There were whispered stories among the elite that he had truly gone mad, or even worse, had gone Socialist.[33] But if the moneyed aristocracy of Delaware was outraged, the general public was delighted with the proposals, and it was the public's applause that rang in the legislators' ears as they assembled in Dover during the first week of January 1917.

The Old Guard's only hope to block this populist movement was to bottle up the bills in committee. To do that, a machine Republican must be chosen as president of the senate, for in Delaware, the presiding officer had the power to appoint all senate committees. Although the Democrats controlled the lower house, in the seventeen-man senate the Republicans were in the majority, ten to seven. Six of these Republican senators were, unfortunately for the reformers, loyal to the old Coleman du Pont machine, and they were pushing for the election of Senator Harvey Hoffecker, one of Coleman's most loyal henchmen. The independents put up Senator John Barnard, and the seven Democrats selected as their candidate Thomas Gormley. There were, in effect, three parties in the senate and, for ballot after ballot, the vote for presiding officer remained the same: 6 votes for Hoffecker, 4 votes for Barnard, and 7 votes for the Democrat, Gormley, who ironically received the most votes even though he represented the minority party.

For three weeks, the deadlock persisted. No one of the three men could achieve the nine votes needed to be elected. Finally on 21 January, Senator Gormley announced that the farce had gone on long enough. If the Republicans could not decide among themselves who the leader of the senate should be, he and his Democratic colleagues would decide for them. On the 41st ballot, the vote was changed to 9 votes for the reformers, 7 votes for the machine, and one Democratic abstainer. The Independent Republican, Senator Barnard, had been elected president, with Democratic votes, and could now select a committee chairman who would ensure full consideration of the tax measures.[34]

The graduated inheritance tax bill was quickly reported out of committee, and once it was on the floor, not even the most die-hard conservative dared to vote against it. It passed both houses unanimously and was speedily signed into law by Governor Townsend.

The far more controversial property equalization tax did not finally become law in the form that Alfred had advocated, however. Two separate bills relating to property and income were introduced, one by the Democrats and one by the Independent Republicans, modeled after Alfred's proposal. The resulting compromise was not an equalization tax on investment income in which the revenues raised would go to the counties, but an across-the-board income tax, with the revenues going to the state. It did provide, however, that all revenue raised by this tax must go to a newly established state school fund and to the state highway commission. Real property owners would no longer bear the full burden of taxation in Delaware, and there would now be additional revenue for needed improvements in public education and state roads. The progressive Republicans had good reason to be satisfied with what they had achieved with their victory in the state legislature. They could now turn their attention to municipal reform.

The mayoralty election to be held in Wilmington on 2 June 1917 would be the first test of the effectiveness of the Independent Republicans in a local election. Alfred was not a resident of Wilmington, since Nemours was situated just outside the city line, and therefore he could not take as active a role as he did in state and national politics. But as leader of the reform Republicans, who were determined to oust the current mayor, James Price, he could not ignore this election.

Alfred was instrumental in persuading his good friend and business associate, William Taylor, to challenge Mayor Price in the Republican city primary to be held on 12 May. Taylor was vice-president of the Delaware Trust bank, which Alfred and William du Pont owned, and was current president of the Wilmington City Council. As Alfred's *Morning News* was the first to point out, "a better candidate than Mr. Taylor could not be put forth."[35]

The machine, however, was not ready to surrender the city of Wilmington without a fight, for Coleman was a legal resident of Wilmington and the loss of his own home base would be the final indignity in the series of defeats he had already endured.

In waging the campaign for Taylor as the Independent Republican candidate, Alfred had to depend largely upon the political skills of E. M. Davis rather than Allee, whose home was in Dover. Davis converted his Committee of One Hundred, which had proved highly useful in the 1916 general election, into the Non Partisan League, which would become a permanent organization, with new and larger headquarters in Wilmington. The League grew rapidly in membership, far exceeding the old Committee of One Hundred. In keeping with its name, the League endorsed both Taylor and John W. Lawson, the Democratic nominee, but it put its major effort into getting the Republican nomination for Taylor.

The primary campaign between Taylor and Price was the 1916 campaign in microcosm. The old machine had learned nothing from its previous defeat. The same rough tactics were used, and the poll officials—all machine appointees—were prepared to manipulate the results to ensure Price's nomination. Even so, the result was surprisingly close. The incumbent Price defeated Taylor by only 132 votes. There were so many known irregularities, so many Taylor votes tossed out by the poll judges as being "spoiled," that Taylor filed a bill of complaint with the Republican City Committee, which by law had supervision over local Republican primary elections. The Committee, controlled by the machine, made short shrift of Taylor's protest. It tabled his complaint, refusing to give it even cursory consideration.[36]

The Independent Republicans had no alternative but to turn to the Democrats, just as they had done in the U.S. senatorial race the previous year. The Non Partisan League was no longer non-partisan and went all out for Lawson, whom Davis preferred in any event, since the Democratic candidate was a fellow labor leader in the railway union.

For this brief and bitter campaign, Alfred acquired a new political ally and behind-the-scenes manipulator, T. W. Jakes, who proved particularly effective in providing Alfred and Davis with inside information on the tactics of the machine. Through his Methodist church contacts, Jakes was also useful in serving as an envoy to the organized black vote of Wilmington, which was traditionally Republican, but given the proper instruction and inducement might be persuaded to vote for a Democrat in this election. "I have taken it upon myself to place some good reliable men on to instruct the colored folks tonight and others how to vote," Jakes wrote Alfred on the eve of the election. "I can use about $50 for special work tomorrow if you think advisable."[37] Jakes was never hesitant in asking for funds for "special" work, and Alfred simply assumed, as he wrote his energetic field worker, that Jakes agreed with him that "an attempt to influence anyone . . . to cast his vote other than for reasons based on the individual's knowledge of what is right, would be manifestly . . . improper."[38] Alfred asked no questions and Jakes provided no accounting as to how voters were influenced to vote for "what is right."

However the votes may have been garnered, enough were obtained from Republicans, both black and white, for the Democratic candidate to give Lawson a narrow victory over the incumbent Republican mayor on 2 June. For the first time in the memory of most voters, Wilmington had a Democratic mayor, and once again the Independent Republicans had given dramatic proof that no Republican could win, even in the bastion of Republicanism that was Wilmington, without the support of Alfred's reformers.

Following the city election of 1917, Alfred du Pont became the acknowledged leader of the Republican party of Delaware. As victor, he was now prepared to offer amnesty to his foes and to make an appeal for party harmony. Coleman du Pont's battered machine was equally prepared, after this defeat, to acknowledge his leadership and to accept reconciliation on his terms, however painful this would be to T. C. du Pont and his friends.

Alfred spelled out those terms in an interview that he gave to the Wilmington *News* on 25 January 1918:

> While the next election is still many months away, I realize that the people of Delaware are thinking a great deal about the matter, and, as one of them, I have no hesitation in expressing my opinion, even at this early date. . . .
>
> The coming campaign presents an opportunity . . . to materially strengthen Republican representation in both branches of Congress. I am quite aware that the chief issue will be the election of a Senator to succeed Senator Saulsbury, and I can see no reason why his successor should not be a Republican, providing the Republicans of the State exercise their rights as voters and do what is manifestly their duty as honest citizens. . . .
>
> In order to accomplish this result, it is vitally necessary that the Republican party in Delaware should put forth its undivided strength in the coming campaign. . . . The nominees must receive their nominations honorably, without fraud or duress. A repetition of the scandals of 1916 can only again end in disaster. . . . If the Republican party in Delaware is to become united and remain so . . . assurances must be given to the people that the next State primaries and convention will be free of fraud, coercion, or any other ill practice, and that delegates duly and honestly elected at those primaries will be seated in the convention.[39]

Two months later when the Republican State Committee met in Wilmington on 19 April, Alfred was asked to give an address to the open assembly of voters that the committee had arranged. This was the famous "harmony meeting" that Allee had been working for weeks to bring about. In his speech Alfred held out the olive branch, but he let the assembled Republicans know that it was not "peace at any price," but peace only if the old machine recognized and accepted the principles that had motivated the Independents since 1916.

> Those principles upon which the Independent Republican elements base their claim for prior recognition have been thoroughly exploited and all the gentlemen present are doubtless familiar with them and recognize their merit. In order, however, to anticipate any uncertainty which may still exist, I will once more define them. The Independent Republicans champion the cause of the individual voter as against any political combination operating adversely to his interests and in suppression of his right to self-government. We are, therefore, adverse to and denounce influence of any nature whatsoever brought to bear on the voters the effect of which will be to detract from the value of his vote as an expression of individual opinion. . . .
>
> The Independent Republicans stand for what is admittedly the right—for clean politics and honest elections, results expressive of the wishes of the people by the selection of officials desired by the majority of the voters to act as servants of the people and not as dictators. These are the fundamental principles of pure democracy. Harmony under these conditions should be a simple matter, as such would not be a difficult one to follow. With these principles as a foundation for further constructive values, I call on each man here assembled to pledge himself to give his undivided support to the cause of unity in the Republican party of Delaware.[40]

The national Republican chairman, Will Hays, who had watched with consternation the civil war that had raged among the Republicans in Delaware for the preceding two years, was overjoyed by the news of Alfred's harmony meeting. The Republicans were making an all-out effort to gain control of the U.S. Senate in 1918 so that they would be in a position to influence any peace treaty that President Wilson might submit to the Senate at the conclusion of the war, and every senatorial election was crucial to the Republican hopes. If the Republicans remained divided in Delaware, the current incumbent Democratic senator, Willard Saulsbury, would surely be re-elected, and his might well be the one vote that would determine which party would control the Senate. But united, the Delaware Republicans could take the seat away from the Democrats. Hays wrote Alfred expressing his pleasure at the efforts for reconciliation: "I have heard with the very greatest interest and appreciation the suggestions for united effort in Delaware. I think this is splendid. I thank you again and again."[41]

Harmony, like motherhood and the American flag, was something machine and anti-machine Republicans alike could embrace as a symbol, but the true test of each side's allegiance to the principle would come when the party began to consider actual candidates for offices—particularly for the top office of United States Senator. Would either side actually support a candidate who had long been conspicuous as a leader in the opposing faction, even though that man was nominated by a fair and honest political process and was clearly representative of the majority opinion of the party?

One reason why T. W. Jakes was such a valuable field lieutenant for Alfred was that Jakes was the brother-in-law of Daniel Hastings, long recognized as Coleman's chief political aide. Frequently, from conversations with his wife's brother, Jakes was able to pick up information about the machine's activities and intentions that proved invaluable to the reformers. Soon after the harmony meeting, Jakes reported to Alfred a conversation with his brother-in-law, in which Hastings was clearly testing Alfred's commitment to party unity:

> He [Hastings] then asked me if T. Coleman du Pont was to allow his name to be used and should happen to get the votes, would Alfred I. du Pont support him. I said I thought he would if he received the most votes, but I said, Dan, T. Coleman hasn't enough money to get enough delegates . . . to nominate him. The talk wound up by his asking me to find out how you feel towards him, said he would like to go to the senate for the honor attached to it, but he would not be fool enough to run if you were against him. I said I would let him know your feelings—but the best thing for you to do is to come out in the open and let the people know where you stand and work for one common cause—Republican success—he like the rest *know that we are it.* I tell you my dear sir as I have told you often before you are the "Big bug" in Delaware politics.[42]

Alfred thanked Jakes for the information he had provided on Hastings. In answer to the question as to how he felt about Hastings, Alfred replied, "I scarcely know the gentleman you refer to, having only had the pleasure of meeting him two or three times. . . . I have always identified him as the closest friend T. C. du Pont had in this state, but of course this is very laregly due to what I have heard. There is no harm in anyone's being a candidate, and of course that does not necessarily mean that he will or will not be considered. So far as I personally am concerned, I shall throw my influence with the man who I believe is the choice of the Republicans and who will look after their interests sedulously."[43]

Once again, great pressure was put on Alfred by the Independents to declare his candidacy for the U.S. Senate, and once again he adamantly refused. Cousin Coleman, to be sure, was making noises about his availability, but this time the word went out from friend and foe alike, as well as from the Republican national headquarters, that he would be ill-advised to permit his name to be presented to the state convention. Coleman quietly withdrew from the race and kept a very low profile throughout the entire campaign. Party unity could survive through the general election only if there was no du Pont running for the Senate. Nor, for that matter, would any conspicuous du Pont follower, be it Alfred's Allee or Coleman's Hastings, be considered. Only someone who had name recognition but not faction affiliation could preserve the very precarious détente that had been established.

Two possible candidates met these exacting criteria: William Heald of Wilmington, who had served as Delaware's Congressional representative from 1909 to 1913, and Dr. Lewis Heisler Ball, who had served in the U.S. Senate from 1903 to 1905, as the compromise choice in the deadlock between the Addicks and the Henry du Pont forces. Alfred personally favored Heald, whom he judged the stronger of the two men, but when Allee began a campaign for Ball, Alfred let it be known that he would find either man acceptable. If Allee wanted his old friend and Senate colleague for the two brief years they had served together to return now for a full term in the Senate, Alfred would not stand in the way.[44]

The 1918 campaign was probably the cleanest—and the dullest—in Delaware's history up to that time. The county primaries were held with a maximum amount of harmony and a minimum amount of corruption, and when the state convention met in Dover on 20 August, it proved to be a love feast, rather than the battle royal that the Republicans had so frequently staged in the past. There was not even a squabble among the triumphant reformers. Former Congressman Heald, aware of the prevailing sentiment among the delegates, courteously withdrew his name from consideration and Dr. Ball was nominated on the first ballot. There was equal harmony in the selection of Dr. Caleb Layton as the Republican nominee for the U.S. House of Representatives. The Republicans were united and that boded ill for the Democrats in November. Alfred issued a statement praising the political harmony in his party: "I am very pleased with the absolute harmony that prevailed and that there were practically no contests for delegates throughout the State. I also wish to express my entire and complete satisfaction with the candidates selected by the delegates.... I have been working hard in politics and mean to work still harder in order to bring success to the Republican party."[45] Coleman issued no statement in celebration of party harmony, but on the other hand, he made no effort to sabotage it. He simply kept quiet, and that in itself was a major contribution to party unity.

The campaign between the Democrats and Republicans that followed the primaries was also abnormally subdued, not so much for a lack of issues or partisanship, but due to the great influenza epidemic that swept the United States in the autumn of 1918. Alfred wrote his brother-in-law, Edward Bradford, on 15 October, "I am afraid the influenza is going to prevent our having any political meetings whatsoever, as one can hardly expect it to decrease to any great extent for the next two or three weeks, at least not sufficiently to warrant the removal of the ban against public meetings. I have been unable to see Mr. Hayes [sic—Will Hays] or Colonel Roosevelt, for that matter, as I have not been in New York since our last interview."[46]

Bradford answered Alfred's letter the following day:

> It is urgently important that you get in touch with Colonel Roosevelt and Chairman Hays as soon as possible. With the epidemic and the Liberty Loan Drive, the campaign has been virtually at a standstill. It is most important that we make up for lost time in the last week or two.
>
> I hope you are all well and escaping the influenza. My friend, Hugh Garland, was buried this morning and I attended his funeral at the Cathedral Cemetery, where there was most convincing and depressing proof of the severity of the epidemic. Burial was being made in long trenches, wide enough to accommodate two coffins, side by side, and there was a continuous procession of funerals getting in each other's way in the narrow drive of the Cemetery.[47]

Hays and Roosevelt were eager to confer with Alfred as Delaware was considered to have one of the most critical and doubtful senate races in the country. Any conference with Alfred, however, would have to be via telephone. Because of Alicia's always somewhat precarious health, Alfred seldom left Nemours during the last weeks of the campaign, for fear that he might bring the infection home to his family. As he wrote Bradford, "I advise you to take all possible precautions against infection. So many good people have died from pneumonia that those few remaining should exercise a little precaution."[48]

The election of 1918, held in the midst of the worst epidemic to sweep the world since the Black Death of the fourteenth century, proved at least one thing. Americans could conduct a political campaign without the usual mass meetings, hand shaking, or baby kissing. The absence of such hoopla may have been an uncomfortable experience for the politicians, but it probably was a welcome relief to the voters.

By election day on 5 November the epidemic had subsided sufficiently to bring approximately the same number of voters to the polls as was usual in an off-year general election. The Republicans won both houses of the state legislature and the few minor state offices that were up for election, but most important, they captured both the House and the Senate seats in the U.S. Congress from the Democrats by a margin of 1500 votes. Party unity had brought the desired results, and Delaware was securely back in the Republican column.

It must not have been easy for Alfred to make an all-out effort to defeat his old friend Willard Saulsbury, one of the few du Pont in-laws whom he truly admired, and to have Saulsbury replaced in the U.S. Senate, where he had achieved a commendable record, by a nonentity like L. Heisler Ball. Alfred, however, justified his position on several grounds. First, he was a Republican partisan, and as the man most responsible for bringing about party unity on his terms, he could hardly do less than wage an all-out campaign for the entire Republican ticket. Second, Alfred had been a consistent critic of Wilson's foreign policy both in peace and in war, and he was genuinely concerned about the kind of peace treaty Wilson might negotiate when the war was terminated. It was very important, Alfred believed, that there be a Republican Senate to serve as a check on the treaty-making power of the executive. Third, although he was generally supportive of Saulsbury's position on most domestic questions, there was one major issue with which Alfred was in violent disagreement with the Senator. Saulsbury was an outspoken critic of women's suffrage, in opposition to President Wilson and most of his colleagues in the Senate. Alfred, perhaps influenced by Alicia's political interests, had since 1916 been a leading proponent for women's suffrage in Delaware, and he had made this a major point of attack in the 1918 campaign against Saulsbury. He wrote Jakes on 25 October that both Ball and Taylor must be more outspoken in support of universal suffrage. "I think if they do not they will have lost a great opportunity to strike a blow to the Democratic ticket. I trust my efforts [to persuade them to do so] will be successful."[49]

For Alfred, these were sufficient reasons to rejoice in Ball's victory. He suffered no guilt pangs over the role he had played in defeating Cousin Willard. When Chairman Hays wrote Alfred a note "to congratulate you again and again and thank you for all you did in Delaware," Alfred could without hesitation respond, "The Republicans here are naturally proud of what they accomplished.[50]" To his hardworking lieutenant in the field, T. W. Jakes, Alfred wrote a note of "thanks for your unswerving, loyal assistance. . . . It is not so much a Republican victory, although from a political standpoint the results are almost immeasurable, but the moral principles upon which our fight was originally begun."[51]

History would certainly corroborate Alfred's statement that the Republican victory in Delaware had produced "results almost immeasurable," although many individual historians would argue that this political victory had not meant as complete a triumph of moral principles as Alfred believed.

Chairman Hays and former President Roosevelt had had good reason, as it turned out, to regard a victory in Delaware as being absolutely essential to the Republicans' gaining control of the Senate. By a margin of one vote, the Republicans achieved the majority necessary to enable them to organize the Senate when it convened on 4 March 1919. That majority in turn enabled Senator Henry Cabot Lodge to become chairman of the Foreign Relations Committee, and in that powerful office to be able to bring about the Senate's rejection of the peace treaty that President Wilson brought back from Versailles.

Had the Delaware Republicans been as divided in 1918 as they were in 1916, had Alfred du Pont not worked as industriously and as effectively to achieve party unity as he did, had only 800 Delawareans cast their votes for Senator Saulsbury instead of Dr. Ball, then Senator Gilbert Hitchcock of Nebraska, Wilson's most loyal lieutenant in the Senate, would have been chairman of the Foreign Relations Committee instead of Henry Cabot Lodge. The Senate hearings on the treaty, which included the establishment of and the United States' participation in the League of Nations, would most surely have been conducted in a much different and far more positive manner of support for both the treaty and the League. Senate acceptance of the treaty would then have been a much more viable possibility. Had that eventuality been realized, then most certainly the history of the twentieth century would have been different. It is by such "for-want-of-a-nail" accidents that the direction of history is altered, producing "results indeed immeasurable"—except by speculation, using that chimerical yardstick, What If.

Following the general election of 1918, Alfred du Pont was at the zenith of his political power in the state. The old du Pont machine was apparently smashed beyond repair; Colonel Henry du Pont was in retirement, working on a genealogy of the du Pont family and finding the past much more pleasant than the present; and T. Coleman du Pont was pursuing paths to glory in New York instead of Delaware. The state belonged to Alfred even though he had lost the du Pont family and the Du Pont Company. His last hope of reversing the stockholders' vote had dissipated when his appeal was rejected by both the Circuit Court of Appeals and the United States Supreme Court in 1919.

Alfred gave no indication following the election of 1918 that he felt that his political goals were accomplished and that he was now ready to retire from the field of battle. Although he sought no office for himself, he was still thoroughly enjoying the game of politics. Difficult as it was, because of his deafness, to be a full participant in the conferences and private confabs that were the essence of political campaigns, he had at his service the dependable ears of such men as Allee, Davis, and Jakes, and he had reached such a position of eminence in the party that he could now command the attendance of his supporters, who would shout out or write down their inside information within the privacy of the library at Nemours.

Politics had also provided Alfred with a precious bond of common interest that united him with his wife. Alicia loved the intrigue involved in the clever maneuvers directed against those du Ponts whom she most despised, and Alfred's many triumphs over the du Pont machine had won for him Alicia's approbation which the defeats in other fields of battle had cost him. To have her look of approval as they clinked champagne glasses in toasts to their political victories on those glorious November nights in 1916 and 1918 was for Alfred ample compensation for all that politics had cost him in both effort and money.

Alfred deserved the congratulatory acclaim that he received during these years, not only for what he had eradicated—the old du Pont machine, and it was that, of course, that most delighted Alicia—but also for what he constructed: cleaner elections than Delaware had ever known, party unity, and above all, economic and social reforms that, as he said, would make "Delaware a better place in which to live."

The Non Partisan League, in particular, became in these years an unofficial ombudsman organization for the general public, which too often in the past had been voiceless and hence impotent in protecting its interests against those in power. Although the League had no legal authority to effect reforms, it could, with Alfred's financial and political support and with E. M. Davis's talent for organization and publicity, force local, state, and even national officials to become attentive and take remedial action. The League carried on successful campaigns against the fare increase of 2¢ that the city streetcar lines tried to impose and the 25¢ monthly tax that the Wilmington Gas Company placed on all consumers' meters. It demanded and got needed street repairs, and it carried on such a highly publicized crusade against "food gougers" and "rent profiteers" who sought to profit from wartime shortages in food and housing that it made a major contribution to the national effort to get Congressional action for price controls.[52]

So successful was the League in winning public support, which at election time could be translated into votes for Alfred's reform movement, that Pierre, who had hitherto kept a very low profile and had been content to leave politics to Coleman, finally decided that he himself must enter the arena. Coleman was using the wrong tactics and consequently was losing the battle. Pierre knew that Alfred couldn't be put down with the old weapons of bribery and poll manipulation, which too often backfired. The fire of reform had to be fought with a diverting fire of reform. Attention must be turned away from such dangerous and—in Pierre's opinion—semi-socialist schemes as the redistribution of wealth through taxation, producer-consumer cooperatives, and state-imposed price controls. Pierre proceeded to organize his own civic betterment organization and to offer his own program of reform.

Using John Raskob as the front man, Pierre created the Service Citizens of Delaware in the autumn of 1918. Unlike the Non Partisan League, it was an elitist organization of business leaders. No one belonging to the League was invited to join, however. Pierre wanted none of the wild-eyed progressives who had enlisted in Alfred's army.[53] Public education—better schools, better teachers, a more modern and relevant curriculum—would be the program that Pierre would offer to the public. In all fairness, it must be said that this was not merely a political ploy on Pierre's part to deflect attention away from Alfred and to provide a less dangerous substitute for his cousin's program. Pierre was genuinely interested in improving an educational program in a state that sorely needed such improvements. In the preparation of the survey on the status of education in Delaware that the Service Citizens brought out in 1919, Pierre became the first and most zealous convert to his own cause. He soon won recognition as the state's leading non-professional authority on public education, and he was even willing to implement his advocacy with his own money. He proposed that the state build one hundred new schools, and he himself would pay for one-half the cost of those buildings.

Because the proposal came from Pierre's Service Citizens Committee, it was naturally assumed by both his friends and his foes that Alfred would oppose it. Alfred's commitment to reform, however, transcended even his enmity toward Pierre. He sent word to his legislative friends via Allee that he wished them to give their support to the proposed new school bill, and he gave orders to Charles Gray, who had become editor of the *Morning News* following the death of Roy Jones in the great influenza epidemic of 1918, that the paper was to give all-out support to Pierre's education bill.[54] As a reformer genuinely interested in educational improvements, Pierre must surely have welcomed Alfred's support, but as an opponent in the game of politics, Pierre found his pleasure at this unexpected endorsement tinged with disappointment. It was hard to carry on a political battle against so supportive a foe. Even with the combined efforts of Alfred and Pierre, however, the school proposal, which involved consolidation of the small rural schools, faced tough opposition. A new school code finally passed the legislature in 1920, and over the next fifteen years Pierre would spend six million dollars of his own money in building new schools—racially segregated, to be sure, but at least equal as to facilities and curriculum.[55]

Alfred had said, when he first entered politics, that he was in the game to stay, and never did he seem more committed to that promise than during what proved to be his last year of participation. He made another strenuous effort to get William Taylor elected mayor of Wilmington in the spring of 1919. What remained of the old du Pont machine, now largely directed by Pierre and Raskob, once again tried to block Taylor. After Taylor captured the Republican nomination over the old organization's opposition in the primaries, the word went out quietly from Raskob that "respectable" Republicans should now vote for the Democratic incumbent. The mouthpiece of what Alfred called "the Market Street gang," the *Evening Journal,* not so quietly endorsed all Republicans running for city office except Taylor. The Philadelphia press, which always reported on Wilmington politics as fully as it did on that of its own city, devoted a great deal of space to the contest. In a lengthy feature story, the *Philadelphia Record* declared, "Victory at the coming city election would make Alfred I. du Pont virtually the political king of Delaware."[56]

On 7 June, the king could claim his throne. In spite of their efforts to persuade Republicans to vote for the Democratic candidate for mayor, Pierre and the *Evening Journal* did not prove as successful in split-ticket "instruction" as were Alfred and Jakes two years earlier. A large number of "respectable" Republicans did vote Democratic, probably for the first time in their lives, but an even greater number of Democrats crossed lines and voted for Taylor, who won by a margin of 1,000 votes. Alfred's party was now the real Republican machine of Wilmington and Delaware.

The du Ponts had always had a proclivity for expansion, and Alfred was prepared to expand his journalistic influence in the state, and perhaps even his own political role. It was not beyond the realm of possibility, now that he had defeated all of his du Pont political foes—Henry, Coleman, and Pierre—that he might reconsider his refusal to accept office for himself. Certain hints he dropped to his closest associates gave them reason to hope. He accepted the presidency of the commission established to build a much needed bridge across the Brandywine on Washington Street, and minor as this post might seem, there was no other position, not even the mayoralty itself, that would attract as much attention or earn as much favorable comment for a job well done.[57] Alfred also clearly indicated that he would be willing to challenge Coleman for the position his cousin still held as Republican National Committeeman from Delaware. In order to get that post now, Alfred would not have to enter into any unsavory bargain with such men as Hirons and Benson.[58]

In February 1919, Alfred entered into serious negotiations with Joseph Bancroft and the latter's attorney, George T. Brown, for the purchase of the Wilmington *Every Evening,* Delaware's leading Democratic newspaper. The acquisition of *Every Evening* would have given Alfred both a morning and afternoon paper and would have provided him with a loud journalistic voice that could drown out almost all press opposition in the state.[59] The sale was never consummated, however, due to a failure to agree on a price. Alfred had to be content with the outright purchase of the Dover *Delaware Leader,* a paper that he had been subsidizing for the past two years.[60]

In three short years, Alfred du Pont had completely refashioned Delaware politics. He had put into office the states' two U.S. Senators, one Democrat and one Republican, he had picked the governor of Delaware and the mayor of Wilmington, he had united the badly divided Republican party, and his reformers controlled both houses of the state legislature. Seldom has a politician ever had so meteoric a rise or so complete a dominion over the politics of any state. He was, as the *Philadelphia Record* had predicted, "the political king of Delaware."

Then, in January 1920, quite abruptly and without previous warning to even his closest associates, the king suddenly abdicated. He designated no one to be his heir, and he left his empire to be squabbled over by his lieutenants and his foes.

Alfred gave no farewell address nor even a public statement to explain his retirement. The chronicler of his political activities, George Maxwell, attributed Alfred's unexpected departure from the political scene solely to the sudden death of Alicia on 7 January 1920. Marquis James, Alfred du Pont's biographer, believed that Alfred was obliged to abandon politics primarily because of the great financial crisis that confronted him in the fall of 1919.[61] Probably both interpretations provide partial explanations for Alfred's self-imposed exile from politics at the height of his power. Certainly it had been Alicia who had first pushed him into the arena, and her continuing interest, support, and applause for his victories had encouraged him to stay. Alicia had wanted revenge over Henry and Coleman du Pont and she got it. She would also have liked to be a reigning queen among the wives of the United States Senators, and in time she might have got that as well. But Alicia did not have time, and with her death, she took with her much of Alfred's compelling motivation to play the game and win.

James is also right in giving emphasis to the fact that in the critical period from 1919 to 1921, when the Internal Revenue Service and a host of creditors were laying claim to his fortune, Alfred no longer had either the time or the necessary pile of chips to stay in the game. Politics, even relatively honest politics, as opposed to the Addicks/Henry du Pont brand, was very expensive. Every one of Alfred's newspapers except the *Morning News* lost money, and the latter, priced at one cent at Alfred's insistence, barely broke even. Alfred in 1920 simply could not afford to play.

Valid as these two explanations offered by Maxwell and James may be, they provide only partial explanations for why a man would voluntarily give up the power that Alfred had so successfully seized for himself. Alfred suggested another explanation in a letter that he wrote to Frank Allee on 19 January 1920:

> Some day soon I would like to have you come up and have a talk with me on the whole situation. I do not like this constant cry for money. It is against my principles to conduct campaigns along these lines. I knew it was necessary originally to spend a good deal of money to throw off the shackles of serfdom which had been thrown around the Republican voters by the old Republican machine, but it is not my intention to continually pay out money to keep them destroyed. I am rather inclined to believe as you do that Republicans in this state do not wish to be free. If they wish to be slaves, they must be slaves to someone other than myself.
>
> I cannot agree with you in the thought that if we beat Coleman du Pont at the Convention for National Committeeman, we will beat him for good. The way to beat Coleman du Pont is to remove, or cut off, the source from which his funds flow. I am convinced they do not come from him, so you can draw your own conclusions as to who is the giver. [Clearly, Alfred believed that Pierre served as Coleman's financial angel.] I am convinced that if this could be done, there would be no further demand for funds in the state, other than the funds necessary to carry on a legitimate campaign. So long as they are able to pit one pocketbook against another, the cry is always for more money. I am not sufficiently interested in the incumbency of the National Committeeman to make any expenditures at all. It is true that we decided to make a fight, but I have taken a somewhat different view of the situation since that decision was made. If every campaign is to be a constant cry for funds, the whole political situation in Delaware generates into a case of who furnishes the most money in order to win, which is not consistent with my ideas of decency.
>
> Next week I will be feeling more like it and will be very glad to talk the matter over with you.[62]

Although written only three days after Alicia's funeral, the letter gives no indication, except for a veiled reference to feeling more like talking with Allee next week, that his wife's death had precipitated Alfred's decision to withdraw from politics. The continuing demand for money was given emphasis by Alfred, but it was not only the expense of politics, which he could no longer tolerate, that was paramount in Alfred's decision. It was the broader issue of buying elections that concerned him, for this practice violated his concept of decency; if continued, it would ultimately doom democracy. Fifty years before there was an organization called Common Cause, Alfred was essentially arguing its basic premise. The increasing cost of elections in America was not only an intolerable burden to the individual contributors but to American democracy itself. Having achieved political dominance, Alfred was now questioning the value of that victory. If he stayed in the game, would he not become as corrupt as the very men he had ousted? Alfred had no desire to be a replacement for Addicks or Henry du Pont.

In a follow-up letter to Allee written five days later, he was even more explicit in stating his disillusionment:

> There is nothing in it but wear and tear, mud-slinging [and] an awful lot of expenditure. . . . I know how hard you work in a campaign and what it must mean to you in both physical and mental stress and the disagreeable personal side of a political controversy. As you say, you are forced to associate with people you would never think of having anything to do with under normal conditions.[63]

Politics, in short, was not a normal condition for an honest man, and Alfred had become sick of its abnormality.

Alicia's brother, who had stood at Alfred's side in every hard-fought campaign, could not at first believe or accept Alfred's decision to take an honorable discharge from the army of politics. Edward tried to goad Alfred into returning to his command post by predicting that "unless something is done . . . immediately the control of the Republican organization will revert to the former corrupt hands with Bazaza & Co. in the saddle. To my mind this would be deplorable in view of the time and substance that has been devoted to the cause, when the full fruits of victory are just within our grasp."[64]

Alfred, however, was not to be lured with so obvious a bait. As far as he was concerned, the "full fruits of victory" had already been garnered. He dismissed Edward's cry of wolf as being a false alarm. He told his brother-in-law that there was no chance of a return for "Bazaza" (Alfred's nickname for Cousin Coleman, first used in their college days, which Alfred said was an old Portuguese African word meaning "wild man"). "Bazaza and Company are just about as near dead ducks in this state as anything above ground could be, and what small portions which happen to be living, and which would be shunned by a turkey buzzard, will be rapidly disposed of within the next week or two, unless I am much mistaken. . . . They [the Republicans] can look to me for no further support."[65] Edward was sure that Alfred was mistaken about Bazaza. Perhaps Alfred himself did not really believe that Coleman's political obituary could now be published, but as far as Alfred was concerned someone else would have to administer the coup de grace.

Having once made up his mind to abdicate his position of command, Alfred lost no time in disposing of his arsenal. All subsidies to downstate newspapers abruptly ceased, and on 24 May 1920, Alfred wrote to Charles Gray, editor of the *Morning News,* to inform him that "I have arranged to dispose of the paper to interests here in Wilmington represented by C. P. Townsend. . . . I wish to take this opportunity to thank you for the excellent work you have done on the paper during your short stay with us and to point out that the physical appearance of the paper has changed many hundred percent. This is not my opinion alone, but the opinion of all who have noted the difference."[66]

Unfortunately for Alfred's political program, however, the *Morning News* would in the immediate future change its editorial policy even more drastically than it had changed its appearance under Gray's editorship. C. P. Townsend proved to be merely a front man for the real purchaser, Pierre du Pont. Not one to allow such an opportunity to pass him by, Pierre happily paid the $195,000 Alfred had asked for Wilmington's leading newspaper.

Before Alfred exchanged his uniform for mufti, there was one final campaign, in the spring of 1920, that he felt he must engage in—the battle waged in the Delaware legislature for the ratification of the proposed Nineteenth Amendment to the federal Constitution, which would give full suffrage rights to women. It is ironic that in this encounter he would enter into an alliance with his old foes, Coleman and Pierre du Pont, who were also supporters of the amendment. Alfred's leading opponent, however, was another old enemy, Mrs. Henry B. Thompson, the grande dame of Wilmington society, who had been especially insistent upon imposing social ostracism upon Alicia after her marriage to Alfred. Mrs. Thompson alone was enough to rekindle Alfred's ardor for battle, even if he didn't have Coleman in the opposing ranks. Alfred conducted this last campaign for Alicia. It would serve as his final tribute to his departed wife.

By May 1920, thirty-five of the necessary thirty-six states had already ratified the amendment. Delaware, which had been the first state to ratify the Constitution in December 1787, could now provide the single vote necessary to extend American democracy to the disenfranchised one half of its population. Alfred had been advocating this action since 1916 and he had tried to make it the leading issue against the Democrats in 1918. He was now determined that Delaware in 1920 would play the decisive role.

In March, Governor Townsend called the legislature back into special session solely for the purpose of considering the amendment. After a week of debate, the lower house flatly rejected the amendment by a vote of 22 to 9. The house then turned its attention to other matters it considered to be more important, such as Pierre's new school code and the question of authorizing additional funds for Alfred's Washington Street bridge, both of which received approval.

Before the senate could follow the house's negative example on suffrage, a motion to postpone consideration for a month was passed in the upper chamber. This delay gave the pro-suffrage forces time to rally their forces. Over the next six weeks Alfred exerted more pressure than he ever had on any other issue to persuade the senators friendly to him to vote for ratification. He had eight sure votes he could count on but he needed to pick up one more vote for it to pass the senate. The noted woman suffragist, Alice Paul of Swarthmore, Pennsylvania, president of the National Women's Party, was delighted to have Alfred's statement of support. She issued a statement following the defeat in the Delaware house. "The temporary setback which the suffrage cause received yesterday in the House has caused many favorable to the cause to realize that a strong battle must be waged. This is what probably actuated Mr. du Pont to announce his stand for suffrage. We cannot thank him too much for his support."[67]

One letter from Alfred to Senator J. C. Palmer of Dover will suffice to illustrate the effort he made to get an additional senate vote:

> I believe that women (I am now referring to the majority) stand for what is best in the home and what is best in the home is best for all. . . . Furthermore, it is not a local issue. . . . It is, therefore, the duty of all members of the Delaware Assembly to view this question purely from a national and party standpoint . . . and, not be influenced by any small or narrow view impregnated with political personal complications. I am expressing my opinion so clearly to the end that you may possibly endorse my views and use your vote in the Legislature to the end that the suffrage amendment will receive the sanction of the Delaware legislature . . . for the ratification of the amendment will place Delaware in the dignified position of fitting the keystone to the arch of universal suffrage and of placing herself in a position where the whole country must bow to her in acknowledgment of their respect.[68]

Alfred's efforts paid off. On the day after he sent his letter to Palmer, the senate ratified the amendment by a vote of 11 to 6—a two-vote margin over the required majority.

A much tougher battle still lay ahead in the House, for here the Democratic opposition, led by Representative J. Edward McNabb, was firmly entrenched. Having already once defeated the amendment in March, McNabb was confident that, in spite of Alfred's pressure, he could do it again.

On 14 May, Alfred, in one of his last acts as owner of the *Morning News,* wrote out a statement and gave orders to Editor Gray to publish it on the front page of the next morning's edition.[69] The statement read in part: "The women of this country will be given the franchise and that very soon. . . . [It] rests with the House of Representatives of the Delaware Legislature to make Delaware the keystone of the arch of universal suffrage and to the members to accord to their native State the honor of completing this step in the evolution of our national civilization." Then, taking a page out of British history, Alfred called upon the people of Delaware to converge upon the state capital in a manner reminiscent of the great public rallies in London for the Reform Act of 1832. "I am urging that every man, be he Republican or Democrat, who is convinced that this amendment to our Constitution should be ratified by the House to proceed to Dover on Monday morning, May 17, and by his presence there give support to his conviction to the end that the enactment of this amendment by the House be assured."[70]

Alfred got his crowd in Dover on 17 May. Delawareans by the thousands crowded into the state capital—and not just the men whom Alfred had called for, but women by the hundreds, including suffragists from across the nation. Alfred had written Bradford, "I am going to Dover Monday to see what can be done towards putting across the suffrage amendment. I would like you to be there too." And then a line that must have startled Edward: "I have arranged for P. S. and T. C. to be on hand. The latter has promised, but of course we do not know what he really will do."[71] T. C. did live up to his promise. He also brought P. S. with him. This certainly was the first and last mass populist meeting Pierre S. du Pont attended.

Had the house voted on the amendment that day, with its chamber packed with spectators and with the largest crowd ever to assemble in Dover overflowing the capital square, ratification very probably could have been secured. McNabb, however, was a master at parliamentary maneuvers. He was able to postpone action until "such time as conditions are more conducive to sober deliberations."

The memorable day of 17 May passed into history. The great crowds dispersed, the fever subsided, but still the house delayed a vote. This postponement gave McNabb and his followers the time necessary to influence a majority of the representatives to revert once again to their usual "political or personal complications." On 2 June, a motion to reconsider the previous action of the House in voting down the ratification of the proposed amendment was itself defeated by a vote of 24 to 10.

By this procedural maneuver, the house was prevented from even taking a second vote on ratification. Delaware's failure to provide the decisive vote of ratification giving suffrage to women made little difference, however. Two months after the Delaware house refused to reconsider its negative vote, Tennessee had the honor of providing "the keystone to universal suffrage" by being the thirty-sixth state to ratify the Nineteenth Amendment. Alfred had the satisfaction of seeing the women of Delaware troop to the polls in November, and by huge majorities, vote a straight Republican ticket. Alfred took special pleasure in the fact that one of the Democratic victims of the women's righteous wrath was J. Edward McNabb. After twenty years of occupying what was regarded as a safe Democratic seat in the Delaware House of Representatives, McNabb paid the penalty for his male supremacist views.

Alfred had to be content with this somewhat delayed victory. He by now had had enough of politics. Someone else could attend to the affairs of state. He must now attend to some very pressing personal affairs.

XI

Exit Alicia, Enter Jessie
1915–1921

T HE FIFTEEN-YEAR BATTLE that Alfred du Pont waged against some of the most powerful members of his family coincided almost exactly with the time span of his marriage to Alicia Bradford. During the extended contest of du Pont versus du Pont, Alicia proved to be both help-meet and hellcat to her embattled husband. It had been in defense of Alicia that Alfred had engaged in battle in the first instance, and it had been largely at her insistence that he had continued the struggle until he finally achieved a measure of success. Alicia's frenzied demand for retribution had driven him to irrational acts of folly against members of his own family, acts that had greatly weakened his own defenses against later attacks by Cole-man and Pierre and had earned him nothing except Alicia's scorn. But it was also Alicia who had provided the motivation for him to enter politics, and his triumphs in that arena had won him those rare and precious instances of her approbation.

In entering what was a second marriage for both of them, Alicia may have taken Alfred "for better," but she was not a woman who could take any man "for worse," especially a man to whom she was not passionately committed. Many cruel things could be and were said about Alicia—about her pride, her vindictiveness, and her contempt for weakness and failures in others. There was one thing, however, that Alfred could never accuse her of having done. She had not entered into their marriage under any false pretenses. She had told Alfred at the time she accepted his proposal of marriage that he could expect only so much from her—her admiration for his successful career, her appreciation for his support and counsel during the troubled years of her first marriage, and her gratitude for his having given her life a new direction and for his being a devoted stepfather to her only child. All of that she could and at times actually did give him, but he could never expect to receive from her the kind of passionate love that she had given to Amory Maddox.

Alfred had been more than eager to accept her on her terms, believing that his complete devotion would eventually win even that which she said she could never give. Perhaps in time, given the right circumstances, Alfred might have succeeded in winning her love. But the right circumstances were never in evidence. On the contrary, their marriage had to be endured under the worst possible circumstances: the ostracism that cut them off from the very society Alicia wanted not only to be a part of but also to rule over; the death, soon after birth, of two children whom Alicia had desperately wanted; and the loss of position that Alfred suffered within both the family and the company. Neither Alfred's blind devotion nor his great wealth could compensate for her losses. "The worse" far outweighed "the better," and their marriage had been doomed to unhappiness from the start. Yet ironically, there was one part of their marriage vows that was observed. They maintained this unfortunate union until death did them part.

It was not be be expected, given her temperament, that Alicia would bear the adversities they suffered with quiet submission. Unable to meet her foes in direct confrontation herself, she was dependent upon Alfred to be her champion in battle, and when he suffered defeat, as he often did, she had no other outlet for her frustration and anger than to turn against the person who had fought her battles for her. The result was frequent rages, yelling tantrums, and scornful denunciations of the one person who truly loved her.

It is not difficult to understand why Alicia should be tagged as the villain in the tempestuous drama that was their marriage. Alfred's friends and foes alike blamed Alicia for everything wrong in the marriage. They believed it had been she who had seduced him and broken up his first marriage; it had been her possessiveness that had caused him to abandon his children; it had been her extravagant tastes that had forced him to waste his fortune on pretentious show palaces; it had been her insistent demand for revenge that had led him to such follies as slander suits against elderly women and shameful attempts to change the name of his only son; it had been her pride that had drive him to court to wrest control of the company from those who were successfully managing it. On and on ran the bill of indictment against the second Mrs. Alfred I. du Pont—a mixture of truths, half-truths, and outright lies. The great natural gifts with which she had been endowed proved to be curses that damned her. Alicia was too beautiful, too intelligent, too strong, too proud for her own good, in the opinion of most of her contemporaries.

No one except Alfred had any appreciation or sympathy for what Alicia herself had suffered during these years of isolation from her family. After the death of her mother, no one was really close to her except Alfred and her brother. Her two sisters were now her most outspoken critics, and only in the last years of his life did Alicia achieve a measure of reconciliation with her father. Nemours was hardly a show palace if there was no one to whom she could show it. Once physically strong and active, she had ruined her health in her repeated futile attempts to have children. Alfred had paid a dear price for this marraige, but so had Alicia, and there were very few who realized that.

It would be a mistake, however, to depict Alfred's and Alicia's life together as one long, sustained period of torment for both of them. In spite of the frequenct quarrels, Alfred never endured the same pain that he had known in his first marriage. Alicia's angry explosions of temper were infinitely preferable to Bessie's cold scorn or indifference. At least there was the heat of temper in Alicia's outbursts, not the frigid rejection of Bessie's locked door.

There were also moments of real companionship, which Alfred had seldom known with Bessie. He and Alicia thoroughly enjoyed their joint efforts in the planning, building, and furnishing of Nemours, even if, upon its completion, it contributed more to isolation than it did to the courtly hospitality for which it was designed. Alicia also shared Alfred's love for his hunting and fishing lodge on Cherry Island. For both of them it was a welcome retreat from the Brandywine, and some of the most companionable moments of their marriage were spent on this small island in the Chesapeake Bay.[1]

There were times, however, when even this secluded spot was not far enough away from Wilmington to be a true escape for Alicia. From the moment of her first visit, Alicia had fallen in love with Paris. The City of Lights had been her asylum of sanity in the darkest moment of her insane relations with Amory Maddox. From then on, the only place in the world where she felt truly alive and well was when living in Paris. In 1908, Alfred agreed to her leasing an apartment on the Avenue du Bois de Boulogne, and this became over the next several years her very special pied-à-terre, her treasured contact with reality.

After the birth and death of her child in 1912, Alicia finally realized that it was hopeless to try to have more children. And as far as she was concerned, the long battle of the Brandywine was also over. She had had enough. In 1913, once she had sufficiently recovered her health to travel, Alicia returned to Paris for what promised to be an extended stay. She would have been content to stay there for the rest of her life.

Alicia was still in Paris on 4 August 1914 when the German army invaded Belgium and began its relentless and apparently irresistible drive toward the French capital. Alfred sent frantic cables begging his wife to leave France while she could still get out. With considerable reluctance, Alicia left her apartment and its valuable furnishings in charge of S. G. Archibald, an American lawyer in Paris who for years had served as the Du Pont Company's European legal counsel. After much difficulty, Alicia managed to get around the rapidly advancing German army and make her way to Le Havre in time to sail on the French line *Espagne,* leaving for New York on 22 August.[2]

For her, homecoming was not an occasion for celebration. Unlike Alfred, she had no illusions that this would be a short war even if the Germans should succeed in capturing her beloved Paris. She feared that she might never see again the Avenue du Bois de Boulogne, which she now regarded as her true home. She returned to find all of the old family controversies as malignantly active as ever, and soon they would be intensified manyfold as Alfred became embroiled in the greatest fight of his life against Pierre for control of the company.

The one area in which Alfred's affairs were prospering was his income. The war years brought to the Du Pont Company the greatest boom in its 115 years of existence. By 1918 it would have sold to its own government and to the Allies more than one billion, five hundred million pounds of explosives, under contracts that grossed the company more than one billion dollars. By 1917 Alfred's personal income from his stock in the company approached $6,000,000 annually.[3] He could now afford to give his wife anything she wanted in the way of material possessions, but with all of his wealth he could not give her the things she wanted most—a healthy child and her home in Paris, far removed from all that she detested in Wilmington.

If Alfred could not give Alicia an escape to Paris, his wealth could provide her with a sumptuous substitute for the Bois de Boulogne apartment. In 1916, Alfred entered into negotiations with Thomas Hastings, senior partner of the architectural firm of Carrère and Hastings, which had designed Nemours, to build Alicia a second home, to be located in Roslyn, Long Island, New York. Alicia named the estate "White Eagle" (no old du Pont place names for Alicia this time). It was a measure of Alfred's devotion to his wife that he agreed to build their second home in the metropolitan area of New York, a place that he had always hated above any other urban center in the country, but which Alicia found the best possible substitute for Paris.[4]

Alicia also found a substitute for the child she wanted and could not herself give birth to. Without any prior consultation with her husband, Alicia made arrangements through a charitable institution in Paris to adopt a French baby. The negotiations were carried out in great secrecy. Alfred had no advance warning of the great surprise that awaited him when he arrived home one evening in January 1916 and found Alicia seated in their bed-room holding in her arms an eleven-month-old girl. Both mother and infant could have served as models for a Renoir painting. Alicia told her amazed husband to come over and kiss their little daughter, Adelaide Camille Den-ise du Pont.

When Alfred had recovered enough from his shock to inquire as to where and how Alicia had obtained the child, his wife simply said she had known the baby's parents in Paris. The father was a French lieutenant who had recently been killed in battle, and the mother had agreed to give her daughter to Alicia for adoption. No further details concerning the identity of the child's parents were offered by Alicia, and Alfred never questioned further. For him it was enough to see a rare smile of contentment on Ali-cia's face as she held once again a baby in her arms.[5]

Alicia had now provided Alfred with two daughters: Alicia junior, now twelve years old, who would always be his very special "little peach," for Alfred also needed substitutes, and little Alicia was exactly the same age as his own Victorine, whom he had not seen since she was four; and baby Denise, who would be for both of them the beautiful, healthy baby they had not themselves been able to produce.

It had taken Alicia considerably less time to obtain a child than it was to take Alfred to provide her with the home she wanted away from Wil-mington. Hastings had given his solemn promise that the house would be ready for occupancy by 1 April 1918 at the latest, but the difficulty in war-time of obtaining building materials of the quality the du Ponts wanted plus a labor strike greatly delayed the construction. As the length of time for completion increased, so did the costs. The original estimate for White Eagle, including the very expensive land in that area of Long Island, had been $400,000. When the final bill was submitted, it totaled $1,102,000. It was considerably more than the little pied-à-terre that Alicia had expected, but Alfred had the money, and Alicia was delighted with her late Tudor style country home, beautifully decorated by Charles of London. They finally moved into White Eagle in July 1918.[6] Alicia unfortunately was to have very little time to enjoy this elaborate substitute for her Paris retreat.

The last few years after Alicia's reluctant return from France were, how-ever, perhaps the happiest and most peaceful years of their marriage. Once again they had the fun of building and furnishing a new home, this time far away from the critical eyes and contemptuous comments of their envious du Pont relatives. They had their two children to cherish, and for the first time they both knew the exhilaration of victory over their du Pont enemies as they worked together in plotting the strategy of the political campaigns.

It was also during these years that Alicia undertook a project that she had often considered but had never taken the time to execute. While living in Paris, she had become particularly fond of the poetry of the seventeenth-century French bishop François de Salignac Fènelon. She now began the translation of his odes into English. She also worked on another book, *The Vineyards of France*. By thus immersing herself in the study of French literature and culture, she found some consolation for having to live in exile from Paris.

If Alicia was more content with her life during these years, Alfred himself was restless. He missed being a part of the Du Pont Company more than he cared to admit. His sister Marguerite's son, Maurice du Pont Lee, was now employed by Du Pont, and he faithfully reported to his Uncle Alfred whatever inside information he had on the company, but reports from a nephew hardly sufficed. It was hard to be only a witness to the great expansion and the biggest boom in the company's history. Alfred kept thinking of the feverish days in the summer of 1898 when he had been responsible for the production of most of the military's ordnance to fight the Spanish. How monumental had seemed then both the task and the achievement, and how puny that effort now seemed with virtually the entire world at war. Alfred longed to be a part of the excitement, to make the contribution he knew he could have made to Du Pont production if he were still a part of the team. He would have been in complete agreement with the sentiment expressed by the poet W. H. Auden:

> I was not looking for a cage
> In which to mope in my old age.[7]

Alfred felt that he *had* been locked in a cage by Cousin Pierre, and he was moping, even though he had, at fifty-two, hardly reached old age.

Suffering from this feeling of impotent frustration, Alfred was easily influenced by Alicia and his lawyer, William Glasgow, to continue to pursue his suit against Pierre for control of the company, although his better judgment told him that it was a futile quest. Following the adverse decision of the stockholders on the company's purchase of Coleman's stock, Alfred yielded to Glasgow's insistent urging that the case be sent to the United States District Court of Appeals to seek a reversal of Judge Thompson's decree that the stockholders' vote must prevail. In October 1918, Glasgow argued the case before the court. It was a masterful presentation for the rights of minority stockholders and should have given Alfred encouragement, but it didn't.

"I have little expectation that the Court of Appeals will in any wise change the findings of the lower court," Alfred wrote to one of his attorneys, H. B. Brown. "I believe that my original decision that it would be unwise to take this litigation to the Court of Appeals will be justified, though I hope to the contrary."[8]

The result of the ill-advised appeal proved far worse even than Alfred's most gloomy predictions. Not only did Judge Joseph Buffington, in presenting the unanimous decision of the three-judge court, deny the appeal, but he went far beyond that. While upholding Thompson's decree that the stockholders' vote was binding, Buffington at the same time denied Thompson's original finding that Pierre du Pont and his associates had been guilty of deliberately violating their trust. The Court of Appeals essentially accepted the opinion that Charles Evans Hughes had provided Pierre as a propaganda piece to influence the stockholders' vote. It gave Pierre and his syndicate a clean bill of health. Even the moral victory that Alfred had won in 1917 was thus summarily snatched away from him by this higher court.[9] Pierre had gained the vindication he had sought, but it hardly satisfied Alfred—nor, curiously, Pierre, either. The latter continued to mull over the question of why he had done what he did, and Alfred continued in his futile quest to undo what Pierre had done. After the disastrous decision of the Court of Appeals, Alfred applied to the U.S. Supreme Court for a writ of certiorari, which the highest court denied the following year. That denial closed the door to any further legal action upon Alfred's part, but the scars of battle and the unanswered questions remained for both Alfred and Pierre to live with for the rest of their lives.

Barred from a role in the Du Pont Company's war effort, Alfred had to make his contribution to the cause in his own way. He could at least use a portion of his war profits in the service of his country. He entered into discussions with Navy officials, including the very energetic Assistant Secretary of the Navy, Franklin D. Roosevelt, as to what the U.S. Navy most desperately needed in order to achieve victory at sea. Alfred had already turned over to the Navy his yacht, the *Alicia,* and his smaller speed boat, the *Petrel,* but the Navy seemed at a loss in knowing how to use for wartime pursuit a yacht built for peacetime pleasure. "My experience with the *Petrel* and the *Alicia* has shown the wisdom of caution in accepting the views of government officials regarding the desirability of acquiring boats," he wrote to S. H. MacSherry, captain of the *Petrel,* now designated the U.S.S.S.P.59. "They expressed a desire for the 'Alicia' and put me to a great deal of expense in getting her in shape for work quickly. Now I find that there is very little probability of their wanting her."[10] Before making any further gifts, Alfred wanted to make sure that he was not simply providing the Navy with more white elephants.

F. D. Roosevelt and the Navy brass told Alfred that what was desperately needed were small craft of great speed, properly armored, which could be used as submarine chasers. Alfred promptly entered into negotiations with the Herreschoff Manufacturing Company, a maritime engineering firm, for the construction of a sub-chaser incorporating some of the mechanical innovations that he and a noted boat designer, A. Loring Swasey, had worked out. The chaser, when completed in the fall of 1917, was 110 feet long, with a draught of only 4½ feet so that it could enter into very shallow coastal waters. Two high-pressure steam engines, which Alfred had given particular attention to, generated 1500 horsepower, giving the vessel the capability of speeds up to 27 knots. It was a model for that time of an antisubmarine destroyer, for which its proud donor was happy to pay the bill of $85,000.[11]

Writing checks, however, was not enough of a contribution to the cause to satisfy Alfred; not even checks amounting to two million dollars for the various Liberty Loan drives. "My war record," he later wrote to his brother Maurice, "consisted in staying at home and trembling in my shoes for fear I would be eaten alive by a Boche."[12] This typical hyperbolic self-denigration does not, of course, describe what Alfred actually felt during the war years. He longed to be a part of the action, either in the Brandywine mills where he belonged, or on the high seas, where because of his physical handicaps, he did not belong. "How I wish I were with you," he wrote Mac-Sherry, who was aboard the *Petrel*. When MacSherry was later transferred to the command of the destroyer U.S.S. *Henley,* Alfred again wrote him:

> It must make you feel like a real man to be in the game and know that you are going to be in the midst of it before long. My miscreants [Pierre et al.] seem tame alongside the Huns. The difference is that my antagonists are cowards and yours are good fighters. I understand the Gang is saying that I am treating them badly and that I should let up on them. What a pity they don't join Barnum's Show, which is due here on the 13th, and add to the clown contingent.[13]

Each man had his own battle to fight, but Alfred would gladly have exchanged both battlefield and enemy with MacSherry.

Denied the participation in the war effort he would have welcomed, Alfred began as early as 1916 to look beyond the war to the equally important question of the reconstruction that must follow. Perhaps then he could, with his wealth and his contacts, find a place for himself that would satisfy his unrealized ambition to lead a meaningful life outside of the Du Pont Company. The new venture, however, would have be his own show. No du Pont cousins would be invited to join him as co-managers this time, and it would not be located in the Brandywine valley. Possibly because of his earlier decision to build a retreat for Alicia on Long Island, Alfred, like Coleman earlier, began to eye the possibility of centering his new venture in New York. This thinking went against all of Alfred's deeply engrained prejudices against the Babylon on the Hudson, but if he was to reach out to a war-devastated Europe, New York was the place to be. Fortunately, New York was big enough to contain both him and Coleman without their rubbing against each other.

The new venture did not emerge full grown, like Minerva, out of Alfred's head. He knew that immediately following the cessation of the fighting there would be a demand for consumer goods from a continent that for four years had subsisted on only the bare necessities of life in order to keep millions of men supplied on the battlefield. There would also be, especially in France and Belgium, which had borne the full brunt of the war on the Western Front, a need for machinery and raw materials to rebuild their industries. The Western Hemisphere, untouched by the destruction of war, had the necessary goods and tools. Somehow, Alfred realized, there must be a way of getting demand and supply together, although initially he could not determine how.

As a first tentative step in the direction of reaching his ambitious goal, Alfred in 1916 entered into negotiations for the purchase of the Grand Central Palace, located directly behind Grand Central Station on Lexington Avenue in New York. For the price of one million dollars, it was a rare real estate bargain, for it contained a half million square feet of space in the very heart of Manhattan, the largest exhibition area in New York. Here Alfred intended to open a world trade center, which he promised would provide the facilities for "stimulating trade between the United States and the Allies, Latin America and the Far East."[4]

It took nearly two years to complete the arrangements for the sale, and it was not until 20 May 1918 that Alfred at last had the gigantic agora in which supplier and buyer could meed face to face.[5] Even then, the Palace was not immediately available for use as a marketplace; Alfred's first tenant was the United States Government, which leased the entire building to serve as a hospital in which to receive wounded American soldiers brought directly from the field hospitals in France to the port of New York.[6]

In the meantime, Alfred was busily engaged in setting up agencies both in New York and in Europe that would facilitate the trade exchanges that he envisioned. With the aid of Charles Dickson, who was vice-president of the Merchants and Manufacturers Exchange (the corporate name for the ownership of the Grand Central Palace), and S. G. Archibald in Paris, Alfred created the Allied Industries Corporation.

It was at this time also that Alfred first met Duncan M. Stewart, a former bank president in Toronto who had come south to explore the greener pastures that New York offered. There is no record of how Stewart and A. I. du Pont first made contact, but it was soon apparent that each saw in the other someone who could be useful to him. Alfred was impressed with Stewart's knowledge of finance, his ingratiating personality, the personal contacts he had with important financial leaders both in Canada and Great Britain, and above all with his energetic drive. Stewart was another Coleman, only more sophisticated.

Stewart, in turn, was impressed with Alfred's millions. Like his new acquaintance, he also had foreseen the rich opportunities that the reconstruction of Europe could offer to the smart entrepreneur, and in very short order, the two men had become close associates in this new venture. Stewart told Alfred that Allied Industries Corporation and the Grand Central Palace were all very well as far as they went, but they didn't really touch the basic problem that lay at the heart of a meaningful reconstruction. If international trade was to be successfully stimulated, then it must be a two-way street. If Europe was to buy goods in Alfred's great marketplace, then it must also have goods to sell. The only way that goods could be produced in France in the immediate future was through the reconstruction of its industries. What was needed was a corporation that would enter into contractual relationships with the French government, both at the national and regional levels, to rebuild the factories and provide the most advanced machinery for both agriculture and industry. Then France could once again offer its textiles, foodstuffs, glassware, and fine porcelains to the world market.

All of this made very good sense to Alfred. It was in line with his thinking, but spelled out in very concrete terms.[17] So in Alfred's Wilmington office in the early spring of 1917 the French-American Constructive Corporation was born, anticipating by thirty years the Marshall Plan that would rebuild Europe after another even more devastating war. Unlike its famed successor in European reconstruction, however, the French-American Constructive Corporation was dependent upon the French government, not the American taxpayer, for its funding, and it was designed to be a profit-making corporation. In both respects, it proved to be a dismal failure.

The corporation was created with high hopes and great expectations, however, and Alfred, the eternal optimist, envisioned a restored and prosperous Europe that would send industrialists, craftsmen, and merchants to his Grand Central Palace offering their goods and in turn buying the products of all the American nations. Swords would indeed be beaten into plowshares. The Allied Industries Corporation bought a controlling interest in the National Tractor Company, located in Cedar Rapids, Iowa, in order to get a reliable supply of agricultural machinery for the farms of Normandy and Brittany.[18]

Stewart at first proved to be all that his self-promotion had proclaimed him to be. He really did know important people and was able to persuade such distinguished and responsible men in the fields of finance and diplomacy as Viscount Furness of Britain, Sir John Carson of Canada, former

American Ambassador to France Myron Herrick, and the New York investment broker William Bonbright to become directors of the French-American Constructive Corporation.[19]

Unfortunately, it was belatedly discovered that Stewart had had an even more interesting financial past than that of which he so frequently boasted. Some of the underlings in the organization, resentful of Stewart's power as general manager and his authoritarian rule, on their own had investigated his Canadian career. To Alfred's chagrin, their investigation revealed the reason for Stewart's sudden departure from Canada. The bank over which he had presided had failed, and there was a suspicion that its failure involved unwise if not illegal activities upon Stewart's part. Stewart was ultimately able to clear himself of all criminal charges, but in the meantime, Alfred had no alternative but to ask in August 1918 for Stewart's resignation as general manager.[20]

With the departure of its able manager, the French-American Constructive Company floundered. Stewart had had the same vision of reconstruction as Alfred, even though Stewart's motivation was largely mercenary. He had been the driving force in negotiating contracts, and he had provided the necessary discipline and order within the organization. After Stewart left, there was no one on the European scene except S. G. Archibald whom Alfred could depend upon or confide in, and in the New York office, Alfred was not sure there was anyone whom he could fully trust, not even his own personal lawyer and friend, Robert Penington.

The whole venture would very possibly have failed even if Stewart had stayed on. The French government was most reluctant and niggardly in providing capital for the reconstruction of Arras and Verdun; the National Tractor Company failed to produce the five tractors daily that it had promised; thousands of pairs of shoes and thousands of pounds of sugar that the Allied Industries Corporation had purchased from Argentina, Brazil, and Cuba piled up in the warehouses. No buyers showed up to purchase this mountain of goods. What did show up with great regularity were the monthly storage bills. These many problems would have presented a Herculean task to any organization, no matter how well conducted, but Stewart's forced resignation doomed whatever small hope for success the venture may have had.

Alfred had taken on far more than he or any other individual could possibly handle. It had taken all of the Great Powers collectively to create this devastation and disruption of world trade. No small private organization could now pick up the pieces and put Humpty Dumpty together again. Given all of his other interests—his political activities, his continuing legal campaign in court against Pierre, his banking interests in Wilmington, and his persistent personal concern about his marriage—Alfred could give only partial and fleeting attention to a venture that would have demanded the complete commitment of the most skilled and experienced entrepreneur. Alfred was a great powdermaker, but he, like Coleman before him, found himself on very dangerous terrain when he ventured into the alien financial world of New York, London, and Paris. In spite of some of the bitter experiences he had had in dealing with Coleman and Pierre, which should have disillusioned him about human nature, Alfred continued to maintain an almost naive trust in others. Consequently, he was continually shocked to find his trust betrayed, through either deliberate malice or clumsy incompetence. Alfred's nephew Maurice Lee gave perhaps the best evaluation of Alfred as a business executive when he said that his uncle had the qualities that are essential for a great leader—vision, imagination, drive, and the ability to work well with others and to inspire their confidence in his leadership. In the fields in which he was an expert by reason of natural aptitude and training, he was a true genius, but, Lee added, like many other great men, Alfred was a poor judge of talent in others. Too often he chose as his associates men who were greatly inferior to him, men incapable of sharing his vision and carrying out the tasks assigned to them.[21] Nowhere was Lee's evaluation of his uncle more fully borne out than in Alfred's venture into postwar reconstruction.

After Stewart's departure, no person other than Alfred had oversight of all the various operations. Control was scattered among several individual companies—the French-American Constructive Corporation, which was itself divided into two separate and autonomous branches in France and the United States; the Allied Industries Corporation; the Merchants and Manufacturers Exchange of New York, in charge of the Grand Central Palace; the National Tractor Company; and the New York Furniture Exchange Company. The venture had become a huge, sprawling holding company without a holder.

In 1919, in a desperate effort to bring some kind of order to this amorphous mass, Alfred put all of the individual companies under the supervision of yet another organization, the Nemours Trading Corporation, which he had set up in 1918.[22] This reorganization came too late to have much effect. It only increased the dissent and inefficiency among the subordinate parts.

The New York offices, in particular, were in disarray. In an attempt to find out just what was going on, Alfred turned to an old friend, his former secretary, Jimmy Dashiell, who had once been so useful to Alfred in keeping him informed as to what was happening within the Du Pont Company's executive office. Dashiell was sent to New York ostensibly to serve as an accountant for the Nemours Trading Corporation, but in actuality to be Alfred's ears and eyes within the company's operations.

Dashiell's reports came almost daily to Alfred's desk in Wilmington. Never one to minimize existing problems or to dismiss office gossip as being trivial, Dashiell presented an ugly picture of waste and outright dishonesty within the company's operations. A single sample of his reports will suffice to illustrate the disorder that prevailed within the Nemours Trading Corporation's offices:

> There has been a continued process of misrepresentation of facts ever since I have been connected with this outfit. We first started off with $500,000 assets which were contained in Beaumont Alexander's suitcase, and which could have been bought for $10.... We charge up salaries as an asset, switch over cash from one month to another to boost a financial statement. We put in a quarter of a million dollars worth of shoes as sales, that in fact *were never shipped*.... When I took charge of the Accounting Department, as I told you, there was not a book in balance. There is no Adjustment Department or Traffic Department. Anyone who knows what to do could have established one in a week's time. We paid $3,500 the other day to Gale Bros. on a shipment of shoes. The shipment was not received, but our contract was cash against bill-of-lading. There is no authenticated person in this office to take this up with the railroad company to recover, although I am making an attempt to do so. As a matter of fact, it is a question whether this shipment was ever made *to be lost*....
>
> These accounts were never started right. The condition we are in is absolutely the fault of Mr. Fay, who had charge.... [and] was put in the office here by Mr. Nixon [Stewart's successor] and tutored by him so that he would do as he told him. The attitude of most every man in this place was to keep me as far away from you as possible. Not because they did not think I knew anything but because they knew I knew too much. I have done nothing since I came on this job but observe human indiscretion....
>
> I am having an audit made of the shoe department by outside auditors to give an impartial report. I am investigating all sorts of exorbitant bills that are nothing but graft.[23]

On and on, page after page, report after report, ran Dashiell's tale of woe, until Alfred no longer knew whom he could trust or what he could do to remedy an increasingly hopeless situation. One thing he did know was that war-devastated Europe was no more in need of reconstruction than were Alfred's own agencies designed to provide reconstruction. It had been a noble plan but an ignoble operation. On 16 December 1919, by action of its board of directors, the French-American Constructive Corporation was formally dissolved, the Paris office was closed, and all further activities became the responsibility of the Nemours Trading Corporation.[24] Deconstruction, not reconstruction, had become the order of the day.

Although Alfred had agreed to the building of their second home in New York largely to give Alicia what she wanted, he had found it to be a convenient location for him as well, once they were able to move into White Eagle in the summer of 1918, for the urgent problems that were pressing in on him in connection with his New York business ventures necessitated his spending several days of each week in New York. Home for him, however, would always be Nemours, not White Eagle, and he returned there as often as possible, even though Alicia prefered to stay in her new home.

The family was together at Nemours for Christmas of 1919, and Alfred didn't allow his business worries to interfere with the holiday activities. Christmas was always a joyous occasion for him. Once again, as so often in the past, he donned a Santa Claus suit, and he delivered gifts to the delighted four-year-old Denise and to the more sophisticated but nevertheless appreciative Alicia junior.

During this holiday season Alicia senior was preoccupied with making preparations for a trip south soon after New Year's Day. She had invited Anna du Pont, Francis I.'s wife, and her eldest daughter, Elise, to be her guests for a trip to South Carolina. Alicia had invitations to attend the annual St. Cecilia Ball, Charleston's most famous and glittering social event. It would be a marvelous escape from winter, from new house problems, and from Alfred's constant business worries. She could hardly wait for the departure date.

Alfred had also made plans for some much-needed rest and recreation while Alicia was consorting with the Southern aristocracy. He intended to take young Alicia, who had never been much further west than the Allegheny mountains, across the country by train to southern California. There they would soak up some sun, drink lots of orange juice, swim in the Pacific Ocean, perhaps even make a quick trip south of the border into Mexico from their planned base in San Diego. Alfred may have mentioned casually to Alicia that while in California he hoped to see some Southern aristocrats himself—his friends from the old hunting days at Harding's Landing, Virginia, the Ball family. Captain Ball had moved to California soon after Alfred had paid his last visit to Ball's Neck in 1906, and had been joined there by most of his family.[25] The Captain was now dead, but Tom, the eldest son, was a lawyer in Los Angeles, the youngest child, little Eddie, was selling autos, or office furniture, or law books—Alfred was not quite sure which—in the same area, and Miss Jessie was an assistant principal in a high school in San Diego. It had been fourteen years since he had seen any of them, and it would be great fun to renew their acquaintance if the opportunity arose.

What Alfred surely did not stress in discussing his travel plans with Alicia was that he had made absolutely certain that the opportunity would arise. Indeed, it was to see the Balls, and one member of the family in particular, Miss Jessie, that had prompted him to plan the trip west. He intended to go directly to San Diego, check into the Hotel del Coronado, where he had a reservation for a suite of rooms for himself, little Alicia, and her maid, and then renew in person his acquaintance with Jessie Ball.

Over the years, Alfred had managed to keep in touch with his old friends from Northumberland County, Virginia. His two best correspondents had been Mrs. Fannie Harding, who frequently turned to him for help when one of her neighbors in that impoverished area of Virginia was in dire financial need, and Miss Jessie Ball, who would recall to him those happy times of hunting and fishing all day and dancing far into the night at the community center.

Jessie had been a better correspondent than he, but when he was on trips he would occasionally send her a postcard or short note telling of his activities. In 1916, he wrote to tell her about the Republican convention he attended as a delegate from Delaware:

> I have just returned from the Republican Convention in Chicago, which I attended more as an agent of Providence than from any desire to be there, and in this way wasted a week, caught a cold, used a good many words which should never appear in print and returned home even more disagreeable than I had hoped. . . . What I really am in need of, and what I have been seeking for the past half century, is a word beginning with "s" and ending with "y," pronounced by some archeologists as sym-pa-thy.[26]

Jessie did not hesitate in providing the sympathy that Alfred needed. The letters they exchanged became more frequent and ever more friendly and warm. It was difficult, however, for Alfred to dismiss as unimportant the twenty-year difference in their ages. He continued to use the salutation "Dear Miss Ball," and to close his letters rather formally with "Sincerely yours, Alfred I. du Pont," as late as May 1919, when he wrote to give her a full account of the adverse Court of Appeals decision and of his political successes in Delaware. In answer to her apparent request for his photograph, he answered:

> I think you have the last picture which I have had taken and which you politely informed me you did not like. I think it is a copy of one I had taken in 1913 or 1914, which was my last effort at self-perpetuation. In any event, I think it much better to let you remember me as you must on the banks of Dividing Creek in 1906, which I believe was the last time we met. My life has not been what is usually termed in fiction as "A bed of roses," and Papa Time with his usual sense of humor has not added to my stock of beauty, which at its best would hardly have been considered in business terms of quick asset. . . .

Please write me again when you feel so disposed and have conde-
scended to overlook my lack of courtesy, which as you well know is no
indication of the facts as they actually exist. . . . I must thank you for the
pretty picture of the U.S. Grant Hotel. . . . With best wishes, I remain,
Sincerely yours,[27]

On Alfred's part, there was still the avuncular cordiality of an old family
friend, but for Jessie it was not enough to remember him as he was on the
banks of Dividing Creek in 1906 nor to continue sending picture postcards
of the U.S. Grant Hotel. By the autumn of 1919 their correspondence had
progressed from "Dear Miss Ball" and "Dear Mr. du Pont" to this letter
from Jessie:

Dearest Heart—Your little, hurried, worried note of the thirtieth is
very dear to me. My first impulse was to write at once—am learning that
impulses are not always to be acted upon—a wise order must be obeyed.
You have enough to worry and trouble you without the additional uneas-
iness of one of my letters falling into enemy hands—hard though it be,
I'll post but one letter each week and that on Sunday as instructed so
you'll know when to expect it. . . . Oh, I want to see you, I must— I
cannot longer bear it. You don't know what the pain is, Dear. You have
never,—thank God—known what it is to worship and adore one and
be through all time separated from the Loved One.—If I could see you,
If only for a little while . . .[28]

Alfred, of course, knew what the even greater pain was to "worship and
adore one" and yet not be separated from that "Loved One" who wouldn't
love back. For the first time, however, he now knew the quite different
experience of loving and being loved in return, even if it must be from a
great distance.

Shortly after New Year's Day, Alfred and Alicia went their separate
ways. On 3 January, Alfred boarded the Pennsylvania train, accompanied
by little Alicia and her maid, and headed west to San Diego. Three days
later, Alicia, Anna, and Elise du Pont headed south to dance at the St.
Cecilia Ball.

From the train, Alfred had sent Jessie a wire telling her the time of his
arrival in San Diego. Their first meeting could only be a polite handshake
in that public place, but Alfred told her he would come over to her hotel
as soon as he and little Alicia were settled in their suite at the Hotel del
Coronado.

When Alfred checked in, the desk clerk handed him a telegram, containing the following message from his secretary, Ruth Brereton:

Following message was send [sic] Mr. Du Pont this morning.
Charleston SC 1220 P Jan 7 1920

Alice [sic] passed away eleven o'clock this morning. Francis and I extend our deepest sympathy. Everything possible was done but she never regained consciousness. Arrangements will be made to take here [sic] to Wilmington. Please advise me your wishes. Anna du Pont.[29]

It must have been difficult at first for Alfred to comprehend this strange message: "Alice"? . . . "passed away"? . . . "never regained consciousness"? Then came the full realization. Alicia, unaccountably, incredibly, was dead. He had no time to wait for further details, only time for a hurried call to Jessie to tell her the unbelievable news and that he was taking the next train back to Wilmington. There would be no reunion after all.

From the train station, which he had so recently left, Alfred sent a telegram to Miss Brereton telling her he was on his way home and asking her to wire further information to him on the train. The follow-up telegram from Anna Du Pont provided a little more information. Alicia had been taken ill while on board the train soon after it left Union Station in Washington. She became increasingly worse during the night, and by the time they reached their destination she was in a coma. She was rushed from the train to a hospital in Charleston, where she died shortly before noon. Her brother Edward was on his way south to bring her body back to Wilmington. No funeral arrangmeents would be made until Alfred arrived home.[30]

It must have been a heart attack. Alfred had no other explanation. For the past two years, Alfred had been concerned about Alicia's health and had had a specialist from Baltimore, Dr. C. B. Gamble, come to both Nemours and White Eagle to give her frequent examinations. It was thought that she was suffering from "improper metabolism," and she was placed upon a carefully regimented diet.[31] By the late spring of 1919, Alicia was eager to return to Paris, but Dr. Gamble advised against it, a decision in which Alfred heartily concurred.[32] Now, Alicia would never again see her apartment on the Avenue du Bois de Boulogne, never again find peace in her beloved Paris.

On the long four-day ride back to Wilmington, there was a great deal of time for thinking—of the regrets Alfred felt that the marriage had never been what he had wanted it to be, of the pain they had both suffered from the marriage, of the happiness he had never been able to buy for her. He must hold on to the memory of those better moments and be grateful for them. White Eagle had been finished in time for Alicia to enjoy it. She had completed her translation of the Fènelon ode, and Brentano's of New York had published it.[33] She had also just completed another book, *The Vineyards of France,* which she had planned to dedicate to Ambassador Myron Herrick, and Alfred would make sure that the book was published.[34] It helped to remember that the last few years had not been as bad as some of the others they had endured.

Alicia's funeral was held at Nemours on 16 January. The dean of the Cathedral of St. John the Divine brought a choir of fourteen boys from New York for the service, and Charles Dubell, who had formerly served as clergyman at the little Episcopalian chapel near the Du Pont powder works at Carney's Point, read the Episcopalian burial service.[35] Knowing how Alicia had felt about having du Pont relatives as nearby neighbors during her lifetime, Alfred decided she would not wish to spend eternity beside them in the family cemetery, so she was buried on the grounds of Nemours.

A few lines form the *Ode of Fènelon* that Alicia had translated might well serve as her epitaph:

> Sheltered from the tempest blind
> Blasting all the Rich and Great
> Underneath these leaves I find
> In all weather a retreat.
> There alone my life begins
> There alone my spirit wins
> Far from books, a deeper truth.[36]

Letters and telegrams of condolence came by the hundred, from old friends, business leaders, politicians, and the powdermen from the mills. As he wrote to his daughter Madeleine in Germany, Alfred even received from his daughter Bessie, with whom he had had no contact for fourteen years, "a very beautiful letter. Unfortunately, owing to the mother, I was unable to respond in the way I should have liked, as it would not have done for me to make any concession at that time, as I realized the rumor which they would cast abroad, and which I understand has been pretty well circulated, and that is to the effect I would return to the mother. . . . What would they say if such a rumor had the slightest foundation for truth?"[37] So Alfred kept his distance from his lost children: not even a death in the family could still old animosities.

One surprising letter that he did answer in kind came from Coleman, but there was no word from Pierre.[38] In a curious way, it was easier for Alfred and Coleman to make some gestures of reconciliation than it would ever be for Alfred and Pierre, who had once been much more closely bound in friendship.

Madeleine, as soon as she heard the news, had wanted to leave on the next ship from Germany to come to him, but he cabled her not to come, and then wrote:

> I am much better alone, as there is no one in the world who can mitigate in the smallest degree a loss such as I have experienced. . . . I cannot see that any good would be accomplished by my burdening other people with even the smallest portion of my misfortune.[39]

There was, however, one person in the world whom Alfred believed could mitigate the loss he had experienced, and on the day after Alicia's funeral, when all the people were gone and Nemours seemed more empty than it ever had during Alicia's frequent absences, Alfred sat down to write a long letter to Jessie, unburdening himself of all the very mixed feelings he had regarding the loss of Alicia. It is probably the only time he ever expressed to anyone else just what his marriage to Alicia had been.

> Dearest little Jessie:
> I wonder if you can possibly estimate what your letter meant to me today—for it came this morning. Just when everything is at its worst & I am nearly crazy with what seems to me my empty life. I made a statement to my wife some time ago, apropos of my unending solicitness [sic] for her welfare & happiness to the effect that in my opinion the bond of misery was far stronger than any bond wrought by happiness and so it has proven to be. I have worked by day and by night without ceasing for her, as I should & though my efforts were not successful my mind was centered on one object & my body strove for but one goal. Now she has gone & while she is happy at last and at rest in paradise, I find myself alone & without reason for existence.
> Mrs. du Pont was a most unusual personality of the strongest type, a woman of remarkable finesse in all her standards—living up to the highest ideals & expecting the same of others. . . . Poor health & a broken nervous system endured by the loss of nearly all her children. Her strong maternal instinct knew no bounds but that of ultimate success which her poor weak body would not grant her. It was her misery that made . . . me what I am. . . . I am wondering if I failed my wife in any way—whether I was wanting in understanding & sympathy. I trust not. This beautiful home (Nemours) built for her by me is of no use to anyone. I am trying to map out a plan for its future. . . .
> Again a thousand thanks for your letter, your love, your sympathy & for you.—Alfred.[40]

Jessie did not intend that Alfred would be alone much longer. She would give him a "reason for existence" and she had ready a map for Nemour's future. She asked and received from the superintendent of the school in which she taught an extended leave of absence. In April, accompanied by her younger sister, Elsie Ball Wright, she came east, ostensibly to visit relatives in Virginia, but in actuality to spend as much time with Alfred as could be discreetly arranged.

Jessie's journey to the East Coast lasted eight months. She made a short visit to Nemours during the Easter season and another for the usual elaborate Fourth of July celebrations and much longer visits to aunts and cousins in Virginia, during which time she wrote passionate love letters to

Alfred and presumably received the same in return.[41] By the time she returned to California in December to submit her resignation, her courtship of Alfred I. du Pont had reached its inevitable and successful conclusion. Jessie would have been happy to have been married at Nemours, but Alfred, still haunted by ghosts of the past, insisted that they be married in California and that they wait the year that propriety decreed.

In the meantime, Alfred gave no hint to anyone of his approaching marriage, not even to his sister and brother. In his letters to them that summer, he spoke lightly of the pleasant visit to Nemours of some old Virginia friends, the two Ball sisters, and their Tappahannock relatives, whom he had entertained for a week in July.[42] As late a 5 January 1921, he wrote to his old friend James Hackett in London that he was "arranging to take a little vacation in the West, beginning the 8th, and will probably be gone a month, and once I get away from my secretary, I doubt if anyone will ever hear from me. . . . I find I am badly in need of a little nervous rest. . . . I want to see what a complete loaf will do for me."[43] No mention, even to pal Jimmy, that before Hackett could get his letter, Alfred would be married.

On 22 January 1921, at the residence of the Episcopalian Reverend Mr. Baker P. Lee in Los Angeles, Alfred and Jessie were married. Jessie's sister Elsie was the matron of honor, and her brother Edward Ball was Alfred's best man. The *Los Angeles Examiner* reported that "immediately following the ceremony, the newly wedded couple left by motor for Santa Barbara. From there Mr. and Mrs. du Pont will make a honeymoon tour of California that will consume three or four months. Then they will return to Delaware to make their home at 'Nemours,' Mr. du Pont's 2,000 acre estate on the outskirts of Wilmington. . . . His fortune is estimated at $400,000,000."[44]

This honeymoon trip was very different from Alfred's previous one with Alicia—no explosions at Fontanet to call him back to grim duty. This would be a very different marriage, to a very different woman.

XII

Hard Times, Good Times
1921–1925

F OR A DECADE and a half, Alfred du Pont had been a highly prized
source of copy for the American popular press. His separation and
subsequent divorce from his first wife Bessie, his quick second marriage to
Alicia, the slander suits he brought against two female relatives, his break
with Coleman and Pierre, and the protracted court fights over the possession of Coleman's stock which followed that break—all of these events were
rich lodes for alert reporters and prying gossip columnists to mine.

By 1921, Alfred should have become inured to sensational news stories
in which he starred, often cast in roles so completely fictitious as to be
unrecognizable to his friends and to himself. Alfred, however, could still be
amazed by the falsehoods that the newspapers presented to the public as
established facts. Surely the *Los Angeles Examiner* had set something of a
record for inaccuracies, in reporting his marriage on 21 January 1921. In a
single paragraph the paper quite gratuitously made him a brother of
Thomas Coleman du Pont and at the same time generously provided him
with a fortune of $400 million.[1] Alfred must have been as glad that the
Examiner's first statement was false as he was sad that the second statement
was equally erroneous.

In actuality, Alfred's assets in 1921 were only about one-twentieth of
what the *Examiner* had attributed to him, and even the greatly reduced
figure of $20 million did not represent a true picture of his financial stand-
ing. For in 1921 Alfred found himself hard pressed by a host of creditors,
including, most ominously, the United States government. Jessie Ball was
not marrying a man as rich as the readers of the *Examiner*'s society columns
believed. Alfred, of course, had no illusions about his wealth, nor had he
attempted to conceal his financial difficulties from his wife-to-be. Indeed,
in the letter which he had written to Jessie on 17 January 1920, the day
after Alicia's funeral, Alfred had not only revealed for the first time to

another person what his marriage to Alicia had entailed, but he also frankly told Jessie of the crisis that had been reached in his business affairs.

> I can't see a map ahead at this time. My business affairs in New York are in an awful mess & the trip to California & back had only added to the mess & I don't think there is any solution save a liquidation under a receiver to save anything. Hopeless incompetence & dishonesty, graft & lack of attention on my part is the cause & I refuse to jeopardize what part of my fortune will remain. As it is I can't get out without an enormous loss of at least a few millions, possibly more. But I am tired of constant worry—So I am going to take the loss & know some peace.[2]

Alfred soon discovered, however, that it was not easy to get out of the "awful mess" he found himself in, and it would be another five years before he untangled his very tangled business affairs.

Alfred had had warnings prior to the moment of showdown that should have prepared him for the inevitable collapse of his export–import business. Jimmy Dashiell's reports of the chaos that prevailed in the New York office of the Nemours Trading Corporation in the spring of 1919 were sagas of woe that no one could ignore. But Dashiell had always been one to find the worst in any situation, and Alfred felt he needed confirmation from outside sources, mainly from his creditors, as to the gravity of the situation. In October 1919, he had written to the president of the Chemical National Bank in New York, H. K. Twitchell, asking if he would "kindly advise me whether your business relations with the Nemours Trading Corporation are entirely satisfactory?"[3] Twitchell was only too happy to comply with this request.

> We avail ourselves of the opportunity offered by the inquiry contained in your letter of the 4th instant to write to you frankly regarding the account of the Nemours Trading Corporation. . . . It has been our opinion for some time that the company was handling too much business on the working capital employed. Not long ago the writer deemed it advisable to notify the company that the then existing line of discount of approximately $700,000 was all that we felt justified in granting until more capital had been paid to the business. Notwithstanding this request, immediate application was made for an additional line of credit extending the amount to $1,000,000. . . .
>
> In our dealings with the company we have endeavored to be very liberal, because of our long and most satisfactory banking relationship with you. We will state to you frankly, however, that in view of the fact that the business of the company is being built up rapidly on the strength of your splendid personal credit, we feel that we are justified in looking to you for a definite assurance that you are in close touch with the company's affairs, that you are in accord with its present policy, and that it

is your expectation to see to it that it is furnished with the necessary additional capital to permit it to handle its banking matters in accordance with usual banking practice.[4]

In subsequent correspondence, Twitchell informed Alfred that the directors of the Chemical National had decided, "in view of more recent information received, . . . we should be furnished with your personal guarantee if we are called upon to continue to grant a substantial line of credit to the [Nemours Trading] Company."[5] He pointed out that his bank has arranged either directly from the Chemical National Bank itself or from among "banking friends of this institution" a total of $3,462,000 in loans to the Nemours Trading Corporation.

> We are led to feel that the merchandise belonging to the Nemours Trading Corporation is not being disposed of to any great extent and that steps are not being taken looking to the liquidation of the indebtedness of the associated companies to this Bank in the near future and the placing of the company in a position to take care of the $1,500,000 collateral loans assigned to other institutions who will doubtless expect payment at maturity. The situation as above outlined has received the attention of the directors of this Bank, who concur in the opinion of the officers that under present conditions no new commitments should be made and that early action should be taken in the matter of the sale of merchandise, even though it is necessary to disregard the question of profits on such sales.[6]

Although Twitchell had spelled out in very precise terms why the Nemours Trading Company was in serious financial straits and what must be done to alleviate that situation, only one phrase in his letter caught Alfred's full attention. Twitchell's statement, "in view of more recent information received not only by the Bank, but by the directors themselves," was a red flag for Alfred. In his state of paranoia with respect to both Pierre and Coleman, he was sure that this cryptic line meant only one thing. Either Pierre or Coleman was attempting to sabotage Alfred's company by spreading the word among New York financial agencies that the Nemours Trading Company was so badly mismanaged that no further loans or extension of existing loans should be granted. Alfred knew for a fact that Pierre and his brothers had considered the possibility of bringing suit against Alfred to prevent him from using the name Nemours for his company, lest it be considered by the general public to be subsidiary of the E. I. de Pont de Nemours Company. They had not actually carried out their threat because their legal counsel had advised them that Nemours was not a trade name that could be protected under the copyright laws. Alfred, however, was convinced that his enemies were now determined to destroy his company by another and more effective means.

Alfred wrote Twitchell to complain that there was

> a disinclination on the part of the New York Banks to make loans to me, or to the Nemours Trading Corporation, either on collateral or personal guarantee. I am unable to reconcile this condition with any normal state of affairs. . . . I am, therefore, forced to conclude that there must be some influence at work against me personally and against my interests, and as no other possible explanation presents itself to my mind at this time, such an influence, should it be a fact, can come only from one of two sources and I thought that you might perhaps be in a position to advise me confidentially as to the actual state of affairs as it is important that I be fully informed in order to handle the situation forcefully and intelligently.[7]

Twitchell promptly answered that "we want to assure you that we do not know of any particular influence at work against you personally, or the interests which you represent. In our opinion, however, there does exist a prejudice against the Nemours Trading Corporation, based upon the business methods employed by a previous management of that company." Twitchell once again detailed the reasons for the reluctance of lending agencies to extend credit to the Nemours Trading Corporation. "You will appreciate that the fact that a large part of the company's receipts have been used in the purchase of merchandise instead of the liquidation of bank loans and also that the company may suffer a severe loss in connection with a large sugar export have served to retard the movement toward a better credit situation."[8] In short, it was unnecessary for Pierre or Coleman to instigate any nefarious schemes for the ruination of the Nemours Trading Corporation. The company was capable of doing that job itself.

Apparently, that was all it was capable of doing. In desperation, Alfred, in October 1919, had fired J. S. Nixon, the successor to Duncan Stewart as vice-president and general manager of the company.[9] Jimmy Dashiell had been urging Alfred to do just that for the past two years, but now it was too late to have any beneficial effect. Nixon's departure simply created a vacancy that no one else could even pretend to fill. It is small wonder that Alfred had written to Jessie in January 1920 that there was nothing to do but "take the loss" and by so doing hope "to know some peace."

The big question was how to determine the amount of the loss. The company's books were in shambles. As Dashiell had earlier reported, liabilities, such as salaries, were listed as assets, and no one, certainly not Alfred, had any real knowledge as to who were the creditors or the exact amount of the outstanding claims against the company.

The easy way out, of course, would have been to declare bankruptcy. In the late spring of 1921, after Alfred's return from his wedding trip, a conference was held to discuss how best to liquidate this sorry business. The meeting was attended by William du Pont, Charles W. Mills, the vice-president of Alfred's and William's Delaware Trust bank, and Robert Penington, Alfred's lawyer, who had played a conspicuous but not very commendable role in the Nemours Trading Corporation's affairs. Alfred made clear that under no circumstances would he consider filing a suit for bankruptcy. As he later reported the discussion to Dashiell, Alfred told his associates, "It was my name that gave the company its credit standing. Whatever his other failings, no du Pont has yet sought the protection of the bankruptcy laws. I am not going to be the first."[10]

Alfred had taken a noble stance, but one that the rest of the assembled company, with the exception of William du Pont, neither understood nor applauded. Either Alfred was a greater fool than they had thought possible, or he was so immensely wealthy that he could afford this extravagant gesture of protecting the family name. Mills pointed out that since no one knew the full extent of the outstanding claims, Alfred stood to lose not only the five million he had already invested in the company but possibly several additional millions on claims that would appear as soon as the word went out that Mr. du Pont was personally assuming all outstanding debts of the Nemours Trading Company. Bankruptcy, Mills argued, was an accepted American business practice. No one's honor was impugned by doing what the laws of the land allowed. It was all part of the risk of being an entrepreneur. But Alfred remained adamant. All legitimate claims would be considered, all valid debts would be paid off. Mills and Penington were instructed to round up all claims against the company and then attempt to negotiate the best possible settlement.

Alfred's two agents had no difficulty in rounding up claims. They poured in from every American, European, and Latin American who had ever had even the slightest business connection with Nemours Trading. Mills also discovered that some of the bank loans, such as that with J. P. Morgan, were carrying an excessively high interest rate of 9% when the current rate during this period of postwar depression was at most only 5 percent. Nixon and Dickson, in order to secure the loans they had desperately needed to keep the company operating, had paid a very handsome premium indeed.

Alfred's second instruction to his representatives—to negotiate the best possible settlement of outstanding claims—was not as easy to accomplish as the first. The company's creditors insisted on being paid in full, and Mills and Penington were singularly ineffective at the bargaining table. Particularly troublesome was the debt owed to the White Shoe Company of Boston for $1,500,000 worth of shoes—shoes that could not be sold in Europe and for which the Nemours Trading Company had already paid out hundreds of dollars in warehouse storage fees. The White Shoe Company refused all efforts at negotiation. They had shipped the shoes to the NTC in good faith; they would not now consider taking the shoes back; they demanded payment in full. Only the storage company could be happy with the resulting stalemate. Clearly, the cost of liquidation, without resorting to banruptcy, was going to be high. Alfred would be fortunate if he managed to salvage a quarter of the fortune that wartime profits had given him.

As if the burden of the Nemours Trading Company were not enough to contend with, Alfred suddenly had dumped upon his already bent back a staggering additional load from the United States Government. In December 1919, he was informed by the Internal Revenue Service that he owed an additional $1,576,015.18 in taxes on personal income for the year 1915. Alfred was stunned. He had always paid his tax assessments promptly and in full, had indeed considered it not only the obligation but the privilege of American citizenship to pay one's taxes. Immediately upon receipt of what he considered to be an erroneous assessment, he returned the statement to H. T. Graham, Collector of Internal Revenue for the Wilmington district, demanding to know why he had been given this additional assessment and pointing out that even if this was a legitimate claim of the government, which he was sure it wasn't, the statute of limitations precluded the reassessment of additional taxes if three years had elapsed since the time the return was due. "I will ask you to note that inasmuch as my return for income received during the year of 1915 was due and was filed on or before March 15, 1916, that the limit of time in which assessment could have been made expired March 15, 1919 and that, therefore . . . no further assessment on income received during the year 1915 can be made subsequent to that date. Therefore, your demand for payment at this time is improper and illegal."[11]

The reason for the additional assessment was promptly explained by the Collector. On 1 October 1915, at the instigation of Pierre du Pont and his Delaware Securities Company, the new E. I. du Pont de Nemours & Company had been incorporated under the laws of Delaware, to replace the former E. I. du Pont de Nemours & Company, incorporated in New Jersey. Stockholders of the old New Jersey company were told to turn in their shares, for which they would be issued two shares in the new company for each share they had held in the old company. Although Alfred, along with all the other stockholders had been informed at the time that this was merely a corporate reorganization and that the two-for-one split could in no way be considered a taxable dividend, the government now belatedly had decided that it was indeed a dividend, and therefore taxable income for the year 1915.

Nor would the government accept Alfred's argument that the legal time period for reassessing his income for 1915 had elapsed. "With regard to the period of limitation," the Commissioner of Internal Revenue brusquely informed Alfred, "information which established that you were liable for additional tax was on file in this bureau prior to March 1, 1919. Therefore since the date of discovery and not the date of assessment operates in fulfilling the requirements of the statue [of limitations], the bureau was within its rights in levying the assessment."[12]

Confronted with this new and totally unexpected claim upon his rapidly vanishing fortune, Alfred turned in desperation to his old friend and former legal counsel, William Glasgow. He asked Glasgow to go over all existing records and attempt to discover the full extent of the company's liabilities for which Alfred had so nobly if somewhat rashly assumed total responsibility. Mills and Penington had failed miserably in their attempts to ascertain the company's indebtedness, and Glasgow could understand why. "I am very troubled," Glasgow wrote Alfred on 15 July 1920. "At nearly every interview I have on the subject some additional liability of yours comes up." He pointed out an obvious fact. Alfred should have set a limit on the extent of his liability before he started accepting outstanding claims, but it was too late for that now. Glasgow also strongly urged Alfred to file suit for bankruptcy. Otherwise the claims, some legitimate, many not, could strip him of everything he owned, including his own home, Nemours. In neat columns, Glasgow wrote down the large outstanding debts that were valid and collectible by court action. There were the several bank loans, including a four-million-dollar loan from J. P. Morgan & Company, with an interest rate of 9 percent, a two-million-dollar loan from Chemical National, and a one-million-dollar loan from Bankers Trust Company of New York. Bonds on various attachment suits had been posted which totaled $1,280,000. Another guarantee of one million dollars had been posted on existing contracts, and Mills had informed Glasgow that it would take at least $600,000 to settle "some sugar matters in France." Then there was the White Shoe Company bill of one and half million dollars for those thousands of pairs of shoes rotting away in warehouses. When the additional tax assessment of $1,650,000, including interest, was added to the list, the columns of debts came to the impressive total of ten million dollars.[13] And these were only the known, legitimate debts. What really worried Glasgow was the fact, as he wrote Alfred two weeks later, "Something new develops every time I take this matter up. Suits now pending against the corporation . . . [involve] an undetermined liability, which may run high." Again Glasgow advised Alfred to consider bankruptcy if he hoped to save even his own home and its furnishings.[14]

Upon receipt of this letter, Alfred hurried up to Philadelphia to confer with Glasgow in his law office. Together, they went over the list of debts and gave them an order of priority in terms of payment. Glasgow insisted that Alfred must have cash in hand to pay the tax assessment. He could, of course, appeal the Collector's ruling, but he must be prepared to pay if the appeal failed, unless he wished to get several years of free housing in jail. The bank loans also must be reduced as rapidly as possible, for the interest rates were excessive and could rapidly multiply Alfred's indebtedness several times over.

Glasgow made no attempt to minimize the problem. His client stood on the brink of ruin, and if he refused to consider bankruptcy, he might well be pushed over the edge. Once again, Alfred reiterated that such a course was not open to him. The loans had been obtained because he had given his word to Twitchell, to Thomas Cochran of J. P. Morgan, and to others. He might not emerge from this fiasco with anything else, but he would emerge with his honor intact. There would be no further talk of bankruptcy. Somehow, with Glasgow's help he hoped, he would work things out. He would liquidate all possible assets, even it it meant selling his entire holding of Du Pont stock; he would pare his living expenses down to the bare necessities; and he would free himself of this burden. Then, providing he was still alive and functioning, he would start all over again. One look at Alfred's face—his big chin stuck out, his one good eye bright with anger—and Glasgow knew that further discussion on the point was useless. He agreed to stand by Alfred whatever the conclusion might be.[15]

It had been an emotionally exhausting session, and Alfred left Glasgow's office feeling utterly drained. Once out on the street, he was not sure he could make it to the Broad Street railroad station. His legs would hardly support him, and he thought he would collapse. Perhaps this was the way his life would end, as it had for his great grandfather, E. I. du Pont, by his suddenly falling and dying on a main thoroughfare in downtown Philadelphia. This would certainly be the easy way out—even easier than bankruptcy and much more honorable. Neither his body nor his spirit would accept such an escape route, however. Too weak even to step to the curb to hail a cab, Alfred somehow managed to inch his way down Broad Street by feeling the sides of buildings with one outstretched hand, like a drunken man. He finally reached the station and the train that would take him back to Nemours, which he could still, at least for the present, call home.[16]

Over the next twelve months, Alfred's financial situation continued to worsen. The entire nation was experiencing the usual postwar economic slump, and no industry was more severely hit by the readjustment to a peacetime economy than was the high explosives industry. The value of Alfred's seventy-five thousand shares of Du Pont stock plummeted from $67,500,000 at its wartime peak in early 1918 to less than seven million in 1921. One could not get the par value of $100 a share on the open market, even if one could find a buyer, which was doubtful. With debts outstanding that amounted to nearly double the value of his single major asset, the prognosis for Alfred's economic recovery was, to say the least, not bright. Any financial counsellor would have called his a terminal case.

Ignoring the fact that the market for the major output of Du Pont production had suddenly evaporated at eleven o'clock on the morning of 11 November 1918, Alfred blamed the precipitous decline in the value of his stock on poor management of the company. Pierre and his financial wizard friend Raskob should have foreseen the imminent cessation of hostilities the moment that the United States entered the war and should have begun planning for a peacetime economy by a diversification of its products, Alfred argued.[17]

This was, of course, a manifestly unfair charge. In the first place, the entire resources of the Du Pont industrial empire were strained to the limit of capacity in an attempt to meet the insatiable demands of the Allied Powers' war machines up to the very moment that the armistice was signed. In the second place, Pierre and his brothers had, as a matter of fact, given what thought and attention they could spare to the question of product and investment diversification for a civilian market even while they were preoccupied with the war. They were able to take over as war contraband the superior German patents for synthetic dyes, and they were already making plans to expand their existing facilities for the promotion of synthetic textiles. Moreover, in the midst of the war, Pierre had invested some $50 million of the company's surplus funds in General Motors stock, which gave the Du Pont Company a controlling interest in the giant automobile firm. Alfred, at the time, had vigorously opposed the move, considering it one more example of Pierre's megalomania, but in the long run even Alfred would have to admit that Pierre's decision to invest in General Motors had been crucial in sustaining the Du Pont Company, and Alfred himself, during the difficult years of postwar adjustment.

In one instance, however, Alfred did have some justification for blaming Pierre and his associates for the financial catastrophe that now confronted him, and that was in respect to the staggering tax bill that the government had presented to him. Shortly after the Supreme Court had ruled, in the case of *Phellis* v. *U.S.,* that the 1915 two-for-one stock distribution made by the reorganized Du Pont Company was indeed a taxable stock dividend, Alfred wrote a letter to his old political ally, J. Frank Allee, in which he explained just how the 1915 reorganization had benefited Pierre and his Du Pont Securities Company but had proved to be a disaster to all other stockholders who were not a part of Pierre's syndicate. It was Alfred's hope that Allee, through his many political connections in Washington, might be able to secure some alleviation from this unwarranted tax assessment.

In connection with the proposed discussion of the whole matter relating to the taxability of the du Pont Delaware stock issued as a dividend to the New Jersey stockholders as of October 1, 1915, the following are the important facts which should be brought out, cited as near as I can remember them, chronologically.

On or about March 1, 1915, Pierre du Pont and his associates ... formed a company known as the du Pont Securities Company for the purpose of financing the acquirement of Coleman du Pont's interest in the New Jersey Company, which, together with their own, gave them an absolute control of the organization. They had no money to pay for the stock and the amount they agreed to pay was roughly $14,000,000. They borrowed $8,500,000 from Messrs. J. P. Morgan & Company. . . . Subsequently, on or about October 1, 1915, this loan was increased to $10,000,000 and a portion of the notes given to Coleman du Pont [were] paid off. . . .

From that moment these people guided the destiny of the du Pont Company entirely along lines as would best conserve the interests of the du Pont Securities Company. It was important that the market value of their stock be kept to a point that would be satisfactory to the bankers as measuring the value of the security behind the loan, and they immediately proceded to do this by purchasing stock for the Company's bonus plan with Company's fund, to the end that this stock was purchased at $260.00 in the middle of March 1915; $400, July 23rd, and $670, July 26th: and so on up. Roughly, $4,000,000 of the Company's funds was [sic] used to purchase stock for the bonus plan.

The Department of Justice having the du Pont Securities Company under investigation in July 1915, on the hypothesis that it infringed the decree of the United States Court involving the du Pont Company under the Sherman Anti-Trust Law, and fearing that the court might order the dissolution of the du Pont Securities Company on the ground that it was in violation of this decree, it became vital in the interests of the stockholders of the du Pont Securities Company that this common stock of the New Jersey Company, which was subject to suspicion be taken out of the said du Pont Securities Company and another form of security supplied as collateral to Morgan and Company for the above mentioned loan.

It was to this end, therefore, that the Delaware Company, whose common stock has now been declared by the United States Supreme Court as income, was created, and the common stock of the Delaware

Company received in payment of a portion of its assets to the Delaware Company was distributed to the stockholders of the New Jersey Company as a dividend. . . .

There never was any need of the Delaware Company, so far as it affected the interests of the stockholders of the New Jersey Company other than those interested in the du Pont Securities Company; nor was there even any need of paying out such enormous dividends, over 200% being paid in 3 years, save that it was needed by the du Pont Securities Company to discharge its obligations.

It is impossible for me to understand how the Surpeme Court could have viewed this translation of values from the New Jersey Company to the Delaware Company except as a plan of financial re-organization, in which the entire capital of the New Jersey Company—that is everything that created wealth—plant, real estate, money, securities and everything else—passed to the new company. . . . Everything which the New Jersey Company had which could be construed as capital passed to the Delaware Company, and I can only account for such a decision by the Supreme Court on the ground that they were entirely unfamiliar with what actually occurred. . . . No stockholder was one whit richer or had one penny greater income on October 1, 1915, than he had on September 30th, the day before. . . .

The result of this litigation had been disastrous to all of the stockholders other than those who by virtue of the incorporation are subject only to a normal tax and who will reap all of the benefit of having their stock assessed at this abnormally high figure and can readily register losses of $20,000,000 thereby, while the others by virtue of the decision of the Court, which I am sure would be changed on review, by the absurdly high basis of assessment many of the stockholders, including myself, will be forced to sell their stock assessed on a basis of $347.50, at much below par in order to pay the tax at a time when all securities are at their lowest ebb. The manifest injustice of the net result of this decision is only too patent and needs no further comment.[18]

For Alfred this was the final ironic conclusion to his protracted battle with Pierre over the ownership of Coleman du Pont's stock—that after having carried the fight to the Supreme Court and having spent a half million dollars in legal fees, only to lose that battle, he was now to be assessed over two million dollars in taxes for a reorganization plan that had enabled Pierre to obtain the stock in the first place. It is not surprising that Alfred should blame his cousin for his financial woes.

Recriminations against Pierre, whether justified or not, may have served as a useful emotional outlet for Alfred, but they did not help to pay the debts that he himself had assumed. That could only be done by selling what assets he could and by a vigorous retrenchment in his style of living.

So began the grand liquidation sale. The first to be disposed of, since they were the most easily convertible into cash, were the Liberty Bonds that he had purchased during the war, worth two million dollars. He was also able to unload on Cousin Willie his entire holdings in the Delaware Trust Company, 2,248 shares at $125 per share.[19] Then there were some minor and generally unfortunate investments that he had made in such highly questionable enterprises as the Great Falls mine for the extraction of gold in Maryland (!) and something called the New York Trap Rock Corporation. Alfred was only too happy to get rid of these stocks, even if what he realized on their sale was a fraction of what his initial investment had been.[20]

From these various sales, Alfred was able to raise nearly three million dollars, which enabled him to pay off three-fourths of the loan that J. P. Morgan and Company had made to the Nemours Trading Corporation. He would now be paying 9 percent interest to Morgan on only one million dollars, and both Alfred and his attorney, Glasgow, could breathe a little easier.

Another piece of property that Alfred was delighted to be rid of was the Long Island estate that had been his gift to Alicia. White Eagle, when finally completed in the summer of 1918, had cost Alfred nearly one and a half million dollars. Now it stood deserted, its curtains pulled and its expensive Charles of London furnishings draped in white muslin dust sheets, for Alfred had never returned to it after Alicia's death. It was for him only a haunting presence of his failed marriage. Except during its planning stage, it had never brought happiness to Alicia, nor had it been a real home to either her or Alfred. Alfred would have been eager to exorcise this ghost even if financial circumstances had not made it imperative to do so.

In the late spring of 1920, five months after Alicia's death, Alfred commissioned his Wilmington lawyer, Robert Penington, to find a buyer for White Eagle. He told Penington he hoped to get one and a quarter million dollars for the estate, but if necessary, he would accept a somewhat lower offer.[21]

Alfred ultimately accepted a much lower offer. Penington proved to be as ineffective in selling White Eagle as he had been in dealing with the affairs of the Nemours Trading Corporation. After nearly a year of waiting for Penington to produce a buyer, Alfred turned the matter over to a New York realtor, Arthur C. Sheridan & Co., who advised a public auction.[22] Alfred reluctantly agreed to this proposal, but only with the stipulation that if no bid approximated a reasonable offer, he could take it off the market.

The auction was held on 23 April 1921. Only a few bids were submitted, and the highest was for $470,000, less than one-third of what White Eagle had cost Alfred. This offer came from the Honorable Frederick K. Guest of London, England, who had had the good fortune to marry the daughter of Andrew Carnegie's partner in the steel business, Henry Phipps of Pittsburgh.[23] Alfred could hardly think that this bid "approximated a reasonable offer," but the Sheridan people advised him to take it, and he did.[24] The year 1921 was not the most propitious time for selling Long Island white elephants, even if they were billed as White Eagles.

The sale of Alicia's Long Island retreat for a bargain basement price did nothing to reduce Alfred's debt load, however, for White Eagle had been given to Alicia, and as part of her estate, it had gone to her daughter, little Alicia. It was for her that Alfred, acting as her guardian, had sold the estate. He himself gained only one monetary advantage from the transaction. He no longer would have to pay the bills he had hitherto always paid for the taxes, insurance, and maintenance of the estate. More rewarding to Alfred was the relief he felt in having disposed of a relic of the unhappy past. As he wrote to his favorite female cousin, Ellen LaMotte, "No, the Long Island place never had any reason to exist, and I am glad to be rid of it."[25] It was more than past debts that Alfred was trying to unload—it was a past life.

Difficult as it was for Alfred to convert his assets into cash sufficient to pay off outstanding claims, it was even more difficult for him to carry out the second part of his new economic program, the retrenchment of his living expenses. Not since his school days at Phillips Andover and M.I.T., when he had had to subsist on the paltry allowance provided him by Uncle Fred, had Alfred ever given much thought to economizing. What had been a comfortable lifestyle in the early years of his marriage to Bessie had developed into the luxuriant extravagance of his most recent years with Alicia. In 1918, neither he nor Alicia could begin to spend their monthly income, however gifted Alicia might be in the art of conspicuous consumption. Now, only two years later, he was solemnly warned by Glasgow that he must pare expenses not only of all fat but of some lean tissue as well. He could no longer afford to spend two hundred thousand a year simply in the maintenance of Nemours. "All I may be able to keep of what I now seem to own," Alfred told Jessie shortly before their marriage, "is Cherry Island."[26] But even he did not really believe that. Much as he loved Cherry Island as an occasional retreat, its simple hunting lodge facilities were hardly an adequate substitute for Nemours, which was his home. It would be the last thing, and then only under the most extreme duress, that Alfred would ever part with. He wanted it to be Jessie's home as well. Together they could make Nemours the warm, hospitable place that it had been designed to be but had never achieved with Alicia as its chatelaine.

Alfred, however, did make what for him was a valiant effort to reduce expenses. The domestic staff at Nemours was drastically reduced, and he wrote to his sister Marguerite that the food bills had been cut to such an extent that it was costing only $1.85 per person per day for meals. "I have closed the greenhouses down also. While there are any flowers, I have instructed Fullerton [the gardener] to send you a few each week. I do not know when I shall be able to open them again."[27] His fleet of limousines were not sold, but all except two were put up on blocks in the Nemours garage and were not used.

These few stringent measures did little to alleviate the situation. Alfred had for so long had so much wealth that he did not have any real concept of economy, and Jessie had always had so little that she found it hard to believe, now that she was Mrs. Alfred I. du Pont living at Nemours, that there was a need to practice the same economy that had always been her way of life, whether living at the edge of poverty on Ball's Neck, Virginia, or subsisting on a teacher's salary in San Diego, California. She wanted the fine things of life that the wife of Alfred du Pont might expect and that her doting husband wanted her to have. So there were frequent trips to New York to shop at B. Altman's and Bonwit Teller. And there were new furnishings that were needed to make Nemours truly her home instead of simply a hand-me-down from the departed Alicia: new rugs and draperies, new chinaware, and little items such as the bill for several hundred dollars from Samuel Kirk and Son, Jewelers of Baltimore, for having the initial "A" removed from all the silverware and replaced with the initial "J"—an important symbolic act for both Jessie and Alfred, but an expenditure that, under the existing circumstances, would hardly seem warranted to Alfred's many creditors.[28]

Nor could Alfred, even in the midst of the financial crisis that was crushing him, turn his back on the continuing needs of his own family and old friends. He instructed the Title Guarantee and Trust Company of New York to extend the $10,000 loan he had made to his actor friend, James Hackett, for another year at no interest, and to continue the annual payment of $1,000 each to his sister-in-law Margery du Pont and her two daughters, Nesta and Charlotte.[29] His brother, Maurice, needed even more aid. "Unfortunately, Maurice has to be looked after," Alfred wrote his sister Marguerite, who was herself always begging him for more money for her church work, "as his investments, as you may know, have not been remunerative; he is the victim of every schemer, as they all know of his access to me, and it keeps me busy helping him out. . . . The amount expended in this way has already reached into several hundred thousand dollars, and I am still contributing $10,000.00 a month in order to purchase certain stocks in which he has invested, the value of which is most exceedingly problematical. I simply tell you all this to let you know that I am constantly beset by appeals of this kind. . . . Of course, not a word of this must go to him [Maurice]," Alfred hastily added. "I am sending you a check for $1,000 to help you out in the present emergency. Would like to send more but my own exchequer is at the lowest ebb that it has been for many years.... I will try to finish your Church up next year somehow and I expect to keep up my regular contributions to your exchequer, but all other contributions will have to be abandoned."[30]

His marriage to Jessie had also brought new family obligations. In the same letter to his New York bank instructing them to pay Maurice's wife and daughters $1,000 each, he also authorized them to send gifts of $5,000 to each of Jessie's two brothers, Tom and Ed.[31] The family circle of donees was widening at almost the same rate that his own assets were shrinking.

Embarrassing as it was to have to refuse the appeal from the president of M.I.T., Richard Maclauren, for a gift to his alma mater, it came almost as a relief to Alfred to receive an appeal to which he could say no. "I cannot tell you how sorry I am not to have been able to participate in this wonderful work," he wrote Maclauren, "but ... I am busy trying to arrange to keep out of jail."[32]

William Glasgow, who was keeping a close check on Alfred's expenditures, must have shaken his head in despair as he looked over his client's account books. A meaningful retrenchment of personal expenditures was clearly not the road to salvation that Alfred would or could take. As new claims continued to flow in and existing assets continued to wash out, the road that Alfred was traveling was straight downhill toward the crash barrier at the bottom.

"My motto, 'Porkum Amat Libertas' or 'Pig Loves Freedom' applies in this case," Alfred wrote his sister at this bleak moment of his life. "A great motto and as I am thinking of getting up a new coat of arms for myself, I think I shall use it. It certainly is more appropriate than *Rectitudine Sto* [the motto on the du Pont family coat of arms] unless it is spelt 'Wreck' instead 'Rec.' As 'I stand here a wreck' is more appropriate than 'I stand through uprightness.'"[33] It was Alfred's concern for uprightness, however, as his lawyer would have been the first to point out, that had reduced him to the financial wreck that he was. Glasgow was sure that it would be a long time, if ever, before Alfred could enjoy the *Libertas* that he claimed every *Porkum Amat*.

By the first of the year 1922, Alfred finally realized that if he was ever to emerge from his burden of debt without losing everything in the process, he would have to dispose of at least a portion of his major asset, the Du Pont Company stock. This was not an easy decision for him. The stock provided him with the collateral that had enabled him to borrow another two million dollars to pay the additional tax assessment on his 1915 income. Much more than that, it remained his last tangible link with the company that he had served as employee, partner, vice-president, and director for over thirty years. But sentimental attachment paid no monetary dividends, and Alfred needed hard cash.

In disposing of his stock, Alfred was confronted with the same problem that he had faced in selling White Eagle. It was a buyer's, not a seller's market. He could consider himself fortunate if he could find any one who would take his stock at its $100 per share par value. Moreover, Glasgow warned him, if he should suddenly offer his large holding of some 75,000 shares of Du Pont stock for sale, he would only further depress an already depressed market.

Nevertheless, since there seemed to be no other alternative. On 28 January 1922 Alfred wrote to his ever ready but seldom helpful lawyer and agent, Robert Penington, to find buyers for his stock.[34] He would like a single buyer, he told Penington, someone of business acumen who, by obtaining so large a block, could demand a directorship in the company, and could perhaps thus bring to its management the sound business judgment and innovative policymaking, which, in Alfred's opinion, the Du Pont Company now sorely lacked.

In Alfred's recent correspondence with the president of M.I.T., Maclauren had revealed that the anonymous benefactor who had provided the institution with five million dollars in matching funds was no other than George Eastman. This gave Alfred his first potential lead. If the Kodak tycoon had the kind of money that allowed him to make such a gift to M.I.T., then he surely had the funds available to purchase Alfred's stock. Eastman, however, was not interested in Du Pont stock. A disciple of Andrew Carnegie's Gospel of Wealth, Eastman was now interested only in disposing of his money, not in garnering more. So, like an obedient retriever, but alas, a retriever with an exceedingly poor nose, Penington went out into the field to return with what game he could find. He made a concerted effort to clamp on to Henry Ford. He went to Detroit and spoke with Ford's personal secretary, but he never was able to get into the Ford inner sanctum, and after a month of letter writing and phone calls, the secretary, Mr. Leibold, coldly informed him that Mr. Ford did not like the du Ponts, who made war products and General Motors automobiles.[35] Penington was equally unsuccessful with the Rockefellers and the Mellons. He couldn't find any of the big birds out in the field waiting to be bagged.

Alfred had tried to ease the pain of selling his stock by asserting, as he did to Marguerite, that it was his lack of confidence in the present management that had prompted his decision to sell. "I am thinking seriously of getting out of the whole shooting match if I can get somebody to buy my stock," he wrote to his sister in March 1922. "I don't see how any business being run as this is being run and how such people as they have on the

Board of Directors can have but one end and that is a receivership. It may
take some time, but unless they change their methods and put some new
blood on their Board, I am afraid that that ending is due within a comparatively short time."[36]

After Penington's failure to return with an outside entrepreneur who
might have introduced some "new blood" to the board and thus have
helped Alfred to save face in selling his stock, Alfred was forced to turn to
the one potential buyer whom he least wanted to have his stock—the Du
Pont Company itself. He felt sure that Pierre and his brother would jump
at the chance to get his big block, the only sizable holding that lay outside
of their control. Although such a sale would make a mockery out of his
professed claim that he was selling out in order to reform the company's
management, Alfred swallowed his pride and asked Thomas Cochran of J.
P. Morgan and Company to negotiate a sale with his cousins. Alfred said
he would sell for $150 a share.[37] How Pierre must have smiled at this offer,
remembering how Alfred had objected to Coleman's suggested price of
$160 seven years earlier. The current price for Du Pont stock on the open
market was slightly below par, at $90. Irénée, Pierre's brother, told Cochran
that he would buy any or all of Alfred's stock at $83 a share, an offer that
Alfred promptly and scornfully rejected.[38]

If there wasn't any big game to be had, even among the du Ponts, then
Alfred would settle for a flock of small birds. This hunt would take longer,
but even so, it would be better than selling out to Pierre. Alfred asked the
New York brokerage house of McBee, Jones and Company to come up with
a proposal for disposing of his stock to the public.[39]

In April 1922, Alfred wrote to Glasgow that he had been discussing
with R. W. Jones "the question of disposing of some of my common stock
. . . in order to pay off the obligations of the Nemours Trading Corporation
at the Bankers Trust Company. . . . In view of the fact that there is no
market for the stock, in Mr. Jones' opinion the market will have to be created by proper advertising. In other words, he will have prepared a dignified
circular, setting forth the history of the du Pont Company for the past ten
years, with a view to showing its earning capacity; send out thirty or forty
thousand of these to the various people throughout the Country with a view
to getting subscriptions to the stock. . . . I should like to have your views.
It seems to me it is the only way to sell it. . . . My feeling is that they [the
Du Pont Company] will continue the 8% dividend throughout the year and
for that reason I think it is the logical time to sell the stock before they are
forced to either curtail or pass the dividend entirely."[40]

Glasgow did not delay in sending his views as soon as he received
Alfred's letter. Glasgow strongly felt that the worst possible way to sell the
stock was by advertising it to the general public. This move could only

depress the market price. If Alfred had to sell, then he must sell quietly and to only one or two persons. Alfred, put on the defensive by Glasgow's negative reaction, responded:

> "In the first place there is no market for this stock; there never was any market for it after the Company stopped buying it for bonus purposes. . . . Nobody here connected with the business wants any more stock. I tried to dispose of it to the people now in control and the best offer I could get was $83.00 a share in deferred payments. . . . I took the matter up with Mr. Cochran, Morgan & Company, and he knows of no one who wants it and has no plans to suggest for disposing of it. I offered it to Morgan & Company themselves, and they would not touch it. . . .
>
> I cannot see how it would have the effect of depressing the value of my stock when it would not be sold below a fixed price. It will not be made public that my stock is for sale, or that there is any large amount for sale. The letter of propaganda will simply state that this stock can be purchased in small amounts at such a price. . . . Not long ago, Dominick & Dominick advertised in the same manner. . . . I presume they got rid of it and I think it was quite a large block; sold by the Colonel [Henry A. du Pont], I think, or his interests, to pay the tax on the du Pont stock. The sale of this stock did not result in depressing the market value . . . as it was offered at the market value.[41]

Glasgow still was not happy with the proposal, but as with the question of declaring bankruptcy, Glasgow's counsel could not dissuade his client once he had made up his mind. The only question Alfred had to resolve was how much he would offer for sale. He vacillated between selling 10,000 shares and selling the entire lot. He finally settled on the compromise figure of 20,000 shares.[42] On 18 May 1922, Alfred with some pride wrote to Glasgow that he had sold his 20,000 shares at par value of $100 a share and now had an additional two million with which to pay off his debts. Immediately after Alfred had sold, the price of Du Pont promptly went up. One reason for this sudden rise was that the Du Pont Company decided that it would, for the first time, list its stock on the New York Stock Exchange and thus make it available to the general public. As usual, neither Pierre not his brothers had informed Alfred that they had this in mind. Had he waited only six more days before selling, Alfred would have received three million dollars instead of two.[43] Alfred wrote Glasgow on 6 June, "A du Pont Director was heard to remark here the other day that it would soon be up to 200. It appears that General Motors is doing unusually well and will resume dividends in September."[44] At least Alfred could console himself with the thought that he had sold only 20,000 shares and not the entire bundle.

Alfred's sale of some of his stock by subscription had not only provided him with some much needed cash but it had also served the less welcome function of broadcasting to the general public the fact that he was in serious financial difficulties. Suddenly, much to Alfred's surprise, offers of help began to arrive from many different quarters. Dr. Miller Hutchison, the akouphone inventor whom Alfred had once hoped would be his partner in business, wrote offering to lend Alfred as much money as he needed, an offer that Alfred most gratefully and politely refused.[45] To Alfred's utter amazement, Cousin Coley also wrote a warm letter of friendship, offering a loan of two million, which Alfred also quite firmly refused. Most touching of all was the appearance at his office of the widow of a former Brandywine mills employee who brought a few dollars tied up in a handkerchief and asked the secretary to take them as a gift to Mr. Alfred, who needed help.[46] These demonstrations of solicitous affection were certainly appreciated, but Alfred would no more consider charity than he would bankruptcy. The only assistance that he did accept from a du Pont cousin was from William. Cousin Willie had invested $300,000 in Alfred's Nemours Trading Corporation. When it was announced that all claims against and investments in the company would be covered by Alfred, Willie told his cousin that under no circumstances would he put in a claim for the return of his investment. With considerable reluctance, Alfred agreed to let his cousin suffer the loss.[47]

One relative who did not come forward with an offer of assistance was Pierre du Pont. Yet, curiously, a story rapidly gained circulation among the du Pont family that Pierre had not only offered but Alfred had accepted a six-million-dollar loan to bail him out of his financial difficulties. Apparently having been told this story by their mother, Alfred's two youngest children, Alfred Victor and Victorine, believed it to be true. Not until much later did they finally become convinced by Jessie du Pont that there was no truth to the rumor. It was as unlikely that Pierre would have come to the aid of his cousin as it was inconceivable that Alfred would accept anything from Pierre even if it were offered. Alfred would have parted with Nemours and would have gone by foot to the county poorhouse before he would ever be a recipient of Pierre's charity.[48]

By the summer of 1922, buoyed up by the addition to the credit side of his ledger of the two million dollars received for the sale of Du Pont stock, Alfred began to feel that the worst of the financial crisis was over. Always the eternal optimist who often mistook the evanescent glow of swamp fire for the light at the end of the tunnel, Alfred, as well as his faithful counsellor, William Glasgow, did have some grounds for encouragement. Not only had three-fourths of Alfred's biggest debt—the four million dollars to J. P. Morgan and Company—been paid off, but the Morgan bank had agreed to cut the interest rate on the remaining one million from 9 to 7 percent. Moreover, partially through the political influence exerted on his behalf by his friends in Washington, J. Frank Allee and Senator Lewis Ball, who owed his seat in the U.S. Senate to Alfred's political support in 1918, the Treasury Department had agreed not to charge the $700,000 in interest that had accrued on the 1915 tax assessment while Alfred was appealing the judgment in the courts.[49]

It now seemed certain, even to the cautious Glasgow, that Alfred would not have to go to jail for nonpayment of taxes, would not have to give up all of his Du Pont stock, and would not lose his beloved Nemours to satisfy the claims of his creditors. They were not out of danger yet, Glasgow warned his client, but such admonitions could not really check Alfred's ever expansive spirit. He began to plan for the building of a new yacht to replace the ill-named and ill-fated *Alicia*. He and Jessie also made arrangements for a holiday trip to Europe. Jessie had never been abroad and was eager to go. Alfred himself had a very special reason for wanting to go to England in the summer of 1922, for Alicia Maddox, who had remained his very special "little pechette," was to be married in London, and Alfred wanted to be there to give his stepdaughter in marriage.

In the year following her mother's death, young Alicia, then eighteen, had begged Alfred for permission to go to France to study voice. The proud stepfather was quite certain that his little peach had the sweetest soprano voice since Melba—it only needed training. So Alfred had readily consented. Accompanied by her chaperone, Miss Tripp, Alicia had gone to Nice to become a student of the voice instructor Professor Jean de Reszke. She had written ecstatic letters about the beauty of Nice and the greatness of her teacher, who "is really a perfectly marvelous teacher and justly deserves the name of being the greatest teacher in the world, which he has here. I have had five lessons with him, and he has produced tones out of me that no one else could ever produce. . . . He is a perfect father to all of his pupils and we all worship him. I am the youngest one he has and it is a great privilege to study with him as he has only a few pupils. . . . His method is considered the hardest in existence and very few stick it out to the end, but I want to and the present bunch of pupils do also. I hope all of this will please you, dear, for I am so happy studying with the 'Maestro' as he is always called." She had added that she was also playing golf every day and was taking Italian lessons.[50]

Alfred replied to her outburst of wild enthusiasm that he was delighted that she was one of "the maestro's" pupils. "I suppose you realize you have a splendid opportunity to make good with your voice under de Reske [sic], but of course it means unfailing work on your part. Otherwise, I am sure he will drop you like the traditional hot cake."[51] Alfred was also "glad to know you are going to take up the dago language again. Keep up your French, also." He was less enthusiastic about Alicia's taking up golf. "Candidly, I do not consider myself old enough to participate in a game quite so sedate and dignified. I prefer chasing the festive bunny through the brairs, brambles, mud and burrs, a couple of hours of which reduces me to a partially solvent condition with the loss of a pound or two."[52] Alfred's prejudice against golf would always remain even stronger than his bias against Italians.

It soon became apparent from Alicia's letters, however, that she had met someone even more marvelous than Professor de Reszke. While on the way to France, she and Miss Tripp had stopped in England for a couple of weeks and during a visit to Oxford Alicia had met a young American Rhodes scholar, Harold Glendening. That brief meeting had progressed rather rapidly via correspondence and a return visit to England at Christmas time from mere friendly acquaintance to a serious romance. Alicia had not returned home to Nemours as soon as her summer vacation began in June 1921, as Alfred had hoped. She and Miss Tripp had made another extended visit to England—mainly to Oxford—and by the early spring of 1922, Alicia told Alfred that she and Harold were planning to marry that summer. Alfred was favorably impressed with the letters that Glendening had written him ("I have concluded that any young man who at his age takes life as seriously as he does should make an excellent husband") but still felt that Alicia, at nineteen, was much too young to make such an important commitment as marriage.[53] Alicia had-made the mistake of emphasizing that Harold wanted and needed her badly, and she felt so much sympathy for him as to make a postponement of a few years impossible.[54]

This was the wrong approach for Alicia to have made. Alfred immediately replied that no one should ever marry out of pity for the other person. He himself had done that, and it had only brought unhappiness to both him and his partner. By this time, interestingly enough, Alfred had apparently convinced himself that he had only married Alicia because he felt sorry for her after her disastrous marriage to Maddox, and not because he had been infatuated with her.

To the charge that she was marrying only out of pity, young Alicia hotly replied:

> It is not pity I have for him, but love. . . . It is he I want and am
> going to marry, for I love him. . . . As to what his father may have been
> [Harold's late father had been a mail carrier in Norwalk, Connecticut],
> it is not the slightest importance to me, as it is Harold I am marrying,
> not his father. He may have been poor but he earned his livelihood in
> a straight and honest way and his ancestors were most probably as good
> if not better than those of most people supposed to be America's social

leaders. . . . Harold stands as well if not better than most boys at Oxford and is invited and received everywhere so that any fears you may have as to his being unequal to accompanying me socially should be dissipated. He has only just come back from visiting Lord and Lady Swaythling [?] at their country place. . . . Hope that your idea about my marrying for pity will have vanished when you have finished this letter.[55]

Alfred's idea about her marrying for pity may not have vanished, but his resistance to the marriage did. Little Alicia, like her mother before her, could always get what she wanted from Alfred. So the wedding date was set for 28 June 1922, and on 17 June, Alfred and Jessie sailed from New York on board the liner *Majestic* to attend the wedding.

Alfred felt that he, as well as his stepdaughter, was about to enjoy a real honeymoon. For a blessed six weeks there would be no conferences with lawyers, government tax officials, or demanding creditors. After the wedding, he and Jessie planned a leisurely trip on the Continent. They would tour France, Switzerland, and Germany, stopping off in Munich so that he could introduce Jessie to his daughter Madeleine, who since her divorce from John Bancroft in 1913 had married a German, Max Hiebler, and by him had had three sons whom Alfred had never seen. It would be a real family reunion, or at least as much of a family reunion as Alfred at that point thought he could ever expect to have. He had not seen his other three children—Beppie, Alfred Victor, and little Victorine—since shortly before his marriage to Alicia fifteen years ago.

Above all, the trip to Europe would give him the opportunity to devote more time and attention to Jessie, which was what would makes this holiday an Elysian excursion for Alfred, who at age forty-eight had at last found his true love and was as giddy as a schoolboy smitten by puppy love. When Alfred married Jessie, he had worried about the twenty-year difference in their ages. Was he too old for her? Would she in time come to resent being tied to a man old enough to be her father? Would she feel eventually that she was more of a nursemaid than a wife? He had written to his faithful confidante, Ellen La Motte, soon after his marriage, apparently expressing some of these fears, for Ellen wrote back reassuringly, "I think it is nice that she is younger than you, for you need a young spirit to match your own. You have always had a boyish, youthful spirit, and thank heavens, you will now have an opportunity to let it go and be as happy and natural and inconsequent as you have meant to be. I think from what you tell me, your marriage is to exactly the right person"[56]

Ellen, as always, had said just the right thing, which is one reason why Alfred always turned to her for advice and comfort. Her assessment of his marriage was correct. Alfred felt younger now than he had at any time since he left M.I.T. at the age of twenty to take his first job in the Du Pont

Company. He couldn't agree with Ellen more—his marriage had certainly been to exactly the right person. Jessie was neither as well-educated nor as sophisticated and aristocratic in bearing as his first wife, Bessie, nor did she have the breathtaking beauty and grace of Alicia. Men would never resent her, as they did Bessie, because of her sharp and sometimes cruel wit, and women would never envy her, as they had Alicia, because of her beauty. Jessie Ball was a pleasantly attractive young farm girl from rural Virginia who had taught school and had had to work hard for everything she ever had, but who now, unbelievably, found herself, by courtesy of marriage, dressed in satin, furs, and jewels, presiding as mistress over one of the great palaces of America. Hers was one more version of the Cinderella story, which has always been the favorite fairy tale of Americans. But the fairy godmother's wand can only change the outer appearance. Jessie would always be what she had been—smart in a way that Bessie, for all of her education, could never be, tough in a way that Alicia, for all of her pride and temper, never was, and far more determined to endure than either of Alfred's other two wives, for they had never known what it meant to strug-gle for survival. It would be Jessie who would prevail. She would be the enduring Mrs. Alfred I. du Pont, not simply the last of Alfred's three wives.

And she prevailed not just becasue she was tougher and smarter and more determined. To regard Jessie as a greedy, ambitious golddigger who had managed to hoodwink Alfred into marriage with her sweet talk in order to take from him his millions, which is the way Alfred's sister, Marguerite, saw her, is manifestly unfair to Jessie and falsely defines the meaning of their marriage. Jessie was truly in love with Alfred and had been since she had first met him when he came by chance to their farm to fish and hunt. Except in the wild fantasy of a schoolgirl's dream, she could not then believe that this love would ever be reciprocal, but now that, miraculously, it was, she could freely and completely give to him what he had never known before. She could give him love in all of its manifestations—physical passion; genuine admiration for him as a person, not just as a representative of a distinguished family name or as a holder of a high office in the com-pany; compassion for his physical disability; and a comfortable companion-ship, whether it was simply walking through the woods at Nemours or sail-ing in high style on their yacht along the Altantic coastal waters. She could be and was his ardent lover, his sharp manager, his protective mother hen, and his pal, and Alfred loved her in each one of these roles, because for her it was not acting, it was real.

So in spite of his financial worries during the first two years of his mar-riage, Alfred was happier than he had ever been before. Small wonder that he felt and acted like a twenty-year-old, for he was experiencing, at nearly age fifty, the fullness of a mutual love that most men, if they are fortunate, experience thirty years earlier. He wrote Jessie love poems with the same ardor, if not with the same mastery of lyricism, that Edgar Allan Poe had addressed his Helen:

> Heart upon heart Sweetheart with lips pressed
>   close to mine,
> The gates of Heaven apart; the golden glare.
> Thy tender arms about me strong, entwine
> The fragrance of flowers beyond compare. . . .
>
> Heart upon heart Sweetheart with lips pressed
>   close to mine,
> Oh God forefend thy love should ever cease
> The bitter waters thou hast turned to wine,
> The Glory of a tortured soul's release.[57]

He flooded his friends with letters singing the praises of Jessie. To his and Jessie's mutual friend, Lucy Maitland, Alfred wrote that Jessie's devotion "makes me feel very much flattered and proud and gives me that unnatural desire to shave twice a day. She surely is a wonderful girl. Just think of living with a girl for nearly two years and never having one word of misunderstanding or anything approaching a misunderstanding. I did not think there was anyone in the world who could live as long without quarrelling with me as, if I cannot find anyone to quarrel with, I fight with myself to keep in trim."[58]

Alfred was particularly eager to get his sister's approval of Jessie and their marriage. Marguerite, always the sharp critic of her younger brother, had never liked either Bessie or Alicia. She had resolutely refused ever to step inside the walls of Nemours, considering it a sinful waste of money, money that she could have put to good use for her settlement house in Washington. Now, however, prompted partly out of a curiosity to meet Jessie, Marguerite consented to visit Nemours for the Easter holiday in 1922.

Jessie made every effort to charm her sister-in-law, for she knew how much Marguerite meant to Alfred and how eagerly he had always sought, usually in vain, his sister's approval. Apparently Jessie succeeded in this first meeting, for when Marguerite returned home she wrote the obligatory thank you note in which she expressed pleasure in meeting Jessie, whom she had found to be a very nice and friendly young woman.

To this apparent acceptance by Marguerite of his latest venture into matrimony, Alfred responded with exuberance:

> I do not wonder you find Jessie such a fine woman. Everything you
> say about her is true but one must live with her from day to day to realize

the full sweetness of her splendid character. She is the most companionable woman imaginable and has a disposition of gold. Any person who can get along with me, meet all my demands, for you know I am pretty finicky, must have a lovely disposition and there has never been the slightest misunderstanding on the part of either of us and each day is fraught with nothing but sunshine and happiness. She is just as you find her always and one must take her or leave her, just as she is. To my understanding, she incorporates nothing but the highest principles and the highest standards. She is so kind to everybody and particularly to poor people, for whom she has the greatest sympathy, that it is a pleasure to see the good she does from day to day.[59]

The last sentence of this encomium to Jessie Alfred had purposefully designed for Marguerite's benefit. That Jessie was particularly kind to poor people "for whom she has the greatest sympathy" should endear her to Marguerite if nothing else did. When it soon became apparent to his sister, however, that there would be no noticeable increase in the largesse that would flow her way for charity work, Marguerite's initial approbation of Jessie rapidly cooled until she came to distrust and then despise the adventuress who had apparently bewitched her brother into serving only her own and the Ball family's selfish interests. But for the brief time that Marguerite remained on friendly terms with Jessie, Alfred was as proud as he had been when as a boy he had by luck bagged a squirrel from the highest tree branch with a single rifle shot and Marguerite had exclaimed, "Good shooting!"

Alfred began to plan all kinds of extravagant trips for Jessie, Marguerite, and him to take together—cruises on his new yacht to Havana, winter holidays in Florida, whatever Marguerite might enjoy. She would accept none of these proposals, however. She thought only of how many more people she might feed in the soup kitchen she had set up in the Georgetown ghetto with what one of the outings cost. In frequent letters to her brother Maurice, and even to her cousin Pierre, with whom *she* had never parted company, Marguerite would decry her brother's wild extravagance at a time when he himself was nearly drowning in debt. Marguerite, of course, placed the entire blame for all this foolishness on Jessie. Marguerite's visits to Nemours became ever less frequent until they finally ceased altogether. Alfred, however, never gave up in his efforts to bring Jessie and Marguerite together in harmonious companionship. His invitations to join them for the various annual holidays—Christmas, Easter, the Fourth of July, and Thanksgiving—were repeated with calendar regularity and were just as consistently declined, probably to Jessie's considerable relief.

If Jessie could not, with all of her best efforts at hospitable charm, win over Marguerite, she could at least provide Alfred with a plentiful supply of her own siblings as substitutes. The Ball children, growing up in the rural isolation of Northumberland County, Virginia, had been, of necessity, a close-knit family. These familial ties had not been broken even when the five children had gone their separate ways as adults. Jessie and her two brothers had joined their parents in California, Jessie as a teacher in San Diego, Tom as a lawyer in Los Angeles, and Ed as a peripatetic salesman of automobiles, office furniture, and law books throughout the whole Pacific Coast area. Jessie's two sisters had both married and remained on the East Coast. Isabelle, the eldest, was married to Addison Baker and lived in Virginia; Elsie, the youngest sister, now divorced, was living with her young son in Baltimore. Following Jessie's marriage to Alfred and her return to the East, Nemours became the new focal center for the Ball family, not only for Jessie's siblings, but for her numerous Virginia relatives as well.

Alfred accepted Jessie's family with open arms. Jessie's favorite aunt, Aunt Ella Haile, quickly became Alfred's favorite aunt as well—the only aunt he whom he genuinely liked. Sister Elsie Wright, with whom Jessie was particularly close, became Alfred's dear "Sister Upright," who enlivened many a holiday occasion, for there was never any problem in persuading sister Elsie to accept an invitation to come to Nemours.

His first two wives having been also his cousins, Alfred now for the first time had a sizable collection of in-laws who had no relationship with the du Ponts. This was a new experience, and he thoroughly enjoyed it. There was no need to be careful of what you said, lest it be spread like entrapping molasses throughout the whole du Pont family. All of the old du Pont family stories he could now tell to a new and much more appreciative audience.

In this totally different milieu, Alfred's attitude toward his own family began to change. The old resentment for having been betrayed by both Coleman and Pierre still persisted, to be sure, and the old pain of family rejection could never be erased, but he no longer had to contend with social ostracism. He had found a new family where he was not only accepted but honored as the ruling patriarch. During the long years of his unhappy marriage to Alicia, he often felt that the only bond they had was their common hatred of their du Pont family. Together they had fed and indeed largely subsisted on each other's paranoia. Every slight to the one was magnified ten times over by the other.

To Jessie, these ancient feuds belonged to a history in which she had no part. She was, of course, totally sympathetic to Alfred's former difficulties, totally defensive of the actions he had taken in every family battle, but she did not dwell on a past that was not hers, nor did she live for sweet revenge that could never be consummated. She helped Alfred put aside that which was done and could not be undone, and this dismissal of the past may have been a major contribution to the success of their marriage.

Soon Alfred was joking instead of choking over his relationship with those whom he called his "delectable cousins." He was greatly amused when Maurice's wife, Margery, told him that she had recently met a woman who had been in Wilmington visiting some of the du Ponts. The woman told Margery "that my fond relatives referred to me as 'A perfect devil'" Alfred wrote Ellen in delight. "Don't you consider that flattering! It is always nice to be perfect and when you have such people, who consider themselves angels, refer to you as 'A perfect devil,' what could be more flattering!"[60]

Now that he felt secure within a new family relationship, he could even enjoy spreading a bit of du Pont gossip himself. He wrote to Marguerite,

> Perhaps I have never told you of one day many years ago, when I boarded a train at Wilmington for Penville [Philadelphia], and when, for lack of a more comfortable and artistic seat, I was forced to ensconce myself alongside a very plump, tow headed lady. I was not surprised shortly after the train left Wilmington to hear dulcet tones make the following inquiry: 'Do you know any of the du Ponts?' Not well, I said. 'Do you know Harry du Pont?' I said, whom do you mean, the one they call Colonel? I know him by sight. She remarked, 'Great boy with the girls!' I remarked, most men are; and with a suitable excuse, and my hand over my pocketbook, said I was going to have a smoke and disappeared out of the car! All I have to say about the Colonel is that he must go in for quantity and not quality.[61]

Alfred could now be amused by the reported peccadillos of his cousins and quite obviously found it pleasant to be the caster, rather than the catcher, of the gossip mongers' missiles. He could even tolerate with a rather genial forbearance a surprising effort that Irénée du Pont, Pierre's younger brother, made in the spring of 1924 to patch up the old family quarrel. Again in a letter to Marguerite, Alfred related Irénée's attempt at détente:

> You will be interested to learn that Irénée was awaiting me in my office the other day when I came in and said he had come to the conclusion that all family feuds should be set aside, or made up, or whatever you do with a feud, and we should side swipe each other's jaws and generally be happy. I told him I had not known there ever was a feud— so far as I was concerned, I had never recognized any—that difference of opinions did not constitute a feud, nor did mud slinging by any one person necessarily make a feud with the person at whom the mud was slung; so when it came to falling on my celluloid shirt front with a view to exchanging the accolade, there was nothing doing! What 'Irony' was really after, was to get me to support Coly for the Senate and he thought he could accomplish this by holding out the offer of complete restitution

and was 'damfounded' to find that there was nothing doing! I told him I considered myself on the most satisfactory relations possible with all my relatives and that I would not have them changed for anything, because as they now stood, I found them most entertaining and, in particular, Harry, whom I described as the only really decent milkmaid the family had ever produced, at which description Irénée smiled. Altogether, it was very amusing and he is coming back again to chew on the old piece of gum once more.[62]

In the Age of Alicia, if such an attempt at rapprochement had even been tried, Alfred would have tossed Irénée out of his office, but in the Jessie era of his life, Alfred could listen with amusement and even agree to another session of "chewing the old piece of gum."

Alfred's newly acquired tolerance toward at least some of his alienated du Pont cousins did not mean that he was ready to abandon his opposition to Coleman du Pont's political ambitions, which he had made patently clear to Irénée at their little summit meeting in Alfred's office. Having repeatedly blocked Cousin Coly's attempts over the past decade to become a U.S. Senator from Delaware, Alfred was outraged by the political maneuver that his cousin had successfully executed in 1921, which had enabled Coleman to slip into the Senate by the back door.

Coleman's gambit was as brazen a piece of Machiavellian politics as anything the little Blue Hen State had ever seen—and Delawareans had seen everything in politics. The game plan was comprised of the following maneuvers that could only be put into operation after Alfred had so obligingly abandoned his political leadership in 1919. First, Coleman had to find a man who would be willing, in exchange for the governorship, to carry out Coleman's orders once elected. Coleman knew just the man for this role, William Denney, whom Coleman had once before pushed for the governorship in 1916, only to have him be rejected by the nominating convention in favor of the reformer's candidate, John Townsend. In 1920, however, with Alfred out of the political game, Coleman was able to so rig the nominating convention that his stooge, Denney, was able to secure the Republican nomination for governor. That being successfully accomplished, it proved easy to get Denney elected, for in the general election of 1920— the year of the Warren G. Harding presidential landslide—every Republican candidate for office in Delaware was assured of victory.

Once Denney was securely ensconced in the governor's chair, the next step in the game plan had to be handled a little more delicately and secretly. The then senior U.S. Senator form Delaware was Josiah Wolcott, a Democrat, who had owed his election in 1916 to the fact that Alfred's insurgent progressive faction had refused to support the Republican Senator, Colonel Henry du Pont, for re-election, and had so divided the Republican party by putting up its own candidate as to ensure the election of Wolcott. In the Senate, Wolcott had proved to be a total nonentity, an embarrassment to both his party and his state. By 1921, even he realized that his days of political glory were numbered.

Very subtle overtures were made to Senator Wolcott by the friends of Coleman du Pont. It did not prove to be too difficult to convince Wolcott that even if he could get the Democratic nomination in 1922, which in itself was highly doubtful, there wasn't a chance of his winning re-election to a second term. Given these circumstances, would Senator Wolcott consider a proposition that Governor Denney and the Coleman du Pont machine were prepared to offer? Would Wolcott consider resigning from the United States Senate in 1921 in order to accept an appointment as Chancellor of Delaware? As Chancellor, Wolcott would hold the highest judicial office in Delaware, comparable to that of Chief Justice of the Supreme Court in most other states. Wolcott would be giving up one year in the Senate, but that most assuredly would be his last year in any event. In exchange, he would get a well-paid state office that carried much prestige and, what was even more attractive, a twelve-year tenure. If Governor Denney could win confirmation of his nomination, Wolcott would be freed of the uncertainties and perils of again having to seek an elective office.

Wolcott found the proposition a most attractive one. He would indeed be happy to resign his senatorial office, providing Denney could deliver on the chancellorship. Wolcott's resignation would then create a vacancy in the United States Senate that would enable the governor to appoint Coleman du Pont as interim Senator until the next general election in the fall of 1922.

Only when Governor Denney announced his surprising nomination of a Democratic Senator to be the next Chancellor of Delaware did Alfred have his first inkling of the nefarious scheme that was afoot. He had been so preoccupied with his own personal affairs—his marriage to Jessie and his financial difficulties—that he had given hardly a thought to state politics since his futile attempt to get Delaware to ratify the Nineteenth Amendment in the spring of 1920. He had taken no real interest in the presidential election of that year, for he regarded both the Democratic candidate, James Cox, and the Republican candidate, Warren G. Harding, as inferior men unworthy of the high office they sought. Alfred wrote to Marguerite, "I never saw a more wishy-washy, back-boneless pair of candidates for the American people to vote for. There is no selection between the candidates. It is purely a question of which party will do the least damage. Unquestionably, the end of both parties is in sight. I remarked a year ago that if I had a pair of ears and $500,000 to spend for speakers, I could put myself or anybody in the President's chair under the banner of the National

Party."[63] Clearly, Alfred was sick of American politics and happy to be out of the game.

The Denney-Wolcott deal, however, reawakened all of his political combativeness and once again renewed his desire to be an active participant in the defense of basic integrity in government. Alfred summoned his old political lieutenants, Frank Allee and Ed Davis, to a council of war. The only hope of keeping Coleman out of the U.S. Senate was to block the confirmation of Denney's nominee for the chancellorship, for obviously Wolcott would not resign his present high office until he had his new position secured. A simple majority in the seventeen-member Delaware senate was all that was needed to either confirm or reject a gubernatorial appointment. Allee told Alfred that they could count on seven votes in the senate against confirmation, which meant they needed to pick up only two more votes to block the appointment. Allee and Davis both believed that this was possible. They carefully scrutinized the available data on each of the seventeen senators and determined who among their supporters would be the most effective in influencing the vote of each individual senator.

They were immensely aided in their campaign by the popular outcry that arose throughout the state against the blatantly corrupt deal. Delawareans were inured to dirty politics, but this deal exceeded the wildest excesses of the "Gas" Addicks era. Even the normally stand-pat Republican press of Wilmington refused to support this latest bit of chicanery on the part of the machine, and much to Alfred's surprise and delight, that past master of political machinations, the aged Colonel Henry du Pont, himself, spoke out against the Denney-Wolcott deal. Alfred wrote to his cousin for the first time in many years, thanking him for his opposition to the nomination of Wolcott and urging him to do even more to block this attempt "to debauch the judiciary of this state." He asked Cousin Henry "if it is possible to do so that you will see as many of these senators personally as convenient during the coming week, or address letters to them expressing your views in connection with this important issue."[64]

To this request, the Colonel replied rather coldly and tersely, "Your letter of the 17th inst. has been rec'd and noted. I am and always have been strongly opposed to the nomination of Senator Wolcott as Chancellor. To this end I am doing and shall continue to do everything in my power in opposition to the same."[65] In short, the Colonel did not need any instructions from Alfred as to how to use his influence. It was an uncomfortable and unaccustomed alliance for both men.

Alfred himself, of course, wrote numerous letters to all the Republican state senators, putting particular pressure on such doubtful senators as Asa Bennett of Frankford. In the end all of these efforts proved to be futile. The du Pont-Denney machine, with its power and money, won over enough Republican votes to provide, along with the Democratic votes in the senate, a majority of ten for confirmation.[66] Asa Bennett, who had once been one of Alfred's young Progressives, cast his vote with the majority. He wrote to his former political mentor to explain his vote, but Alfred would accept no excuses. "I have no comments to make in connection with your act in Dover in voting for the confirmation of Wolcott," Alfred responded in cold fury. "That you will regret it to the end of your life and probably afterwards is a foregone conclusion. I am sorry for you but you have chosen and there is no going back as you will find out."[67]

To the brave seven who dared to stand up to the machine, Alfred wrote a quite different note:

> As every citizen of Delaware lies today under a debt of gratitude to each Senator who ... stood strongly against the confirmation of Mr. Wolcott as chancellor, I take this opportunity to thank you personally for the splendid work which you accomplished, and for the unflinching manner in which you discharged the trust placed in you by the people who elected you to the State Senate, and which confidence in you has been so strongly vindicated, in contrast to the craven parasites who voted against the interest of the State. ... The mere fact that your fight was not successful only lends additional brilliancy to your efforts and to the encomiums which you are bound to receive from your fellow citizens.[68]

Coleman at last had what he had been actively seeking since he first signed up with the Colonel du Pont machine in 1906, a seat in the United States Senate. Alfred, however, was determined that his cousin would not sit in that seat long enough to warm the cushion. In the fall of 1922 Coleman would have to face the voters, for Governor Denney's shameful appointment was only valid until the next general election. Somehow, some way, Coleman would have to be defeated in that election.

In the meantime, when he wasn't in Colorado for reasons of his continuing poor health, Coleman enjoyed immensely his life as a Senator in the nation's capital. With his wealth, his talent for palaver, and his expansive Babbitry boosterism, he was a natural for the Washington of the Harding era. He quickly became a part of the White House Ohio gang and was a frequent diner at the White House, where he regaled Harding and his cronies with an inexhaustible supply of off-color jokes and his old sleight-of-hand card tricks. He was an occasional big winner and, what was more important to his political standing, an even more frequent big loser in President Harding's nightly poker games. In disgust, Alfred wrote to his old friend, Mrs. John Sebree:

> It is impossible to keep track of the political situation today. ... The whole situation seems to be hopeless, what with seating a man like New-

berry and with Bazaza [Coleman du Pont] dining at the White House and holding Mrs. Harding's fin underneath the unbleached muslin tablecloth, the administration would seem to be taking on a load that it will be difficult to carry through the elections of the coming fall! . . . It is all very amusing, standing on the pinnacle, as Moses used to do in the desert, when he delighted in watching his children . . . setting up the embryonic heifer and giving it a coat of radiator bronze in order to hand him a gold brick when he descended.

Much to the delight of Mrs. Sebree, who was a staunch Democrat and had for years been carrying on a political debate with Alfred, her old Republican friend now found a certain sympathy and respect for her hero, Woodrow Wilson:

> He [Wilson] has nothing but my sympathy. I think his ideals were all right and, had he remained free from the influence of Mrs. Wilson, nee Mrs. Norman G., he might have carried out his ideals in consistency, but with that wily vamp hanging on his rack, he really could not be himself. At any rate, an idealist like me or Wilson should return to the farm and get near to nature, . . . remain in the turnip patch, the natural abiding place for idealists.[69]

Alfred, however, had no intentions of remaining in his "turnip patch" or of continuing to view the sorry political scene in Washington from a "pinnacle in the desert" like Moses. He had decided as soon as Coleman was appointed to the Senate that he would once again enter politics.

Alfred's re-entry into Delaware politics was the one thing Coleman feared, even though it did not seem likely that his cousin would be able to do much to harm him. The progressive element of the Republican party had largely evaporated after Alfred had left the scene in 1919. Alfred had given up his newspapers, and the Wilmington *Morning News,* which had once been such a strong voice of reform in Delaware, had become, under Pierre du Pont's ownership, the loyal drum-beater for the Coleman du Pont machine. Alfred quickly realized that there wasn't the faintest chance of his taking control of the Republican party in 1922 as he had so effectively done in 1916. There was no hope of his being able to block Coleman's nomination for a full term to the Senate. The only alternative was to become a Democrat, and Alfred was prepared to make this supreme sacrifice for the cause. Mrs. Sebree, after years of trying to convert Alfred to her party, was pleased to receive the following note from Alfred in September 1922:

> We have quite a decent Democratic candidate for United States Senator. . . . I hope he will be elected. Were I to change my political creed, or rather keep the same political creed and change the name of my party,

I could condense my views in one short phrase. I am a Democrat because
I am tired of being a Republican.[70]

Mostly, Alfred was tired of being in the same party with Coleman du Pont.

He became greatly alarmed when he heard that Coleman was not only
rigging the Republican nominating convention to ensure his own nomination but that he was also, through Wolcott and his Democratic friends,
attempting to manipulate the Democratic machine into nominating as its
candidate for the Senate Bessie G. du Pont's old drinking companion and
Alfred's very particular bête noire, Andrew Gray. The prospect of a senatorial contest that would pit Coleman against Andrew Gray was so frightening that this alone would have driven Alfred into the Democratic camp.
It was essential that the Democrats nominate someone whom those people
who were outraged as he was by the Denney-Wolcott "dirty deal" could
vote for in good conscience. As Alfred wrote to Ellen La Motte, "It was
T. C.'s plan to have Andrew Gray the Democratic candidate and then ask
the people which pill they preferred. Poor old Andrew is about on his last
legs, I should judge; but in any event, his reputation is so bad that nobody
wants him in any capacity."[71]

So Alfred moved into the Democratic camp, bringing with him what
remnants of the Republican progressive faction he could round up. Alfred
would have preferred to have the Democrats nominate Henry Ridgely of
Dover. "He would have made the best Senator," Alfred wrote Ellen, but
when the push for Ridgely failed, Alfred was content to settle on Thomas
Bayard, whose family name would be of immense help, for the Bayards had
been leaders of the Democratic party in Delaware since the early nineteenth
century, and he himself, as Alfred wrote Mrs. Sebree, "is a good man, in
no way controlled by the 'gang' here. He has no great personality [but] may
have some undeveloped ability. His record is clean and it was necessary to
have some decent man for whom those against Coleman and his methods
could vote."[72]

Even so, the prospects for defeating Coleman du Pont in 1922 were
not very bright. Delaware was now considered to be a safe Republican state.
Coleman was ready to spend any amount of money to ensure his election,
and as Alfred wrote Ellen, "the gang here . . . thought that failure was absolutely impossible. Please remember they had every daily newspaper, both
Republican and Democratic and which they brazenly used in the interest
of T. C. They had all the prestige and influence supposed to go with the
du Pont Company. They brazenly lined up their employees and tried to
make them sign up in T. C.'s interest or else walk the plank; they had all
the money of course they needed; but it turned out to be all without
avail."[73]

What Coleman and the other Republican leaders had not realized was the extent and intensity of the disapproval that his cleverly rigged deal with Governor Denney and Senator Wolcott had aroused among the general public. Not only did Coleman fail to win election for a full term in the Senate, he could not even win the short term, which would have allowed him to stay in the Senate until the new Congress convened in March. He was out of office as soon as the votes were tallied. In high good humor, Alfred wrote to Mrs. Sebree:

> Delighted to have your congratulations on our new senator. This winds up my political efforts for sometime to come. I think I made a good clean job of it this time. It was not only a case of my delectable cousin T. C. being worked off the slate, but I wanted the whole gang, including the bunch in the hotel [Pierre du Pont and Raskob] repudiated by the public vote and that was accomplished.... He [Coleman] is disposed of for all time now and is going to take his spite out on the state by moving to New York! 'Don't make me laugh—me lip is cracked'....
>
> The situation has been an open sore in this poor little state for the last twenty years and I feel much relieved at having finally disposed of it. I can now retire on my laurels for the rest of my days.[74]

Alfred obviously was convinced that the final chapter of Coleman du Pont's sorry political career had been written. As he had done so often in the past, however, Alfred underestimated the pertinacity and determination of his cousin. If Alfred was ready to retire on his laurels, Coleman was not ready to retire in his latest defeat. The ex-Senator from Delaware moved to New York, but he very carefully kept his legal residence in Wilmington. Two years later, in 1924, Coleman was back on the Delaware hustings again, as ebullient and as optimistic of victory as ever. It had been at Coleman's insistence that Irénée du Pont had called on Alfred in the spring of 1924 to seek an end to the family feud. Alfred had curtly dismissed the offer of reconciliation and had let Irénée know that under no circumstances could he ever support Coleman for any office. On the other hand, Alfred showed no inclination in 1924 to lead the charge against Coleman once again. He made a sizable contribution to Allee's Non Partisan League, but in doing so he stated that "I wish it implicitly understood that this money is to be used purely in the interest of securing a clean, honest election, and can only be used for legitimate expenses, such as hiring automobiles and ... any other work to the end that anybody will be prevented from using money illegally."[75] Alfred himself, however, was now ready, out of boredom and disgust with Delaware politics, to climb Mt. Pisgah and watch with cold detachment the worship service of the golden calf that was taking place below. As he had written earlier to Mrs. Sebree, "I have presented the gentleman [Coleman du Pont] to the voters absolutely, I presume, in an unvarnished condition and as an undecorated personality and if they want that kind of representation in the United States Senate, they can have it."[76]

Apparently that was what the voters wanted, for that is what they got in 1924. Coleman had no difficulty in brushing aside the claims of incumbency that the Republican Senator, Lewis Ball, had for a nomination to a second term. Coleman was nominated by a rigged Republican state convention, and with Alfred maintaining a strictly hands-off posture, Coleman was elected overwhelmingly for a full term in the Senate in the fall election. Against only nominal Democratic opposition and with the broad coattails of President Coolidge to ride upon, any Republican in Delaware could have been elected to the U.S. Senate in 1924, even a man with as tarnished a political past as T. Coleman du Pont. Shortly after this election, Alfred in turn moved out of Delaware. Unlike Coleman, however, he made it a point to transfer his legal residence to another state. He had no interest in ever again being a political leader or even a voter in the sovereign state of Delaware.

As for Coleman, it would be death this time, not Alfred, which would prevent him from enjoying a full six-year term in the Senate. By the summer of 1928, Coleman was desperately ill with cancer of the lungs and larynx, but he carefully waited until after the general election in November 1928 before submitting his resignation from office. In that way, the governor of Delaware could appoint his successor, and the governor, as instructed, chose Coleman's loyal henchman, Daniel Hastings, as his replacement. Even on his deathbed, Coleman du Pont was still capable of wheeling and dealing. It would be a long time before his ghost would be exorcised from the body politick of Delaware.

Unlike Alicia, Jessie was neither an instigator nor a sustainer of Alfred's interest in politics. She was only too happy to have him retire from the poltical scene in Delaware. She wanted him to detach himself completely from all the past that had caused him so much grief and pain. Alfred was now out of the company, he was out of politics, and soon he would be out of debt. It was time for him, with her help, to start a new life. Bury the dead past, which needed burying for it stank to high heaven, and become a part of the living present.

This kind of exhortation was true enough and needed to be said as only Jessie could say it, but Polyanna generalities would not suffice. Jessie was smart and loving enough to realize that she had to offer Alfred something more than clichés if she was to succeed in turning him around to face the future.

There were three great losses in Alfred's life that no amount of inane cheerfulness could either trivialize or erase. First, there was the loss of his hearing. Then there was the loss of his children—the three whom he had lost to Bessie by divorce and the two little infants whom Alicia had given him for only a brief moment of life. Finally, there was the loss of his career as the nation's greatest black powder maker. Jessie knew that each of these three deprivations presented a threat to their mutual happiness, and each had to be dealt with in its own way, and dealt with all the ingenuity, force-fulness, and determination that Jessie could command. Fortunately these were qualities that she had in an almost inexhaustible supply.

First on the agenda was Alfred's deafness—a disability that, in terms of isolation from society, can be even more devastating than blindness, partic-ularly for someone as fond of and gifted in music as was Alfred. By the time of their marriage, Alfred had abandoned hope of finding a treatment that might alleviate his condition. At the time that his good friend Miller Hutch-ison, by means of his akouphone, had provided Alfred with one brief mirac-ulous period of restored hearing, he had warned his patient that it would be only a temporary relief. Alfred's condition was degenerative and could never be remedied, and with that grim prognosis, Alfred had given up chas-ing after illusory cures.

Jessie considered this defeatism a foolish attitude. Alfred couldn't be any worse off than he was. He might as well, therefore, try out every pos-sible treatment or invention, no matter how fraudulent it might prove to be. So within three months after their marriage, Alfred was once again undergoing treatment and writing hopeful letters to friends similarly afflicted to urge them to visit a Dr. Rice of New York, who had developed a new machine for "treating middle ear deafness" and which had "caused a distinct improvement, and I expect to improve further as time pro-gresses."[77] There was no improvement for Alfred as time progressed, but undaunted, Jessie kept coming up with new leads—a Dr. J. D. Edwards in St. Louis and a Dr. Curtis Muncie in Brooklyn.[78]

These efforts to find a cure would continue for the remainder of Alfred's life, for Jessie would never give up hope and Alfred would never cease trying, no matter how great the cost in both money and pain these treatments might exact. It was so wonderful to have for the first time a wife who cared enough to seek a cure for him—who neither ignored nor made fun of his affliction as Bessie and Alicia had done—that Alfred was willing to suffer through each new trial and accept the inevitable disappointing failure that would follow. Jessie even tried to learn and then to teach Alfred lipreading, but he claimed he was too old and his eyesight too poor to master that art. Jessie also encouraged him to try a new machine that Ellen

La Motte, in London, had heard reports by which one could pick up sound vibrations through one's hands. "I do not believe much in hearing with your feet or your hands or anything," Alfred wrote Ellen, "except what the Lord gave you to hear with or eavesdrop; because of course everybody with two ears is an eavesdropper and one with one ear is a voluntary hearer and one with no ears is one who minds his own business. I read somewhere one time that if you wrapped up the head of a goose in a bag so that he could not beathe, he would breathe through his wings! I tried this and the goose died! However, I believe in following up everything even if there is only one chance in a thousand of success."[79]

Jessie never found a cure for his deafness, but the chase after each rumored rainbow of hope continued, and the search provided a remedy for Alfred's despair. That in itself was a major victory for both Jessie and Alfred.

With equal determination, Jessie sought to assuage Alfred's grief over the loss of his children. This loss had been further accentuated within two years of their marriage by his stepdaughter Alicia's apparent alienation from her father. Her marriage to Harold Glendening had proved to be a disaster. Alfred's initial fear that Alicia was too young to accept the responsibility of marriage proved to have been well founded. Alicia had too much money to spend and too little opportunity to spend it in the places and ways she desired. She had given up what she believed would have been a promising musical career for marriage. Now she found that that choice had been a mistake, and she was bored—bored with Oxford, bored with housekeeping, bored with a husband who spent all of his time studying.

Alfred was inclined to believe that much of the blame for her unhappiness rested with her husband. Alfred had been surprised when Harold suddenly abandoned his graduate work in chemistry and began the study of law. It seemed that this young man intended to become a perpetual student, living off his wife's income and never accomplishing anything. Alfred got news of the Glendenings through his cousin Ellen La Motte, who lived in London and saw Alicia frequently. After one of Ellen's letters informing Alfred that Harold had been ill, he responded:

> Sorry to learn of Alicia's husband being laid up in bed, but I have an idea he beds easily and is relaxing himself in the feathers of matrimonial indolence. Who was the wiseacre who remarked "that unearned wealth turned many a fine young man into a dangerous fool"? I think it was Carnegie, or some other authority on the subject. Unquestionably, the Almighty Dollar is an Almighty Curse when placed in the hands of people who know not its value and who are thereby turned into spendthrifts and loafers. Personally, I have no respect for any man, or anything

used as stuffing for a man's suit, who will live off his wife. It is impossible
for a woman to respect a man who does this and unless the woman does
respect her husband, the end is in sight and in a Damn sight.[80]

The end of Alicia's marriage was in sight before the young couple had
celebrated their first anniversary. Not even the birth of a son could save the
marriage. It only further exacerbated their marital difficulties, for Alicia was
even less prepared for motherhood than she had been for marriage. In the
summer of 1924, she returned to the United States, leaving her husband
and child in England, and went to Reno for a divorce.

Alicia did not come to Nemours to see Alfred and Jessie. Alfred wrote
his old friend Lucy Maitland, who had been a close friend of his wife Alicia,
and continued to take a keen interest in little Alicia, that after obtaining
her divorce, his stepdaughter was now living in New York. "She has taken
on a new line of friends in Reno and New York which has resulted in
cutting herself almost completely from Jessie and me. I have not heard from
her for a long time."[81] Perhaps ashamed because the marriage that her stepfather had counselled against had not worked out, perhaps not wishing to
hear Alfred say, "I told you so" and Aunt Jessie's little homilies on the
responsibilities of being a married woman, Alicia had abruptly broken relations, and Alfred had lost yet another child.

Alfred, to be sure, still had his foster daughter, Denise, now nearly ten,
for whom he had great affection, but there had been a special relationship
between Alfred and little Alicia. Alicia had come into his life just at the
moment that he had lost his other three children. She was the same age as
his own little Victorine and quickly became the substitute daughter upon
whom he could lavish all of the affection he had once given to his own
children. Denise, however, had suddenly appeared in his home without any
advance knowledge. She had been his wife Alicia's discovery and a balm for
the grief she had felt over the deaths of three previous children. Alfred
found it difficult to believe that Denise belonged to him as well. After Alicia's death, Alfred had left Denise to the care of nurses and governesses
and had given very little personal attention to this child. Jessie, however,
had accepted Denise as her own child and had taught the little girl to call
her mother, not Aunt Jessie.

Fortunately, Denise was by nature a very happy, outgoing child who
apparently adapted easily to the many changes she had experienced in family relationships during her short life. She was sent off to boarding school
at a very young age, and during vacations, when she came back to Nemours,
she and Alfred became much closer than ever before. Together they would
take long rambles across the fields and through the woods along the Brandywine in pursuit of the ever prevalent but always elusive rabbit. When she

was away at school, Alfred's letters to her became more frequent and more endearing as he told of the latest adventures of the family dogs, Monk and Yip, with skunks and porcupines and with the many cats on the estate, whom both Alfred and the dogs detested. Something resembling a true parental-filial relationship began to develop between Alfred and Denise, which they both needed and wanted.

This, however, as Jessie realized, was not enough. Alfred should have more than Denise, home for summer vacation, to satisfy his need for a family of his own. Jessie's own family—and that included not only her siblings but also her many aunts, uncles, and cousins in Virginia—had always been so closely bound together in genuine companionship that she found it difficult to imagine how anyone could be so alienated from his family, and particularly from his own children, as was Alfred. The fact that he had lived in the same relatively small town with three of his children and had not to the best of his knowledge set eyes upon any one of them for over fifteen years was almost beyond Jessie's comprehension.

If somehow Jessie could manage a reconciliation between Alfred and his children, it would not only give her husband much happiness, which was Jessie's main objective, but it would also, she was convinced, further endear her to her husband. Jessie had heard enough of the old family gossip to know that one of the most severe charges that many of the du Ponts had made against Alicia was that she had deliberately and maliciously insisted upon the severance of all ties between Alfred and his three youngest children. This was a false accusation, for it had been Alfred's own decision to stop seeing his children. Nevertheless, the story had not only ensured that Alicia would never be accepted by most of the family but it had also driven one more wedge into their marriage. Had Alfred been willing to continue the same joint custodial relationship he had had with his children prior to his marriage to Alicia, the whole situation might have been different. Now, after all these years of separation, what a tremendous boon it would be to Alfred's and Jessie's marriage—and to Jessie personally—if she could reunite Alfred with his children.

Jessie got her first opportunity to achieve family reconciliation in the summer of 1923. Bill Scott, Alfred's old friend and former associate in the black powder department, was visiting Nemours, and he mentioned to Jessie in private that Alfred's daughter, Bessie, was in town with her two young children on a visit to her mother, Bessie G. du Pont. Scott told Jessie that little Bessie—Beppie, as she was always called—had been Alfred's special favorite. It had been Beppie who had come to the mills each day to bring Alfred his lunch box, and it had been Beppie who had played violin duets with her father in the old music room at Swamp Hall. Bessie was now married to a young Washington attorney, Reginald Huidekoper, the nephew of Scott's and Alfred's hunting and fishing companion, Wallis Hui-dekoper. What a great thing it would be for Alfred, Scott told Jessie, if he could see what a fine young woman and mother little Beppie had become. Jessie thought so too. She asked Scott if he thought he could arrange such a meeting, and Scott was only too happy to serve as the intermediary, for it had been he who had fired the shot that had nearly cost Alfred his life and did cost him an eye. Scott would do anything that might help to atone for the dreadful accident he had caused.

Jessie relayed Scott's information to her husband, and to her immense relief, Alfred, with tears in his eyes, said he would very much like to see Beppie again and meet his grandchildren. So it was quickly arranged. Scott saw Bessie, and she agreed to visit her father in his office, and she would bring her children with her.[82]

It proved to be a very difficult meeting. Alfred was delighted with his little grandson, Henry, age four, and his nine-month-old granddaughter, Ann, whom he bounced on his knee like any proud grandfather, but beyond discussing the children, neither father nor daughter knew what to say to each other. Too much time had elapsed, too much pain was still present for both of them. Bessie was shy and uncertain, Alfred grew stiff and formal. After a half hour, Bessie left, wishing she had never come. She did write Jessie a sweet note of thanks for having arranged the meeting. "It is impossible for me to tell you how deeply I admire you for the spirit you have shown, and the great help you have given. Thank you."[83]

When Jessie showed her husband this note, Alfred, who had been miserable about the failure of his reunion with his daughter, hurriedly wrote Bessie a letter of apology. Although this letter has not been preserved, Alfred was able to put into writing what he had not been able to say to Bessie when they met. He apparently told her of his love for her and apologized for his seeming coldness. His only excuse was that the meeting had brought back so many painful memories that he had retreated into an aloof indifference that he in no way felt. He asked for forgiveness, and hoped that she would come again to see him.

Bessie answered:

> Daddy dear:
> Your letter has made me very happy. I left your office on Tuesday disappointed and discouraged. I thought from your sending me the message to come to you and bring my babies that you must have some feeling of affection for me left and yet your manner showed none. You were lovely about the children but then they *are* two precious little people and no one could think otherwise. I knew the interview was hard for you, after our fifteen years of separation, but no harder for you, Dad, than for me. I had definately [sic] made up my mind that I was not

wanted. Our meeting seemed only to renew old heart aches and not to obliterate them—but your affectionate letter has changed all that and I will go to see you when I am here again in the autumn. When you are in Washington, or if you come to New Bedford during the summer, won't you come to my house. My husband and I want you. Thank you, again, for showing in your letter the gentle father I remember so well when I was a little girl.

Your affectionate daughter, Bep[84]

Bessie did return for a second visit with her father in the autumn, and this time there was no wall of reserve, crowned with sharp spikes of fear and pain, to separate them. It was a true reunion, and Bessie became once again Alfred's favorite child—and Jessie's as well.

Reconciliation with Alfred's two youngest children took longer to accomplish. They had been so young when Alfred last saw them, Alfred Victor only seven and Victorine only four, that they were but shadowy memories to their father, and he to them. As small children, the two had never heard his name mentioned in their mother's presence, and for them he became a legendary figure of romance and mystery. As they grew older, they had been able to glean some information about his activities, for Alfred was frequently featured in the news stories of the day. They learned which of the large automobiles parked on the Wilmington streets belonged to him, and Victorine and Alfred would frequently wait to catch a glimpse of him as he came out to the car. Once Victorine had followed him for several blocks, not daring to speak but finding pleasure in simply being near him.

For Alfred, there had not been even surreptitious contacts with his children. He knew that his son, against his vigorous opposition, had left the Hallock School, where Alfred had placed him, and had gone to Lawrenceville preparatory school and from there to Princeton. He also knew, through his nephew Maurice Lee, that Alfred Victor was employed in the Du Pont Company, but he did not know that his son had served in the Marine Corps during the war until Beppie told him at their first meeting. He later read in the paper in the summer of 1924 an announcement of the marriage of his son to Marcella Miller of Denver, Colorado.

He knew even less about Victorine. This was pointedly brought home to Jessie in the late summer of 1925 when she and Alfred attended the annual Wilmington Horse Show. A bevy of young girls passed in front of their seats in the grandstand, and Alfred, in glancing up, muttered to Jessie, "There goes some of the family spawn," for he could recognize some typical du Pont features, including the prominent du Pont nose, on several of the young faces. A friend sitting on the other side of Jessie whispered in her ear that it was indeed a group of du Pont cousins, and, pointing to one tall young blonde, he told Jessie that that was Victorine du Pont, Alfred's own daughter. Jessie hurriedly pointed her out to Alfred, who responded in gruff emabarrassment, "Well, I don't like the way she's dressed." That was all he said, and Jessie didn't think this an opportune moment to press the issue of a father-daughter meeting.[85]

It was Victorine herself who took the initiative at a reconciliation. Of Alfred's four children, this youngest daughter was the most like him both in appearance and personality. She was the strongest in will, the most forceful and forthright in expressing herself, and a natural leader in any group in which she found herself. Had she been Alfred's son instead of his youngest daughter, she might well have forced herself to the top of both the du Pont family and the company, for she had the brains, the toughness, and the determination to have successfully run that difficult course, but like so many of her female forebears—her great-grandmother Margaretta, and her Aunt Marguerite, to name but two—she would find that her gender was an insuperable barrier in that male supremacist society. She would have to be content in dominating her restricted world of women's clubs and charity balls.

It is not surprising then that Victorine, soon after her marriage to Elbert Dent, a Philadelphia attorney, in the early spring of 1927, boldly decided that twenty years was long enough to continue playing a hide-and-seek relationship with her father. She wanted to show off the big, handsome Princeton graduate to whom she was happily married, and, above all, she once again wanted a father in place of a myth.

So she wrote Alfred a short note asking him if she could bring her husband to Nemours to meet his father-in-law. She also mentioned that Alfred Victor and his wife, Marcella, would also be in town and she would like to bring them as well. With Jessie's encouragement, Alfred replied that he would be delighted to see both young couples at Nemours, and he picked the evening of 22 April 1927 as the date. "That's the night of Irénée's big party," Alfred told Jessie. "It will give the young people an excuse for leaving early."[86]

Alfred's reunion with his two youngest children proved to be not as difficult as had been his meeting with Beppie four years earlier. This was due in part to the presence of Jessie and the spouses of his two children. Mainly, however, it was due to his two youngest children not having been as central in his affections as Beppie had been. Therefore, it was not as painful an experience for Alfred to meet them face to face once again.

Marcella du Pont later recorded in her diary that "the atmosphere was strained at first but got more lively later." Alfred showed the young people through Nemours, that fabled palace of which they had previously seen only the high wall with its jagged glass topping—"to keep du Ponts out," as Alfred was reputed to have said. Now these very special du Ponts were in, and much to everyone's surprise, the evening passed quickly and easily. The four young people did not get to Irénée's big party until well after eleven PM.

From then on, Alfred du Pont's children and their families knew that the gates of Nemours were open to them. The chasm of twenty years' separation could never be filled in, of course, but it could be and was bridged. Jessie justifiably took credit for being the chief architect and engineer of that bridge, and Alfred, during the remaining eight years of his life, would be much indebted to her for the heavy traffic that flowed across it.

XIII

The Way to a Fountain of Youth in Florida 1925–1930

O
N 12 MAY 1924, Alfred du Pont celebrated his sixtieth birthday. Except for his deafness and the loss of an eye (rather large exceptions, as Alfred would readily admit) he was in excellent health and had all of the vigor and stamina of a man of forty. Alfred, in fact, had become something of a fanatic in the observation of a strict health regimen. Having smoked his last cigar on the night of his great political victory over the du Pont machine in 1916, he from then on preached against the use of tobacco, especially cigarette smoking, fifty years before the United States Surgeon General made such strictures accepted medical doctrine. Alfred was not a teetotaler with respect to alcohol. He and Jessie both enjoyed a whiskey highball before dinner as well as a glass of wine with their meals, and Alfred, foresightedly, had laid in $100,000 worth of spirits, wine, and beer just prior to the date that the Eighteenth Amendment went into effect. Alfred confidently expected that this supply would last him until the end of his life, even with frequent gifts of brandy and Scotch to his sister Marguerite and his brother Maurice.[1] Anything in excess of his own moderate consumption, however, he considered to be a sure sign of alcoholism, which could only lead to an early grave.

Above all, Alfred was a great believer in daily exercise and a very sparse but nutritionally healthful diet. Most heart attacks, Alfred was convinced, were caused by too much food and too little exercise. He himself favored walking, swimming, and the "daily dozen" form of calisthenics so widely advocated but so little practiced in the 1920s. For those who demanded competition in their exercise to make it interesting, he favored tennis, not golf, which he put on the same level of silly sedateness as croquet, a game fit for only nonagenarians. One turned to a doctor only as a last resort, and for most people, Alfred was sure, it would be the *last* resort. If their own bad habits didn't kill them outright, doctors would.

Like most health fanatics, Alfred was a proselytizing missionary who quickly bored his friends and relatives with his unsolicited sermons. Both Marguerite and Maurice were subjected to constant health tips from their brother. Alfred was particularly worried about Marguerite's frequent consultations with doctors and her expectations of an early death. "I wish you would stop talking about dying and, also, dying alone," he scolded. "I do not like this frame of mind which you get into. It is the result of living entirely too much alone and would cure itself if you would only come to Nemours and associate with your family. Any person that can skip up those ... steps of yours at sixty millimeters a minute and not have palpitation of the gore pump ought not to talk of dying alone. To [sic] much pill dispenser prognostication. Leave docs alone (they are no good) and come up and see me."[2]

In the summer of 1924, Jessie and Alfred discovered the Kellogg Sanitarium in Battle Creek, Michigan. They both became converts to the cornflake king's austere regimen of eating only food made of vegetables, nuts, fruits, and grains—preferably Mr. Kellogg's own breakfast cereal products. Upon their return to Nemours, Jessie enthusiastically endorsed the Kellogg "cure," and wrote to John Harvey Kellogg that she hoped "it will be in our power to spend a few weeks out of each year there."[3]

Kellogg, delighted to have such distinguished and wealthy guests express their appreciation of his sanitarium, responded that he was convinced that man, like other primates, was naturally herbivorous, for monkeys prefer vegetables to meat, and raw vegetables to cooked vegetables. Alfred, who shared some of Jessie's enthusiasm for the sanitarium, wrote back:

> Before visiting your wonderful institution, I was not a strong believer in the Darwinian theory, but due to the health habits acquired during my stay in Battle Creek, I find that I no longer have the slightest desire for "cow" and so far have been quite contented with the "grass" diet. This leads me to believe that I am undoubtedly descended from a monkey, as friend Darwin insists. However, with all desire to be fair and give the devil (cow) his due, I must confess, sotto voce, that I can see no difference in my general health as between cow and grass, but being perfectly satisfied with the diet of watercress and raw tomatoes, I propose to stick to it until I become as strong and festive as you.[4]

Kellogg was not sure how he should respond to this letter. Had he won another convert, or was Alfred making fun of him and his treatment? In any event, Kellogg did not wish to become involved in the science vs. religion controversy, for this was the time of the Scopes "monkey trial," and he did not wish to cause any offense. So he answered, rather cautiously,

"In my opinion none of us is descended from monkeys, but undoubtedly our ancestors were on very friendly terms with the ancestors of higher apes, and in the long ago prehistoric times when, according to Elliot, there were no carnivorous animals, lived and thrived upon fruits . . . and other products of the forest. . . . We have no teeth especially adapted to the eating of meat, and . . . [by] the art of cooking we disguise some of its most offensive features and by long use have cultivated a liking for its particular flavors. But that we are better off without it there is no room left to doubt."[5]

Alfred never became a complete convert to vegetarianism. He continued to enjoy an occasional meat entrée, but he did become a vociferous advocate of the Kellogg sanitarium for his relatives. He wrote to Marguerite urging her to stop seeing her "pill pusher," Dr. Greer, and try Kellogg instead:

> I am a great believer in the principle advocated by all sane physicians, and that is there is but one physician of any merit and that is old Dame Nature, but in order that she may effect a cure, she must be given some show.
>
> The trouble with most people is . . . that they go through life on the principle advanced by Mr. Glasgow, who, when I requested him some two or three years ago to give up rum, coffee and tobacco, cried out in alarm, what on earth would there be left to live for! If you and Maurice would accept my invitation to go up to see Dr. Kellogg and his vegetarian monkey and remain there about three months, it would not only give me great pleasure and happiness, but you would both return, I am confident, if not altogether cured, certainly in much better health. My heart, my home and my purse are wide open to you both, but the trouble has always been that neither of you would heed my advice or injunctions.
>
> I have been trying to get Maurice to come and live with me at Nemours and give up the life of Broadway and a diet of "hot dogs" made chiefly of cats and rats, but you might as well take a drink of brandy and make eyes at the hitching post, for all the response I get.
>
> It is not yet too late to take the matter in hand scientifically and energetically and go where your body will be treated as a delicate machine and not like an old worn out wheelbarrow which is the treatment accorded to most bodies at home.[6]

Marguerite hastily declined his invitation to go to the Kellogg sanitarium, all expenses paid, saying that her poor old "wheelbarrow" was beyond repair. Alfred answered, "You are wrong in your assumption that the old wheelbarrow is too far gone to be put back into shape. One advantage of the wheelbarrow is its extreme simplicity as a mechanical organism. It is never beyond repair; and what you believe to be the matter with your gore pump, and which is more probably due to toxemia, can be readily corrected by a proper diet and a careful daily regimen. I sincerely urge you once more to do this for my sake and for Maurice's sake and for the sake of all those who love you."[7]

Marguerite, however, refused to abandon her "pill dispenser" for Alfred's much touted sanitarium. She would never agree with her brother that the Kellogg "grass" was greener than Dr. Greer's multi-colored pills. Alfred had no luck in changing Maurice's habits either. He couldn't even get his brother to drink more water. "Maurice is absolutely intractable, like yourself," Alfred wrote Marguerite, "and won't even make the concession of drinking water. . . . Unless an adequate supply is poured into the human map, it is only a question of time before you dry up and blow away in dust. Maurice, so far as I can see, drinks no water except what he gets in coffee, or an occasional ration of grogg [sic], and my efforts to get him to drink the normal ten glasses a day have been as 'bread cast upon the waters.'"[8]

For all of his efforts to interest his relatives in the benefits of a visit to the Kellogg sanitarium, the only customer that Alfred succeeded in sending to Battle Creek was Maurice's daughter, Charlotte, the oldest of Maurice's three children. Charlotte had always been a strangely withdrawn and unresponsive child. After the death of her brother Gerald in France in 1917, she withdrew even further into her own private world. She became greatly overweight and completely unresponsive to her parents and her younger sister Nesta.

Alfred was sure that the Kellogg treatment was just what his niece needed. Once Charlotte lost some weight, she would have a new pride in herself and would emerge from her shell. Because Maurice and Margery did not know what else to do with their strange, sullen child, they accepted Alfred's offer to pay all expenses, and poor Charlotte was shipped off to Battle Creek in September 1924.

From the moment Charlotte entered the place, she hated it, and so informed her parents and uncle whenever any of them made an occasional visit. She did perforce lose weight while eating only raw vegetables and drinking skim milk, but she also seemed to lose what little interest in life she had previously had and retreated even further into herself.[9] All in all, one could hardly judge Alfred's attempt to bring health to his family as one of the great successes of his life. Those who might have benefited by accepting his counsel, Marguerite and Maurice, simply ignored him, and the one person who was forced to accept his prescription, Charlotte, would have been better off if she too could have refused his help.

Alfred himself, however, seemed to offer living proof that the careful observation of his rules for good health worked. Except for an occasional cold, he was never ill. He could walk for miles through the woods and fields along the Brandywine or along the beaches in Florida without tiring, and neither the humid heat of the Delaware valley in summer nor the cold blasts of northeasters off the Atlantic in winter bothered him. True, he spent almost as much time with doctors as did his sister with her favorite physicians, seeking in vain for the magic elixir or machine that might restore his hearing—a fact that Marguerite was not at all hesitant to point out to him, but he insisted that he did it largely to please Jessie.[10] He himself was convinced that most medical practitioners were "quacks or no damned good" and most people would do well to avoid them like the plague.

Obviously a man as robust and energetic as Alfred would not be ready at the age of sixty for a quiet retirement from all business activity. Jessie's great fear was that, without some meaningful pursuit to occupy his time, Alfred would become bored with his life, perhaps even bored with her. She could encourage him to pursue each new lead for the treatment of his deafness and thus keep alive a hope that his hearing might be restored, she could have even greater success in restoring his children to him, but she knew that unless Alfred were given a satisfactory substitute for his third great loss—the loss of a career— he could never be truly content with the new life they were building together. Jessie began work on this problem even prior to their marriage.

In the fall of 1920, during her lengthy visit to the East Coast, Jessie introduced Alfred to her cousin Robert Harding. This young man had designed a machine that could peel, core, and prepare tomatoes for canning in one simple operation. It still needed some further refinement before it was ready to be marketed, but Harding was convinced that his machine could revolutionize the tomato-canning industry. He needed capital to perfect and market his invention; Alfred needed a new field of interest. If the two men could form a partnership, it would be, Jessie believed, a most happy conjunction of resources that would satisfy both men's needs.

Always intrigued by any new technological advance, Alfred was easily persuaded to join forces with Harding, and soon the two were engaged in working out the numerous technological problems involved in handling by machine alone the somewhat delicate fruit of the tomato plant. In a long letter to Ellen La Motte, Alfred explained both the need for a tomato-canning device and how the Harding machine met that need:

> No, the tomato skunter is not my own invention, at least not altogether. The idea was brought to me by one of Jessie's cousins, a Southern boy, by the name of Harding. . . . I do not know whether you are familiar with the ordinary hand peeling method used in the tomato factories all over the world and which, if once witnessed, would make the average gizzard forever turn against any dish containing canned tomatoes. As you know, the season is short, about six weeks, and the

amount of labor required to remove the skin and cores from several million bushels of tomatoes is tremendous. The result is that the factories have to go to the cities and secure any old thing they can get as they are all bidding for this excess labor at the same time, it makes it that much more difficult to get and the prices paid are always excessive and the character of the labor power generally confined to Negro women, Poles and Italians, and the sight of these damsels removing the cuticle and epidermis from the tomats [sic] with their dirty hands, often bleeding from cuts inflicted by their own knives, is not conducive to appetite on the part of the observer for some time. This new machine, the principle of which was presented by young Harding to me and which we have jointly developed, takes the tomatoes from the tank of water into which they have been dumped from the farmers' baskets, placed them stem side uppermost on a wire conveyor belt in perfect alignment of six rows, carries them through the steam scalder and pushes them out of the skin. . . . The tomato pulp at the moment of its being pushed out of its skin is separated from the tomato water . . . and in to a conveyor which carries it directly to the filling machine, in which it is again mixed with as much of the tomato water as is desired; from there is goes into the can, which is immediately sealed; it is then cooked, labelled and boxed.

The manifest advantages of this process over the old process are:

First, the tomato is not touched by human hands between the farmer's basket and the completed container.

Second, [in] the old hand peeling method . . . there was no means of attaining a standard water content, but with a machine skunter it is very simple and any desired standard of water content can be guaranteed. . . .

The machine will pack an ordinary can of tomatoes, containing 20 ounces, in a second and so far as I can see, all the labor required to keep the machine in operation will be about half a dozen, . . . whereas, by the old method, it required anywhere from 50 to 75 women, maybe 100. . . .

I am confining its uses for the next year to Delaware where I can observe the operation of the new machine much better than if I were to put it out broadcast. . . . Of course, it is going to be a little difficult and maybe quite expensive to let people know that the product can be had, as it is going to be difficult to get the cooperation of either the wholesalers or retailers for the reason that they know at once that the terms, prices, etc., are absolutely in the hands of the owners of the patent, and, as my object will be to see that the housewife gets the benefit of the reduction in price, which is bound to ensue, I cannot expect much cooperation from the retailers. However, this will work itself out even if I have to organize a special selling company to handle this product."

This was precisely the kind of enterprise, combining profitmaking with public service, that appealed to Alfred. He was soon writing enthusiastically to Marguerite, "The more I go into this tomat [sic] machine, the more I realize its possibility. . . . One food distributing house in Pittsburgh when shown some cans of our product and having the process explained to them, immediately stated that they would be very glad to underwrite 500,000 cases for next season. This one order would call for a million baskets of tomats; so it behooves us to get busy and see the Hayseeds with a view to getting the necessary number of acres planted next Spring."[12]

In her eagerness to help Alfred find a new career, Jessie did more than introduce her husband to a kinsman whose invention could arouse Alfred's interest. She also brought forth another family member to assist Alfred in the project, her younger brother, Edward Ball. At her invitation, Ed Ball took a leave from his furniture company and came east to offer his services to his brother-in-law in the marketing of his machine.

Little Eddie was only eleven years old in 1899 when Alfred came to Ball's Neck on his first hunting expedition. With his short stature, round cherubic face, and bright little eyes, the boy more resembled a small Irish leprechaun than a mortal child as he went hopping all over the Ball farm, tending with hyperactive energy to his multitude of money-making enterprises: setting crab traps in the Chesapeake, hunting for black walnut logs in the nearby woods, and serving as game flusher for rich hunters like the du Pont party.

It was soon apparent to Alfred that the bright little elfin creature was the spoiled darling of both his parents and his three older sisters. At the age of eleven the boy had persuaded his father to let him drop out of the local country school so that he could devote all of his time to his many business affairs. Eddie was in constant pursuit of the rainbow's end where lay the pot of gold, and for that quest, now that he had learned the rudiments of the three Rs, he saw no need for Latin conjugations or Euclidean theorems.

After the hunting accident in 1904, Alfred was not to see the boy again until seventeen years later when Alfred married Jessie Ball and asked Edward to serve as best man at the wedding.

Although little Eddie was now a man of thirty-two, Alfred had no difficulty in recognizing his brother-in-law to be, for Edward had changed very little in physical appearance from the lad Alfred had gone hunting with two decades earliers. He was not much taller than he had been as a boy, and there was the same intense look of determination on his small cherubic face. He was still the busy little leprechaun, and as if to give further emphasis to his elfin Irish appearance, he now sported a tie pin in the shape of a shamrock, set with a small emerald stone.

Edward had led as busy and varied a life as a man as he had pursued as a boy on the shores of the Chesapeake. Moving to California with his parents in 1906, he had at various times been a prospector for gems in the Sierra Nevada, had owned a grocery store in partnership with a Texas cowboy, had been an automobile dealer in Los Angeles, a traveling representative of a law book publisher, and now was employed by Barker Brothers, the largest furniture store on the West Coast.[13] Whatever occupation he undertook, however, was for him simply one more way to further his frenetic search for the end of the rainbow. It was not until his sister Jessie asked him to come east to help Alfred that Edward would at last be on the right path leading to the pot of gold.

When Eddie arrived in Wilmington on temporary leave from Barker Brothers, it was ostensibly only for the purpose of helping in the organization and marketing of the Clean Foods Products Company, which Alfred and young Harding had recently formed. Jessie, however, had more long-range plans in mind for both her husband and her brother. If she had her way—and she usually did—Ed would stay on to become Alfred's assistant, confidant, agent, and general factotum for all future enterprises in which her husband might become involved.

Perhaps Ed Ball, as well, had had this role in mind for himself from the very beginning. In any event, he welcomed the confidential information that his sister gave to him regarding Alfred's current financial status. Jessie apparently told her brother everything—Alfred's tax problems, the one-million-dollar loan still outstanding with J. P. Morgan, and that troublesome debt of one and a half million owed to the White Shoe Company for those thousands of pairs of shoes that the Nemours Trading Company had never been able to sell.

Ed Ball was fascinated by his sister's revelations, and immediately saw in Alfred's difficulties his own great opportunity. Ball did not share the enthusiasm of his brother-in-law for the Harding tomato machine. He would, of course, take over the financial and sales departments of the Clean Foods company, for that, after all, was the reason he had been asked to become Alfred's employee at a salary of $500 a month. But this little enterprise which, for some unaccountable reason, his brother-in-law found fascinating, was to Ed Ball a mere sideshow with no promise of having a very long or profitable run. Alfred, who had once been vice-president and general production manager of the vast Du Pont Powder Company, and who had on his own attempted to reconstruct the export–import business of war-devastated Europe, surely had a brighter and more important future than tending to the peeling and coring of tomatoes. Ball was sure that there was no pot of gold to be found in the tomato venture—there wasn't even a very colorful spectrum to this particular rainbow. It looked all red to Ed.

If, however, with the information obtained from his sister, Ed Ball could extricate Alfred from some of the old financial obligations that still enmeshed him, then there would be the opportunity for Alfred to move on to bigger and better things, and in the process of cutting the old fetters, Ed could prove himself to be indispensable to his new employer. Ball, to be sure, had never seen a Wall Street banker, let alone tried to negotiate with one, but in his cocky self-assurance he imagined that these fabled captains of finance would probably prove to be no more difficult to deal with than the many small-town, tight-fisted lawyers upon whom he had been able to force his law books. At least it was worth the try. There was nothing to lose, and there could be that pot of gold to gain.

Later, in relating the story to Marquis James, Ed Ball boasted that without informing either Alfred or Jessie, quite on his own initiative, he made an offensive end run against the defense of the Wall Street financiers, and immediately scored a touchdown. Many years earlier he had learned, when selling law books and cash registers, that one didn't waste time trying to interest the underlings in the outer office in one's product. One went directly to the senior partner in the firm. Ed Ball took the train from Wilmington to New York. Presenting himself as Mr. Alfred I. du Pont's business manager, he breezed into the inner sanctum of high finance, the office of George F. Baker of the New York First National Bank. To Baker and his son, George Jr., who was also present, Ball candidly presented the facts of Mr. du Pont's current financial situation. Ball could present quite a heroic tale of Alfred's noble efforts to pay off all of his debts in full, rather than to declare bankruptcy as most men would have done. His employer, Ball told Baker, had succeeded, at great sacrifice to his own fortune, in meeting most of the claims against the Nemours Trading Company, including three-quarters of the four-million-dollar loan the company had borrowed from J. P. Morgan. At the present moment, Mr. du Pont was in the process of developing and marketing a new machine that would revolutionize the vegetable canning industry. It seemed grossly unfair that he should be hampered in his promising new enterprise by old debts carrying an usurious rate of interest. To put the matter briefly and bluntly, which is precisely what Ed Ball did, Mr. du Pont would like to borrow from the First National one million dollars at a reasonable rate of interest in order to pay off the Morgan loan, which bore the absurd rate of 7 percent interest. After thirty minutes of conversation, Ball left Baker's office with a promissory note for one million dollars at 4 percent interest for Alfred to sign.

Following his major success in cutting the interest rate nearly in half on Alfred's one outstanding bank loan, Ball was now ready to tackle the old White Shoe Company debt, which was seemingly an insoluble problem. Negotiations had been under way for the past three years between the company's lawyers and Penington and Glasgow, representing Alfred's interest, but with no success. The shoe company wanted the one and a half million dollars for the shoes, which the Nemours Trading Company had ordered but had never paid for. It refused to settle for anything less than the full amount. As Ball later told James, the deadlock presented a challenge to his skill in sharp bargaining that he found irresistible. This time Ball set off with his employer's blessing, although Alfred had little hope that his brash young brother-in-law could succeed where such a respected lawyer as Glasgow had failed.

According to his own account, Ball spent two days with the White Shoe Company's lawyers in Boston. Although there is no record of what was said in the deliberations, one can surmise that in this instance Ball presented a quite different picture of Alfred's current financial situation than that which he had given to George Baker. This time it was important to portray du Pont as having his back against the wall. There might be no other recourse open to Alfred but to file for bankruptcy, for Ed was not afraid to threaten the creditors with bankruptcy, even if Alfred's scruples ruled out such action. If the White Shoe Company wished to realize anything on its long overdue bill, it had better settle for what it could get now, rather than wait and get nothing.

Whatever arguments Ball may have used, they apparently had an effect that the legal arguments of Glasgow and Penington had not had. At the end of the two-day discussion, the shoe company agreed to cut the claim in half—$750,000. If the company's lawyers expected Ball to jump at this generous concession, however, they were disappointed. Ball had only begun to haggle. Saying that he would take their offer to Mr. du Pont for his consideration, but also stating that he was certain his employer would not be able to accept it, Ball returned to Wilmington. A month later, he was back for more discussion and received a second offer from the company, a settlement for $630,000. Take it or leave it, the lawyers said. Ball left it and again returned home. Alfred, who had been following the proceedings with amazement, would have been willing to settle on these terms, but again Ball insisted on holding out for a better deal. He was making progress, but the White Shoe Company was now obviously in retreat and would make an even better offer. Three weeks later, Ball was asked by the company's lawyers to come back to Boston for further discussions. This time the bill was cut to $490,000, less than one-third of the original claim. As far as Alfred was concerned, the bargaining process had gone on long enough. He could stand the strain no longer, even if Ball was thoroughly enjoying himself and would have liked yet another round of negotiations. Alfred promptly sent off a check for $490,000, and with that the last of his big debts had been paid off. The Nemours Trading Company was now finally buried, and there was much rejoicing in the Nemours household as the last obsequies for a past disaster were observed. Little Eddie had been most impressive in his role as undertaker, and in the process had proved himself to be indispensable, just as he had hoped might be the case.[14]

There still remained the troublesome present to deal with, however, before Ed Ball could walk with his boss into greener pastures where the rainbow touched down and turned the ground to gold. There still remained Robert Harding's pesky tomato machine. Reluctantly but manfully, Ed Ball took up the task of marketing the machine and buying up the tomato crop from local gardeners for processing in the summer of 1923.

Alfred meanwhile was pursuing the dream of revolutionizing the canning industry with the same naive hopefulness with which he had once searched for gold in Maryland and radium in Colorado. He bought into the Piggly Wiggly Stores, one of America's first nationwide grocery chains, in order to get that company to stock his canned tomatoes "untouched by human hands," and he exhorted his two associates, Harding and Ball, to go all out in promoting the Harding machine with major canned food processors throughout the country.[15]

Outwardly, Ed Ball gave every appearance of being as committed to tomatoes as Alfred was. He worked industriously all that summer supervising the processing of the bumper crops that the local tomato growers had produced, and he used his most persuasive sales talk to urge major canners to adopt the machine. But Ed's heart was never in the venture. He may even secretly have hoped that the machine would fail to do the job for which it was designed, but if so, he was to be disappointed. To Alfred's great delight, it easily cored, peeled, and canned in a fraction of the usual time all of the tomatoes that Ed had managed to contract for. The Clean Foods Company seemed to have made a most auspicious debut.

Not only did the Harding machine work—it worked too well. As Alfred du Pont's biographer, Marquis James, wrote, "The Harding-du Pont tomato peeler and canner was a classic example of the machine age outdoing itself."[16] It was soon apparent that only thirty of these machines could can the nation's entire yearly pack of tomatoes, even if one could persuade every canner in the country to purchase the machine. There simply was not a large enough market to make the continued production of the machine profitable.

For Ed Ball this surprising development proved just as useful as if the machine had not worked at all. At the beginning of the venture, Robert Harding had informed Alfred that there were several thousand companies engaged in canning tomatoes throughout the country who would be potential customers for the machine. Ball had done a little investigating on his own and presented Alfred with a list showing only 325 canners in the nation. Even if one could persuade every canner to buy one of the Harding machines, the market would soon be saturated. Ball said that it was his sad duty to inform Alfred that Robert Harding had lied to him about the potential market. Alfred called Ed Ball's cousin in, and young Harding had to admit he had misrepresented the facts.[17] An unhappy and disillusioned Alfred wrote to Ed Ball's older brother, Tom, "I find that . . . Harding does not tell the truth, so I have relegated all the statements, which I had supposed to be true and which he has made to me regarding the machines in the past, to the shelf of uncertainties, if I might so charitably characterize them. I am glad to have Ed here, as he is intelligent and up-to-snuff and an absolutely reliable character, and anybody that puts anything over on him has to get up very early in the morning and remain up until the next morning."[18]

No one got up earlier than Ed Ball. Having disposed of Robert Harding as a possible rival influence, Ball then persuaded Alfred to get out of the tomato business altogether. He counseled Alfred to write off the $250,000 he had already invested in the company, use it as a business loss for tax purposes, and turn his attention and talent to a greater purpose. It proved not too difficult to persuade Alfred to rid himself of tomatoes. In April 1924, Alfred again wrote Tom Ball, "Ed and I have given up the tomato peeling machine, chiefly for the reason that it did not hold out sufficient inducement in the way of remuneration; secondly, for the reason that there seemed to be an antagonism to this product among retailers. It appears that they have no interest in the sanitary condition of the pack but they have an interest in purchasing from the small packers unlabelled products, very often below cost, putting on their own labels, indicating that the contents are some wonderful fancy pack and selling at high figures."[19] Alfred might have added a third reason—that Ed Ball was determined to get Alfred "to give up the tomato machine" and to direct his attention elsewhere.

Having successfully extricated Alfred from what he considered an unprofitable venture and one, moreover, in which Harding, and not he, would have played the supporting role, Ball forced the issue of his own future with Alfred du Pont by politely informing his brother-in-law that with his work with the Harding Tomato Machine company now finished, it was time for him to return to his job with the Barker Brothers furniture store in Los Angeles. He could see no reason for staying on. Besides, Ball added, almost parenthetically, he was earning $18,000 a year at Barker Brothers, but only $6,000 a year with Alfred.[20]

At this opportune moment, Jessie spoke to her husband. Alfred should make every effort to keep her brother in his employ, for in Ed Ball he at last had an associate whom he could trust implicitly. "Let Eddie be your two good ears," she counseled.[21]

It took no great persuasion upon Jessie's part to convince Alfred that he did indeed need Ed Ball. There was the settlement of the J. P. Morgan loan and the long-standing dispute with the White Shoe Company which his brother-in-law had handled with consummate skill and in so doing had saved Alfred well over a million dollars. Alfred took his brother-in-law on a short yacht cruise, and the matter was quickly settled. Ed Ball would immediately submit his resignation to his Los Angeles employers and would accept a permanent position with Alfred I. du Pont at a salary of $25,000 a year.[22] No detailed job description was asked for by Ed Ball or given by Alfred. It was simply assumed that Ed was at Alfred's service in whatever capacity was needed. Eddie would indeed be Alfred's two good ears, and his two good eyes, and his quick sharp tongue as well. Alfred thought he had at last found the faithful aide whom he had long sought—an agent who could drive a sharp bargain, a confidant who could be trusted. "Ed is surely a nice boy," Alfred wrote to Tom Ball. "Being a Ball, he is naturally a gentleman and excuse me from living with anybody less. He is a little pig-headed—another Ball feature, (also a du Pont feature, I being the exception)—so it is necessary to bat him over the head with a club once in a while to make him amenable to reason; but he has a well-balanced cabeza and is a fine, loyal, hard worker, and is as tenacious as a bull dog on a tramp's pants, all qualities appealing most strongly to me."[23]

Jessie was even more pleased with the newly formed association between her husband and brother. She may have been unable to find a doctor who could restore Alfred's hearing, but she had provided him with a pair of ears. More than that, she was convinced she had given her husband an astute counselor—someone who could be the Richelieu in the new empire that she was sure Alfred would now create. There were others close to Alfred within the du Pont family, however—most notably his sister Marguerite—who did not hail Ed Ball as a second Richelieu. Marguerite was reminded of a quite different royal counselor when she saw Ball moving into the center of the inner circle—another advisor whose name also began with the letter R—not Cardinal Richelieu, but the monk Rasputin, would be a more apt comparison in Marguerite's judgment, for here was someone who had hypnotized her brother into complying with what she saw as the Ball family's evil designs.

In the previously quoted letter to his brother-in-law, Tom Ball, in which he had announced that he and Ed were planning to give up the tomato-peeling machine, Alfred had boasted, "We are engaged at present in much more interesting and remunerative work."[24] This was an empty claim, however—nothing but a posture of bravado upon Alfred's part. In actuality, he didn't at that moment have the slightest idea where he should next direct his attention and resources. Since 1915, when he had been forced out of the Du Pont Company, Alfred must have felt that he had been drifting without either sail or compass—not a very comfortable situation for an old, seasoned yachtsman to be in. Twice, in high enthusiasm, he had set his course for a new land but had failed to make a successful landfall. He had first been dashed against the rocks on the Nemours Trading Company voyage, and more recently he had been becalmed in a Sargasso Sea with the Harding tomato machine.

Throughout these post Du Pont Company years of trial and frustration, he had been receiving regular reports on the old company that he had once served so well and which, in spite of his bitterness, he still loved. These accounts of the current successes of E. I. du Pont de Nemours Company were sent to him by his nephew, Maurice du Pont Lee, Marguerite's younger son, who in 1923 was working in the company's rayon plant in Buffalo. Of all his du Pont relatives, Maurice Lee was the one who had over the years remained the most loyal to Alfred, and it was he who kept his uncle abreast of recent Du Pont developments. He wrote regularly, reporting in full on the many diversified products with which the company was now engaged: synthetic fabrics, synthetic dyes (whose patents had been expropriated from the Germans as contraband of war), and new enamel paints for automobiles. It was an impressive list. Gunpowder and high explosives became in these post World War I years only a minor part of the Du Pont Company's production, and the old powder mills on the Brandywine had at last been closed and were falling into ruins. Alfred was not there to protect them, but even if he had still been the director of the company's production, he was no longer sure that he would have tried to save them.

It was with mixed feelings that Alfred read his nephew's letters. He appreciated Maurice's keeping him informed, but he also envied his nephew for being a participant in the chemical revolution. How very much Alfred would have enjoyed taking a leading role in the transformation of the consumer market that his company was bringing to the industrial world. As early as 1910 Alfred had urged the company to modernize and diversify its production. At that time he had tried to provide his cousin, Francis I. du Pont, with the opportunity to devote his time, talent, and laboratory facilities to a scientific quest for more creative uses of cellulose than that of destructive explosives. But Pierre and Coleman had been interested primarily in the assured profits that smokeless powder could bring, and Francis was given orders to direct his research only toward the perfection of high explosives. Now the company, in the years immediately following the war that had been fought to end all wars, had made an abrupt about-face. It had built a great research center in the suburbs of Wilmington that in time would rival the best laboratory facilities of any major university in the country, and would employ hundreds of bright young Ph.D. chemists, who would devote all of their time and a great deal of the company's profits to research on new uses for cellulose.

Alfred, however, was only a spectator to the revolution in chemistry that he had envisioned. So far, his one contribution to this brave new world of technology was a machine that could peel and can tomatoes, but unfortunately did that task so quickly and so well that nobody wanted it! Napoleon Bonaparte, sitting in lonely exile in Elba in the early spring of 1815 while the great powers assembled in Vienna to transform the map of Europe, could have been no more frustrated than was Alfred du Pont, sitting in lonely exile at Nemours in the spring of 1924, so close to but so far removed from all that interested him.

Alfred desperately needed a new direction to his life, but he could not expect his recently acquired associate to provide him with the compass and map. Ed Ball may not have been the Rasputin that Marguerite thought he was, but neither was he the Richelieu that Jessie believed him to be. He was, as Alfred himself had said, "a fine, loyal hard-worker." He was a good lieutenant, but he was no brilliant field marshall of strategic planning. Alfred would have to mark out his own path. Only he could fix his future.

It was quite by chance that Alfred stumbled upon the proper road to travel, and Jessie deserves the credit for serendipitously providing the right environment in which her husband could at last successfully end his search for a new career. Jessie's long years of residence in San Diego had made her particularly sensitive to cold. Alfred might be a man for all seasons, but Jessie could not become easily re-acclimated to the damp, penetrating chill of the Chesapeake winters she had known as a child. Early in their marriage, she had persuaded Alfred to go to Florida during the last dismal weeks of January when the weather was at its most tiresome oppressiveness and the winter stretched ahead interminably.

The du Ponts made a leisurely cruise southward in their yacht, *Nenemoosha,* stopping at Charleston, Savannah, and Jacksonville before they finally berthed their vessel in a marina at Miami. The further south they traveled, the happier and more animated Jessie became. If she couldn't have San Diego, then Florida was surely the best substitute on the Atlantic Coast. In some ways Florida was even better than California in 1922—or at least more exciting, for the du Ponts had begun their annual pilgrimages south just as the Florida land boom was beginning its spectacular expansion. Jessie found a wonderful exhilaration in the speculative excitement, which crackled like an electrical charge along the Gold Coast from Palm Beach to Miami. This was what Jessie imagined the Sacramento Valley must have been like in 1848, or in her own lifetime, the Klondike in 1898. Here was the same heady excitement of instant fortunes being made and the spectacle of new millionaires tossing away what they already had as they frantically grabbed for still more.

Jessie, a gambler by nature, was already deeply involved in the bull market speculation on Wall Street in these halcyon days of stock exchange prosperity. She could hardly resist placing a few chips on the giant roulette table that was southern Florida, where everyone's number seemed to be coming up a winner, while the wheel kept spinning faster and faster. In 1923, against Alfred's advice, Jessie bought two large lots on Miami Beach for $33,000. To Jessie it seemed a good investment, for she believed there was no way in which she could be a loser. The best possible result of the purchase would be that she could persuade Alfred to build her a new home on these lots—a home that she herself would plan and that would be truly her own, in contrast to Nemours, which she had merely married into. At the very worst, should she fail to persuade her husband to establish a permanent residence in Florida, she could at least sell the lots in this crazily expanding market and turn for herself a tidy profit.[25]

While Jessie was thus busily engaged in real estate negotiations or happily shopping in the many exclusive little dress and jewelry boutiques that were springing up all over Miami with the same rapidity and abundance as realtors' offices, Alfred sat in a rocking chair on the wide verandah of the Royal Palm Hotel, content merely to observe the wild scene that passed before him: men in white Palm Beach suits and floppy Panama hats pushing their way through the crowds, women proudly wearing their newly acquired sable fur stoles and perspiring under the hot Miami winter sun, and everywhere the street vendors hawking their $50,000 lots in El Portal or Coral Gables as if they were selling hot dogs or roasted chestnuts. For only $1,000 down, one could get a "binder," or short-term option, to buy a $50,000 lot that in another week might be worth $75,000. It was a capitalistic extravaganza in its most flamboyant form, with the buyers as eager to buy as the sellers were eager to sell.

All of this Alfred observed with the cold, analytical detachment of an anthropologist studying a Polynesian tribe performing an exotic harvest dance. He found the passing scene entertaining and instructive but also deeply disturbing. As he told his brother-in-law after one such afternoon of crowd watching, "Ed, it is the craziest thing I ever saw. These people are on the brink of the precipice right now; and they talk about the good times only getting started. . . . They'll all go broke. No other result is possible. The aim is to get something for nothing. It never works—for long."[26]

But while it lasted, it was quite a show. Each winter that the du Ponts came south, from 1922 through the winter of 1925, they found the tempo more allegro, the sound ever more fortissimo. The old money of the Rock-efellers, the Flaglers, and the Whitneys at Palm Beach mixed uneasily with the brand-new money of the Ohio hardware merchants and the Michigan druggists who had flocked to Biscayne Bay. The story of the Chicago house-wife who had inherited a tract of land near St. Petersburg originally worth $400 that she sold in 1925 for $450,000 was told over and over again to further fan the flames of avarice. One didn't even have to go to Florida to play the game. A farmer in Iowa or a grain elevator operator in North Dakota could buy ten acres of palmetto scrub in the interior of Florida for $100 from a mail-order real estate agency and then wait for the golden tidal wave that he had been assured would eventually engulf even his remote and quite worthless tract.

Alfred was not surprised when one day he encountered his cousin Cole-man du Pont in the main dining room of the Royal Palm. Coleman would, of course, be here. He was never one to miss any great barbecue, and Flor-ida in 1925 was a feast of such extravagance as to put Lucullus to shame. Cousin Coley was in a most affable mood, for he had finally succeeded in winning an election for a full six-year term to the United States Senate. No thanks were due Alfred, to be sure, for his electoral triumph, but on the other hand, Coleman could be and was grateful that Alfred had not rallied any strong opposition to his candidacy. So it was a smiling Coleman who greeted his cousins effusively, and over lunch showed Jessie some of his famous card tricks and gave Alfred a few hot, inside tips on the best buys in Florida real estate. Neither of his companions, however, was much inter-ested in what he had to offer.[27] From the very beginning of their visits to Florida, Alfred had expressed some concern that Miami might be too crowded for comfort, but now, with Coleman's arrival, he knew for sure that it was.

In January 1926, Alfred and Jessie came back to Miami for what proved to be their last winter sojourn in this Babylon on the Biscayne. The sun was just as warm, the breezes just as balmy as on the previous occasions, but there was now a pronounced chill in the air that could be felt on Flagler Street and in the causeways leading to Miami Beach. The crowds were thinner, and sellers outnumbered buyers by a ratio of at least three to one. There was a great deal of brave talk about "a wholesome readjustment of prices," and "a healthy stabilization of land values," but clearly the frenzied St. Vitus's dance was turning into a lugubrious *danse macabre*. The "binder boys" had been pushed over the precipice, the lights were dimming, and the party was nearly over.

This was no surprise to Alfred. The only remarkable thing was that the dance had lasted as long as it had. Jessie's two lots on Miami Beach were put up for sale. She was fortunate enough to sell them for $165,000, and thus realize a nice little profit of 500 percent on her $33,000 investment only weeks before the bottom fell completely out of Florida real estate.

There had never been any real possibility, of course, that Jessie would have the opportunity to use the two lots as a site upon which to build a home. Alfred had been willing to visit Miami for a couple of months each winter in deference to her wishes, but he never would have consented to making this tinsel town his permanent residence. This did not mean, however, that Jessie had no chance of persuading Alfred to make Florida his home. Fortunately, Miami and the equally popular Tampa Bay region on the Gulf side were only small parts of the state, and Florida had a very special appeal for Alfred, as it did for Jessie, although for a different reason. He did not need the state for its climate, and he took no special pleasure in its speculative promise, as did his wife, but in a curious way he felt that Florida, like the Brandywine, was a part of his parental heritage. It has been his mother's favorite retreat after the Civil War. She had made frequent trips to Florida to escape all that so greatly distressed her in Wilmington. He could remember that when he was a small boy, upon Charlotte's return from one of her many prolonged absences, he would snuggle up on his mother's lap and listen in wide-eyed wonder as she told him of the sunny land of flowers and palm trees, of rivers with unpronounceable names filled with real alligators, and of strange, great winged birds who fed only on snails and who had to sit and dry their feathers after each dive into the water for their food. He had vowed at that time that some day he too would go to Florida and see these wonderful sights for himself—perhaps even make his home there as his mother said she wanted to do.

When Charlotte Henderson du Pont made her occasional escapes from her home on the Brandywine, she usually fled to the Putnam House in Palatka, Florida. This imposing hotel was located near the St. Johns just at the point where the stream widened to become the beautiful broad avenue of water that flowed north to Jacksonville and the Atlantic Ocean some sixty miles away. From the verandah of the Putnam, Charlotte could view the river with its steamboats and pleasure crafts and find a peaceful contentment that she never knew when viewing the Brandywine from her porch at Swamp Hall. In the decade of the late 1860s and early 1870s, Palatka and St. Augustine, directly east on the coast, were about as far south as the northern tourist had yet ventured. Miami at that time was an unknown and insignificant little clearing in the Everglade swamps some three hundred miles further south, and it would be another twenty years before Henry Flagler, with his Standard Oil millions, would build his railroad and his great hotels to lure Yankees southward to the very tip of Florida and to the many little keys that lay beyond.

So Alfred, when he thought of Florida, thought of the northern part his mother had known—the land of the St. Johns River, the long, white-sand, coastal beaches that stretched from Amelia Island to St. Augustine, and the one community in all of Florida large enough to be called a city—Jacksonville. This was the land his mother's stories had taught him to love—not the crazy boomtown of Miami.

One of Alfred's business associates, W. T. Edwards, remembered years later how Alfred would frequently take him on weekend trips on the *Nene-moosha* up the St. Johns as far as Palatka, where they would anchor for the night. "On more than one occasion," Edwards would recall, "he asked me to go for a walk thru the town. On every occasion we would stop at a now vacant lot. . . . He would remove his hat and stand in revery [sic] of silent memories for a few moments, then say, 'My mother used to stop in an old Hotel which once stood here and was afterward burned.'"[28]

No, Miami had nothing from the past or of the present to hold Alfred as a permanent resident—but Jacksonville could be a quite different story. Here was a town where a Yankee could feel at home—the most northern city in spirit anywhere in the Deep South. In Jacksonville men wore proper blue serge suits in the winter, not silky, white pongee outfits that looked like pajamas and felt like nudity. Here the banks were solid stone structures. They looked like banks, not like the Alhambra pleasure palaces that passed for banks in Miami. One could walk down Laura Street in the center of Jacksonville and, except for a few palm trees in the little park nearby, imag-ine oneself on State Street in Boston or Market Street in Wilmington. If there was one town in all Florida where some semblance of financial con-servatism and responsible management still prevailed in the boom and bust days of the 1920s, surely it must be Jacksonville. Any place in Florida where Alfred du Pont would consider living would have to be north of the 30° parallel, for below that line lunacy set in, and only money-mad or just plain mad Yankees would care to settle there.

The year 1926 was a strange time for anyone to consider investing in Florida. The hurricane that struck the peninsula during the second week of September not only took 243 lives and destroyed millions of dollars worth of property, but it also blew away the last flimsy hopes of those who had bought undeveloped tracts of sand and swampland in the belief that they could make a fortune. By the winter of 1926, Miami had begun to resemble a ghost town. Many of the hotels were closed for lack of guests, and hun-dreds of real estate offices remained boarded up, for there had been little point in taking down the storm shutters after the hurricane had swept through. The Great Depression came to Florida three years before the Wall Street crash of 1929.

Alfred had been vindicated in his Cassandra-like warnings of doom, which he had been pronouncing since he and Jessie had first come to Miami in the winter of 1922. Like all clairvoyants, he took a certain satisfaction in having been proved right. Furthermore, Alfred was convinced that it had been all to the good for Florida, and quite possibly for him. The hurricane had blown away the last of the speculators, and that was a very good thing. But the land was still there and so were the sea and the sun—and that was even better. In the wreckage that was now Florida, there was an opportunity for salvage—not for those who had speculated in paper profits, but for those who were interested in a true reconstruction by creating real wealth out of the state's rich natural resources. Here might be the opportunity he had been seeking, the new career he so desperately wanted to fill the void in which he had been living since 1915.

Jessie enthusiastically encouraged these as yet vague entrepreneurial musings of Alfred, and her eagerness to begin a new life in Florida would probably have been enough to induce Alfred to take the big step of severing his formal affiliation to his native state of Delaware. Ironically, however, it was Pierre du Pont who would provide the final push to Alfred's becoming a legal resident of the state of Florida.

By 1924, Pierre's only official position within the industrial world he had so long dominated was that of chairman of the board of E. I. du Pont de Nemours Company. Five years earlier, he had turned the presidency of the company over to his brother, Irénée, in order to accept the presidency of General Motors, which greatly needed his executive talents. Once the giant automobile firm was on sound footing, Pierre had relinquished that office as well and had returned to Wilmington in 1923. His duties as chairman of the board were not very demanding, however, and he had ample time to devote to the civic activities that interested him, especially the reforms in the state public education system that he had long espoused. Better schools, of necessity, demanded more public funding, and Pierre turned his attention to tax reform in the state. In 1925 the governor appointed him Tax Commissioner, and, with his usual efficiency, Pierre took up the task of improving the system of tax collection in Delaware. In so doing, Pierre was carrying out much of the program that Alfred himself had advocated when he had been the leader of the progressive wing of the Republican party in the state. Now, however, Alfred was not at all happy at the prospect of having his cousin in a position to examine personally the accounts of Alfred's various property holdings and other investments. When one of Pierre's deputies called on him to inspect his account books, Alfred hesitated no longer over the question of expatriation. He took immediate action to remove himself from Pierre's official scrutiny by organizing the St. Johns River Development Company, incorporated in the state of Florida, to which he transferred all of his property holdings, with the single exception of his estate, Nemours. He then coldly informed the Tax Commissioner's office that as he was no longer a resident of the state of Delaware and his real property was now assigned to a company incorporated in Florida, neither Commissioner Pierre du Pont nor any of his deputies had any legal right to pry into his financial affairs. The big step had been taken. Alfred du Pont had become a Floridian, although his and Jessie's legal residence was as yet only a suite of rooms in the Mason Hotel in Jacksonville. But Pierre du Pont would surely understand that, because for years his legal residence had been a suite in the Hotel du Pont in Wilmington.[29]

To I. D. Short, his old political ally in many Progressive battles of the past, Alfred felt some word of explanation was necessary. Alfred denied that he was moving to Florida to escape the Delaware income tax that he and Short had long advocated; rather he was determined to escape Pierre du Pont. "It was not for the purpose of serving the people that he [Pierre] desired this office," Alfred wrote Short, "but for the purpose of striking me personally as, under the law, he would have access to my books and thus familiarize himself with my private affairs."[30] The old paranoia was still there. Jessie was quite right. It was high time to leave Delaware and the unhappy past behind him.

To another political cohort, Jacob V. Hill, Alfred explained his move within a broader political context. "I shall always be exceedingly interested in the welfare of little Delaware, though at present I have become a resident of Florida, for the reason that until Delaware cleans house and restores the government of the state to the hands of the people and we have once more our regular Republican and Democratic parties instead of one party, namely the du Pont party, I do not consider it a fit state in which to live."[31] Clearly, Alfred himself had no intention of trying once again to help Delaware clean its political house. His native state was no longer the center of his universe. It had now shrunk to "little Delaware," and Alfred was ready to move onto the larger stage of Florida which awaited him.

As a first step in assuming the new role that he had assigned himself, Alfred set up a corporation which he named Almours Securities, Inc. To this holding company, he assigned all of his assets—the Du Pont Company stock plus other investments in stocks, bonds, and real estate. The company was chartered in the fall of 1926, under the corporation laws of Florida, with capital assets of $34,000,000. The name that Alfred chose was symbolic of his desire to redeem past errors with present plans for future success. He kept the suffix "mours" from the old ill-fated Nemours Trading Company, but to it he added the prefix of his own nickname, "Al," indicating his determination that his new company would be under his direct supervision and management. This time there would be no blind delegation of authority, no avoidance of responsibility for sound administration that had doomed his earlier venture. He would do for post-boom Florida what he had failed so miserably in doing for postwar Europe.

With the creation of Almours Securities, Inc., Alfred had in effect converted his entire fortune into liquid assets, so that any portion of his wealth could be readily used for investment in whatever areas he deemed most advantageous for the rehabilitation of Florida. He was ready for action, but there still remained the question of where he should make his first move. There was so much that needed to be done in this depressed state and, once the word was out that Alfred du Pont had moved himself and his entire fortune to Florida, so many proposals were made as to where he should invest his wealth, that he could have easily dispersed the entire capital of Almours within a month. Each eager promoter who came hat in hand to see the great man promised with most eloquent rhetoric that if Mr. du Pont would only accept this proposition, he could double his money within a year. Alfred, however, was not interested in doubling his money. He was interested in the reconstruction of Florida. As he told Ed Ball and William T. Edwards, a Baltimore businessman whom Alfred respected and had persuaded to join him in the Florida venture, "Boys, we are now in Florida to live and work. We expect to spend the balance of our days here. We have all the money which may be necessary for any reasonable cause in our efforts to help Florida grow and develope [sic]. We will find means to this end. Our business undertakings should be sound but our primary object should not be the making of more money—we have enough money for ourselves and any of these needs. Thru helpful works, let's build up good will in this State and make it a better place in which to live. In my last years I would much rather have the people of Florida say that I had been of help to them and their State,—than to double the fortune I now have."[32] As Edwards said in the manuscript of his reminiscences that he wrote for Jessie du Pont, "This declaration, made in serious thought, made a profound and everlasting impression upon me."[33] Edwards claimed that he had remembered Alfred's statement, word for word, exactly as it was spoken at the moment that the three men began their great Florida venture.

In surveying the scene of devastation that was post-boom, post-hurricane Florida, Alfred decided that the most pressing need was to restore some measure of confidence in Florida's banking institutions. At the height of the land craze in 1924, Alfred had been appalled to learn how deeply involved the bankers of the state were in promoting and underwriting the speculative mania. Loans had been made to both land companies and individuals with blue-sky hopes as the only collateral. Now the skies had turned black, the borrowers had fled, the loans were uncollectible, and banks were closing all over the state. As Edwards later reported: "Many large towns . . . had not a single banking institution left, many had only one or two and were therefore handicapped for customary means of carrying on the orderly pursuit of business. Checks on Florida banks received dubious glances everywhere outside of the state. The entire financial credit of the state was a matter of national comment and skeptical concern. Consequently anything like a speedy upturn in business in Florida was utterly impossible. Communities were crying for banking but nobody was answering the call."[34]

It was into that chaotic situation that Alfred decided to move. In the early spring of 1927, he instructed Ed Ball to begin purchasing a small amount of stock in each of the three leading banking institutions in Jacksonville. This would enable Alfred, as a stockholder, to receive the annual reports of these banks, so that he might better assess their financial soundness and the possibility of obtaining a controlling interest in one of them.

With a few shares and the annual reports in hand, along with the gossip he was so adept at garnering from the streets, Ball was able to report to Alfred on the current banking situation in Jacksonville. Of the three major institutions in the city, the biggest and most powerful was the Atlantic National. It was largely the creation of one man, Edward Wood Lane, who was, as Ball reported, "Florida's version of J. P. Morgan." Lane had come to Jacksonville from Georgia twenty-five years earlier, and backed by the fortune he had made with prudent investments in some of Henry Flagler's various enterprises, he had made of the Atlantic National the most powerful financial institution in Florida. Most of its stock was held by him and his family. Although this bank was the soundest of any bank in Florida, with deposits of $26,000,000, there was no possibility that Alfred might move in and take over the Atlantic National to make it the center of his proposed bank chain for Florida.

Much the same situation prevailed in respect to the city's second largest banking institution, the Barnett National. It, too, was largely the personal possession of one man, old Bion H. Barnett, a former Indianian, who had come south in the 1890s. Until Lane moved onto center stage, Barnett's bank had been the major financial institution in the state, and among old Floridians, it was still the most respected. As with the Atlantic National, there was little chance of Alfred du Pont's taking over the Barnett.

That left one other possibility—the Florida National. It was only slightly smaller than the Barnett as measured by the amount of deposits, but it was not nearly as sound financially as its two rivals. It was, as Ball told Alfred, "loaded with slow notes at best," It had one great advantage, however. Unlike its two rivals, it was not the private fiefdom of one man. Its stock was widely distributed throughout the city and state and was available for purchase. Here was a bank that Alfred could take over, but, Ed Ball warned his brother-in-law, it might prove to be a very dangerous venture, for the Florida National was far more deeply involved in bad loans to land speculators than either of the other two Jacksonville banks. Alfred, however, was determined to have a bank in Jacksonville as the center of his Florida operations, and he gave the order to Ball to keep on buying its stock until Almours had the necessary 51 percent control.[35]

While Ball was quietly buying up Florida National stock, Alfred read an item in the Jacksonville *Journal* on 18 April 1929 to the effect that the directors of the Florida National and the Barnett National had voted to merge the two institutions. This surprising action may have been largely motivated as an attempt to bolster the financial structures of both banks, and especially that of the Florida National, the weaker of the two, but Alfred, suspected that canny old Bion Barnett had got word of the Almours Securities Company's surreptitious purchases of Florida National stock. This was his way of blocking the entry of a new and formidable rival into the Jacksonville banking circle. Whatever the reason might be for this merger, Alfred saw it as a direct threat to his attempt to establish a financial base of operations in Florida, for if this consolidation should occur, the Barnett stockholders would have the majority of the shares, and Almours could never obtain the controlling interest of any Jacksonville bank.

By this time, Ball had acquired enough Florida National stock to give Almours Securities a very strong voice in that bank's affairs. So Alfred sent his lieutenant around to see Arthur Perry, the president of Florida National. Perry had just finished writing a letter that was to go out to all of the stockholders of Florida National urging them to approve of the recommendation of the board of directors for a merger with Barnett National at a joint meeting of the shareholders of both banks, which would be held within a fortnight. Ball read a copy of this letter which Perry said Almours Securities would be receiving in the next morning's mail. Ball then said that by coincidence he had in his pocket a letter that Mr. du Pont planned to send out to all the Florida National stockholders urging them to vote against the merger, as such a consolidation would only benefit the Barnett stockholders. If Perry sent his letter, Ball asserted, then most assuredly du Pont would send his. After some heated discussion, Ball prevailed. Neither letter would be sent. A few days later, the plan of a merger was abruptly scrapped by the directors of both banks. Apparently, Perry realized and had been able to persuade the Barnett officials that a merger could not be accomplished because of du Pont's strong opposition. The Florida National would have to remain an independent institution. Alfred could now proceed with his plans to obtain a 51 percent controlling interest of the bank, which in a few short weeks he accomplished.[36]

The officers of both banks were more than a little disgruntled by Alfred's successful efforts to thwart their plan for a merger, but small-town bankers throughout the state, who had long been resentful of Jacksonville's domination of Florida finance had seen the proposed merger as but one more step in reinforcing that power, vigorously applauded Alfred's bold stand. In the opinion of C. D. Dyal, the president of the Daytona Bank & Trust Company and who would later serve Alfred well as president of the Florida National in St. Petersburg, blocking the merger was "one of the greatest, if not the greatest single thing Alfred I. du Pont ever did for Florida. It restored confidence when confidence, above everything, was needed. It marked the dawning of a new day in Florida."[37]

Once Alfred had his bank, however, it became his responsibility to make sure it survived, and the summer of 1929 provided a real test of the Florida National's ability to endure. Summer has always been a difficult time for Florida banks, even in the heyday of the land boom prosperity. In late March when the winter tourists pack their bags and, like migratory snow geese, head north, they also close out their bank accounts that they opened during their winter sojourn. In a few short weeks, all Florida banks suffer a serious drain on their deposits, and they can only hope that they will have enough liquid assets in reserve to meet the demands of their regular year-round customers. If they have also made shaky, unsecured short-term loans that they cannot collect, as most Florida banks had done during the boom days from 1923 to 1925, then they can be in serious trouble.

With each passing winter after the 1926 crash, the situation of cash reserves on hand had worsened, and every bank faced the prospect of surviving the summer with grim foreboding. The winter of 1928–29 had been the poorest tourist season since the end of the war, and by April 1929 more than eighty Florida banks had closed. Even such seemingly sound institutions as Atlantic National were trying to build up cash reserves through borrowing to tide them over the summer.

On 12 July 1929, the Citizens Bank in Tampa closed its doors, and the dreaded run on bank reserves was under way. Alfred and Jessie were in Europe at the time. At Jessie's insistence, they had gone to the famous spa at Carlsbad, Czechoslovakia, so that Alfred could be treated by yet another specialist in audiology of whom Jessie had heard wonderful reports. There Alfred received a frantic cable from Ed Ball informing him that the Florida National was experiencing serious drains on its cash reserves. Ball, who had been left with general power of attorney over the holdings of Almours Securities, asked Alfred's permission to borrow $15 million from New York banks, using the Du Pont stock as collateral. Alfred cabled back his assent. "You're on ground. Use own judgment but pull our bank through."[38] Ball acted quickly. The stock certificates were sent off to New York by express, and the loan was secured. When the bank opened its doors on 25 July, the long line of anxious depositors that Ball had anticipated had already formed. The tellers had been carefully instructed to greet each depositor with a broad smile and an immediate payment of the funds requested, along with the cheery announcement, "Here's your money, sir (or madam). You know, of course, that Mr. du Pont has just deposited $15 million of his own money in this bank which is at our disposal." By noon, the run was over, and by mid-afternoon, many rather sheepish-looking depositors were back in line, asking that their accounts be reopened.[39]

Having successfully weathered this test of both his nerve and his material resources, Alfred now felt in a strong enough position to make a public pronouncement of his basic philosophy of sound banking principles. "I have been associated with banks all my life," Alfred told the press in one of the few interviews that he himself initiated, "and I have come to the conclusion that they are public trusteeships. Their primary object should be safe custodianship of the public's money entrusted to them—not the making of money for their shareholders. If all Florida bankers had had this in mind in the past, there would have been no failures. I have never borrowed a penny in my life without giving what I knew to be proper security. I felt that the money that I would obtain was the public's money, not the banker's money"[40] If this sounded gratuitously sanctimonious to those ruined bankers who had been forced out of business, or if it seemed that this multimillionaire was finding smug satisfaction in rubbing salt into their bleeding wounds, nevertheless the soundness of his principles could not be denied. Banking is a public trust, and that trust had been grossly violated during the great land boom. Now, unfortunately, both the bankers and the general public alike were suffering the consequences.

Over the next two years, Alfred proceeded to how the bankers of Florida just what he meant by banks being "public trusteeships." Florida National banks were established in five other communities: in Daytona Beach on the Atlantic coast, in St. Petersburg on the Gulf coast, and in three interior communities, Orlando, Lakeland, and Bartow. Alfred also applied for a charter to open a bank in Tampa, but when the Comptroller of Currency, the federal officer in charge of issuing new national bank charters, informed Alfred that he was reluctant to issue another charter for that city inasmuch as the two existing national banks there were putting up a good fight to survive but could not withstand the kind of competition that du Pont's bank might present, Alfred readily agreed with the Comptroller's decision. "It must be no part of our policy to compete with other banks when the general situation will not be helped by it," he told Ed Ball, who was always more than eager to expand. "We must help the other banks who are doing a sound banking business, not hinder them."[41]

There was no oversupply of sound banking business in Miami, however; consequently, in August 1931 Alfred opened his seventh and last bank in that city. In order for him to get the charter, the federal Comptroller required Alfred to supply enough cash to pay off the depositors of Third National Bank of Miami, which was in the process of liquidation. It cost Alfred $652,000 to get his charter, but he was determined that in the very center of what he regarded as Florida's lunatic fringe it was incumbent upon him to provide at least one institution that followed sound banking practices. His would be a bank that would make no unsecured loans and would keep enough cash in reserve to meet any demands made by its depositors.

Alfred had never viewed his entry into banking as an end in itself. Banking was to serve as a base from which he could extend his operations into many other areas of economic development for the rehabilitation of Florida. One region of Florida that had a special fascination for Alfred was the panhandle, which extended for over two hundred miles from Pensacola on the Alabama border to Apalachee Bay, where the coast turns abruptly southward to form the long peninsula of Florida. This panhandle was to the Deep South what Appalachia was to the south of Kentucky, Tennessee, and West Virginia—isolated, impoverished, and neglected by the rest of the nation. The little inland towns of Bonifay and Blountstown and the small coastal fishing villages of Apalachicola and Carrabella were as remote and as different from Miami as Jumping Branch, West Virginia, and Yancey, Kentucky, were from New York City. Once, in the distant, pre-Civil War past, the panhandle to be sure, had been the main center of what economic life then existed in the Florida territory. Down its short but navigable rivers had come cotton bales from Alabama and the yellow pine timber from its own forests. Apalachicola and Port St. Joe had been major seaport towns on the gulf at a time when Tampa and Fort Myers were but small outposts on the fringes of the Seminole Indians' territory. The first state constitutional convention had met in Port St. Joe, and the town had had grandiose dreams of being to Florida what Mobile and New Orleans were to Alabama and Louisiana. The deadly scourge of yellow fever, however, had decimated the population in the 1840s, and the Civil War and the Northern blockade in the 1860s had finished the job of economic ruination. The panhandle had sunk into obscurity, and except for Tallahassee, the state capital, it was ignored by the rest of the state and was terra incognita to most of the nation. Not even Henry Flagler, in all of his ambitious planning for his adopted state, ever deigned to look westward or to point his magic wand in the direction of the panhandle. But for Alfred du Pont this region, with its great pine forests, its natural harbors in close proximity to the major shipping lanes that led from Latin America to New Orleans and Mobile, and its undefiled beaches, which on St. George and Santa Rosa islands were superior to anything that Miami Beach could offer, represented a frontier of great potential. The panhandle, Alfred believed, might finally justify its name. It could, indeed, be the handle with which to lift the flat pan that was Florida out of the economic depression into which the state had sunk.

In the fall of 1926, Alfred began to purchase land in western Florida. During the first year, he acquired 66,000 acres of mostly timberland, for which he paid $808,000. Over the next eight years, through 1934, he purchased an additional 253,000 acres, paying only a little over two dollars an acre as land prices continued to tumble downward. These additional land purchases were centered largely in Liberty and Gulf counties and also included one thousand city lots in the old and now largely deserted town of Port St. Joe, five miles of waterfront property in that port town, and control of the Apalachicola Northern Railroad and St. Joseph Telephone and Telegraph Company. The railroad was a spur line that ran north from Port St. Joe to River Junction, Florida, where it connected with the major trunk line of the Louisville and Nashville Railroad. It had once been an important railroad link between the cotton belt in the north and the busy port at St. Joseph Bay, and Alfred was determined that it would once again see a revival of its antebellum traffic.[42]

Alfred's various activities in setting up new banks and buying up timberland received the maximum amount of publicity in the Florida press. That anyone should be investing a fortune in Florida in the years immediately after the great debacle of 1926 was indeed news, and depressed Florida desperately needed that kind of news. Banner headlines announced "DU PONT MILLIONS COME TO FLORIDA," and, as always, the press greatly exaggerated the number of millions that Alfred had. Most of the papers asserted that Almours was a two-hundred-million-dollar corporation, and Alfred made no effort to correct the figure to its true amount. It suited him just fine if the businessmen and politicians of his adopted state, and particularly his rivals in the banking field, thought that he had that much financial clout. When Alfred I. du Pont spoke, the Lanes and the Barnetts, and even the really big money at Palm Beach, listened.

Inevitably, comparisons were drawn between Florida's first big promoter, Henry Flagler, and its latter-day patron. Both had played important roles in the building of a major corporation in America, Flagler in Standard Oil, du Pont in the company whose name he bore. Both had given up executive positions in their respective businesses and, at an age when most men might consider retirement, had come to Florida to build an entirely new career. There were strikingly similar circumstances in their personal lives as well. Like Alfred, Flagler entered into two marriages that ended unhappily before late in life he found real happiness with a third wife considerably younger than he. Both men even suffered from the same physical handicap, loss of hearing, which made their personal and business relations with others difficult.

Alfred was never pleased, however, to have the press hail him as a second Flagler. He always insisted that he perceived his role as benefactor to his adopted state as being quite different from that of his predecessor. Flagler had come to Florida to build a playground for the American rich—great hotels, supervised beaches, big cities with expensive shops, and a railroad line that would bring, in special Pullman cars, the playboys of the north to the sun and sea of this semi-tropical extension of the American mainland. Alfred had come to Florida to build an economy for the producer, not the consumer: railroads and highways to carry wood pulp and citrus fruit to northern markets, not to transport northern tourists to luxury hotels; banks to finance new industrial and agricultural enterprises, not to make unsound loans to profit-seeking land sharks. There was none of the Flagler glamour in Alfred's vision of a new Florida. It was neither Baroque apartment houses on Miami beach nor Spanish haciendas in Coral Gables that he was interested in financing, but paper factories in Port St. Joe, warehouses for citrus growers in Orlando, and fencing and feedlots for cattle ranchers in Bartow.[43] No, it was not Alfred du Pont, but Walt Disney, some forty years later, who was the true successor to Henry Flagler.

In the fall of 1927, Alfred sent Ed Ball out to explore the panhandle and to inspect the lands that had already been purchased in those remote western counties. Ball was away for several weeks, and he must have felt that his mission was very similar to that of Lewis and Clark, whom Jefferson had commissioned in 1803 to explore the newly acquired Louisiana territory. The panhandle was nearly as much of a wilderness to Ball, as he tried to make his way westward in his 1926 Chevrolet over the nearly impassable trails that passed for roads, as were the upper reaches of the Missouri river to Jefferson's two explorers. When he finally got back to civilization in Jacksonville, Ball could report to his chief that the land they now owned was indeed there, but the major problem was trying to get in to find it. What the panhandle needed most were improved roads, and so at that moment there was created the Gulf Coast Highways Association, with Alfred I. du Pont president, W. T. Edwards executive vice-president, and Edward Ball and the great circus impresario, John Ringling, on the list of honorary vice-presidents.

The main purposes of the association were to serve as a lobbying agency to get both state and federal funds to aid in the construction of a major national highway on the panhandle, and to provide a central coordinating organization to promote the issuance and sale of local municipal bonds for the construction of bridges as well as hard-surfaced feeder roads for the main arterial highway. Alfred provided the money for the enterprise, and Ball and Edwards provided the legwork and the sales pitches to persuade the little isolated communities to support this grandiose project. It was a difficult task, for the people of the panhandle were as suspicious of outsiders as the West Virginian mountain folk were leery of "revenuers." Rural Floridians were not interested in having new taxes imposed upon them, and it was not easy to explain that bond sales did not mean the same thing as property taxes. Alfred's grand vision, which was graphically portrayed on the letterhead of the Gulf Coast Highways Association's stationery, was to build what he called "The Florida Loop," a paved highway that would follow the thousand-mile circuit from Jacksonville across the panhandle to Pensacola, down the Gulf coast and around the tip of Florida to Miami, and then up the Atlantic coast, terminating in Jacksonville. It would be longer than a road from New York to Saint Louis, and it would have to be built over innumerable bays, estuaries, and swamps. This would certainly make Cousin Coley's much touted ninety-mile highway down the center of level little Delaware look pretty puny indeed. There was, of course, no possibility that Alfred would ever live long enough to see his dream accomplished, but during the remaining eight years of his life a start was made from Pensacola along the Gulf coast to Panama City, with the aid of federal funds during the first phase of the New Deal. Above all, Alfred had given the state of Florida a vision of the future, and his Gulf Coast Highways Association kept that dream alive until it could become a reality under the Eisenhower administration some thirty years later.[44]

Alfred also became an active promoter of three proposed waterways that he was convinced would be of immense benefit to the economy of Florida. One was the extension of the Atlantic Inland Waterway, a project that had been long under-way to provide an inland, protected water system stretching from Boston southward to Miami. Before Alfred's death in 1935, the waterway was complete. One could travel by boat from Massachusetts to the southern tip of Florida without ever having to leave protective land cover. There was a touch of Flagler in this enterprise, to be sure, for it was of special benefit to yacht and pleasure-boat owners heading south for the winter, but the waterway was heavily utilized by commercial shippers and fishing trawlers as well.

During the first days of the New Deal, Alfred's highways association also pushed the extension of the Gulf coast inland waterways through his beloved panhandle area of Florida. This extension would provide a protected water passage from Corpus Christi in southern Texas, past New Orleans to St. Mark's Bay east of Apalachicola. In June 1933, Alfred received the welcome news that the War Department in Washington had included the last stretches of the Gulf coast inland waterway along the panhandle of Florida in its recommended projects on the National Recovery List of rivers, harbors, and canal projects.[45]

A third and even more ambitious proposal that Alfred continued to push throughout his life was for a canal across the peninsula of Florida. Although he was amenable to any route that could win the support of the federal government, he had in mind using the St. Johns river south past Palatka to the interior lake region of central Florida, and then to construct a series of canals between the lakes to form a waterway across the hundred-mile width of the peninsula to either the Waccasassa river or the Suwannee river and then down that river to the Gulf of Mexico. This proposed waterway, Alfred argued, would be to the Gulf of Mexico and the Atlantic seaboard what the Panama Canal was to the Pacific and Atlantic oceans in Central America, and it could be built for a fraction of the cost of the famed Panama Canal. Alfred had little success in promoting the project, however. There was no Theodore Roosevelt in Washington to push for this waterway, and even among Floridians, especially those in the port cities south of the twenty-ninth parallel, there was little interest in, if not outright hostility to, the plan. Sixty years later the project remains an unrealized but still much discussed dream of the gulf coast shippers in Texas, Louisiana, and the Florida panhandle.[46]

In promoting sound banking, inland waterways, and the much needed hard-surfaced roads in Florida, Alfred du Pont was attempting to provide his adopted state with the basic services of finance and transportation that are essential to a sound economy. Having spent the major portion of his business career as general manager of production for the E. I. Du Pont Company, however, Alfred was under no illusion that the initial projects that he had undertaken would in themselves be an answer to Florida's economic problems. Service industries can never be anything more than buttresses to a viable economy. There is little need for banks unless there are industries to finance; there is little need for highways and canals unless there are goods to move. The Florida economy had always been too service-oriented, and valuable as tourism had been in producing revenue for the state, it was a shaky foundation upon which to build an entire economy. Tourism was too seasonal and too dependent upon the economic prosperity of the rest of the nation. Yet Nature had given this long peninsular appendage of the United States a climate that permitted a twelve-month growing season. Here were rich lands to be tilled, lush pastures for grazing, and millions of acres of pine forests waiting to be harvested. Surely Florida could offer the nation more than hotel rooms on the Gold Coast and marinas for yachts in its bays. Simply providing winter-weary Northerners who could afford it a sunny retreat would not be enough to lift the state out of the deep depression into which it had fallen after the Great Bust of the 1926.

Alfred was still intrigued with the idea of revolutionizing the food canning industry even though his recent flier into that territory with his miracle "tomato skunker" machine had turned out to be a short, rough flight that had ended in a crash landing. As he looked at the highly productive citrus groves of central and southeast Florida and saw mounds of fruit rotting on the ground because of a lack of adequate marketing facilities, he deplored the high cost of this waste. Even worse was the sight of substandard and overly ripe fruit being shipped off to spoil, unsold, on the fruit stands of northern cities. Why couldn't the juice of this fruit be canned? Alfred was once again fired with enthusiasm for the possibility of providing good, wholesome food to the American family at a price that it could afford.

Alfred commissioned his newly employed assistant, W. T. Edwards, to undertake as his first task a careful study of the entire citrus industry of Florida. Alfred posed some basic questions that he wanted Edwards to answer. How was the industry presently organized? Were there a few major producers with whom Almours Securities might negotiate? Had the large producers formed any kind of cooperative associations, and if not, would they be amenable to such a proposal? What would their reaction be to using substandard fruit for canned juice? The present practice of letting this marginal fruit rot, or worse, shipping it off to market, did nothing but damage the reputation of the industry and further curtail the consumer's interest in Florida's oranges and grapefruit.

As he outlined the course that Edward was to follow in his investigation, Alfred waxed eloquent as to what could be done to build up Florida's citrus industry. In 1927, the public had not yet realized the benefits of Vitamin C, and few people thought of orange juice as being as essential to their breakfast menu as coffee and cereal. To most Americans the orange was a luxury item, not quite as exotic as the pineapple, but still a fruit to be purchased for special occasions, a good stocking-filler on Christmas Eve. Alfred told Edwards that this fruit would have to be popularized, and in particular, an appetitie must be stimulated for its juice. Through price and quality control a market could be created. Just as the Coca-Cola Company had boosted its product through advertising, the same could be done for canned orange juice with far less effort and far greater benefit to the health of the American people.[47]

With this rousing pep talk ringing in his ears, Edwards set about to educate himself and his employer on the prevailing conditions in the Florida citrus industry. At the same time, the two men hoped to educate the growers to an appreciation of the possibilities Alfred envisioned for their product. What Edwards discovered, however, both as pupil and as teacher, was not very encouraging. The citrus growers of Florida were as widely dispersed and as little organized as were the tomato growers of Delaware and New Jersey. Two years of investigation, which included talking with hundreds of growers, packers, and shippers, revealed that no one was interested in either quality control or cooperative marketing. Nor could Edwards discover any enthusiasm for investing in or even investigating the possibility of a market for canned orange juice. The only thing that counted, Edwards reported, was volume. "The packing house, of which there are scores, is interested in volume, because they got a certain sum per box no matter what the fruit brings on the market. The Exchange or marketing agency, of which there are scores, is interested in forwarding the highest possible volume to market, because it gets so much per box for every box. Each selling agency is primarily interested in volume. Each vies with the other to control the grower, the packer and the market."[48] So substandard fruit continued to be packed and shipped north, and the fruit that was too poor for even the packer to accept continued to rot in the groves or was fed to hogs. As Edwards later recounted, "Mr. du Pont's plan to rescue the Florida citrus industry had to be abandoned with the hope that some future event might make the plan workable."[49] That future event that both Alfred and Edwards so hopefully awaited would come a decade after Alfred's death, with the perfection of a technique for quick-freezing the concentrated juice of the orange, the grapefruit, the lemon, and the lime. A quarter of a century after that process was developed, orange juice became as basic an ingredient in the average American's diet as bread and milk. As had been true in other instances, Alfred's imagination ran ahead of technology.

Alfred had not had the authority which was necessary to implement his ideas for a diversified usage of cellulose when he had been the Du Pont Company's vice-president in charge of production, nor did he own the groves which would have allowed him to carry out the reforms he proposed for the citrus industry, but Alfred did own hundreds of thousands of acres of southern pine lands. It was in this direction that he now turned his attention after his failure to interest the orange growers in his proposals. His beloved panhandle had an inexhaustible source of wood products—not just for naval stores and turpentine, but wood pulp for paper and cellulose for whatever purposes modern chemistry might discover. Alfred was suddenly fired with enthusiasm for a new mission. Here was an opportunity for one of the most impoverished sections of the country to lift itself into prosperity by providing a basic natural resource for the new synthetics that the chemists were creating. But native Floridians, upon hearing Alfred expound upon his plans for the timberlands of the panhandle, scoffed at his ideas. To them it was not a realizable opportunity but simply another of those idealistic visions which had so often in the past led Alfred to engage in quixotic and futile crusades. Du Pont was tilting at windmills again, they said, for everyone in the South knew that the southern yellow pine had too high a resin content to make it useful for anything other than orange crates, turpentine, or firewood. "You can't turn a sow's ear into a silk purse" was the conventional wisdom that was accepted without questioning.

It is the obligration of science, however, to question the prevailing wisdom. Chemists, in particular, since the day when their medieval alchemic forebears had searched for the magical philosopher's stone, have always been ready to turn their hands and their brains to the task of transforming the base into the precious, be it lead into gold, sow's ears into silk purses, or southern pine wood into newsprint. Science was even then at work to turn Alfred's latest romantic vision into reality at the very moment that he was reluctantly abandoning his earlier dream of producing gold from the orange juice of the Florida citrus farmers. In Savannah, Georgia, a young chemist, Dr. Charles H. Herty, was already far along with his experiments on the southern pine in a laboratory which had been provided him with funds from the Georgia legislature and the Savannah Chamber of Commerce.

Dr. Herty's experiments proved conclusively what he had long suspected—that the high resin content of the southern pine, which made it unsuitable as a source of pulp for paper, was contained only within the heartwood of the mature pine. This heartwood does not begin to form until the tree reaches an age of twenty-five years. Younger trees produce a pulp that is as free of resin as that of the Canadian and Swedish spruce trees from which most American newsprint was at that time made. Herty ascertained that the optimal age for harvesting the southern pine is fifteen years.

After many trials using varying cooking conditions and various sizes of burrs on the pulp wheel, Herty's laboratory was able to produce a paper that had "a marked velvety feel, required less ink for printing and was more pliable than the average commercial newsprint."[50] Suddenly, a dazzling prospect for the Southern economy had been opened up. By 1930 the United States used approximately three million tons of newsprint annually, of which over two-thirds had to be imported from Canada and the Scandinavian countries. Yet the South had a forestland that could provide all the newsprint the country could possibly use. Alfred's illusory castles in Spain were beginning to take the shape of actual paper mills in the panhandle of Florida.

As early as 1931, rumors of a possible breakthrough in the adaptation of the southern pine for commercial use in the manufacture of paper had reached Alfred, and he began to lay plans for the quick exploitation of this development. He asked Edwards to open negotiations with some of the major paper companies in America for the establishment of paper mills along the Gulf coast of Florida. Edwards reported in September 1931 that several companies had indicated an interest and that he was urging them to consider two places in particular, old Port St. Joe on St. Joseph Bay in Gulf County, and the small fishing village of Carabelle on St. George Sound in neighboring Franklin County.[51] In the meantime, Alfred told Ed Ball to go out to the panhandle and buy up more timberland, particularly in Liberty and Gulf counties.

At first Alfred envisioned his role in the development of this new industry as being largely that of supplier of the basic raw material to the established paper companies that he hoped to attract to the Gulf coast. He sent Edwards in the fall of 1931 to make arrangements with contractors to cut the timber on the land held by Almours Securities, and with barge companies to ship the cut lumber to whatever location the paper companies might select. Edwards reported back that Southern Kraft Corporation, which manufactured the strong cardboard used in cartons and which was already in operation in Panama City, was interested in purchasing all of the young pine timber that the du Pont interests could provide. On 6 September Edwards had gone with a Captain Brown, director of the Towboat and Lighterage Company of Jacksonville, to meet with G. P. Wood, a timber-cutting contractor in Sumatra, Florida. The three men had gone over the Almours property in western Florida. Wood had agreed to do all of the cutting, hauling, and delivering of the lumber to the barges for $2.50 a cord. Captain Brown, in turn, agreed to deliver the lumber to the Southern Kraft Corporation in Panama City for one dollar a cord. "The Southern Kraft Corporation, as you know," Edwards wrote Alfred, "pays us $5.25 per unit from which has to be deducted the normal unloading cost of 15¢ per unit, leaving a net of $5.10. Deducting $3.50 per unit for cutting and freighting to Panama City, as per above contracts, we would have a net of $1.60 per unit. . . . I was thoroughly satisfied with the officials of the Southern Kraft Corporation, and I think they are now perfectly willing to play ball with us and encourage us to keep in business in a very large way, which would, of course, be more satisfactory to them than having a large number of contracts with small operators."[52]

Before these contracts could be signed, however, Alfred's concept of his proper place in the new enterprise had expanded along with his holdings of timberland. Why be content with being merely the supplier of raw materials? Alfred was a manufacturing man. Why shouldn't he build his own plant in the now nearly deserted village of Port St. Joe, which Edwards had assured him had the best natural harbor, outside of Pensacola Bay, on the entire panhandle coast?

Ball and Edwards both enthusiastically supported this more ambitious undertaking, for no one knew at that point where the pathway that science had recently opened into the hitherto neglected pine forests of Georgia, Alabama, and Florida might lead. From the laboratories in Savannah and the experimental station of the Hercules Powder Company in Alfred's own native ground of Wilmington, Delaware, new reports were emerging that suggested even more exciting possibilities for the use of southern pine than simply for newsprint or cardboard boxes. Royal Rasch, a research chemist in Herty's laboratory in Savannah, reported in the *Manufacturers Record* of November 1934 that recent experiments with southern pine had produced cellulose fibers through the viscose process with a degree of purity that would make it suitable for the manufacture of rayon and transparent wrappings.[53] Another chemist, E. F. Puckhaber, employed by the Hercules experimental station, had produced an alpha cellulose content of over 97 percent from the southern pine.[54] Over 70 percent of the nation's production of rayon came from textile plants that were located in six southern states. These plants had been built in the South because of the availability of cheap labor, but the savings in labor costs had been partially offset by the costs of transporting spruce lumber from Canada. Now, however, if the recent research findings proved to be commercially sound, the southern mills would have all of the raw material they needed in their own back yard.[55]

Upon reading these reports, Alfred lost no time in writing to his nephew, Maurice Lee, informing him of these development and asking him to send him a sample of the cellulose fiber that the Du Pont Company was currently using in its production of both rayon and cellophane, in order that Herty might compare the quality of the Du Pont cellulose with that which he was producing in his laboratory from the southern pine.[56] How wonderful it would be if the Florida panhandle had the potential to support new plants for synthetic fabrics and transparent protective wrappings that would be equal or superior to those of E. I. Du Pont de Nemours. That would give Cousin Pierre and his two brothers something to brood over.

Alfred even had a name for the product that he hoped could be produced from southern pine cellulose and that could be entered into competition with the Du Pont Company's much touted cellophane. It would be called "glassophane," a name that would emphasize its qualities of transparency and protectivity.[57] As far as Alfred was concerned, no armistice had been declared in the long war of du Pont vs. du Pont. The battle had only shifted to a new front.

Once again, Alfred let his imagination turn wishes into fast horses upon which he could ride into a triumphant future. It would be a long time, if ever, however, before the Du Pont Company would find its rayon and cellophane plants threatened by southern pine research. In the meantime, the immediate prospects of developing a southern paper industry were bright enough to dazzle any entrepreneur in these dark days of the Great Depression.

Alfred had made a real find in W. T. Edwards. Had he had both Edwards and Ball on his payroll in 1917, the whole history of the Nemours Trading Corporation might have been different. Edwards, in particular, had so impressed Alfred with his competence and thoroughness in the preliminary investigations of both the citrus fruit and timber industries in Florida, that Alfred was now ready to give Edwards the major responsibility for the further development of the wood-pulp enterprise, while Alfred and Ed Ball concentrated their attention on keeping their seven Florida banks afloat at a time when banks all over the nation were going under.

Edwards undertook his assignment with his usual zest and industry. He had been delighted with Alfred's decision to build his own paper mill at Port St. Joe. There had been a clear indication from Alfred, as expressed by his interest in Herty's research, that the mill would be for the production of newsprint, and Edwards proceeded with plans for a Port St. Joe mill based on that assumption.

In the fall of 1932, Edwards secured an option to purchase the assets of five small land companies that had holdings in the St. Joseph Bay area. The properties included not only town lots and five miles of waterfront in the nearly deserted village of Port St. Joe but also the ownership of the Apalachicola Railroad and the St. Joseph Telephone and Telegraph Company, which served all of Gulf and Franklin counties. The asking price for all of these holdings was $381,000.

Alfred advised Edwards to hold on to the option but to delay consummation of the deal during the current state of national economic crisis. The following year, in July 1933, soon after the inauguration of Franklin Roosevelt, the federal government announced plans to establish additional national forests in the southeastern section of the United States, to be used as conservation and federal works project areas. Edwards immediately contacted the Federal Forest Service in Washington, and with Alfred's authorization, offered to sell to the government 186,400 acres of the Almours Securities timberland holdings for $450,000. Their offer was accepted by the government, and with the U.S. Treasury check in hand, Edwards was able to buy up the Port St. Joe area with all of its port facilities, railroads, and telephone lines and still have a cash reserve of over $60,000 to make needed improvements on both the railroad and the communication lines.[58]

It was a neat little transaction. Edwards had more than earned his $25,000 annual salary even if he lived to be a hundred years old.

Edwards now had the location, the port facilities, and the railroad and wire communications upon which to build a first-class newsprint mill. And the goddess Fortuna continued to smile upon him. In the summer of 1934, he was given the inside track with the Southern Newspaper Publishers' Association. This powerful organization, composed of the major newspaper publishers of the southeastern United States from Mississippi to Georgia and from Tennessee to Florida, was holding its annual meeting in Asheville, North Carolina. The major topic on the agenda was a consideration of how best to utilize Charles Herty's successful research in producing high-quality newsprint out of southern pine. Naturally, these publishers were excited about the prospect of obtaining all of the 800,000 tons of newsprint that they used annually from timberland close at hand.

Prior to the meeting in Nashville, the Association had created an ad hoc newsprint committee to which it invited interested parties to submit proposals for the construction and operation of a newsprint mill. The committee would then consider the proposals and make its recommendation to the Association as to which proposal should be accepted. The Association would then enter into a contractual relationship with the company that had submitted the most satisfactory proposal for a specified amount of newsprint on a trial basis for the year 1936.[59]

In fear that the annual meeting would be overrun with eager, would-be mill entrepreneurs, the committee had stipulated that under no circumstances would it receive any representative or delegation at its Nashville meeting. Only written proposals would be accepted and considered. As soon as Edwards heard of this development, he had gone to Atlanta to talk with Clark Howell, the distinguished editor/publisher of the *Atlanta Constitution*. Howell was so interested in the plans already under way for the building of a newsprint mill at Port St. Joe, which Edwards submitted to him, and so impressed with the fact that the plans had the financial backing of Almours Securities, that he told Edwards he would personally submit the proposal to James G. Stahlman, the publisher of the *Nashville Banner,* who was chairman of the newsprint committee.

Shortly thereafter, Edwards received an invitation from Stahlman to come to Nashville to meet with the committee and present the Almours proposal. Apparently, the committee was willing to waive its own rule of receiving no representative when that agent had the backing of the du Pont millions. Edwards hurried to Nashville on 28 July and spent the next two days in discussion with the committee.

"We were the only organization from the United States or Canada which was permitted to attend any of the committee meetings . . . or to contact the individual members for any discussion of the subject," Edwards, with understandable excitement and pride, wrote to Alfred. "The committee had received so many importunities from paper manufacturers, paper mill executives and promoters that they had been compelled to adopt a rigid rule disbaring [sic] any organization getting its representative before the committee. As stated above, we however, were taken into the family under an agreement that the fact should be kept confidential and that no part of our discussion be divulged for a period of thirty days or until our plans had matured."[60]

Edwards further reported that the Association was prepared to offer Almours a contract for an order of 100,000 to 125,000 tons of newsprint for the year 1936 "and this would increase from the members of the association during each year of the life of the contract by at least 50%. . . . [The] SNPA is exceedingly anxious to establish this industry in the South and to be certain that it is southern owned, southern controlled and operated. . . . The Committee is aware that the northern mills and the Canadian mills are violently opposed to the establishment of a newsprint mill in the south. These northern mills know that newsprint can be manufactured cheaper in the south." The SNPA was fearful that northern interests would attempt to move into the South, take control and then boost prices in order to protect their northern mills. "The SNPA therefore, want to be certain that they are sledding with strictly southern interests who have at heart not only the interest of the newspaper business of the south but the future development of the south. They regard the initiation of newsprint manufacturing in the south as the greatest single industrial development which has ever been taken in the south."[61]

In his presentation of the Almours proposal, Edwards could confidently assure the committee that their interests were identical to those of his employer. Mr. du Pont had moved south and had established Almours Securities for the primary objective of being of assistance to the southern economy and particularly to the economy of his adopted state of Florida. He even hoped, as soon as it was feasible, to employ only native residents of the Florida panhandle in his Port St. Joe mill. This gave some indication of how strongly Mr. du Pont felt about keeping this industry a southern enterprise.[62]

Alfred was delighted with the results of Edward's negotiations with the Southern Newspaper Publishers' Association in Nashville. He seemed particularly pleased that the proposed mill at Port St. Joe would be built for the production of newsprint. Not only would this mean that he would be a true pioneer in an entirely new industry for the South, but also it would bring him into direct contact once again with a profession that had always had a strong attraction for him, even as a schoolboy at Phillips Andover.

He had often said that if his name had not been du Pont and if he had not liked black powder as much as he did, he would have become a journalist. His brief venture into newspaper ownership during the years of his political leadership in Delaware had further whetted his interests in journalism. Now, once again, in the autumn of his life, he had the opportunity to be so engaged, as a producer of newsprint if not of the printed word. At the same time he would be providing jobs for the people of the panhandle. What could be better?

Alfred wrote to Maurice Lee on 1 September 1934 with great enthusiasm:

> You might be interested to know that we are seriously considering starting a newsprint mill in Florida. We believe we have practically solved the manufacture of the necessary cellulose from Southern pine, of which there is almost an unlimited quantity in the state, reproducing itself ... [in] from seven to ten years. ... The first mill will cost Five Million Dollars.[63]

With strong backing from Alfred and with assurance of a contract from SNPA, Edwards feverishly began to make plans for the construction of the newsprint mill at Port St. Joe. Time was of the essence, for Edwards wanted to make sure that by 1936 the company was in a position to fulfill its contract for at least 100,000 tons of newsprint. Engineers were consulted and architectural plans were commissioned. Edwards was determined that theirs would be the most modern, most efficient newsprint mill on the North American continent, for the entire publishing world would be watching.

Within a month, however, the entire scenario was rewritten, and it was Ed Ball who was the rewrite man. On 22 September, Ball wrote to Alfred, who was at Nemours for the fall season:

> Last week we carried the paper people through West Florida for three days. ... Each and every one of them was very favorably impressed with West Florida. Dr. Herty, who was with us, said that we had enough wood in that territory to take care of all of the imports that the United States makes, which is two million tons per annum.

And then Ball produced the new script that he wanted to be adopted.

> In addition to the newsprint people, we had a kraft paper man with us, who is most enthusiastic. On checking up, it appears that kraft offers possibilities of much larger profit than newsprint and it is my thought that we should have further conversations with a craft [sic] outfit before definitely making commitments on the newsprint. ...
>
> So far, I have ascertained this much: newsprint sells ordinarily at approximately $40.00 per ton ... and the cost of newsprint, according to Dr. Herty ... is $31.00 per ton.

On the other hand, kraft paper cost $35.00 to $38.00 per ton to produce and sells currently at $75.00 per ton. . . . There is a much larger margin of profit, and it is my idea, from what I have learned so far, that we should weigh carefully the advantages to be gained in both newsprint and kraft before going ahead on the newsprint. . . . We are having another meeting Wednesday evening in New York. Edwards and myself will leave Tuesday night, arriving in New York late Wednesday afternoon. We will report to you the results of this session so as to receive your guidance and instructions before making any definite commitments.[64]

It must have been a spirited discussion that took place when Ball and Edwards reported to Alfred at Nemours after their visit to New York, with Edwards arguing the case for a newsprint mill and the commitments he felt that had already been made to the SNPA, and Ball even more vehemently speaking for kraft paper. Alfred had told both men at the beginning of their Florida venture, "our primary object should not be the making of more money." Making more money may not have been Alfred's primary object, but it most certainly was Ed Ball's. To Ball, the $28 difference in profit per ton was an irrefutable argument in favor of kraft paper over newsprint. Suppose Herty's research was faulty. Why should anyone gamble five million dollars on an untried product when one had a proven product with a ready market at hand, and one that would in addition pay a four times greater dividend?

In the end, of course, it was Ed Ball who prevailed, as he almost always did. It was he who provided the guidance and instructions for the new venture in papermaking. In early October, Alfred wrote another letter to his nephew, Maurice Lee. "The newsprint business does not look very promising at the present time. . . . We are now looking into the kraft situation."[65]

It was left to the frustrated Edwards to explain to the Southern Newspaper Publishers' Association as best he could that Mr. du Pont had decided not to enter the newsprint field after all, but to go into the production of kraft paper instead. The sudden about-face did, of course, demand some explanation, and the word was circulated within the paper industry that Mr. du Pont had not been satisfied with the quality of newsprint that southern pine produced. It absorbed too much ink. Herty's research, of course, had proved just the opposite, and subsequent developments within the industry showed Herty to have been correct.[66] Since neither Alfred nor Ed Ball cared to divulge the real reason for the change in their plans, the public, including Alfred du Pont's biographer, were left with an incorrect impression. Marquis James could only write in explanation, "Mr. du Pont would have preferred to make newsprint because he loved to pioneer and because of the social significance of giving the South a new industry. But when at length he came to the conclusion that the time was not ripe to invest $5,000,000 in an attempt to produce newsprint in the South, he switched to kraft with what must have appeared to outside observers as startling abruptness."[67]

It appeared to be startling abruptness to some inside observers as well. Bill Edwards, worn out by his efforts to secure a newsprint mill at Port St. Joe, and sick with the recurrent ulcer attack that had most certainly been exacerbated by his confrontation with Ball, was sent by his doctor to a health resort in Asheville, North Carolina. Alfred wrote solicitous letters to him weekly, urging him to take as long a sick leave as necessary in order to be restored to good health because both the company and Alfred needed him.[68] There were times, however, when Edwards wondered just how much he was needed—or wanted—as long as Ed Ball sat at Alfred's right hand.

During these busy years from 1926 through 1934, while Alfred was establishing a bank chain, buying up timberland, and attempting to build up a citrus fruit and paper industry for Florida, Jessie has been busy on a project of her own. As soon as the decision had been made to establish a permanent residence in Jacksonville, she had begun planning a new home. This would be her home. She would design it, furnish it, name it. By Nemours standards, it would be modest, with only seven master bedrooms and a limited domestic staff. There would be nothing French about this place—no Louis Seize rococo pretentiousness, but rather a Spanish hacienda graciousness—pink stucco exterior with wrought-iron balustrades shaded by great old live oaks festooned with Spanish moss. The estate would be called Epping Forest, the name given to the Virginia plantation of Mary Ball, George Washington's mother, with whom Jessie proudly claimed an ancestral relationship. The name did not exactly fit either the geographic location, the indigenous flora, or the architectural style of the house, but it certainly was no more inappropriate to Jacksonville, Florida, than were Nemours, Chevannes, or Montchanin in Wilmington, Delaware, and it was a name that belonged to her family history not to the du Pont history.

A site was selected a few miles south of Jacksonville on the St. Johns at a point where the river is at its widest—nearly three and a quarter miles across. Here Alfred could have his boathouse and his own marina. The highly respected local firm of Marsh and Saxelbye was commissioned as the architects to build the home according to Jessie's specification, and by the spring of 1927 the house was far enough along for the du Ponts to move in. In March 1927, Alfred wrote to Maurice Lee, "Our shack is pretty well completed but it seems difficult to drive the workmen out so that it can be cleaned. The electric fixtures, most of which were ordered from Spain, have just been put in place. . . . I have a hunch that you and your dear wife will approve of it when you see it. . . . The watchman caught a sixteen pound catfish off the end of my little wharf, and likewise killed a sixty-seven inch water moccasin, so if you come there is no reason why you should not live a more or less active life."[69]

Both Alfred and Jessie had certainly proved that even without catching catfish or killing water moccasins, one could lead a very active life in the deep South. Now nearly seventy, Alfred had never been more busy or more actively engaged in living than here in Florida, where most men his age came to doze away the remaining years of their lives in beach chairs. One of Henry Flagler's biographers commented that in "this work of transforming neglected beach and swamp into one of the most luxurious playgrounds in the world he found his second youth."[70] Although Alfred disliked being compared with Flagler, this certainly was one more common denominator the two men shared, for in transforming the neglected pinelands of Florida into a thriving economy, Alfred had found his own fountain of youth.

XIV

The Way to a New Deal
1930–1933

I N THE YEARS from 1926 through 1929, while Alfred I. du Pont was devoting his attention and his financial resources to the reconstruction of depressed Florida, most of his fellow capitalists were concerned only with how they might best realize the greatest profit for themselves from what appeared to be the biggest economic boom in the nation's history. In those bright zenith days of Coolidge prosperity, industrialists, financiers, stockbrokers, and even small-time, penny-ante investors were crowding up to the board to feast upon the Great Barbecue that the stock market, as their genial host, was serving up on Wall Steet.

Pundits and historians, always eager for a simplistic label, have hailed these years as "America's coming of age." The country had now donned long pants and felt it had every right to be cocky as it swaggered its way into adulthood among the nations of the world. America had entered the Great War as a debtor nation; it had emerged as the world's single great creditor nation. The old Western powers, Great Britain and France, were now our debtors, not only for the money we had lent them and which we were now clamoring to have returned with interest, but also for our having accomplished what they had been unable to do in three long years of bloody carnage—defeat the German Wehrmacht. At the moment when the weary Allies were entrapped in the mud of futility that was the Western Front, unable to push forward, unwilling to surrender, Uncle Sam in shining armor had arrived on the scene and had slain the evil dragon. That mission now accomplished, we wanted only to retire into well-earned isolation, tend to our own garden, refurnish our house with the luxurious trappings we could now afford, and let the rest of the world pay tribute to us.

For the first time in its history, the United States now had to be taken seriously as a major force in Western civilization. The Old World, to its great surprise, discovered in the 1920s that there was an American culture worthy of recognition. Even before the American elite had found it respect-able to do so, Europe had snatched American jazz from the brothels of New Orleans and Memphis and had carried its improvised sounds and its erotic rhythms across the Atlantic to delight the café society of Paris and London. The lost generation of European youth found in the writings of F. Scott Fitzgerald, Ernest Hemingway, and Eugene O'Neill a dialogue which best expressed its own brittle, cynical disillusionment as it traversed the postwar wasteland which had been mapped out by the American poet T. S. Eliot.

American intellectuals still felt obliged to make the required pilgrimage to the cultural mecca that was Paris, but now they went not as Mark Twain's innocents abroad but as George Gershwin's Americans in Paris—not in supplication but in the brash self-assurance of conquerors. And when they arrived in that fabled City of Light, they quickly found their way to the city's most glittering salon for artists, writers, and philosophers, where reigned, incredibly enough, not some latter-day French Voltaire or Madame de Staël, but an American from Allegheny, Pennsylvania. At 27 rue de Fleu-rus, surrounded by her Gauguin, Braque, Matisse, and Picasso paintings, Gertrude Stein sat enthroned, as one of her biographers has said, like "'a great Jewish Buddha,' calmly authoritative, arbitrating—and imposing her views upon—every manner of aesthetic dispute."[1] Now for the first time, Europe could appreciate what Walt Whitman had meant when he wrote that America would sound its own unique "yawp across the roof tops of the world," and the whole world would be forced to listen.

It was not, however, the aesthetic criticism which Stein issued as fiat from her salon; nor was it the Fitzgerald stories of the "beautiful and the damned;" nor even the vibrant dances and songs of Josephine Baker, the American black from St. Louis, Missouri, who now reigned supreme at the Follies Bergère, that impressed most Europeans in the late 1920s with the importance of America. It was simply America's wealth, which Henry Ford's automobiles, John D. Rockefeller's oil, Andrew Carnegie's steel, and the du Pont family's chemicals had produced. There was also, and perhaps especially, the unreal paper wealth that Wall Street was creating out of little more than ticker-tape numbers and the gullibility of an avaricious public. Even the sophisticated were now ready to believe what the poor peasant immigrants from Sicily and Greece had once believed: America's streets were paved with gold—or at least one street in New York City was. Never mind the midwestern farmer who had still not recovered from the 1921 collapse of agricultural prices; or the eastern European immigrants who had found that on Hester and Mulberry in New York, the streets were paved with garbage, not gold; or the black sharecroppers in Alabama, or the red-neck crackers in Panacea, Florida, whose poverty made a cruel irony out of the town's name. The America that Europe was interested in was Samuel Insull's, E. F. Hutton's, and George F. Baker's America. America the Bountiful—of thee they sang in 1928.

From his self-imposed exile in Florida, where little evidence of Coolidge prosperity could be found, Alfred du Pont watched the Wall Street phenomenon with the same disbelief and foreboding that he had earlier witnessed the Florida land boom of 1924. He was convinced that he was a spectator of a repeat performance of the one-act tragedy that he had viewed from the verandah of the Royal Palm Hotel in Miami, only now the play was on national tour. Inevitably, Alfred was convinced, there would be the same abrupt conclusion and a quick curtain to close the show. All through the year of 1928 and the spring and summer of 1929, stock prices continued to rise, and Alfred's mounting pessimism kept abreast of their advance.

In this second performance of frenzied speculation which the decade of the 1920s had produced, however, Alfred was something more than a disinterested observer. Except for Jessie's little fling with Florida real estate, the du Ponts had not been involved in the Florida land boom, but in the Wall Street extravaganza, Alfred was a major participant, whether he liked it or not. The 55,000 shares of Du Pont stock that he had managed to retain through the hard years had now multiplied ten times. There had been two Du Pont stock splits since 1924, each one resulting in three and one-half new shares for one previously held share. These divisions, plus the additional shares that Alfred had purchased on the open market, had increased his holdings to 582,000 shares, making him second only to Pierre du Pont in the amount of Du Pont stock owned. After each stock split, the value of the new shares continued to rise, so that by the spring of 1929, with Du Pont shares selling at about $200, Alfred's wealth had grown to what seemed to him the unreal figure of nearly $120 million.[2] Thus, in five short years, without any effort on his part, Alfred had risen in a vertical, phoenix flight upward from the ashes of insolvency to the pinnacle of affluence.

To Alfred, however, the unexpected largesse which Wall Street had bestowed upon him was unreal. It had no firm structural support. It was made of paper which the first gust of wind would blow away. With respect to physical assets, earnings, and dividends paid, it was as absurd for Du Pont to be selling at $200 a share as it was for General Electric to be at $400, and Adams Express—a great favorite of the speculators—to be at $750. "It isn't worth that, Ed," he told his brother-in-law. "Everything is too high. Let's sell a little and keep the cash on hand."[3]

Ball, however, reacted quite differently to the bull market. The rainbow's end he had been pursuing all of his life had finally been reached. The arc had touched down right on top of Almours Securities, where Alfred's precious Du Pont stock was deposited. Having found the pot of gold, like all real leprechauns, Ball wanted more. When the respectable brokerage house of Cassatt & Company of Philadelphia suggested that Almours, with its sizable capitalization, should go on the market by offering $50,000,000 in preferred stock to the public, which Cassatt offered to handle at par value, Ball was eager to seize this opportunity of increasing the assets of Almours by nearly 50 percent. But this time, Ball did not win out. "No," Alfred told his eager partner, "there is going to be no Almours preferred. I'll tell you, Ed, all hell is going to break loose. We don't know when, but it's coming. We'll have enough trouble looking after what we've got with having $50,000,000 of other people's money."[4]

To his son, Alfred Victor, who was employed at the Du Pont rayon plant in Buffalo and, like everyone else, was tempted to play the stock market, Alfred sounded his usual note of caution:

> In the matter of investments a young man like you should be guided, in my opinion, by the dividend return principally. Of course, the asset or book value of a stock is always to be taken into consideration, but I would not get into the habit of "playing" the market which is only another term for gambling. No one ever made a dollar on the New York stock exchange that did not take that dollar from someone else.[5]

Alfred's warnings of impending catastrophe were as little heeded by his friends and relatives in 1928 as his strictures against buying Florida land had been in 1924. The market kept rising, and bulls and lambs continued both to gamble and gambol together in the bright sunshine of prosperity with never a bear in sight. And Alfred, willy-nilly, continued to make more and more dollars, at least on paper, from the very boom he was decrying.

Alfred and Jessie returned from their extended stay at the Carlsbad spa in Czechoslovakia on 15 October 1929—just in time to witness the crash of the Wall Street paper tower. All hell had indeed broken loose—and sooner than even Alfred had expected. In a few short months, the assets of Almours Securities, which consisted largely of Alfred's half million shares of Du Pont stock, plummeted from a high of $150 million down to less than $50 million. On the evening of Black Thursday, 29 October, Alfred wrote his son:

> Du Pont was selling at 80 and the ragman was looking for certificates. Candidly, I think Du Pont is not worth much more than par [$100 a share]. I never did. . . . Of course, the pyramid house of cards could not last forever, for when one is paying 10 or 12 percent interest on money he borrows to purchase securities which return only 3% or less [in dividends], it is only a matter of time when the collapse must come. . . . A good many people are richer in experience; also brand new Diesel yachts in New York can be had for a song.[6]

But no one on 29 October 1929 felt like singing and new Diesel yachts went begging. The fat bulls had been struck by a lightning bolt, and the little lambs who had wandered into the marketplace were now shorn.

Ed Ball greatly feared that at least three of his and Alfred's associates in Almours Securities were among the shorn. The day after Black Thursday, Ball wrote to Alfred to express his concern about the effect that the market crash had had on Henry Dew, Jessie's cousin, whom Alfred had only recently brought into the Florida enterprise;[7] on James Bright, the president of Alfred's bank in Lakeland; and on Willard Hamilton, president of the Orlando bank. Like so many other investors, these three had bought stock on a 10 percent margin, and now the brokers were demanding either payment in full or the surrender of their stock certificates. Ball asked if Alfred would be willing to rescue these three by having Almours take over their accounts and meeting the margin calls of their brokers. Their holdings were in good stocks, Ball assured Alfred, "and will naturally go up when conditions have again become normal." As might be expected, Ball's motive in suggesting this help was not entirely altruistic. "Other than helping out these boys who are connected with you, the thought occurred to me that if they should be called for margin and unable to meet it, that it might become known in the community and the community in turn might feel that your interests were affected." If the public should suspect that Alfred himself was in financial trouble, it could, of course, result in dangerous runs on all of Alfred's banks, and Almours Securities itself might be shaken.[8]

Alfred promptly gave his consent, but only on the condition that Ed Ball himself received the same assistance. Ball was happy to report that "while I have a few tooth and claw marks of the bears on various parts of my anatomy, I have not been seriously inconvenienced . . . and believe that my account will get through without any need of assistance."[9]

Conditions, however, did not "again become normal" as quickly as Ball had so confidently predicted, and soon he, as well as Jessie, along with many other relatives and friends needed Alfred's help. As Alfred later told his cousin Francis I. du Pont, he had in all "spent about Two Million and a Half in caring for the interests of others, and those accounts still stand and probably will stand for a long time to come before they can be liquidated without inviting enormous losses."[10]

The Great Crash of '29 had reduced Alfred du Pont's fortune to only one-third of what it had been six months earlier, but he regarded most of what he had lost as mere fantasy anyway. He still had his 582,000 shares of Du Pont stock, and although they were now worth only $80 a share— if one could find a buyer—they were at least all his. No brokerage house had a claim on them, and he was still solvent. Naturally, he felt an obligation to assist those close to him who were not as fortunately situated as he, although he could and did call them damn fools for believing they could end up winners in that gambling casino on Wall Street.

There were others—many thousands of others—however, who were not the authors of their own financial distress and who stood to lose more than just money invested in stocks. These were the people upon whom the sun of Coolidge prosperity had never shone, who had always had to scrabble for what little they had, and who now, with the collapse of the American economy, faced the frightening prospect of homelessness and actual starvation. To these people, too, Alfred felt a very real sense of responsibility.

Of the many stories that had been told of Pierre Samuel du Pont I and of the family's origins in France, a favorite among five generations of du Pont children, because it had all of the romance of medieval chivalry, was the story of the time that Pierre Samuel, soon after the death of his wife, had called his two sons together and, asking each to kneel, had touched the shoulder of each with his sword, saying with great solemnity, "Remember, my sons, with great privilege must also come great responsibility." Over the succeeding one hundred and fifty years some of the du Ponts had not only remembered and repeated this legend but had even accepted the moral of the story as a personal charge.

Two of Eleuthère Irénée du Pont II's children in particular seemed especially impressed with their great-great-grandfathers concept of noblesse oblige and took its message to heart. Marguerite du Pont Lee devoted her life, at the expense of her marriage and what money she had, to the care and instruction—both practical and moral—of the poor in the nation's capital. Although her brother Alfred frequently complained about her incessant demands for more money for her settlement-house work, he almost always complied with her requests, for he too was sensitive to the social obligations attached to privilege.

Alfred's and Marguerite's views on who should be the major beneficiaries of their philanthropic bequests and how these charitable programs should be administered differed greatly, however. Marguerite gave her attention to young men and women whose physical and spiritual welfare was placed in jeopardy due to poverty, ignorance, and bad habits. She opened a school in her settlement house to teach young women homemaking skills—cooking wholesome food, planning a nutritional diet, sewing sturdy, sensible clothes, and nursing and first aid that a nonprofessional could perform in the home. The school offered young men instruction in useful trades—carpentry, plumbing, and metal work.

Alfred, on the other hand, believed that young adults should be able to take care of themselves if they were industrious and took proper care of their health. If they failed, it was their responsibility, and society need not provide for them except in the most extraordinary circumstances of natural or man-made disasters. Alfred's concern was for those at the two extremes of the age spectrum—the very young and the very old—as well as for those who because of physical handicaps were unable to accept responsibility for their own well-being.

Marguerite sought publicity for her good works. Public attention given to her projects meant more support from those who could afford to sponsor her programs. She believed in organization and planning, and she prepared for her campaigns against poverty and sin with all of the scrupulous attention to long-range strategy and immediate tactical details that a General Clausewitz might give in preparing for the defense of the Prussian state. For Marguerite, charity was something more than alms. It must be didactic as well as benevolent. The recipients of her good works must expect to receive, along with hot meals and baths, solemn sermons on temperance, thrift, and women's rights, as well as heavy doses of Episcopalian theology.

Alfred preferred to give anonymously and spontaneously—a thousand-franc note handed to an amazed beggar child on a street in Beauvois, France, or a hundred dollar bill suddenly thrust under the blanket of a baby who lay whimpering in the arms of its shabbily dressed mother was the form of charity that Alfred practiced. He detested philanthropy that had an ulterior purpose, the kind of giving that had strings attached to self-righteous piety. One day when he and his good friend Dr. Charles Hanby were walking down Market Street in Wilmington, they were approached by a black man who asked for a quarter to buy a hot meal. Alfred handed him a five dollar bill. Hanby was as amazed as was the beggar. "Why did you do that?" Hanby demanded. "You know he'll only get drunk on that." Alfred answered that he hoped the man would, since the poor fellow probably had found no other fun in life than what was contained in a bottle of cheap whiskey.[11]

Alfred had not only accepted Pierre Samuel du Pont's dictum that along with social privilege also came social responsibility, but in a very marked way, he was the only one of Pierre Samuel's many descendants who fully subscribed to his great-great-grandfather's basic economic philosophy. In many respects, Alfred du Pont was the last genuine French physiocrat. He truly believed in the principles of laissez faire and a self-regulating economy. Government, he believed, should stay out of business and not attempt to regulate it, because regulation prevented the operation of the natural laws of a free economy. There should be no statutes that set maximum hours of work or minimum wages. (Many of his du Pont cousins would, of course, heartily agree with that.) But when Alfred spoke of laissez faire, like the French philosophers of the eighteenth century and the Manchester liberals of the early nineteenth century, he really meant "let the economy alone to do as it naturally would." If there should be no governmental restrictions

on the economy, there should also be no governmental favors that artificially supported certain segments of the economy—no protective tariffs, no
subsidies, no government-sponsored monopolies through long-term fran-
chises—in short, no welfare state for those who already fared well. Here,
of course, Alfred parted company with most of his fellow industrialists and
his cousins who had long dominated the Du Pont Company's affairs.

Alfred went even further in his allegiance to a free economy. Not only
should government be denied the right either to restrict or to foster the
economy artificially through law, but capital itself must refrain from those
activities that sought to eliminate competition through trade associations
and unfair, under-the-table practices. There should be no secret rebates, no
pooling agreements to fix prices and divide the market. Competition was
the natural regulator and promoter of a free economy. Without competition
there was economic despotism. As a young man, Alfred had opposed the
Gun Powder Trade Association that Uncle Henry had established shortly
after the Civil War in an effort to bring order—his order—to the explosives
industry and to limit, if not eliminate, competition. Later, Alfred had been
an outspoken critic of Cousin Eugene's policy of buying up newly built
powder plants before they could become established and offer real competition to the Du Pont Company—a practice that Cousin Coleman tried to
continue during his presidency. Nor was Alfred ever an enthusiastic supporter of Cousin Pierre's program of building corporate conglomerates and
international cartels. The profits that the Du Pont Company earned—and
they were large—should be put back into the business in the form of new
plants and equipment and, above all, into research for new products. The
profits should not be used to buy out old rivals like Laflin & Rand or Hercules. The Du Pont Company should become bigger by becoming better,
not by becoming a monopoly. Competition, as he said repeatedly in letters
to both his subordinates and his board of directors, is the necessary fertilizer
to industrial growth and progress. If Du Pont couldn't compete effectively
with such rivals as Laflin, & Rand or Nobel, then it was not fit to survive.
Far more than either John D. Rockefeller or Andrew Carnegie, who so
loudly proclaimed himself to be Herbert Spencer's disciple, Alfred du Pont
accepted the Social Darwinist belief in competition.

Alfred was an industrialist, a producer of manufactured goods, but even
so, he remained a physiocrat, believing that a nation's true wealth ultimately
lay not in its gold reserves and certainly not in the paper wealth which the
stock market engendered, or even in the value of its manufactures, but
rather in the land and in the natural resources which that land contained.
Wealth should be measured by the richness of the land's top soil and by the
abundance of its underground minerals and fossil fuels. The Du Pont Company was itself as dependent upon the land as was the farmer or the miner.

Alfred's program for the reconstruction of Florida and his unflagging faith
in its future were based upon land—not playground land for the promoter
and the speculator, but tillable land and timberland for the citrus grower
and the forester. In 1932, at the bottom of the depression, Alfred confidently told the press:

> The production of wealth—the taking of money out of the ground
> is the really important thing to be considered in times like this. That is
> where Florida holds a preeminently important place in the sun, because
> the industries of the state are wealth producing. Crops are stabilized.
> They are money-making crops. There lies the salvation of Florida, and
> that is why Florida stands well out in the list of states calculated to
> recover rapidly from depression.[12]

In one important respect, however, Alfred du Pont, like Pierre Samuel
du Pont I before him, was not a true physiocrat. Both men were consciously
aware of the social responsibility that came with privilege, and this knowledge moderated their concept of a free economy, unrestricted by anything
but natural law. Both men envisioned an ideal community—Pierre Samuel
dreamt of a Pontiana in the Piedmont area of Virginia and Alfred Irénée
planned for a new Port St. Joe in the Florida panhandle. They both envisioned communities populated by free men who would work as individuals
but who would consciously strive for the common good—communities that
would take care of their own. Both men, although they would deny the
term, wanted the Utopian socialist ideal—"from each according to his ability, to each according to his need." Competition would exist, but it would
be the competition of the relay runner, seeking to contribute the most to
his team, not the competition of the robber baron seeking to amass the
most wealth for himself.

Neither man realized his dream of the ideal community, although
Alfred might have created something resembling his plans for Port St. Joe
had he lived another five years. Both Pierre Samuel I and Alfred had to
accept the fact that their planned communities would at best serve only as
models, not as solutions to the basic problem of the maldistribution of
material wealth and services in society. For a long time, Alfred clung to the
idea that the only way to achieve belance between capital and labor was
through education, not governmental interference in the economy.

In 1918, in response to Professor Irving Fisher, whose pamphlet
"Wealth and War" Alfred had just received, Alfred wrote: "In my opinion
the solution of the problem of industrial discontent, if so it can be termed,
lies in the proper distribution of wealth between capital and labor in years
to come, based on some economic principle which will be satisfactory, and
which can be defended on the grounds of science and fairness. There can

be no other possible solution. This will gradually result in the elimination of the line between capital and labor, and each laboring man will become a capitalist to a certain extent. When this condition arrives, all interests will be common.""[3] To Samuel Swett, who had written to him expressing his concern over the Bolshevik revolution in Russia, Alfred responded, "[The] solution of the misunderstanding between what you term 'Capital' and 'Labor' is rather a simple one and can only be brought about by a fair distribution.... A plan looking for a more rapid elimination of this line [between Capital and Labor] is most desirable. This plan must look purely to the education of those who are at present in ignorance of the very basic principle of economics.... In the past, Capital received a larger share of the wealth earned by its utilization than it was justly entitled to. This has been the basis of all Labor vs. Capital disputes.""[4]

Alfred was still the physiocrat, still the loyal disciple of Turgot and Pierre Samuel du Pont, who had believed that the great natural laws of political economy would result in a human society as orderly and benevolent as that which prevailed in the physical universe that Newton had revealed.

Alfred had done what he could to live by this philosophy while an officer in the Du Pont Company. He promoted the establishment of a pension plan for all Du Pont employees who stayed with the company until retirement, not just for those who had been incapacitated or for the families of men who had been killed in mill accidents—provisions that Eleuthère Irénée had set up following the first explosion in the Brandywine mills. Alfred was also a strong proponent of a profit-sharing plan for all employees who had been with the company for five years—not just stock bonuses for top officials. Such plans, if widely adopted, Alfred was convinced, would "rapidly obliterate" the dividing line between Capital and Labor which he had discussed with Swett.

In the meantime, it was the responsibility of men of substance to take care of those who could not care for themselves and were not the beneficiaries of the enlightened capitalism that he envisioned for the near future. As early as 1918, Alfred had corresponded with Miss. E. R. Keim regarding children who needed help. As Secretary of the Delaware State Fair, Keim had contacts throughout the state and frequently heard of children whose parents could not provide them with needed medical attention or even such basic necessities as food and clothing. Alfred found her reports accurate and her requests not excessive, so he began a practice of sending an occasional stipend, accompanied by instructions for her to use her best judgment in how the money should be distributed.[15]

Sometime in the late 1920s a teacher in Lincoln City, Delaware, Mrs. Laura Walls, wrote to Alfred describing the poverty she saw daily in downstate Delaware, particularly among people who, after a lifetime of hard work, were forced into retirement by old age and illness. These people had never benefited from any pension plans of an enlightened capitalism and very often had no children who could or would give them any support. They lived on the razor's edge of starvation and homelessness; frequently, their only alternatives were the notorious county almshouses or suicide. Many preferred the second option, Walls reported. Walls undertook investigations of the most desperate cases for Alfred and reported to him the amount each person needed. Alfred was impressed not only with the thoroughness and care of her investigation of each case, but also with her compassion—a compassion carefully controlled by the toughness of her appraisals.[16]

As the monthly reports came in, Alfred began to understand the magnitude of the problem. Here were cases of desperate need that private pension plans could not help and so numerous as to make private charity like his inadequate. Alfred slowly came to the realization that there *was* a place for government in meeting the problem of the proper distribution of wealth. It would be this realization that would motivate Alfred to re-enter the Delaware political arena that he had forsworn.

Since January 1920, when he had voluntarily and unexpectedly abdicated his throne of political power within the Republican party of Delaware, Alfred du Pont had scrupulously maintained his self-imposed exile from politics. Once bitten by the political bug, however, as Alfred had been in 1916, few politicians are able to slough off its infection. For most former political leaders, there is a recurrent fever that causes a temperature rise with biennial regularity. The year 1928 provided the first real test of Alfred's resolution to abjure all political involvement. It had been relatively easy for him to ignore both the presidential election of 1920, in which he considered neither major candidate fit for the office, and the presidential contest of 1924, which, of course, had been no real contest at all. The presidential election of 1928 was a different story, however, and Alfred quickly demonstrated that he was no exception to the general rule that once infected there is a lifetime susceptibility to political fever. Soon after the two major parties had selected their presidential candidates, Alfred was as badly smitten as he had been in 1916 when Cousin Coley had made his absurd attempt to get the Republican presidential nomination.

With Alfred, there was a special reason for his recurrent bouts of political involvement: his continuing feud with his family. By 1928, Alfred had removed himself both physically (for at least six months of the year) and legally from the Delaware scene. He had severed relations with his old political allies, and he evinced little interest in or knowledge of state politics in Delaware. It is possible that he might have sat out the election of 1928 as he had the two previous presidential campaigns, had not Cousin Pierre's involvement in the campaign served as a compelling inducement for Alfred to enlist on the opposing side.

Pierre du Pont's conspicuous role in the campaign of 1928 came as a great surprise both to his family and to the nation. Hitherto, Pierre had always avoided the political limelight, leaving that to Cousin Coley. He never sought an elective office and made few political pronouncements to the press. He was, of course, a substantial contributor to the Republican party both at the state and national level, and Alfred was always convinced that Pierre was the major financial supporter of the Coleman du Pont Republican machine in Delaware. Early in the 1920s, however, Pierre became convinced that the national prohibition amendment that had become a part of the United States Constitution in 1920 represented the most flagrant violation of civil liberties in the history of the American republic. In addition to being an unwarranted invasion of the state into the personal lives of individual law-abiding citizens, it had also proved to be unenforceable, giving rise to the most sinister manifestation of organized crime that the country had as yet experienced. Pierre had hoped that by 1928 the Republican party would recognize that the so-called "noble experiment" had proved to be a most ignoble failure and would acknowledge that fact in its national platform. The Grand Old Party, however, had once again re-iterated its staunch support for prohibition and had reaffirmed its determination to enforce vigorously what had so far been unenforceable.

Two weeks after the Republican convention in Kansas City, Missouri had nominated Secretary of Commerce Herbert Hoover for the presidency, the Democrats met in Houston. The Democratic platform attempted to straddle the prohibition issue by refusing to reaffirm its approval of the Eighteenth Amendment while at the same time stating that it favored a strict enforcement of the law as long as it was in effect. The delegates, however, chose Alfred E. Smith, the governor of New York, as the party's presidential nominee, and Smith promptly announced that as president he would personally seek a repeal of national prohibition.

Smith's announcement would in itself have probably been enough to win Pierre du Pont's support, but his support became a certainty when Smith, to the great surprise of party regulars, chose John J. Raskob, the former treasurer of the Du Pont Company, as the new national chairman of the Democratic Party and his personal campaign manager. Raskob, who had become chairman of the finance committee of General Motors, was Pierre du Pont's closest friend and business associate. Pierre had first employed Raskob as his personal secretary while at the Johnson Steel Company in Lorain, Ohio. When Pierre returned to the Du Pont Company in 1902, he brought Raskob with him to Wilmington. During their long years of association the two men had developed a father-son relationship. Raskob, like Pierre's younger siblings, called his mentor and patron "Dad," even though there was only a nine-year difference between the two men's ages.

In the wake of Smith's bold opposition to prohibition and his preferential treatment of Raskob, Pierre publicly announced that he was deserting the Republican party. He would cast his vote for Alfred E. Smith. Pierre's unexpected and quite atypical public stance shook the political terrain with earthquake force. To the Democratic high command, Pierre's conversion was little short of a miracle. To have this pillar of Republican Protestant affluence suddenly become the champion of a lower East Side, New York, Roman Catholic Democrat was a transformation that ranked along with Saul's conversion on the road to Damascus or the pagan Emperor Constantine's acceptance of a flaming cross in the sky as the new sign that must prevail throughout the Roman empire. To the shaken Republican leadership, Pierre's defection must have called to mind different historical analogies—Judas Iscariot, Marcus Brutus, and Benedict Arnold, to name but a few.

No one was more surprised by Pierre's remarkable political metamorphosis than Alfred. Ironically, Alfred himself had long been rather favorably disposed toward the New York governor, who during his four terms in office had demonstrated a remarkable talent for executive leadership and had successfully implemented a program of progressive legislation that had placed New York in the forefront of social reform. Alfred, moreover, had had a special fondness for Irish Catholics that dated back to his earliest childhood on the Brandywine, when his playmates had been the McMahons and the Dolans, and the only church in which he had felt at home had not been the Episcopal Christ Church of the du Ponts, but St. Joseph's, where genial Father Kelly had ministered to his Roman Catholic flock. Although he could not share Smith's love for New York City, Alfred did greatly admire Smith's successful climb from the slums of the Bowery to the governor's mansion in Albany. Smith's life story was the true American, self-made-man success story, which always had a strong appeal for Alfred.

The Republican candidate, on the other hand, elicited no similar admiration or affection from Alfred. In spite of the heavy pro-Hoover propaganda that both his sister Marguerite and his close friend William Glasgow had laid upon him for ten years, Alfred had always regarded Hoover as a cold, impersonal technocrat who lacked the vitality and warmth of a Jefferson, Jackson, or Lincoln so essential for truly great presidential leadership. Because he knew it would irritate Marguerite, who had been beating the drum for Hoover since 1920, Alfred always referred to her hero as "your Mr. Hoo Hoo."[7] Nor was Alfred a great supporter of national prohibition, which Hoover and his party so vigorously defended.

It is difficult to say what Alfred's stand might have been in 1928 had there been no personal family issues involved. It is conceivable that Alfred might actually have voted for Smith, for he had not refused in the past to give his support to Democratic candidates in Delaware when he felt the political situation warranted it. At the very least, it is probable that Alfred would not have taken a conspicuous role in opposition to the New York governor. But Smith's decision to name Raskob as his campaign manager and Pierre's widely publicized statement of support for the Democratic candidate could have no other effect than to drive Alfred du Pont into the Hoover camp as an active and vociferous supporter. Rumors were rife that if elected, Smith planned to name John Raskob as his Secretary of Treasury and Pierre du Pont as the new ambassador to Great Britain. The prospect of his two most detested enemies holding such high offices, and particularly even the remote possibility of seeing Pierre in formal morning attire presenting his credentials to His Majesty King George V at the Court of St. James's, was more than Alfred could bear. He came roaring out of his self-imposed political isolation to pick up the banner for Herbert Hoover.

Alfred saw no need to offer aid to the Republicans in Delaware. That state, in spite of Pierre's and Raskob's defection, was so solidly under the control of Coleman's machine that Alfred's support was not needed. Alfred decided he would play a much bolder and more dramatic role. He would use his influence and his wealth to move his adopted state into the Republican camp. Florida had not given its electoral vote to any Republican presidential candidate since the disputed election of 1876, when, in the final days of Radical military occupation, Florida's four electoral votes had been awarded to the losing candidate in the popular vote by the special Congressional Electoral Commission, and Florida's votes, along with the equally contested electoral votes of South Carolina and Louisiana, had ensured the election of the Republican nominee, Rutherford B. Hayes. Florida had never forgotten nor forgiven this "stolen election." For fifty years it had been firmly embedded within the Democratic Solid South. It seemed inconceivable that native Floridians would ever vote for a Republican for any office, especially that of the presidency. In 1928, however, the Democrats were offering to their loyal Southern constituents a Roman Catholic, and if there was anyone that the Florida "cracker" considered more of anathema than a Republican in the White House, it was His Holiness, the Pope in the Vatican. Moreover, no state was more supportive of national prohibition than Florida. Southern Baptists supported it on moral grounds; white supremacists, as long as they could obtain their own illegal "moonshine," favored it in order to keep liquor from the blacks; and Miami rum-runners, who were making huge profits out of their illicit trade with Cuba, were as ardent advocates of prohibition as were the members of the Anti-Saloon League. In leading a campaign against a Democratic candidate who was a Roman Catholic "wet" from New York City, Alfred was not pursuing the unobtainable, as Smith, Raskob and company believed.

Before leaving for a trip to Europe with Jessie in mid-summer of 1928, Alfred put his brother-in-law in charge of the Democrats for Hoover Committee that he had succeeded in establishing in several Florida counties, including Dade County, where Miami was located, and several of the panhandle counties where the rural population would be most hostile to Smith. Alfred told Ball to play up prohibition but to avoid religion. Rum, not Rome, was the issue to be stressed.[18]

Ball, like most of his family, was a life-long conservative Southern Democrat, and he was certainly a more appropriate person to head an organization of Democrats for Hoover than was Alfred. He was willing to emphasize the issue of Rum as Alfred had instructed him to do, but unlike his employer, he had no scruples against using a little religious bigotry as a means of winning Florida's electoral votes for Hoover. He quickly entered into a political alliance with one of Florida's most blatantly anti-Catholic demagogues, former Governor Sidney J. Catts, the Ku Klux Klan's favorite politician.[19]

Fiery crosses lighting up the black night sky in Florida became the angry symbols of fear and hate in the ugly campaign of 1928, as the very visible white-sheeted, white-masked foot soldiers of the Invisible Empire of the Imperial Grand Dragon marched against Al Smith and his alleged allies, the Pope, northern Jewish liberals, and the few courageous blacks who might dare to exercise their constitutional right to vote.

It was not a campaign that Alfred du Pont would have taken pride in sponsoring had he been on the scene to witness it. Nor was Ed Ball's success in promoting the cause of Democrats for Hoover one that he himself later remembered with satisfaction. Years after the event, Ball asserted that his efforts on behalf of Herbert Hoover in 1928 represented the greatest mistake of his life. "If Smith had only been elected in 1928," Ball told this writer, "we would never have had Franklin Roosevelt in 1932, and so would have been spared that great catastrophe of the New Deal."[20]

In accepting responsibility for Hoover's victory in Florida in 1928, however, Ed Ball was taking too much credit—or blame—upon himself. Florida would have undoubtedly been carried by Hoover, as were four other states of the old Southern Confederacy, if Alfred du Pont had not spent a penny or Ed Ball had not enlisted a single campaign worker on Hoover's behalf. The President-elect, however, was properly grateful for Alfred's support and sent a telegram of appreciation immediately after the election.[21] There were even rumors that Alfred would be offered a cabinet position in the new administration, but if such tentative feelers were actually sent Alfred's way, he most assuredly rejected them. He was much too preoccupied with his plans for revitalizing Florida even to consider moving to Washington, especially to serve a president whom he personally disliked.

Alfred did, however, regard Hoover's Florida victory margin of 42,000 votes as his own personal triumph over his enemies in Wilmington. On the day after the election, he wrote to his niece Dorothy Lee, "I've just licked Pierre and Raskob and made Florida Republican and I am reeking with gore."[22] Significantly, it was not Smith's defeat that Alfred celebrated, but rather the defeat of Pierre and Raskob.

Having once again plunged back into national politics, Alfred found it relatively easy to re-enter the turbulent waters of Delaware politics. His return to the political stage in his native state, where he no longer even had a vote, was not motivated, however, simply by a desire to flex his muscles against Pierre and company again. Alfred was motivated by his determination to push for progressive social legislation that he felt was long overdue in Delaware, namely adequate benefits for the indigent elderly.

In the summer of 1928 Alfred had hoped that the Republicans in Delaware would nominate the progressive state senator and Alfred's old political ally, I. D. Short, for the governorship, and he had reestablished contact with E. M. Davis to urge Davis to work for Short's nomination.[23] That effort had proved unsuccessful, and Alfred was certainly not pleased but hardly surprised when the Coleman du Pont machine easily achieved the nomination of Coleman's son-in-law, C. Douglass Buck, for governor at the state convention in July.

Alfred was most pleasantly surprised, however, to discover that Buck as governor proved to be no subservient tool of the Coleman du Pont machine. He was his own man, with his own views, which were not entirely in accord with those of either his father-in-law or Cousin Pierre, but were in harmony with Alfred's. Buck soon let it be known that he, like Alfred, was favorably inclined toward a bill introduced in the Delaware legislature that provided old-age pensions for those whose income was below the poverty level. Alfred was delighted to discover this unexpected and influential ally, and from his winter base in Florida, he wrote to Senator Short urging him to push the pension bill through the legislature and to send it to the governor's desk, where it would undoubtedly receive favorable attention. Alfred wrote Short: "This would be an excellent Bill, if properly compiled to the end that old people be left in their own homes on a small pension, rather than be forced into the alms house, and I doubt whether it would cost any more in the end, if the investment represented by the cost of public institution be taken into consideration. It seems to me that it would be a fine thing if this Bill could become a law."[24]

The national Fraternal Order of the Eagles, which had been advocating state pension bills throughout the United States since 1920, pushed hard for the enactment of the Delaware bill. Indeed, one of its members, Representative Edward Glenn, the former president of the Wilmington Eagles Aerie No. 74, had introduced the bill in February 1929 in the lower house, where it passed by a vote of 21 to 10. Unfortunately, in the Senate the bill was defeated by nearly the same two to one margin.

Although disappointed by the result, Alfred was not surprised. He felt there had not been a careful enough study made of just how many people might be eligible for benefits under the provisions of the bill and what the total annual cost to the state would be. One could hardly expect the legislature to sign a blank check.[25] Alfred himself was prepared to make such a study, and in the most practical manner possible. He called in Laura Walls and asked her if she would resign her teaching position to become director of his own personal pension program. He told her that over the next two years he was prepared to pay, out of his own pocket, pensions to all Delaware citizens over the age of 65 whose eligibility for such assistance Walls would determine. The eligibility criterion was that the persons be without any other means of support, from either their own or their children's resources. When the legislature next convened in 1931, Alfred du Pont could then present accurate figures as to the number of people needing such assistance and the exact amount such a pension plan would cost the state. Was Walls willing to undertake such an assignment? Indeed, she was both willing and able. She already had a list of 1,200 eligible persons whose names she had obtained through her own investigations and from ministers and priests throughout the state. So began a program that is surely unique in the history of American social welfare—a statewide old-age pension plan financed by a single individual. Only in a state as small as Delaware and only with an individual so philanthropically inclined and with sufficient wealth to act upon his eleemosynary impulses could such an experiment be carried out.

Alfred's original intention was to remain the anonymous sponsor of the program. He gave Walls instructions to give the pension plan the widest possible publicity so that every eligible recipient would hear of it, but at the same time, he wanted no publicity for himself. When, however, speculation was soon rife throughout the state as to the identity of the unknown benefactor, and when Alfred heard rumors to the effect that it had to be a du Pont, possibly Pierre, he decided to step forward and claim the credit. Before leaving for Europe, he wrote to Walls:

> There is no harm in my name being made known. In this case the facts are better than untrue rumors and speculations. The hardest part of your work will be in separating the worthy cases from the others. . . .

It is going to be a strenuous task, but I feel that it is one that you are fitted to tackle very successfully. Your plan is well thought out and should succeed. You have the assistance of the ministers and some of the best thinking people of the State. Please accept my deepest appreciation for your splendid assistance in this fine work for to you must go the credit.[26]

Walls was delighted to be free to make such a public announcement. She wrote to Alfred:

The above announcement [in the press] is glorious news for the neglected aged of 'Dear Old Delaware' that you have planned to come to their aid and make their sunset bright and happy. This worthy class of people have been in the blackground [sic], unnoticed and uncared for, needing food, medicine and clothing. Some of the greatest battles for country, State and righteousness have been fought and won by this class of most worthy fathers and mothers. They have not been praying for a full loaf of bread but a few crumbs. The God of Israel has heard their cries, and you are the angel of mercy that is coming to their assistance. No greater work has ever been done, or can be done by any individual for Delaware than this.[27]

Soon after Alfred established the private pension program for Delaware, he and Jessie sailed for Europe. Jessie had insisted that he endure another month-long session with the noted audiologist, Dr. Mueller of Carlsbad, and Madeleine was eager to have her father visit her and her sons in Munich. Alfred had little reason to believe that a second round of the doctor's painful treatments would prove to be any more successful than those of the previous summer, but he readily consented to the trip, as he had his own reasons for visiting Europe in this summer of 1929. He wanted to observe firsthand the administration of some of the social welfare programs, particularly the old-age pension plan in Germany. This program had been inaugurated, surprisingly, forty years earlier during the chancellorship of the arch-conservative Prince Otto von Bismark.

The du Ponts returned home in the late fall. Alfred's hearing was no better than before, but his understanding of how a state could successfully administer a nationwide pension program was certainly enhanced. He was more determined than before to push such a program through the Delaware legislature, and if it required his reentry into the Byzantine machinations of that state's politics, he was prepared to make that sacrifice.

The first step in Alfred's plan was to secure the reelection of Governor Buck, who was outspokenly committed to a statewide pension plan. In offering his support to Buck, Alfred, of course, had to accept the entire Republican ticket, including the party's candidate for the United States

Senate, Daniel Hastings, Coleman's old crony and Alfred's political foe of long standing. Alfred found himself in the anomalous position of having to rally the remnants of his old progressive wing of the party in support of the Coleman du Pont machine, but he did not waver in doing what he regarded as his duty. Holding his nose tightly, he bravely swallowed the entire Republican ticket, Hastings and all. The bitter dose was somewhat sweetened by the knowledge that Pierre and Raskob were forced to gulp down what to them was an equally unpalatable offering. Their recent political conversion meant that they had to campaign for Thomas Bayard, a former Democratic Senator, whom Alfred and his insurgents had put into office in 1922 over their strenuous opposition. Bayard had been defeated in the Hoover landslide of 1928, but he was making another try in 1930 against Pierre's and Raskob's former pal, Hastings. Politics supposedly make strange bedfellows, but never had either Pierre or Alfred been required to share covers wtih such an uncongenial companion as each was forced to do in this particular election.

As for Coleman, who lay dying of cancer of the throat in his manor house on the Eastern Shore of Maryland, this little game of musical beds, which he no doubt watched with fascinated delight, provided a much-needed comic relief to alleviate his final hours of suffering. Coleman could no longer speak, but he was still able to write, and on 28 October he scrawled out his last note to Alfred, thanking him for his unexpected support of Dan Hastings and telling his cousin how much it meant to him personally that Alfred and his son-in-law, Governor Buck, had become good friends and close political allies. Coleman signed his letter "Your affectionate cousin," and this time he really meant it.[28]

On 4 November, the entire Republican ticket won an easy victory over the Democrats. Even Hastings managed to obtain an 8,000-vote margin over the highly respected Bayard. Once again Alfred had trounced Pierre and Raskob—and this time on their own home turf. "We wiped up the floor with Bayard and your friend P. S.," Alfred joyfully announced to Marguerite.[29]

One week later, Coleman du Pont died. He had lived long enough to savor this last political victory and to know that the unlikely combination of Coleman du Pont machine politics and Alfred du Pont progressivism was an invincible force.

For Alfred, the death of Coleman marked the end of a long, tempestuous association, beginning with their youthful adventures in Boston and ending with an unnatural political alliance after having apparently been irreparably shattered in 1915. Too much food, too much liquor, too many cigarettes—in Alfred's opinion, Coleman had killed himself. The only surprise was that it had taken him so long to complete the job. Yet, even with the bitterness and scorn that Alfred had long felt toward his cousin, their old camaraderie could never be totally dissipated. The du Pont family circle would be considerably duller without Coleman's boisterous presence, and although he always exuded an air of P. T. Barnum showmanship and parlor magic sham, behind his tomfoolery lay real substance, which Alfred recognized. "He possessed an original mind," Alfred told his wife. "But for his shortcomings in other directions, Coly could have been one of the great men of his time. The trouble was that you just could not trust him."[30] Alfred had good reason for not trusting Coleman. He also had equally strong reasons for both loving and hating his cousin, and this ambivalence would remain as a troublesome legacy.

One reason for Governor Buck's landslide victory over his hapless Democratic opponent was that he had made the old-age pension bill a major issue in his campaign. Not only had this won him Alfred's enthusiastic support, but also it brought the old people of the state out to the polls in unusually large numbers. As early as August, at the beginning of the campaign, the governor had announced that he was appointing a special commission to study the question of old-age pensions and to prepare a bill that could be presented to the legislature as soon as it convened in January 1931.

Alfred, at first, questioned the need of such a commission. Soon after he read the announcement in the *Evening Journal,* he wrote to Buck that although he didn't "know exactly what you have in mind, I should like to point out that at the present time, as far as the 1,200 pensioners which I now have on my list are concerned, . . . there is no room left for further study, no information to be procured that has not already been placed on file. . . . It is possible you may have other indigent ones in mind over and above the class for which I am caring, in which case perhaps it will be necessary to amplify my data to a certain extent; but so far as the old people on my list are concerned, no possible further study of the situation would be anything but a loss of time."[31] In addition to the files that he had on the pension cases that he was funding, Alfred also had the detailed information that he had gathered on state pension plans in Europe during the previous month.

It was not difficult, however, for Governor Buck to persuade Alfred of the political wisdom of appointing a commission to go through the formality of a study that would satisfy the legislature, even though Alfred already had all of the necessary data at hand. Alfred quickly assented to the proposal, particularly when Buck informed him that he would be appointed chairman of the study commission. Alfred urged Buck also to appoint John Rossell, a Wilmington lawyer who had worked with Alfred on the Washington Street Bridge commission, and in that service had earned Alfred's respect. The governor accepted this suggestion and in addition appointed the Rev. Charles L. Candee, a Methodist minister in Wilmington who had been particularly helpful in soliciting ministerial support throughout the state for the pension proposal.[32]

It was a committee with which Alfred could work in harmony, and he immediately set about to accomplish the commission's task as quickly as possible. In a letter to his two fellow commission members, Alfred reviewed the cases of all of his private pensioners in a report that Laura Walls had prepared. He then outlined three basic items for the commission's consideration:

> *First Item:* Is it desirable to help these old people and provide them with the necessities of life . . .?
>
> Assuming that we agree on this point, namely, that it is not only proper from a humane standpoint but as viewed by people professing Christianity and following the precepts and teachings upon which it is based;
>
> *Second Item:* Should the care of these old people devolve upon the state, or upon one or more individuals who are willing to furnish the funds?
>
> In the event that it is agreed that it is proper for the state to assume the responsibility for the care and maintenance of these old people, which has been generally conceded to be proper among most of the civilized nations of the world:
>
> *Third Item:* Then comes the more complicated analysis of the various means to attain this end.
>
> The maintenance of the old people in institutions has been the usual means adopted for many centuries and is still in vogue in most countries of Europe.
>
> Personally, I have always felt that if some plan could be worked out for maintaining these old people in the environment in which they have been accustomed . . . among the people they have known, this would be the ideal arrangement. The results achieved in the past twelve months by our efforts would seem to show this to be true; but then we must remember that the so-called alms houses in Delaware are not by any means attractive and nothing is done to make them attractive for the inmates, beyond giving them a scant maintenance; and we have come across a number of cases where this maintenance was barely sufficient to keep them alive and without comfort or medical attention.[33]

Alfred concluded his letter with the statement that in his and Mrs. Walls's opinion it would be less expensive to to maintain the indigent old people in their own homes than it would be to place them in institutions, although he admitted that the institutions could not "be abolished altogether, because there are some pensioners who would prefer an institution, properly run." He also hoped that eventually the state might establish "a contributive, or self-supporting system" available to all persons upon retirement, similar to the pension program in Germany, but that, Alfred conceded, would take a great deal more study and time to work out. The immediate problem at hand was to supply relief to those who had no other resources at their command.[34]

The other two members of the commission were in complete agreement with Alfred on all three items that he had brought to their attention. By 12 December 1930, Alfred was able to submit a comprehensive report on the scope of the problem and the costs involved in providing state pensions to the elderly indigent. The report also contained a draft of a bill that could be introduced in the legislature at its opening session. The bill as proposed would provide a maximum of $25 a month to all persons who had been residents of the state for five years, who were sixty-five years of age or older, and who had no other source of annual income, although no limitation was placed upon the amount of real property the applicant might possess at the time of his or her application. This latter provision was to ensure that a person owning a home would not be denied a pension, providing that assistance was needed for the other necessities of life.[35] The Governor was delighted with the expeditious manner in which the commission had carried out its assigned task and expressed his thanks to Alfred for "an excellent report."[36]

When the legislature convened in the first week in January, the governor was ready with the report and the proposed bill. Representative William Virden, another member of the Wilmington Eagles Aerie 74, introduced the bill in the lower house, and on 23 January the Committee on Appropriations reported favorably on the bill. The House rules were suspended on that same day, and the House then approved of Alfred's bill by a vote of 31 to 2. Four days later, the Senate passed the bill by a vote of 14 to 1. The passage of this measure set a new record for the Delaware legislature, both in speed and in the near unanimity of enactment.[37] It was a great political victory for Alfred du Pont.

Abraham Epstein, executive secretary of the American Association for Old Age Security, in the association's journal, extended "hearty congratulations to Alfred I. du Pont for his vision, generosity and humanity," to which Alfred replied that "I am always naturally pleased to be successful in any of my endeavors—in this one, in particular, I think I am enjoying the greatest happiness in my life."[38]

Alfred and the other two members of the governor's commission had been in agreement from the first not only that old-age pensions must be provided by the state, but also that the existing county almshouses had to be abolished and replaced by a single, modern state welfare home for those who needed or preferred institutional care. Alfred, however, felt that this second reform should probably not be presented to the legislature at the same session as the pension bill. "In connection with the other matter which we have under consideration, namely, the Bill looking towards the abolishing of the alms houses, it is not at all clear to me that it will be a wise thing to ask the Assembly at this session to pass such a Bill as would solve this problem in a manner creditable to the state, " he advised Governor Buck at the opening of the legislative assembly. "Anything quite so revolutionary, the members of the Legislature will I am confident approach with caution. . . . I had hoped to be able to sidetrack this particular problem until the next session of the Legislature, two years hence, which would give us ample time to formulate something that would meet with general approval, through a campaign of education prosecuted in the meantime. Owing to the lack of accurate data obtainable from the records of the different alms houses, the problem will take more time than originally contemplated for an accurate, cohesive and comprehensive plan to be worked out, and one that will reflect credit upon you, the people of Delaware and your commission."[39]

Yet, when the governor and the other proponents of old-age welfare assistance all seemed eager to present the entire package to the legislative assembly in a single session, Alfred leaped on the bandwagon, and became an ardent advocate of immediate consideration of a central welfare home to replace the existing wretched almshouses.

It was one thing to get the Assembly to accept the pension plan for the indigent elderly. This measure carried an annual price tag of only $200,000. For a state that still had, even in the depression year of 1931, a treasury surplus of $13 million, it was an inexpensive way to demonstrate what Alfred called "a true Christian concern for humanity"—and also (not so incidentally) to purchase the good will of the electorate. No one, therefore, was especially surprised by the lopsided favorable vote. The only surprise was that there were even three legislators who would dare to vote against the pension plan.

A new welfare home to replace the existing alms houses was quite a different matter. In the first place, the bill came much higher—$500,000 as a minimum initial outlay for the facility itself, plus additional annual maintenance costs that could not be determined in advance. All that the supporters of the measure could say in respect to operating costs was that they should be less than the cost of maintaining separate almshouses in each county. Then there was the problem that the elimination of the present facilities would mean a loss of jobs as well as the loss of local county control and political patronage. As Alfred had pointed out to Governor Buck in his letter cautioning against the introduction of the proposal, a new state facility would mean that "the investment in New Castle County [which included the city of Wilmington] would have to be scrapped. . . . I don't believe the Legislature will sanction it."[40]

The establishment of a pension plan, however, made it even more imperative to replace the almshouses. Given the conditions that prevailed in the almshouses, few people—even those who greatly needed institutional care—would elect to go to them if they were receiving a stipend sufficient to provide them with the necessities of life in their own homes. Alfred stressed this point in a letter to one state senator. "The approval of the Old Age Pension bill tacitly carries with it that acknowledgement of the debt owed to stricken humanity and there can be no logical reason for accepting one Bill and disapproving the second; and I am confident, when viewed from this stand point, the Legislators will give this Bill their unanimous approval as they did the former."[41]

If Alfred could see no logical reasons for accepting the pension bill while rejecting the state welfare home bill, the opposition, led by Senator J. Austin Ellison, certainly could. As Alfred himself acknowledged in a letter to John Rossell, "I am advised that Ellison stated he was unable to withdraw his opposition to this Bill for the simple reason that he was fighting for his political existence and the people now in charge of the poor houses were his friends and working for him."[42]

Ellison organized a group of Republicans throughout the state to fight the bill. This hastily formed Republican Active Association proved to be very active indeed, and it began to attract so much publicity that Governor Buck wrote to Alfred in despair, "Opposition from this source, coupled with a lack of interest on the part of those who [sic] I hoped we could count on, may defeat the bill in the Senate where it was introduced. As I have said, I doubt if it can be passed, but insofar as I can, I will assemble as much strength for it as I am able to."[43]

It was now Alfred's turn to reinforce the governor's support of the bill.

> The consensus of opinion of all seems to be that the passage of this Bill is only possible, conditional upon the united efforts of all of us to the end that it be passed, which union of strength alone can offset the disintegrating influence, purely political, of those who would be adversely interested. The enactment of this Bill into a Law ... is a wonderful opportunity for you to crown your administration by the most remarkable step forward in the cause of humanity, brotherly love and Christian endeavor—such an opportunity that comes to few in their lives. The people are almost a unit in their desire to have this Bill passed and the feeling seems to be that a successful termination of our efforts as well as yours to this end is very largely a question of your own personal interest and determination to see it through to a successful conclusion.[44]

Having very early committed himself to the support of the state welfare home bill, Governor Buck now had no choice but attempt to see the measure "through to a successful conclusion" and to grab the crown that Alfred had promised would be his. No less than Ellison, Buck was fighting for his political existence. Alfred, too, was more deeply involved in this issue than he had been in any issue since his fight for women's suffrage in 1920. He kept a steady pressure on every state senator, including even Ellison, threatening them with defeat in their next election if they dared to stand against the will of the people.[45] Alfred did not himself go to Dover to lobby for the measure, as he had done in 1920 for the Nineteenth Amendment, because he felt his total deafness made such a trip useless, but he was well represented by the other two members of the Welfare Commission, Rossell and Candee.

As Rossell later reported to Alfred, "It was a battle royal. When Dr. Candee and I arrive in Dover yesterday forenoon, there was evidence of demoralization in our ranks.... The Sussex [County] members of the House were reported as against the Bill, and inroads appeared to have been made on the Democratic members.... I was accorded the privilege of the floor, and never before do I believe I spoke with greater ernestness."[46]

Nor apparently with greater effect. To the surprise of both sides, the bill passed the Senate by the wide margin of 12 to 5. It was then rushed to the House. The rules were suspended and the roll-call vote gave 24 votes in the affirmative, 10 in the negative. Two members who had voted against the measure then changed their votes to affirmative in order that they could later move for a reconsideration of the measure. But, according to Rossell, the bill's supporters had anticipated this parliamentary ploy. As soon as the vote was tallied, one of the House members rushed the document to Governor Buck, who was in his office awaiting the outcome. Before there was an opportunity to make a motion to reconsider, the governor had signed the bill, and it was law. Buck dashed off a telegram to Alfred, "State Welfare Home bill passed house this afternoon."[47]

The governor had won his laurel crown, and Alfred had won the sweetest victory of his entire political career. Delaware now stood in the forefront of every state in the union in welfare legislation for the elderly. Thirteen other states had passed pension bills prior to Delaware, but no other state had replaced county poorhouses with a central welfare home, where, as Rossell told Alfred, "proper institutional care will be given to indigents."[48]

Governor Buck followed up his telegram with a congratulatory letter to Alfred:

> I think the passage of the Welfare Home bill is one of the Legislature's greatest achievements. Efforts to defeat the Bill continued from

the time it was reported by the committee until it was brought to me half an hour after its passage in the House, when efforts to recall it would have been futile. Credit for enacting this piece of splended legislation belongs to you and I congratulate you on such an achievement.[49]

Jessie du Pont, who had long made it a practice to read Alfred's more important pieces of mail and to add her own editorial comment where appropriate, wrote at the bottom of this letter, "Reckon he knows now who is the most powerful & constructive politician in Del."[50]

The fight for the state welfare home was Alfred's last great political battle. It was also his most unalloyed victory, for it had been solely a contest for a humanitarian principle, untarnished by the vindictiveness of a family feud.

It was fortunate that Alfred and Governor Buck had pushed for the adoption of the entire old-age welfare program at the time that they did, for the Spring of 1931 was the last possible moment in which they could have argued with any degree of plausibility that Delaware could afford such legislation. The $13 million surplus in the state treasury, to which they had pointed as tangible evidence of their claim, would within the year have many other pressing demands placed upon it. As the Great Depression deepened, it was no longer just the elderly and infirm who were suffering. Able-bodied men walked the streets looking for jobs that no longer existed while their wives and children lined up before emergency soup kitchens to get what meager rations private charity and municipal and state relief funds could provide.

Never had any society plunged from such a dizzy height of prosperity to such a dismal pit of depression within so short a time. In his speech accepting the Republican nomination for the presidency in San Francisco in August 1928, Herbert Hoover had said "We in America today are nearer to the final triumph over poverty than ever before in the history of any land. The poor house is vanishing from among us. We have not yet reached the goal but given a chance to go forward with the policies of the last eight years, and we shall soon with the help of God be in sight of the day when poverty will be banished from this nation."[51] Twenty-one million Americans, comprising a majority of the electorate in forty of the forty-eight states, had believed and had voted for Hoover. But in October 1929, eight months after his inauguration, Hoover's predicted triumphal march against poverty had come to an abrupt halt, and now two years later what had been a Wall Street panic had turned into a cataclysmic economic collapse. God's help seemed to be singularly lacking, and the "policies of the past eight years," of which Hoover had so proudly boasted in 1928, had proved to be a disaster. On only one point had the Republican presidential candidate proved to be a true prophet. In 1931–32, the county poorhouses were indeed vanishing from the land, but only because local governments could no longer maintain them. Every house in the United States, however, had become a poorer house, and for hundreds of thousands of Americans there were no houses at all that they could call home.

After the stock market crash, which he had long considered to be inevitable and perhaps even desirable, had worsened into a real depression, Alfred du Pont's initial reaction to the unprecedented economic crisis had been essentially the same as that of President Hoover. Both men believed the crisis was largely psychological. People had panicked when the stock market broke. They had stopped buying anything but the essentials of life. This drop in demand for consumer goods meant a cut in production, with an inevitable loss of jobs, which in turn meant a further cut in production and even more unemployment. In order to live, people had been forced to withdraw their savings from the banks, and many weaker banks that could not withstand the loss of deposits had closed their doors. These failures had prompted runs on stronger banks and further closings. Thus the nation had been plunged into a major depression. Yet everything needed for a strong economy was still in place. The fertile land was still here, the most modern and efficient steel and auto plants in the world were only lacking customers, the people still wanted and needed the goods and services, and their demands could stimulate production and provide employment. America, in short, was not a victim of some catastrophe externally imposed by the Four Horsemen of the Apocalypse, but rather was suffering from a self-imposed national psychosis. What was needed was the restoration of confidence in the basic strength of the American economy.

It was on the question of how to restore that confidence that the two men parted company. Ralph Waldo Emerson had once said that in America there were always only two parties, and no matter what political label they might bear, be it Federalist or Whig, Republican or Democratic, they were in reality either a party of memory or a party of hope. Hoover was a staunch adherent to and presided over the party of memory. During the great crisis of the Depression, he sought to restore confidence by recalling the virtues of the American past: the courage of the westward moving pioneer, the small-government philosophy of Thomas Jefferson, the self-reliance of the self-made man, the generous charity of the next-door neighbor, the unregulated, unfettered enterprise of the laissez-faire capitalist. He would cure the Depression by praising American business, by sermonizing on the Christian duty to help one's neighbor in distress, and by scolding those who wanted something free from the government, like the veterans who wanted a bonus. Above all, it was important for him, as President of the United

States, to keep a stiff upper lip and to reassure the nation that by holding to the traditional values that had made America great, it would survive the greatest crisis since the Civil War. Hoover was to make an unintentionally cruel cliché out of the statement that "prosperity is just around the corner." The difficulty was that no one as yet could even find the corner to go around. All that could be seen was a road going straight down hill.

Nowhere did Hoover better exemplify his allegiance to the party of memory than in his speech accepting the GOP's nomination for a second term, which he delivered in August 1932. "The solution of our many problems which arise from the shifting scene of national life," he told the nation at the very nadir of the Depression, "is not to be found in haphazard experimentation or by revolution. . . . It does not follow, because our difficulties are stupendous . . . that we must turn to a State-controlled or State-directed social or economic system in order to cure our troubles. That is not liberalism; it is tyranny." Hoover reaffirmed his faith in a high, protective tariff, in restricted immigration, in insistence upon payment of war debts in full, and stated that "the first necessity of the nation . . . is to reduce expenditures on government, national, state and local. It is relief of taxes from the backs of men which liberates their powers. . . . This must be done." Finally, he promised, "Come what may, I shall maintain through all these measures the sanctity of the great principles under which the Republic over a period of one hundred and fifty years has grown to be the greatest nation on earth. . . . I have but one desire, that is . . . to see the principles and ideals of the American people perpetuated."[52]

To Alfred du Pont, Hoover's program did not provide the means by which confidence in the American economy could be restored. The party of hope had always had a greater appeal to Alfred than the party of memory. In this great crisis especially, foolish prattling about "the sanctity of great principles" provided no new jobs and fed no starving people, and a false optimism based on no substance other than inane clichés could only further aggravate, not alleviate, the national psychosis. It was high time for innovative experimentation, which could be rational, not "haphazard," as apparently Hoover thought all experimentation must be. Otherwise, there most assuredly would be the revolution that both men feared. "Millions of men going around with empty dinner kettles. . . . One can never tell what social upheaval may be imminent," Alfred wrote his nephew Cazenove Lee in October 1930. "Hoover, in my opinion, is not handling the situation with the force it needs. The emergency is much more drastic than we were facing during the war. . . . It is a national emergency of major importance and he should have at his disposal, by act of Congress, no less than five billion dollars to tide over the situation. . . . He would not have to spend one billion, I am confident, as the mere knowledge that he had it at his disposal would inspire the necessary change of confidence to change the whole psychology of the situation."[53]

This letter was written in 1930, two years before the bottom had been reached, at a time when most financiers and politicians were still mumbling about "a necessary correction in market values," or at worst "a temporary set-back in economic growth." By 1932, when Congress, through the Reconstruction Finance Corporation Act, finally provided the $5 billion that Alfred proposed in 1930, Alfred himself had moved far beyond that suggested remedy. The "trickle down" policy that Hoover had finally been persuaded to accept in 1932 might earlier have been sufficient to restore confidence in the economy and to provide jobs at the bottom by pouring money in at the top, but now it came too late. What was needed in the summer of 1932 was direct, immediate relief for the millions who were hungry, naked, and homeless. A slow trickle down of the manna given to the top echelon was neither fast enough nor sufficient to keep people from starving or freezing during the coming winter. What was needed, Alfred told his son-in-law, Elbert Dent, was for the government to borrow $3 billion by selling bonds, just as it had done in 1917 to finance the war, and with this fund to start a program of public works all over the country. A public works program would be far better than simply providing a dole, as Great Britain was doing for its unemployed.[54] But in his acceptance speech of 1932 President Hoover said that he would continue to oppose what he called "pork-barrel nonproductive works which impoverish the nation," and he vetoed the Garner-Wagner relief bill which would have provided direct relief to individuals and would have established a system of public works throughout the country, on the grounds that it was "impractical," "dangerous," and "damaging to our whole conception of governmental relations."[55]

The campaign of 1932 placed Alfred in a most uncomfortable situation. He was determined to use what political influence he had within the state of Delaware to secure the reelection of Governor Buck. To achieve this goal, he was even willing to work along side his cousin, Lammot du Pont, but he refused to give a cent to the Better Government League, which Lammot had organized to give financial support to the entire Republican ticket. Alfred considered the League to be simply a group of wealthy Wilmington industrialists, including many of his own relatives, who were once again trying to buy an election.[56] Alfred refused to make any financial contributions to the campaign in 1932, not even to the one candidate he was strongly supporting. As he explained to I. D. Short, "I refused to contribute to the campaign fund for Governor Buck, as I don't believe in spending money for campaign purposes, and I believe that Governor Buck should be renominated and reelected upon his record; and I have decided to make no contributions of a political nature this coming Fall. . . . The many hungry people invite prior consideration."[57]

Alfred was willing to join forces with the League to secure the election of the man who had been such a staunch champion of the social legislation Alfred had espoused.[58] His one great fear, as he told his old ally, T. W. Jakes, was that "so many Republicans will want to vote for Roosevelt and, fearing to cut the ticket, they will vote the whole Democratic ticket. I am quite prepared to see Delaware go for Roosevelt but . . . [if the Democratic nominee for Governor] is elected, it will be for the above reason. A lot of voters are afraid to cut a ticket, for fear it will be thrown out."[59]

Whether Alfred himself was ready "to cut the ticket" by voting for a Republican for Governor and a Democrat for President must remain open to conjecture. It is difficult to believe, judging by the comments he made to his friends and relatives before and immediately after the election, that he cast his vote for Hoover. As early as the summer of 1931, he had clearly given up any hope that the incumbent President would provide the leadership the crisis demanded. He wrote to his new German son-in-law, Madeleine's recently acquired third husband, "Our Government is purely an organization of political parties, in preference to the interests of the people. We lack a leader, a man of courage at the head."[60]

If Alfred did indeed split his ticket in 1932, then he must have been pleased with the results, for the immensely popular Governor Buck easily defeated his Democrat opponent by almost the same percentage of votes that Roosevelt received nationally over his Republican opponent. Although Hoover had no coattails upon which incumbent Republican office-holders could ride to victory, Buck's coattails were long enough to carry Hoover to victory in Delaware although by less than three thousand votes.

The letters that Alfred wrote immediately after the election do not read as if he were grieving over having backed a losing horse. To Ellen La Motte, he wrote: "The Republican party . . . had gotten to think that nobody else in the world could run the country. . . . They . . . [got] their natural good licking which they all deserved."[61] He invited Hoover's Secretary of War, Patrick Hurley, who was Alfred's good friend, to come to Epping Forest to rest his voice after all that "hoo-hooing for Hoo-Hoo."

Jessie's older brother, Tom Ball, who was considerably to the right of Calvin Coolidge in his political views, wrote a frantic letter to Alfred immediately after the election, predicting the end of civilization in the United States and probably in the Western world. In Tom Ball's opinion, Hoover had been bad enough, for Ball always regarded him as a crypto-radical ever since Hoover had provided emergency famine relief to the Bolshevik Russians, but compared with Roosevelt, Hoover was a pillar of conservatism. Ball predicted that America would soon experience a revolution that would put the French Reign of Terror to shame.

Alfred tried to calm his distraught brother-in-law:

> Well, the Election turned out about as most people had expected it would. . . . The Republican Party has been so inflated with its own estimation of its own value and the utter impossibility of any other Party or Parties being able to run the country, or do anything for that matter, that it was about ready for a "damn good licking." . . . As to the ultimate outcome of the Democratic Party's attempt to do what the Republicans have wholly failed to do, that is purely a matter of surmise; however, as a large majority of the people voted for a change, believing that a change would bring better times, the mere fact that so many people believe the change will bring better times is about one-half the battle won; just as the bad times were to a very large extent the result of a mental condition. . . .
>
> Had any Democrat in the country headed the ticket this Fall, the Party would have won; it probably would have won without any one. . . . Those who know Roosevelt personally say he is a very pleasant person. . . . I don't look for anything radical; the platforms of the two major parties are practically identical. Of the two, I prefer the Democratic platform, as it is cleaner cut on the matter of Prohibition, and I am hoping for a 'beer and light wine bill' from the 'Lame Duck Congress' and a full repeal as soon as it is possible. . . .
>
> So far as the business of the country is concerned, the Democrats have some little sense and they will not antagonize. . . . There are just as many good business men in the Democratic Party as there are in the Republican Party and such men realize that capital and labor must go hand in hand if prosperity is to return.[62]

Jessie also felt obliged to send a note of sisterly advice to her brother. "The only thing to do is to get heart and soul in back of the Administration and pull America out of the chaos into which she has fallen by virtue of the debauch of the last six years. We, the citizens of America, are responsible for it, not the governing officials. To my way of thinking this is one of the greatest faults of the human race, always looking for someone on whom to put the blame, rather than saying, 'I was a d--n fool and I know it. I will try not to make the same mistake again.'"[63] It is doubtful that brother Tom was consoled by Alfred's calm acceptance of the results of the election or inspired by his sister's counsel. Tom continued to threaten to move to Canada and let his two Pollyanna relatives in Florida have their precious New Deal. To Tom, they must have seemed like Louis XVI and Marie Antoinette, sitting in Versailles while the canaille prepared the tumbrils to carry them to the guillotine.

In the meantime, the dreadful interregnum between Hoover's defeat on 7 November 1932 and the inauguration of Roosevelt on 4 March 1933 had to be endured. Most Americans had thought it impossible for conditions to get any worse than they were in the summer of 1932, when the veterans had marched on Washington, only to be driven out by General Douglas MacArthur's tanks, and desperate farmers in the Midwest had attempted a food strike in order to raise agricultural prices at the same time that millions of Americans could not afford food at any price. But conditions did get worse—much worse—during the four months of waiting for the new administration. Winter alone would have ensured that, since in addition to the need for food there were the cruel demands that winter makes for clothes, fuel, and shelter. The incumbent government was now the recumbent government, lying paralyzed by defeat and incapable of any action. There were new runs on the banks that were still trying to function, which threatened the destruction of even the soundest financial institutions. In terms of human suffering, these were undoubtedly the worst four months in the nation's history.

Despairing of any remedial action from the defeated President and the lame-duck Congress, Alfred could no longer sit complacently in luxurious comfort and do nothing about the situation. Although he knew it would be about as meaningful as providing a single bowl of soup to an orphanage of hungry children, Alfred decided in January 1933 to inaugurate his own miniature emergency relief agency in both Delaware and Florida. To assist him in Delaware, he once again turned to his trusted and competent co-worker in social welfare, Laura Walls, whom he had initially hoped would be named director of the Old Age Welfare Commission in Delaware, but who had been turned down because she was judged to be too partisan a Democrat. Consequently she was available to help Alfred, and he employed her as his own personal director of welfare relief. Through her, he provided a statewide pension plan for retired and destitute teachers who, because of age and some small personal resources, were not eligible for the regular state pension plan.[64] He also bought up a large supply of blankets and winter overcoats and asked Walls to distribute them to those most in need of winter garments and bedding.[65]

Alfred's most ambitious program of emergency relief, however, was undertaken in Florida, where he and Jessie had gone during the first week in January. He utilized the services of Jessie's cousin Henry Dew to engage a fleet of pick-up trucks with drivers. Each morning, Dew would lead this caravan down the streets of Jacksonville, picking up men, both black and white, who were sitting idly on street benches but were eager to work. The men were taken to parks throughout the city to clean the walks and grounds and to plant shrubs and flowers. At the end of each day, every man was paid $1.25. Alfred's relief program cost him over $400 a day, but he thought it a small price to pay to feed over three hundred men and their families.[66]

Alfred's most immediate personal, depression-inflicted problem was to keep his own chain of seven banks open and solvent as the banking crisis worsened in the first two months of 1933. Several small banks in St. Louis had closed early in January, which touched off a bank run throughout the state of Missouri. The next wave of closings and runs swept over California later in the month. On 13 February a major bank in Detroit ceased operations, and the governor, not waiting for the kind of statewide runs that had occurred in the other two states, promptly closed all the banks in Michigan. The final collapse of the American economy appeared to be at hand. Most bankers throughout the East angrily denounced what they regarded as the governor's rash and much too precipitous action. Alfred, however, startled his own bank presidents by applauding the governor's action and announcing that if Hoover had any backbone or sense at all, he would promptly close all of the banks in the country, pass a strong banking bill that established strict regulations on loans, and send out examiners to every bank, who would then determine if a bank was sound enough to reopen for business. He believed that the hysteria would then pass—for the people would know that if a bank was open, their money was secure—and the runs would promptly stop. Alfred's associate, W. T. Edwards, could hardly believe his ears. "To me this theory was astounding, because it was so unorthodox. We had been taught that a bank should stay open until the last dollar was gone. Mr. du Pont explained that if the banks were allowed to remain open the hysteria of the people and the fright of the bankers would bring about the doom of both—and the country."[67]

Such unorthodoxy was, of course, as inconceivable to President Hoover as a proposal to nationalize all the banking institutions would have been. It was left to the state governors to take such drastic action if conditions continued to worsen. And worsen they did. The governor of Maryland closed all banks in his state, and with that action, the panic swept the entire East Coast. In some instances, mayors, with doubtful legal authority, closed all the banks in their cities when the governors failed to act. On Thursday, 2 March, one-quarter of a billion dollars were withdrawn from the banks that were still open across the nation. On Friday, 3 March, the same amount again was withdrawn. It looked as if the last dollar would disappear from every bank and savings institution in the country. At midnight of that day, Governor Herbert Lehman closed all the banks in New York, and by morning the chief executives of Illinois, Iowa, Missouri, and Minnesota followed suit. At twelve noon on the following day, Saturday 4 March, the long, terrible interregnum was over. Franklin Roosevelt took the oath of office and was now President of the United States.

On that same Saturday afternoon and Sunday morning, in Jacksonville, an emergency meeting of the officers of the three largest banks was held. Florida's governor had still not acted, and the immediate question each of the three banks faced was whether or not to open on Monday morning. Alfred did not attend the meeting because of his deafness but sent as his representatives George Avent, president of the Jacksonville Florida National, and of course, Ed Ball. Alfred sat in his office and had Ball report to him periodically on the progress of the meeting and to relay his position on the issues raised by the conference.

Both Ed Lane of the Atlantic National and old Brion Barnett of Barnett National insisted on keeping their banks open. They were as orthodox in their banking views as Hoover. Alfred sent instructions that he favored closing the banks and keeping them closed until they saw what kind of emergency banking bill the new administration would produce. The Florida National would, however, stay open if the other two banks did. Alfred was outvoted two to one to have the banks stay open.

The next issue was over the question of what percentage of a depositor's holdings the banks would allow to be withdrawn, for all three banks were in agreement that they could not possibly allow a full withdrawal or they would be forced out of business within an hour of opening. Lane favored allowing a withdrawal of only three percent of a depositor's total holding, while Barnett argued for ten percent. Again Alfred stunned his banking colleagues. Ed Ball reported to the conference that although Mr. du Pont, as they all knew, favored an immediate closing, if they were determined to stay open he believed that a three percent withdrawal allowance, or even a ten percent allowance, would be the worst possible plan of operation. Restrictive limits on withdrawals would only increase the panic and would bring out customers who might otherwise stay away. Anything less than twenty-five percent would be unacceptable.

"Have you got the money to do that?" Ed Lane shouted.

"You're damn right," Ed Ball boasted. "We're bringing more funds in all day."[68]

This was pure bluff. There was no way that Alfred could have liquidated his assets in Almours Securities that weekend in time to meet the Monday morning opening. If the anticipated crowd of depositors were actually allowed to withdraw twenty-five percent of their deposits, the Jacksonville bank would have been forced to close by noon and Alfred's other banks would have quickly followed. But Lane and Barnett could not be sure what cards Alfred was holding in this grim game of financial poker. All they could think of was the $200 million that the Florida press had so conveniently manufactured as the reputed assets of Almours Securities. If the Florida National actually had that kind of money in reserve and all three banks opened on Monday with a twenty-five percent allowance for withdrawal, then the Atlantic and Barnett banks would be closed within a few hours, and Alfred would be the one remaining monarch of the banking capital of Florida. On the other hand, Ball had been known to bluff before. Should they call Alfred's hand, or should they quietly withdraw and agree to a closing? It was hard to read the faces of their opponents with Alfred sitting down in his office and Ed Ball sitting like a smug little leprechaun on a pot of gold. Neither Lane nor Barnett could decide which course to follow, so they talked of other matters. What news was there from other banks in the state? What news was there from Washington? On and on the discussion dragged. In disgust, Alfred went back to Epping Forest at five on that Sunday afternoon, but he stayed up late listening for bulletins over the radio. At midnight the news he had been waiting for came. President Roosevelt had just signed a proclamation closing every bank in the nation. He assured the American public that there would be an immediate survey of every bank in the land. Banks would reopen as soon as they could demonstrate financial stability, and when a bank opened its customers could be assured that their money was safe and need not be withdrawn.[69]

The poker game with Lane and Barnett was over. Alfred went to bed confident that there was now a real leader in the White House. "Roosevelt had the courage to handle the situation with brass knuckles," he wrote to his brother Maurice. "I say Hurrah!"[70]

During the next one hundred days Alfred had many other occasions for saying "Hurrah for Roosevelt." The Emergency Banking Act passed by Congress in only four hours on 9 March provided desperately needed immediate aid to banks that were allowed to reopen. The more permanent provisions of the Glass-Steagall Banking Act of 1933 also met with Alfred's approval, except for the provision that established a Federal Deposit Insurance Corporation to insure all deposits up to $2,500. "I am utterly opposed to any form of Federal guarantee of deposits, which, even to a layman, must show that such a law would put a premium on bad banking," Alfred wrote to his son in April.[71] Nevertheless, the Federal Deposit Insurance, more than any other feature of the Banking Bill, restored the public's confidence in banks. Deposits flowed back into all of Alfred's Florida banks. Once again, people believed banks to be safer than the mattresses on their beds.

Alfred also hailed the National Recovery Act as a major step toward bringing together labor and capital, as well as for setting standards for both production and wages. The major accomplishment of these first three months of the New Deal was in providing immediate relief to millions in desperate want. The Civilian Conservation Corps, put into operation in April, the Federal Emergency Relief Administration, which had $500 million at its disposal, and above all the Civil Works Administration, which not only provided relief but employment, were precisely the measures that were needed. In late March, Alfred gave one of his rare interviews to the press: "The depression," he told the surprised reporters, "is over. . . . Now the sound banks are open again, doing a bigger business than before. The people are confident. Fear is gone from their minds. President Roosevelt forced the cards down on the table for a showdown and the people judged for themselves. They are now assured that banks permitted to reopen have the stamp of federal approval and all fear is gone. . . . Since his inauguration, Mr. Roosevelt has made the finest showing of any President since Lincoln. He has more fully realized the needs of the country than anyone else and has acted wisely."[72]

To his brother-in-law, Tom Ball, who had viewed all of these new alphabetical agencies with horror, Alfred wrote, "We are in for a complete change in our industrial and financial situation. We might as well face that fact. The upper income tax brackets will undoubtedly be eliminated in a comparatively short time and every effort made leading to a more equable distribution of national income. This on the assumption that it will increase purchases, which of course means increased production. Unquestionably, we are in for Government management of industry and a distribution of earnings along lines tending to this much desired condition; and then in the future, all surplus earnings of corporations, instead of being used to expand production unduly, will be utilized in raising wages and salaries, so as to increase the purchasing power of the masses."[73] Alfred, however, would set a limit on executive salaries. The top "banking salary should be Fifty Thousand Dollars," he told his nephew Maurice Lee. This was precisely the salary that Alfred was paying Ed Ball.[74]

That a du Pont could accept not just with equanimity but with apparent approval the redistribution of wealth, which Alfred predicted was now inevitable, was to Tom Ball simply incredible. Alfred du Pont had come a long way from his inherited laissez faire philosophy, but then so had the nation by the Spring of 1933.

XV

The Patriarch and His Family:
Last Days 1934–1935

S PRING had never been so welcome as it was in March 1933. The long, dreadful winter of discontent had at last ended, and now both nature and the nation showed new life. It was a moment when all America, from Wall Street to Main Street, looked to Pennsylvania Avenue, for the leadership that was desperately wanted and the remedial action that was desperately needed. In the ensuing three months—the first one hundred days of the New Deal—the people were to get both. Hastily written bills were rushed through Congress in record time with scarcely a dissenting vote. The alphabet suddenly took on a new prominence as one acronym after another became part of the American vocabulary: NRA, AAA, FERA, CWA, CCC, TVA—and they all spelled relief and reform to a people who had long despaired of obtaining either from their government.

Never before had the nation been so united in support of its elected officials, never had the general will been so manifest as it was in the spring of 1933. Alfred du Pont's publicly voiced approbation was only one small contribution to the chorus of millions who were shouting "Hurrah for Roosevelt." From the moment he had announced in his inaugural address that "the only thing we have to fear is fear itself," Franklin Roosevelt had struck the right note, as far as Alfred was concerned. For years, Alfred had been saying precisely the same thing to anyone who would listen to him. The only affliction America suffered from was fear—"nameless, unreasoning, unjustified terror which paralyzes needed efforts to convert retreat into advance," as Roosevelt so aptly stated it. Banish fear and you banish unemployment, bread lines, bank runs, and foreclosure sales. Fear could not be banished, however, simply by keeping a stiff upper lip and prattling empty clichés about prosperity being "just around the corner." The crisis demanded "action—and action now."[1]

Alfred du Pont had been waiting to have these words uttered for the past three years. Although, because of his deafness, he could not hear them pronounced over the radio and would thus be immune to the magic of Roosevelt's voice, which charmed the nation, he could read the inaugural address in the newspapers, and he liked what he read. He liked even more the action that the President had promised—not action in some unspecified future, but action now. Most of the program of the first months of the New Deal had Alfred's unqualified support. The government gave jobs, not degrading, unearned charity in the form of a dole; youths were given the opportunity to get off the streets and out of freight cars to find constructive employment in the Civilian Conservation Corps in building park shelters and nature trails; banks were allowed to reopen only after they had been inspected and had received the government's stamp of approval; and industrial leaders were asked to draw up codes to establish national standards for quality of production and humane conditions of labor. To Alfred all of these emergency relief and reform measures were but the concrete enactments of his frequently expressed proposals for the reconstruction of the American economy. It was almost as if the new administration had been privy to his correspondence with friends and relatives and had then accepted his counsel. Only the Federal Deposit Insurance feature of the Glass-Steagall Banking Act provoked criticism from Alfred, for he regarded this measure as having the dangerous potential of rewarding bad banking practices.

Alfred's honeymoon with the Roosevelt administration did not last beyond the first year, however. In 1934, he wrote his nephew Maurice Lee, "I do wish Roosevelt would let up on his experiments and scrap all his different NRA's and PDQ's and let business take care of itself once more. If he does not he will rapidly accept Socialism as part of our National system. . . . The people of this country are rapidly drifting into a condition where they are willing to accept from the Government instead of living by the sweat of their brow."[2] This view sounded like the pre-Depression, laissez-faire Alfred of old, and although he kept most of his criticism of the New Deal confined to letters to his closest relatives, he did openly attack the Wagner National Labor Relations bill when it was presented to Congress in 1934. He wrote a very blunt letter to Senator Hastings telling him that he expected the Delaware Senator to use "every proper effort . . . to defeat this Bill."[3]

Although his initial ardent admiration for the new administration cooled down considerably, Alfred never made as open a break with Roosevelt as did Cousin Pierre, who became one of the founding fathers of the ultra-conservative American Liberty League in 1934. In his infrequent press interviews, Alfred continued to express confidence in the administration and a personal admiration for Franklin Roosevelt. Alfred was the first person in Florida to make a contribution to the Roosevelt Birthday Ball, a gala event to be held on 30 January 1935 that would serve as a kick-off for the annual March of Dimes drive in the fight against poliomyelitis, an enterprise that had Alfred's whole-hearted support.

Even as late as March 1935, just one month before his death, Alfred responded to two hysterical letters from his old friend Sterling Joyner, who was convinced that Roosevelt's policies were little more than carbon copies of Stalin's Five Year plans: "I have your two letters under date of March 19th, and I am afraid that you have been unduly influenced by listening to pessimists in discussing the general political affairs of the world and this Nation in particular. I see no eminent [sic] threat to our country."[4]

Indeed, by the spring of 1935, Alfred's support of the New Deal had been revived by the administration's proposed Social Security program that sought to establish a national unemployment insurance and an old-age pension plan that would be self-supporting through contributions from both employers and employees. The proposal closely resembled the German social welfare program that Chancellor Bismark had established fifty years earlier and which Alfred had been advocating for the United States ever since he had given it careful study while on a trip to Germany in 1929.

Moreover, Cousin Pierre and John Raskob were waging a fierce campaign against the New Deal under the banner of the American Liberty League, a slick, Madison Avenue propaganda organization that they had created and largely financed. In Alfred's opinion, anyone whom Pierre and his buddy "Raskal" were attacking could not be all bad. Unfortunately, Alfred did not live long enough to be able to savor Roosevelt's rejoinder to Pierre du Pont and his Liberty League friends. "They are unanimous in their hate of me," Roosevelt told a wildly enthusiastic crowd at Madison Square Garden on 31 October 1936, "and I welcome their hatred."[5] Only someone who himself had long been a recipient of that same hatred could fully appreciate and enjoy Roosevelt's defiance. That statement alone would have forever endeared Franklin Roosevelt to Alfred.

Although the various New Deal emergency relief programs had done much to alleviate suffering in the United States, many people, particularly the elderly, were still in dire need of assistance in 1934–35. Even if the Social Security proposal became the law of the land—and Alfred sincerely hoped it would—a whole generation of retired and disabled persons would not be covered by the national insurance program and would continue to be dependent upon other sources for assistance. The welfare pension program that Alfred and Governor Buck had succeeded in getting the legislature to establish in 1931 would remain an essential program for thousands of Delawareans for many years.

When the Delaware Old Age Welfare Commission was created Alfred had hoped that Laura Walls would be named executive director of the program. Alfred had implicit confidence in her ability to determine who were the truly needy and in her managerial skills in distributing the pensions. He had pulled every political string that he could grab to secure her appointment. When, however, she was denied that appointment for what Alfred considered grossly unfair partisan reasons, it became imperative for him once again to become involved with the distribution of old-age pensions in Delaware. Six months after the Delaware Old Age Pension program was established he wrote his daughter Madeleine, "I had hoped by this time to have gotten rid of my 1600 pensioners in toto . . . but owing to the inefficiency of the new Welfare Commission, or rather their lack of experience, only about one thousand, so far, have been cared for. This leaves quite a number who have to be looked after from time to time to prevent starving."[6]

Alfred was never entirely rid of his pensioners during the remainder of his life, for Laura Walls continued to find needy cases that the Welfare Commission either could not or would not on her recommendation place on its pension list. Walls remained on Alfred's payroll and her recommendations regarding any family in dire straits continued to reach Alfred's desk. She never asked for much, and Alfred could be certain that every case was deserving. Each month produced a new list:

"Mrs. Stella Hayman, Georgetown, Del. This woman's house was burned a week or two ago, and they lost everything they had. They have 7 children, and in greatly need of assistance. $20.00 would be a Godsend to them."

And: "Mrs. Lida Stevenson, Fredrica, Del. Mrs. Stevenson hates to ask for help from anybody. She and her children have gone hungry last winter. She even used bread wrappings to grease her griddle (waxed paper). I gave her some white potatoes. She and four children have been making out some way on $1.25 per week. $15.00 would be a Godsend to them."[7]

These and many more "Godsends" were always forthcoming. In addition to the special individual gifts, Alfred also sent funds on a regular monthly basis to Laura Walls for distribution to needy cases that the Delaware State Aid Society and the three-state Delmarva Supply Ministerial Aid had recommended for assistance and with which Walls concurred.[8]

Jessie and Alfred also jointly contributed $600 monthly to a private charity, The People's Settlement, to provide school lunches for children whose parents could not afford the hot lunches that the Wilmington Public School district provided at a minimal cost.[9] Jessie was an enthusiastic supporter of all of Alfred's various contributions to the needy, and in general, she concurred with his political views regarding the New Deal, although many of her own family, especially her brother Tom, would have felt quite at home in Cousin Pierre's Liberty League. In February 1935 Jessie was shocked to receive a letter from Tom in which he attributed the current depression to Franklin Roosevelt. She replied tartly:

> Tom, I am going to take up your letter. . . . Please remember that this depression did not start in 1933, as you seem to think. It started before the New Dealers came into power. It was precipitated by the sudden drop in the market in October 1929. . . . I certainly do not approve, and I don't want you to interpret the above as approval or endorsement of most of the things that are being done. Think what we need are some far-sighted business people, and the chief officials recognize this. You know, Tom, and know it from your personal experience that people don't listen to advice. When you were here last year both A. I. and Eddie advised strongly against taking on any liability both for you, for me and for them. Did you heed that advice? Now just as individuals are, nations are. The pendulum has swung far too much to one side. Of course, we have a government by a mob. In our form of government it can be no other way. . . . However, a nation must have a popular ruling body. . . . Wherein [sic] the great populace was ruled, it was very unfair and in many instances inhuman treatment, where poverty reigned supreme. . . .
>
> During Mr. Hoover's Administration he [Alfred] advocated a Five Billion Dollar bond issue just after the collapse in 1929 for public works to stem the tide and as his remark then was, "It will save America many times that many billions and prevent a great period of depression"; but the Republicans did not see fit to follow it, as they did not deem it wise, but looking back, they think it was.
>
> Forgot to mention that we had a depression from 1921 to 1923, which was also during a Republican Administration, but going back through history you will find this has always been true. It is not a partisan thing.[10]

Jessie could have been even more forceful in refreshing Tom's memory as to when the Great Depression began if she had reminded him that it was in 1932, a year before Roosevelt took office, that she had had to cut by ten percent his monthly allowance of $500 that he had been receiving for the previous five years from her and Alfred.[11] Tom undoubtedly resented Jessie's contradiction of his views on politics, and her history lesson on business cycles would in no way convince him that depressions were "not a partisan thing." In his mind, they would always be calamities created by Democrats whenever they got into office. Suffering Jessie's and Alfred's political heresies, however was something that Tom had to accept along with his monthly allowance check, for much as he decried welfare payments made by the government to the destitute, Tom eagerly accepted the dole he received from his benefactors.

When, after a bitterly contested political battle, the bill to establish a central welfare home to replace the county almshouses finally passed both houses of the Delaware Assembly in April 1931 and promptly received the governor's approval, Alfred felt that he had won the greatest political battle of his life. Even though that battle had been won, however, the campaign, unfortunately, was far from over. Like many of the old-age pensioners for whom Alfred still had to provide funds, the problem of the central welfare home remained with him for the rest of his life.

The first major issue to arise was the choice of an architect for the new home, to be built in Smyrna, Delaware. Here Alfred took a very personal interest, for it was an issue which would involve his own son, Alfred Victor, who had only recently become an architect.

After attending Yale University for one year, young Alfred Victor, had served as a private in the Marine Corps from September 1918 until November 1919. Upon his release from the service, he returned to Yale to study at the Sheffield Scientific School. Leaving Yale after only one more year of study, he worked at the Buick plant in Flint, Michigan, until his mother was able to persuade Cousin Lamont to find a place for her son in the Du Pont Company. In November 1922, Alfred Victor began work at the Du Pont Laboratory in Repauno. He was later transferred to the Louviers Chemistry laboratory and then in February 1925 to the Barksdale plant.[12]

From the first Alfred Victor was not happy working for the company whose name he bore. He was well aware of his deficiencies in education in comparison to the bright young engineers whom the company continued to bring in from the best technical schools. One year at Sheffield, where he had done only moderately well in his courses, could not be compared to an engineering degree from M.I.T. or Rensselaer Polytech. His name had got him his job, but on the job it proved to be a burden, not a boon. His peers resented him, and his superiors were hesitant to give him the intensive, on-the-job instruction and criticism he needed. In November 1926, Alfred Victor wrote to Pierre to inform his cousin that, after talking the matter over with his mother, he had decided that it would be better if he worked for a different company. "I feel that being a du Pont in the Du Pont Company is more detrimental than helpful while under men who never hope to attain high positions and feel they must struggle to hold their places they have, and when one's only chance of promotion lies in their hands." Pierre answered that he had consulted with Lammot and that both he and his brother felt Alfred Victor should try to stick it out. "During my younger years I suffered many pangs of disappointment concerning possibility of advance. I am happy that I stuck to the job."[3]

Pierre had himself quit the Du Pont Company after suffering his "many pangs of disappointment" and had returned only at the invitation of Alfred and Coleman to be part of the triumvirate in the ownership and management of the company. No such glorious future seemed likely for Alfred Victor, and in April 1927 he sent his formal resignation to J. W. McCoy, assistant general manager of the Explosives Department.[14]

The du Ponts had always been reluctant to let go one of their own, and so Maurice du Pont Lee rushed forward with an offer of a job in his rayon plant in Buffalo in order to keep the only son of the eldest son of the eldest son of the eldest son of the Founding Father bound to his familial heritage.

By this time Alfred I. had been reconciled with his son and was delighted to learn that Maurice had found a place for Alfred Victor in the rapidly expanding rayon branch of the company. "Of course, in any new industry, such as the Rayon," Alfred wrote his son, "there is a greater opportunity . . . than in an industry such as High Explosives, which in this country is over forty years old. . . . The Rayon industry is much more interesting from every standpoint, though the element of danger which entered into one's life in the powder business always gave it additional spice; but you have before you almost a clear field and, with the stimulus which independence of thought and action engenders, so it is a case of matching your own brain power and vitality against the field."[15]

Alfred Victor had the brain power, but his lack of real interest in the Du Pont Company itself greatly diminished any vitality he might have exhibited on the job. In offering Alfred Victor a job Maurice had assured him that "the Rayon Company was managed as a separate entity from the du Pont Co.—its methods and standards quite different and its policy not dictated by Wilmington."[16] Once on the job, however, Alfred Victor discovered that the same situation prevailed at the rayon plant as at Barksdale. He was resented because of his name and for his close ties with Maurice Lee, and he was ridiculed for his lack of technical training. Finally, after much hesitation and a great deal of trepidation, Alfred Victor wrote a long letter to his father to explain the situation and to justify his decision to leave the Du Pont Company permanently. He told Alfred that he had had two offers for other positions, one with an insurance company and the other in an investment house, "but I felt pretty sure that they would have capitalized somewhat on my name." He then told his father what he would really like to do. "All my life I have had it in the back of my mind. I enjoyed draughtsmanship in school and have since often dallied with drawing plans, as amusement. I believe, taking it all in all, that now I am convinced I made a mistake in re-entering duP. Co work. Architecture may prove to be my calling. Also, because of the income I have, thanks to your provision, I am in a position to study it and work hard at it before getting monetary returns, and meanwhile Marcella [his wife] & I can live comfortably."[17]

Alfred Victor had no idea how his father would react to his leaving Du Pont and breaking the last primogenital tie with the company, but to Alfred Victor's immense relief, Alfred gave an enthusiastic response:

> I am not in the least surprised at what you write. . . . The situation as I view it now is that efforts are being made by Wilmington to put you at a disadvantage. This is clear to me. . . . If this premise is a correct one, then regardless of where you work in the Rayon Co. you will never get a square deal. . . . There is nothing more distasteful than working under any sort of a handicap. Your vocation should breathe freedom of thought and action, results all dependent on your own efforts as an individual, leading to hope and encouragement. . . .
>
> I am pleased to know you have a natural trend towards architecture. It's a fine profession and to a man who makes good even in a moderate way offers a decent living, the prospects for great things to those who excel are ever present. . . . As you are a free agent and young, you should succeed in a profession selected as a mature man. . . . Please remember— that I worked 18 years for the old du Pont Co. and never a thank did I get—and I hope you will be spared eating potatoes out of the same barrel.[18]

No son could have asked for a warmer paternal blessing before starting out on a new venture. And so it was quickly arranged. Alfred Victor left the Du Pont Company on 1 October 1928. After a cram course in the French language at a school in Philadelphia, Alfred Victor and Marcella sailed for France on December 29th.[19]

While enrolled in the Department of Architecture at the Ecole des Beaux Arts in Paris, Alfred met a young Italian architect who was looking for a partner, preferably an American, who had money and contacts in the United States. It was a most felicitous union. Gabriel Massena had the professional training and experience that Alfred lacked, and Alfred had a seemingly assured position in America that Massena was eager to obtain. In March 1929, they formed a partnership, and Alfred immediately gave the new architectural firm of Massena and du Pont its first commission—to design a formal sunken garden for Nemours.[20]

Alfred was delighted with the drawings that Massena and Alfred Victor sent him:

> The plans submitted by you are a distinct improvement over Mr. Hastings' [Thomas Hastings, the architect for Nemours who had also submitted plans for a formal garden] and embody many of the features I tried in vain to make Mr. Hastings adopt. . . . The stairways are much

more artistic ... and the whole effect is more open, lighter and more graceful—and much cheaper.[21]

Receiving this kind of encouragement, the two young architects hurried to America before Alfred Victor had completed his course of study at the Beaux Arts in order to build the sunken gardens, complete with marble stairways, statuary, medallions, and a small, domed rotunda bearing the same name, the Temple of Love, and in the same style as Marie Antoinette's little trysting place at the Petite Trianon.

Alfred also commissioned Massena and his son to draw up plans for a small non-sectarian church to serve as a filiopietistic, if not entirely appropriate, memorial to his deistic French ancestor, Pierre Samuel du Pont de Nemours.

There was a limit, however, to the work that Alfred himself could provide for his son and Massena. They completed the work on the gardens, but before they could undertake the commission for the church, the demands that providing for the needy placed upon Alfred's greatly reduced income forced him to abandon the church project.[22] Alfred Victor and Massena were obliged to look elsewhere for commissions, which were not easy to come by in the depression. Their only real opportunity lay in government projects, both state and federal, which the various New Deal programs were authorizing. Alfred used all of the political clout he had with Senator Daniel Hastings, which was not much, and with local politicians to obtain architectural commissions for Massena and du Pont in the building of post offices and new schoolhouses in Delaware, but he was thwarted at almost every turn. He was convinced that in each failure he could see the hand of Cousin Pierre, who was still carrying a grudge against Alfred Victor for having left the Du Pont Company and for becoming reconciled with his father. All of these commissions went to Pierre's favorite local architects, Betelle and Martin.[23]

The one political plum that could not be denied to the Alfred I. du Pont family, however, was the building of the new welfare home at Smyrna. With Governor Buck's backing, Alfred was assured of obtaining this commission for his son, and in the late fall of 1931, the members of the Welfare Commission selected Massena and du Pont to be their architects. The Commissioners never had any reason to regret the choice that had been forced upon them.[24]

Work on the central welfare home began in the early summer of 1932, and the main building was completed one year later. For a public institution designed to house the elderly indigent, it was a remarkably graceful and pleasing building. It truly looked like a home—a welcoming, red-brick

Georgian country estate—rather than a forbidding state asylum for the destitute. The Wilmington *Star* waxed lyrical over its design:

> Fronted with a monumental georgian [sic] colonnade, the administration and hospital building is the hub of the group of buildings which will eventually be added to the home. . . . The great, carefully proportioned columns of the main building, back of which is a beautifully carved Georgian doorway with broken pediment, form the focal point of the scene, and the axis of the grounds. Through this portal, the line of vision carries directly through the grounds to the twinkling waters of the lake. . . . Amid clean and restful surroundings, those who have greeted the December of life with nothing which might rightfully be called a home will spend their declining years in comfort. The benevolent hospice of a place to go is theirs forever.[25]

The dedication of the welfare home on 11 October 1933 was a proud day for both father and son. In addressing Governor Buck and the other guests assembled for the occasion, the chairman of the Old Age Welfare Commission, the Rev. Charles Candee, paid special tribute to Alfred I. du Pont, whose persistence in the political arena had made the home possible and to his son and his partner, Massena, who had turned a legislative act into an architectural reality. "We feel," said Candee, speaking for the Commission, "that the architects have been signally successful in combining and harmonising permanency and fireproof construction with that which is architecturally and artistically beautiful. One thing which we feel certain will impress all of our visitors today . . . is the attractiveness and the beauty of the different colors that are on the walls . . . thus making more homelike and beautiful these rooms, instead of finishing them in the usual institutional white and we are told at practically no more than $15.00 additional expense."[26]

Alfred and Alfred Victor were never closer as father and son than when they listened to these words of shared praise. Alfred Victor wrote his father two days after the ceremony:

> Dr. Candee's fine tribute to you . . . made me very happy indeed, but it couldn't be high enough to cover *really* the very great movement forward in civilized charity that had its beginnings in your dream come true. Except for your work, generosity and untiring pressure until the necessary laws were enacted this undertaking would never have been realized. . . .
>
> When this community is complete, the required units added and all of the indigent aged of the state housed and cared for there—the other states must follow until the wonderful inspiration conceived by you is

nationally recognized as the proper method of providing for the old people. And to you is due the credit of the entire scheme.

I want this to be waiting for you on your return, so that as soon as you are home you will realize how very deeply I feel about it all.[27]

The community was not complete with the opening of the central hospital and administration building on that bright October morning in 1933. Still awaiting construction were eight "guest pavillions," a physical maintenance and central heating plant, and a medical laboratory.[28] Massena and du Pont had drawn up the plans, the Commission had given them approval, and the staff was insistent upon the necessity for these additions—all that was lacking was the necessary funding. During the next two years Alfred wrote many letters to state legislators and the governor in an attempt to persuade the state to apply for a federal grant that would cover one-third of the $500,000 required to complete the state welfare home project and to appropriate the remaining $350,000 from the state treasury, or if necessary to supply for a federal loan for that amount.[29]

With the central building completed, however, the welfare home no longer headed Governor Buck's list of priorities. As Alfred wrote Candee, "My feeling is that the Governor is not behind us in our desire to secure the total sum of $500,000, for, were the total sum appropriated, it might encroach upon what the Board of Education believes to be its interests."[30]

The issue dragged on for another year. Alfred lowered his expectations and pushed for only $300,000 in appropriations and loans. Finally, in desperation, he offered to make a gift of one-half that amount if the legislature would appropriate the remaining $150,000.[31] The problem was still unresolved at the time of Alfred's death.

Laura Walls, who could always look to Alfred as the patron of the poor, must have often wondered what would happen to her neediest cases when Alfred died. There was no one else to whom she could turn with the assurance that her urgent requests would be granted.

Her fears were not unfounded. Only three days before his death, Alfred sent out what proved to be his last check to Laura Walls, which he postdated 30 April 1935. Unfortunately, however, Walls was not able to distribute this last payment to the needy. On 29 April 1935, only a few hours after his death, Alfred's secretary, M. E. Brereton, apparently upon orders from Ed Ball, sent a telegram to Walls ordering her not to "cash check of April thirtieth until further notice."[32] There is no record that any further notice was ever given. The little "Godsends" that Laura Walls had for so long counted on abruptly ended. The deserving destitute of Delaware had lost their benefactor.

No father could have been more supportive of a son's career than Alfred was during the first years of the Massena and du Pont architectural partnership. Anyone outside of the immediate family who witnessed Alfred's untiring efforts to secure commissions for his son would have seen their relationship as a model of familial harmony, of generational cooperation, and of mutual admiration. Alfred's letters to both Massena and Alfred Victor—and to others, as well, whom Alfred was trying to induce to become clients—are filled with praise for the two young architects' work. And with good reason, for Massena and du Pont were as good as their major employer and chief publicist said they were.

More intimate friends of the family, however, knew that a barrier existed between father and son that had never been breached, although neither man quite understood why. Perhaps Alfred saw in his son a reminder of the most inexcusable act of his life, when in blind vengeful fury against his former wife, he had attempted to change, in the most public method possible, the name of his son. Alfred Victor, in turn, must have been reminded of the pain he had suffered as a result of the action undertaken by his father for no reason that Alfred Victor could ever understand.

Jessie's attitude toward her stepson and his wife, Marcella, must also have affected the relationship. Alfred had become increasingly sensitive to and influenced by the opinions that Jessie either overtly or by innuendo expressed to him about the family. Of Alfred's four children, Alfred Victor was her least favorite. She suspected him to be—as indeed he was—the most closely bound to his mother, the much hated Mrs. B. G. du Pont. Although Jessie had never met her predecessor, Alfred had long ago convinced Jessie that his first wife was nothing less than an ogress who had turned his own children against him. Jessie was also influenced by her brother's ill-disguised hostility toward Alfred's only son. In Alfred Victor, Ed Ball saw the only potential rival who could conceivably replace him at Alfred's right hand, should this father-son reconciliation business be carried too far.

During Alfred's lifetime, Jessie never openly expressed hostility toward his son, but she did indirectly attack him through his wife, by spreading the word among the family that Marcella had accused her of trying to prevent Alfred from being too closely allied with his son. With admirable if not very diplomatic directness, Marcella met this attack head on. In May 1934 she wrote Jessie:

> We dined with the Dents shortly before we dined last with you at
> Nemours. After dinner Victorine or Elbert—I can no longer remember
> which one—out of a clear sky said, "What is the trouble between you
> and Jessie?" It was as if someone had thrust a dagger in my heart. Both

Alfred [Victor] and I had been noticing what we considered coolness in your associations with both of us for a long time; but being conscious of no wrong-doing, and hoping you were not allowing trouble-makers to fill you with "he does this" and "she says that"—we tried to disregard it and regain your warm friendliness by patience. . . . My reply was "So far as I am concerned there is nothing wrong. I feel as I have since first knowing Jessie. If there is a difference in feeling it is on Jessie's side." . . . Why didn't you come direct to me the very first moment some "kind friend" of mine tried to make you dislike and distrust me? . . .

I do *not* think that you are trying to separate Pop [Alfred] from his children. I think you are contributing to their closer union by inviting them and the "in-laws" repeatedly for all kinds of visits and by general thoughtful affection on all sides. . . . [W]hat I really feel is that you have had a difficult job to do and have done it miraculously well. . . .

Until you can come to lose your prejudice against me, Jessie, perhaps it is better for me to remain away from you. . . . It will be less difficult for Alfred [Victor] for he will not see my unhappiness so much as under the existing state of affairs and perhaps some day you will get over the smart of these purported feelings that I do not possess—and we can come together again in all enthusiasm. I shall be waiting for such a time, Jessie, whether it comes now or not till much later, and I am sure that Pop will not miss me if you do not, and that, on the other hand, the moment that you are genuinely convinced of my integrity and affection, he also will be happier.[33]

The moment when they all could "come together again in all enthusiasm" would never come, and although there were occasional invitations to Nemours after the open confrontation between Jessie and Marcella, the chilled and mannered politeness of the two women only further raised the barrier between Alfred and his son.

Initially, Jessie had favored Alfred's second daughter, Bessie, over Alfred's other three children. Bessie was the first of his estranged children to become reconciled with her father, and Jessie, who had skillfully arranged their meeting, could take full credit for instigating what eventually proved to be a successful reunion. The reconciliation was a major victory for Jessie, and this alone made Bessie a very special person to her stepmother. Jessie also knew that Bessie—little Bep—had once been the closest to her father of all of his children, and upon meeting Bessie, Jessie could well understand why. Little Bep had developed into a woman of beauty, grace, and a quiet poise that her two sisters lacked. Bessie's husband, moreover, was a Huidekoper. His uncle Wallis Huidekoper had been Alfred's hunting companion and had done much to make Alfred's six months of exile in South Dakota bearable. Alfred was predisposed to like any Huidekoper, and Bessie's Reginald more than lived up to Alfred's expectations. The two men shared the same views on politics, and Alfred quickly came to have great respect for Huidekoper's intelligence, his even, sober temperament, and his success as a lawyer. Ever more frequently, Alfred turned to his son-in-law for advice, particularly in legal matters that involved his own family, since he knew that he could trust both Reginald's discretion and his sound judgment. It is hardly surprising, then, that Bessie and her husband were favorites, although they never asked for any special favors and never made any demands, as the others in his family did, upon Alfred's pocketbook. They were proudly independent, and the affection they showed for Alfred and Jessie was real, never fawning.

In the winter of 1930, when Jessie was suddenly struck by a stomach abscess and it appeared for a time that she would not survive, Bessie and Reginald were Alfred's constant companions and support. Jessie later wrote to Maurice's wife, Margery:

> Bessie and Reggie have just been in to see me. They have been perfectly lovely all through my illness and untiring in their affection and willingness to assist A. I. in any way possible. As soon as they heard I was going to be operated on they came right over to the Hospital, and sat up the greater portion of that night ... which was a great comfort to him. Then they would return every few days, and call up two and three times a day. I think they are true blue!
>
> Victorine and Elbert have likewise been very attentive and affectionate in every way, so has everyone, but Bessie's and Reggie's stand out of course. All of Bessie's early love and devotion to her Father, I think, is as keen to-day as it was in those years.[34]

Over the next few years, however, this warm, close relationship slowly became more distant and strained. Alfred continued to be in close contact with his son-in-law, Reginald, but Jessie's view of Bessie inexplicably changed. Perhaps Bessie was too independent and didn't ask for enough favors, or perhaps she, like her brother, remained on too friendly terms with her own mother. She resolutely refused to be drawn into the controversy between Marcella and Jessie and to become the purveyor of stories against her sister-in-law. Although there was no immediate open break between Bessie and Jessie—that would come only after Alfred's death—Bessie was replaced by her sister Victorine at the center of Jessie's affection. Victorine and Elbert lived close to Nemours, while the Huidekopers, living in Washington, D.C., were outside the immediate family circle. Victorine made it a point to stay in close contact with Jessie, stopping in for tea and sharing the family gossip.

Jessie's changing attitude toward Bessie and Victorine inevitably affected Alfred's own relations with his daughters. He wrote a letter to Madeleine in which he revealed the changed situation:

> I think your understanding of Victorine is practically correct. I have been in closer touch with her recently than with any other of your sisters [there was, of course, only one other sister—Bessie] and I find her [Victorine] most normal in every way, without any bad habits. She has herself thoroughly under control and is improving her husband rapidly. She is very companionable, very quiet, self-reliant, doesn't have to be entertained at all, and always sweet and affectionate. She is, as you say, more du Pont than any of the others, saving yourself, but much more so than Bep and Alfred, who of course, is a pure Gardner. . . . Confidentially, she doesn't seem to be able to get very near Bep, as she puts it, she is afraid of her.[35]

Alfred's characterization of Victorine is precisely what he earlier and perhaps more accurately would have ascribed to Bessie. Most of Victorine's friends would have been amused by her being described as "very quiet," and they would have found it difficult to imagine Victorine being afraid of anyone—least of all her sister Bep. As for poor Alfred Victor, there could be no more damning charge leveled at him by his father than to be called "a pure Gardner."

Then there was Madeleine, Alfred's first born—big, exuberant Madeleine, a woman of strong loves and hates—but mostly love, a Whitmanesque character who wore her heart on her sleeve and sought to embrace the entire world in an all-encompassing bear hug. Madeleine had opted to go with her father at the time of her parents' divorce, for she hated her mother with an intensity that equalled her father's, and this greatly endeared her to him.

Two months after Alfred's marriage to Alicia, Madeleine, at the age of twenty, married John Bancroft, the scion of the Bancroft family, whose wealth and social position in Wilmington predated that of the du Ponts. Madeleine's first marriage was brief and stormy. On the couple's wedding trip to Germany, Madeleine met a handsome young German ne'er-do-well who was as much attracted to her wealth as she was to his good looks and his courtly, hand-kissing European manners. According to later testimony given by her husband, Madeleine's acquaintance with the German youth, Max Hiebler, rapidly progressed from hand-kissing to something considerably more ardent. Bancroft hurried back to Wilmington while his wife remained in Paris. Nine months later Madeleine gave birth to her first son, who was christened John Bancroft III, although Madeleine's husband had considerable doubt as to the child's true parentage. When in the following year, however, Madeleine, back in Germany, again became pregnant, her husband no longer had any doubts about his wife's infidelity. He filed for a divorce, and in the messy contest that followed, the press had a grand time with the lurid details the aggrieved husband so obligingly provided in his testimony.[36] Bancroft got his divorce and custody of the first child.

Alfred, who had recently been subjected to the same scandal mongering by the press and public, should have been able to empathize with Madeleine, but both he and Alicia were righteously outraged by his daughter's conduct. For a brief time, he severed all relations with Madeleine. Following the birth of her second son, to whom she mischievously gave the first name Bayard, after another distinguished Delaware family, Madeleine returned to Germany with her second son. Two years later, she finally married her German lover, and had two more sons by him, Benno born in 1914, and Alfred in 1915.

It was not possible for Alfred to remain alienated from his daughter for long, no matter how shocking her conduct might be to him. During the war years from 1914 to 1917, Madeleine wrote frequent, heart-breaking letters about the food shortages and the desperate economic plight in which she and her three young sons found themselves. Alfred sent money and parcels of food and clothing until America's entry into the war in April 1917 closed all further lines of communication.

In the immediate postwar years of revolution followed by the cataclysmic inflation of the early 1920s, Alfred managed to keep in touch with Madeleine and through various business and diplomatic contacts kept her supplied with enough funds to feed and house her family. Her handsome Lothario of a husband had become a brutal, wife-beating tyrant once he had discovered that Madeleine was not a rich woman but was dependent upon what small allowance her father could manage to send her. Hiebler was reluctant, however, to consent to the divorce she wanted, for he clung to the hope of her eventual inheritance.

Through this time of *Sturm und Drang,* Madeleine kept her sense of humor and her zest for living. In her own careless, loving way, she was a good mother to her three sons. Her first child, John Bancroft III, she was not to see again from the time he was one year old until he was a young man of twenty.

There was never any suggestion from Alfred or Jessie that Madeleine should return to America with her sons, nor did Madeleine ever express any such desire. Munich was her home and her children's Vaterland. Difficult as things were for her in Germany, Madeleine found life there infinitely more pleasant than would have been her existence in the Brandywine valley among all the du Pont and Bancroft gossip mongers.

On his and Jessie's first trip to Europe together in 1922, Alfred saw Madeleine for the first time since 1914, and he was introduced to three grandsons whom he had never before seen. He found Madeleine to be the same exuberant and demonstrative daughter who had been his companion in exile when he left Swamp Hall to live at Rockland Farm. She had changed only physically. She now looked like just what she in fact was: a heavy, jolly German hausfrau, beloved by her sons and by all of her neighbors on Prinzregentstrasse, Munich.

Alfred was surprised and delighted to find that, in spite of a bad marriage and all the difficulties of war and economic collapse, Madeleine had been able to raise three very handsome, intelligent, and well-behaved sons. Alfred had to concede that, all things considered, Madeleine had managed her life surprisingly well. After she finally succeeded in getting a divorce from Hiebler by buying him off, it seemed that she would at last achieve the peace, security, and freedom she had never before known, and which Alfred devoutly wished her to have.

Madeleine, however, was always capable of springing a new surprise. Early in November 1930, Alfred received a letter from Hermann Ruoff, of Mannheim, Germany. The name on the envelope seemed vaguely familiar to Alfred, and then he recalled that Ruoff was the name of Madeleine's closest friend, of whose death the previous summer Madeleine had written her father.[37] He further recalled that Madeleine's letter had mentioned Frau Ruoff's youngest son, Hermann. This recollection, however, in no way prepared Alfred for the content of Herr Ruoff's letter.

> I wrote you nearly eight years ago my first letter to ask your help in finding a position in the States. At that time the conditions were too poor to realize this plan.
>
> Today I have a new—and more important wish, to beg your consent to marry your daughter Madeleine. . . .
>
> I made the acquaintance of Madeleine in 1913 and since this time she has been the best friend of our family. If you will look back in your correspondence to the spring of 1921, you will find among others a letter from my mother, which will tell you more what Madeleine has meant to us, than many words from me. The time has been long enough to judge each other's characters. I know fully well that Madeleine has had to face hard and difficult propositions and I propose to prove to her, as well as to those who care for her, that she has in me a faithful comrade and a good friend to her boys.
>
> I am younger than Madeleine—the greatest fault I have, she says— but the beginning was not easy. Almost a boy, I was a soldier and nearly two years in the trenches. In 1919 I went back into business and was three years later confidential assistant with power of attorney. I began my own business in 1924 as partner in an old and esteemed firm, founded 1883 and since then have made good headway. . . .
>
> I only mention this to prove to you that my affection for your daugh-

ter is based in ideal principals, [sic] and that the material, fortunately, does not have to be considered.

To be [sic] your daughter a loyal and honest husband, to respect and to protect her person, I will promise you with heart and hand! If you will give your consent, which I know also means everything to Madeleine, we have in view to marry soon, quietly.[38]

Hermann's letter was quickly followed by one from Madeleine:

My Boy is all right—and I know you will like him too. . . . I fear his letter will have mistakes—but he wanted to write you all to himself—so be indulgent. . . .

We will not be married before April—but then very quietly—and let people "explode" while we are away a few weeks.

Think of *me* having a husband who calmly says: "You leave your pocketbook at home those four weeks, my girl—for once you are going to learn that some other man can take care of you"—!

That certainly is a new experience, Dad.

Well I am tired—and happy and am going to bed—and cuddle up—and hope I'll have as much luck as you.

"There is luck in odd numbers—" only #1 was no good to either of us. The only luck I had there was you!! For your #1 I mean![39]

Both to Hermann's quite formal politeness and to Madeleine's joyful gushing, Alfred replied quite stiffly. To Hermann he wrote:

Naturally, I am somewhat surprised at the news [your letter] brings and I have very little to say on the subject, beyond the following:

Madeleine, of course, is old enough to decide a matter of this kind for herself. If I disapproved of the alliance, it would probably make no difference to her—she would proceed according to her own desires. . . .

I am not a strong advocate of married life as a means of securing happiness for the man or for the woman, and particularly is this view of mine true when the difference of thirteen years, which I judge is the difference between your age and hers, exists as an additional setback. However, you have made your own arrangements, I presume, regardless of my advice or wishes, and your request for my approval is purely a matter of policy and entirely superficial.[40]

To his daughter, Alfred was more pointedly severe:

I do not know the young man, and, if I did, being deaf and unable to speak his language, I could not arrive at any well balanced opinion as to what will be the outcome of such an alliance as you propose. Everybody must look out for themselves in this world. You are free, white and twenty-one, and, while I do not say you are fully able to judge for yourself in a matter of this kind, I can say you ought to be.

You are passing through what the poets call "The dangerous age," when ladies are prone to break loose and do all sorts of damn foolish things—not so with the men—they make up these messes somewhat later on. Thank the Lord, I have never made a mess of that kind without knowing what I was doing. My first marriage, as you know, was a boy's marriage, made on the impulse of chivalry, with the outcome you also know. My second marriage was made very largely for the purpose of protecting a sick woman, who was being persecuted by every human being who had every right to do the opposite. I failed in my first marriage and I failed in my second.

Jessie and I get along like a house afire, because she recognizes my rights as I do hers. . . . It is true that there was even greater disparity in our ages than there should have been, but it is not an unheard of thing for a man to marry a girl nineteen years his junior.

My fear is that you will not be able to keep your young husband interested after ten years have elapsed, although on that point I may be entirely wrong. My one object is your happiness and my solicitude is for your welfare alone. . . .

Now that I have blown my face off about marriage in general, I want you to know that your happiness will be mine and anything that makes you happy will make me happy also. You have nothing but my prayers and best wishes for the future.[41]

Madeleine replied that she had "often heard about the so-called 'dangerous age'—but I fear I am not there yet. Maybe there is a touch of 'manliness' in my makeup. . . . and I hope I have wisdom enough to avoid them as you have this time. Hermann happens to be the very rare sample of a European who allows a woman her opinion—and her rights. . . . I do not worry for a moment of having my own personality crushed—which so often happens here. . . . I hope to the Lord that I will have brains enough to keep my husband interested in me in ten years time!"[42]

Following that rejoinder, Alfred gave his blessing—or what had to count for a blessing—to the union: "I am sorry that I cannot be with you on the date of your wedding, as I am professionally a flower girl and with the right colored wig and the right length of skirts, I am sure I would do you credit. Of course, I carry nothing but sun flowers. Of course, we will be hoping and praying for you, particularly praying and I am trying to get in some advance work in order that my prayers may possibly be answered."[43]

Alfred must have been successful in his advance work on prayers, for this time Madeleine had chosen the right man. As for her father, #3 was her lucky number, and Madeleine was one of the few du Ponts of her generation of whom it could be said, "they were married and lived happily ever after." And in spite of his deafness and the language barrier, Alfred did come to know and to like and respect Hermann through the correspondence they established. One can judge his changing attitute by the salutations that Alfred used. They progressed in cordiality from the very stiff "My dear Sir" of the first letter, to "Dear Mr. Ruoff" after the marriage, and finally to "My dear Hermann," as indeed he was.

Prior to her third marriage, Madeleine had bought a large Biedermeier style house in the little resort town of Herrsching-am-See, fifteen miles southwest of Munich. Here she and her family were to spend most of their time. This was particularly true after Hermann joined them, for he, being a skilled carpenter, was able to renovate the old house and turn it into a showplace on the lakefront. Madeleine gave up her large apartment in Munich, and a small flat sufficed as the family's city base. In spite of the depressed economic conditions and the political turmoil that marked the last days of the Weimar Republic, these years were the most peaceful and happy period of Madeleine's life. Her relationship with her father and Jessie had never been more secure or more stable. With an ocean separating them and with her marital difficulties at an end, she and her father were able to establish through their correspondence a warm rapport that was even further enhanced by Madeleine's over-zealous support of Jessie in any dispute between her stepmother and Madeleine's siblings. Alfred also thoroughly enjoyed writing to and receiving letters from his three German grandsons, who, at their mother's urging, were faithful correspondents. Alfred actually felt closer to them than to any of his American grandchildren, partly, of course, because they were much older than Bessie's and Victorine's children. He was happy to accept his role as grandfather, sending them unsolicited advice on careers, marriage, and politics. Hermann and two of the boys, Bayard and Benno, were greatly alarmed over the increasing strength of National Socialism in Germany during the early 1930s, a sentiment with which Alfred was in total agreement. In April 1932, following the German presidential election, he wrote to his grandson Benno, "Delighted to see that poor old Pop Hindenberg [sic] came through successfully and consigned our mutual friend Hitler to his logical resting place in Hell. . . . The repudiation by the people of Germany of such radicals as Hitler will do more to help the situation in Europe than anything else."[44]

Unfortunately, Hitler was not to remain long in what Alfred regarded as his "logical resting place." He moved into the Chancellor's chair in January 1933, bringing with him the Hell to which Alfred had hoped the aging President of the republic had permanently consigned him. The new Chancellor had only one ardent admirer in the Hiebler-Ruoff family—Madeleine's youngest son, Alfred. He and his grandfather entered into a lively exchange of letters regarding *Der Fuehrer* of the Third Reich. Young Alfred wrote his grandfather two months after Hitler took office:

> We have now in Germany a very excited time, since Adolf Hitler is the German Chancellor. But, I do not know how you think about him. I am so happy he got Chancellor. . . . Our country was full of giddiness, everybody tried to overtake the others. And we Germans do not need Sovjet-Russia!! We do not need 'Marxisten'! And I am very happy that Adolf Hitler cleans our country of those people. It would be very interesting for you to see those Marxisten-people we have. There is only one word! Terrible!! But they are now very quiet. Their leaders are now in captivity. Many of them fled! And it is since years the first time that the German Reichstag has a national majority! 44% of the German people are Hitlerianer. And, I guess you still know it, I am too one![45]

Alfred replied to this enthusiastic outburst, "I don't know what to make of your friend Hitler. We don't get all the facts here, but I was so disappointed in his attitude towards the poor Jews, that I have come to the conclusion that he is one of these men that is too small for a big job and too big for a small job. If this analysis of the gentleman is correct, I am afraid that he will not last long as Dictator and Chancellor of Germany."[46]

To his daughter Madeleine, Alfred was even more blunt in his criticism of the new Nazi regime and its leader. Madeleine had written her father, "We have a political explosion (inasmuch as three parties are represented in the family) almost every 24 hours," with young Alfred as the one "ardent Hitlerite." Alfred clearly sided with the majority against his namesake: "I don't like at all the way the Nazis have been attacking the Jews. It isn't a question of whether you like them or not. I have too many in my own family to be desperately in love with them; at the same time, everybody must admit they are inoffensive citizens and do boom business in their particular line. I think that, considering Hitler's origins, he is not going to measure up to his job. I don't think he handled the Jew situation as a big man would have or could have. He ought to be above attacking them."[47]

Most of the correspondence between Alfred and his grandsons, however, was not of such heavy political stuff. Alfred wrote about his dogs, his yacht, and his various business activities in Florida, and the boys wrote about their grades in school, their small sailboat named for their grandfather's beloved yellow mongrel, Yip, and the Buick that Alfred had bought Madeleine and which was the family's pride and joy.[48]

After their marriage, Madeleine and Hermann Ruoff made two trips to the United States before Alfred's death in 1935. In his biography of Alfred du Pont, Marquis James asserts that on the Ruoffs second trip to Wilmington in 1934 "Alfred urged them to settle here. . . . Alfred sent the Ruoffs on a coast-to-coast tour of the United States in the hope that they would perceive the wisdom of throwing their lot and that of the boys with the land of Madeleine's birth." But, according to James, the Ruoffs "refused to desert the Fatherland. . . . With a heavy heart Alfred accepted the fact that the fates of his daughter and of three manly grandsons were in the hands of Adolf Hitler."[49]

James's only source for this story of Alfred's attempt to rescue the Ruoffs from "the hands of Adolf Hitler" is apparently an interview with Jessie du Pont several years later. Certainly there is no evidence in the du Pont papers that Alfred ever urged the Ruoffs to make their home in America. At least two members of the family—Hermann Ruoff and his stepson Bayard—would have been only too happy to accept such an invitation. There is, however, a great deal of evidence in the family correspondence to the effect that Bayard pleaded with his mother and his grandfather to be allowed to come to the United States and claim his United States citizenship. He was told quite bluntly that he was not wanted. Alfred did provide Bayard with a temporary job in his bank in Jacksonville, which allowed the young man to come over in 1933 on a one-year visitor's visa. Jessie was determined, however, that Bayard should not stay and ask for his citizenship papers, because she feared that this would resurrect all of the old scandal about his illegitimate birth and the details of the divorce suit brought against Madeleine by her first husband, John Bancroft. Upon the expiration of his temporary visa, Bayard reluctantly returned to Germany in 1934, but only after expressing openly both to his grandfather in Florida and to his mother in Germany his determination to return to the United States to claim the American citizenship, which he thought was rightfully his. Jessie, greatly alarmed over this declaration, at once wrote to Madeleine begging her to keep her son in Germany. "A. I. is in no condition to have all the scandals of the past published. He is entitled to protection now," Jessie emphatically asserted. She also wrote to Bayard himself, expressing the same sentiment.[50]

Madeleine, always eager to be helpful when she thought her father's well-being was at stake and knowing full well that Bayard's return would forever alienate her from Jessie, somehow managed to keep Bayard in Germany. Her son was to pay a very high price for respecting Jessie's hypersensitivity to the "scandals of the past." After being obliged to serve in the Hitler Youth Work Corps—the Freiwillige Arbeitsdienst, he was conscripted into the Germany army.[51] Bayard was destined never to fulfill his vow to become an American citizen. He died in February 1945, fighting on the side of the Nazis, whom he despised, against the country he loved and had wanted to claim as his own. Ironically, it would be his younger brother Alfred, the erstwhile ardent Nazi, who would survive the war and would eventually come to the United States, establish his American citizenship, and settle in New Jersey.

After Alicia's death in 1920, Alfred had willingly assumed the guardianship of her two children. Although he never legally adopted either Alicia's own daughter, Alicia Maddox, nor the French orphan child, Adelaide Denise, both children remained as dear to him as the four children he himself had sired. Between him and Alicia Maddox, in particular, there would always be a very special attachment. He was the only father the child ever knew, and she adored him. She in turn had served as Alfred's surrogate daughter in the unhappy first years after his divorce from Bessie when he had cut himself off from all contact with his own three children. By her very presence at Nemours, little Alicia had also provided one constant bond of affection between Alfred and her mother, Alicia.

In the years after his second wife's death, Alfred continued to give most of his parental affection to little Alicia. Even though he could only agree with all of the criticism that Jessie directed against this foster child whom she had never liked—Alicia's immaturity, her emotional instability, her failure to accept the responsibility that marriage and motherhood entailed, and her extravagant life-style—still Alicia always remained Alfred's darling Pechette, his one remaining tie to an unreciprocated past love which Alfred, try as he would, could never forget, and which Jessie, try as she did, could never obliterate. He supported Alicia throughout all of her unhappy romantic entanglements—her first marriage to Harold Glendening, her divorce and the bitter child-custody fight that ensued, and her second, equally unfortunate marriage to George Kent. Alfred accepted and tolerated all of these impulsive acts with a forebearance he would never extend toward his own children.[52]

Knowing how dear Alicia was to Alfred, Jessie felt obliged to remain on cordial terms with her, no matter how difficult she found the relationship to be. Her occasional letters to Alicia always bore the salutation "Dearest little Alicia," and concluded with "Dearest love and best wishes," but even someone far less perceptive than Alicia could read between the lines Jessie's true feelings. From her temporary exile in Panama where she had fled with her son during the custody fight with her ex-husband, Alicia had written to Jessie announcing her intention to return to the United States with her son. She asked Jessie to recommend a good private school in California where she could place the boy and where he would be safe from recapture by his father. Jessie was only too happy to give advice about schooling, along with some unsolicited counsel for Alicia:

> Of course, in California, you are out of the jurisdiction and away from the worry and trouble of that New York party [i.e., Harold Glendening]. . . . I trust that your stay in New York will be short, as I imagine that if Harold finds you are there, he will cause you serious

trouble. . . . It has just occurred to me that it might be within his power to have you arrested on a charge of kidnapping. . . . Don't concentrate on little Alan's strong will and temper. Both are excellent traits, if properly guided. I feel sure that within the next six months you will get him in the right environment. You know so little about children. I do not mean that in any sense to injure you, but being an only child and a child growing up without the association of other children, naturally you wouldn't know what most youngsters are like.[53]

In Jessie's opinion, Alicia was the unfortunate victim both of bad heredity (the product of a worthless cad of a father and a selfish vixen of a mother) and of bad environment (a child raised by uncaring servants and pampered by a much too indulgent foster father). It was small wonder that Alicia was what she was—a flighty, irresponsible child who would never grow up.

If Alicia was a hopeless case, Jessie could at least do something about the second foster child who had been bequeathed to Alfred by his late wife. Little Denise was not quite five when Alicia Bradford du Pont died, and she was not yet six when Jessie married Alfred and was perforce obliged to assume the responsibility of raising this child. Jessie, of course, did not know the heredity with which she might be contending in undertaking this task. Alfred had only the vague story of the child's origins that Alicia had told him—something about a French lieutenant father who had been killed in battle in 1914 and a young widowed mother who had placed the baby in an orphanage. Jessie might not be able to assess Denise's heredity, she could most decidedly control the child's environment, henceforth. Jessie was convinced that she could produce the model young woman that she wanted and expected Denise to be. Nurses and governesses were chosen with great care, discipline was strict, and indulgence or pampering was never considered to be an essential ingredient of parental love.

Although Denise's early childhood must have been lonely, with Alfred and Jessie frequently away for extended yachting and motor trips, the child nevertheless seemed to adapt well to this regimen, and her love for the only parents she ever knew was genuine and very meaningful both to her and to them. Five other children called Alfred father, but only Denise called Jessie mother.

At the age of fourteen, Denise was sent to a girls' school in Broadstairs, Kent, England. With understandable pride in the effectiveness of her parental guidance, Jessie wrote to Alfred's other foster daughter, Alicia, "She [Denise] has grown into the loveliest little girl I ever saw—thoughtful and sweet in every way. I know she is going to be a great little pal and companion and a joy to all of us. She is very clannish, very devoted to her family and is beginning to realize that there is nothing in life without love—real love—not love based on selfishness and that no one, regardless of who the person is—is going to put up with a display of temper."[54] If Alicia was perceptive enough to make a personal application of Jessie's little homily on the importance of unselfish love and a controlled temper, so much the better.

Alfred was equally proud of Jessie's handiwork. He wrote to Madeleine, "Denise returned on the 22nd [of December 1930, from her school in England], looking in fine condition, much improved in every way. She certainly is going to turn out to be the most creditable young woman and bring the greatest possible happiness into the lives of those who know her. She is returning on the 16th of January. Yesterday she went to New Haven to visit some people she knows there and was very much impressed with her own importance in travelling alone. She is very well able to look after herself and gets along splendidly with all people, particularly older people, to whom she is very sweet always."[55]

Jessie did not consider her parental duties at an end just because she had shipped Denise three thousand miles away to a boarding school in England. She issued a steady flow of letters reprimanding Denise on everything from sloppy penmanship to excessive expenditure on the rare occasions when the girls were taken to London for a dinner and the theatre:

> I notice, Love, that you are not as careful as you should be . . . in your letter writing. Two have commented on the number of misspelled words and the use of plural subjects and singular verbs. It creates a bad impression.[56]

And again:

> Incidentally these parties are very, very expensive. Just to give you an idea of what one evening in London cost, including theatre, supper, taxis, etc.—$40.00. Now Father and I do not spend money like this, Dear, and you were not entertaining. Had you had some friends to entertain, it might have warranted it, but certainly not just for yourself. I never spend that much on myself in one evening, and have never done it, and that is not all, I never expect to do it.[57]

Alfred always left the disciplining of Denise to Jessie. His letters were filled with praise for "such a nice letter came to me, which reached me on my return from Florida," and conveyed the news that Denise really wanted to have—about the family's three dogs and their continuing battles with the skunks, rabbits, and cats at Nemours:

> I wish you could have seen them [the dogs] this morning. As I was walking down to the pumping station, the Three Graces, running ahead

... failed to observe the greenhouse cat (new), which was sunning itself in the gutter by the side of the road, and so interested were they in getting down the road, each one ahead of the other, that they passed within six inches of the cat (which I think entitles the cat to at least two or three months more of life) without seeing it. The joke was, they all saw the cat when it was too late—clamped the brakes on, one dog rolling over the other in an attempt to stop, while I reached down and scooped up the cat. I tossed the cat to a nearby gardner, who tried to secrete it under his coat, while the three wretched hounds pranced around and tried to pull his clothes off! It was with difficulty that I saved the gardner, plus the cat! You certainly would have laughed your nut off, had you seen the mixup.[58]

In the spring of 1930, Denise spent Easter vacation in southern France, accompanied by a chaperone, Mrs. Ronald Sutherland, whom the headmistress of Denise's school, Mary Wolseley-Lewis, had selected. Shortly after Denise's return, Jessie was surprised and greatly alarmed by a letter from Miss Wolseley-Lewis informing her that, while Denise was in France, Benno Hiebler and a young friend had appeared on the scene. "Mrs. Sutherland did not allow them to be alone together and herself treated their relationship as perfectly natural," the headmistress reported. "But she thinks I ought to tell you that there is a certain amount of infatuation on both sides, that the boy talks openly as though they were likely to be betrothed and that Denise seems to be under that impression." Furthermore, Miss Wolseley-Lewis felt that Mrs. du Pont should know that Mrs. Hiebler "wrote Denise a letter (which she [Denise] showed to Mrs. Sutherland) in which she used rather strong expressions about the boy's great attachment to Denise and begged her not to disappoint him in anything he might ask."[59]

Greatly agitated, Jessie responded immediately to the headmistress, thanking her profusely for relaying this information:

I was shocked at the contents of the letter that Mrs. Hiebler wrote to Denise. Please see that Denise and Benno meet no more. Mr. du Pont and I are both absolutely opposed to any such ideas being fostered regarding Denise. She is just a little girl and we want no such silly notions put in her head regarding betrothals and matrimony. The child has not the slightest idea what it means, but no doubt if Mrs. Hiebler and Benno could have Denise off to themselves they might quickly educate her. Then again we had at no time considered an alliance with a German for our daughter.

Please, Miss Lewis, accept my deepest appreciation for writing me regarding this matter. I shall not mention it to Denise, of course, nor to Mrs. Hiebler, but shall endeavor to see that Denise forget [sic] the boy entirely in her summer vacation while she is in school with you.[60]

Poor Madeleine—in spite of all her efforts to curry favor with Jessie, she had with her usual impulsiveness and ineptitude once again committed a social blunder that Jessie was certain to find unforgiveable. Her clumsy attempt to effect an alliance between her son and Denise cooled relations considerably between Nemours and Herrsching-am-See for some time.

Jessie might endeavor with some measure of success to see that Denise forgot "the boy entirely" but she could not ensure that the boy would forget Denise. In the following year, Benno appeared once again, at the place where Denise was staying in Paris during her spring vacation, and it took all of Mrs. Sutherland's considerable skill as a duenna to make sure that the two young people did not meet except in her presence.

After this second rendezvous, Jessie decided that, much as she liked Miss Wolseley-Lewis's boarding school, it would be far better to have an ocean separating her foster daughter from her youthful and overly zealous German admirer. For the next school year, Denise was enrolled at the Emma Willard School in Troy, New York. Thus in addition to the "past scandals" surrounding Bayard Hiebling's origins, Jessie had another good reason for making sure that the Hiebling-Ruoff family remained in Germany, a safe three thousand miles away. Jessie wanted no part of du Pont scandals—past, present, or future.

It could never be said that Alfred du Pont's two secretaries, M. E. Brereton in Wilmington and Irene Walsh in Jacksonville, did not fully earn their monthly paychecks, for Alfred was an energetic and industrious correspondent. His total deafness ruled out the telephone as a means of communication for him. He could not even depend upon ordinary conversation with the business associates whom he encountered daily in his own suite of offices. Written memos and letters had to substitute for the customary personal appointments and conferences that most businessmen utilized.

In addition to the heavy burden of business correspondence that Alfred's handicap imposed upon his hardworking clerical help, there was also the very extensive personal correspondence which Alfred conducted not only with the members of his immediate family—his children and grandchildren—but also with such favorites among his extended family as his cousin, Ellen La Motte, his two nephews, Maurice and Cazenove Lee and their families, Jessie's two sisters and her brother Tom in California, and even some of Jessie's more distant kin, such as her favorite aunt, Ella Haile, and her two cousins, the Tyler sisters in Tappahannock, Virginia. Alfred might dictate as many as ten or twelve letters to relatives or friends in a single day, and these were not brief notes of greeting, but three- or four-page epistles in which Alfred not only detailed his and Jessie's recent activities but also pronounced judgment on the great issues of the day—

Franklin Roosevelt's New Deal, the current condition of the nation's economy, the necessity for increased world trade, which in his opinion, could
only be achieved by cancelling all war debts and reparations, Hitler's Third
Reich and the danger it posed to world peace, and the current low moral
standards that prevailed among Delaware politicians.

Alfred especially enjoyed writing to young people, giving his grandnephews, grandnieces, and grandchildren a running account of his three
dogs and their ever running battles with each other and with all other livestock, both domestic and feral, on the Nemours estate. These letters were
also the soap boxes from which Alfred could lecture his youthful correspondents on proper health habits, urging them to exercise more, eat less, and
abstain from "booze and coffin nails." He also repeatedly warned them—
particularly the young girls with whom he corresponded—on the dangers
inherent in matrimony. Some thirty-five years before Betty Friedan, Alfred
decried the same feminine mystique which assigned to young women the
role of getting a man—any man—and becoming his slave for life. His letter
to the daughter of his old friend Dr. Charles Hanby is typical of the many
strictures he issued about marrying in haste and repenting at leisure:

> If I were a young woman and had sufficient brain power and physical
> energy to put on my own clothes and go to work and earn enough food
> to keep my stomach from becoming deflated, I would like to see the
> man who could inveigle my slats into his embrace. This is not against
> marriage as an abstract proposition. Marriage is more or less a necessary
> evil and may result in a most enjoyable comradeship; but most girls grab
> the mule by the wrong end, and the husband must be some beautiful
> pink face bunch of slickum . . . without any fixed source of revenue, and
> as a rule entirely without any appreciation for a woman of merit. . . .
>
> I do not know whether you have ever thought of inviting some Wee
> Willie to share your breakfast table; while he sits and reads the news-
> paper and smokes a cigarette, you pour out his coffee for him and slash
> the hen fruit over the edge of the cup for him. My advice to you, if in
> moments of temporary mental aberration you have such ideas, is to put
> them aside and go in for something really interesting. Married life is not
> interesting; it is prosaic, horribly so, hard work, Hellishly hard, thankless
> work, not a damn thank you. Not a nice way to wind up a picture but
> more truth than fiction. . . . Do everything you can to maintain a very
> high standard of physical excellence, which always reflects to a large
> extent one's mental condition, but above all, seek interesting occupa-
> tions . . . and, as a final injunction to anyone seeking happiness and con-
> tentment, cultivate your sense of humor.[61]

The youthful recipients of his counsel were always polite and wise
enough not to argue with him, whether or not they ever accepted his

advice. One correspondent, however, was not at all hesitant in sharply contradicting his pronouncements and in return telling Alfred just what he should or should not be doing. The liveliest exchange of letters Alfred had was with his sister Marguerite du Pont Lee. Marguerite had always been a liberated woman, and she needed no man—particularly no younger brother—to tell her how to conduct her life. Although she had married at the youthful age of nineteen a man twelve years her senior, she had never allowed matrimony to interfere with her major interests in life, which were many and varied and did not include simple housekeeping but did include water color painting, the collection, writing, and publishing of "true" ghost stories, and above all, attending to her social welfare work in the nation's capital. As far as she was concerned, her husband, Cazenove Gardner Lee, had fulfilled his one necessary function when he had sired her two sons. Beyond that, he was only a shadowy background figure in her life. She hardly noticed when, after eighteen years of marriage, he quietly departed from their home and took up permanent residence in a rooming house in Washington, where he lived until his death in 1912. Marguerite had no desire to dignify—or worse yet, to sensationalize—this inconsequential act of separation by filing for a divorce.

Like her grandmother, Margaretta Lammot du Pont, whom she so closely resembled, Marguerite was another strong-willed, strong-minded du Pont woman who had assumed the role of grande dame of the family, the one who took most seriously the family motto, *Rectitudine Sto,* and the solemn injunction of her great-great-grandfather Pierre Samuel du Pont de Nemours that "no privilege exists that is not inseparably bound to a duty." Duty was Marguerite's favorite word, and she saw herself as the very embodiment of William Wordsworth's ode to that particular virtue, for she thought herself to be the self-appointed

> Stern Daughter of the Voice of God! . . .
> Who art a light to guide, a rod
> To check the erring, and reprove;[62]

To the best of her ability, considering her limited resources, she did, with her Georgetown settlement house and school for adult training, provide "a light to guide," and with equal vigor and enthusiasm she used her tongue and her pen as "a rod to check the erring, and reprove," particularly her brother Alfred. In spite of her deep faith in her Episcopalian God, Marguerite found it hard to accept the inscrutable workings of His Providence, which gave so much material wealth to her brother to waste on yachts and mansions and harebrained business ventures while she had to beg for a few crumbs from that Lucullan feast to feed the hungry and clothe the naked.

As Alfred grew more wealthy, Marguerite grew more imperious in her demands. In 1930, Marguerite wrote Alfred asking if he would give her a small lot in Washington that he had purchased several years before so that she could use it as a recreational ground for her settlement house. He had answered that inasmuch as all of his assets were now a part of the Almours Corporation he could not on his own deed over the property to her without proper compensation to the Almours Corporation. She responded with her usual tartness:

> I want to make clear to you that when I requested a gift of the 50 × 150 feet of land at the Nation's Capitol [sic], I was under a misapprehension as to true conditions. Had I understood conditions as they *are,* of course I would never have made the request. I *knew* that you had purchased that property and paid for it with your own money. I was *not* aware others had a lien upon it. In your letter of July 28th you elucidate by saying in part—"Far from your belief in my ability to produce this piece of mud by raising a wand, sine cash expenditure, the property belonged to a Corporation and it was therefore among its assets and it is impossible to remove any of these assets without proper compensation, unless I desire to spend a certain portion of my remaining days in jail for embezzlement." That your brother-in-law has the power to throw *you* in jail, should cause food for thought, and 'tis surely a far cry from driving a grocery wagon in California to assigning a multi-millionaire to a dungeon in Delaware!—[63]

Marguerite was in error when she referred to Ed Ball as a former grocery wagon driver, and Jessie, the assiduous reader of and commentator on Alfred's correspondence, did not fail to point out to her husband his sister's mistake. "He never drove a grocery wagon in his life!" was her sharp marginal notation on this letter.[64] Marguerite unfortunately had randomly selected almost the only occupation that Ed Ball had not pursued during his busy life. Nevertheless, Marguerite got the property she wanted, for Alfred, in spite of his usual protests, once again acceded to his sister's request and sent her the deed for the lot after he paid the Almours Corporation the assessed value of the land.

Marguerite was as defensively protective of her one brother, Maurice, as she was offensively aggressive toward her other brother, Alfred. Putting more trust in the old adage that "God helps those who help themselves" than she did in the beatitude that "the meek shall inherit the earth," she constantly demanded from Alfred support not only for her own charitable enterprises but also for her meek brother Maurice and his family. When on one occasion Alfred asked Marguerite bluntly what more she thought he ought to be doing for Maurice than he was already doing if "you stood in my shoes," Marguerite had a ready answer: "Had I your tremendous income I would make a trust fund of five hundred thousand dollars for my only brother's benefit, for his life, the *principle* to revert to Nesta [Maurice's daughter] at his death."[65] Alfred's response, that Nesta could not be entrusted with such a large fortune, was a blunder, as he should have realized, since he had implied that a woman could not handle money as well as a man. Marguerite lost no time in setting him straight.

> Your criticism is that did 500,000 fall into Nesta's keeping, she would soon be in the poor house; the inference being that men, not women, have the god-given acumen necessary unto successful invest-ment of money; which reads a bit poorly, this side, just now, inasmuch as four prominent bankers, on account of "poor investments" were last heard of by me shackled together . . . headed for Atlanta! . . . Of course if a lack of investing gray-matter is your fear, the Trust *could* extend from Maurice to include the years she is due to remain on this sphere of experience. Do not for one moment suppose that any of your "ben-eficiaries" in whose veins course the same blood as in your own, are either dissatisfied with the benefits received, nor are they passing their time indulging in irridescent hopes for the future. They do not count their chickens before they are hatched. . . . I hope in the future you will realize that none of your blood relations . . . are planning attacks upon your strong box. *This* I *can* say, and say *most emphatically*—did I possess an income flirting around the three million mark, and felt that 2400 gross was all I could afford to give yearly to my own brother, *Hell would freeze over* before *any one* with one drop of blood alien to that coursing through my veins would get *one copper cent* from my pocket, so help me God! I quote from your letter—"I again desire to make this statement. If there is any 'good thing which this world has to offer' to its incum-bents which either your or Maurice have not, or which you could not have if you so wished, I shall be happy for the information." In reply there is one thing as certain as are taxes and death: of these "good things" from Maurice you will *never* hear. I am fashioned in a coarser mould. All my life, I have always wanted, and shall always want, until the earth covers me in the Congressional Cemetery, *Money* and *plenty* of it! Not for myself, but in order to lighten life's burden for many of my friends, and to bring the light of pleasure into many places now very dark. *It was not to be!* With love, Affectionately your sister
>
> Marguerite du Pont Lee[66]

Marguerite was fully aware of the fact that Jessie read and offered writ-ten comments on Alfred's incoming mail, and she was not hesitant in let-ting Alfred know that she knew: "You write to me at considerable length, and very confidentially about Maurice and his intimate personal affairs, knowing no one other than myself sees what you have to say. You forget that you are not similarly situated; and therefore it is not possible for me to go into private and intimate family matters with you in a letter. . . . As I said to a married man whom I favor with an occasional letter, and whose wife opens mine: 'My dear boy, a family letter cramps my style!' I do know that business men in N. Y. and other places say 'There is much Mr. Alfred I. du Pont should know, but there is no way to approach him *confidentially.*' Relatives and acquaintances have said la meme chose. How much I could tell you, astonishing! How much vastly diverting!"[67] If letters like this were the products of a style cramped by Marguerite's knowing that Jessie was a diligent reader of all of her husband's correspondence, then Alfred and Jessie dreaded to think what might be produced by an uninhibited Marguerite.

Marguerite had for so long billed herself as the moral conscience of the du Ponts—the rigid rod always quick to chastise the erring—that it was inconceivable for any member of the family to imagine that any scandal could ever be attached to her name. Alfred, to be sure, had frequently teased his sister about her many ministerial "boy friends," but this had been done in jest as a response to her frequent requests for contributions to her several church projects.

In the mid 1920s, however, Marguerite began to concentrate all of her attention upon one young Episcopalian minister, the Rev. William T. Reynolds, whom she had first met when he was an impoverished seminary student. She had been so impressed with his religious zeal and his commitment to social service that she had helped finance his education, and upon his graduation and ordination, she had secured for him the pastorate of Grace Episcopal church in Georgetown, of which she was a member. The Rev. Reynolds soon exhibited, at least to Marguerite's satisfaction, remarkable powers of healing by the laying on of hands, a talent rare within the conservative Episcopal diocese in Washington and one usually attributed to the more radical fundamental Protestant ministry. Marguerite became increasingly aggressive in her demands upon Alfred for financial assistance for her protégé so that he might extend his ministry of healing those suffering illnesses of body and soul.

With a great deal of grumbling, Alfred acceded at least partially to his sister's requests. He was convinced that Reynolds was just one more charlatan "imposing on Marguerite's good nature," as he stated in a letter to his brother Maurice. "There is nothing I would not do for Marguerite. . . . This you know well enough. But to give Marguerite more money than she actually needs for her own comfort and expense merely means that she gives it away to someone else and there is no limit to the sum which may not be disposed of in this manner. Parson Reynolds will take everything he can lay his 'flippers' on, and in spite of the fact that I contribute over $4,000.00 a year in order that his stomach may not collapse entirely, he still had the nerve to try and tap me for an additional $2,500.00 for the specific purpose of paying his debts and advancing the salary of his chauffeur. Now, if you can beat that for ecclesiastical nerve, I would like to have a look at it."[68]

Marguerite had been the supplicant for so many other ministerial enterprises over the years that even though this particular pastor had shown more gall in making his demands than had the others, Alfred never suspected that his sister's interest in the Rev. Mr. Reynolds was anything but eleemosynary. It therefore came as a great shock when Alfred in 1928 learned that Marguerite was embroiled in a nasty slander suit brought against her by Reynolds and a wealthy real estate operator, Mrs. Maude E. Ford. It was revealed that the good reverend, dissatisfied with the yearly allowance that Marguerite had provided him, courtesy of Alfred du Pont, had found another patroness whose purse strings were not held by a tight-fisted relative. It also became apparent that Marguerite's love for her pastor was far more libidinous than it was Christian agapeic. Believing that he had jilted her for a younger and more attractive woman, Marguerite had struck back with characteristic aggressiveness. At the very fashionable St. Paul's Episcopal Church in Washington Circle, to which the Reverend Mr. Reynolds had been invited to conduct one of his famous healing services, Marguerite had suddenly arisen from her pew and in a loud voice had proclaimed to the startled congregation that the great healer was intoxicated. She then had pointed an accusatory finger at Mrs. Ford and denounced her as the temptress who had offered up something a bit stronger to the officiating minister than sacramental wine. Needless to say, the healing service was abruptly turned into a wounding tumult. Shortly thereafter, the Reverend Mr. Reynolds brought suit for $100,000 in damages to his reputation, and Mrs. Ford, more modestly, asked for $50,000 compensation.

Reynolds subsequently dropped his suit and, pleading poor health, resigned his ministry at the Grace Episcopal church. Mrs. Ford, however, was made of sterner stuff, and she persisted in her suit. Alfred tried to counsel Marguerite on what should be her proper response to the legal contest:

> As I understand it, the woman is suing you for libel and anything aside from this issue cannot be aired in court. Mrs. Ford's real estate transactions are not pertinent to the issue; nor has Mr. Gardner's [W. Gwynn Gardiner, Mrs. Ford's attorney] character anything to do with it.... Whether Mrs. Ford is a real lady or otherwise is of no interest to the Judge. What interests the Court is what you said about Mrs. Ford.... On the other hand, Mrs. Ford, in order to collect damages, must be able to show them—in other words, that the statements purported to have been made by you would actually cause her damage, which can be measured in dollars and cents.... As I understand it, you do not deny having made the statements. Your position would be that

they were not made for the purpose of injury and not intended to be heard by the public or brought to the public's notice. . . . Nothing, as I see it, can be gained by your going on the witness stand. . . . Your lawyer, if you employ a lawyer, should have a free hand, not restricted by you.

You are not conversant with legal procedure or you would not have made the remark [you did] about Gardner [sic]. He is not being tried, and, therefore, he is immune from personal attack. . . . The same can be said of Mrs. Ford. It is not she who is being tried and any attempt to make insinuations derogatory to the lady's character will not get very far.[69]

Wise counsel, but Marguerite, as Alfred should have known, found it totally unacceptable. Only binding, gagging, and incarceration could have possibly kept Marguerite from testifying when the case finally came to trial in the late fall of 1930. The witness stand was for Marguerite precisely the pulpit she wanted from which she not only proudly boasted that she had made the statements she was charged with but also hurled further defamatory accusations against Mrs. Ford's character and that of her lawyer. Marguerite even produced two witnesses who testified that they had been in the company of Reynolds and Ford when both were highly intoxicated.

Mrs. Ford, in turn, produced a bundle of love letters that Marguerite had written to Reynolds over the years. These billets-doux were quite as extraneous to the issue at hand as were Marguerite's personal attacks upon the character of the plaintiff's lawyer, but nevertheless they impressed the jury and delighted the press, for they were vintage Lee epistles. They showed Marguerite at her most uninhibited best, since they were obviously written in blissful ignorance that they would also be shared by a third party, like those written to Alfred. All of the letters were signed "Your Stormy-Petrel," and with this closure, at least, the jury and the general public could find no fault.

On the 17 December 1930, the jury found that Mrs. Lee had indeed slandered Mrs. Ford by her remarks publicly voiced at a church service and had damaged the lady's reputation to the extent of $4,000.[70] On the same day that Marguerite received this unwelcome news, she also received Alfred's annual Christmas check of $1,000. She promptly sent off her brand of a thank-you letter to her brother:

Thank you for your Christmas check. Yes, I can 'put it to good use for Christmas.' Mrs. Ford was awarded 4,000—the usual result when a self support (?) woman is off set by a Capitalist (?) I must pay my lawyers 2500, costs and find a 4000 bond as we have noted an appeal. I am told that a *great deal* of surprise is expressed in legal circles that I was not represented by more experienced counsel. I feel sure the *splendid* Judge wondered were those *boys,* with whom he continually became irritated, the *best* du Pont money can obtain! The Judge does not know du Ponts. You will recall long ago, my reply to you saying I "must have a smart lawyer," that I was employing all I could afford to pay. You had ample funds to salvage your brother in law, when gambling in stocks he got squeezed; you have ample funds for automobiles, houses, jewels, fine clothes and trips to Europe for your sisters in law . . . but when your only sister was faced by a slander suit for 50000 you did not even put your lawyers at her service, nor offer *her* any financial assistance. I may be wrong, but to me the tragedy lies not in the suit, but in the brother. I thank you for your present.

Affec. your sister
Marguerite du Pont Lee[71]

To Marguerite's complaint that because she did not have a competent lawyer the outcome had somehow become Alfred's fault, her brother replied rather gently, "I am sorry you haven't got a half decent lawyer, but, as the Illinois miner remarked to me one day down in the depths of mother earth, apropos a conversation on powder, 'Mr. du Pont, I have always used du Pont powder and God knows the best is none too good.'"[72]

Alfred was more forthright in stating his feelings about this unhappy incident to Maurice:

> Marguerite's case was lost, not by poor handling but, for the simple reason that she had no case. The Ford woman accused Marguerite of making certain statements derogatory to her character in public. Marguerite admitted she did it; thus the suit was settled before they went into Court. It was merely a question of how much damages she had to pay. . . . Marguerite should know, by this time, that one cannot utter scurrilous statements about people with impunity. . . .
>
> Marguerite notified me in a letter, upbraiding me and making untrue statements about Jessie, Ed Ball and other members of her family. . . . She is naturally angry at losing a case, which was already lost before it went into Court, and I am the natural one upon whom she could vent her anger. However, this is not a new experience for me, so I really don't mind.[73]

It is difficult to believe that Jessie, had she been answering as well as reading Alfred's incoming mail, would have been as forebearing as he in responding to Marguerite's attacks. Her sister-in-law's ill-tempered outburst against Alfred was, however, a small price to pay for the satisfaction of seeing the family's most censorious critic herself publicly censured. Here was one du Pont scandal that Jessie had not been averse to having aired in the press. Marguerite's day in court, moreover, helped to counterbalance a minor court suit that Jessie herself was confronted with at this time. In December 1929 the ex-wife of Jessie's brother, Tom Ball, brought suit in the Duval County Circuit Court of Florida against her former sister-in-law for having maliciously interfered in her marriage with the purpose of alienating her husband's affection. Mary Ball asked for one million dollars in damage. Fortunately, Jessie had at her disposal the best legal service Alfred could obtain, and the suit was finally settled out of court in 1934 with a payment of a paltry one thousand dollars to the plaintiff.[74] Even more fortunate for Jessie was the fact that the entire affair was kept out of the newspapers. Marguerite never had the satisfaction of knowing that Jessie had also had her own legal problems with an angry woman who had sought to obtain a piece of the du Pont money.

In spite of Marguerite's sharp words directed against her brother and her constant demands that he do more for her charities and for his own family, Alfred never took umbrage at her sharp criticism. Her explosions were simply her way of blowing off steam and, as he wrote to Maurice, "this is not a new experience for me, so I really don't mind." Marguerite, in turn, dealt with his frequent lectures on proper health habits, his cynical remarks about her good works, and his teasing gibes about her parsons by simply ignoring them. The sibling rivalry of their childhood on the Brandywine persisted into old age. So also—and this was of utmost importance to both of them—did the affection that had bound them together since the death of their parents. They would always be dependent upon each other, even while boldly asserting their independence and their disdain for the other's views and life style. Theirs was an alliance that no one—not even Jessie—could ever understand, let alone sunder.

A by-product of their attachment was the very close bond that existed between Alfred and Marguerite's two sons—Cazenove, Jr., and Maurice du Pont Lee. In many ways Alfred had always been the two boys' surrogate father, inasmuch as their own father had played an insignificant role in a household dominated by Marguerite. Alfred served as the boys' male role model in their youth—the heroic figure of the Brandywine black powder mills, inventor and production chief *par excellence,* who had single-handedly saved the Du Pont Company for the du Pont family; the progressive reformer who had cleaned up Delaware politics; the man who had dared to lay his future on the line by challenging Cousins Pierre and Coleman. Even defeat in that struggle had added to his stature, for the halo that the rebel martyr earns in a just cause will always outshine the laurel crown of the undeserving victor. Marguerite might be an outspoken critic of Alfred's shortcomings, but to her two sons he would always be the valiant hero of the du Pont family. Especially was he so to Maurice, who, at Alfred's invitation, had joined the Du Pont Company in 1908 after obtaining his degree in mechanical engineering at Cornell University.

When the great battle of 1915 between Pierre and Alfred for the control of the company started, Lee was assistant manager of the Smokeless Powder division in Wilmington. He was one of the favored members of the du Pont family who were invited to exchange their stock in the Du Pont Powder Company for stock in the newly created Du Pont Securities Company, which was worth 50 percent more than the old stock. Maurice took advantage of the offer, but his loyalty to Alfred could not be purchased. When the time came, after the court decision, for the stockholders to vote on whether the company would purchase the Coleman du Pont stock that Pierre had grabbed for himself and a few associates, Maurice, as he later wrote in his memoirs, "was the only bonus beneficiary of the Du Pont Company, so far as I know, who voted for him [Alfred du Pont]. . . . Alfred was asking for nothing for himself—only for his employees. On the evening of the election I got into a crowded elevator in the Du Pont Building with Irénée du Pont. In a loud voice, he turned to me and said, 'Maurice, I am sorry for you,' and I replied, 'Thank you very much, Irénée,' From then I knew I could not expect very much advancement in the Company. In spite of this feeling I decided not to resign and by hard work tried never to give cause for criticism."[75]

Over the next thirty-five years that Maurice was an employee of the Du Pont Company, he continued to pay the penalty for having sided with Alfred in the great family battle, but he never regretted having made that decision. He wrote in his memoirs:

> On January 10, 1916, the Board of Directors did not re-elect Alfred du Pont as Vice President and member of the Finance Committee of the Company. He was thus separated from the Company he had once saved from extinction as a du Pont property and which he had served longer and with as little self-interest as any du Pont then living. This cut him to the quick—to think that his family and every other employee of the Company (all of whom owed him their positions and wealth) should treat him so when the court of the land had said that he was correct in his contention. . . . [I]s this what is called gratitude and fair play?[76]

In the years that followed Alfred's ouster from the Du Pont Company, Maurice continued to demonstrate an unflagging loyalty to his uncle. His frequent letters kept Alfred informed of all the new product developments, such as synthetic fabrics, dyes, paints, and cellophane, as the company diversified its production. Much as Alfred appreciated this information, it did not provide the company's second largest single stockholder with the complete picture of the Du Pont administrative policies and operations that he deserved but with which Maurice, unfortunately, could not provide him. Late in 1930, Maurice decided to do what he could to remedy that situation. He first called upon Ed Ball and broached the subject of getting Alfred back on the Board of Directors. Maurice told Ball that if Alfred was amenable to the idea, he (Maurice) would present the idea to the powers that be. If because of his deafness Alfred felt that he could not himself serve effectively on the Board, then Maurice was willing to suggest that Alfred appoint another person to represent him on the Board.

Ball was clearly excited by the suggestion, but apparently immediately suspected that Maurice was angling for the appointment himself, for that suspicion was what he reported to Alfred. Immediately after Maurice's conference with Ball, Alfred wrote to his nephew:

> I had a talk with Ed Ball and he referred to a suggestion made by you that I take some step looking to representation on the Board of the du Pont Company and that you represent me in that capacity.
>
> This would be an ideal arrangement but I doubt if it is practical. Don't forget that the Tenth and Market gang hate me worse than the Devil hates Holy Water (I am the Holy Water). (They are the Wholly Devils!) and unless they took to it kindly, the mention of your name as a possible representative of my interests as a stockholder would not redound to your benefit. I trust you see the point.
>
> If there is any way of feeling out Lammot, for instance, as to just how such a proposition would be considered without mentioning your name, it might be well to follow it up, provided they all did not die in a fit at the mere thought, in which case I should naturally like to have a Release and not be responsible for their funeral expenses.
>
> My private opinion is that they would consider it a ruse on my part and that I would use such knowledge as I would acquire by virtue of such representation for the purpose of further attacking them in the Courts. A guilty conscience needs no accuser.
>
> I have often given thought to such a plan but have always put it from me for the reason that they are all too small and too guilty to desire representation on the Board by anyone they cannot directly control.[77]

It was several weeks before Maurice had the opportunity "of feeling out Lammot," but on 27 February, Maurice did get an appointment with Lammot and broached the subject. The next day he reported the results of that interview to Alfred:

> I have considered all you said on the subject quite carefully and finally concluded the best thing to do was to go direct to Lammot myself which I did yesterday.
>
> Told him I had been sitting around for some twenty two years doing chores for the Company and had formed some conclusions of my own. If he saw fit to use the suggestion please not to mention my name to you since I did *not want you* to feel I was butting in on your affairs. Did so in order that he would not think you had sent me up there.

I told him I thought you should have a representation on the board, some one who would have the liberty of talking over with you the Companies [sic] affairs and mentioning interesting items to you. That I thought it would be a most graceful act and would without question be to the benefit of the Company, the community and the State. He said that you were most always kept informed of important questions such as stock issues etc. and that *he had never thought of giving you representation* but it was worth considering provided some one of ability, common sense and level headed were the representation. . . .

Have an idea I will hear more on the point later. If he speaks or writes to you do not intimate that I have had any conversation with you on the subject. Have had some fun working the thing out any how whether anything comes of it or not.[78]

Maurice also took care to set Alfred straight about his own reason for promoting the proposal: "Ed [Ball] made one mistake in his presentation to you. I did not put myself or in fact anyone else up as a proper candidate but merely asked his opinion as to whether you would benefit by representation. However, all this is immaterial."[79]

If the question as to who would represent Alfred on the board was really immaterial to Maurice, it certainly was not to Ed Ball. The latter remained suspicious that Lee had an ulterior motive behind his offering his service as mediator in the negotiations. Ball was determined that if anyone should represent Alfred on the Board, it would be he. Better to have no representation than to have a du Pont move into the inner circle in the superintendence of Alfred's affairs.

To Maurice's letter, Alfred responded rather cautiously:

Your interview with Lammot was interesting in a way and I want to thank you for the affectionate thought which was behind your desire to interview him on the subject in question. I am afraid that while he was naturally desirous of being polite to you, that he has not the slightest intention of giving the matter any further consideration. . . .

His statement that I was always kept informed of important questions is, to put it barely, an exaggeration. He has never informed me about anything, except when the recent stock was sold to stockholders at 80, when they were naturally desirous of getting my co-operation and the pourboire of a million and a half dollars, in the way of a subscription to the new stock. That, I think, is the limit of confidence to which they have treated me.

I think he told the truth when he said that he had never thought of giving me representation and I am confident he will never think of it again. . . . They will never divulge one iota of the Company's doings, if they can prevent my receiving it; however, you have my thanks and gratitude for taking the matter up, regardless of what the results may be.[80]

If Alfred was confident that Lammot would never think of the matter again, he himself continued to think of it. After doing some calculation on the extent of the Almours Company holdings in du Pont stock, Alfred wrote a memo to Ball stating that the current holdings were "now 582,000 shares—5.26 percent of the total 11,065,710 shares. This should entitle us to 1.9462 directors on the Board."[81] Alfred would, of course, be happy to settle for one director, but clearly now the ball was in his court. Maurice had done as much as he could in getting the game started.

After waiting fifteen months for an overture from Lammot, Alfred finally decided to broach the subject himself. He did it somewhat indirectly, by writing to Lammot to ask for information on several topics: the amount of life insurance the Du Pont Company carried, if any, for its officials, directors, and employees; a list of all of the stockholders of the Du Pont Company; the extent to which "the salaries of the higher salaried officials have been reduced . . . as compared to the reductions made in the salaries of the lower paid officials," and finally, any definite policy the company may have adopted "relating to the encroachment upon the Company's surplus in the interest of the Common Stockholders." "It seems to be a propitious time," Alfred wrote, "for all stockholders, particularly the larger ones in the Du Pont Company to take a more personal interest in its affairs and to seek proper information. With this end in view, I expect from time to time to ask you certain questions which you can answer or not as you see fit, but such questions as I shall ask I am confident you will agree, as would any impartial one, are proper and pertinent."[82]

Lammot responded quickly and courteously to this letter, providing Alfred with complete information on all of the questions he had raised and expressing pleasure "to note that you are taking a personal interest in these matters." Lammot then pointed out that he had noticed at the time of the annual meeting that the Almours Corporation had not sent in its proxy vote on the large number of shares that it owned, and which Lammot understood, Alfred controlled. "At the time I had in mind writing you inquiring whether any dissatisfaction with the management had caused the withholding of this proxy, but concluded I had better not single out any small group of stockholders to address such an inquiry. I would appreciate it if you would let me know whether there was any dissatisfaction on your part, or whether the failure to execute the proxy was merely an oversight, perfectly natural in case of stock being registered in the name of a corporation."[83]

This reference to Alfred's failure to vote his shares gave Alfred precisely the opening he had been waiting for. He thanked Lammot for answering so promptly and "quite fully the queries which I submitted to you." As for withholding his proxy votes, Alfred minced no words in providing an answer:

The proxy of Almours, Inc. at the Annual Meeting was withheld simply for the reason that the sending of it would suggest personal approval of the Management. As I had no means of judging as to the business discretion or acumen exercised by the Management, I could not honestly submit it. The earnings of a Company do not always reflect the character of its management, despite the fact that most stockholders, I assume, consider satisfactory earnings as a reflection of good management. You have some good men on your Board of Directors ... [but] there are others who, I judge, are without value from an advisory standpoint. It has, however, in the past, or at least while I was connected with the Company, been the policy to have large holdings of stock represented on the Board of Directors, for the purpose of keeping the large interests directly informed on all matters of importance. To what extent this policy has been changed, I am without information.

Separate and apart from any obligation which rests upon the du Pont Company, due to the fact that its corporate entity as of today is due entirely to my foresight and determination as exercised in 1903, and furthermore that the prosperity of those who have profited thereby rests upon the same foundation, I assume that the Almours holdings represent one of the largest on the stockholders' list, and it would have been, therefore, logical to presume, ere this, personal representation on the Board of Directors would have suggested itself as a proper act. I have no time which I could allot to personal service but perhaps a representative, carefully chosen for his business ability, would be in a position to give me a clear insight into the Company's policies and management, and in addition it might be a distinct addition to the personnel of your Board. Such an arrangement would permit of my voicing my approval through the medium of annual proxy.[84]

Lammot had asked for a reason for Alfred's having withheld his proxy votes, and he had received an answer in no equivocal terms. The issue of Alfred's representation on the Board could no longer be evaded. Lammot answered, "In your letter you refer to the matter of representation on the Board for your large holdings. I would like to take this up with you at a later date and shall try to make an appointment at a time satisfactory to you."[85]

Three weeks later the two men met in Alfred's office. To facilitate matters, Lammot handed Alfred a written memorandum in which he suggested some possible alternatives for securing representation for Alfred on the Board. One means would be that which had been arranged with the French minority stockholders at the time that the Du Pont Company had bought the French rights to the manufacture of rayon and cellophane. Since Mr. Edmond Gillet, the representative of these French interests, lived in Paris and was unable to attend the monthly meetings of the Board of Directors in Wilmington, the company sent him a "monthly résumé of the matters taken up at Board meetings and he seemed highly appreciative of that action. I wonder if a similar résumé would be helpful to you." Or as another possibility, "Is it not possible that there is at the present time some member of the Board with whom you are in sufficiently close contact and in whom you have sufficient confidence, to feel that he is already your representative. . . . I would be glad to speak to any one or more of the Directors whom you might designate and point out to them the desirability of conferring with you frankly on the Company's affairs.

"If neither of the above suggestions seems to meet the situation, I would be glad to consider any other suggestion you have. You have suggested that a representative chosen for his business ability might be a distinct addition to the personnel of the Company's Board. Did you have anyone in mind whom you would suggest? . . . Might I add that your taking this matter up is not only timely, but entirely in order. I think if I were in your position, I would feel about it as you have expressed."[86]

Neither of Lammot's two suggestions were acceptable to Alfred. Like the American colonists of 1775, Alfred was not about to buy the English concept of "virtual representation." A monthly résumé would not suffice, and certainly there was no member of the "Tenth and Market street gang" in whom he had "sufficient confidence" to feel that he "is already your representative." "No participation without representation" was Alfred's motto. Lammot's hesitant invitation to suggest a person who might serve as his representative was, however, something Alfred could seize upon with alacrity, for he most certainly did have "someone in mind." He immediately sent in the name of Edward Ball, and this proposal was just as quickly rebuffed by Lammot.

Some time later, in response to an inquiry from Alfred's biographer, Marquis James, regarding Alfred's attempt to place Ball on the Board of Directors, Lammot wrote:

> I think he [Alfred] mentioned Mr. Ball, but he was not known to any members of the du Pont Company management—at least well known—and his lack of manufacturing experience seemed to me to make him unsuitable. We had, of course, several people in mind who might have served, if that had been satisfactory to Alfred, but it was not my feeling that I should suggest the names to him; and I did not do so.[87]

Maurice Lee made one last effort to secure representation for Alfred on the Board. Since the Du Pont officers would not accept Ball, Maurice urged his uncle to propose the name of Cousin Emile du Pont, whom Maurice believed "undoubtedly would be an acceptable person to all concerned, but this suggestion was not carried out."[88] For Alfred it had to be Ball or noth-ing. So it was nothing—no representation on the Board for Alfred, and no proxy votes from Alfred for the company. The deadlock continued for the remaining three years of Alfred's life.

Lack of representation on the Board, however, did not mean silence on Alfred's part when the management pursued policies that he considered wrong. In December 1934, he wrote to Lammot to inform him that there was "considerable unfavorable criticism of the Company's act in declaring extra dividends when the men down the line have not had the full losses to their salaries restored. I have been told that the reduced salaries have not been returned to their original plane by 10%, and this fact is being unfa-vorably commented upon—and it is something I can understand. I thought perhaps you and the other Directors, who are responsible for these extra dividends, were unaware of the comments engendered thereby, and I feel it my duty, as a large stockholder, to bring it to your attention."[89] On or off the Board, Alfred never ceased to present and defend the interests of the company's employees and to continue to be a prickly burr underneath those who rode in the seat of power.

Alfred by nature was no social recluse. He was genuinely interested in people, and he especially liked young children who listened to his stories and didn't monopolize the conversation with long discourses he could not hear. He also enjoyed company when taken in small doses in an informal setting. During most of his adult life, however, there had been little place within his daily routine for entertaining. During the years of his marriage to Bessie, he had been much too busy meeting the demands of his job to give much time or thought to any social activities except for the happy evenings he spent with the Tankopanicum Musical Club. Most of Bessie's friends, moreover, he did not like and did his best to avoid. Later, after his divorce and his marriage to Alicia, when he might have had the time, the money, and the inclination to entertain, he and his unhappy wife were ostracized in Wilmington. He became inured to a life of lonely isolation, while Alicia sought refuge in Paris.

His social isolation was to come to an abrupt end when Jessie became mistress of Nemours. Naturally gregarious, she had always managed, even with very little money, to have an active social life in Virginia and Califor-nia. The quick transition into the world of wealth that her marriage had produced was for her a dazzling opportunity to enjoy to the fullest the pleasure of open-ended hospitality that she had always craved. Like an orphan suddenly let loose in a toy shop, Jessie revelled in the spaciousness of Nemours, with its great drawing room, dining room, and terraces, obviously designed with lavish entertaining in mind. She invited back into Alfred's life his children along with the grandchildren he had never known. She also brought her own large company of siblings, in-laws, and cousins into the household as guests. Seldom was there a week at Nemours or at Epping Forest when there were not at least two or three houseguests.

Alfred was never able to adjust to such crowds of people in his home. In the world without sound in which he had to live, he must often have found the whole scene to be similar to watching a silent movie without benefit of subtitles. In a letter to Marguerite, written from Jacksonville, where he had gone by himself for a couple of days to attend to business, he wrote:

> I am only here for two days as I must perforce be back at "Nemours" on Friday to help Jessie entertain her week-end "crush." The lady visitors enjoy the prattle and the rum which are dispensed; the men likewise and, in addition, my cigars and cigarettes. So great has been the recent consumption of tobacco that I perforce may seek out cheaper brands as I cannot afford to dispense cigars sealed in two's at $1.00 per to tobacco consumers who are as thoroughly satisfied with an old pipe or a cheap cigarette. Of course, you know all about this, as all of your friends, particularly Parsons, are hopeless addicts to the weed.[90]

The best times for Alfred were the occasions when he and Jessie had Epping Forest all to themselves. "We usually go to bed about half-past nine," he wrote Ellen La Motte, "unless we go out to dinner, which is not often, and I wish it could be a trifle more seldom, as I have about as much use for social functions as a dog would have for wax legs in Hell."[91]

In addition to the demands placed on him by his many business ventures in Florida, there were domestic tasks that he assigned to himself: supervising the construction of the formal gardens at Nemours, designed by his son and Gabriel Massena, and the constant improvements he made on the house and grounds at Epping Forest. Alfred spent many happy months constructing a giant waterwheel at Epping Forest. In the summer of 1930, he proudly reported to his sister:

> I am at present very much involved in a waterwheel, which I am building, to pump water into the water pipes at Epping Forest, for nothing. I happen to have an artesian well, which exerts a small pressure at the surface, and I am using this donation by the good Lord to effect economies and incidentally produce much better water. . . . This waterwheel is 20 feet in diameter, made entirely of brass, with the exception of the main shaft, runs on ball bearings, and is a thoroughly reliable wheel, running only when people open a faucet, otherwise remaining quiescent, in complete physical relaxation. It has been rather a strain on

my reliance and confidence in my own ingenuity, as many a head has been shaken as the wheel gradually developed. Strange to relate, however, when the water was turned on, it rotated majestically and all four pumps went to work with the wheel.[92]

A half century later, Alfred's waterwheel would still be working as silently and efficiently as the day he first opened a faucet in the house and pure artesian spring water gushed forth.

Next to being alone with Jessie at Epping Forest, Alfred found his greatest relaxation aboard his yacht. These cruises almost always provided Jessie with yet another opportunity to entertain guests, but the space available on a yacht, even one as luxurious as *Nenemoosha II,* limited the number who could be accommodated, much to Alfred's satisfaction.

There were a few occasions when Jessie went on trips without Alfred, and he took advantage of those times by taking short runs on the yacht with only the crew and at most two or three companions. In 1931, Jessie, much to her delight, was invited by Governor and Mrs. A. J. Montague of Virginia to accompany them on an official visit to an international conference in Rumania. Alfred, left alone at Nemours for the longest period of time since his marriage to Jessie, found the unusual emptiness of the place not as pleasant as he had imagined it would be. He wrote to Jessie's sister, Belle Baker, that he was planning a short run down the Chesapeake to Yorktown, providing he could find the time. He assured his sister-in-law that he planned to take "no Governors on my yacht. A Governor is no different from anybody else, only as a rule, they are all swelled up like a bag of wind with pompousness, attributed to their political emolument . . . but, if I do go, I shall expect you and Victorine to be my guests and possibly Ed—nobody else—and don't you dare bring anybody else on that boat without my permission. I am trying to get a little bit of rest and I am sure you and Victorine can entertain each other."[93]

The times when he had his home and his yacht to himself, devoid of guests, were rare, however. Much more typical was the situation he described to Ellen La Motte: "Mr. and Mrs. Glasgow are with us and have been here about a month. . . . Maurice is also with us for a short time. . . . We are expecting company from now on, which will keep our little Spanish home quite full."[94]

His deafness made conversation almost impossible, and Alfred of necessity became a great "people-watcher." He wrote Ellen, "It is really refreshing to watch the people, which I do every opportunity, sometimes sitting in the park nearby [to his bank in Jacksonville] to rest from my labor and again in Liggetts Drug Store where I surround my mid-day meal, with consideration both for my 'gizzard' and my pocketbook. Incidentally, he disports quite a nice line of detective stories at 75¢ each, which under normal conditions bring $2.00."[95]

Alfred was an avid reader of mystery stories and was surely one of the Crime Club's best subscribers, taking every month's selection along with several of the Club's alternate selections.[96] People-watching, reading, and long walks with his dogs, who left all the talking to him, apparently were adequate substitutes for conversation. When the director of the Kinzie Institute of Speech Reading in New York offered to come to Nemours to teach Alfred lipreading, he answered that he was getting along very well without that special skill. "Not being of a garrulous nature, I am quite satisfied with meditation and good reading and an occasional billet doux from someone who is sufficiently interested in me to scribble on a pad."[97]

Jessie was definitely sufficiently interested in him to scribble on a pad short, telegraphic-style messages. Some of the many notes she wrote to him daily can still be found among his personal papers.

Even if Alfred was not "of a garrulous nature," Jessie most certainly was, and she found the necessary outlet for conversation in her guests. Alfred, therefore, tolerated the steady stream of visitors to Nemours and Epping Forest, as long as he was not expected to be an active participant in the gatherings and was allowed to go his own way: to putter about in the gardens, to read in the library, or simply to sit by himself in a crowded room and watch with wry amusement the antics of the passing show.

Most of the guests were either relatives or business acquaintances, particularly men who Ed Ball believed might prove useful in promoting the fortunes of the Almours Company. Jessie was usually present at these business conferences, relaying memos of the conversation to Alfred to which he could respond, for she took a keen interest in every detail of the company's operations. She sometimes felt, however, that the quantity of the guest list at Nemours and Epping Forest did not offer the quality or variety she would have liked. Bankers and paper manufacturers were certainly important in her and Alfred's life, but as she on one occasion wrote to Ed, "It is good for everyone to get out with people on something than a business line once in a while. Our contacts are too strictly business. No doubt, you realize this just as much as I do."[98] Jessie must not have known her brother as well as she thought she did, for the need to extend one's social horizons beyond the narrow world of business was the last thing Ed Ball would ever realize. He found it difficult to imagine why his sister, who had such a good, practical head on her shoulders, would waste time and money entertaining people who could not in some way further their business interests. Ed would also have liked to apply this qualification to all relatives—both du Ponts and Balls; for in his opinion most of them were simply leeches to whom, were he in charge of the social calendar, he would not even give a passing nod, let alone free meals and lodgings.

Alfred's concern about Jessie's social activities was somewhat diffferent from that of his brother-in-law. He frequently worried that the strain that Jessie was under in conducting a perpetual open house would prove too much even for her seemingly unlimited energy and physical stamina. In December 1929, when Jessie became seriously ill, Alfred's worst fears were at last realized. On doctor's orders, Jessie was forced to cancel the plans for the usual elaborate Christmas festivities at Nemours, and Alfred hurried her off to Epping Forest, where for this season at least there would be no guests. Alfred wrote to Jessie's Aunt Ella Haile, "My dear Jessie is much better, but oh, she is so thin it worries me just to look at her. . . . The more I think of her recent illness, the more I am convinced that it was purely a nervous breakdown. She has given the limit and her poor little body just refused to stand any more. I hope this will be the means of knocking a little selfishness into her, because unless she cares for herself a little, the break-down will surely be repeated."[99]

The breakdown was repeated in early January, and it proved to be more serious than simply a case of nervous exhaustion, as Alfred later reported to his and Jessie's mutual friend, Lucy Maitland Francis:

> She got along fairly well for awhile; but in January she was taken seriously ill again and after playing fast and loose with some local pill dispensers, it became evident that she would have to go North for treatment.
>
> They took the poor girl up there on a stretcher and arrived in Baltimore on January twentieth. After keeping her under observation for a week, the Doctors determined there must be some serious case of local infection and . . . they determined to see what was the matter. The operation was a very serious one—removing one or more abscesses . . . where the Doctors believe they had been living for six months or more. . . . For a week none of us expected that she would come through, nor did the Doctors . . . but about that time a turn for the better came and since then she has been fighting her way along the road to recovery.[100]

Jessie attributed her survival to her sister Elsie: "The night after my operation, . . . when I was supposed to be sinking very rapidly, and calling for cracked ice every few minutes, . . . Elsie told them to put some whiskey on the cracked ice and give it to me. The nurse said she couldn't do it without the doctor's orders, so then she went and got the doctor's orders. He said, 'Yes, give it to her, nothing can hurt her now.' Who knows but that it was the whiskey that turned the tide for me. You know we have all been good single-handed drinkers. It's a good thing Dad and Mother trained us right in our youth, so as to take the right amount and not too much."[101] Whether Jessie's remarkable revival could be attributed to the whiskey or, as Alfred preferred to think, "Your love for me . . . [which] gave the needed strength to win,"[102] she was out of danger within two weeks and by mid April had sufficiently recovered to travel to Florida with Elsie to join Alfred at Epping Forest. Alfred had hoped that Jessie's close brush with death would convince her that she must, as he wrote to Denise, "be careful of herself . . . for once in her life. She has never taken care of herself, as you know, but [is] always looking out for other people's interest."[103] In the realization of this hope, however, Alfred was to be sadly disappointed. By summer, Jessie was once again hostess to a full house.

Jessie's restless energy and her constant need for other people and new experiences were character traits that Alfred never understood or appreciated. If Nemours in the summer or Epping Forest in the winter was not booked full with guests, then Jessie was eager to be off to some distant place where she would herself be a guest. Alfred himself enjoyed short runs down the Chesapeake or along the inland waterway of Florida in his yacht, as well as occasional weekends at his Cherry Island camp. He could even find pleasure in more extended motor-trips into New England in the autumn, but generally he would have been content to remain at home, with no guests present, and enjoy the unusual peace.

What Alfred particularly disliked were extended trips abroad. After his first trip to Europe with Bessie in 1889, he had had no great desire ever again to leave the continental limits of the United States. He had, to be sure, derived a vicarious pleasure in observing Jessie's delight in seeing Europe for the first time when they went abroad in the summer of 1922 to attend Alicia's wedding in London, but as far as Alfred was concerned, seeing Europe once should be enough to satisfy any good American. It was not enough to satisfy Jessie, however. She would have taken a trans-Atlantic trip every year if Alfred had been willing. She did manage to get him to Europe for two successive summers in 1928 and 1929 to receive treatments from the noted Carlsbad audiologist, Dr. Mueller, but those painful and, as Alfred had predicted, futile ventures were hardly experiences that would further stimulate Alfred's interest in European travel.

Jessie's trip to Rumania without Alfred in the summer of 1931 simply whetted her appetite for seeing more of the Old World. She had repeatedly assured Alfred that she would never again take a long journey out of the country without him, but that promise did not mean that she would be content to stay at home.[104]

After a great deal of wheedling, Jessie succeeded in persuading Alfred to take her on a Mediterranean cruise in the winter of 1932. They went from Naples to Athens and on to Egypt where they took a river boat, the *S. S. Arabia,* from Cairo down the Nile to the Aswan dam. Jessie, of course, was fascinated by all that she saw and wrote rhapsodical letters to friends back home of "the starry skies on the desert" and "the donkeys and camels just as it was in the days of the Bible. . . . A. I. was a splendid pal to travel with."[105]

Alfred, of course, was anything but "a splendid pal to travel with." The weather was the coldest ever recorded. They viewed the Acropolis as best they could through the swirling snow of a blizzard. The river boat had no heat. Alfred caught a cold and spent most of the trip huddled morosely in his cabin, cursing his stupidity for ever having agreed to this damn fool expedition. In the diary that Jessie recorded on the back of picture postcards, she gave a much more honest account of the trip than in the glowing letters she sent to her friends. "Trip has been an awful disappointment—'A. I.' miserable all the time. . . . Oh, so happy to start home . . . it hurts to have him so unhappy on our first trip alone."[106] Jessie surely realized by then that this was their last trip abroad, for no one, not even she, could ever again induce Alfred to leave the United States.

The only bright moment for Alfred in the whole miserable affair occurred shortly before they intended to leave for home. On a walk through a park in Cairo, Alfred found a little black-and-white mongrel bitch lying in the bushes. The little dog was emaciated, too weak to stand and obviously near death. Alfred scooped the dog up and carried her back to his suite in the Shepheard Hotel. A veterinarian was called in, and Alfred immediately made other arrangements for their return trip, as he refused to leave until the dog was sufficiently recovered to permit them to take her back with them. As with his beloved Yip, Alfred had acquired another waif to adopt and to cherish.[107]

If it was a lucky day for the little dog when Alfred took his stroll through a Cairo park, Alfred in turn believed his foundling more than reciprocated the good fortune he had bestowed upon her. He named her Mummy, and he told everyone aboard the ship that she was a reincarnated Egyptian of great importance—possibly Cleopatra herself. He was only half facetious when he wrote to a friend:

> Mummy, as a mascot, is proving a wonderful success. She calmed the storms on the Ocean while we were on the *Aquitania,* much to the relief of the Captain, who I think was on the point of becoming seasick. On our arrival in Jacksonville, the Captain of my yacht advised me that his wife had a little baby—something they had been hoping for for many years but never expected . . . so much for Mummy!
>
> Two days after I picked Mummy up in Cairo, I was visiting a junk shop and cast my eyes on what appeared to be a mass of copper discs cemented together. Inquiring of the owner, he told me they were taken from an ancient tomb. . . . Evidently, some water, containing either an

acid re-action or perhaps chloride, percolated into the jar and the resultant chemical action cemented them together, so it was necessary to break the vase in order to get this ball of green coins out. He told me they were bronze and didn't amount to anything. However, I paid him his price, a few dollars, and took them home. On taking them apart and getting them cleaned up, they turned out to be coins issued by Ptolemy I, made about 300 years B.C., known as penta-drachmae, and apparently made of a sheet of gold with a sheet of silver on either side. . . . I am sending a few of them to different museums to have them interpreted. . . .

About a year ago, I purchased from a man . . . a painting. It was discovered on investigation, to have been superimposed on another painting. The removal of the top layer of paint disclosed the most beautiful Madonna and Child by Murillo, painted about 1680, in perfect condition and worth of course an enormous sum. Mummy did this!

I have one or two more jobs for Mummy and then I will give her a vacation before I put her to work again.[108]

Alfred had scored over his sister Marguerite. Let her collect Virginia ghosts if she wanted them. Only he had a reincarnated Egyptian mascot who could produce wonders as great as those of Aladdin's genie.

The Murillo painting that Alfred mentioned in his account of Mummy's magic powers was in reference to a portrait, purported to have been painted by Jacques David, of Pierre Samuel du Pont's wife Marie Le Deé, holding her infant son Victor in her lap. Alfred had acquired the portrait from the heirs of the noted art collector Ferdinand Danton. The Danton family had asked $25,000 for the painting, but because they were in desperate financial straits, they were obliged to accept Alfred's offer of $1,000. As Alfred later told Louise Bayard:

He [Van Eaton Danton] wanted a fabulous price but compromised ultimately for $1,000.00; considering the frame alone was worth four or five hundred dollars, I figured I could not be let down very badly if the thing turned out to be a chromo originally issued by the Atlantic and Pacific Tea Company. I had an expert, in the form of a house painter, come and look at it and we decided that the whole picture was not painted by the same so-called artist. Further examination developed the fact that it was superimposed on an old canvas; and five hundred dollars worth of attention by an expert in removing the top coat of paint showed the most wonderful Murillo "Madonna and Child" you ever cast your peepers on, absolutely without blemish, worth almost anything.[109]

Alfred believed that in all likelihood "the Murillo was given Pierre Samuel du Pont, probably by Louis XVI, or one of his wealthy French friends and that in order to get it out of the country he had the painting changed and the picture of his wife put on top of the Madonna and Child."[110]

This painting was only one item in a very large collection of Pierre Samuel du Pont de Nemours memorabilia that Alfred acquired from the Danton family, including jewelry belonging to Pierre Samuel du Pont's mother, Anne Alexandrine Montchanin; marble busts of Pierre Samuel and Marie Le Deé du Pont, reputed to be the work of Jean Antoine Houdon; watches, vases, and a bird bath given to Pierre Samuel by Lovis XVI; a sketch of Thomas Jefferson by Gilbert Stuart; and an exceptionally fine collection of du Pont family portraits by the French artist Robert Le Fevre.[111]

At the time that Alfred was negotiating for the purchase of these du Pont family heirlooms, he received a form letter that Cousin Pierre had sent out to all members of the family warning them not to purchase the jewelry that Van Eaton Danton claimed had belonged to Marie Le Deé du Pont. Pierre said that this box of jewelry had first been offered to him, but Pierre had noted that the individual pieces had been inscribed "Marie le Deé du Pont de Nemours." He advised the family that as "she died in 1785 and the 'de Nemours' was not used until 1789, I do not think it possible that this jewelry could have been marked while in her possession. . . . While I was interested in the jewelry at first, I am so doubtful of it that I have ceased negotiations. . . . Therefore, if you should be approached on the subject, I advise a most careful search before making purchases, if you are inclined to do so."[112]

As Alfred had already agreed to pay $20,000 for the jewels by the time he received Pierre's letter, he felt it necessary to justify his purchase, and at the same time to give his cousin a little lesson in the family's history:

> My dear Pierre:—
> I think you are in error in your conclusions, as set forth in your circular letter of July 23rd. . . . I am quite familiar with their history and believe them to be genuine for the simple reason that they cannot be otherwise. . . . I have had some of them examined by different experts and they all agree. . . .
>
> Your lack of belief in their genuineness apparently rests in the thought that Marie le Deé died four years before the title "de Nemours" was affixed to the du Pont name. In this premise, I believe you are wrong. The Patent of Ennoblement given to Pierre Samuel du Pont by the King was signed at Versailles in the month of December, 1783, and so one may logically assume that he used the title from that date, and that Marie le Deé did likewise. This date is two years previous to her death, instead of four years subsequent. I have in my possession the original Patent of Ennoblement given to du Pont de Nemours by Louis XVI; . . . so you see the use of "de Nemours" by Marie le Deé was not only proper but well within reason, and the little trinkets undoubtedly belonged to du Pont de Nemours and members of his family. It is true

that they are without material value, but . . . their sentimental value is
beyond measure.

> Yours very truly,
> Alfred I. du Pont.[13]

Alfred might also have added that, quite apart from their sentimental
value, the acquisition of these jewels had provided him with a rare opportunity to correct the mistake in family chronology made by know-it-all
Pierre, the self-appointed family historian. That alone must have been
worth considerably more to Alfred than the $20,000 which he had paid for
Marie Le Deé's "trinkets."

That Alfred had become an avid collector of family relics this late in his
life must have struck many of the du Pont family, including Pierre, as decidedly odd and uncharacteristic of the man. Alfred had always made fun of
ancestor worship. He regarded those in his family who proudly emblazoned
the du Pont coat of arms upon everything from limousine doors to match-
book covers as simply foolish advertisers of the fact that they themselves
had accomplished nothing and subsisted only on the past glory of their
forebears. Unlike them, Alfred was not overawed by the name he bore.
When an acquaintance in Jacksonville wrote to inquire about a du Pont
living in France in 1848, thinking it might be one of Alfred's ancestors,
Alfred replied curtly, "The du Ponts in France are the equivalents of Smiths
in America. . . . My direct ancestors came here from France in 1800. It
would be quite a relief if most of their descendants returned."[14]

Alfred had always been willing to acknowledge the historic significance
and greatness of his great-great-grandfather, Pierre Samuel, and of that
man's son, Eleuthère Irénée, but of their descendants, one could count on
the fingers of one hand the few who were distinguished in their own right:
his great uncle Boss Henry, his uncle Lammot, his second cousin Francis
Gurney, and the latter's son, Francis I. du Pont. Conspicuously absent from
Alfred's list of du Pont notables were Cousin Pierre Samuel II and Cousin
Coleman du Pont, although the latter might have made the du Pont Hall
of Fame, Alfred believed, if only he could have been trusted.[15] As for all
the other du Ponts (himself excepted, of course), they had nothing going
for them except their name. "I only hope that Pierre Samuel du Pont de
Nemours cannot witness the net result of his youthful folly," he wrote Louise Bayard, "for would he not curse the day when he first set eyes on his
Annie Morato alias his first wife, Marie le Dée! The 'family' surely has gone
to seed."[16]

Marguerite Lee held a similar view about the decline of the du Pont
family, but as might be expected, she looked back not to a great founding

father but to a noble founding mother. In commenting upon a sketch of the du Pont family that the director of the Historical Publications Society of Philadelphia had prepared and sent to her for review, she wrote: "My dear Mr. Du Bin, take it from one who knows, and moreover is not camouflaging: we are today, small potatoes in the bin of Life. The Aristocracy and Strength flowing to us from the veins of Pierre Samuel du Pont's great mother, Anne Alexandrine de Montchanin, are now visible only in an occasional cleft in some fool's chin."[117]

In Marguerite's opinion, the family's history had been one of decline and fall ever since the daughter of the aristocratic Montchanin family had foolishly married the lowly Parisian watchmaker Samuel du Pont in 1737. She had been as surprised as anyone else in the family by Alfred's belated interest in collecting and memorializing the family, and she had no great regrets when Alfred had to abandon the idea of building the chapel on the Nemours estate that was to serve as a monument to their founding father, Pierre Samuel du Pont de Nemours.

Three months before his death, Alfred wrote to Marguerite to inform her that although there would be no monument to Pierre Samuel (perhaps Alfred feared the general public might confuse the original P. S. du Pont with the current P. S.), he was "working on the plans for a Memorial Tower in memory of our father and mother. . . . It will have a carillon, and I have about selected the following inscription . . . on one of the small bronze tablets to be placed in the entrance-hall base:

> 'This musical tower was erected by Alfred I. du Pont as a token of love and affection to the memory of our parents, Eleuthère Irénée du Pont and Charlotte Sheppard [sic] Henderson; also to lovers of music, he believing that great happiness will accrue to all by virtue of its construction.' Massena & du Pont, Architects 1935.

"The tower is to be constructed of pink and white marble, with the new Brandywine blue rock terrace and steps surrounding it.

"I thought you might be interested to see this. What I would like to know is whether I have our mother's name spelled properly. The word 'Sheppard' is spelled so many different ways, that I thought the quickest way to ascertain the facts in the case would be to appeal to you."[118]

Marguerite was less than thrilled by this announcement. It was her belief that far greater happiness would accrue to all if Alfred, instead of constructing this pretentious bell tower, would contribute these funds to her charitable work in Washington. At least he had included their mother as well as their father in the memorial, even if he didn't know how to spell their mother's name, and for that Marguerite could be thankful. She didn't like the proposed inscription at all—it was far too wordy and gave too much prominence to the donor of the tower. Marguerite added a comment on the letter,"I suggested as an inscription simply 'Erected to the memory of Eleuthère Irénée du Pont and Charlotte Shepard Henderson his wife 1935.'"[119]

Marguerite and other members of the family should not have been surprised, however, by Alfred's sudden interest in memorializing the past. On 12 May 1934, Alfred celebrated his seventieth birthday. He had reached the age when most people become acutely aware of their own imminent mortality, and at that time it is natural to reach out for some hold on immortality, either through a strengthening faith in eternal salvation, which for Alfred could never be an abiding certainty, or by leaving for future generations some evidence of one's existence. It is a time to gather together the relics of the past, a time to collect one's papers, to write one's memoirs, or, if one has the means, to build monuments in pink and white marble. This desire to deprive death of its sting and the grave of its absolute victory is a far greater spur to philanthropic giving than even the lure of tax deductions—a fact that is fully appreciated by every church and college in the land.

Alfred, to be sure, with his many business enterprises in Florida, was now more involved with present and future planning at seventy than he had been at the age of fifty, when he was summarily ousted from the Du Pont Company management. Busy as he was, however, he still had time to look at the calendar and to make a reckoning from the actuarial statistics. Even if he had wanted to ignore the passing of time, there was always Marguerite to remind him that he was now an old man. In the summer of 1934, she wrote him another begging letter, this time on behalf of their cousin, Will Gardner, who was dying of cancer.

> What is to become of the wife and *five* little children? They are your mother's own kin. . . . You and I are going to 'put on immorality' in much less than *one year*. This is a *fact,* and might as well be faced bravely, however *unpalatable.* My heart, poor at birth, is simply *worn out. That's that!* Yours, stricken in old age, finds itself in even *worse shape! That's that.* To take money into the cemetery, no matter how ardently we may wish to do so, a means has never yet been found! What difference does it make to *us* what becomes of it when we stand at the pearly gate pleading our cause with Peter? For *God's sake be a sport* at last! Come across with an adequate check to make these children comfortable this last year you and I are destined to cumber the ground.[120]

Alfred was not at all appreciative of this sisterly reminder of his mortality. He answered with an unusual irascibility: "To be perfectly frank, a man in Will Gardner's position, at sixty years old, has no business having babies. The knowledge that he is without any self-control, when practically an old man, does not appeal to the sympathetic nature. . . . If he had had one or two children, it was the maximum he should have undertaken to bring into this world, which is already overstocked with anaemic children, who by reason of their heritage, are without value.''[121] Having thus blasted Marguerite with this Scroogeian "Bah, humbug," Alfred then proceeded with his usual acquiescence to send to Gardner's widow a sizeable check that paid for all of the medical bills and funeral expenses. He then made arrangements to send her a monthly check for the support of the five children.[122]

Jessie's response to Alfred's newly acquired interest in historic preservation and highly selective ancestral veneration was considerably more enthusiastic than Marguerite's. Her delight in the Danton collection had been particularly sweetened by the knowledge that Pierre had hastily and mistakenly rejected it, enabling Alfred to purchase the relics at a bargain counter price. While supportive of her husband's acquisition of du Pontiana, she was determined that her own distinguished ancestry should not be forgotten or slighted. The Balls, after all, were landed gentry in Virginia when the du Ponts were mere petit bourgeois shopkeepers in Rouen. Jessie was a bit weary of hearing so much about Pierre Samuel's friendship with Thomas Jefferson when it should be remembered that George Washington's own mother was a Ball. Jessie deeply resented the scarcely veiled intimations of the du Pont clan, particularly the sneers of Marguerite, that Alfred had married below his station when he took as his wife the sister of "a grocery wagon driver."

So Jessie herself took up a bit of family commemoration. She became one of the most generous patrons of the Lee Society of Virginia's program to purchase and restore the ancestral Lee home at Stratford. She regarded her sizeable contributions as nothing less than a solemn obligation owed to her own family, for the Balls had been neighbors and kin to the Lee family a good century and a half before Marguerite du Pont had acquired her Lee connection, of which she was so proud, by marrying Cazenove Lee.

Jessie's major coup in the battle of genealogy with the du Ponts came in 1932 when she purchased the ancient estate of Ditchley in Northumberland County, Virginia. This tract of land had been part of the large grant given by King Charles I to Colonel Richard Lee prior to Lee's migration to Virginia in 1647. The Colonel's great-grandson, Kendall Lee, had built a lovely Georgian brick mansion on the Ditchley estate in 1752, and it was subsequently sold to Kendall Lee's nephew, James Ball. In 1930, Jessie heard from her Northumberland cousins that Ditchley was for sale, and she at once set about acquiring it. Fearing that if the owners knew that she was a prospective buyer the price would be immediately inflated, she asked her old friend, Judge Joseph Chinn, to negotiate the sale for her. "As you know 'Ditchley' originally was a Lee home, but has been a Ball home for close to two hundred years," she wrote Chinn. "While I have never been there, I understand the house is in very good condition and that there are sixty acres of land with the house, a portion of which is on the water. . . . I am willing to give anywhere from Ten to Fifteen Thousand dollars for it."[123] When Chinn informed her that the asking price was $20,000, she considered that excessive, and told him to try to get a better deal. Nevertheless, she assured Chinn, "'Ditchley' is a nice old place and *I want it.*"[124] What Jessie wanted, she usually got. After two more years of negotiation, Chinn purchased the estate for $18,000.[125]

Jessie never spent much time at Ditchley, but she had the satisfaction of possessing her own ancestral home, which was a half century older than the du Pont mansion at Eleutherian Mills. She even graciously offered to let Marguerite investigage the legendary Lee ghost that was reputed to haunt its upper hall.

Enjoyable as Alfred found the business of collecting du Pont relics, of finding a Murillo painting underneath a family portrait, and of building the imposing carillon tower that would be the highest structure in the state of Delaware, unfortunately other, less pleasant tasks had to be attended to in preparing for one's final exit. When his old friend Olive Seiss wrote him expressing concern about making her burial arrangements, Alfred responded that he too was "very much concerned about my own resting place at the present moment." He told her that he had a "dislike [for] the thought of being buried in the du Pont cemetery, almost totally devoid of friends and surrounded by my enemies, so I am developing a plan looking to the creation of a private cemetery and chapel on my estate at Nemours, limited to one hundred occupants. The idea has worked out very beautifully indeed, but I lack encouragement—no one seems to like the idea of being buried anywhere and, when I approach the subject, they always shy off. Should my cemetery materialize within the next five years, of course, I shall be glad to offer you a resting place therein, where you may secure the association of none but good people, nice people, and all white."[126] His wife Alicia and their two infants were already buried on the Nemours grounds, and Alfred even toyed with the idea of disinterring the remains of his parents from the du Pont family's exclusive cemetery, where poor Charlotte lay "almost totally devoid of friends," and bringing them to the even more highly selective graveyard that he envisioned at Nemours.

An even more important and pressing problem than the final disposal of one's body was the disposal of one's possessions. An estate as large as

Alfred's, which even in the depressed market of the early 1930s amounted to nearly $40 million, presented a considerable problem to Alfred and a rich prospect for the Internal Revenue Service.

As a first step, Alfred in 1926 offered to each of his four children 2,000 shares of Du Pont preferred stock. In characteristic fashion, Marguerite informed her son Maurice about the offer, of which Alfred had told her in confidence:

> A. I. gave me some interesting information which you must not repeat. He has given each of his four children $200,000 of stock and they "very willingly" released his estate from further obligations. He told them it did not mean that he would not remember them in his will; but they could not claim anything. Glasgow drew up the releases which seemed to give satisfaction, I suppose upon the admirable theory that "a bird in the hand is worth two in the bush." A new will is being drawn, to be ready for his signature at the end of Nov. This means there will be no one to contest the Will and therefore bequests, of which he said, "there are *many*" will be paid. You & Kits [Cazenove, Jr.], and I suppose Nesta, are to receive $15,000 a year.... Each of the six children [Marguerite's grandchildren] gets $100,000, in trust until they are 30. At that date, they get the principal. I hear no mention of Maurice and myself, and as I wrote *him,* we are presumed to have removed ourselves to another world in the meanwhile; which supposition I fancy is correct.... It is a great comfort to know A.I.'s will cannot be contested. He said he had rumors of contests, and therfore this [was] the cheapest way to settle matters.[127]

The new will drawn up by William Glasgow that Marguerite mentioned to Maurice proved not to be Alfred's Last Will and Testament. Over the next nine years of his life, Alfred continued to rethink how best to provide for those who were dependent upon him and at the same time to make sure that the bulk of his estate would eventually be used for philanthropic purposes. He had a special dread of postmortem litigation, for he knew of too many instances in which rich estates had been nearly totally consumed by legal fees. After Glasgow's death in 1930, Alfred decided to write a new will and asked a Jacksonville lawyer, Raymond D. Knight, to prepare it. The will was finally completed to Alfred's satisfaction and was signed on 19 November 1932.

Under the terms of this document, Jessie was to receive Epping Forest in fee simple, along with all of his personal possessions, including furniture, art objects, silver, jewelry, automobiles, and the yacht. She was to have the use of Nemours as a home during her lifetime, but it would belong to his estate, not to her.

In addition to the annuities he had already provided for his sister and brother under a special trust fund established in 1929, he designated grants and annuities for certain other beneficiaries. These bequests included 5,000 shares of Almours Securities to each of his four children. His foster daughter Denise was to receive an annuity starting at $5,000, which would increase to $30,000 per year after she reached the age of thirty.

Jessie's two sisters, Isabel Baker and Elsie Bowley, and her brother Thomas each were to receive 1,000 shares of Almours stock, but Edward Ball was to receive the same amount as Alfred's children, 5,000 shares. Maurice Lee was to receive a life annuity of $15,000. There were also annuities of $1,200 to each of his secretaries and to his butler, Thomas Horncastle, and smaller annuities to several other servants at Nemours. After all of these annuities had been paid, the remaining income from his estate was to go to Jessie Ball du Pont during her lifetime.

Upon Jessie's death, the income from his estate was to be used by the Nemours Foundation for charitable purposes. He specified that the Nemours mansion and grounds were to be maintained for the public's enjoyment. The remaining income was to be used for the treatment of crippled children and for the care of the indigent elderly. It was this Foundation which held Alfred's major interest and which he was most concerned to protect.

The will provided that the Nemours Foundation need not begin its charitable work until after Jessie's death unless she should wish to build a hospital on the Nemours grounds prior to that time. Alfred named as trustees of his estate the Florida National Bank of Jacksonville, Jessie Ball du Pont, Edward Ball, and John F. Lanigan, a Miami businessman and friend of Ed Ball.[128]

Six months after this will was signed, Alfred had Knight prepare a codicil guaranteeing Jessie $200,000 each year before any of the other annuitants received a penny: "Having become impressed, after a study of my said Will and deliberate consideration of world conditions, that I have not in my Will properly protected the interests of my wife, Jessie Ball du Pont, against possible eventualities, and desiring that she shall receive, to commence at the time of the date of my death, an annuity of Two Hundred Thousand Dollars, or such part thereof as shall be available each year out of current income . . . before any of the [other] annuities . . . are paid either in whole or in part . . . and that such other annuities should be second, subject and subordinate to the said annuity to her."[129] One can be certain that this was one piece of "interesting information" regarding his estate that Alfred did not pass on to Marguerite. Had she been informed of this codicil, Marguerite would no longer have had "great comfort in knowing that A. I.'s will cannot be contested." She would have been convinced that his will had already been successfully contested prior to his death.

If Alfred's lawyer believed that this codicil finally completed a will that satisfied his client, he was to find out differently. Alfred was soon to have further thoughts on the disposition of his estate.

In early February 1934, during one of those rare moments when there were no guests at Epping Forest, Jessie persuaded Alfred to take her on a run aboard their small boat, the *Gadfly,* down the inland waterway. En route, they made a brief stop so that Alfred could have a walk along the beach. He suddenly experienced a shortness of breath and pain in his chest. As Jessie later reported to Reginald Huidekoper, "Needless to say we returned to the yacht at once and attempted no further exercise. When we reached Miami, we went to the doctor there and had the usual tests, etc. They pronounced his condition A-1 with all organs in perfect order, but in order that the heart might have the opportunity to regain its full strength, they ordered absolute rest of several weeks. Hence we started to Jacksonville on Sunday. Here we have a very good doctor, who ordered A. I. to bed for some length of time."[30]

Jessie was not unduly alarmed. She wrote a friend, "We are rather inclined to think it has been more of a nervous break than anything else, which will correct itself over a period of weeks. . . . Never having been sick before in his life, it has come a trifle hard [sic] with him than with the average person. . . . The only manifestation which could be attributed to the heart was the shortness of breath . . . and as you say, all of us get out of breath as the years advance."[31]

Marguerite, however, upon getting Jessie's report of Alfred's illness, immediately suspected the worst. She dashed off one of her gloom and doom letters to Jessie. "I of course am sorry to grasp the tenure of your letter concerning Alfred's sudden attack; but I cannot say that it was altogether a surprise. You must remember Alfred is an old man. He is reaching in May the allotted three score and ten. I have had much experience with heart conditions and am sorry to say your letter sounds far from promising. It contains many dubious aspects—to a knowing eye. . . . We both know how invariably Doctors 'jolly' their patients; and those anxiously watching their condition. The breathing was probably occasioned by a slight thrombosis."[32] This, of course, was precisely what Jessie did not want to hear. It was difficult to believe that any one still so active in the pursuit of both business and pleasure as he could be either sick or old, and she refused to accept Marguerite's gloomy prognosis.

If Jessie was eager to brush aside her sister-in-law's dire alarms, Alfred himself was somewhat more concerned. It may have been an apprehension that he was more seriously ill than either Jessie thought or he would admit to her that prompted him to consider his will once again. Whatever his motivation, in the fall of 1934 he informed Knight that he wanted to make a new will that would involve a few substantive changes from his present will. First, he wanted to add several more names to his list of annuitants, for he kept thinking of people who should be included, especially Laura Walls and Ellen La Motte. How could he possibly have omitted his favorite cousin from his list of beneficiaries? Second, he wanted to make a change in the designated trustees of his estate. In lieu of Ed Ball's friend, John Lanigan, he wanted his son-in-law Reginald Huidekoper as a trustee. Surely there should be at least one representative of his own du Pont family among the directors of the Nemours Foundation. Of even greater importance were two other changes he wished to make. He had apparently given second thought to the codicil that gave Jessie first claim upon the income of his estate. Since in a bad year this first claim provision could mean that all of his other annuitants would receive nothing, Alfred wanted Jessie to share whatever income the estate produced with the other annuitants on a pro-rated basis. Finally, Alfred wanted all of the income from the estate, over and above that needed to pay the specified annuities, to go directly to the Nemours Foundation for charitable purposes rather than to Jessie for her use during her lifetime.

However surprised Knight may have been by the changes that Alfred wished to make, it is not the business of a lawyer to question the testamentary provisions desired by a client. Knight informed Alfred, however, that Florida had recently enacted a law that required a six-month "waiting period" before charitable provisions in a second will could supersede a previous will. This requirement was designed to prevent "death bed conversions" in which a dying person suddenly disinherits legitimate heirs in favor of a convincing "converter" who has had undue last-minute influence.

Alfred was outraged. He told Knight that had he had "the question of residence under consideration at the present time, I should certainly object . . . and refuse to become a resident until it was made more within the bounds of reason."[33]

Alfred was prepared to use what political infuence he had in the state to get the law changed. He wanted an amendment added to the legislation to the effect that a new will would not supersede an earlier will within a six-month period "unless proper or adequate provision has been made by the testator for members of his immediate family." Knight, however, said that such an amendment had no chance of passage since it was much too vague as to what "proper or adequate provision" meant. Knight did think that the legislature might be amenable to passing an amendment reducing the waiting period from six months to ninety days.[34]

Alfred was willing to accept this compromise amendment. He wrote to Knight, "I think the period of ninety days is a wise one and as far as my own individual interests are concerned, will make my [new] Will executed

on 15th of January legal on the 15th of April, at least such is my understanding of the proposed amendment."[35] It was almost as if Alfred had the
same psychic powers that Marguerite claimed for herself and that he could
foresee the date of his death. Knight had to inform Alfred that the proposed amendment would "not accomplish what you understand that it will
accomplish. Your Will of January 15th will not become effective on April
15th even though the amendment becomes law on that date. When the
Will was drawn the six months statute was in force and effect and therefore
. . . the new Will would not become effective for a period of six months
and then only in case you survived that period. . . . I can conceive of no
way to make your Will of January 15th effective before July 15th."[36]

Knight told Alfred that he would have to be satisfied with a second
codicil to his first will that would allow him to make a change in the trustees
and would permit him to add additional annuitants whom Alfred wished
to remember. What Alfred could not do by codicil, however, was to take
income from any beneficiary named in the first will and give it to a philanthropic organization. It was precisely that particular transition of funds
that the Florida law was designed to prevent. It would take a new will to
deprive Mrs. du Pont of first claim on the estate for her annuity and to give
all residual income from the estate, after the annuities had been paid, to
the Nemours Foundation rather than having that income go to Jessie du
Pont during her lifetime as the present will provided. If Alfred wanted these
latter changes to be valid, he would simply have to live until 15 July when
the new will would go into effect.[37] Alfred could push the matter no further. He had to be content with a second codicil to his first will to secure
some of the changes he had wanted in the event he should die during the
six-month waiting period.

The du Ponts went south earlier than usual in the winter of 1934–35,
forgoing the usual Christmas festivities at Nemours. Alfred was preoccupied
with writing his new will, and he was also pushing the Port St. Joe paper
mill project with great vigor. Neither Ed Ball nor W. T. Edwards had ever
seen their boss exhibit such a sense of urgency about any previous project.
It was as if he had set for himself a deadline that was inexorable and must
be met.

Denise came down to Epping Forest for the Christmas holidays, bringing with her one of Jessie's second cousins, Elizabeth Ball, who was Denise's classmate at Sweet Briar College. The house was filled with young
people home from college, for Jessie was eager to introduce Denise into
Jacksonville society. All of this excitement Alfred found very tiring, and it
was a considerable relief to him when the young girls departed for school
a week after New Year's. Jessie wrote to her sister Elsie, "Don't know just

what I am going to do about Denise's spring vacation as I notice the strain is very hard on A. I. You know he was looking so well before coming down here. He looks very tired and worn now, though we have done nothing in the way of social life, except the New Year's Day reception, which by the way was a great success. . . . 'Yours truly' spent all day Sunday making thirty gallons of egg nogg. Made none too much . . . by actual count we had five hundred and twenty-seven guests.''[38] Alfred might well have considered 527 guests something a little more than "doing nothing in the way of social life," but because the reception was for Ed Ball and his new wife, Ruth Price Ball, Alfred took the whole affair in good humor. He was delighted that Ed had finally married, and Alfred liked the bride—a beautiful young woman who was manager of the gift and tea shop that Ed had opened just outside the entrance to George Washington's home at Mount Vernon— one of the many little enterprises that Ball managed to conduct out of his hip pocket. Alfred wrote his niece that his role at the reception was to serve as "fist shaker No. 1. I was supposed to cripple fists, so they would do less injury to Jess, Ed Ball and his wife . . . and others down the line. . . . By seven o'clock, all our guests had went [sic] and peace was once more restored.''[39]

The peace that descended upon Epping Forest after the holidays did not seem to bring peace to Alfred, however. Jessie, for the fist time, became worried about his health. She wrote Elsie's husband, General A. J. Bowley, "Have never seen A. I. drive as hard as he has since he came to Florida this time, and he is beginning to show it very plainly. He looks very tired and is very, very nervous. For this reason, we are leaving on Monday or Tuesday on the yacht to be gone, I hope, the rest of the month.''[40]

The du Ponts did not start on their yacht trip as soon as Jessie had hoped, for Alfred couldn't leave until the new will had been drawn up and signed. Nor were they able to be away as long as Jessie had planned, for after a few days cruising down the inland waterway Alfred became ill, and they hurried back to Epping Forest. Jessie again wrote the Bowleys, "This spell was very much different from any he has had before, but no sign of thrombosis or any heart condition. In fact they haven't found the occasion for it, but I am in hopes he will soon be all right.''[41] The doctors diagnosed his affliction as nervous exhaustion, but Ellen La Motte, who had been a registered nurse and who now hurried down to Florida to see if she could help out, quickly ascertained that he had the old fashioned "grippe." Under her care, Alfred quickly recovered.

He continued to push himself on the St. Port Joe development, how- ever, and he also was involved in planning a major social event of his own, something that truly surprised Jessie. He invited all three of his children living in the United States, along with their families, plus his two nephews and their families and Jessie's sister Elsie, to "a family excursion to Miami Beach, March second to March sixteenth . . . [I] want you to help me advertise Florida and incidentally have a good time."[42] The whole party was to be housed at the Pancoast Hotel in Miami Beach, with all expenses, including travel, to be paid by Alfred. It was to be a grand family reunion, for, as Alfred wrote his nephew Maurice, "as Time passes, blood ties become stronger."[43] It was also to be, although none of them except possibly Alfred realized it at the time, his farewell to the family.

The excitement of the reunion proved to be a little more than Alfred had bargained for, and before the two-week Miami Beach excursion had ended, Alfred and Jessie quietly departed, Alfred pleading pressing business matters as his excuse. Back in Jacksonville, he again resumed his heavy work schedule while Jessie entertained Secretary of War George Dern and his wife. On 10 April, Jessie persuaded Alfred to take a couple of days' rest and to accompany her, the Derns, and Ellen La Motte on a short trip up the St. John's river on the *Gadfly*. As always, Alfred thoroughly enjoyed steaming up the St. Johns. He wrote Maurice Lee's wife:

> The St. Johns River [was] at its best, that is, complete with alligators, turtles and snakes. . . . We had two cars meet us at Sanford . . . and we returned to Johnsonville [Jacksonville] via Ocala and the Silver Springs, at which place Jess and Ellen and "Durn yer" and "Gol Durn yer" [the Derns] spent three hours in the boats, while I sat on a bench and was interviewed by local satellites, politicians and reporters; had one bottle of coca cola and a five cent plug of "Baby Ruth" . . . and cast an occasional eye over the abbreviated costumes of the ladies and gent swimmers as they passed silently in review. . . . I think I am gaining strength daily, though the visit to Silver Springs on Friday seemed to give me a temporary setback, due to the convention of nudists which I judged was being held that day—professional jealousy, I suppose![44]

In addition to his heavy business correspondence, Alfred continued to turn out letters to his family: to Alicia, congratulating her on finally getting custody of her son; to Madeleine, on the demerits of Der Fuehrer; to Maurice Lee on business conditions as they affected the Du Pont Company; and to Cousin Francis I. asking if he would serve as one of five trustees of "La Societe du Carillon de Nemours," which would oversee his proposed bell tower at Nemours. "It will not entail any great amount of work and there will be no salary to go with it—entirely a work of love."[45] Francis I. hurried down to Florida to accept this appointment and to see his old mentor and ally once again. And, of course, Alfred sent a letter to Marguerite, telling her that doctors were no good. They couldn't find "the cause of a man's headache when he has been kicked in the back of the head by a

Government mule," and enclosing yet another check for the Will Gardner family.[146]

On Wednesday, 24 April, Alfred as usual put in a busy day at the office. The Port St. Joe project was progressing, but not as fast as he wanted, and he was concerned about housing plans for the many new employees that would be needed. He returned home very tired but insisted on helping two plumbers who had come to install a new water heater. No one could fiddle around with his precious waterwheel except under his supervision.

The next morning Alfred was quite ill. A heart specialist, William D. Stroud, was summoned from Philadelphia, and Alfred Victor came down with him by plane.

Two days later, on Saturday, 27 April, Alfred seemed greatly improved. He talked local politics with Ed Ball, urging him to push for an additional bridge across the St. Johns in Jacksonville. Dr. Stroud, believing the crisis had passed, returned to Philadelphia.

On Sunday, Alfred sat up in bed, read the Sunday comics, argued with his young Catholic nurse about some of the differences between the Catholic and Protestant Bibles, and all in all, seemed well on the road to recovery. In the evening, Jessie wore the evening dress Alfred particularly admired when she brought him his dinner. She then went down to join her sister Belle, Addison Baker, and Alfred Victor for her dinner. Halfway through the meal, Jessie was summoned by a nurse. Alfred apparently had had another heart attack. A local doctor was hastily summoned, and an oxygen tent was wheeled in. Alfred muttered, "So it's come to this." An hour later, he lost consciousness. With Jessie and Alfred Victor standing close by his bed, Alfred regained consciousness about eleven o'clock and murmured, "Thank you doctors. Thank you nurses. I'll be all right in a few days." Those were his last words. At 12:22 A.M. on Monday, 29 April 1935, he died.[147]

Marquis James titled his biography *Alfred I. du Pont: The Family Rebel.* If Alfred du Pont was indeed the family rebel, circumstances, not his own aspiration, placed him in that role. He preferred to regard himself as the family conservator—the conservator of the old mills on the Brandywine when Coleman wanted to tear them down; the conservator of his employees' interests when his associates sought new and cheaper labor; the conservator of the company itself for the du Pont family when his elders wanted to sell out; above all, the conservator of the family honor in the bitter battle of the Three Cousins. If "family rebel" is not an appropriate sobriquet for Alfred I. du Pont, then perhaps "family fighter" is a term that everyone, including his enemies, could accept. Alfred's earliest boyhood ambition was to be a great prizefighter, and his youthful hero was John L. Sullivan, America's legendary boxing champion, whom Alfred knew and loved.

That eloquent aficionada of the manly art of boxing, the contemporary American novelist Joyce Carol Oates, wrote in her book *On Boxing* that "At its greatest intensity, it [boxing] seems to contain so complete and so powerful an image of life—life's beauty, vulnerability, despair, incalculable and often self-destructive courage—that boxing *is* life and hardly a game." That statement provides a fair summation of Alfred's life. His story was, as Oates said of boxing, "love commingled with hate [which] is more powerful than love. Or hate."[48] Alfred could both love and hate intensely, and from that mixture he drew the source of his power. He once wrote his niece Dorothy Lee, "Perhaps *I* am a trifle different from most men—God knows I look it; but I have an ingrained fundamental principle in life and that is never to admit defeat. My maxim is Never apologize; never explain. The time to call on your resources and spend your surplus is when the bricks are coming at you the thickest."[49] Alfred fortunately had the necessary resources in large reserve for the many battles he had to fight.

Of all of the many members of the du Pont clan, the only one who had Alfred's complete respect was the founding father of the family, Pierre Samuel du Pont de Nemours. The great French physiocrat once wrote, "I am too free, too proud, too frank, too assured in my conscience for any evil principle to sway me."[50]

His great-great-grandson might have been too modest to say that of himself, but Pierre-Samuel's self assessment can well serve as Alfred Irénée du Pont's epitaph.

Epilogue
The Will and the Way
1935–1985

O N 24 APRIL 1935, one day before his fatal heart attack, Alfred du Pont wrote what proved to be his last letter, addressed to his and Jessie's dear friend, the Reverend Baker Lee. Alfred wrote of the happiness he and Jessie had found in Florida, and concluded with the statement, "I think you are probably right in thinking that Jess and I will spend more time at Epping Forest than at Nemours. It is so nice and quiet down here and the people in Jacksonville are so nice and genuine. Did I make any remark of being distant from my relatives? If not, I should have included that also."[1]

Alfred did love Epping Forest, almost as much as Jessie did, but he remained a Yankee from the Brandywine who had found himself late in life transplanted to the banks of the St. Johns. Nemours would always be his true home. And when he contemplated the prospect of a final resting place, he knew it could only be back in his native soil. Not, however, in the Sand Hole Woods graveyard where lay four generations of the du Pont family. He had informed Jessie that he wished to be placed in a crypt at the base of the great carillon tower that he had commissioned his son and Gabriel Massena to construct on the grounds of Nemours.

The time had now come when Alfred's request must be faithfully satisfied. On 30 April, Jessie took the body of her husband in a private railroad car on a final trip northward to Wilmington. A few days later, Jessie wrote identical letters to her step-daughter Madeleine in Munich and to her brother Tom in Los Angeles, telling them of Alfred's last illness, his death, and the funeral arrangements.

> We came to Wilmington, arriving here on Wednesday, May first. The services here were conducted on Thursday, May second, at three o'clock, with Bishop Cook and Reverend Dubell, a lifelong friend of A. I.'s. The boy choir from the Cathedral of St. John the Divine in New

York with four adult voices chanted part of the service, and sang two hymns selected by Alfred [Victor], namely "Abide With Me" and "Onward Christian Soldiers." It was all very dignified, and I think just as A. I. would have wanted it.

He is still in the Morning Room, but a temporary crypt has been constructed and will receive him tomorrow. Reverend Dubell will hold the services at ten o'clock in the morning. We will then all return to Florida in the afternoon. Will be down there about a month, and come back up here, where we shall endeavor, and I trust that God will give me strength to carry on his work as he wished it.[2]

The Wilmington *Morning News,* which had once served as Alfred's principal medium for the promotion of political reform in Delaware but which, since its acquisition by the Du Pont Company, had been his unrelenting critic, pushed aside for the moment the du Pont family feud and carried a detailed account of its former owner's last rites:

> The simple Episcopal burial service, imbued with more solemnity and richness by the participation of a choir of twenty voices, sounded through the halls of Nemours yesterday afternoon as nearly a thousand persons paid final tribute to Alfred I. du Pont.
>
> Filling several rooms on the first floor of Mr. du Pont's home outside Wilmington, with many standing, friends and former associates of his social, business and political career, at the conclusion of the brief service, silently filed past the casket, . . . which had been placed in the morning room [and which] was covered with white orchids, gardenias, roses and lilies. . . . In the conservatory sat a large number of men who had worked with Mr. du Pont years ago in the old powder mills on the Brandywine and many close associates. The other rooms were reserved for members of the family, relatives, and out of town friends.[3]

Jessie was undoubtedly surprised to read further on in this same newspaper account that among the other members of the du Pont family in attendance was Pierre S. du Pont. Jessie, who had never met Cousin Pierre, was unaware at the time that, just as the services began, this least expected visitor had slipped quietly through the front entrance and had stood inconspicuously at the rear of the overflow crowd.

The surprised few who did note his presence among the standees surely must have speculated on what Pierre's thoughts were as he made this, his first and only visit to Nemours. Did Pierre still retain some affection for the man who in their youth had been his closest friend? By his presence at Alfred's funeral, was Pierre now expressing regret for the rupture in their relationship that had turned Alfred into his most implacable foe? Or had Pierre come to the service only out of a sense of family duty—not out of respect for the memory of his onetime mentor, who had completely changed the course and fortune of Pierre's own life? These were, of course, questions which most probably would never be asked of him, and even if asked, would certainly never be answered by Pierre. Perhaps he could not possibly have answered such questions, for a relationship as convoluted as his and Alfred's could only have produced in Pierre a jumble of emotional responses—love coupled with hate, remorse ameliorated by success, and nostalgia terminated by curt dismissal. Of only one thing could Pierre be certain. The case of du Pont vs. du Pont had now finally ended with the closing of the flower-draped casket in the Nemours morning room.

On Wednesday, 8 May, after a brief interment service in the morning, Jessie departed for Jacksonville.[4] She had dutifully taken Alfred back to his home at Nemours, and she was eager now to return to what she regarded as her own home, Epping Forest. There are always many duties which death imposes upon the survivor—letters to write, papers to sign, and most pressing of all for Jessie, a will to be filed and an estate to be settled. "It is my wish to have A. I.'s children here while the will is being read," Jessie wrote her brother Tom soon after arriving in Florida. "I realize that it is not necessary, yet I deem it a wise and fair measure, as they should be the first ones to know the disposition of his property."[5]

On the following Monday, 13 May, Jessie and Alfred's children, with the exception of Madeleine, who was in Germany, gathered in Ed Ball's Jacksonville office for the reading of Alfred's will. Although it had been Raymond Knight who had actually drawn up the will in 1932, for an occasion as important as this, Knight's senior law partner, Henry P. Adair, claimed for himself the honor of presiding at this solemn ritual announcing "the disposition of Alfred's property."[6]

A will, however, is much more than a prosaic legal document prepared for the sole purpose of disposing of property. A will also serves as a personal summation of the testator's life. It is a memoir, and usually the only autobiography the deceased leaves to posterity. The beneficiaries of the will—as well as biographers—can learn much from the document, for the testator is in effect saying, "Find here in the listing of my material possessions, some evidence of my accomplishments. And in the naming of my beneficiaries I present those persons who were meaningful to my life." A will is an expressed remembrance of loves enjoyed—and occasionally of hates suffered. A will is also a last reach for immortality, for in providing for family, friends, and good works, it seeks to insure that some substance of the departed person's life will endure and be cherished.

The actual reading of a will is seldom an inspired performance, but in spite of its being a dull droning of ponderous legal phrasings, this particular form of public reading never fails to captivate its audience. Alfred's 1932 will, with its two codicils of 1933 and 1935, was a lengthy document. It took Adair over an hour to complete the reading, but he never lost the attention of his auditors. Forty pages, after all, could hardly be considered excessive in providing Alfred's widow and his children with a summary statement of his life, for Alfred had succeeded in encapsulating within these pages much of what had been meaningful to him and about him.

Here was his expression of love for Nemours and his desire to share that home with others: *My Trustees are specifically instructed that it shall be their duty first, to care for the mansion and grounds and gardens surrounding "Nemours" in order that they be maintained for the pleasure and benefit of the public.* Here too was his pride in ancestry, or at least pride in a few ancestors: *The "Nemours Foundation" shall be created and maintained as a memorial to my great, great grandfather, Pierre Samuel du Pont de Nemours, and to my father, Eleuthère Irénée du Pont de Nemours, and a proper tablet shall be erected in the present mansion to so indicate.* And here were his social welfare concerns: *My said Trustees are hereby to pay over . . . to the said corporation . . . the net income of my said estate, subject to the annuities and legacies hereinabove mentioned, for the purpose of maintaining the said Estate of "Nemours" as a charitable institution for the care and treatment of crippled children, but not of incurables, or the care of old men or old women, and particularly old couples, first consideration, in each instance, being given to beneficiaries who are residents of Delaware.*

Here also he remembered those persons who had been bound to him by kinship, friendship, and above all, affection: gifts of Almours stock to Jessie's siblings, and annuities to his foster daughter Denise, his nephew Maurice Lee, his two nieces Charlotte and Nesta du Pont, his secretaries and his servants at Nemours, his six grandnieces and grandnephews— Alfred's Pirate Gang who had romped with him through the woods of Nemours in hot pursuit of rabbits and had raised the Jolly Roger flag on his yacht—his favorite cousin Ellen La Motte, his old political allies T. W. Jakes and George Maxwell, his trusty dispenser of old age pension funds Laura Walls, and a few old friends who had long been dependent upon Alfred's generosity for their sustenance.

In the listing of beneficiaries there were a few notable omissions which Alfred had taken care to explain: *I make no provision in this my Will for Alicia Amory Moddox Glendenning, [sic] the daughter of my former wife . . . for the reason that I consider she is amply provided for under the Will of her mother. . . . I make no provision in this Will for my sister, Mrs. Marguerite du Pont Lee . . . nor for my brother, Mr. Maurice du Pont . . . for the reason that ample provision has been made for them in a certain Trust Agreement.*[7]

Alfred's children, who must have listened to the reading of the will with rapt attention, heard nothing in the terms that greatly surprised them. They were well aware of the fact that in 1926 when they had signed waivers of any filial rights to the estate in exchange for 8,000 shares of Du Pont stock, they had relinquished any legal claim to be beneficiaries. But they also hoped and believed that their father would not omit them entirely from his will, and indeed, he had not. Each was to receive 5,000 shares of Almours common stock, valued at approximately $750,000.

Nor could Alfred's children have been surprised to learn that the principal beneficiary of his will was Jessie Ball du Pont. It had been apparent to all who knew Alfred that Jessie held first place in his affection. If there was anything unexpected in the bequest to his wife, it was not in how much but rather in how little she was to receive in fee simple. Only Epping Forest and his personal effects were to go to Jessie outright, along with a first claim annuity of $200,000. The major portion of his estate, however, estimated at $40 million, would belong to the Nemours Foundation, the income from which Jessie would have during her lifetime. It would be hers to use but not to own or to disperse to others.

The will did not specify a precise date for the establishment of the Nemours Foundation. There was an implication that the foundation might be created prior to Alfred's death, but if not, it was to be established as soon as possible after his will was probated. Alfred did name in the will the trustees who would administer the foundation: Jessie Ball du Pont, Edward Ball, Reginald Huidekoper, and a representative of the Florida National Bank in Jacksonville.

Undoubtedly some of the du Pont family were both surprised and pleased that most of Alfred's estate had not gone directly to Jessie in fee simple, but Jessie herself was neither surprised nor disappointed by the terms of her husband's will. Alfred's decision to create the Nemours Foundation had long been known to her, and it had had her complete approval. Her own personal wealth was now great enough to provide for any beneficiaries she might wish to remember in her will, and even if her concern for crippled children and old people had never been as keen as her husband's, nevertheless she respected his philanthropic interests. What was important to her was the foundation itself. It would be a continuing memorial to Alfred. She accepted it as a sacred trust, and for as long as she lived, she would be the zealous keeper of that eternal flame to his memory.

Although Alfred's will had contained few surprises for the family assembled in Ed Ball's office, Alfred's children must surely have been puzzled by one enigmatic reference Alfred had made in the second codicil, dated 15 January 1935. "I had considered executing a new Will at this time which would immediately revoke the aforesaid Will and Codicil . . . but my attorneys have called my attention to Section 20 of Chapter 16103, Act of 1933, Laws of Florida, relating to charitable devises and bequests . . . rendering such bequests invalid unless the instrument making them was duly executed at least six months prior to the death of the testator.''[8] What could this mean? There was no elucidation forthcoming from the lawyer. The family was never to know the details of the new conditional will to which Alfred had referred, nor were they to know that this will had actually been placed in the hands of his lawyers with instructions for it to supersede his previous will if he lived until 15 July 1935.

The second will would have in no way affected his children's inheritance, for it would neither increase nor diminish the bequests that they were to receive as outright gifts. But for the heirs named as annuitants, the new will would have been much more beneficial, since it did not stipulate that Jessie should have first claim on $200,000 each year before any other annuities could be paid. It would have prorated the annual income from the estate among all the annuitants, including Jessie, up to the full amounts stipulated in the will. In prosperous times, this distinction in the two wills would have had no real significance, because Alfred's estate would then provide enough income to pay all of the annuities in full and still leave a healthy surplus in reserve. But in the depressed year of 1935, however, economic prospects were at best uncertain. It could well be that there would be precious little pie remaining for the other annuitants to share after Jessie had received her very generous slice.

An even more important distinction between the two wills was that, under the terms of the second will, the Nemours Foundation was to be created immediately after his death. All of his property except Epping Forest and his personal effects were to be transferred directly to the foundation, which would not only hold Alfred's estate in fee simple but would have the annual income paid directly to it for use in philanthropic missions. Jessie would not have the use of that income during her life, but in compensation for that loss her annuity was to be increased from $200,000 to $475,000. She could also lease the Nemours mansion and grounds for the bargain price of $500 a year. Had Alfred lived and had this will become effective on 15 July 1935, it would ultimately have made a great difference in Jessie's annual income. She would have received $475,000 a year instead of the $15 million a year she received by the late 1960s.

Fortunately for Jessie's relationship with Alfred's other beneficiaries, the existence and details of the second, conditional will remained Knight's and Adair's carefully guarded secret. If Alfred's sister, Marguerite Lee, had ever got wind of the second will, her letters to the estate's legal counsel would certainly have been acerbic. Even without benefit of this knowledge, her letters regarding the management of the estate were sharp enough to cut deeply into the hide of the most callous recipient.

Following Adair's reading of the will, the three executors—Jessie du

Pont, Edward Ball, and Reginald Huidekoper—immediately agreed to retain Henry Adair as attorney for the estate. The family had conducted its business with seeming good grace. Alfred's children had accepted the terms of the will with apparent gratitude for what they themselves had received, and also for Bessie's husband Reginald's having been named as one of the executors of the estate.

In the spirit of the familial harmony that prevailed, Jessie invited all of them to return to Epping Forest with her for lunch. On the ride out to her home, Jessie sat in the back seat of one limousine between her two step-children, Victorine and Alfred Victor. According to Alfred Victor's later account of their conversation, Jessie almost immediately turned to Alfred Victor and said that it was her intention to begin at once to implement a part of the Nemours Foundation's charitable enterprises. She wanted to build a hospital for crippled children on the grounds at Nemours as soon as possible. She then said, "Alfred, of course you will be the architect." Alfred was delighted to hear this. "That is wonderful," he replied. "Of course, Alfred!," Jessie added, "your father would not have it otherwise." Alfred assured her that he and his partner, Gabriel Massena, could begin work on the preliminary plans as soon as the carillon tower was completed, and with that the matter seemed settled.[9]

Determined to create the Nemours Foundation as soon as possible, Jessie within the following week called together the other two members designated in the will as trustees of the foundation plus W. H. Goodman, head of the Trust Department of the Florida National bank, who had been chosen by that institution to be its representative on the board.

The disposition of an estate can frequently evoke dissension, even among closely knit families, and Alfred's immediate family was anything but harmonious. There had been suspicion and tension between the du Ponts and the Balls from the moment that Jessie entered the family, bringing in her train her siblings as well as numerous Virginia cousins. Alfred had welcomed them all with hearty enthusiasm. He was far more comfortable with the Balls and their Harding cousins than he had ever been with his own du Pont kin—even those who were the closest to him.

There remained, however, a bond of deep affection between him and his two surviving siblings, dating back to their childhood. Neither Marguerite's continuous carping nor Maurice's quiet but equally disturbing aloofness could ever sunder this bond. By the sheer strength of his affection for both his own and his wife's kin—as well as his remarkable patience in turning quite literally a blind eye and a deaf ear toward the ill-concealed discord that surrounded him, he had been able to hold together the family he loved during his lifetime. With his departure, however, there was left no center to hold, and it was inevitable for things to fall apart. Alfred had been the centripetal force for family unity. His will, unfortunately, proved to be an equally powerful centrifugal force for disintegration.

The family troubles began when Marguerite, supported by her son, Maurice Lee, wrote to Goodman to demand that he provide her with a list of all of the annuitants and the amount each was to receive annually. By a vote of three to one (with Huidekoper in the minority) the trustees refused to give out this information. Only after Marguerite and Maurice Lee threatened to take the matter to court did the trustees yield. Adair sent her the list, and in return received a scorching letter:

> I am in receipt of your letter of the 8th, and would remark before going further that it is of quite a different nature from half a dozen received sometime since from the Trustee, who cried aloud to High Heaven that the matter was strictly confidential, and could not be divulged. Of course, I know my rights in the matter (inasmuch as Mr. Goodman was dealing not with a stupid Ball, but with a smart du Pont), and that you would get an opinion from the Court; and that if you did not, I would. I knew the provisions of the Trust, and the names of the annuitants long before writing to the august Trustee; but Mrs. J. B. decreed I should not have the information from the Trustee and I was bound I should. The Trust was originally made by Mr. du Pont for his own kin; his words to me were "Jessie has her own money and can take care of her own." This before her tremendous crash in the stock market. The 15% reduction [in the annuities as originally set in 1927] was necessitated by Thomas Ball—afflicted, poor man—and a preacher by the name of Lee, being foist for support upon Mr. du Pont's relatives. . . . I might remark in passing, this holy man of God violated the laws of his church when he undertook to bind a man and a woman in holy ties of "staggered concubinage." . . . Mrs. J. B. du Pont cannot browbeat *me*.
>
> What saith Shakespeare? "The robb'd that smiles steals something from the thief."[10]

Marguerite, unlike Alfred, had continued to maintain a friendly relationship with Pierre du Pont. She sent her cousin a copy of this letter, marked "Pour vous amuser." Pierre was indeed amused by the intramural squabble of Alfred's family. "Your letters are always welcome and your style is inimitable," he wrote.[11]

Marguerite had long regarded Jessie as being bent upon depriving Alfred's blood kin of their proper share of his wealth. She blamed her sister-in-law for frivolously wasting Alfred's income on yachts, a second mansion in Florida, and expensive furs and jewels, thus denying Marguerite of the funds she desperately needed to carry out her noble charity work. Now with Alfred's death, in her position of power as chief executor and trustee of his estate, Jessie would, Marguerite was convinced, cheat Alfred's own family of even the pitifully small amounts he had allotted them in his will.

Marguerite was also angered that Reginald Huidekoper had been named as an executor of the will and a trustee of the foundation, rather than her own son, Maurice, to whom Alfred had at one time promised these assignments. To Marguerite this was conclusive evidence that the fine Machiavellian hand of Jessie was busily at work to achieve her nefarious ends. Such suspicions boded ill for future family relations.

Alfred's brother Maurice held the same views of Jessie as did Marguerite, although he did not engage in writing angry letters to the trustees. Always somewhat resentful of his older brother's success and embarrassed by having to be bailed out of his own unsuccessful ventures by Alfred, Maurice had nevertheless maintained a close relationship with his brother and a correct posture of friendship with Jessie during Alfred's lifetime. He seldom accepted Alfred's repeated invitations to visit Nemours and Epping Forest, however, and he found real pleasure in his brother's company only on those rare occasions when they went off alone on a fishing trip to Cherry Island or a sail down the Chesapeake. Now with Alfred's death, Maurice saw no reason for keeping even a pretense of kinship, not to mention friendship, with Jessie. In 1939, soon after she had engaged Marquis James to write a biography of Alfred, Jessie wrote Maurice a friendly letter in which she asked her brother-in-law to grant James an interview:

> My dear Brother Maurice:
> It has been a long time since I have heard from you and longer still since I have seen you—a fact I have long regretted deeply . . .
>
> I am leaving tomorrow for Florida then to California. Will be back the middle of October when I hope to have the pleasure of seeing you. In the meantime I am writing to know if you would see Mr. Marquis James, who is collecting material for a very interesting biography of A. I. He is anxious to get from you experiences of your childhood days and the many pranks . . . , which I have heard you and A. I. so often reminisce over as the audience would be convulsed with laughter. . . .
>
> With love to you, Margie and Nesta, and hoping your health is very, very good, I am
>
> > Affectionately yours,
> > Jessie Ball du Pont.[12]

Maurice in reply sent a very curt refusal to see James. Maurice was insulted that he had been asked only for information about "childhood pranks," and in spite of repeated requests from James himself and even from Maurice Lee, Maurice steadfastly refused to meet James.[13] Thus ended all personal contact between Jessie and her brother-in-law.

Jessie could not decide which was the more exasperating—Marguerite's vitriolic communications or Maurice's chilly rebuff of any communication. In both instances, however, Jessie was painfully aware that their animus was directed against her, not the other trustees. Both Marguerite and Maurice would agree with that autocrat of the breakfast table, Oliver Wendell Holmes, when he wrote, "Man may have his Will—but woman has her Way!"

Other members of the family soon realized, however, that the real directing force in the management of Alfred I. du Pont's estate was not Jessie, but Ed Ball. Jessie, to be sure, was nominally *prima inter pares*. She could enforce her will upon the others and could veto proposals of any of the other three. Increasingly, however, it was Ball who suggested and Jessie who concurred, with the other two of necessity either accepting or finding themselves ignored. It was Ed's decision to expand greatly the number of banks from the seven acquired during Alfred's lifetime to thirty by the time of Jessie's death in 1970. It was also Ed's decision to push full steam ahead with the development of the Port St. Joe Paper Company, which Alfred had only begun at the time of his death. And it was Ed who sought to postpone or, if possible, to evade entirely the obligation of meeting the annuities and outright grants that Alfred's trusts and will had imposed upon the estate.

During the first several years after Alfred's death, Jessie was so preoccupied with getting the Nemours Foundation established and the hospital for crippled children built on the Nemours grounds that she left the more mundane day-to-day operations of the estate to her brother. It seems fair to assume that Jessie would have greatly preferred to have preserved some semblance of family unity after Alfred's death if that had been possible. Although she had always found Marguerite difficult, Jessie had genuinely liked Maurice and especially Maurice's wife Margery. And she did want to keep close contact with Alfred's children, whom she prided herself on having brought back into the fold. More important to her, however, than continuing personal relations with Alfred's family was the preservation of Alfred's memory through his cherished foundation. If the two concerns came into conflict, as they sometimes did, then family ties had to give way.

Ed Ball, on the other hand, was never disturbed by such conflicting interests. He had very little sentimental attachment to anyone, even to his own kin. Ultimately, he would break relations with both his sister Elsie and his sister Belle. Needless to say, if he cared little for his own siblings, he had no affection whatsoever for his du Pont in-laws. Ed Ball's major interest was to hold tightly to the pot of gold that had finally come his way and, by his own shrewd efforts, to increase its contents.

Ball had no difficulty justifying his position. Most of Alfred's kin were, in Ed's eyes, simply drones in the du Pont hive. They wanted to partake of the honey without working for it. Maurice du Pont, for example, was an incompetent bungler who had failed in every enterprise he had ever undertaken, and as for Marguerite, anything given to her would immediately be passed on to drunken bums or fawning preachers. Nor could Ed Ball see any particular talent in Alfred's children except that of spending their father's hard-earned money.

It had been Jessie who had stepped in when Alfred had reached the lowest point in his life—deposed from the company he had saved for the family and burdened with debt from his unsuccessful ventures in New York finance. It was she who had given a new direction and meaning to his life. And who had stood at Alfred's right hand and had guided him to a successful conclusion of the new course that Jessie had set for him? Edward Ball, of course.

In Ed Ball's view it was only proper that the estate should be governed by the two people who had done the most to preserve and to increase it during the last fifteen years. Jessie and Ed owed nothing to any du Pont except Alfred himself, and the most appropriate way that they could serve Alfred's memory was to increase the estate manyfold in the years that were left to them. Given this line of reasoning, it proved to be not too difficult to convince Jessie that her brother Ed was pursuing the right way.[14]

The first clear indication of what Ed Ball's "way" would mean to the Ball–du Pont relationship came in November 1938, when Almours Securities, which contained the entire assets of the Alfred I. du Pont estate except for the Nemours property in Wilmington, was abruptly dissolved and its capital, $50 million, was liquidated.

The first official news that the twelve annuitants had of the dissolution, which was to have an immediate and adverse effect upon their annual income, was the notice that they received on 26 December 1938. They were informed that as of 1 January 1939, they would receive no further annuity payments until such time as the assets of Almours Securities could be reinvested and income had been received by the estate from the new investments. The annuities would then be prorated among the recipients on the basis of that income. Henry Adair's only explanation for this totally unexpected move was that the heavy state and federal inheritance taxes that had been levied against the estate had forced the trustees to take drastic measures to raise capital to meet the tax obligations.

Fortunately for the well-being of the trustees, Marguerite du Pont Lee was no longer on the list of annuitants to receive this notice, for she had died in October 1936. Had she been alive to protest this action, even her remarkable vocabulary would not have been extensive enough to convey her feeling of outrage.

It was her son, Maurice Lee, who took it upon himself to become the spokesman for the other annuitants. In a somewhat more temperate letter than his mother would have written, Maurice immediately brought the issue directly to Jessie:

> The sudden and complete cessation of all monthly payments to the trust beneficiaries, without any warning whatsoever, has placed Uncle's brother, his brother's wife and two children in such straits that I, from my meager resources, am compelled to come to their assistance to the extent of $2,500 per month. This I am going to continue so long as necessary or until my funds are exhausted. . . . It is unbelievable but true that in less than four years after Uncle's death, leaving an estate of more than $50,000,000., his only brother should be in such a position, while at the same time an office building, a factory and an institution costing millions are being created by his estate. . . .
>
> I am going to do all in my power to make every year that remains to Uncle Maurice as comfortable and happy as I can, knowing that my mother and A. I. would so desire. Should my resources fail in the doing, I am certain that a personal appeal to Pierre will not go unanswered.[15]

If Maurice had expected his letter to bring remedial action from Jessie, he was disappointed. She replied politely but stiffly, "It is rather difficult for me to understand why you directed your letter of January ninth to me rather than to the Trust Department of the Florida National Bank, as the content of the letter is pertaining to the trust that was set up at the bank ten years or more ago by Mr. Alfred I. du Pont—a trust over which I have no jurisdiction and in which my interest is nil, unless there be a remainder from the income, after all annuitants receive their annuities."[16]

Maurice had shared the contents of his letter to Jessie with his cousin, Francis I. du Pont, who had continued to remain on friendly terms with both the Balls and the Alfred du Pont family. Francis I. was appalled at the sentence in Maurice's letter in which he mentioned that it might be necessary to ask Pierre to come to the financial relief of Alfred's brother. Marguerite, to be sure, had often in the past effectively used this ploy of threatening to go to Pierre to get funds for her favorite charities in order to get money from Alfred, but Francis was sure such tactics would not work with Jessie. "I don't believe you realize what a serious mistake you made when you mentioned Pierre the way you did in the letter you read to me, and I am going to recommend that you do the best you can to repair this damage," he counselled Maurice. "I should think you ought to open your letter with a humble apology for having written that. . . . If you do, it may possibly soften her feelings a little and she may come to look upon you as somebody whose advice would be very welcome.

"I do not think the world contains many women who would have so laboriously and so assiduously used her energy to carry out the wishes of her husband, and for this reason, I feel certain that she has every desire to do the right and proper thing in regard to these trusts. . . . It seems to me that the only thing Jessie could do would be to get the Court to sanction some act on the part of the Estate which could be equivalent to this guarantee. In my opinion, she is the person to do it, and no one else."[17]

Although Maurice Lee did not hold as generous a view of Jessie as did Francis, he nevertheless accepted his cousin's advice. He wrote Jessie as apologetic a letter as he could under the circumstances, and she graciously received his apology. "Your letter of February fourth apologizing for the one of January ninth came not as a surprise to me. Being the gentleman you are . . . I was certain that upon deliberation you would send forth another letter, which I was glad to receive and gratefully accept."[18]

Jessie had suggested that Maurice offer to the Florida National Trust Department any proposals he might have as to how the recently liquidated trust funds could best be handled in order to provide for the annuities that Alfred had established. Maurice promptly presented to Jessie, not to the Trust Department, two possible options for how the annuities could be paid: "The executors can apply to the Court for permission to transfer from the estate to the trust sufficient additional funds so the trust can be carried out in complete detail; or they can apply to the Court for permission for the estate to guarantee that trust and its fulfillment. The latter method is much to be preferred since it would give complete protection to the annuitants as well as to the minor beneficiaries in years to come, by providing for any shrinkage in principal, which is almost certain to occur over a long period of time where it is necessary to sell and reinvest securities."[19]

Jessie promptly turned these suggestions over to Henry Adair for his advice, and after some deliberation, Adair gave his opinion: "The conclusion is (a). That the Executors and Trustees have no duty, moral or legal, to pursue either of the courses suggested by Mr. Lee; (b). That the Executors and Trustees have no power to pursue either of such courses; and (c). That no court has the power to authorize the Executors or the Trustees to follow either of the suggestions made by Mr. Lee or to do anything except administer the Will and Codicils according to their terms."[20] Ed Ball was, of course, delighted with the opinion offered by the estate's legal counsel. Other than paying the inheritance taxes on the estate (which could have been handled in a different way), the major motivation for the dissolution of Almours Securities and the liquidation of its assets was to provide Ball with the funds he needed for the ambitious plans he had in mind—the acquisition of additional banks in Florida, the purchase of fifty thousand more acres of forest lands in the Florida panhandle, and the completion of the Port St. Joe paper mills.[21] Supported by his own counsel's opinion that it would be unwise, if not illegal, to petition the court to set aside capital to pay in full the annuities, Ball simply ignored Maurice's suggestions.

For the next six months, Maurice out of his own pocket provided Uncle Maurice, Aunt Margery, and their daughter Nesta the annuities they were no longer receiving from the trust fund. Only the second daughter, Charlotte, still a patient in the Kellogg Battle Creek clinic where her Uncle Alfred and Aunt Jessie had placed her many years earlier, received her full $8,500 annuity.

In August 1939, W. H. Goodman of the Florida National informed Maurice Lee that the annuity trust fund had been reinvested and was now earning income sufficient to pay 60 percent of the original annuities.[22] Over the next several years the annuities were gradually increased, and by 1941 the beneficiaries were receiving 86 percent of the full amount of their annuities: "None of the shortage was ever made up," Maurice wrote in his unpublished memoirs.[23]

Another trust created by Alfred's will provided for the payment of $100,000 to each of Marguerite's six grandchildren and Alfred's own six American grandchildren, born to Bessie and Victorine prior to the time of Alfred's death. These grants were to be paid to each recipient upon reaching the age of thirty. "The first of these amounts should have been paid July 14, 1944," Maurice Lee wrote in his memoirs, "but, in spite of much correspondence and discussion, was not paid until late in 1964, when it was twenty years overdue, and was only paid then after one beneficiary and I went to court in Jacksonville with our lawyer and demanded an accounting. Five have now been paid, up to the end of 1966, and the others will follow at intervals."[24] It is hardly surprising that the Lee and du Pont families were to harbor over an entire generation considerable resentment toward the man whom they held to be largely responsible for the failure to fulfill promptly the obligations that Alfred's will had placed upon its executors.

Personal resentments directed against Ed Ball never caused him to lose any sleep. He was accustomed to family squabbles and he relished court suits, whether brought by or against him. He expected to lose some, to be sure, but during his long and actively litigious life he managed to win far more than he lost. Alfred once said to his wife, "Don't let anyone minimize Ed Ball's ability to you. I have worked with him for years; I know his ability, his integrity and his devotion to you and to me and his loyalty to a trust."[25] Some might question Ball's integrity, but no one could doubt his loyalty to the preservation of the estate of Alfred I. du Pont, which it was his destiny to hold and to augment over the next forty-six years. For Ball, increasing Alfred's estate became an end in itself, an ark of the covenant that he believed to have been created between him and the only man he ever completely admired. He was as dedicated as any religious zealot to protect that ark and to make it ever more splended. It pained Ball more to disperse any of the fortune to anyone, even to Jessie, than it would had it been his own money, and he lived as parsimoniously as a man in his position could—two small rooms in a hotel directly across the street from the Florida National Bank, daily lunches at the Walgreen drugstore counter, the least expensive car available on the market. The only luxury he allowed himself was Wild Turkey bourbon. Each evening in his small hotel sitting room, in the company of his closest subordinates, he would lift his glass and offer the same toast—"Confusion to the enemy." Ball, during his long life, would have many enemies. They were always confused by this powerful little man, and they were usually defeated.

In her own, quite different way, Jessie Ball du Pont was equally dedicated to the preservation of the trust that she believed Alfred had bequeathed to her. As social as her brother was antisocial, she was determined to keep Nemours and Epping Forest as beautifully and expensively maintained and as full of guests as they had been during her husband's lifetime. The house parties and cruises down the Chesapeake and formal dress for dinner each night, even if she dined alone, continued. She welcomed all of the family, both Balls and du Ponts, into her home, and she kept alive what remained of the life she had known with Alfred.

In the spring of 1936, the carillon tower was at last completed. It was an impressive monument, rising 210 feet to be the highest structure in Delaware, and as the Wilmington *Star* reported, it was "the largest reinforced concrete tower in the world," built to carry in its apex, "the carillon of 31 bronze bells, which have been successfully cast."[26] Alfred Victor was now as eager as Jessie to turn to the building of the children's hospital.

Jessie, however, apparently had had second thoughts about the promise she had made to Alfred Victor a year earlier. The first hint of her change of mind came in May 1936, shortly after the dedication of the tower, when, in conversation with her stepson, she told him that Alfred had had to correct the structural drawings that Massena and du Pont had made for the tower by inserting cross bracings and that Alfred furthermore had been highly critical of the firm's structural engineer. Alfred Victor was dismayed by Jessie's comments and immediately wrote her a note saying, "There must be some misunderstanding here as he never, to the knowledge of anyone in our office, criticised [sic] or asked revisions of any of our structural drawings either on the Tower or any other projects which we have executed.. . . . I am sure you will realize that if a rumor were to get about whereby any of our structural drawings were reported not to be up to snuff, it could do our office untold harm."[27]

Jessie, however, refused to retract her statement that Alfred I. had been dissatisfied with his son's work. Alfred Victor was told that if his firm were to get the commission for the hospital it would only be on the condition that Massena and du Pont accept an outside consultant who would work closely with them on the project.

Both Alfred Victor and the highly sensitive Massena found this condition both insulting and damaging to their reputation. Alfred Victor appealed to the one du Pont family member of the trustees, his brother-in-law Reginald Huidekoper, who did his best to convince the other trustees that Massena and du Pont should have the commission with no strings attached.[28] The most Reginald could achieve was to get the commission for Alfred Victor, but with an outside consultant chosen by the trustees. Massena and du Pont had no alternative. Major private projects were hard to come by in the Depression, and once the carillon tower job was finished, there was no other work in the offing. The firm accepted the commission on Jessie's terms, but with considerable reluctance, and the whole project got off to a bad start.

The next hurdle that confronted Alfred Victor was that of cost. Ed Ball, who had never wanted Alfred Victor to get the commission, informed the architects that costs must not exceed $700,000, even though Alfred's will had provided a maximum of one million dollars for the construction of any institution at Nemours. Alfred Victor and Massena were not free to pick contractors who they thought would do the best job but instead open bidding was required. The lowest initial bid was for $740,000, which necessitated an entire new set of plans and specifications. "I understand from Alfred [Victor]," Maurice Lee reported to Uncle Maurice, "that the specifications are now being cut in such a manner that he is very fearful of the success of the venture. I do not know how true this is—it is merely his view. The architects, I believe, are getting somewhat discouraged with the many changes and are about ready to give up."[29] There must have been times when Alfred Victor and Massena, desperate as they had been to get the job, regretted that they had succeeded in obtaining the commission. The hospital did get built, however, and was officially opened in 1940. Although somewhat smaller than Alfred Victor had originally planned, it was, like the Welfare Home in Smyrna, another architectural triumph for Massena and du Pont. Even Jessie expressed her complete satisfaction with what the two architects had accomplished with a very limited budget. It would be, however, the last commission that Massena and du Pont would obtain from Alfred Victor's father's estate.

Jessie was equally fortunate in her choice of the man who was to serve as the first medical director of the hospital. After consulting with several of the most distinguished and knowledgeable specialists in the field of orthopaedic medicine, she selected Dr. Alfred R. Shands, Jr., who was then a professor of surgery in charge of orthopaedics at Duke University. She could not have made a better choice. Dr. Shands had received his medical training at Johns Hopkins, and after serving as instructor in surgery at George Washington University, he had been invited to help set up the Duke University medical school. Author of what is still today considered to be the classic text in the field, *The Handbook of Orthopaedic Surgery,* Shands had the highest respect of the entire profession. For the next thirty-two years, until his retirement in 1969, Alfred Shands gave distinguished service and leadership to the Alfred I. du Pont Institute as administrator, teacher, and surgeon, making the institute a leading center in the research and practice of orthopaedic surgery. During his tenure, nearly 25,000 crippled children were treated at the hospital, and through that treatment were given the opportunity to lead productive lives.[30] Jessie could take great satisfaction in having successfully fulfilled one part of Alfred's vision of a better society.

The succeeding decades were as satisfying for Ed Ball in augmenting the estate's fortune as they were for Jessie in dispersing a portion of that fortune in the good works Alfred's will had called for. For the next thirty-five years, the most operative word within the administration of the Alfred I. du Pont Estate was expansion. During that time, the estate acquired twenty-three more banks in Florida, making it one of the largest banking institutions in southeastern United States. It bought up nearly one and a half million acres of forest land in Florida and southern Georgia to provide the four billion pounds of wood chips that the paper mill at Port St. Joe needed to produce each year 1,250 tons of kraft paperboard and 500 tons of bleached pulp. By 1960, the latter enterprise had expanded far beyond the little fishing village of Port St. Joe. There were container plants as far north as Massachusetts, as far west as Texas, and as far east as Dublin, Ireland.[31]

In 1961, the estate acquired the bankrupt Florida East Coast Railway Company, built by Henry Flagler in the 1890s. Ball took over as chairman of the Board of Directors of the company and within five years had turned a failing enterprise into one of the few highly profitable railroad lines in the country. He did this by eliminating the passenger service and greatly reducing the number of employees. When the unionized workers went out on strike, Ball successfully defied both the unions and the federal government and won a court decision that made the Florida East Coast Railway the only non-unionized railroad line in the nation. The railroad was, in Ball's own words, "my most cherished possession."[32]

Ed Ball's power in Florida and the nation, as well as among the trustees of the estate, expanded as rapidly as did the estate's holdings. In the first years after Alfred's death, one of the trustees, Reginald Huidekoper, had been somewhat of an obstructionist in Ball's otherwise smooth operations. Huidekoper had protested when Almours Securities had been dissolved and had sought unsuccessfully to provide an adequate reserve so that the annuities could be paid in full.[33] He had also intervened, this time successfully, to obtain the architectural commission for the hospital to Alfred Victor over Ball's opposition. True, Huidekoper was almost always voted down by a vote of three to one, but still it was a nuisance to have his single voice raised in opposition. It was, therefore, a considerable relief to Ball when Elbert Dent, Victorine's husband, replaced Huidekoper when the latter died in 1943. With Dent, Ball knew he would have no difficulty, for when the board of trustees was first established, Dent had written an extravagantly flattering letter of appreciation for Ed's many years of faithful service to Alfred du Pont and had added as a postscript, "Don't let 'Beppy Heidi' [Bessie Huidekoper] and 'Maury' Lee annoy you. Their bark is worse than their bite."[34] After Dent replaced Huidekoper, there was no one in an official capacity even to transmit "Beppie's" and "Maury's" bark to the other trustees.

From 1943 until 1966, Edward Ball reigned supreme within his fiefdom with no one except Jessie (and she very rarely) to gainsay him. As the estimated value of Alfred's estate passed the one-billion-dollar mark, Ball was clearly the most powerful man in Florida, a state which was expanding almost as fast as Ball's own empire. There was no effective opposition to him either politically or economically throughout the Sun Belt. His wish was the state government's command, in the legislative halls, the governor's mansion and the Supreme Court chambers in Tallahassee. It would seem that the only enemy whom he had not yet confused and with whom he would still have to contend was death itself, and there were even some in Florida who feared that the old man might successfully evade or at least indefinitely postpone that final encounter.

In the mid-1960s, however, some cracks began to appear in the power structure that Ball had erected. He might control Florida, but not even his financial resources were great enough for him to control the federal government, and within Congress, he had some very powerful enemies, most notably Senator William Proxmire of Wisconsin and Congressman Claude

Pepper from Ball's own state whom the ultra-conservative Ball had infuriated by his defiance of the National Labor Relations Board.

There were other old-time New Deal Democrats in Congress who were also angered by Ball's successful war on the unions. In retaliation for the Florida East Coast Railway's tactics of union busting, these opponents in 1966 were able to muster enough strength to pass a law which prohibited charitable trusts from engaging in both banking and nonbanking enterprises. This measure was clearly directed specifically against the Alfred I. du Pont estate, and for once Ball's will could not prevail. He did not have the influence in the federal courts that he had in the Florida state courts. Ball was given five years to divest the estate of its controlling interests in either its banks or its nonbanking enterprises. Given those choices, Ball had no difficulty making a decision. The banks had to go, for under no circumstances would the estate give up its paper mills, its forest lands, and its railroad. The law did allow charitable trusts to hold up to 24.9 percent ownership in banks while still engaged in other business activities, however, presumably on the somewhat questionable assumption that it took at least a 25 percent ownership to control a bank. Over the next five years, the estate did reduce—and how painful all of this must have been to Ed Ball—its 59.6 percent ownership of the Florida National Bank group down to the required maximum of 24.9 percent ownership. Inasmuch as Ball personally owned 6.4 percent of the bank shares and Jessie owned 4.5 percent, however, there were many both within and outside of Congress who doubted that Ball had really lost control of the banking empire that Alfred du Pont had built.[35]

The next attack came in 1969, when Congress passed a law regulating charitable trusts and foundations. To keep their tax-exempt status, such institutions would have to disburse *all* income received from their holdings to the charitable enterprises for which they were incorporated except for necessary operating expenses and a reasonable surplus to be held in reserve. Ball countered this attack by reclassifying the Nemours Foundation as a hospital, but the Interval Revenue Service was not favorably impressed by this ploy and continued to keep a careful check on the income and disbursements of this tempting plum.

In 1970, Jessie Ball du Pont died, and under the terms of Alfred's will all of the income from his estate was now to go directly to the Nemours Foundation for its use. The plum now looked ripe for the picking. The income did go to the foundation, but Ball seemed exceedingly reluctant to let much of it go any further than its resting place in the Florida National Bank of Jacksonville. In 1974, Alfred's and Ed's old nemesis, the *Wilmington News-Journal* carried a blistering story by one of its brightest young investigative reporters, Curtis Wilkie, under the headline "Nemours Foundation Stockpiles $35 Million." After several weeks of painstaking investigation Wilkie discovered that out of an annual income of $15.2 million in 1972, only $3.3 million was given to the philanthropic enterprises for which the foundation had been created—$500,000 for the maintenance of the Nemours estate itself and $2.8 million to the Alfred I. du Pont Institute hospital. The remainder, nearly $12 million, was simply added to the surplus.[36]

As if these outside attacks from Congress, the IRS, and the media were not enough, Ed Ball suddenly came under fire where he least expected it—from what he could only regard as quislings within the hitherto subservient board of trustees. Elbert Dent died in 1964 and was succeeded on the board by William Mills. Mills, a lawyer by training and a banking expert, had become president of the Florida National in Jacksonville. It had been at Ed's insistence that Mills had been added to the board, although Jessie herself would have preferred to have Dent's younger son, Alfred Dent, replace his father. Ball, however, felt he could count on Mill's unswerving loyalty, and moreover, Ball wanted to keep Dent in reserve as a possible successor to Jessie on the board. As usual, Ball had his way.[37] It was a decision that Ball would live to regret.

Mills had always had the highest regard for Mrs. du Pont. He believed her to be completely dedicated to Alfred I. du Pont's ideals and he never had any reason to question either her motives or her integrity. Mills did not hold the same view of her brother. As long as "Miss Jessie" was alive, Mills never questioned how the income of the estate was allocated, for after all, the will had explicitly provided that Alfred's wife was to have the use of the income from the estate for as long as she lived. Any contribution made to charity was a decision which she alone could make. Mills also knew that much of Mrs. du Pont's annual income did actually go to philanthropies of her own choosing—colleges, churches, and scholarships for deserving and needy young people.

Mills was a Yankee, born and bred in Maine, with all of the ingrained integrity with which "Down-Easters" are legendarily endowed. Even before Jessie du Pont's death Mills had become increasingly concerned that there had been little discussion or planning among the trustees as to how the income from the estate would be utilized after she died. Following her death, he was at once ready to bring the issue squarely before the board.

In this venture, Mills to his surprise found an unexpected ally in the newest member of the board, young Alfred Dent. Dent had been chosen as Jessie's successor on the board of trustees in 1970 on the assumption by Ball that he would be as amenable and subservient as his father Elbert Dent had been. Young Dent, however, proved to be not the lightweight playboy and sycophantic tool that Ball had wanted on the board. With Mills's backing, Dent quickly demonstrated that he was not only willing to stand up to the old man, but also eager to speak his own mind.

For the first time ever, Ball faced the prospect of not being in control of the board of trustees of the estate. This was for him an intolerable situation requiring immediate action. Although there was no authorization for such a move under the terms of the will, Ball increased the size of the board to seven by appointing three of his subordinates as additional members: J. C. Belin, president of the St. Joe Paper Company, Tom S. Coldewey, retired vice-president of the paper company, and W. L. Thornton, president of the Florida East Coast Railway. These were men of demonstrable business ability, but their most admirable quality in Ball's view was their complete loyalty to him. With the enlarged board now balanced 5 to 2 in his favor, Ball at the age of eighty-seven was now prepared to take on all enemies, both within and outside of his empire.

Mills and Dent brought suit against Ball for having violated the terms of Alfred du Pont's will, which had clearly specified a board of only four trustees. Neither of the plaintiffs was too sanguine, however, as to how this suit would fare in a Florida court. At the same time, in a suit highly unusual in the history of American jurisprudence, the two sovereign states of Delaware and Florida brought suit against the estate for not having properly used its income for the welfare of the people as specified in Alfred du Pont's will.

Thus beleaguered on all sides—even by the sovereign power of his own state of Florida, Ball was forced to yield ground. Plans were drawn up for a multi-million-dollar addition to the Alfred I. du Pont Institute hospital at Nemours. A second children's hospital was also to be built in Jacksonville, Florida, and for the first time the Nemours mansion was at last opened to the general public for its "pleasure and benefit," eight years after Jessie's death.

On 24 June 1981, Edward Ball was forced to yield to yet one more implacable foe. He died at the age of ninety-three in a New Orleans hospital. Even in death, Ball sought to emulate his patron and hero. He left nearly all of his personal fortune of some $75 million for the treatment of "curable crippled children in Florida."

Forty-six years after the death of the man whose last will and testament had provided the script for this family drama, the play was at last over. All of the actors, both those in supporting and those in non-supporting roles, had made their final exit from the stage—Marguerite Lee and her son Maurice; brother Maurice du Pont; all four of Alfred's children; Alfred's beloved Jessie; and, finally, Ed Ball. If in this drama it had been Ed Ball's lot to be cast in the role of the villain, at least no one could say he had not given his all to the part assigned to him.

With Ball's exit, the family squabbles that had for so long enmeshed Alfred I. du Pont both in life and in death should have at last mercifully ceased. Such was not to be the case. In October 1985, *Forbes* magazine carried a story that a reader would more likely expect to find in the *National Enquirer* instead of a respected business journal. In a major article of that issue, Christine Donahue related a story that Alfred Dent had given to her.[38] Donahue took seriously the information given to her, for Dent was the grandson of the late Alfred I. du Pont and for fifteen years a trustee of the du Pont estate. Dent's charge was bluntly and shockingly stated: Jessie Ball du Pont, at the instigation of her brother Ed, had murdered Alfred I. du Pont by poison. Dent had no difficulty in providing a motivation for his allegation. Jessie and Ed had been determined to prevent Alfred's second will from going into effect on 15 July 1935, for if the first will prevailed, Jessie would get the use of the entire income of the estate for as long as she lived, and Ed Ball would get to manage that estate to advance his own fortunes and his economic and political power in the state of Florida.

Dent's only source for this remarkable tale was Ed Ball's former wife, Ruth Price Ball, who had divorced her husband in 1941 after a nine-year marriage, which in the testimony she gave in the divorce proceedings Mrs. Ball had described as being one of sheer hell. After her divorce, Ruth Ball married again, this time happily, but she had continued to live in fear of her former husband. She was certain that he knew that she had damning information regarding Alfred's death. Only after Ball's death did she dare to tell the story to Alfred's grandson.

Dent went to Mrs. Ruth Ball Morrell's home in St. Louis. He found her to be even in her mid-eighties a remarkably beautiful and intelligent woman. She told a convincing story. At least it convinced Dent, whose long-standing hatred of Ed Ball had in no way been softened by Ball's death. Dent found a match for that hatred in Ed Ball's former wife, and she in turn found a most sympathetic and appreciative audience in Alfred Dent. Aside from her own assessment of her former husband's character and her intuition, which had been honed by long years of frustrated hatred directed against Ed and Jessie, Ruth Morrell had little or nothing to support her charge except for a letter written by Jessie to Ed Ball shortly after Alfred's death, which Ruth Ball had found in her husband's files and had kept for the past fifty years. In that letter, Jessie mentioned the sleepless nights she had endured since Alfred's death and had then added one sentence that to both Ruth Morrell and Alfred Dent seemed damning: "I can't eliminate the thought from my mind, that there was a 'slip' somewhere."

This statement was too vague to be of value to a prosecuting attorney without any other corroborative evidence. "A 'slip' somewhere" could refer to anything, for it is not unusual for a grieving widow to brood over her husband's last hours and to blame herself for not having done more to prolong his life. Jessie may have felt that she had "slipped up" in not insisting that the Philadelphia heart specialist stay on at Epping Forest until Alfred was clearly out of danger. Or perhaps she should have placed Alfred in a hospital as soon as he took ill, or the doctors should have prescribed a different medication. Any of these possibilities might have been the "slip" Jessie had in mind when she wrote her note to Ed Ball, as certainly any defense attorney worth his fee would quickly bring to the attention of a jury. To Ruth Ball Morrell and to Alfred Dent, however, this statement meant only one thing—the cry of a guilty conscience.

After telling Mrs. Morrell's story to a few intimate friends, Alfred Dent decided to go public with it in the summer of 1985, and he brought it to the attention of the *Forbes* editorial staff. What Dent hoped to accomplish by this attempted exposure remains as much of a puzzle as the story itself. Even if it were true, what had been done could not at this late date be undone or even punished. The money that Jessie had been given by Alfred's only valid will could not now be taken away; the long years of power that Ed Ball had enjoyed could not now be recalled. Only the satisfaction of setting the record straight—no small matter in itself, to be sure—would be the reward for this revelation, but that reward could only be realized if indeed the story were true.

Since only a simple, very ambiguous note can be offered as evidence to substantiate the story, judging the validity of the charge must finally depend upon an evaluation of character of the two persons who have been accused by Morrell and Dent. Although some people in both Wilmington and Jacksonville would have no difficulty in accepting Ed Ball in the role of Iago for this drama, there are few who believe that Jessie du Pont should be cast as Lady Macbeth. For Jessie to take such a part, strong and determined a woman as she may have been, would have made a mockery out of the most meaningful and genuine theme of her life—her love for Alfred du Pont.

Needless to say, Ruth Ball Morrell's story aroused once again all of the old family gossip about Alfred on the fiftieth anniversary year of his death. It was a singularly unfortunate and inappropriate way by which to commemorate that event. But gossip is ephemeral, and this story proved to be a ten-day wonder. It was quickly brushed aside as the public's attention turned to other tales of more contemporary alleged wrongdoings of the very rich.

As for Jessie and Alfred, they were beyond knowing or caring as they lay side by side within Delaware's highest monument.

Marguerite du Pont Lee had always been fond of concluding her noteworthy epistles with a quotation from Shakespeare. One of her favorite citations, which she frequently used in her begging letters to obtain funds for her charitable works, was:

> What saith Shakespeare—"How far that little candle throws his beam,
> So shines a good deed in a naughty world."

Alfred had often been irritated by his sister's pat quotations. With this citation, however, Alfred might have been pleased, for here could be found the beauty of poetic truth. He and Jessie had also held little candles of light in a naughty world.

Notes

Foreword

1. Henry Adams, *The Education of Henry Adams,* Sentry Edition (Boston: Houghton, Mifflin Co., 1961), p. 3.

2. There has been no consistency among the members of the du Pont family as to the form in which they have written their name. Pierre Samuel generally used Du Pont, his son Irénée used du Pont. Others have used Dupont, DuPont, and duPont. Alfred Irénée usually used the form duPont, but because his great-grandfather, the founder of the company, and most of his descendants have used du Pont, that is the form that will be used in this text. The name of the company, however, is always given as Du Pont, except when the full name E. I. du Pont de Nemours is used.

3. James Truslow Adams, *The Adams Family* (New York: The Literary Guild, 1930), p. 8.

Prologue: Part I

1. For further accounts of Samuel du Pont, Anne Montchanin, and the childhood of Pierre Samuel du Pont, see Pierre Jolly, *Du Pont de Nemours,* translated by Elise du Pont Elrick (Wilmington: The Brandywine Publishing Co., 1977), Chaps. I and II, and William H. A. Carr, *The du Ponts of Delaware* (New York: Dodd, Mead, 1964), Chap. II.

2. Quoted in Carr, *op. cit.,* p. 11.

3. Much of the following material on the youth of Pierre Samuel du Pont up to the time of his marriage is taken from his own memoir, *L'enfancie et la Jeunesse de Du Pont de Nemours Raconties par Lui-même,* as related in Jolly, *op. cit.,* Chaps. II and III.

4. Quoted in Jolly, *op. cit.,* p. 20. Voltaire's classical allusions refer to the river Pactolus, which had gold nuggets in its sandy bottom, and the stream of Permessus, favored by Apollo and the Muses.

5. Quoted in Jolly, *op. cit.,* p. 21.

6. *Ibid.,* p. 34.

7. *Ibid.,* p. 37.

8. *Ibid.,* p. 69.

9. Quoted in Ambrose Saricks, *Pierre Samuel Du Pont de Nemours* (Lawrence: University of Kansas Press, 1965), p. 214.

10. P. S. du Pont de Nemours, *Philosophie de l'Univers,* quoted in Jolly, *op. cit.,* pp. 159–160.

11. Quoted in Carr, *op. cit.,* p. 42.

12. Victor du Pont to Citizen Minister Talleyrand, 6 and 21 July 1798, quoted in Jolly, *op. cit.,* pp. 224–26.

13. The story of the du Pont family's migration to America has been told many times. See Jolly, *op. cit.,* Chap. XVI and Carr, *op. cit.,* Chap. VI.

Prologue: Part II

1. Pierre Samuel du Pont (hereafter referred to as PSdP) to Madame Germaine de Staël, quoted in Jolly, *op. cit.,* p. 237.

2. PSdP to Jacques Bidermann [1801], quoted in Carr, *op. cit.,* p. 52. See Carr, *op. cit.,* Chap. VII, pp. 50–56 and Jolly, *op. cit.,* Chap. XVI, pp. 215–41 for further information on the du Ponts' various business projects in America.

3. Carr, *op. cit.,* pp. 53–54.

4. Henry Seidel Canby, *The Brandywine,* The Rivers of America Series, (Exton, Pennsylvania: Schiffer, Ltd., 1969), pp. 3–4.

5. PSdP to Germaine de Staël, 19 Pluviôse, year VIII, in James F. Marshall, editor and translator, *de Staël—du Pont Letters* (Madison: University of Wisconsin Press, 1968), p. 16.

6. Thomas Jefferson to PSdP, 25 April 1802, in Gilbert Chinard, *The Correspondence of Jefferson and du Pont de Nemours* (Baltimore: The Johns Hopkins Press, 1931), pp. 46–47.

7. PSdP to T. Jefferson, 30 April 1802, *ibid.,* pp. 48–54.

8. PSdP to T. Jefferson, 12 May 1802, *ibid.,* pp. 55–58.

9. PSdP to T. Jefferson, 3 March 1803, *ibid.,* pp. 69–70.

10. PSdP to T. Jefferson, 12 May 1803, *ibid.,* pp. 72–74.

11. T. Jefferson to PSdP, 1 November 1803, *ibid.,* 79–80.

12. T. Jefferson to Secretary of War Henry Dearborn, 23 July 1803, *ibid.,* pp. xlii–xliii.

13. Quoted in Carr, *op. cit.,* p. 68.

14. Eleuthère Irénée du Pont (hereafter referred to as EIdP) to Charles Dalmas, 20 June 1814, quoted in John D. Gates, *The du Pont Family* (Garden City: Doubleday & Co., Inc., 1979). pp. 40–41.

15. EIdP to PSdP, February 1808, quoted in Carr, *op. cit.,* p. 80.

16. Peter Bauduy to French investors, June 1814, quoted in Carr, *op. cit.,* p. 96.

17. PSdP to Francoise du Pont, June 1815, quoted in Carr, *op. cit.,* p. 99.

18. PSdP to Hyde de Neuville, the French Ambassador to the United States, 3 August 1816, quoted in Jolly, *op. cit.,* pp. 341–45.

19. PSdP to T. Jefferson, 26 May 1815, quoted in Chinard, *op. cit.,* pp. 216–19.

20. Quoted in Jolly, *op. cit.,* p. 332.

Chapter I

1. *Delaware State Journal,* 4 November 1834.

2. See association agreements for 1837 and 1851 in Longwood Mss., Group 10, Series A, Box 8, Eleutherian Historical Library, Wilmington (hereafter referred to as EHLW).

3. A favorite story in the du Pont family, told to me by Maurice du Pont Lee in an interview in 1974.

4. See letter from Daniel Lammot to his daughter Margaretta Lammot du Pont, 26 May 1829, Acc. 761, Box 3, EHLW.

5. Victorine Bauduy, "The Brandywine Manufacturer's Sunday School," unpublished and undated mss, Longwood Mss, Acc. 918, Group 10, Series A, Box 9, EHLW; *Eleutherian Mills—Hagley Foundation News Letter,* Vol. 10, No. 3, Winter 1982, p. 1.

6. Marguerite du Pont Lee to Pierre S. du Pont, 15 September 1935, Longwood Mss., Group 10, Series A, File 328, EHLW.

7. Gates, *op. cit.,* p. 49.

8. See Carr, *op. cit.,* p. 135, for an account of this accident.

9. *Ibid.,* p. 148.

10. *Ibid.,* pp. 152–53; Gates, *op. cit.,* p. 51.

11. Sophie du Pont's account of Alexis I. du Pont's death, exhibited in the "Mourning Glory" Exhibit at the Eleutherian Historical Library in October 1980 (Catalogue No. W9-39866).

12. See Anthony F. C. Wallace's *Rockdale* (New York: W. W. Norton & Co., 1980), pp. 430–35, for an excellent statement on what he calls "the drama of salvation and death."

13. Francis Gurney Smith to Joanna Smith du Pont, 3 September 1857, "Mourning Glory" Exhibit at the Eleutherian Historical Library in October 1980.

14. R. Page Williamson to Mary P. Williamson, 7 October 1852, Correspondence of R. Page Williamson, Acc. 1698, EHLW.

15. RPW to MPW, 11 November 1852, RPW Correspondence, *loc. cit.*

16. RPW to MPW, 14 October 1852, *loc. cit.*

17. RPW to MPW, 28 October 1852, *loc. cit.*

18. RPW to MPW, 25 March 1853, *loc. cit.*

19. RPW to MPW, 28 October 1852, *loc. cit.*

20. RPW to MPW, 4 November 1852, *loc. cit.*

21. See RPW's letter to MPW, 9 June 1853, *loc. cit.,* for a full account of Gilpin's party.

22. RPW to MPW, 7 October 1852, *loc. cit.*

23. For a genealogy of the Cazenove family, see "Autobiographical Sketch of Anthony-Charles Cazenove, Political Refugee, Merchant Banker, 1775-1852," edited by John Askling, *The Virginia Magazine of History and Biography,* Vol. 78, July 1970, pp. 295–307; and Bessie G. du Pont's manuscript genealogy, Box 51, Longwood Mss., EHLW.

24. "Note Found in an Old Bible Belonging to Mrs. Sophie du Pont, the wife of Admiral du Pont," reprinted in B. G. du Pont, *A New England Family and Their French Connections,* pp. 66–67.

25. Askling, ed., *op. cit.,* p. 304.

26. See Sophie du Pont's diary entry for January 1837 in which she mourns the

death of Charlotte Cazenove Shepard: "the strongest most cherished tie of friendship that had bound my heart was broken." Mss. copy of Sophie du Pont's diary, EHLW.

Chapter II

1. For one family member's account of Charlotte's inherited insanity, see the mss. by Henry A. du Pont, undated but probably written in 1923. Henry Algernon du Pont, Series A, Box A, 1922–23, 1923 folder, EHLW.

2. Marguerite du Pont Lee to Major Edwin N. McClellan, Historical Section, U.S. Marine Corps, 18 May 1932, papers of Maurice du Pont Lee, EHLW; autobiographical sketch by Maurice du Pont Lee, MdP Lee Papers, Box 6, EHLW.

3. See letter from Mary Sophie dP to Victorine dP Kemble and Paulina dP, undated but probably in the summer of 1860, Acc. 761, Box 5, Children of Alfred Victor dP, Out File, 1849–1892, EHLW.

4. Lammot's dealings with the British government in 1861 over the saltpeter shipment became a favorite family legend. See the mss. account by Bessie G. du Pont, dated 25 March 1907 in Group 10, File 918, Box 7, EHLW.

5. See letters of Lammot dP to Mary Belin, dated 31 May 1864 & 4 August 1864, Acc. 761, Box 9, Mrs. Lammot dP File, EHLW.

6. See Askling, *op. cit.,* p. 305, fn. 46, for an account of Cazenove Henderson's participation in the Battle of Bull Run.

7. Margaretta dP to Victorine dP Kemble, Thursday, 18 October [1860], Acc. 761, Box 2, Mrs. Alfred Victor du Pont, Out File, EHLW.

8. Partnership agreements of the E. I. du Pont de Nemours Company, in File 1203-51, Group 10, Series A, Box 8, EHLW.

9. Harold B. Hancock, "The Income and Manufacturer's Tax of 1862–1872 as Historical Source Material," *Delaware History,* Vol. 14, October 1971, p. 256. See also the *Delaware Republican* for 30 March 1865, p. 3, for a listing of incomes reported to the federal government for the year 1864.

10. Joanna Smith dP to Fanny dP Coleman, 1 July 1863, Acc. 918, Group 10, Series A, Box 9, Folder W18a, EHLW.

11. Margaretta dP to Bidermann dP, 3 July 1865, Acc. 761, Box 2, Mrs. Alfred Victor dP Out File, EHLW.

12. Margaretta dP to Fred dP, 5 May 1873, Acc. 761, Box 3, Mrs. Alfred Victor dP, Out File, EHLW.

13. Charlotte dP to Lammot dP, 18 August 1875, Acc. 384, Box 8, Folder 16, EHLW.

14. Charlotte dP to Lammot dP, 31 August 1875, Acc. 384, Box 8, Folder 16, EHLW.

15. Margaretta dP to Fred dP, 22 April [1870], Acc. 761, Box 2, Folder marked Mrs. Alfred Victor dP to Alfred Victor dP, Jr., EHLW.

16. Margaretta dP to Mary Lammot, 4 January 1871 and 29 March [1871], Acc. 761, Box 2, Mrs. Alfred Victor dP, Out File, EHLW.

17. Margaretta dP to Fred, 4 February [1872], Acc. 761, Letters of Mrs. Alfred Victor dP to her children, EHLW.

18. Marquis James, *Alfred I. du Pont: The Family Rebel* (Indianapolis: Bobbs-Merrill, 1941), p. 31.

19. *Ibid.,* p. 20.

20. *Ibid.,* p. 22.

21. Margaretta dP to Mary Lammot, 3 October 1877, Acc. 761, Box 2, Mrs. Alfred Victor dP, Out File, EHLW.

22. Louisa G. dP to Sophie dP, 19 August 1877, Winterthur Mss., 7/B, Box 18, EHLW.

23. *Ibid.*

24. Margaretta dP to Mary Lammot, 3 October 1877, Acc. 761, Box 2, Mrs. Alfred Victor dP, Out File, EHLW. Marquis James is in error when he states in his biography of Alfred I. du Pont: "Eleuthère stood by his wife, beginning an estrangement which never ended, the son refusing to see his mother when on his deathbed." James, *op. cit.,* p. 22. On the contrary, Margaretta was with Irénée almost daily during the final five months of his life when he was confined to his bed.

Chapter III

1. Quoted in James, *op. cit.,* p. 47.

2. Bessie G. dP's notes from her mss., Acc. 384, Box 49, EHLW.

3. The actual appraised value of the estate was $498,871.41. The will was filed 8 November 1877, Will Record II, 517 ff., New Castle County Court House, Wilmington. See James, *op. cit.,* p. 26.

4. Marquis James gives a colorful account of the confrontation of the children with their elders at Swamp Hall, based upon Alfred's recollection of the event as related to his wife, Jessie Ball du Pont. See James, *op. cit.,* pp. 22–24.

5. AIdP to Sterling Joyner, 10 November 1928, Alfred I. du Pont Papers (hereafter referred to as AIdPJ), located at the time of this writing in the offices of the Alfred I. du Pont Estate, Florida National Bank of Jacksonville, Florida. These papers were filed in a rough chronological order, alphabetized by the last name of the correspondent to Alfred I. du Pont or by the addressee to whom AIdP was writing. I have not attempted to give file classification since the papers will be reorganized and assigned new classifications when they are given to Washington and Lee University, as stipulated in Mrs. Jessie Ball du Pont's will.

6. See Sophie dP to Camille Bidermann, 27 September 1878, Acc. 761, Box 17, Bidermann Family, EHLW.

7. Sophie dP to Margaretta dP, 23 October 1877, Acc. 761, Box 8. Alfred Victor du Pont In File, EHLW.

8. See Sophie dP's letter to Margaretta dP, undated, but written some time in September 1879. Sophie wrote, "I am anxious to hear if both my dear Alfreds have gone off to Andover today!" Acc. 761, Box 8, EHLW.

9. AIdP to Mrs. Estelle Bakewell-Green, 16 January 1931, AIdPJ.

10. Frank H. McIntosh, Bronxville, New York, to Mrs. Alfred I. dP, 22 July 1939, Mrs. Jessie Ball du Pont Papers, Jacksonville (hereafter referred to as JBdPJ).

11. AIdP to Mary Victor dP, 18 October 1879, JBdPJ.

12. AIdP to Fred Victor dP, 1 September 1980, Acc. 761, Box 5, Children of Alfred Victor dP File, EHLW.

13. J. C. MacKinnan, Registrar of M.I.T., to Marquis James, quoted in James, *op. cit.,* p. 49.

14. See James, *op. cit.,* pp. 50–54, for a lively account of the A. I. du Pont–J. L. Sullivan friendship.

15. *Ibid.,* p. 52.

16. A. I. dP to Margaretta dP, 20 March [1884], Acc. 761, Box 8, Alfred Victor dP Family In File, EHLW.

17. See the letter from Marguerite dP to Pierre S. dP, 15 September 1935, Longwood Mss., Group 10, Series A, File 328, EHLW.

18. Typed mss. of the Henry Miller-A. I. dP interview, undated, in the AIdPJ.

19. See William S. Dutton, *Du Pont: One Hundred and Forty Years* (New York: Charles Scribner's Sons, 1942), pp. 81–84, for a more complete account of the development of "B" blasting or soda powder; Lammot du Pont, Improvement of Gunpowder, Letters of Patent No. 17,321, dated 19 May 1857, U.S. Patent Office.

20. Quoted in Dutton, *op. cit.,* p. 109. See *Ibid.,* pp. 107–41, for a brief but excellent summary of the development of the high-explosive industry in America.

21. *Ibid.,* pp. 119–21; P. S. dP II's notes, Longwood Mss., Group 10, Series B, File 1203–38, EHLW.

22. Dutton, *op. cit.,* pp. 121–23; P. S. dP II's mss. on history of the company, dated 8 April 1948, Longwood Mss., Group 10, Series B, EHLW.

23. See Dutton, *op. cit.,* Chap III "Repauno," pp. 126–41 for an excellent summary of how the Du Pont Company became involved in the manufacture of nitroglycerine products.

24. *Ibid.,* pp. 138–39; Francis G. dP to Alexis I. dP, Jr., 1 April 1884, Longwood Mss., Group 10, File 918, Box 3, EHLW.

25. Lammot also anticipated by some twenty years Frederick W. Taylor's work method studies. See Norman B. Wilkinson's provocative article, "In Anticipation of Frederick W. Taylor: A Study of Work by Lammot du Pont, 1872," *Technology and Culture,* Vol. 6, No. 2, Spring 1965, pp. 208–21.

26. James, *op. cit.,* pp. 55–57.

27. *Ibid.,* pp. 61–65.

28. *Ibid.,* pp. 76–78.

29. See Henry A. dP's letter to W. W. Laird, 15 February 1923, in which HAdP criticizes Bessie G. dP's history of the Du Pont Company for not doing justice to the accomplishments of his father. Winterthur Mss., Series A, Box 10A, 1923. See also Dutton, *op. cit.,* p. 140.

30. Dutton, *op. cit.,* p. 157.

31. See Alfred I. dP's letters to his cousin Francis G. dP, especially AIdP to FGdP, 18 April 1889, Acc. 504, Allan J. Henry Collection, Box 1, Series A, EHLW.

32. See AIdP's interview with Henry Miller, AIdPJ; also his letter to E. H. Batson, Washington, D.C., 8 June 1931, AIdPJ.

33. AIdP, Hotel des Deux Mondes, Paris to FGdP, 18 April 1889, Francis Gurney du Pont Papers, 54, D. 10, Box 1, EHLW; AIdP to FGdP, 1 May 1889, Acc. 504, Series A, Allan J. Henry Collection, Box 1, EHLW.

34. See James, *op. cit.,* pp. 81–83 for a full account of AIdP's trip to France, Britain, and Belgium in 1889. Also see letters of AIdP to FGdP, 10 April, 11 April, 18 April, and especially 1 May 1889, Acc. 504, Allen J. Henry Collection, Series A, Box 1, EHLW.

35. AIdP to FGdP, 11 April and 10 April, Acc. 504, Allan J. Henry's Collection, Series A, Box 1, EHLW.

36. The source for this account of Fred du Pont's being offered the presidency of the Du Pont Company can be found in a letter that Marguerite dP Lee wrote to PSdP

10. AB to LdP, 6 April 1892, CP,EHLW.

11. AB to LdP, 19 January 1892, CP,EHLW.

12. AB to LdP, 12 November and 30 September 1892, CP,EHLW.

13. AB to LdP, 19 January 1892, CP,EHLW.

14. AB to LdP, 17 August 1892, CP,EHLW.

15. Alicia Bradford du Pont, as quoted by Mrs. Francis I. du Pont in an interview with Marquis James. James, *op. cit.,* p. 193.

16. AB to LdP, 19 November 1893, CP,EHLW.

17. See the letter from "Big-boy," to "Skritzen" [Alicia Bradford], 8 December 1896, JBdPN.

18. AB to LdP, 19 November 1893, CP,EHLW.

19. Information on the Maddox family provided by John Beverley Riggs, Manuscript Curator of the Eleutherian Mills Historical Library, Wilmington, 8 July 1977.

20. Eleuthera du Pont Bradford to Mary Pauline du Pont, 2 July 1897, temporary acc. no. W8-41143, EHLW.

21. Marquis James to JBdP, 11 March 1940, Misc. File, JBdPJ.

22. For an excellent, although not the sole example of such writing, see Leonard Mosley, *Blood Relations: The Rise and Fall of the du Ponts of Delaware* (New York: Atheneum, 1980), pp. 158–62.

23. Edith Goldsborough to Alicia Amory du Pont Maddox, 10 March 1921, Misc. File, JBdPJ. Written one year after Alicia du Pont's death, this letter by Mrs. Goldsborough to young Alicia Maddox, at the latter's request, relates in considerable detail Alicia Bradford Maddox du Pont's relations with both Alfred du Pont and her husband George Amory Maddox. It is a refutation of Judge Edward Bradford's widely accepted version of those relationships and the subsequent gossip in the du Pont family.

24. Louisa d'A. du Pont Copeland to PSdP, 9 February 1900, P. S. du Pont Papers, quoted in Mosley, *op. cit.,* p. 160.

25.*Ibid.*

26. Judge Edward Bradford to HBdP, 21 March 1902, Misc. Papers, JPdPJ.

27. *Ibid.*

28. Edward Bradford to TCdP, 10 March 1902, copy in Edward Bradford's letter to HBdP, 21 March 1902, Misc. Papers, JBdPJ.

29. See letter of J. S. Dashiell to Jessie Ball dP, 10 March 1941, Misc. Papers, JBdPJ.

30. See letter of Edith Goldsborough to Alicia du Pont Maddox, 10 March 1921, *loc. cit.*

31. A collection of Eleuthera du Pont Bradford's letters to her daughter was carefully saved by Alicia. They are in the Nemours Mansion archives and were made available to this writer by Ruth Linton, Curator of the Nemours Mansion.

32. Eleuthera Bradford to Alicia Maddox, no date but sometime in late July 1903, *loc. cit.*

33. Eleuthera Bradford to Alicia Maddox, no date but sometime in August 1903, *loc. cit.*

34. Eleuthera Bradford to Alicia Maddox, 30 July [1904], *loc. cit.*

35. Eleuthera Bradford to Alicia Maddox, 23 August 1904], *loc. cit.*

36. See the letter from Eleuthera Bradford to Alicia Maddox, 8 September 1904, *loc. cit.*

37. See the letter from Edith Goldsborough to Alicia Maddox of 10 March 1921,

loc. cit., in which Mrs. Goldsborough tells Alicia of her grandmother's expressed desire to have Alfred and her mother marry.

38. James, *op. cit.,* pp. 184–86.

39. AIdP to Jessie Ball, 3 December 1903, JBdPJ; quoted in James, *op. cit.,* p. 187.

40. James, *op. cit.,* pp. 188–89.

41. Madeleine du Pont Ruoff to Jessie B. du Pont, 30 March 1937, JPdPJ.

42. Deposition of Rebecca V. Abbott in divorce case. *A. I. du Pont* v. *B. G. du Pont, loc. cit.* See also letter of R. V. Abbott to Jessie B. dP, 27 June 1935, JBdPJ.

43. Madeleine dP Ruoff to JBdP, 30 March 1937, *loc. cit.*

44. See letters of Eleuthera dP Bradford to Alicia Maddox, undated, but written in November and December 1904, JBdPN.

45. Madeleine dP Ruoff to JBdP, 30 March 1937, *loc. cit.*

46. Statement of Madeleine du Pont Ruoff in her letter to Jessie BdP, 30 March 1937, *loc. cit.*

47. Edith Goldsborough to Alicia du Pont Maddox, 10 March 1921, *loc. cit.*

48. A copy of this trust agreement between Alfred I. du Pont and Bessie G. du Pont, dated 26 September 1905, in the JBdPN.

49. James, *op. cit.,* p. 191.

50. Eleuthera Bradford to Alicia Maddox, 2 August 1905, *loc. cit.*

51. Newspaper clipping [*N.Y. Herald,* 9 December 1906?] in AIdPJ.

52. *Alfred I. du Pont* v. *Bessie G. du Pont,* suit for divorce, deposition of Marguerite du Pont Lee, R. V. Abbott, and T. W. Huidekoper, Circuit Court records, 7th Dist. of South Dakota, Sioux Falls, to be found in Box 4, AIdP Papers, Nemours (hereafter referred to as AIdPN).

53. James, *op. cit.,* pp. 195–96.

54. AIdP to G. A. Maddox, 18 May 1907, Records of the E. I. du Pont de Nemours & Company, Acc. 1075, Series II, Presidential File #33, EHLW.

55. AIdP to TCdP, PSdP, F. L. Connable, George Patterson, all dated 18 May 1907. Acc. 1075, File #33, EHLW.

56. James, *op. cit.,* p. 199.

57. Wilmington *Morning News,* 16 October 1907.

58. James, *op. cit.,* p. 201.

Chapter VIII

1. This is Alfred's account as to what words were exchanged in this meeting as Alfred later told it to two members of the du Pont family and as he is quoted in James, *op. cit.,* pp. 201–2.

2. Telegram and letter from Leighton Coleman to Edward G. Bradford, Jr., both dated 17 October 1907, Misc. Files, JBdPJ.

3. Edward J. Bradford, Jr., to Leighton Coleman, 18 October 1907, copy in the Misc. files, JBdPJ.

4. AIdP to Alfred Victor dP, 27 March 1907, in the papers of Alfred Victor du Pont, Miami, quoted in James, *op. cit.,* p. 198.

5. Madeleine dP Ruoff to JBdP, Xmas 1936, Misc. files, JBdPJ.

6. AVdP as told to Marquis James and as quoted in James, *op. cit.,* pp. 108–9.

7. Victorine du Pont Dent as told to Marquis James, *ibid.,* p. 216.

8. See Mosley, *op. cit.,* p. 162.

9. George Q. Horwitz to PSdP, 14 June 1907, PSdP papers, Group 10, Series A, File 407, EHLW.

10. PSdP to G. W. Horwitz, 17 June 1907, PSdP papers, Group 10, Series A, File 407, EHLW.

11. G. Q. Horwitz to PSdP, 20 June 1907, PSdP Papers, Group 10, Series A, File 407, EHLW.

12. AIdP to G. Q. Horwitz, 21 April 1909, PSdP Papers, Group 10, Series A, File 407, EHLW.

13. See AIdP's letter to T. Bayard Heisel, 25 May 1914, which describes these arrangements, AIdPJ.

14. See letter of PSdP to G. Q. Horwitz, 23 January 1908, PSdP Papers, Group 10, Series A, File 407, EHLW.

15. G. Q. Horwitz to BGdP, 28 May 1909, PSdP Papers, Group 10, Series A, File 407, EHLW.

16. See correspondence between AIdP and Carl Hagen, the Du Pont Company Foreign Sales representative in London who negotiated the sale, especially the letter quoted from Hagen to AIdP, 30 October 1913, AIdPJ.

17. See correspondence between G. Q. Horwitz and PSdP, September–October 1908, PSdP Papers, Group 10, Series A, File 407, EHLW.

18. *Town and Country,* 28 June 1913, pp. 21–23.

19. JBdP to Marquis James, 19 November 1940, Misc. files, JBdPJ.

20. James, *op. cit.,* p. 227.

21. The newspaper stories quoted are taken from James, *op. cit.,* pp. 228–29.

22. Possibly a reference to the gossip about Alicia being spread to the Bidermann branch of the family living in Paris.

23. AIdP to Frank C. Jones, 2 November 1909, AIdPJ.

24. See letter from AIdP to Frank C. Jones, 6 April 1910, describing Alicia's despondency following the death of their son, AIdPJ.

25. Quoted in James, *op. cit.,* p. 229.

26. *Ibid.,* p. 230.

27. *Ibid.*

28. Jessie Ball du Pont later denied that Alfred and Alicia lived in isolation as Marquis James reported in his biography. See her letter to James, 17 March 1940, in Misc. Files, JBdPJ. From all available accounts, however, James was correct in his interpretation of their life at Nemours.

29. Marquis James (as well as subsequent biographers who have used James as their source) was incorrect in stating in his biography that Alfred's and Alicia's first child was born on 9 January 1912 and the second child on 21 October 1914. James also had the order of the children wrong. The son, Samuel, was born first and the daughter Eleuthera second. See James, *op. cit.,* pp. 246 and 253. The official Du Pont Genealogy is correct for the dates and order of these children. Substantiating letters, particularly from Alfred to F. C. Jones, confirm the genealogical record as given.

30. AIdP to Thomas Bayard Heisler, 15 January 1913, AIdPJ.

31. AIdP to T. Bayard Heisler, 15 January 1913, AIdPJ.

32. Alfred I. du Pont as quoted in the Wilmington *Morning News,* 7 February 1913; also quoted in James, *op. cit.,* p. 250.

33. *New York Times,* 26 February 1913.

34. *Philadelphia Press,* 27 February 1913.

35. James, *op. cit.,* pp. 250–51.

36. *Ibid.,* p. 251. James does not give the source of the statement.

37. Interview by this writer with a du Pont cousin who asked that she not be identified by name.

38. Jessie Ball du Pont attributed her husband's actions, particularly in the effort to change his son's name, to "the hidden woundings of a sensitive spirit." It is as generous an interpretation as can be given. See JBdP's letter and suggested revision of the account (which James did not accept), to Marquis James, 1 August 1941, Misc. files, JBdPJ.

Chapter IX

1. This file of the A. I. du Pont-Connable correspondence covers the period from July through November 1906 and is catalogued under Acc. 1599, Box 1, EHLW.

2. See Frank Connable letter to AIdP, 13 July 1906 and subsequent correspondence over the next several weeks. Acc. 1599, Box 1. EHLW.

3. AIdP to Connable, 5 August 1906, *loc. cit.*

4. *Ibid.* Also, Connable to AIdP, 17 August 1906, *loc. cit.*

5. Connable to AIdP, 18 August 1906, *loc. cit.*

6. AIdP to Connable, 24 August 1906, *loc. cit.*

7. AIdP to Connable, 4 August 1906, *loc. cit.*

8. AIdP to Connable, 13 July 1906, *loc. cit.*

9. AIdP to Connable, 20 November 1906, *loc. cit.*

10. Connable to AIdP, 24 November 1906, *loc. cit.*

11. AIdP to Connable, 22 November 1906, *loc. cit.*

12. AIdP to TCdP, 23 August 1906, copy in the Connable file, *loc. cit.*

13. See AIdP's letter to Connable expressing his delight that "the Brandywine is running again," 29 October 1906, *loc. cit.*

14. *Ibid.*

15. AIdP to Connable, 4 August 1906, *loc. cit.*

16. AIdP to Connable, 24 August 1906, *loc. cit.*

17. AIdP to Connable, 1 September 1906, *loc. cit.*

18. See letters of AIdP to Connable, 24 July, 1 September, and 17 November 1906, *loc. cit.*

19. AIdP to Connable, 24 July 1906, *loc. cit.*

20. AIdP to Connable, 5 August 1906, *loc. cit.*

21. AIdP to Connable, 24 July 1906, *loc. cit.*

22. AIdP to Connable, 30 October 1906, *loc. cit.*

23. AIdP to Connable, 23 July 1906, *loc. cit.*

24. AIdP to Connable, 1 September 1906, *loc. cit.*

25. AIdP to Connable, 27 July 1906, *loc. cit.*

26. AIdP to TCdP, 18 September 1906, Records of E. I. du Pont de Nemours and Company, Series II, Part 2, Presidential file, Acc. 1075, File 33 "Resignations 1902–07," EHLW.

27. James Dashiell to Mrs. JBdP, 31 July 1939, Misc. File, JBdPJ.

28. *Ibid.*

29. *Ibid.*

30. See PSdP's chapter entitled "Coleman Sells Out," from his manuscript history of the Du Pont Company, PSdP Papers, Series B, File 54, EHLW. Pierre is also quoted in C&S, *op. cit.,* p. 306.

31. Even Marquis James appears to have been influenced by "only a black powder man" theme, although he does give emphasis, as few others do, to Alfred's interest in advanced chemical research. See James, *op. cit.,* pp. 236–37.

32. For an excellent account of this suit, see C&S, *op. cit.,* pp. 261–300.

33. *Ibid.,* p. 278 and p. 281.

34. AIdP's testimony before the U.S. Circuit Court, District of Delaware in the case of *The United States of America* v. *E. I. du Pont de Nemours & Company, Testimony,* Vol. I, p. 253 *passim.* See also James, *op. cit.,* pp. 230–32.

35. Senator J. Frank Allee quoted Taft's description of Coleman to Alfred in his letter to AIdP, dated 13 July 1921, AIdPJ.

36. See James, *op. cit.,* p. 244. James obtained his information regarding this meeting from Jessie Ball du Pont and her brother, Edward Ball, as it was related to them by Alfred some ten years after the event. Alfred may have been correct as to what was said at this meeting, but apparently he was mistaken as to where this meeting took place. It was not a meeting which he had had Senator Frank Allee of Delaware arrange for him at the White House, but was instead an accidental encounter with Wickersham at the court hearings. For an accurate account of this meeting, see C&S, *op. cit.,* p. 282 and p. 660 (footnote 96).

37. For a complete listing of the allocation of companies required under the Final Decree, see C&S, *op. cit.,* p. 662, footnote 134.

38. *Ibid.,* pp. 290–91.

39. *Ibid.,* p. 306.

40. Executive Committee minutes of 18 January 1911, Meeting #239. See also C&S, *op. cit.,* pp. 303–9 for an account of the reorganization of 1911, presented particularly from Pierre's and Coleman's point of view.

41. AIdP to TCdP, 27 January 1911, Records of the E. I. du Pont de Nemours Co., Acc. 472, EHLW.

42. AIdP to the Executive Committe, 27 January 1911, *ibid.*

43. TCdP to AIdP, 3 February 1911, *ibid.*

44. C&S, *op. cit.,* p. 307.

45. AIdP to TCdP, 7 February 1911, *ibid.*

46. E. S. Bader notebook, quoted by Marquis James with the permission of Bader, James, *op. cit.,* p. 234.

47. Interview of S. F. Mathewson with Marquis James, quoted in *ibid.,* pp. 235–36.

48. Interview of M. R. Hutchison with James, *ibid.,* p. 236.

49. The story of the farewell ceremony was reported in the Philadelphia *Public Ledger,* 11 June 1911, also quoted in James, *op. cit.,* pp. 238–39.

50. Elizabeth Dorman and Ed Bader in an interview with Marquis James. Quoted in James, *op. cit.,* p. 239.

51. For a simple and readable account of these early developments in the use of cellulose nitrate, see Dutton, *op. cit.,* especially the chapters "Cellulose Marches," pp. 153–56, and "Cellulose Marches Again," pp. 295–301.

52. Henry Adams, *op. cit.,* p. 450.

53. Wilmington *Morning News,* 11 October 1913.

54. See correspondence between AIdP and James Force, his agent in Denver, Col-

orado, 15–20 May 1914, regarding terms of the sale, AIdPJ. See also Maxwell Interview, AIdPJ.

55. TCdP to PSdP, 24 August 1914, quoted in C&S, *op. cit.,* p. 318. C&S gives a full account of the many differences arising between Coleman and Pierre du Pont, see especially pp. 314–19.

56. TCdP to PSdP, 27 August 1914, and PSdP to TCdP, 28 August 1914, both letters quoted in *ibid.,* p. 319.

57. See testimony of Pierre S. du Pont in the suit of *Philip F. du Pont, et al.* v. *Pierre S. du Pont, et al., U.S. District Court, District of Delaware, No. 340 in Equity,* Final Hearing, Vol. II, p. 617 (hereafter referred to as *dP vs. dP*) as quoted in C&S, *op. cit.,* p. 328. See C&S, *op. cit.,* pp. 322–58, and James, *op. cit.,* pp. 254–96, for the two best, although contradictory, accounts of this famous du Pont controversy. This writer is indebted to both sources for the following account of the controversy.

58. AIdP to PSdP, 14 December and 21 December 1914, reprinted in *dP vs. dP,* Final Hearing, Vol. I, pp. 165–66; AIdP to William dP, 14 December 1914, AIdPJ.

59. Finance Committee Minutes, Special Meeting of 23 December 1914, #56, as reprinted in Testimony of PSdP, *dP vs. dP* Final Hearing, Vol. VIII, p. 17.

60. Finance Committee's report to Board of Directors on action taken at the Special Meeting of 23 December 1914, reprinted in *ibid.,* Vol. VIII, p. 19.

61. Testimony in the Kraftmeier affair given by Pierre in *dP vs. dP,* Final Hearing, Vol. IV, p. 315, quoted in C&S, *op. cit.,* p. 332.

62. PSdP to TCdP, 4 January 1915, reprinted in *dP vs. dP,* Final Hearing, Vol. IV, p. 295, also in C&S, *op. cit.,* p. 331.

63. PSdP to TCdP, 9 January 1915, reprinted in *dP vs. dP,* Vol. IV, p. 298, also in C&S, *op. cit.,* p. 331.

64. This conversation is given as related in the testimony of Alfred I. du Pont and William du Pont, *dP vs. dP,* Final Hearing, Vol. II, pp. 166–68, Vol. III, pp. 1–7, and Vol. IV, pp. 58–59, and also in James, *op. cit.* pp. 259–60.

65. AIdP to TCdP, 16 February 1915, in *dP vs. dP,* Vol. III, p. 18.

66. TCdP to PSdP, 13 February 1915, as quoted in C&S, *op. cit.,* p. 333.

67. TCdP to AIdP, 19 February 1915, as quoted in James, *op. cit.,* pp. 260–61.

68. Testimony of PSdP, *dP vs. dP,* Final Hearing, Vol. II, p. 763, quoted in C&S, *op. cit.,* p. 333.

69. Telegram from PSdP to TCdP, 20 February 1915, reprinted in C&S, *op. cit.,* p. 334.

70. Data presented as exhibits in *dP vs. dP* suit. See also C&S, *op. cit.,* pp. 334–35, 338. Agreement of 19 March 1915 between Du Pont Securities Company and Irénée du Pont, *et al.* Defendants' Exhibit No. 45, *U.S.* v. *Du Pont, General Motors et al., U.S. district Court for Northern District of Illinois* (1957).

71. *Wilmington Star,* 28 February 1915.

72. Testimony of AIdP as to this conversation, in *dP vs. dP,* Final Hearing, Vol. III, pp. 16–19; quoted in James, *op. cit.,* p. 269.

73. Philadelphia *Public Inquirer,* 1 March 1915.

74. Copy of this telegram is in the Testimony of William du Pont, *dP vs. dP,* Vol. IV, pp. 60–61.

75. Testimony of PSdP and Francis I. du Pont, *dP vs. dP,* Vol. VII, pp. 26–28 and pp. 89–92, 134. See also James, *op. cit.,* pp. 269–70.

76. PSdP to Board of Directors of E. I. du Pont de Nemours Powder Company, 5 March 1915, quoted in C&S, *op. cit.,* p. 340.

77. James, *op. cit.*, p. 273.

78. Minutes of Board of Directors, 10 March 1915, as reported in James, *op. cit.*, pp. 273-75.

79. James, *op. cit.*, p. 275.

80. E. I. du Pont de Nemours & Co., Annual Report, 1918, p. 30; C&S, *op. cit.*, pp. 359-60.

81. Johnson's reason for taking the case (as given to Alfred) was later told to Marquis James by Jessie Ball du Pont. See James, *op. cit.*, p. 229 and footnote 1, p. 555.

82. Minutes of the Board of Directors, 10 January 1916, see C&S, *op. cit.*, footnote 90, pp. 667-68.

83. Philadelphia *Public Ledger*, 14 March 1916, quoted in C&S, *op. cit.*, p. 345.

84. John H. Frederick, "Johnson, John Garver," in the *Dictionary of American Biography* (New York: Charles Scribner's Sons, Centenary Edition, 1946), Vol. X, p. 107.

85. For this accounting of shares, see James, *op. cit.*, pp. 306-7; see also AIdP to Howard D. Ross, 12 October 1917, AIdPJ, in which Alfred tallies up the votes for Ross to explain how he was defeated by the stockholders.

86. Argument of G. S. Graham, *dP vs. dP*, Vol. XII, pp, 159-60.

87. Argument of John G. Johnson, *dP vs. dP*, Vol. XII, p. 76; quoted in James, *op. cit.*, p. 291.

88. James, *op. cit.*, p. 287.

89. Judge Thompson's Opinion, *dP vs. dP*, Vol. XVI, pp. 8-29.

90. *Ibid.*, p. 90. Quoted in C&S, *op. cit.*, pp. 350-51.

91. Thompson's Opinion, *dP vs. dP*, Vol. XVI, pp. 70-71; see also Carr, *op. cit.*, p. 275.

92. Thompson's Opinion, *dP vs. dP*, Vol. XVI, p. 88ff.

93. *Ibid.*, p. 83. Quoted in James, *op. cit.*, p. 303.

94. Irénée dP to A. Felix dP, 20 April 1917, quoted in C&S, *op. cit.*, p. 352.

95. *Ibid.*

96. Account given by Mrs. Bessie Deery Gallagher to Marquis James, James, *op. cit.*, p. 305.

97. Judge Thompson's Opinion, *dP vs. dP*, Vol. XVI, pp. 111-12.

98. AIdP to William Glasgow, 24 April 1917, AIdPJ. See also letter of PSdP to B. Clark, 14 September 1917, cited in C&S, *op. cit.*, p. 354.

99. Wilmington *Morning News*, 13 April 1917.

100. C&S, *op. cit.*, p. 353.

101. James, *op. cit.*, p. 307.

102. Printed copy of letter to TCdP to Directors of E. I. du Pont de Nemours & Co., 11 May 1917, quoted in C&S, *op. cit.*, pp. 353-54.

103. PSdP to TCdP, 8 March 1915, quoted in C&S, *op. cit.*, p. 326.

104. TCdP to PSdP, 17 March 1915, quoted in C&S, *op. cit.*, p. 317.

105. *Ibid.*, p. 337.

Chapter X

1. Quoted in John A. Munroe's *History of Delaware*, Chapter 10, "Addicks and Du Pont," p. 10/4. An unpublished manuscript at the time that I read it, Munroe's history subsequently was published by the University of Delaware Press, in 1980. The pagination is that given in the manuscript.

2. *Ibid.*

3. *Ibid.*, p. 10/5. See also Carol E. Hoffecker, *Delaware: A Bicentennial History* (New York: W. W. Norton & Co., 1977), pp. 187–91; Henry M. Canley, Jr. "J. Edward Addicks: A History of His Political Activities," unpublished senior thesis, Princeton University, 1932.

4. Munroe, *op. cit.*, pp. 10/5–6; *Biographical Directory of the American Congress, 1774–1971* (Washington, D.C.: U.S. Govt. Printing Office, 1971), "Du Pont, Henry Algernon," pp. 888–89; "Kenney, Richard Rolland," pp. 1225–26.

5. Munroe, *op. cit.*, pp. 10/5–8; *Biographical Directory of the American Congress,* "Du Pont, Henry A.," *loc. cit.;* "Kenney, Richard R.," *loc. cit.*

6. Munroe, *op. cit.*, p. 10/7–8.

7. *Biographical Directory of the American Congress,* "Du Pont, Henry A.," *loc. cit.*

8. James, *op. cit.*, p. 254.

9. *Ibid.*, p. 282, *New York Times,* 16 and 31 March 1916.

10. For the terms of the sale, see letter of E. M. Hooper to AIdP, 11 December 1911, and receipt for sale signed by Hooper, dated 16 December 1911, AIdPJ.

11. *Delaware: A Guide to the First State,* Federal Writers' Project of the WPA (New York: Viking Press, 1938), Part I, The Press, pp. 134–35.

12. James, *op. cit.*, p. 284.

13. *Ibid.*

14. See correspondence of AIdP and J. Frank Allee regarding the investment. Letters dated 2 February, 15 April, 27 May, 1 July, and 1 August 1912, AIdPJ.

15. AIdP to J. F. Allee, 12 November 1912, AIdPJ.

16. Quoted in an unpublished manuscript "The Political Activities of Alfred I. du Pont," by G. T. Maxwell, Part I, p. 10, AIdPJ. Maxwell in 1916 was a newspaper man in Dover who served as a stringer for the Wilmington *Morning News.* This manuscript, written sometime in the early 1930s, was intended to be a part of a definitive biography of Alfred du Pont. Maxwell, who had good reason to be grateful to Alfred for the many favors given him, including many loans and an all expense paid course of instruction at the Chicago Art Institute, was always highly supportive of his benefactor. In this manuscript he overdid himself in waxing rhapsodic over Alfred's greatness as a man and as a politician. Maxwell's manuscript, to say the least, is heavily biased, but it contains information on Alfred's political activities that is not available anywhere else and is therefore a valuable source. Maxwell never completed his biography, however, and the chapters he did write were never published.

17. *Ibid.*, also James, *op. cit.*, pp. 284–85.

18. Quoted in Maxwell, *op. cit.*, Part I, p. 18.

19. Wilmington *Morning News,* 20 May 1916.

20. Maxwell, *op. cit.*, Part I, pp. 23–24.

21. *National Party Conventions, 1831–1971* (Washington, D.C.: *Congressional Quarterly,* 1976), Part I, pp. 48–49; Part II, p. 29.

22. See Maxwell, *op. cit.*, Part I, pp. 64–84 for a very complete account of the county primaries and state Republican convention of August 1916.

23. *Ibid.*, Part I, pp. 86–87.

24. See *ibid.*, Part I, pp. 93–97, for an account of the Progressive convention in September.

25. *Ibid.*, Part I, p. 99.

26. Quoted in *ibid.*, Part I, pp. 105–7.

27. James, *op. cit.*, p. 296.

28. Wilmington *Morning News,* 25 November 1916.

29. Public statement of AIdP, quoted in Maxwell, *op. cit.,* Part II, pp. 20–22.

30. *Ibid.,* p. 20.

31. E. M. Davis to AIdP, 12 January 1918, AIdPJ; see also the letter of E. M. Davis to AIdP, 15 April 1920, in which he lists the monthly subsidy due to four of the papers, AIdPJ; also James, *op. cit.,* p. 287.

32. Maxwell, *op. cit.,* Part II, pp. 40–41.

33. James, *op. cit.,* p. 299.

34. Maxwell, *op. cit.,* Part II, pp. 27–32.

35. Wilmington *Morning News,* 15 April 1917.

36. Maxwell, *op. cit.,* Part II, pp. 66–70, provides a very complete account of the Wilmington primary campaign of 1917.

37. T. W. Jakes to AIdP, 1 June 1917, AIdPJ. There is in the Alfred I. du Pont Papers in Jacksonville a very large file of correspondence between Jakes and Alfred on political activities in Wilmington at this time.

38. AIdP to Jakes, 20 January 1917, AIdPJ.

39. AIdP interview in Wilmington *Morning News,* 25 January 1918, quoted in full in Maxwell, Part II, pp. 91–93.

40. "Address by Mr. Alfred I. du Pont to the Republican State Committee," 19 April 1918, Wilmington, Delaware. Copy of this broadsheet, printed by the Republican State Committee for mass distribution, in AIdPJ.

41. Will H. Hays, Republican National Committee chairman, to AIdP, 16 March 1918, AIdPJ.

42. T. W. Jakes to AIdP, 11 June 1918, AIdPJ.

43. AIdP to T. W. Jakes, 12 June 1918, AIdPJ.

44. Maxwell, in his account says that AIdP originally favored Heald but at the last moment was convinced by Allee that Ball was more loyal to their reform movement. See Maxwell, *op. cit.,* Part II, p. 150.

45. Quoted in Maxwell, *op. cit.,* Part II, p. 152.

46. AIdP to Edward Bradford, Jr., 15 October 1918, AIdPJ.

47. Bradford to AIdP, 16 October 1918, AIdPJ.

48. AIdP to Bradford, 15 October 1918, AIdPJ.

49. AIdP to T. W. Jakes, 25 October 1918, AIdPJ; see also Alfred's letter to Jakes of 10 October 1918, AIdPJ, stressing the same issue.

50. National Reopublican Committee Chairman Will Hays to AIdP, 27 November 1918 and AIdP to Hays, 4 December 1918, AIdPJ.

51. AIdP to T. W. Jakes 6 November 1918, AIdPJ.

52. See Maxwell, *op. cit.,* Part II, pp. 81–90 for a full account of the Non Partisan League's activities in behalf of civic reform.

53. *Ibid.,* pp. 145, 148.

54. AIdP to J. F. Allee, 8 March 1919 and AIdP to Charles E. Gray, 19 January 1920, AIdPJ.

55. Munroe, *op. cit.,* pp. 186–88.

56. *Philadelphia Record,* 20 April 1919.

57. See Maxwell, *op. cit.,* Part III, especially pp. 134–40.

58. See AIdP's letter to J. F. Allee, 19 January 1920, AIdPJ, which indicates he had been interested in the post.

59. See the correspondence between AIdP and Joseph Bancroft and AIdP and George Brown from February until 3 April 1919, regarding the sale, AIdPJ.

60. See AIdP's letter to Edward Bradford, authorizing him to make the purchase of the *Ledger* for $4,500, AIdP to Bradford, 17 February 1919, AIdPJ.

61. Maxwell, *op. cit.,* Part III, pp. 131–34; James, *op. cit.,* pp. 335–36.

62. AIdP to J. F. Allee, 19 January 1920, AIdPJ.

63. AIdP to J. F. Allee, 24 January 1920, AIdPJ.

64. Edward Bradford to AIdP, 6 April 1920, AIdPJ.

65. AIdP to Bradford, 7 April 1920, AIdPJ.

66. AIdP to C. E. Gray, 24 May 1920, AIdPJ.

67. Quoted in Maxwell, *op. cit.,* Part III, p. 145. Maxwell gives quite a full account of the women's suffrage battle in Delaware, *ibid.,* pp. 140–66.

68. AIdP to Senator J. C. Palmer, 4 May 1920, AIdPJ.

69. AIdP to C. E. Gray, 14 May 1920, AIdPJ.

70. *Ibid.*

71. AIdP to Edward Bradford, 14 May 1920, AIdPJ.

Chapter XI

1. See AIdP's letter to Cazenove G. Lee, 13 May 1912, telling of Alicia's fondness for Cherry Island, AIdPJ.

2. See letter of S. G. Archibald to AIdP, 24 August 1914, AIdPJ. See also the correspondence of S. G. Archibald with AIdP, covering the period 24 August 1914 to 11 November 1915, regarding Archibald's having custody of Alicia's apartment including a complete inventory of all the furnishings she was obliged to leave behind, AIdPJ.

3. See an address of Colonel E. G. Buckner, Vice President in charge of Military Sales, "The Relations of du Pont American Industries to the War" delivered at the General Sales Convention, Atlantic City, New Jersey, 19 June 1918, copy in Maurice du Pont Lee papers, Acc. 1452, Box 6, Folder 1, EHLW; see also James, *op. cit.,* pp. 285–86, for figures on Alfred's personal income.

4. See correspondence between Alfred I. du Pont and Thomas Hastings (1916–1918), regarding the construction of White Eagle, AIdPJ.

5. James, *op. cit.,* p. 313, for information concerning Denise that James obtained from Jessie Ball du Pont.

6. See various letters in the Thomas Hastings–AIdP correspondence, especially the following: Hastings to AIdP 17 July, 4 October, 10 October, 3 November, 8 November, 12 November 1917, 21 May, 17 June, 29 June 1918; AIdP to Hastings, 8 November and 12 November 1917; AIdP to C. J. Charles, 4 October and 10 October 1917; Charles to AIdP, 8 October 1917. Letter of AIdP to Hastings, 16 November 1920 gives figures on final cost, AIdPJ.

7. From W. H. Auden's "Master and Boatswain," *Selected Poetry of W. H. Auden* (New York: Modern Library, 1958), p. 86.

8. AIdP to H. P. Brown, 6 October 1917, AIdPJ.

9. *Opinion of Judge Joseph Buffington, Du Pont vs. Du Pont; In the United States District Court of Appeals for the Third District.* pp. 32–51. See also James, *op. cit.,* pp. 322–25.

10. AIdP to S. H. MacSherry, 11 May 1917, AIdPJ.

11. Letters of AIdP to MacSherry, 11 May and 25 June 1917, also to T. T. Nelson, Aide for Matériel, Fourth Naval District, 20 April 1918, AIdPJ. See also Maxwell, *op. cit.,* Part II, p. 52, and James, *op. cit.,* p. 315. James is in error in confusing the sub-

marine chaser Alfred built for the navy with his own *Petrel,* which he had earlier leased to the navy. See also AIdP's letter to Briggs Ordnance Co. of New York, 27 February 1917, regarding the order of guns for the submarine chaser, AIdPJ.

12. AIdP to Maurice dP, 10 December 1930, AIdPJ.

13. AIdP to MacSherry, 15 April 1917 and 8 May 1918, AIdPJ.

14. Quoted in Maxwell, *op. cit.,* Part II, p. 183; see also James, *op. cit.,* p. 309.

15. New York *World,* 21 May 1918.

16. See letter of AIdP to Charles C. Dickson, 7 June 1918, AIdPJ.

17. See the long letter which Duncan Stewart wrote to AIdP on 10 October 1918 AIdPJ, at the time of his forced resignation from the company. This letter provides the only information available as to the origins and purpose of the French-American Constructive Company.

18. See the correspondence between AIdP and officials in the National Tractor Company, dated from April to October 1918, especially 5 June, 28 August, 30 September, 2 and 3 October 1918, AIdPJ.

19. Duncan M. Stewart to AIdP, 10 October 1918, AIdPJ.

20. AIdP to Stewart, 19 August 1918, AIdPJ.

21. Maurice du Pont Lee in an interview with this writer, June 1974.

22. See letter of AIdP to Dickinson W. Richards, New York lawyer, 11 July 1918, AIdPJ, regarding the incorporation of the Nemours Company.

23. J. L. Dashiell to AIdP, 26 November 1919, AIdPJ.

24. Letter of Robert Penington, Secretary of the French American Constructive Corporation to AIdP, 23 December 1919, enclosing a copy of the minutes of the Board of Directors for 16 December 1919, AIdPJ.

25. There is one letter extant, from Thomas Ball in San Diego, California, to his daughter Jessie, 4 June 1908, telling her what she needed to do to get a certificate to teach in California, written just prior to her joining her parents. In the personal papers of Jessie Ball du Pont, Hazel Williams's file, Jacksonville.

26. AIdP to Jessie Ball, 14 June 1916, from Jessie Ball du Pont's personal papers, quoted in James, *op. cit.,* p. 329.

27. AIdP to Jessie Ball, 8 May 1919, JBdPJ.

28. Jessie Ball to AIdP, 8 October 1919, personal papers of JBdPJ.

29. Telegram, Ruth Brereton to AIdP, 7 January 1920, AIdPJ.

30. Telegram, Anna dP to AIdP, 8 January 1920, AIdPJ.

31. See AIdP's letter to Dr. C. B. Gamble, 18 February 1919, AIdPJ.

32. See AIdP to Gamble, 19 May 1919, AIdPJ.

33. See correspondence between AIdP and Temple Scott and Arthur Brentano of Brentano's of New York, 12 June and 5 December 1917, 15 December 1919, AIdPJ.

34. Correspondence of AIdPJ with C. J. Herold of Brentano's, 28 May and 1 June 1920; letter of AIdP to Myron Herrick, 2 February 1920, AIdPJ.

35. James, *op. cit.,* pp. 329-30.

36. Alicia Bradford du Pont, *Ode of Fenelon* (New York: Brentano's, 1917), p. 11, quoted in James, *op. cit.,* p. 314.

37. AIdP to Madeleine Hiebler, 7 December 1920, AIdPJ.

38. James, *op. cit.,* p. 330.

39. AIdP to Madeleine dP Hiebler, 23 February 1920, AIdPJ.

40 AIdP to Jessie Ball, Saturday 17 [January 1920], AIdPJ.

41. See letters from Jessie Ball to AIdP written in the spring and summer of 1920, personal papers, JBdPJ.

42. AIdP to Marguerite dP Lee, 10 July 1920, AIdPJ; see also AIdP to Floride Harding Adams, inviting her to come up to Nemours while "Miss Jessie Ball and Mrs. Wright" were there, 24 May 1920, AIdPJ.

43. AIdP to James Hackett, 5 January 1921, AIdPJ.

44. *Los Angeles Examiner,* 23 January 1921.

Chapter XII

1. *Los Angeles Examiner,* 23 January 1921.

2. AIdP to Jessie Ball, Saturday 17 [January 1920], *loc. cit.*

3. AIdP to H. H. Twitchell, 4 October 1919, AIdPJ.

4. Twitchell to AIdP, 6 October 1919, AIdPJ.

5. Twitchell to AIdP, 9 October 1919, AIdPJ.

6. Twitchell to AIdP, 28 November 1919, AIdPJ.

7. AIdP to Twitchell, 27 December 1919, AIdPJ.

8. Twitchell to AIdP, 29 December 1919, AIdPJ.

9. See AIdP's letter to Francis I. dP, 28 October 1919, AIdPJ, informing him of Nixon's forced resignation.

10. J. L. Dashiell to Marquis James, quoted in James, *op. cit.,* 332.

11. AIdP to H. T. Graham, 1 January 1920, AIdPJ.

12. Commissioner of Internal Revenue to AIdP, undated copy, in AIdPJ.

13. William Glasgow to AIdP, 15 July 1920, AIdPJ.

14. Glasgow to AIdP, 5 August 1920, AIdPJ.

15. James, *op. cit.,* p. 334; see also Glasgow's letter to AIdP, 5 August 1920, AIdPJ.

16. The single source for this account of Alfred's near collapse in Philadelphia is Jessie Ball du Pont. See James, *op. cit.,* p. 334.

17. Alfred dP argues this point repeatedly in letters to his sister, Marguerite dP Lee, his cousin, Francis I. dP and his nephew, Maurice dP Lee during this postwar period. AIdPJ.

18. AIDP to J. Frank Allee, 31 January 1922, AIdPJ.

19. See AIdP's letter to William dP, 30 June 1924, acknowledging receipt of a check for $281,000 for his shares in the Delaware Trust.

20. See letters of AIdP to J. H. Eckerley, vice-president of the Great Falls Mine Properties, 28 June 1920, and from Robert Penington to AIdP, 25 April 1922, in regard to the sale of the stock in the Great Falls Mine and the New York Trap Rock Corporation. AIdPJ.

21. AIdP to Robert Penington, 11 May 1920, and Penington to AIdP, 15 May 1920, AIdPJ.

22. AIdP to Penington, 1 March 1921, and Penington to AIdP, 14 April 1921, AIdPJ.

23. Penington to AIdP, 26 April 1921, AIdPJ.

24. A. C. Sheridan to Penington, 19 April 1921, and AIdP to Penington, 25 April 1921, AIdPJ.

25. AIdP to Ellen LaMotte, 23 May 1921, AIdPJ.

26. Quoted in James, *op. cit.,* p. 339, with Jessie B. du Pont as the source.

27. AIdP to Marguerite dP Lee, 29 March 1922, AIdPJ.

28. See the correspondence between AIdP and Samuel Kirk & Son, Baltimore in the spring of 1922, AIdPJ.

29. AIdP to Title Guarantee & Trust Co., NY, 21 August 1923 and 13 December 1922, AIdPJ.

30. AIdP to Marguerite dP Lee, 31 August 1923, AIdPJ.

31. AIdP to Title Guarantee and Trust Co., 13 December 1923, AIdPJ.

32. AIdP to Richard Maclauren, 2 December 1919 and 20 January 1920, AIdPJ.

33. AIdP to Marguerite dP Lee, 10 September 1920, AIdPJ.

34. AIdP to Robert Penington, 28 January 1922, AIdPJ.

35. See the correspondence between Robert Penington and AIdP, 22 February 1922 until 17 April 1922, regarding the Henry Ford negotiations, AIdPJ.

36. AIdP to Marguerite dP Lee, 29 March 1922, AIdPJ.

37. Thomas Cochrane to AIdP, 11 April 1922, AIdPJ.

38. AIdP to Cochrane, 8 April 1922, AIdPJ.

39. See correspondence of R. W. Jones of Jones, McBee, Jones and Company and AIdP in April 1922, AIdPJ.

40. AIdP to Glasgow, 24 April 1922, AIdPJ.

41. AIdP to Glasgow, 28 April 1922, AIdPJ.

42. AIdP to R. W. Jones, 8 May 1922, AIdPJ.

43. James, *op. cit.*, pp. 349–50.

44. AIdP to Glasgow, 6 June 1922, AIdPJ.

45. See AIdP's letter, 11 April 1921, to Miller R. Hutchison, AIdPJ.

46. James, *op. cit.*, p. 335.

47. *Ibid.*

48. The story of Pierre's having offered to help Alfred was told to Marquis James by Jessie B. du Pont. See James, *op. cit.*, p. 335.

49. The account of Sen. Ball's intervention in Alfred's behalf regarding the interest charges was published in the Philadelphia *Public Ledger,* 16 September 1928.

50. Alicia Maddox to AIdP, 20 November 1920, AIdPJ.

51. AIdP to Alicia Maddox, 11 December 1920, AIdPJ.

52. AIdP to Alicia Maddox, 26 November 1920, AIdPJ.

53. AIdP to Alicia Maddox, 29 March 1922, AIdPJ.

54. Alicia to AIdP, 8 March 1922, AIdPJ.

55. Alicia to AIdP, 27 April 1922, AIdPJ.

56. Ellen La Motte to AIdP, 26 February 1921, AIdPJ.

57. AIdP to Jessie dP, July 1921, personal papers of JBdPJ.

58. AIdP to Lucy Maitland, 31 August 1922, AIdPJ.

59. AIdP to Marguerite dP Lee, 28 April 1922, AIdPJ.

60. AIdP to Ellen La Motte, 7 July 1923, AIdPJ.

61. AIdP to Marguerite dP Lee, 30 October 1922, AIdPJ.

62. AIdP to Marguerite dP Lee [no date, but sometime in spring of 1924], AIdPJ.

63. AIdP to Marguerite dP Lee, 29 October 1920, AIdPJ.

64. AIdP to Col. Henry A. dP, 17 June 1921. Correspondence between Henry A. du Pont and Alfred I. du Pont, Winterthur Mss., Series A, EHLW.

65. Henry A. dP to AIdP, 18 June 1921, Winterthur Mss., Series A, EHLW.

66. Wilmington *Evening Journal,* 1 July 1921.

67. AIdP to Asa Bennett, 1 July 1921, AIdPJ.

68. AIdP to J. Carey Palmer, John M. Walker, James W. Robertson, Charles D. Murphy, J. Frank Allee, Jr., I. D. Short, and J. G. Highfield, Jr., 5 July 1921, AIdPJ.

69. AIdP to Mrs. John Sebree, 25 January 1922, AIdPJ. The man Newberry mentioned in this letter was Truman H. Newberry, who had defeated Henry Ford in the

1918 Senatorial election in Michigan, but who was later indicted and convicted of having violated the Corrupt Practices Act. The conviction was overturned by a 5 to 4 vote in the U.S. Supreme Court, and the U.S. Senate by a vote of 46 to 41 in January 1922 allowed him to take his seat in the U.S. Senate.

70. AIdP to Mrs. John Sebree, 11 September 1922, AIdPJ.

71. AIdP to Ellen La Motte, 14 December 1922, AIdPJ.

72. AIdP to Mrs. John Sebree, 11 December 1922, AIdPJ.

73. AIdP to Ellen La Motte, 14 December 1922, AIdPJ.

74. AIdP to Mrs. John Sebree, 11 December 1922, AIdPJ.

75. AIdP to J. F. Allee, 11 October 1924, AIdPJ.

76. AIdP to Mrs. John Sebree, 6 January 1922, AIdPJ.

77. AIdP to Otho Newland, 16 May 1921, AIdPJ.

78. See letter from Jessie dP to Dr. J. D. Edwards of St. Louis, 26 November 1924, and AIdP to Paul Belin, regarding Dr. Muncie's treatment, 11 December 1922, AIdPJ.

79. AIdP to Ellen La Motte, 9 January 1923, AIdPJ.

80. AIdP to Ellen La Motte, 14 December 1922, AIdPJ.

81. AIdP to Lucy Maitland Francis, 2 February 1926, AIdPJ.

82. James, *op. cit.,* pp. 370–71.

83. Bessie dP Huidekoper to Jessie dP, 13 June 1923, Misc. Papers, JBdPJ.

84. Bessie dP Huidekoper to AIdP, 15 June 1923, Misc. Papers, JBdPJ.

85. James, *op. cit.,* pp. 372–73; 375–76. Much of this material on the separation of Alfred from his two youngest children Marquis James obtained in interviews with Victorine du Pont Dent and Alfred Victor du Pont.

86. James, *op. cit.,* pp. 375–78. Marquis James's source for this meeting was Victorine dP Dent. Alfred Victor du Pont, after reading James's book, wrote the author that he regretted James had not shown him the manuscript prior to its publication. "Your description of our evening there is almost entirely, apparently, according to her [Victorine's] vision of it. . . . My memory of it would give a different impression." Alfred Victor dP to Marquis James, undated but some time soon after the book's publication in 1941. This letter was given to the Eleutherian Mills Historical Library by John Ness. Alfred Victor does not say in the letter what his impression of the meeting was.

Chapter XIII

1. See AIdP's letter to his daughter Madeleine, 6 May 1921, in which he tells of laying up a stock of liquor, AIdPJ.

2. AIdP to Marguerite dP Lee, 10 July 1920, AIdPJ.

3. JBdP to John H. Kellogg, 22 September 1924, AIdPJ.

4. John H. Kellogg to Mr. and Mrs. AIdP, 30 September 1924; AIdP to Kellogg, 7 October 1924, AIdPJ.

5. Kellogg to AIdP, 14 October 1924, AIdPJ.

6. AIdP to Marguerite dP Lee, 1 October 1924, AIdPJ.

7. AIdP to Marguerite dP Lee, 11 October 1924, AIdPJ.

8. *Ibid.*

9. See Alfred's letter to J. H. Kellogg of 22 September 1924, in which he makes arrangements for paying all expenses for Charlotte du Pont's indefinite stay at the Battle Creek sanitorium, AIdPJ. These papers contain several letters during the period

1924 to 1935 concerning Charlotte's health, her unresponsiveness to treatment, and her frequent and pitiful appeals to leave.

10. See Alfred's letter to his secretary, Mollie Brereton, 22 December 1924, AIdPJ.

11. AIdPJ to Ellen La Motte, 22 December 1921, AIdPJ.

12. AIdP to Marguerite dP Lee, 1 September 1921, AIdPJ.

13. For Edward Ball's career prior to becoming Alfred du Pont's assistant, see James, *op. cit.*, pp. 358–59. See also two biographies of Edward Ball, Leon O. Griffith, *Ed Ball: Confusion to the Enemy* (Tampa: Trend House, 1975), and Raymond K. Mason and Virginia Harrison, *Confusion to the Enemy: A Biography of Edward Ball* (New York: Dodd, Mead, 1976).

14. Concerning Edward Ball's role in helping to liquidate Alfred du Pont's outstanding debts in 1923, the sole source for the account of Ball's negotiations with George Baker and the White Shoe Company is in an interview which Ball gave to Marquis James in 1940, the details of which can be found in James, *op. cit.*, pp. 357–58.

15. See William dP's letter to AIdP, 24 May 1922, regarding the purchase of "Wiggly Piggly" [sic] stock, AIdPJ. Alfred invested in Piggly Wiggly stock, to be sure, not only to place his tomato cans on the chain store's shelves, but also because Piggly Wiggly of Delaware had rented some property from Alfred and William for one of its stores and agreed to use the Delaware Trust Company bank for all of its deposits.

16. James, *op. cit.*, p. 356.

17. Interview of this writer with Edward Ball, 3 January 1975.

18. AIdP to Thomas Ball, 19 July 1923, AIdPJ.

19. AIdP to Thomas Ball, 22 April 1924, AIdPJ.

20. Interview of this writer with Edward Ball, 3 January 1975.

21. Jessie B. du Pont's advice to her husband to have her brother serve as his "two good ears" was told to me by several persons in Jacksonville who had heard Jessie tell of this advice given to her husband. Apparently it was a favorite story of Mrs. du Pont.

22. Interview of this writer with Edward Ball, 3 January 1975. Also see James, *op. cit.*, p. 359.

23. AIdP to Thomas Ball, 15 October 1924, AIdPJ.

24. AIdP to Thomas Ball, 23 April 1924, AIdPJ.

25. James, *op. cit.*, p. 396.

26. These remarks of Alfred du Pont were quoted by Edward Ball to Marquis James in an interview. See James, *op. cit.*, pp. 392 and 396.

27. *Ibid.*, p. 395.

28. W. T. Edwards, a handwritten manuscript of his reminiscences of his association with Alfred I. du Pont, written at the request of Jessie B. du Pont, undated, in JBdPJ. (Hereafter referred to as Edwards *Reminiscences*.)

29. James, *op. cit.*, pp. 396–97; 401.

30. AIdP to I. D. Short, 11 April 1927, AIdPJ.

31. AIdP to Jacob V. Hill, 30 April 1927, AIdPJ.

32. Edwards *Reminiscences, loc. cit.*

33. *Ibid.*

34. *Ibid.*

35. James, *op. cit.*, pp. 404–5.

36. The details of how Alfred du Pont and Edward Ball blocked the attempted merger were given to Marquis James in an interview with Ball. See James, *op. cit.*, pp. 430–32.

37. Quoted in *ibid.,* p. 432.

38. Text of AIdP's cable in *ibid.,* p. 434.

39. Interview of the writer with Edward Ball, March 1977; also Dave Howell, "The Man Who Rehabilitated Florida," *Orlando Sentinel,* 29 December 1963.

40. Statement issued by Alfred I. du Pont on 18 February 1930 and reprinted in many Florida papers. Copy in AIdPJ.

41. Quoted in James, *op. cit.,* p. 445.

42. See the letter from James G. Bright to Marquis James, 15 February 1941, which gives in detail all of AIdP's land purchases in Franklin, Bay, Walton, Liberty and Gulf Counties in west Florida. AIdPJ.

43. See AIdP's letter to Frank Jennison, 28 June 1930, in which he states "Florida is richer in natural resources than Mr. Flagler ever dreamed." AIdPJ.

44. See Edwards, *Reminiscences, loc. cit,;* also an article, D. B. McKay, "The Men Who Put Florida On Wheels," in *Tampa Tribune,* 28 December 1958, plus correspondence of W. T. Edwards to AIdP in AIdPJ.

45. AIdP to Judge Mansfield, Columbus, Texas, telegram, 19 June 1933, and Mansfield's answer 20 June 1933, informing AIdP the projects were recommended. AIdPJ.

46. See correspondence regarding the trans-peninsular canal between AIdP and W. T. Edwards in the 1930's, in AIdPJ; also the official endorsement by the Gulf Coast Highways Association for the project, and Edwards *Reminiscences,* AIdPJ; also AIdP to Walter Coachman, Chairman of the Florida State Canal Commission, 14 November 1931, AIdPJ. There is also correspondence in the papers of Mrs. W. T. Edwards, Jacksonville (hereafter referred to as WTEJ), regarding the canal: James T. Daniels, manager of the Jacksonville Chamber of Commerce to W. T. Edwards, thanking him for his efforts to see a canal and P. C. Kelly to W. T. Edwards, 14 April 1933 with a statement on earlier efforts to get a Florida canal, beginning with a law passed by Congress in 1826 to study the possibility of a canal across Florida.

47. A paraphrase of Alfred du Pont's "pep talk" to Edwards is given in Edwards *Reminiscences,* JBdPJ.

48. *Ibid.*

49. *Ibid.*

50. Charles H. Herty, "Southern Pine for White Paper," *Scientific American,* May 1934, pp. 234–36.

51. See W. T. Edwards to AIdP, 16 September 1931, AIdPJ.

52. W. T. Edwards, Memorandum to Mr. du Pont and Mr. Ball, 14 September 1931, in WTEJ.

53. Royal H. Rasch, "Pine Cellulose," *Manufacturers Record,* November 1934.

54. See correspondence between AIdP and Robert Kloeppel of Jacksonville, Florida, the uncle of E. F. Puckhaber, regarding the latter's research at the Hercules Experimental Laboratory, 21 November 1931, AIdPJ.

55. Rasch, *op. cit.*

56. AIdP to Maurice dP Lee, 24 November 1931, AIdPJ; see also the letter of Charles Herty to AIdPJ, 16 November 1934, thanking him for the sample of cellulose "as used by the du Pont Company," AIdPJ.

57. See Edward Ball's letter to AIdP of 22 October 1933 regarding "glassophane" as a possible competitor to Du Pont cellophane, AIdPJ.

58. See W. T. Edward's letter to Jessie Ball du Pont, dated 10 July 1935, copy in WTEPJ, and his letter to Marquis James, dated 15 February 1941, in JBdPJ.

59. *New York Times,* 1 June 1934.

60. W. T. Edwards, "Memorandum for Mr. Du Pont Regarding Activities in Behalf of Proposed Paper Mill," undated but sometime in August 1934, pp. 2–3, WTEJ.

61. *Ibid.,* pp. 3, 5–6.

62. James, *op. cit.,* p. 528.

63. AIdP to Maurice dP Lee, 1 September 1934, quoted in James, *op. cit.,* p. 526.

64. Edward Ball to AIdP, 22 September 1934, AIdPJ.

65. AIdP to Maurice dP Lee, 4 October 1934, quoted in James, *op. cit.,* p. 526.

66. This explanation was still given by Edward Ball and his associates at the St. Joe Paper Company forty years after the decision to go with kraft paper had been made. It was repeated by Mr. J. J. Belin, Mr. Ball's associate in the Alfred I. du Pont Estate, in an interview with this writer in February 1977.

67. James, *op. cit.,* p. 527.

68. See letter of AIdP to W. T. Edwards of 8, 20, and 26 October 1934, WTEPJ.

69. AIdP to Maurice dP Lee, 17 March 1927, AIdPJ.

70. Holland Thompson, "Henry Morrison Flagler," *Dictionary of American Biography,* Vol. III, (New York: Charles Scribner's Sons, 1931), p. 452.

Chapter XIV

1. Douglas Day, "Gertrude Stein," in *Notable American Women,* Vol. III (Cambridge: The Belknap Press, 1971), p. 356.

2. See James, *op. cit.,* pp. 421–42; also the correspondence between AIdP and Maurice dP Lee, regarding the 1929 Du Pont stock split: AIdP to Lee, 12 November 1928, and Lee to AIdP, 16 November 1928, AIdPJ.

3. AIdP, quoted by Edward Ball and Jessie BdP, in James, *op. cit.,* p. 422.

4. Quoted in *ibid.,* p. 421.

5. AIdP to Alfred Victor dP, 8 January 1928, quoted in *ibid.,* p. 420.

6. AIdP to Alfred Victor dP, 29 October 1929, quoted in *ibid.,* p. 440.

7. See the letter from Henry W. Dew to AIdP, 25 June 1929, expressing his pleasure in accepting Alfred's offer, AIdPJ.

8. Edward Ball to AIdP, 30 October 1929, AIdPJ.

9. AIdP to Edward Ball, 1 November 1929, and Ball to AIdP, same date, AIdPJ.

10. AIdP to Francis I. dP, 7 December 1931, AIdPJ.

11. These stories of Alfred's spontaneous acts of charity were told to Marquis James by Charles Hanby and Jessie BdP. See James, *op. cit.,* pp. 422–23.

12. *Orlando Evening Reporter Star,* 20 April 1932.

13. AIdP to Irving Fisher, 11 June 1918, AIdPJ.

14. AIdP to Samuel Swett, 27 May 1918, AIdPJ.

15. See letters of AIdP to E. R. Keim, particularly those of 28 February 1918 and 5 February 1919, AIdPJ.

16. James, *op. cit.,* p. 425.

17. There are numerous letters in the AIdPJ during the 1920s from AIdP to Marguerite dP Lee in which he refers to Herbert Hoover, as "Mr. Hoo Hoo."

18. Interview with Edward Ball, 3 Jan. 1975; see also AIdP's letter to E. M. Davis of 7 July 1928 urging Davis to stress prohibition but not the religious issue in Delaware, AIdPJ.

19. James, *op. cit.,* pp. 408–10.

20. Interview with Edward Ball, 3 January 1975.

21. Herbert Hoover, telegram to AIdP, 7 November 1928, AIdPJ.

22. AIdP to Dorothy V. Lee, 7 November 1928, quoted in James, *op. cit.,* p. 410.

23. AIdP to E. M. Davis, 7 July 1928, AIdPJ.

24. AIdP to I. D. Short, 21 February 1929, AIdPJ.

25. AIdP's comment as reported in the Wilmington *Morning Star,* 8 June 1930.

26. AIdP to Laura Walls, quoted in the [Dover] *State Sentinel,* 17 July 1929.

27. Walls to AIdP, quoted in *ibid.*

28. TCdP to AIdP, 28 October 1930, AIdPJ.

29. AIdP to Marguerite dP Lee, 11 November 1930, AIdPJ.

30. Comment of AIdP to his wife, JBdP, quoted in James, *op. cit.,* p. 427.

31. AIdP to C. Douglass Buck, 2 September 1930, AIdPJ.

32. AIdP to Buck, 17 September 1930; Buck to AIdP, 18 September 1930, AIdPJ.

33. AIdP to John S. Rossell and the Rev. Charles L. Candee, 4 October 1930, AIdPJ.

34. *Ibid.*

35. AIdP to Buck, 12 December 1930, AIdPJ; for details of the proposed bill, see the [Newark] *Delaware Ledger,* 13 March 1931.

36. Buck to AIdP, 19 December 1930, AIdPJ.

37. A full report on the legislative history of the pension proposal appeared in the *Delaware Ledger,* 13 March 1931.

38. Abraham Epstein in *Old Age Security,* March 1931; AIdP to Epstein, 3 February 1931, AIdPJ.

39. AIdP to C. Douglass Buck, 5 January 1931, AIdPJ.

4.0. *Ibid.*

41. AIdP's letter to Senator Keith which the latter read in the senate chamber and entered into the senate's journal. Later quoted in a news story in the Wilmington *Evening Journal,* 25 February 1931.

42. AIdP to John H. Rossell, 23 March 1931, AIdPJ.

43. C. D. Buck to AIdP, 19 March 1931, AIdPJ.

44. AIdP to Buck, 3 April 1931, AIdPJ.

45. See AIdP's letter to Senator E. B. Griffenberg, 3 April 1931, as a sample of the letters he sent to every state senator. AIdPJ.

46. J. S. Rossell to AIdP, 17 April 1931, AIdPJ.

47. Buck, telegram to AIdP, 10 April 1931, AIdPJ.

48. Rossell to AIdP, 17 April 1931, AIdPJ.

49. C. D. Buck to AIdP, 16 April 1931, AIdPJ.

50. *Ibid.,* JBdP's marginal notation.

51. Acceptance speech by Secretary of Commerce Herbert C. Hoover, San Francisco, 11 August 1928, reprinted in A. M. Schlesinger, Jr., ed., *History of American Presidential Elections* (New York: Chelsea House, 1971), Vol. III, p. 2683.

52. Acceptance Speech by President Herbert C. Hoover, Washington, 11 August 1932, reprinted in *ibid.,* pp. 2792–2805.

53. AIdP to Cazenove G. Lee, 10 October 1930, quoted in James, *op. cit.,* pp. 473–74.

54. See Elbert Dent's letter to AIdP, 15 May 1931, in which Dent praises Alfred's suggestion and urges him to get "some U.S. Senator to introduce such a bill in the Senate." AIdPJ.

55. See Alfred Link, *American Epoch* (New York: Alfred A. Knopf, 1955), pp. 372–73.

56. AIdP to Lammot dP, 3 June 1932, AIdPJ.

57. AIdP to I. D. Short, 9 July 1932, AIdPJ.

58. See the correspondence between AIdP and Lammot dP from June to September 1932, AIdPJ.

59. AIdP to T. W. Jakes, 5 November 1932, AIdPJ.

60. AIdP to Hermann Ruoff, 2 September 1931, AIdPJ.

61. AIdP to Ellen La Motte, 10 November 1932, quoted in James, *op. cit.*, p. 481. It is in this letter to La Motte that Alfred mentions his invitation to Hurley.

62. AIdP to Thomas Ball, 19 November 1932, AIdPJ.

63. JBdP to Thomas Ball, 1 December 1932, AIdPJ.

64. See the lengthy file of correspondence between AIdP and Laura Walls regarding a Teacher's Retirement Plan, which she had drawn up and had presented to the legislature. The bill had passed the legislature and had been signed by the governor, but was declared unconstitutional by the Attorney General on grounds that it depended in part on its funding by drawing funds from the School Education Fund. It was at that point that Alfred du Pont paid for pensions for teachers in need out of his own pocket. Letters to and from AIdP and Laura Walls, May–September 1933.

65. See AIdP's letter to Laura Walls, 4 January 1933, AIdPJ; see also AIdP's letter to T. W. Jakes, 26 September 1933, asking him to make purchases of blankets and overcoats for distribution by Walls during the coming winter, indicating that Alfred continued the program even after the New Deal relief was in operation. AIdPJ.

66. James, *op. cit.*, p. 474.

67. Edwards *Reminiscences,* JBdPJ.

68. These direct quotations are given in James, *op. cit.*, p. 485. Although no source is given by James, presumably these quotations are from an interview with Edward Ball.

69. *Ibid.,* pp. 485–87.

70. AIdP to Maurice dP, 7 May 1933, AIdPJ, quoted in James, *op. cit.*, p. 486.

71. AIdP to AVdP, 13 April 1933, AIdPJ, quoted in James, *op. cit.*, p. 499.

72. AIdP interview, printed in the *Orlando Morning Sentinel,* 21 March 1933.

73. AIdP to Thomas Ball, 6 July 1933, AIdPJ.

74. AIdP to Maurice dP Lee, 25 October 1933, AIdPJ.

Chapter XV

1. Franklin D. Roosevelt's First Inaugural Address, 4 March 1933, reprinted in Henry Steele Commager, ed., *Documents of American History* (Englewood Cliffs, N.J.: Prentice-Hall, 1973), Vol. II, pp. 240–42.

2. AIdP to Maurice dP Lee, 4 May 1934, AIdPJ, quoted in James, *op. cit.*, p. 508.

3. Quoted in James, *op. cit.*, p. 508.

4. "Alfred du Pont Gives $1,000 to President's Birthday Ball," *Jacksonville Journal,* 4 January 1935; AIdP to Sterling Joyner, 26 March 1935, AIdPJ.

5. Quoted in A. M. Schlesinger, Jr., *The Politics of Upheaval* (Boston: Houghton, Mifflin Co., 1960), p. 639.

6. AIdP to Madeleine dP Ruoff, 16 September 1931, AIdPJ.

7. Monthly list, dated 14 February 1935, Laura Walls to AIdP, AIdPJ.

8. See correspondence between Laura Walls, AIdP, and his secretary, M. E. Brereton, of various dates: 16 May 1934 and 14 February and 20 February 1935, AIdPJ.

9. See AIdP's letter to Sarah W. Pyle, director of the People's Settlement, Wilmington, 27 February 1934, AIdPJ.

10. JBdP to Thomas Ball, 22 February 1935, JBdPJ.

11. Memo of Mary Shaw, Jessie BdP's secretary, giving revised figures of family allowances, attached to a letter to Addison Baker, Jessie's brother-in-law, dated 26 March 1932, JBdPJ.

12. Personnel report on Alfred Victor du Pont while working at the Barksdale plant, dated 4 September 1926, in the Alfred Victor du Pont papers in the possession of his widow, Mrs. Henny Christiansen du Pont, Miami, Florida (hereafter referred to as AVdPM).

13. AVdP to PSdP, 26 November 1926; PSdP to AVdP, 7 December 1926, AVdPM.

14. AVdP to J. W. McCoy, 19 April 1927, AVdPM.

15. AIdP to AVdP, 7 November 1927, AIdPJ.

16. Quoted by AVdP in a letter to AIdP, 18 July 1928, AVdPM.

17. AVdP to AIdP, 18 July 1928, AVdPM.

18. AIdP to AVdP, 3 August 1928, AVdPM.

19. AVdP to AIdP, 10 October and 1 November 1928, AIdP to AVdP, 9 November 1928, AVdPM.

20. See correspondence between AIdP and AVdP regarding the partnership and the commission for the sunken gardens at Nemours, 11 and 16 May 1929, AVdPM.

21. AIdP to AVdP, 16 May 1929, AVdPM.

22. See AIdP's letter to Edward Ball, 12 March 1931, stating that in view of the economic crisis he was postponing indefinitely any plans for the building of the church, AIdPJ; also M. E. Brereton to W. & J. Sloan, Co., Washington, D.C., 19 June 1931, informing that company that all work on the church had been "suspended indefinitely," AIdPJ.

23. See correspondence of AIdP with Secretary of Treasury Ogden Mills, 11 August and 7 December 1932, AIdPJ; AVdP to Senator Daniel Hastings, 20 January 1933, re post office building in Wilmington, AVdPM; PSdP to AVdP, 17 August 1931, and AVdP to PSdP, 22 March 1933, AVdPM.

24. See correspondence AIdP to Governor Buck, 13 August 1931, and Buck's answer expressing the hope that Alfred Victor and Massena would get the commission, 18 August 1931, AIdPJ.

25. *Wilmington Sunday Star,* 15 October 1933.

26. Dedication Speech, dated 11 October 1933, copy in AVdPM.

27. AVdP to AIdP, 13 October 1933, AIdPJ.

28. *Wilmington Sunday Star,* 15 October 1933.

29. See correspondence of AIdP with Reps. Julian T. Robinson, James Latcham, and William Scott, 21 November 1933; to Sen. Jacob V. Hill, 22 November 1933; to Sen. E. B. Griffenberg, 25 November 1933; and to Charles Candee, 25 November 1933, among many letters pushing for additional funding for the welfare home. AIdPJ.

30. AIdP to Charles Candee, 25 November 1933, AIdPJ.

31. See AVdP's letter to AIdP, 21 January 1935, AVdPM.

32. M. E. Brereton, telegram, to Laura Walls, 29 April, 1935, AIdPJ.

33. Marcella dP to JBdP, 16 May 1934, copy in AVdPM.

34. JBdP to Margery F. dP, 24 March 1930, AIdPJ.

35. AIdP to Madeleine dP Ruoff, 24 October 1931, AIdPJ.

36. See Mosley, *op. cit.,* pp. 205, 213–14; James, *op. cit.,* p. 217.

37. Madeleine dP Hiebler to AIdP, 7 August 1930, AIdPJ.

38. Hermann Ruoff to AIdP, 4 November 1930, AIdPJ.

39. Madeleine dP Hiebler to AIdP, 11 November 1930, AIdPJ.

40. AIdP to Hermann Ruoff, 14 November 1930, AIdPJ.

41. AIdP to Madeleine dP Hiebler, 21 November 1930, AIdPJ.

42. Madeleine dP Hiebler to AIdP, 14 December 1930, AIdPJ.

43. AIdP to Madeleine dP Hiebler, 10 February 1931, AIdPJ.

44. AIdP to Benno Hiebler, 13 April 1932, AIdPJ; see also Hermann Ruoff's letter of 22 May 1932 to Alfred in which Ruoff was equally outspoken in his criticism of Hitler and the Nazi party. AIdPJ.

45. Alfred Hiebler to AIdP, 23 March 1933, AIdPJ.

46. AIdP to Alfred Hiebler, 5 April 1933, AIdPJ.

47. Madeleine dP Ruoff to AIdP, undated but sometime in March 1933, and AIdP to Madeleine, 25 April 1933, AIdPJ.

48. See numerous letters 1930–1935 between AIdP and his grandsons Bayard, Benno, and Alfred, AIdPJ.

49. James, *op. cit.,* pp. 515–16.

50. JPdP to Madeleine dP Ruoff, 5 April 1934 and JBdP to Bayard Hiebling, 14 April 1934, JPdPJ. See also Elbert Dent to JBdP, 5 April 1934, and Victorine Dent to JBDP, 14 March 1934, enclosing a copy of the letter she sent to Madeleine urging her sister to keep Bayard in Germany, JBdPJ.

51. Bayard Hiebling to AIdP, 3 December 1934, telling of his experiences in the Hitler Youth Work Camp near Chiemsee, Bavaria.

52. See the numerous letters of AIdP to Alicia Maddox Glendening and her letters to him from 1921–1935 in AIdPJ.

53. JBdP to Alicia Maddox Glendening, 25 March 1929, JBdPJ.

54. *Ibid.*

55. AIdP to Madeleine dP Hiebler, 31 December 1930, AIdPJ.

56. JBdP to Denise dP, 20 December 1929, JBdPJ.

57. JBdP to Denise dP, 2 March 1931, JBdPJ.

58. AIdP to Denise dP, 4 November 1932, AIdPJ.

59. Mary Wolseley-Lewis to JBdP, 6 May 1930, JBdPJ.

60. JBdP to Mary Wolseley-Lewis, 20 May 1930, JBdPJ.

61. AIdP to Margaret Hanby, 26 March 1929, AIdPJ. See also Alfred's letter to the Rev. Dr. Charles Dubell 21 February 1930, AIdPJ regarding Alicia's marital difficulties in which he states that Alicia was much too young when she married. "I am a great believer in the thought that married life is only possible when mutual admiration and respect forms the foundation of love."

62. William Wordsworth, "Ode to Duty," in *The Complete Poetical Works of William Wordsworth,* ed. by Andrew J. George (Boston: Houghton, Mifflin Co., 1904), p. 319.

63. Marguerite dP Lee to AIdP, 7 September 1930, AIdPJ. The italics are Marguerite Lee's.

64. *Ibid.* JBdP's marginal notation.

65. Marguerite dP Lee to AIdP, 25 July 1929, AIdPJ.

66. Marguerite dP Lee to AIdP, 25 August 1929, AIdPJ.

67. Marguerite dP Lee to AIdP, 25 July 1929, AIdPJ. The italics are Marguerite Lee's.

68. AIdP to Maurice dP, 23 February 1928, AIdPJ.

69. AIdP to Marguerite dP Lee, 21 April 1930, AIdPJ.

70. For accounts of the slander suit, see Washington *Times,* 17 December 1930, and Washington *Herald,* 18 December 1930.

71. Marguerite dP Lee to AIdP, 17 December 1930, AIdPJ.

72. AIdP to Marguerite dP Lee, 7 April 1930, AIdPJ.

73. AIdP to Maurice dP, 24 December 1930, AIdPJ.

74. See the memorandum of Alfred's lawyer, Henny P. Adair, to Edward Ball, 13 January 1934, and the legal document which Mary Ball signed in settlement of the suit, found in AIdPJ.

75. Manuscript of the Maurice du Pont Lee Memoirs in the Maurice du Pont Lee Papers, Box 6, EHLW.

76. *Ibid.*

77. AIdP to Maurice dP Lee, 31 December 1930, Maurice dP Lee Papers, Box 10, EHLW.

78. Maurice dP Lee to AIdP, 28 February 1931, AIdPJ. The italics are Maurice Lee's.

79. *Ibid.*

80. AIdP to Maurice dP Lee, 8 March 1931, AIdPJ.

81. AIdP to Edward Ball, 4 February 1931, AIdPJ.

82. AIdP to Lammot dP, 28 June 1932, AIdPJ.

83. Lammot dP to AIdP, 5 July 1932, Administrative Papers of the Du Pont Company, Acc. 1662, Box 21, C12, Treasury Dept., EHLW.

84. AIdP to Lammot dP, 8 July 1932, AIdPJ.

85. Lammot dP to AIdP, 19 July 1932, Administrative Papers of the Du Pont Company, Acc. 1662, Box 21, C12, Treasury Dept., EHLW.

86. Lammot dP to AIdP, 19 July 1932, AIdPJ. A copy of this letter is in Acc. 1662, Box 21 with a notation at the top "Handed AIdP, 8/5/32," EHLW.

87. See Marquis James to Lammot dP, 3 March 1941, and Lammot du Pont to James, 11 March 1941, Administrative Papers of the Du Pont Company, Acc. 1662, Box 8, EHLW.

88. See Maurice Lee's statement on the efforts to get representation for Alfred du Pont on the Board of Directors in Maurice du Pont Lee Papers, Box 10, EHLW.

89. AIdP to Lammot dP, 1 December 1934, Administrative Papers of the E. I. du Pont Company, Acc. 1662, Box 45, EHLW.

90. AIdP to Marguerite dP Lee, 23 July 1930, AIdPJ.

91. AIdP to Ellen La Motte, 28 January 1929, AIdPJ.

92. AIdP to Marguerite dP Lee, 12 September 1930, AIdPJ.

93. AIdP to Mrs. N. Addison Baker, 30 September 1931, AIdPJ.

94. AIdP to Ellen La Motte, 28 January 1929, AIdPJ.

95. AIdP to Ellen La Motte, 23 July 1930, AIdPJ.

96. See AIdP's letter to his secretary, Irene Walsh, 15 July 1931, AIdPJ.

97. AIdP to Cora Kinzie, 2 April 1931, AIdPJ.

98. JBdP to Edward Ball, 18 October 1930, JBdPJ.

99. AIdP to Ella Haile, 17 December 1929, AIdPJ.

100. AIdP to Mrs. Alex Francis, 12 March 1930, AIdPJ.

101. JBdP to Carrie and Will Martin, 11 April 1930, JBdPJ.

102. AIdP to JBdP, 17 February 1930, AIdPJ.

103. AIdP to Denise dP, 7 April 1930, AIdPJ.

104. See JBdP's several letters to Alfred dP written during her trip to Rumania, September and October 1931, JBdPJ.

105. JBdP to Will Martin, 14 April 1932, JBdPJ.

106. JBdP's postcard diary, undated entries during January and February 1932, JBdPJ.

107. See AIdP's letter to the Rev. Dr. Baker P. Lee, 7 March 1932, telling of finding the dog of Cairo, AIdPJ; also the certificate of the veterinarian, Dr. W. L. Harrison of Cairo, Egypt, dated 15 February 1932, attesting to Mummy's good health, JBdPJ.

108. AIdP to the Rev. Baker P. Lee, 23 May 1932, AIdPJ. See also AIdP's letter to Mrs. Cazenove Lee, Jr., 5 March 1932, giving a similar account of Mummy's magical powers, AIdPJ.

109. AIdP to Louise Bayard, 18 May 1932, AIdPJ.

110. AIdP to Madeleine dP Ruoff, 18 May 1932, AIdPJ.

111. See correspondence and receipts of sale between AIdP, Van Eaton Danton and William T. Stewart, New York art dealer and agent for the Danton family, 1928–1931, AIdPJ.

112. PSdP II to AIdP, 23 July 1930, AIdPJ.

113. AIdP to PSdP II, 30 July 1930, AIdPJ.

114. AIdP to James G. Bright, 7 December 1932, AIdPJ.

115. James, *op. cit.,* p. 427.

116. AIdP to Louise Bayard, 23 September 1931, AIdPJ.

117. Marguerite dP Lee to Alexander Du Bin, 17 May 1936, a copy in the papers of Maurice dP Lee, Acc. 1452, Box 14, EHLW.

118. AIdP to Marguerite dP Lee, 12 January 1935, in the papers of Maurice dP Lee, Acc. 1452, Box 14, EHLW.

119. *Ibid.,* Marguerite dP Lee's added note.

120. Marguerite dP Lee to AIdP, 30 July 1934, copy in the papers of Maurice dP Lee, Acc. 1452, Box 14, EHLW.

121. AIdP to Marguerite dP Lee, 9 July 1934, papers of Maurice dP Lee, *loc. cit.*

122. See telegram from Marguerite dP Lee to Maurice dP Lee informing him of Alfred's contribution to the Will Gardner family, papers of Maurice dP Lee, *loc. cit.*

123. JBdP to Judge Joseph Chinn, 22 September 1930, AIdPJ.

124. JBdP to Chinn, 2 October 1930, AIdPJ.

125. See Certification of Incorporation of Ditchley, Inc., dated 1 August 1932; letter of Edward Ball to JBdP, congratulating his sister on the purchase, 19 July 1932; and newspaper clipping from the Richmond *Times-Dispatch,* 28 August 1932, giving a history of the estate, in JBdPJ.

126. AIdP to Mrs. Ralph Seiss, 4 June 1930, AIdPJ.

127. Marguerite dP Lee to Maurice dP Lee, 5 October 1926. See also the letters of appreciation from all four children to AIdP in AIdPJ. See James, *op. cit.,* p. 374.

128. The Will and Testament of Alfred I. du Pont, dated 19 March 1932 as printed in a pamphlet entitled "In Re the Estate of Alfred I. du Pont, Deceased," prepared and printed by the Trustees of the Alfred I. du Pont Estate, pp. 1–22, and containing in addition to the Will, the Codicil dated 4 March 1933; the Second Codicil dated 15 January 1935; and the Order of Probate dated 24 May 1935, as the same appear of record of the Court of the County Judge, Duval County, Florida, Book 5, pp. 42–63.

129. Codicil of 4 March 1933, p. 1, in *Ibid,* Book 5, pp. 64–67.

130. JBdP to Reginald Huidekoper, 12 February 1934, AIdPJ.

131. JBdP to Will Martin, 8 March 1934, AIdPJ.

132. Marguerite dP Lee to JBdP, 12 February 1934, JBdPJ.

133. AIdP to Raymond Knight, 7 February 1935, AIdPJ.

134. AIdP to Raymond Knight, 23 January 1935, and Knight to AIdP, 30 January 1935, Alfred I. du Pont papers in the possession of Alfred dP Dent, Wilmington (hereafter referred to as ADW).

135. AIdP to Knight, 13 March 1935, ADW.

136. Knight to AIdP, 16 March 1935, ADW.

137. Knight to AIdP, 11 January 1935, ADW.

138. JBdP to Elsie Bowley, 3 January 1935, JBdPJ.

139. AIdP to Dorothy Lee, 4 January 1935, AIdPJ.

140. JBdP to Gen. A. J. Bowley, 12 January 1935, JBdPJ.

141. JBdP to Bowley, 28 January 1935, JBdPJ.

142. See copies of the telegrams sent to Alfred Victor dP, Elbert Dent, and Reginald Huidekoper plus letters to the Maurice Lee and Cazenove Lee families, 23 January 1935, AIdPJ.

143. AIdP to Maurice dP Lee, 6 February 1935, AIdPJ.

144. AIdP to Geraldine Lee, 15 April 1935, AIdPJ.

145. See letters of AIdP to Alicia, 7 February and 16 April 1935; to Madeleine Ruoff, 26 March 1935; and to Francis I. dP, 3 April 1935, all in AIdPJ.

146. AIdP to Marguerite dP Lee, 8 March, 1935, AIdPJ.

147. The details of Alfred du Pont's last illness and death are taken from James, *op. cit.,* pp. 540–42. James's sources were Alfred Victor du Pont, Mr. and Mrs. N. Addison Baker, Edward Ball, and of course Jessie Ball du Pont.

148. Joyce Carol Oates, *On Boxing* (New York: Dolphin Doubleday, 1987). This particular quotation is taken from a review of that book by Christopher Lehmann-Haupt in the *New York Times,* 2 March 1987, p. 17.

149. AIdP to Mrs. Cazenove G. Lee, Jr., 25 March 1930, AIdPJ.

150. Quoted in Elizabeth Fox-Genovese, editor and translator, *The Autobiography of Du Pont de Nemours* (Wilmington: Scholarly Resources, Inc., 1984), Introduction, p. 50.

Epilogue

1. AIdP to the Rev. Dr. Baker P. Lee, 24 April 1935, AIdPJ.

2. Jessie BdP to Madeleine Ruoff and Thomas Ball, 7 May 1935, JBdPJ.

3. Wilmington *Morning News,* 3 May 1935.

4. Jessie BdP to Thomas Ball, 7 May 1935, JBdPJ.

5. *Ibid.*

6. See memo from Alfred Victor dP to his sister, Victorine dP Dent, undated but written sometime in 1937 for a description of the reading of the will, AVdPM.

7. The Will and Testament of Alfred I. du Pont, dated 19 November 1932, *loc. cit.* See also Chap. XV for a further discussion of Alfred du Pont's will.

8. Second Codicil, dated January 15th, 1935, In Re the Estate of Alfred I. du Pont, Deceased, *loc. cit.*

9. See deposition of Alfred Victor du Pont regarding this oral agreement; undated, but written sometime in 1937, in AVdPM.

10. Marguerite dP Lee to Henry Adair, 11 August 1936, copy in Longwood Mss., Group 10, Papers of P. S. du Pont, File 328, EHLW.

11. PSdP to Marguerite dP Lee, 17 September 1936, Longwood Mss., Group 10, File 328, EHLW.

12. JBdP to Maurice dP, 13 September 1939. Copy of the letter is in the Maurice dP Lee Papers, Acc. 1452, Box 10, EHLW.

13. See copies of correspondence between Marquis James and Maurice dP, which Maurice apparently gave to his nephew, Maurice dP Lee, and also the correspondence between the latter and Maurice dP, regarding the James biography. Maurice dP Lee Papers, Acc. 1452, Box 10, EHLW.

14. These views regarding the estate and the du Pont family were expressed to this writer by Edward Ball in an interview in September 1974.

15. Maurice dP Lee to Jessie BdP, 9 January 1939, Papers of Maurice dP Lee, Acc. 1452, Box 11, EHLW.

16. JBdP to Maurice dP Lee, 26 January 1939, Papers of Maurice dP Lee, Acc. 1452, Box 11, EHLW.

17. Francis I. dP to Maurice dP Lee, 30 January 1939, Papers of Maurice dP Lee, Acc. 1452, Box 11, EHLW.

18. JBdP to Maurice dP Lee, 8 February 1939, Papers of Maurice dP Lee, Acc. 1452, Box 11, EHLW.

19. Maurice dP Lee to JBdP, 7 February 1939, Papers of Maurice dP Lee, Acc. 1452, Box 11, EHLW.

20. Henry P. Adair to JBdP, 28 March 1939, copy in Papers of Maurice dP Lee, Acc. 1452, Box 11, EHLW.

21. See several newspaper accounts of Ball's projects for expansion: *Miami Daily News,* 28 February 1936; Jacksonville *American,* 19 June 1936.

22. W. H. Goodman to Maurice dP Lee, 11 August 1939, Papers of Maurice dP Lee, Acc. 1452, Box 11, EHLW.

23. Unpublished memoirs Maurice dP Lee in the Papers of Maurice dP Lee, Acc. 1452, Box 11, EHLW.

24. *Ibid.*

25. Quoted in *The Estate of Alfred I. du Pont and the Nemours Foundation,* published by The Estate of Alfred I. du Pont (Jacksonville: 1974).

26. *The Delmarva Star* [Wilmington], 5 January 1936.

27. AVdP to JBdP, 19 May 1936, AVdP Papers, Acc. 1508, EHLW.

28. See AVdP's letter to Huidekoper, 1 October 1936, AVdP Papers, Acc. 1508, EHLW.

29. Maurice dP Lee to Maurice dP, 28 March 1939, Papers of Maurice dP Lee, Acc. 1452, Box 11, EHLW.

30. See *The Estate of Alfred I. du Pont and the Nemours Foundation,* pp. 52–63.

31. *Ibid.,* pp. 31–42.

32. *Ibid.,* pp. 43–50.

33. See Reginald Huidekoper's letter to Henry Adair, 19 November 1938, urging that a reserve for the annuities be established; copy in JBdPJ.

34. This letter is quoted in Maurice dP Lee's memoirs in the Maurice dP Lee Papers, Acc. 1452, Box 10, EHLW.

35. "Florida Banker at Bay," *New York Times,* 14 June 1971.

36. Curtis Wilkie, "Nemours Foundation Stockpiles 35 Million," *Wilmington News-Journal,* 19 March 1974.

37. Interview with Hazel Williams, January 1989.

38. Christine Donahue, "The Death of Alfred I. du Pont—a Postmortem," *Forbes 400,* 28 October 1985. This writer also interviewed Mrs. Morrell in St. Louis in March 1983, and she told the same story. She was obviously sincere in her belief that the story she told was true.

Bibliography

Primary Sources

The collected papers of Alfred I. du Pont, which at the time of my research (1974–1982) were housed in a storeroom at the Florida National Bank, Jacksonville, Florida. Subsequently, the papers were transferred to Washington and Lee University in Lexington, Virginia, in accordance with the provision for their disposal stipulated in the will of Jessie Ball du Pont.

Additional, much smaller collections of Alfred I. du Pont papers at the Nemours Mansion in Wilmington and in the possession of Mrs. Ruth Price Ball Morrell, Edward Ball's former wife, which she and Alfred du Pont Dent, Alfred I.'s grandson, kindly made available to me.

The papers of Jessie Ball du Pont, also at the Florida National Bank in Jacksonville and at the Nemours Mansion.

The papers of Edward Ball, in the Florida National Bank in Jacksonville and in the possession of Mrs. Ruth Morrell.

The papers of Alfred Victor du Pont, Alfred I.'s son, which at the time of my research were in the possession of his widow, Henny Christiansen du Pont, and were later transferred to the Eleutherian Mills Historical Library in Wilmington, Delaware.

The papers of W. T. Edwards, in the possession of Mrs. W. T. Edwards in Jacksonville, Florida.

Letters of Eleuthera du Pont Bradford to her daughter, Alicia Bradford Maddox, in the Nemours Mansion, Wilmington, Delaware.

The voluminous and invaluable holdings of the Eleutherian Mills Historical Library in Wilmington, Delaware, for other members of the du Pont family, including in particular, courtesy of Hagley Museum and Library:

> The Longwood Manuscripts: Papers of Pierre Samuel du Pont de Nemours I; papers of Eleuthère Irénée du Pont; papers of descendents of the sons of du Pont de Nemours; papers of the E. I. du Pont de Nemours & Co.; notes and collections of Bessie G. du Pont; papers of Pierre S. du Pont II.

The Henry Francis du Pont Collection of Winterthur Manuscripts: Papers of the du Pont family, 1588–1785; papers of Pierre Samuel du Pont de Nemours I; papers of Eleuthère Irénée du Pont; papers of the daughters of E. I. du Pont; papers of the sons of E. I. du Pont; papers of Henry A. du Pont; papers of Sophie Madeleine du Pont; miscellany.

Records of the E. I. du Pont de Nemours & Co., 1801–1972.

The Eleuthera Bradford du Pont Collection.

Papers of Francis Gurney du Pont, including the Allan J. Henry Collection.

Papers of Irénée du Pont.

Papers of John J. Raskob, which are restricted and are used here with permission of the Raskob Estate.

Papers of Francis and Louise du Pont Crowninshield.

Papers of Maurice du Pont Lee, including papers of Marguerite du Pont Lee.

Du Pont family miscellany.

Oral Interviews

Kennard Adams

Rebecca Harding (Mrs. Charles) Adams

Edward Ball

Joseph Ball

J. J. Belin

Eleanor Boden

Alfred du Pont Dent

Mrs. E. Paul du Pont

Henny Christiansen du Pont

Mrs. W. T. Edwards

Elise du Pont Elrick

Bill Frank

Mrs. A. Duer Irving

The Rev. Alexander Juhan

Maurice du Pont Lee

The Rt. Rev. Arthur R. McKinstry

William B. Mills

Ruth Price Ball Morrell

Mrs. T. C. Raver

Dr. A. R. Shands, Jr.

S. D. Stoneburner

Irene Walsh

Hazel Williams

Denise du Pont Zappfe

Newspaper Files Consulted

Chicago *Daily News*

Delaware Ledger

Delaware Republican

[Dover, Del.] *State Sentinel*

Delaware State Journal

Jacksonville *American*

Los Angeles Examiner

Louisville Commercial

Miami Daily News

New York Herald

New York Times

New York *World*

Orlando Sentinel

Philadelphia Press

Philadelphia *Public Inquirer*

Philadelphia *Public Ledger*

Philadelphia Record

Richmond *Times-Dispatch*

Tampa *Tribune*

Washington Herald

Washington Times
[Wilmington] *Delmarva Star*
Wilmington *Evening Journal*

Wilmington *Morning News*
Wilmington Sunday Star

Published Letters

Gilbert Chinard, ed., *The Correspondence of Jefferson and du Pont de Nemours,* Baltimore: The Johns Hopkins University Press, 1941.

James F. Marshall, ed. and transl., *de Staël-du Pont Letters,* Madison: University of Wisconsin Press, 1968.

Official Records

Court of County Judge, Duval County, Florida

Philip F. du Pont, et al. v. *Pierre S. du Pont, et al.,* U.S. District Court, District of Delaware

Du Pont v. *Du Pont,* U.S. District Court of Appeals for the 3rd District

Letters of Patent, *U.S. Patent Records,* U.S. Patent Office

U.S. v. *Du Pont,* U.S. Circuit Court, District of Delaware

U.S. v. *Du Pont, General Motors, et al.,* U.S. District Court for the Northern District of Illinois, 1957

Unpublished Manuscripts

Henry M. Canley, Jr., *J. Edward Addicks: A History of his Political Activities,* senior thesis, Princeton University, 1932.

George T. Maxwell, *The Political Activities of Alfred I. du Pont.*

John A. Munroe, *History of Delaware* (later published by the University of Delaware Press, 1980.)

Secondary Sources Consulted and Cited

Canby, Henry Seidel, *The Brandywine,* The Rivers of America Series, Exton, Pennsylvania: Schiffer, Ltd., 1969.

Carr, William H. A., *The du Ponts of Delaware,* New York, Dodd, Mead & Co., 1964.

Cazenove, Anthony-Charles, "Autobiographical Sketch," John Askling, ed., *The Virginia Magazine of History,* July 1970, pp. 295–307.

Chandler, Alfred D., Jr., and Salsbury, Stephen, *Pierre S. du Pont and the Making of the Modern Corporation,* New York: Harper & Row, 1971.

Delaware: A Guide to the First State, Federal Writers' Project, New York: Viking Press, 1938.

du Pont, Bessie G., *E. I. du Pont de Nemours and Company, A History, 1802–1902,* Boston: Houghton, Mifflin Co., 1920.

Dutton, William S., *Du Pont: One Hundred and Forty Years,* New York: Charles Scribner's Sons, 1942.

The Estate of Alfred I. du Pont and the Nemours Foundation, privately printed, Jacksonville, 1974.

Fox-Genovese, Elizabeth, ed. and transl., *The Autobiography of Du Pont de Nemours,* Wilmington: Scholarly Resources, Inc., 1984.

Gates, John D., *The du Pont Family,* Garden City: Doubleday & Co., Inc., 1979.

Hancock, Harold B., "The Income and Manufacturer's Tax of 1862–1872," *Delaware History,* October 1971.

Herty, Charles H., "Southern Pine for White Paper," *Scientific American,* May 1934, pp. 234–236.

Hoffecker, Carol, *Delaware: A Bicentennial History,* The States and the Nation Series, New York: W. W. Norton & Co., 1977.

James, Marquis, *Alfred I. du Pont: The Family Rebel,* Indianapolis: Bobbs-Merrill Co., 1941.

Jolly, Pierre, *Du Pont de Nemours,* translated by Elise du Pont Elrick, Wilmington: The Brandywine Publishing Co., 1977.

Link, Arthur, *American Epoch,* New York: Alfred H. Knopf, 1955.

Mosley, Leonard, *Blood Relations: The Rise and Fall of the du Ponts of Delaware,* New York: Atheneum, 1980.

Rasch, Royal H., "Pine Cellulose," *Manufacturers Record,* November 1934.

Saricks, Ambrose, *Pierre Samuel Du Pont de Nemours,* Lawrence: University of Kansas Press, 1965.

Schlesinger, Arthur M., Jr., *The Politics of Upheaval,* Boston: Houghton, Mifflin Co., 1960.

Wallace, Anthony F. C., *Rockdale: The Growth of an American Village in the Early Industrial Revolution,* New York, W. W. Norton & Co., 1980.

Wertenbaker, Charles, "Du Pont," *Fortune,* November 1942.

Wilkie, Curtis, "Nemours Foundation Stockpiles 35 Million," *Wilmington News-Journal,* 19 March 1974.

Wilkinson, Norman B., "In Anticipation of Frederick W. Taylor: A Study of Work by Lammot du Pont, 1872," *Technology and Culture,* Spring 1965, pp. 208–21.

Standard Reference Works Consulted

Biographical Directory of the American Congress, 1774–1971, Washington, D.C.: U.S. Government Printing Office, 1971.

Dictionary of American Biography, New York: Charles Scribner's Sons, 1931.

Documents of American History, Henry Steele Commager, ed., Englewood Cliffs, N.J.: Prentice-Hall, 1973.

History of American Presidential Elections, A. M. Schlesinger, Jr., ed., New York: Chelsea House, 1971.

National Party Conventions, 1831–1971, Washington, D.C.: Congressional Quarterly, 1976.

Notable American Women, Cambridge, Mass.: The Belknap Press, 1971.

Index